Position-Sensitive Gaseous Photomultipliers:

Research and Applications

Tom Francke
Myon, Sweden

Vladimir Peskov
CERN, Switzerland

A volume in the Advances in Chemical and Materials Engineering (ACME) Book Series

ENGINEERING SCIENCE REFERENCE
An Imprint of IGI Global

Published in the United States of America by
Engineering Science Reference (an imprint of IGI Global)
701 E. Chocolate Avenue
Hershey PA, USA 17033
Tel: 717-533-8845
Fax: 717-533-8661
E-mail: cust@igi-global.com
Web site: http://www.igi-global.com

Library of Congress Cataloging-in-Publication Data

Library of Congress Cataloging-in-Publication Data

Names: Francke, Tom, 1964- | Peskov, Vladimir.
Title: Position-sensitive gaseous photomultipliers : research and applications / by Tom Francke and Vladimir Peskov.
Description: Hershey PA : Engineering Science Reference, [2016] | Includes bibliographical references and index.
Identifiers: LCCN 2016000650| ISBN 9781522502425 (hardcover) | ISBN 9781522502432 (ebook)
Subjects: LCSH: Photon detectors. | Photons. | Photoelectric multipliers. | Position sensitive particle detectors.
Classification: LCC QC787.P46 F73 2016 | DDC 539.77--dc23 LC record available at https://lccn.loc.gov/2016000650

This book is published in the IGI Global book series Advances in Chemical and Materials Engineering (ACME) (ISSN: 2327-5448; eISSN: 2327-5456)

British Cataloguing in Publication Data
A Cataloguing in Publication record for this book is available from the British Library.

All work contributed to this book is new, previously-unpublished material. The views expressed in this book are those of the authors, but not necessarily of the publisher.

For electronic access to this publication, please contact: eresources@igi-global.com.

Advances in Chemical and Materials Engineering (ACME) Book Series

J. Paulo Davim
University of Aveiro, Portugal

ISSN: 2327-5448
EISSN: 2327-5456

Mission

The cross disciplinary approach of chemical and materials engineering is rapidly growing as it applies to the study of educational, scientific and industrial research activities by solving complex chemical problems using computational techniques and statistical methods.

The **Advances in Chemical and Materials Engineering (ACME) Book Series** provides research on the recent advances throughout computational and statistical methods of analysis and modeling. This series brings together collaboration between chemists, engineers, statisticians, and computer scientists and offers a wealth of knowledge and useful tools to academics, practitioners, and professionals through high quality publications.

Coverage

- Thermo-Chemical Treatments
- Fracture Mechanics
- Toxicity of Materials
- Polymers Engineering
- Electrochemical and Corrosion
- Coatings and surface treatments
- Ceramics and glasses
- Sustainable Materials
- Heat Treatments
- Nanomaterials

IGI Global is currently accepting manuscripts for publication within this series. To submit a proposal for a volume in this series, please contact our Acquisition Editors at Acquisitions@igi-global.com or visit: http://www.igi-global.com/publish/.

The Advances in Chemical and Materials Engineering (ACME) Book Series (ISSN 2327-5448) is published by IGI Global, 701 E. Chocolate Avenue, Hershey, PA 17033-1240, USA, www.igi-global.com. This series is composed of titles available for purchase individually; each title is edited to be contextually exclusive from any other title within the series. For pricing and ordering information please visit http://www.igi-global.com/book-series/advances-chemical-materials-engineering/73687. Postmaster: Send all address changes to above address.

Titles in this Series

For a list of additional titles in this series, please visit: www.igi-global.com

Research Perspectives on Functional Micro- and Nanoscale Coatings
Ana Zuzuarregui (CIC nanoGUNE Consolider, Spain) and Maria Carmen Morant-Miñana (CIC nanoGUNE Consolider, Spain)
Information Science Reference • copyright 2016 • 511pp • H/C (ISBN: 9781522500667) • US $215.00 (our price)

Emerging Research on Bioinspired Materials Engineering
Mohamed Bououdina (University of Bahrain, Bahrain)
Engineering Science Reference • copyright 2016 • 441pp • H/C (ISBN: 9781466698116) • US $215.00 (our price)

Petrochemical Catalyst Materials, Processes, and Emerging Technologies
Hamid Al-Megren (King Abdulaziz City for Science and Technology, Saudi Arabia) and Tiancun Xiao (Oxford University, UK)
Engineering Science Reference • copyright 2016 • 539pp • H/C (ISBN: 9781466699755) • US $225.00 (our price)

Applying Nanotechnology to the Desulfurization Process in Petroleum Engineering
Tawfik A. Saleh (Department of Chemistry, King Fahd University of Petroleum and Minerals, Saudi Arabia)
Engineering Science Reference • copyright 2016 • 555pp • H/C (ISBN: 9781466695450) • US $235.00 (our price)

Quantitative Structure-Activity Relationships in Drug Design, Predictive Toxicology, and Risk Assessment
Kunal Roy (Jadavpur University, India)
Medical Information Science Reference • copyright 2015 • 703pp • H/C (ISBN: 9781466681361) • US $295.00 (our price)

Handbook of Research on Recent Developments in Materials Science and Corrosion Engineering Education
Hwee Ling Lim (The Petroleum Institute, UAE)
Engineering Science Reference • copyright 2015 • 493pp • H/C (ISBN: 9781466681835) • US $325.00 (our price)

Processing Techniques and Tribological Behavior of Composite Materials
Rajnesh Tyagi (Indian Institute of Technology (BHU), Varanasi, India) and J. Paulo Davim (University of Aveiro, Portugal)
Engineering Science Reference • copyright 2015 • 464pp • H/C (ISBN: 9781466675308) • US $225.00 (our price)

Handbook of Research on Diverse Applications of Nanotechnology in Biomedicine, Chemistry, and Engineering
Shivani Soni (Alabama State University, USA) Amandeep Salhotra (City of Hope National Medical Center, USA) and Mrutyunjay Suar (KIIT University, India)
Engineering Science Reference • copyright 2015 • 820pp • H/C (ISBN: 9781466663633) • US $395.00 (our price)

IGI GLOBAL
DISSEMINATOR OF KNOWLEDGE
www.igi-global.com

701 E. Chocolate Ave., Hershey, PA 17033
Order online at www.igi-global.com or call 717-533-8845 x100
To place a standing order for titles released in this series, contact: cust@igi-global.com
Mon-Fri 8:00 am - 5:00 pm (est) or fax 24 hours a day 717-533-8661

To Kerstin and Tatiana

Table of Contents

Preface

EXPLORATION OF PHOTOELECTRIC AND PHOTOIONIZATION EFFECTS IN GASEOUS PHOTODETECTORS

The photoelectric emission was discovered a long time ago. In1887 Hertz observed that sparks can be produced between two electrodes if the gap is irradiated by ultraviolet (UV) radiation. His discovery attracted a great attention and was continued by several investigators, e.g. Hallwach, Righi et al. It was proven that a clean surface of zinc, if charged negatively, loses its charge when exposed to the UV light, whereas if this electrode is charged positively, it retains its charge. Later detailed studies were performed by Stoletov (in the period between 1888 and 1891) who observed the direct proportionality between the intensity of the incident light and the induced photocurrents.

It is remarkable that these early works were made before the discovery of the electron. Only after many other investigations, including the one performed by Thomson, it was realized that the observed photoelectric effects can be explained in terms of electron emission induced by light. In the framework of classical physics, the photoeffect was interpreted as transfer of light energy to the metal, which lead to the increase of the kinetic energy of its electrons and subsequent thermal emission. According to this explanation the photoeffect should depend on the light intensity. However, Lenard discovered experimentally that below a certain light frequency there is no electron emission regardless how intense the light is. Only above the threshold frequency the photocurrent was proportional to the light intensity.

A quantum theory of this effect was developed in 1905 by Albert Einstein, who found a mathematical relationship between the energy of photon E_v=hv (where v is the light frequency and h is the Plank constant), the photocathode work function φ and the maximum energy of the released photoelectrons E_k,

$$E_k = E_v - \varphi.$$

For this ground breaking theory, which well explained the experimental data, Albert Einstein was awarded by a Noble prize in physics in 1921.

Although the importance of this theory is difficult to overestimate, until 1929 the practical role of the photoeffect was very limited (mainly it was exploited in photometry) since the quantum efficiencies of the known materials were quite low. In 1929 an Ag-O-Cs photocathode was introduced, which had much higher quantum efficiency and was sensitive not only to UV, but also to visible and infrared light (Koller, 1930; Campbell, 1931). Cs-based photocathodes immediately found many applications, one of the most impressive at this old time was the reproduction of sound from film. The most important, however, was the development of some photoemissive devices: vacuum photomultiplier tubes, iconoscopes etc. This lead to a technical revolution, for example appearance of electronic television cameras capable of operating at low light fluxes.

Later some other highly efficient photocathodes were introduced including cesium, antimony, and multialkali. This opened the way for developing various vacuum photodetectors, e.g. photodiodes and photomultipliers having quantum efficiencies for UV, visible or infrared photons close to 30%. Photomultiplier tubes combined photocathodes with a dynode amplification structure exploiting secondary electron emission. The multiplication factor of such dynode system may reach values above 10^5 enabling detection of single photoelectrons. Unfortunately, the position resolution of these detectors was determined by the size of photocathodes which varied from 1 to several cm.

For a long time vacuum photomultipliers were really unique. There were practically no other devices capable to compete with them. The situation changed dramatically at the end of 1970 when for the first time position-sensitive gaseous photodetectors were introduced (Séguinot, 1977; Bogomolov, 1978). In these detectors three new principles were combined:

1. Gas photoionization,
2. Avalanche multiplication, and
3. Segmented electrodes allowing to measure a 2-dimensional position of the absorbed UV photons.

The gas photoionization is based on the fact that photons with energy higher than the ionization potential of the gas ionizes the atoms or molecules so that free photoelectrons and positive ions are created (see Chapter 1 for more details). Since the ionization energy of atoms and molecules is larger than their work function, $E_i > \varphi$, the first designs of gaseous photomultipliers had high sensitivity only for far ultraviolet light (VUV), i.e. photons with energy between 5 and 25 eV.

In most gaseous detectors, the primary (photo)electrons experience a avalanche multiplication process. This happens in a high enough electric filed so that the electron can gain sufficient kinetic energy for impact ionization (see Chapter 2). As a result of the avalanche multiplication the collected charge on the segmented anode can reach value as high as10^6 electrons (in some cases even more) making it is easy to detect photoelectrons electronically.

Avalanche electrons approaching segmented anodes induce signals on them as well as on segmented cathode. From the measured profile of these signals the avalanche position can be determined with high accuracy, often below 0.1 mm. The typical example of such a design is the multiwire proportional chamber (MWPC), considered in Chapter 3. Photosensitive MWPCs immediate found applications in high-energy physics and astrophysics, mainly in detection of Cherenkov light, and in plasma diagnostics (see Chapter 15).

The next breakthrough happened when the MWPC was combined with solid photocathodes, e.g. CuI (Peskov, 1980; Francke, 2004) allowing to extend the sensitivity of such detectors to longer wavelength, up to 350 nm. For a decade the photosensitive MWPCs were the only position-sensitive detectors of UV and VUV radiation.

Later, gaseous detectors were combined with photocathodes sensitive to visible light (Ag-O-Cs, cesium, antimony and multialkali) (see Chapter 7). These detectors had quantum efficiencies comparable to vacuum photomultipliers making them attractive also for measurements in the visible range of the spectrum.

By the end of 1980-ies the first vacuum position-sensitive detectors appeared on the market. They were microchannel plates and multianode photomultipliers (see Introduction and Chapter 8 for details).

In the 1990-ies, solid state photodetectors operating in avalanche mode started to be used in some experiments. Although only relatively small size matrix ($3x3mm^2$) of these detectors are commercially available (see Chapter 8) they are considered as very promising.

So, what is the role of gaseous photomultipliers among other detectors?

The main advantages of gaseous detectors are:

1. They can be built with very large sensitive areas,
2. They have low cost and relatively simple designs,
3. They are practically insensitive to magnetic fields,
4. Some of their designs are capable of performing measurements in narrow spectral intervals without the use of monochromator or filter.

THE CHALLENGES

It was not an easy task to develop gaseous photodetectors capable of competing with commercial devices.

The main challenges were:

1. To find photosensitive liquids/gases having at the same time the lowest possible ionization potentials and high enough vapor pressure, and select among them those which are not extremely chemically aggressive so that they can be used in practice.
2. To find ways of handling them, which includes storage, transfer technique, cleaning etc.
3. To find detector materials compatible with the photosensitive vapors.
4. Develop large area gaseous photomultipliers and associated electronics.
5. Developing solid photocathodes sensitive from UV to visible light and withstanding contact with gases.
6. Develop a technology of installing these photocathodes inside the gaseous detectors.
7. To find an optima gas amplification structure capable of operating at high gas gains, and being combined with solid photocathodes.
8. Develop not only gas-flushed detectors, but also sealed ones.

Let us briefly comment on these challenges. It is clear that various compounds having low ionization potentials can be found through the search in the literature or through personal contacts with chemists specialized on the subject. Unfortunately, most of these substances are chemically aggressive which is directly related to their low ionization potential. The outer electrons are weakly bound to the molecule. This is one of the difficulties. Another one is to select from them those which have high enough vapor pressures. Note that as a rule these substances are either liquid or solid. To use them in practice is not a trivial task since their vapor pressure is usually rather low, which leads to low light absorption and as a consequence to low practical quantum efficiency. The next difficulties were how to clean these substances from impurities capable of absorbing free electrons, which also reduce the quantum efficiency, and how to transfer them into the gaseous detectors. This required development of a special technology. More over one have to select proper materials for the detectors and to perform long-term tests to be sure they will not be damaged due to contact with the chemically active vapors.

Another great task was to construct and assemble large area gaseous photodetectors, flushed with mixtures containing low ionization potential vapors. One of the problems was to find the way to suppress the photon feedback caused by the avalanche light emission. The feedback reduces the maximum achievable gains. One also had to develop special low noise electronics.

The next challenge was to combine gaseous detectors with solid photocathodes which allowed to extend their sensitivity to longer wavelengths. This required studying known photocathodes, used in vacuum devise, in gas atmosphere. The problem was (and still is) that most of these photocathodes degrade fast if tiny impurities of oxygen or water are present in the gas mixture. They are also easily damaged under ion bombardment.

Traditional gaseous detectors, for example MWPCs combined with highly efficient photocathodes sensitive to long wavelength, suffer from photon and ion feedback problems which becomes much stronger than in the case of photosensitive vapors. The challenge was to develop appropriate amplification structures capable efficiently suppressing both photon and ion feedback.

Finally, the most difficult challenge with regard to the gaseous detectors sensitive to visible light was the development of sealed detectors. It requires finding appropriate materials which does not outgas anything during long times. For commercial applications only sealed gaseous photomultipliers may represent any interest as a continuous gas flow is troublesome.

SEARCHING FOR SOLUTIONS

The search for gases with low ionization potential started immediately after position sensitive photodetectors were discovered (Séguinot, 1977; Bogomolov, 1978). It resulted in that several promising liquids with low ionization potentials were identified and tested, e.g., trimethylamine, triethylamine, ethylferrocene tetrakis (dimethylamino) ethylene (TMAE), see Chapter 3.

Much more difficult was to find ways to handle these liquids. For example, TMAE, which was the most attractive due to its low ionization potential, is chemically aggressive, it has low vapor pressure at room temperature and requires cleaning prior the use in detectors. These serious technological problems were overcome through the hard works of a quite large community of specialists: physicists, engineers, chemists etc. The problem TMAE's chemical aggressiveness was solved by a careful selection of detector materials. The problem of low efficiency to VUV photons was solved by heating the entire detector, including the TMAE injection system, to temperature up to 50-60°C. To clean the TMAE, a special setup was built at CERN serving various experiments at CERN and elsewhere.

Of course, during the development of the TMAE-flushed MWPC some unexpected small and big problems were revealed. One of them was the accumulation of polymer depositions on the anode and cathode wires leading to a slow degradation of the detector performance (a so called aging effect). A special research program was launched to attack this problem. Another problem was the photon feedback. This effect was diminished by a special geometry of detectors, by optimization of the gas mixture and by developing low noise electronics.

As a result of all these measures, large detectors were built and successfully used in various experiments at CERN and elsewhere (see Chapters 10-13). Some of them operated in time projection mode allowing to obtain 2D and 3D images of Cherenkov photons (see Chapter 10). Gaseous detectors filled with photosensitive vapors allowed obtaining important results in high energy physics, astrophysics and plasma diagnostics experiments.

The next important step in the evolution of gaseous photomultipliers was the development of gaseous amplification structures combined with solid photocathodes which allowed to extend their sensitivity beyond the ionization threshold of TMAE.

Note that solid photocathodes can be divided in two main categories:

1. Sensitive only to VUV and UV light, for example CuI, CsI.
2. Sensitive to UV and visible light e.g., SbCs, bialkali.

As a rule, photocathodes belonging to the first category are more robust. They even withstand a short contact with air. This is why UV sensitive photocathodes were the first to be used in practice.

One important problem to overcome was the photon feedback, another one the photoelectron back diffusion. The first is caused by secondary photons produced in the avalanche process in the detector illuminating the photocathode and producing secondary pulses. The second is due to photoelectrons extracted from the solid photocathodes due to diffusion, which return back to the cathode reducing its quantum efficiency. Both of these problems were solved by the gas optimization. It turned out that one of the best gases for this purpose is methane, CH_4.

More complications arose and were overcome when constructing gaseous photomultipliers sensitive to visible light.

These photocathodes degrade fast if the gas contains even small traces of oxygen, water or some other chemicals. For this reason, photocathodes sensitive to visible light is unpractical to use in gas-flushed mode. They are acting as getters for impurities continuously coming with the gas and are eventually irreversibly damaged. Therefore, the best approach is to use them in sealed mode. The development required finding means for photocathode manufacturing, development of sealed gas chambers filled with ultraclean gases made of materials with very low outgassing properties, development of gas amplification structures capable suppressing photon and ion feedbacks etc.

Although the technology of manufacturing of solid photocathodes is described in some publications and books (see for example Sommer, 1968) there are many important details which physicist dealing with gaseous photomultipliers discovered and solved via their own experience.

To make visible photocathodes more robust for contact with gases it was tested to cover them with a thin protective layer, having some transparency for photoelectrons created from the photocathode, but strongly reducing the gas diffusion through the film (see Chapter 7). This approach, indeed, provide remarkable protection, but the price to pay was the loss of some photoelectrons in the layer and thus decrease of detection efficiency especially in long wavelengths.

The success of the gaseous photomultipliers sensitive to visible light was impossible without developing gas amplification structures capable suppressing photon and ion feedbacks. In first prototypes glass capillary tubes operating in avalanche mode were used. Later they were replaced by a more sophisticated structure consisting of several hybrid hole-type structures allowing to reach higher gas gains than with capillary plates, close to10^6 (see Chapter 7).

Above only a few examples of challenges that has been overcome are presented. One can imagine that in reality many more problems arose and were solve before gaseous photomultiplier became mature devices capable competing with other types of photodetectors.

ORGANIZATION OF THE BOOK

The book describes the development of these detectors, as well as practical tips how to make them work in the best possible way.

In Chapter 1 two main principles of conversion photons to photoelectrons used in gaseous photodetectors are considered: gas photoionzation and a photoelectric effect from solid photocathodes. There are also descriptions of liquid photocathodes which offer lower ionization thresholds compared to the corresponding vapors. The great success of solid photocathodes covered with an adsorbed layer of some photosensitive vapors is described in detail. The thin layer of an absorbed photosensitive vapour reduces the cathode work function and as a result it extends the photosensitivity threshold towards long wavelengths and also enhance their quantum efficiencies, sometime by a factor of two.

Chapter 2 describes the process of electron avalanche multiplication of photoelectrons allowing to reach electrical signals sufficient for detection single primary electrons. Electron avalanche multiplication can advantageously be used in position sensitive photomultipliers to with high efficiency detect single photons with respect to position, time, direction and energy.

In Chapter 3 position-sensitive gaseous photomultipliers filled with photosensitive vapors are described. The ultimate goal is not only to detect photons with a high efficiency, but also to determine their position with an accuracy often down to a few micrometers. In many measurements it is important to know when the photon was detected with a precision of less than a nanosecond. The detectors should have excellent stability. Their performance should not degrade with time.

Chapter 4 is dedicated to an exotic photosensitive compound, liquid photocathodes which were studied intensely for a number of years around 1988. The motivation was driven by the fact that the threshold energy for photoionization in liquids is generally lower than in gases owing to the polarization energy of the positive ion and the lowering of the electron conduction band energy. Moreover, it was observed that the gaseous compounds used as photocathodes while absorbed on surfaces, makes them photosensitive even if the thickness of this layer was extremely thin, just a few monolayers. This opened up the dream of a well-defined photosensitive layer. In a photosensitive detector with a gaseous photoelectric converter it is difficult to know where each individual photoelectron is actually emitted. The conversion volume has to be made thick enough to allow efficient conversion of the incoming photons and this smears the position resolution, and reduces the time resolution. A thin layer photocathode would eliminate this smearing in space and time. Furthermore, the gas system might be simplified, or even removed with such a liquid cathode. Chapter 4 summarize the results of these studies, which led to the important development related to solid photocathodes in gaseous detectors.

Chapter 5 describes the early work on UV sensitive solid photocathodes for gaseous detectors. A solid photocathode is advantageous over gaseous and liquid photocathodes due to its well defined emission point of the photoelectrons. Gaseous detectors with solid photocathodes are simpler to manufacture, are sensitive to longer wavelength and have better time resolution than detectors filled with photosensitive gases. They should be able to operate at high gas gains and detect single photoelectrons. Finally, compared to vacuum photomultipliers they have little sensitivity to magnetic fields and can measure the coordinate of photon conversion and can be made with large areas and arbitrary shapes. All this leads to that gaseous photomultipliers with solid photocathodes open a new page in photosensitive detection technique. Chapter 5 reviews the early development of solid photocathodes and the road towards the modern photocathodes which have revolutionized the photosensitive detector technology.

In Chapter 6 UV-sensitive CsI and Cs_2Te photocathodes are described. The development of CsI and Cs_2Te solid photocathodes present the breakthrough in photosensitive gaseous photodetectors we all had been waiting for, and put so much effort in finding. It was actually going back to basic. CsI had for a long time been used in vacuum photomultipliers. Recent studies proved to also work well in gas flows, and can even be exposed to air for shorter times. Hence, it is rather easy to handle compared to other photocathodes. For example, it can easily be mounted inside avalanche gaseous detectors in a glove box, and even in air. Due to these advantages, the CsI-based photon detectors represent the most effective solution concerning cost and performance for large detector area applications in relatively low rate (or low occupancy) experiments. The main application of CsI-based photodetectors is as Cherenkov ring imaging detectors. For example, large area CsI-MWPCs are used in the ALICE (~11 m^2) and COMPASS (5.8 m^2) experiments at CERN.

In Chapter 7 the latest generation of gaseous photomultipliers, sensitive up to visible region of spectra are described. They represent the most exciting and promising developments which may open the possibility of building large-area positon sensitive gaseous photomultipliers. Laboratory and commercial prototypes of such detectors were developed and extensively tested. Typically they consist from a sealed envelope housing a cascaded hole-type structure combined with reflective or semitransparent photocathodes: SbCs or bialakali. Sealed gaseous photomultipliers sensitive to visible light now have high quantum efficiency comparable to commercial vacuum photodetectors and are stable in time. In contrast to vacuum devices, they have little sensitivity to magnetic fields, which is an important advantage allowing them to be used in many applications.

In Chapter 8 a review of alternative position-sensitive vacuum and solid state detectors is presented. This includes microchannel plates, multi-anode photomultipliers, solid–state avalanche detectors and hybrid detectors combining in one design two or more detection principles or sensors.

These types of detector appeared after the development of gaseous photomultipliers filled with photosensitive gases, they start to conquer the market of photo-sensors and applications, and represent now a serious alternative to gaseous detectors.

Chapter 9 describes the history and physics of Cherenkov light, the light emitted in a medium as a charged particle traverses it with a speed greater than the phase velocity of the light. The most important feature of this light is that it is emitted in a cone at a specific angle to the track, depending of the velocity of the incoming particle. This means that by measuring this angle one can determine the velocity of relativistic particles with good accuracy. Combining this information with information about the particle's momentum, the mass of the particle is known and, since the mass of each particle is unique, the particle is identified.

The Cherenkov effect has been used to identify particles ever since its discovery, and is becoming an even more important part of particle physics experiments of today, since new technology has provided better and better equipment to detect the faint light. These detectors are known as Cherenkov detectors.

Chapter 10 describes the history of development of practical devices exploiting the Cherenkov effect and used for particle identification. Slowly, starting from the first small prototypes, bigger and more sophisticated devices were built and used in many experiments. A real breakthrough happened after introduction of Ring Imaging Cherenkov detectors (RICH). However, a full scale implementation of these new detectors would be impossible without a complimentary work, the search for vapors having low ionization potential and compatible with gaseous detectors such as a multiwire proportional counter. As a result of these efforts two useful vapors were identified: Triethylamine and Tetrakis(dimethylamino)ethylene.

After solving the technical problems associated with the chemical aggressivety of tetrakis(dimethylamino) ethylene the latter was successfully used in such a giant detector as DELPHI (Detector with Lepton, Photon and Hadron Identification) RICH as well as in many other detectors.

Note that most of RICH detectors are quite complicated to construct and operate. Chapters 11-14 describe in more detail examples of such detectors and their performance.

Chapter 11 is dedicated to the NaF RICH detectors used in CP LEAR (charge parity violation studies at Low Energy Antiproton Ring) experiment at CERN. In this particular RICH design the NaF radiator is used together with a quartz window separating the drift volume from the photosensitive MWPC flushed with a gas mixture containing tetrakis (dimethylamino) ethylene vapors. The use of these vapors as photo converter allowed to achieve at the same time good separation power and high detector efficiency. To avoid having a large conversion region the detector has to be heated to 40° - 50°C. In the MWPC the pad readout was used to give a non-ambiguous two-dimensional reconstruction of the hit pattern.

The separation power of the NaF RICH counter is superior to conventional methods like time of flight, dE/dx, and threshold Cherenkov counters, etc. It is fast and compact and represents only about 20% of a radiation length, and can therefore be used as an efficient particle identifier in the medium energy range ($p_{proton} < 5$ GeV/c).

In Chapters 12 and 13 examples of RICH detectors used in astrophysics balloon born experiments are given.

In Chapter 12 the Cosmic AntiParticle Ring Imaging Cherenkov Experiment (CAPRICE) is described. It used for the first time a RICH detector combined with magnet spectrometer that is sensitive to unit charged particles (previously, the Chicago University group had flown a gas-RICH detector sensitive to helium and higher charges).

The aim of the experiment was to measure the flux of antimatter, mainly antiprotons and positrons, and light isotopes in the cosmic radiation. This RICH detector used a NaF solid radiator, tetrakis (dimethylamino) ethylene vapor as photo-converter and cathode pad readout in the photosensitive MWPC operated at low gain. Besides particle identification on an event-by-event basis it efficiently rejects multi-particle events and albedo particles.

In Chapter 13 the next generation of the CAPRICE experiment is described. The science objectives were to measure the antimatter component in the cosmic rays in the energy range 2–50 GeV and to study the cosmic ray composition in the atmosphere in the energy region between a few hundred MeV and 330 GeV. The main goal of the gas-RICH detector was to identify antiprotons in a large background of electrons, atmospheric and locally produced negative pions and muons. Because of the large number of detected photoelectrons and the low noise level in the MWPC, a good velocity identification was possible by the induced signal on the pad plane originating from the Cherenkov light. By combining the RICH detector information on Cherenkov angle with the rigidity information in the magnetic spectrometer and the identification capability of the imaging calorimeter, particles were fully identified over large energy ranges.

Chapter 14 deals with the latest generation of RICH detectors based on a CsI photocathode coupled to a gaseous detectors. Large area CsI-MWPC were/are successfully used in several experiments at CERN and elsewhere. For example, an active area of the CsI RICH operating in ALICE (A Large Ion Collider Experiment) experiment is about 11 m^2. All these detectors have shown to be efficient and stable over long periods of time.

After the invention of micropattern gaseous detectors CsI- based photo-sensors gained new momentum and large area CsI-gas electron multipliers made of Kapton were implemented in the PHENIX hadron blind detector. A so-called CsI-thick gas electron multipliers manufactured from the standard printed circuit will be installed in 2017 at the COMPASS upgrade RICH.

In Chapter 15 the main applications of gaseous photomultipliers beyond RICH detectors are described. They include applications in spectroscopy, plasma diagnostic, astrophysics, flame detection, readout gaseous and solid scintillators, and cryogenic detectors. Their advantages are described and compared with alternative techniques.

Some photosensitive substances (e.g. tetrakis (dimethylamino) ethylene vapors, CsI, etc.) were/are used also in other type of gaseous detectors, e.g. light imaging chambers and in gaseous multipliers exploiting a secondary electron emission effects. As follow from this chapter gaseous photomultipliers are unique for some applications.

THE PLACE OF GASEOUS PHOTOMULTIPLIERS IN APPLICATIONS AND AMONG ALTERNATIVE DETECTORS

Introduction of gaseous photomultiplier was an important step in evolution of photosensors. They were the only position-sensitive detectors capable detecting single UV photons. At this time the only other devices were small non-position sensitive vacuum photomultipliers. Moreover, since they were mainly operated at pressure of 1 atm, there was no special mechanical constrains on the window size and the chamber wall strength. This enabled large area detectors. None of the existing detectors at this time could compete with UV sensitive gaseous photomultipliers.

With time appeared commercial position-sensitive vacuum and solid state detectors, but they had / have relatively small sensitive areas. Hence, the UV gaseous photomultipliers remained to be leaders in cost-effectiveness for large area coverage applications. They have also other important features: insensitivity to magnetic fields (up to 4 T), low cost, the possibility to choose construction materials with low radioactive background etc. For these reasons they were employed in the largest RICH detectors ever built, the DELPHI RICH and the SLD CRID, successfully operated between 1987 and 2000.

Later, the development of large-area CsI photocathodes made another breakthrough in Cherenkov photodetection allowing the transition from the photoionization in gas to photoelectron extraction from solid photocathodes in respectively faster MWPCs. Nowadays the majority of large RICH systems presently in operation in high energy physics are still based on such a technology. The necessity to increase the counting rate capability for future applications (for example for the COMPASS RICH) has produced a further evolution of gaseous photomultiplier and all new developments are based on micro-pattern gaseous counters, combining fast detector operation with low cost per unit of surface.

Promising results have also been obtained from R&D on micro-pattern photomultipliers using photocathodes sensitive to visible light. Commercial prototypes were build and their preliminary tests demonstrated good performance, so these detectors also may have a great feature. We hope that this book will be useful for students, engineers, researches, professors and technicians.

REFERENCES

Bogomolov, G.D., & Dubrovski, Yu, V., & Peskov, V.D. (1978). Multiwire gas counter for coordinate measurements in the VUV region. *Instruments and Experimental Techniques*, *21*, 778–782.

Campbell, N. R. (1931). The photoelectric emission of thin films. *Philosophical Magazine*, *12*, 173–185.

Francke, T., & Peskov, V. (2004). Photosensitive gaseous detectors and their applications. *Nuclear Instruments & Methods in Physics Research. Section A, Accelerators, Spectrometers, Detectors and Associated Equipment*, *525*, 1–5.

Koller, R. L. (1930). Photoelectric emission from thin films of caesium. *Physical Review*, *36*, 1639–1647.

Peskov, V. (1980). *Complex/multiparameter plasma diagnostics in UV and X-ray regions*. (Doctor of science thesis). Kapitza Institute for Physical problems, Moscow, Russia.

Séguinot, J., & Ypsilantis, T. (1977). Photo-ionization and Cherenkov ring imaging. *Nuclear Instruments and Methods*, *142*, 377–391.

Sommer, A. H. (1968). *Photoemissive materials: preparation, properties, and uses*. New York, NY: Wiley.

Acknowledgment

This book is based on our lifelong experience of developing gaseous detectors. During all these years we have had the opportunity to meet and work with a large number of very skilled and fascinating persons with a common interest in gaseous detectors of all kinds.

We hope that this book reflects the hard work and interesting discoveries made by all these persons. We are grateful for all the knowledge they have shared with us and the memorable moments we have had together at various research centers; CERN, Fermilab, P.L. Kapitza Laboratory, New Mexico State University, Royal Institute of Technology, Université de Strasbourg and many other interesting and creative scientific environments.

We would like to take the chance to specially thank some of those who have supported and encouraged us through the years: David Anderson, Amos Breskin, Per Carlson, Paolo Fonte, Robert Golden, Paolo Martinengo, Antonio Di Mauro, Eugenio Nappi, Anton Oed, Panagiotis Pavlopulos, Paulo Pichichi, Fabio Pietropaulo, Francois Piuz, Fabio Sauli, Jacques Séguinot, Steve Stochaj, Thomas Ypsilantis, Antonio Zichichi and many others.

None of the work would have been possible without the professional hands of Michel Bosteels, André Braem, Jacob van Beelen, Roger Bouclier, Eric David, Bob Hull, Daniel Lacroix, Rui de Oliveira, Roy Park, Miranda van Stenis, and many more of the skilled engineers.

Finally, we are especially grateful to Martin Suffert and Georges Charpak who introduced and trained us to the world of light, gases and electric fields.

Introduction

FUNDAMENTALS AND MAIN DESIGNS OF SINGLE PHOTON DETECTORS

This book is dedicated to gaseous photomultipliers—gas filled devices able of recording single ultraviolet (UV) and visible photons with a high position resolution. Nowadays, such detectors are mainly used in various research areas as electronic imaging devices. Combined with computers they are capable treating and storing imaging information of UV-light. Serious efforts are ongoing currently aiming to commercialize some of these designs. This will allow them to greatly expand their applications into various fields in the nearest future.

Before we identify the place for gaseous detectors among other types of ultraviolet and visible photons sensors, it will be useful to give a short historical review of the developments in the field of single photon detection.

There are various detectors for single photons. Their design depends on photons energy, the required sensitivity.

The first electronic detector of single photons (visible and UV) was the photomultiplier tube (PMT). The first prototypes of the PMTs were developed around 1930 by Kybetsky and his group, according to (Lubsandorzhiev, 2006 and Summer, 1957). This detector became a commercial product after the famous work of Zworykin and his team (Zworykin, 1936). In this device, primary electrons are produced by photoelectric effect between the photons and the cathode. These primary photoelectrons are attracted towards a second cathode where secondary electron emission of electrons occur which are attracted towards a third cathode and so on. The specially designed dynode system is capable of multiplying the single initial photoelectron by a factor of 10^3-10^6. This multiplication makes it possible to with a high efficiency detect single primary electrons, and hence single photons.

For a long time vacuum photomultiplier tubes were the only devices capable of detecting single ultraviolet and visible photons. Typically, these photodetectors had sensitive areas between 5 and 50 cm^2 and they were not position-sensitive.

The next major step appeared in the form of television (TV) tubes and image intensifiers. The TV-tubes was also developed based on Zworykin's efforts. They were the main imaging detectors of visible and infrared light at this time. However, due to various technical reasons they were not capable of recording electronically individual photons, and they were not sensitive to UV light.

In the 1950's another single photon UV detector appeared, the Geiger gas counter (Friedman, 1951). Although this type of electron multiplier had been invented already in 1908, and was widely used for the detection of high energy particles and X-ray photons, it was the first time it was exploited to detect UV photons. The first photosensitive Geiger gas counter was also based on the principle of photoelectric effect between the UV photons and a metal cathode. Later the metallic cathodes were replaced by photoionization of vapors with a low ionization potential, offering much higher sensitivity for ultraviolet light (Chubb, 1955).

In both these detectors, primary electrons are created from the UV photons. In the photomultiplier tube it is via photoelectric effect in a metallic cathode, and in the Geiger gas counter it is via photoionization of a photosensitive gas. In the Geiger gas counter, each primary photoelectron trigger a short (~1 ms) burst of corona discharges in the strong electric field applied in the gas, and as a result in that up to 10^8 secondary electrons are created and collected on an anode wire (see Chapter III for details). This results in a large amplitude electrical pulse which is fed into the input electronics and allows efficient detection of single photoelectrons and hence single photons. These detectors were not position sensitive either.

A new page in the development of photosensitive detectors was written after the invention of the position sensitive gaseous photomultiplier in the late 1970ies. This device is operating in avalanche mode in a gas mixture at a pressure of 1 atmosphere (Séguinot 1977, Bogomolov 1978). These detectors also exploited photoionization of vapors having a low ionization potential, but had two important new features:

1. They were multiwire proportional chambers (MWPC). These are detectors having a planar geometry with a multiwire anode instead of a single wire as used in the cylindrical Geiger gas counter.
2. In contrast to Geiger gas counters they operated at sufficiently low voltages between the anode and the cathode electrodes to still allow the primary photoelectrons to trigger Townsend avalanches, but not to trigger a corona discharge. Depending on the applied voltage, each primary photoelectron can create between 10^4 and 10^6 secondary electrons. Even with the electronics of that time, they could be detected with efficiency close to 100%. At these conditions avalanches are well localized in space and induce signals on the anode wires in the vicinity of the avalanche. This enables the position of the avalanche, and hence the initial photon, to be determined in the direction perpendicular to the anode wires with an accuracy of better than a few mm. The initial position sensitive gaseous photomultipliers could not determine the position along the wires.

In the early work, Séguinot used benzene vapors which are sensitive to UV light with wavelength $\lambda < 135$ nm (Séguinot, 1977), whereas Bogomolov successfully used toluene vapors which are sensitive to UV light with wavelengths $\lambda < 146$ nm (Bogomolov, 1978).

The quantum efficiency of these detectors defined as the number of created photoelectrons per incident photon reached up to 30% -40% in a narrow UV wavelength region, which is comparable to the efficiency of the best PMTs (Séguinot, 1977).

The long awaited first position-sensitive photomultiplier had been developed! For many years these photosensitive MWPCs were the only position sensitive detector capable detecting single UV photons with high efficiency. At atmospheric pressure there is no serious mechanical constrains on the window between the detector and the outside. Hence, the detector size and the active area of this photomultiplier can easily be several square meters per detector. Another important feature of the photosensitive MWPCs, compared to vacuum PMTs, was their low sensitivity to magnetic fields which are often used in high energy physics experiments.

Photosensitive MWPCs immediately became a very active area of development. This development is still going on. Various new multiplication structures have been developed, including the recently invented micropattern detectors (Francke 2002). The development of new photosensitive compounds has increased the sensitivity to single photons, comparable to the PMT, and has extended the sensitive spectrum into the visible range of spectra.

By the end of the 1980ies, microchannel plates (MCPs) were successfully combined with high efficiency photocathodes allowing the detection of singe photons with a position resolution close to or better than the photosensitive gaseous detectors (Anashin, 1995, and references therein). The microchannel plates had been invented already in the 1960ies and were used to detect charged particles (Wiley, 1962). In these photosensitive microchannel plates electron multiplication occurs due to the secondary emission of electrons on the inner walls of glass micro-capillary tubes. This type of detector also has a low sensitivity to magnetic fields. However, the active area of such device is small, around 10 cm^2, and it is very difficult and expensive to produce MCPs with larger active area.

At the same time photonics companies started producing the first vacuum multi-anode PMT which offered a position resolution of few mm (Salomon, 1992).

Some work was also done on so called hybrid photodetectors which used a new principle of generating secondary electrons (Ambrosio 2003). In this device, the primary photoelectrons are accelerated in vacuum to high kinetic energies and hit either a solid state detector or a scintillator. In the first case many secondary electron-hole pairs are produced in the solid state detector allowing detection of primary photoelectron with a high position resolution. In the second case a bust of scintillation light is produced and can be recorded by other means, for example with small PMTs.

The rapid progress in solid state technology offers new possibilities for detecting single UV and visible photons. One of the most exciting among them is the so called SiPM. This is a solid state detector exploiting a Geiger type of discharge in the p-n junction (Renker 2006). The quantum efficiency of such device is about 20%, similar to vacuum PMTs. The multiplication factor in Geiger mode is $\sim 10^6$, hence each primary photon produces $\sim 10^6$ secondary electron-hole pairs. This detector can be made with high granularity as all semiconductor devices. The individual pixel size can be as small as 25 x 25 μm^2. Similar to the MWPCs they are insensitive to magnetic fields. However, their price per unit of area is quite high which restricts their applications. The use of electron avalanches in semiconductor devices is a blooming field of development. For example, avalanche charge coupling device (avalanche CCD) for imaging visible and UV photons are under intense development and may end up soon as a commercial device (Holl 2006).

In this book we will do a detailed comparison between various types of photosensitive detectors, and analyze their advantages and disadvantages for different applications. Despite great competition coming from other type of photosensitive detectors, in many applications UV sensitive gaseous photomultipliers offer a simple and cost effective solution allowing one to build large area devices capable operating in strong magnetic fields.

In the last chapters we will also describe and exemplify a wide range of applications for various gaseous photomultipliers, from high energy physics and astrophysics experiments to plasma diagnostics and spectroscopy. From this it will be clear why position sensitive detectors of single photons are so important in science, technology, industry and our daily life.

REFERENCES

Ambrosio, C. D., & Leutz, H. (2003). Hybrid photon detectors. *Nuclear Instruments and Methods in Physics Research*, *501*, 463–498.

Anashin, V. V., Berschastnov, P. M., & Golubev, V. B. (1995). Photomultipliers with microchannel plates. *Nuclear Instruments and Methods in Physics Research, 357*, 103–109.

Bogomolov, G. D., Yu, V., Dubrovski, V., & Peskov, V. D. (1978). Multiwire gas counter for coordinate measurements in the VUV region. *Instruments and Experimental Techniques, 21*, 778–782.

Chubb, T. A., & Friedman, H. (1955). Photon counters for far ultraviolet. *The Review of Scientific Instruments, 26*, 493–498.

Francke, T., Peskov, V., & Rodionov, I. (2002). Novel position-sensitive gaseous detectors with solid photocathodes. *IEEE Transactions on Nuclear Science, 49*, 977–983.

Friedman, H., Lichtman, S. W., & Byram, E. T. (1951). Photon counter measurements of solar X-ray and extreme ultraviolet light. *Physical Review, 83*, 1025–1030.

Holl, P., Andritschke, R., & Eckhardt, R. (2006). A new high-speed, single photon imaging CCD for the optical. *IEEE Nucl. Sci. Conf. Record, 3*, 1589–1594.

Lubsandorzhiev, B. K. (2006). *On the history of the photomultiplier inventions.* Arxiv.org/pdf/physics/0601159

Renker, D. (2006). Geiger-mode avalanche photodiodes, history, properties and problems. *Nuclear Instruments and Methods in Physics Research, 567*, 48–56.

Salomon, M. (1992). New measurements of scintillating fibers coupled to multi-anode photomultipliers. *IEEE Transactions on Nuclear Science, 39*, 671–673.

Séguinot, J., & Ypsilantis, T. (1977). Photo-ionization and Cherenkov ring imaging. *Nuclear Instruments and Methods, 142*, 377–391.

Summer, W. (1957). *Photosensitors: A treatise on photoelectric devices and their application to industry.* London: Chapman Hall.

Wiley, W. C., & Hendee, C. F. (1962). Electron multipliers utilizing continuous strip surface. *Nucl. Sci. IRE Trans., 9*, 103–106.

Zworykin, V. K., Morton, G. A., & Malter, L. (1936). The secondary emission multiplier - A new electronic device. *Proceedings of the IRE, 24*, 351–375.

Chapter 1
Conversion of UV and Visible Photons to Photoelectrons

ABSTRACT

The operation of most of gaseous photomultipliers is based either on gas photoionization or on photoelectric effect from solid photocathodes. There have also been attempts to use liquid photocathodes which offer lower ionization thresholds compared to the corresponding vapors. A great success has been achieved with solid photocathodes covered with adsorbed layers of some photosensitive vapors which reduce the cathode work function and as a result extend the photosensitivity threshold towards long wavelengths. It also enhances their quantum efficiencies sometime on a factor of two. The main physic mechanisms of interactions of UV photons with gases as well as with liquid and solid photocathodes are described in detail in this chapter. This basic knowledge is important when designing and using gaseous photodetectors.

1. PHOTOIONIZATION OF GASES

Photons passing through a gas volume may experience absorption if their energy $E_\nu > E_{thr}$ where E_{thr} is a characteristic energy which depends on the particular gas. When absorption occurs, the intensity of the light beam along its propagation line will be attenuated. For example, if N_{ph} monochromatic photons enter the gas volume having a thickness of l then the number of photon exiting this volume, $N_{phex,}$ will be:

$$N_{phex} = N_{ph} \exp(-\sigma_{abs} N_{gas} l), \tag{1}$$

where σ_{abs} is the absorption cross section and N_{gas} is the number of the light absorbing atoms or molecules per cm^3.

As will be shown later, absorption plays an important role in operation of gaseous detectors as it affects their efficiency.

DOI: 10.4018/978-1-5225-0242-5.ch001

If the photon energy E_v is higher than the ionization potential of an atom or a molecule, E_i, (typically $E_i > E_{tr}$) it can photoionize the gas. In a general case this a complicated phenomenon which has many energy transfer channels and even may end up in the creation of few photoelectrons per absorbed photon.

In this chapter, for practical reasons, we will mainly be interested in the process leading to the emission of photoelectrons by photons with an energy E_γ in the range $E_{shell} > E_v > E_i$, where E_{shell} is the energy required to liberate an electron from the inner shell of a given atom.

Schematically this process can be represented in the following way:

$$h\nu + M = M^+ + e^-, \tag{2}$$

where $h\nu$ is a photon with energy $E_\gamma = h\nu$, (h is the Plank constant and ν is the frequency of the light), M and M^+ are the molecule and the molecular ion, respectively, and e^- is a free electron.

Even in the relatively narrow energy interval $E_{shell} > E_v > E_i$ the photon energy dissipation may occur via several mechanisms, leading to the fact that the ionization efficiency even in the case of full light absorption is often below 100%.

In applications such as gaseous photodetectors two parameters are especially important, E_i and the quantum efficiency (QE) defined as the number of created photoelectrons per incident photon. For historical reasons some academic authors prefer to use terms photoionization efficiency or photoionization yield, which are in fact the same as quantum efficiency.

Benzene was the first photosensitive vapor successfully used by Séguinot in his first photosensitive MWPC (Séguinot, 1977). One of the early measurements of the quantum efficiency of benzene vapors C_6H_6 and its isotope C_6D_6 (η_H and η_D, respectively) are presented in Figure 1. In the same figure are also shown the photo absorption cross sections for the same vapors. As can be seen, in the case of benzene vapors the photoionization starts at $E_\gamma > 9$ eV corresponding to wavelengths shorter than 138 nm. The quantum efficiency at high enough photon energies reaches 60%. The deuterium isotope has slightly higher quantum efficiency at all wavelengths than normal benzene. Please also note that the absorption cross section has no correlation with the quantum efficiency.

The quantum efficiency of the toluene vapors used in the early work by Bogomolov (Bogomolov, 1978) gave similar results (Carver, 1964).

Séguinot and Bogomolov demonstrated for the first time that high quantum efficiency and a rather good position resolution can be achieved simultaneously with a single device (Séguinot, 1977, Bogomolov, 1978). These results triggered a great interest to gaseous photomultipliers. A lot of work by various teams focused on searching gases and vapors having small E_i in order to extend the spectral sensitivity interval of these detectors. The great progress in this direction will be described further in Chapter IV.

2. LIQUID PHOTOCATHODES

It is well known that the ionization threshold of a non-conducting liquid E_{liq} is less than the ionization potential the corresponding vapor E_i (see for example Schmidt, 1997). This can be expressed as:

$$E_{liq} = E_i + P_+ + V_0 + E_{val}, \tag{3}$$

Figure 1. The quantum efficiency of benzene C_6H_6 (η_H) and its isotope C_6D_6 (η_D), and the absorption cross sections, σ_H and σ_D, measured for C_6H_6 and C_6D_6, respectively as a function of photon wavelength (Person, 1965). As can be seen, in the case of benzene vapors the photoionization starts at $E_\gamma > 9$ eV corresponding to wavelengths shorter than 138 nm. The quantum efficiency at high enough photon energies reaches 60%. The deuterium isotope has slightly higher quantum efficiency at all wavelengths than normal benzene. Please also note that the absorption cross section has no correlation with the quantum efficiency.

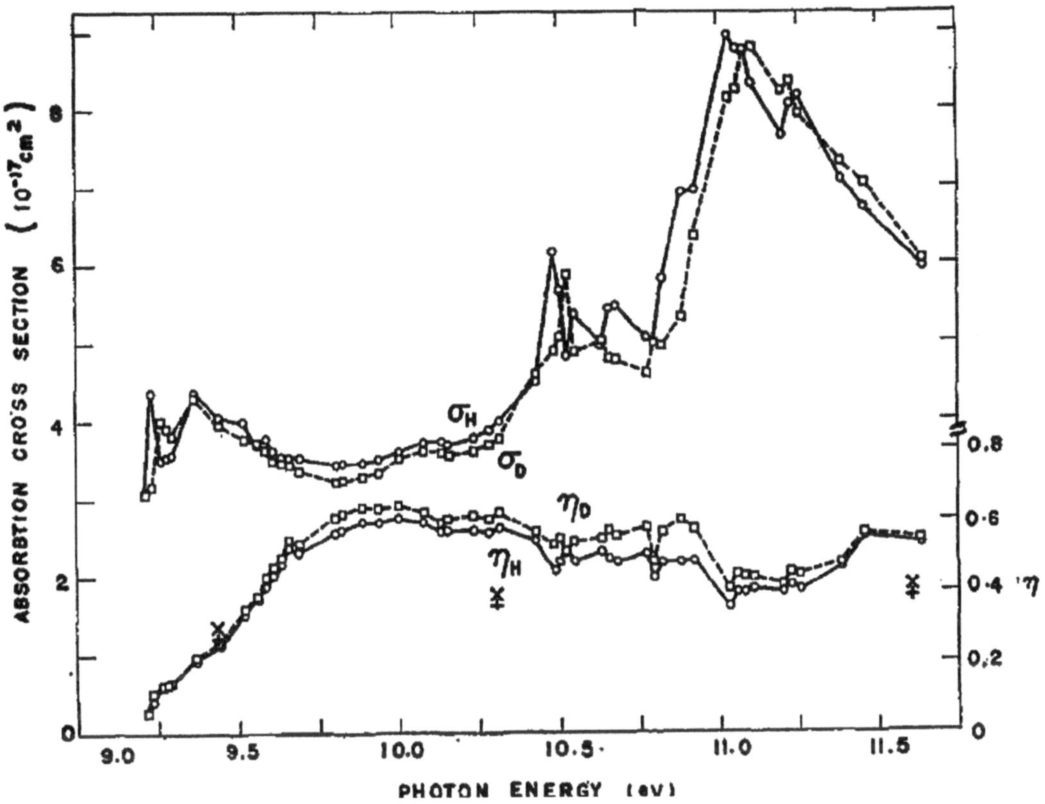

where P_+ is the polarization energy of the positively charged ion in the surface of the liquid (often referred to as surface tension), V_0 is the electron affinity of the liquid or the energy of the conduction band with respect to the outside vacuum, and E_{val} is a broadening of the levels of the valence electrons due to intermolecular forces.

Schematically the energy diagram of a non-conductive liquid compared to the corresponding gas is presented in Figure 2.

To our best knowledge the idea to exploit the fact the $E_{liq} < E_i$ in a practical detector came up by David Anderson. His famous paper (Anderson, 1982) triggered a chain of investigations of various liquid photocathodes (Peskov, 1988, Séguinot, 1990). The physics of photoelectron extraction from a liquid surface is well understood now.

One can consider two main practical cases for extracting photoelectrons from a liquid surface:

Figure 2. Simplified energy diagram of a non-conductive liquid compared to the corresponding gas. E_i is the ionization potential the corresponding vapor. P_+ is the polarization energy of the positively charged ion in the surface of the liquid, V_0 is the electron affinity of the liquid or the energy of the conduction band with respect to vacuum and E_{val} is a broadening of the levels of the valence electrons due to intermolecular forces.

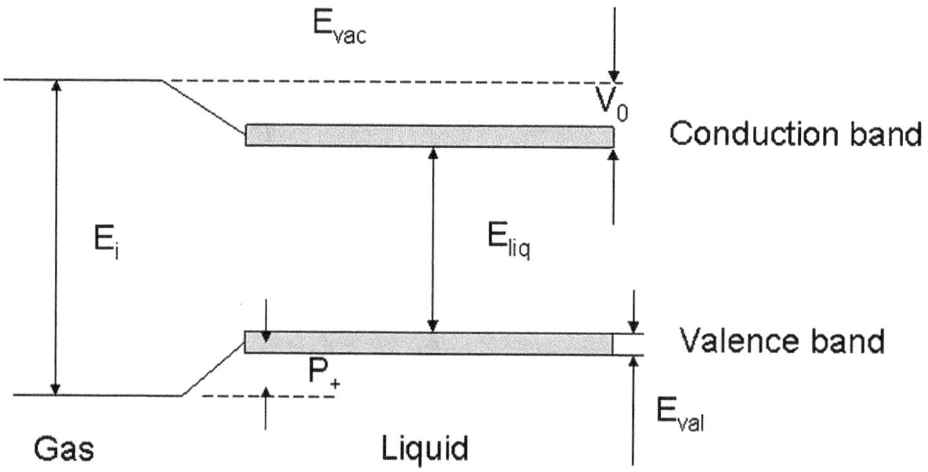

1. UV light illuminates the top layer of the liquid, or
2. UV light illuminates the bottom layer of the liquid via a UV-transparent window.

In the first case, photoelectric absorption occurs in a thin top layer of the liquid liberating a photoelectron in the liquid. If the photoelectron energy $E_k = h\nu - E_{liq}$ is higher than is the affinity V_0 it has some probability to escape from the liquid into the gas phase. An important factor facilitating the photoelectron to escape is if an electric field is applied across in the liquid-gas interface (Peskov, 1988). If so, the photoelectron gains additional kinetic energy while it drifts in the strong electric field and increases its probability to escape from the liquid into the gas phase above it.

In the second case, all photoelectrons will be created at the bottom layer of the liquid close the inner surface of the window. If no electric field is applied across the liquid, they will recombine with the positive ions. However, if a sufficiently strong electric field is applied over the liquid, electrons will be able to drift some distance inside the liquid before recombining. The length of this drift depends on the liquid and the level of impurities which efficiently trap electrons. If the liquid is clean enough some of the photoelectrons will reach the top surface of the liquid, and if their kinetic energy at that moment is larger than the affinity of the liquid, $E_k > V_0$, they will escape into the gas phase above the liquid (Bolozdynya, 1999).

3. SOLID PHOTOCATHODES

3.1 Metallic Photocathodes

Metallic photocathodes were widely used in studies of the photoelectric effect around 1900 and later, and even exploited in some early photosensitive devices. Einstein was the first to explain the photoelectric effect theoretically in 1905, for which he was awarded by the Nobel Prize in Physics in 1921. He established a relation between the maximum kinetic energy of a photoelectron E_{kmax}, the photon energy, and the metal work function φ:

$$E_{kmax} = h\nu - \varphi, \tag{4}$$

Metals usually have rather low quantum efficiency. There are three main reasons for this:

1. Metals have high optical reflectivity so only a fraction of the light penetrates to the inside of the metal and can cause photoelectric effect.
2. The photoelectrons rapidly lose their kinetic energy by collisions with other electrons in the metal and are recaptured in the metal's conducting band. The created photoelectrons have at the beginning a kinetic energy nearly equal to the absorbed photon energy. Actually, the initial kinetic energy of the photoelectron is $E_k = h\nu - E_b$, where E_b is the binding energy of the electron in the atom. Most photoelectrons emitted by ultraviolet and visible light are emitted from the valence band or the conduction band in the case of metals, where the binding energy E_b is low. Hence, one can with good accuracy in most cases assume that the initial kinetic energy of the photoelectron, E_k, is equal to the photon energy, $h\nu$.
3. The work function of metals is relatively high. Typical values are between 2 eV (Cs) and 5 eV. This means that only photons with energy higher than the work function are capable of extracting a photoelectron from the surface of the metal.

Metal's quantum efficiency increases with photon energy and can be between a few to ten percent in the ultraviolet region. This is because ultraviolet light has less reflectivity than visible light, and also because the created photoelectrons have higher kinetic energy so also electrons from the deeper inside the metal can contribute to the photoelectric effect.

3.2 Semiconductor Photocathodes

The intense studies of PMTs and various photocathodes performed in 1930 and later, demonstrated that much higher quantum efficiencies, compared to metals, can be obtained with semiconducting photocathodes. The physics behind this is the following. A photoelectron can be created inside a dielectric material if $E_\nu > E_g$, where E_g is the energy difference between the top of the valence band and bottom of the conductive band, see Figure 3. The "hot" photoelectron created inside a semiconductor will lose its excess energy via phonon scattering, a process in which the energy loss per interaction is much smaller than in the case of electron-electron collisions which occurs in metals. The photoelectron can escape the semiconductor if its kinetic energy is higher than $E_g + E_a$, where E_a is the electron affinity (similar to V_0 in liquids or the work function φ in metals). As a consequence, photoelectrons from deeper regions of the semiconductor can reach the surface with sufficiently high energy to also escape from the surface, whereas in metals only photoelectrons created close to the surface contribute to the photoemission. This is the main factor contributing to the high quantum efficiency of semiconductor photocathodes.

As a rule of thumb semiconductors with $E_g > E_a$ have higher quantum efficiency than those with $E_g < E_a$. This is because in the second case an electron created in the conduction band has sufficient energy

Figure 3. Energy diagram of the energy bands in an ideal semiconductor at low temperature; a photoelectron can be created inside a dielectric material if $E_v > E_g$, where E_g is the energy difference between the top of the valence band and bottom of the conductive band. The photoelectron can escape the semiconductor if its kinetic energy is higher than $E_g + E_a$, where E_a is the electron affinity.

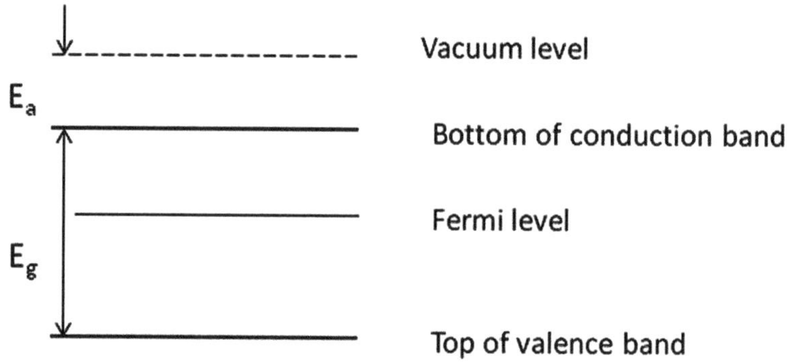

to raise a second electron from the valence the conduction band and then none of them may have sufficient energy to escape from the solid to the gas phase.

The diagram in Figure 3 represents the ideal case of a semiconductor. In real materials and at room temperatures, contributions from defects and impurities are unavoidable. These affect the electrical conductivity, the quantum efficiency and the photoemission threshold energy. Photoemission from defect levels may be caused by photons with energy less than $E_g + E_a$. Another possible effect associated to the defect levels is a so-called "band bending", which happens in for example p-type semiconductors with n-type surface state when the internal electric field is formed. In this case electrons should overcome only the effective electron affinity which is less than E_a.

There are of curse many other fine effects which contribute to the photocathode sensitivity and operation (see for example Pertsev, 1968).

Due to the higher quantum efficiency, in most of the modern single photon detectors only semiconducting photocathodes are used today.

3.3 The Effect on the Quantum Efficiency by Absorbed Layers on Solid Photocathodes

Early studies of the photoemission from metallic photocathodes in vacuum revealed that the quantum efficiency depends on the composition and the thickness of any absorbed layer of another material on the surface (McDaniel, 1964).

In the case of gaseous detectors this effect was also observed (Anderson, 1882, Peskov, 1988). Some absorbed molecules may form a double electric layer on the metallic or semiconducting surfaces which increases the QE and extend the photosensitivity threshold towards long wavelengths (Astruc, 1988, Charpak, 1988). In the case of a thick enough absorbed layer the latter may also act similar to a liquid photocathode. These effects will be considered in more detail in Chapter V where also its practical consequences will be discussed.

CONCLUSIVE REMARKS

As follows from this chapter, two types of photosensitive substances have high quantum efficiency for ultraviolet light: gases with small ionization potentials and semi conductive photocathodes. Highly conductive solid materials, like metals, normally have low quantum efficiency. For this reason modern gaseous photomultipliers are mainly using photosensitive gases or solid semiconducting photocathodes.

REFERENCES

Astruc, D., Charpak, G., & Mine, P. (1988). Organometallic compounds as photocathodes in gaseous detectors. *Proceeding. Int. Workshop on liquid-sate electronics.* Berlin: Hahn-Meitner Inst.

Bogomolov, G. D., Yu, V., Dubrovski, V., & Peskov, V. D. (1978). Multiwire Gas Counter for Coordinate Measurements in the VUV Region. *Instruments and Experimental Techniques*, *21*, 778–782.

Bolozdynya, A. I. (1999). Two-phase emission detectors and their applications. *Nuclear Instruments & Methods in Physics Research. Section A, Accelerators, Spectrometers, Detectors and Associated Equipment*, *422*(1-3), 314–320. doi:10.1016/S0168-9002(98)00965-6

Carver, J. H., & Mitchell, P. (1964). Ionization chambers for the vacuum ultraviolet. *Journal of Scientific Instruments*, *41*(9), 555–557. doi:10.1088/0950-7671/41/9/308

Charpak, G., Peskov, V., & Schmidt, W. F. (1988). Liquid or solid photocathodes based on TMAE or TMPD: physics and application. *Proceeding. Int. Workshop on liquid-state electronics.* Berlin: Hahn-Meitner Inst.

McDaniel, E. (1964). *Collision phenomena in ionized gases.* Wiley.

Person, J. C. (1965). Isotope Effect in the Photoionization Efficiency for Benzene. *The Journal of Chemical Physics*, *43*(7), 2553–2555. doi:10.1063/1.1697162

Pertsev, A. N., & Ruzin, V. A. (1968). O*n a change in the spectral sensitivity of a photocathode during a change in the potential. Journal of Applied Spectroscopy*, *8*(5), 517–518. doi:10.1007/BF01103244

Peskov, V., Charpak, G., Mine, P., Sauli, F., Scigocki, D., Séguinot, J., & Ypsilantis, T. et al. (1988). Liquid and solid organic photocathodes. *Nuclear Instruments & Methods in Physics Research. Section A, Accelerators, Spectrometers, Detectors and Associated Equipment*, *269*(1), 149–160. doi:10.1016/0168-9002(88)90872-8

Schmidt, W. (1997). Liquid state electronics of insulating liquids. CRC Press.

Séguinot, J., Charpak, G., Giomataris, Y., Peskov, V., Tischhauser, J., & Ypsilantis, T. (1990). Reflective UV photocathodes with gas –phase extraction: Solid, liquid and absorbed thin films. *Nuclear Instruments & Methods in Physics Research. Section A, Accelerators, Spectrometers, Detectors and Associated Equipment*, *297*(1-2), 133–147. doi:10.1016/0168-9002(90)91359-J

Séguinot, J., & Ypsilantis, T. (1977). Photo-ionization and Cherenkov ring imaging. *Nuclear Instruments and Methods in Physics Research*, *142*(3), 377–391. doi:10.1016/0029-554X(77)90671-1

KEY TERMS AND DEFINITIONS

Absorbed Layer: A monolayer or, more often, a layer consisting from several atomic/molecular layers attached due to the atomic/molecular forces to the surface of solid substance which can be a metal or a dielectric. This layer affect the work function or affinity respectively a metallic or a semiconducting solid photocathode.

Ionization Threshold: The minimal energy of photons (or correspondingly a maximum photon's wavelength) capable cause a photoionization process in the gas or liquid.

Liquid Photocathode: A thin layer of photosensitive liquid formed on the surface of a metallic cathode and serves as a photosensitive media.

Photoelectrons: Electrons created in the detector volume either due to the photoinization of the gas or extracted from the solid photocathode.

Photoemission: Emission of electrons from the any type of photocathode; gas, liquid or solid.

Semiconductor Photocathodes: Photo-convertors made of semiconductors, i.e. materials for which electrical conductivity values are between the conductors (such as metals) and insulators, e.g. glass.

Work Function: The minimum energy needed to move an electron from the cathode into vacuum or gas.

Chapter 2
Electron Multiplication and Electron Multipliers

ABSTRACT

This chapter describes the process of electron avalanche multiplication, which is the only method known to detect the smallest possible electrical signal, a single electron at rest. Electron avalanche multiplication can advantageously be used in position sensitive photomultipliers to with high efficiency detect single photons with respect to position, time, direction and energy.

1. ELECTRON AVALANCHE MULTIPLICATION IN GASES

Gaseous photomultipliers can detect singe photoelectrons due to avalanche multiplication in a gas where a large number of secondary electrons are created from a single photoelectron. Avalanche multiplication is an interesting phenomenon discovered a long time ago by Townsend (Townsend, 1901; Townsend, 1903a; Townsend, 1903b). In 1908 Rutherford and Geiger for the first time exploited electron avalanche multiplication in a practical device to detect alpha particles (Rutherford, 1908).

Let us first consider electron avalanche development in a uniform electric field. In this field a free electron is liberated, created by external radiation e.g. by photoelectric effect between a single photon and a gas which the electric field permeates in. If the electric field is strong enough, the single primary electron can gain sufficient kinetic energy E_k from the electric field to experience inelastic collisions with atoms or molecules in the gas.

If $E_k > E_i$, where E_i is the ionization potential of the gas atoms or molecules, a secondary electron can be created ionization of the gas in these collisions. This secondary electron, as well as the initial electron, gain further kinetic energy in the electric field, and after drifting some distance dx in the gas and again gaining a kinetic energy $E_k > E_i$, they collide with the gas atoms and produce further secondary electrons through ionization. An electron avalanche is triggered. This process can be described by the differential equation:

$$dn_e = \alpha\, n_e(x)\, dx, \quad (1)$$

DOI: 10.4018/978-1-5225-0242-5.ch002

where n_e is the number of secondary electrons and α is a coefficient often called a first Townsend coefficient.

The number of secondary electrons grows exponentially with the distance x:

$n_e = \exp(\alpha x)$, (2).

The value

$$A = \exp(\alpha x), \tag{3}$$

represents the multiplication factor or the gas gain. As can be seen the avalanche just increases and increases as long as it can propagate.

The Townsend coefficient, α, depends on the electric field strength, E, and the gas pressure, p. There are several parametric descriptions of the first Townsend coefficient (Aoyama, 1985). Most of them satisfactorily describe the experimental data in some ranges of E/p, with correctly determined parameters. The generalized form of the reduced first Townsend coefficient, α/p, is given by the expression:

$\alpha/p = A \exp(-B\ p/E)$, (4),

where *A* and *B* are parameters depending on the gas and the E/p range.

In the avalanche cloud, electrons drift two-tree orders of magnitude faster than the positive ions due to the much higher mass and larger interaction cross section of the ions (the electric field strength is the same for electrons and ions).So, during the avalanche development ions can be considered to a good approximation to be stationary. This is illustrated in Figure 1. First, an exponentially growing number of electrons are created and move fast towards the positive anode. They induce a fast and short pulse in the readout electrodes. When the electron pulse is gone, a cloud of positive ions slowly move towards the cathode producing a long and slow pulse.

At certain conditions a free electron can be extracted from the cathode during the recombination of the ions, triggering a second avalanche in the same place as the initial one.

In early devices the avalanches were generated in coaxial cylindrical electric field geometry. In these detectors a metallic cylinder served as a cathode with a thin anode wire in the center. In such geometry, the electric field strength is inversely proportional to the radius r:

$E = V/r \ln(c/a)$, (5),

where V is the voltage applied to the detector, a and c are the radius of the anode wire and the cathode cylinder, respectively. In this geometry, primary electrons created inside the detector volume drift toward the anode wire and in the region of strong electric field, close to the anode wire, they initiate Townsend avalanches.

At low voltage applied across the detector electrodes the amplification factor, A, can be expressed as:

$$\ln A = \int_a^c \alpha\, dr, \tag{6}$$

Figure 1. The development of a Townsend avalanche; the electrons move 100-1000 times faster than the heavier ions so the avalanche process is separated into two distinctive events separated in time. First, an exponentially growing number of electrons are created and move fast towards the positive anode. They induce a fast and short pulse in the readout electrodes. When the electron pulse is gone, a cloud of positive ions slowly move towards the cathode producing a long and slow pulse.

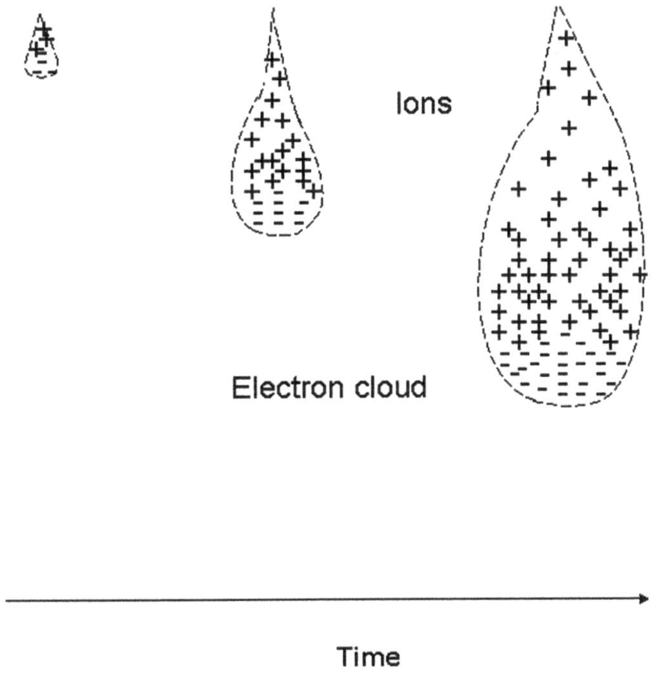

Later in this book will describe in more detail other multiplication structures which are widely used in modern designs of gaseous photodetectors (Francke and Peskov 2014).

2. SECONDARY PROCESSES DURING AVALANCHE DEVELOPMENTS

The picture of the avalanche development given above is simplified. It quite well describes the avalanche development at low gas gains, typically at A $<10^2$-10^3. At higher gas gains secondary effects start contributing to the amplification process.

The critical gain at which secondary processes appear depends on many factors, for example the gas composition. Most gaseous detectors are filled with a mixture of noble gases with molecular gases often called "quenching gases". The quenching gases are added to reduce the probability and effect of secondary processes.

The secondary processes are associated with the creation of electrons triggering new electron avalanches, not belonging to the original avalanche. These avalanches can be created in the gas itself, or from the cathode. They are created by de-excitation of atoms in the gas, or emitted when the ions recombine at the cathode.

In conventional gaseous detectors used to detect charged particles or X-ray photons the generation of secondary avalanches is less important. However, in photosensitive detectors they can be essential, cause feedback pulses and strongly disturb the normal operation of the detector. The reason is that in photosensitive detectors the probability of the creating additional avalanches is much higher than in normal gaseous detectors used for charged particle or X-ray detection.

2.1 Creation of Secondary Electrons by De-Activation of Excited Atoms and Molecules

The energy required to ionize an atom or molecule, E_i, is larger than the energy required to excite atoms or molecules in the gas without ionizing them, E_{exc}. Hence, during the avalanche multiplication many more excited atoms or molecules will be produced for each secondary electron.

These excited atoms and molecules then de-excite which often results in photon emission. The typical de-excitation time is between 10^{-9} and 10^{-6} sec. Some of these photons have high enough energy to ionize the surrounding gas or cause photoemission of electrons from the surrounding surfaces and electrodes. These secondary electrons can under certain conditions trigger secondary avalanches, often far away from the original avalanche.

Excited atoms and molecules can dissipate their energy via several processes. In general most of them are quite complicated and may include a chain of consequent steps. For example, de-excitation of some molecules may lead to creation of excited fragments (Fonte, 1991). For the sake of the simplicity we will consider only two processes here:

1. Emission of photons capable of ionizing the surrounding gas or causing a photoelectric emission of electrons at the detector cathode.
2. Collisions with surrounding atoms and molecules leading to ionization.

2.1.1 Photoionization of the Gas

Some excited atoms and molecules emit ultraviolet light with high enough energy to ionization the surrounding gas. Schematically this process can be described as:

$$M_1^* = M_1 + h\nu, \tag{7}$$

$$h\nu + M_2 = M_2^+ + e^-, \tag{8}$$

where M_1 and $M_2$are atoms or molecules in the gas and e^- is a free electron. The created free electron may in turn trigger another avalanche, called a feedback avalanche.

2.1.2 Photoelectrical Emission of Electrons from the Walls and the Electrodes

Ultraviolet and even visible photons emitted from the avalanche can cause photoelectric emission of electrons from the detector cathode. For a photon with energy E_ν the probability of photoelectric emission from the cathode is proportional to the quantum efficiency of the cathode in the gas, QE_{gas}:

Figure 2. Photoelectron extraction efficiency (the function f{(E/p),E$_\nu$}), from a CsI photocathode into various gases and gas mixtures at atmospheric pressure; the wavelength of the light is 185 nm. The sharp rice of some curves at E>1.75kV/cm is due to the gas amplification and not related to the f{(E/p),E$_\nu$}. The photoelectron extraction efficiency requires certain electric field strength to flatten out. At weaker electric fields the probability is high that the photoelectron will scatter back onto the cathode and be lost before the electron can drift freely into the gas (from Azevero, 2010).

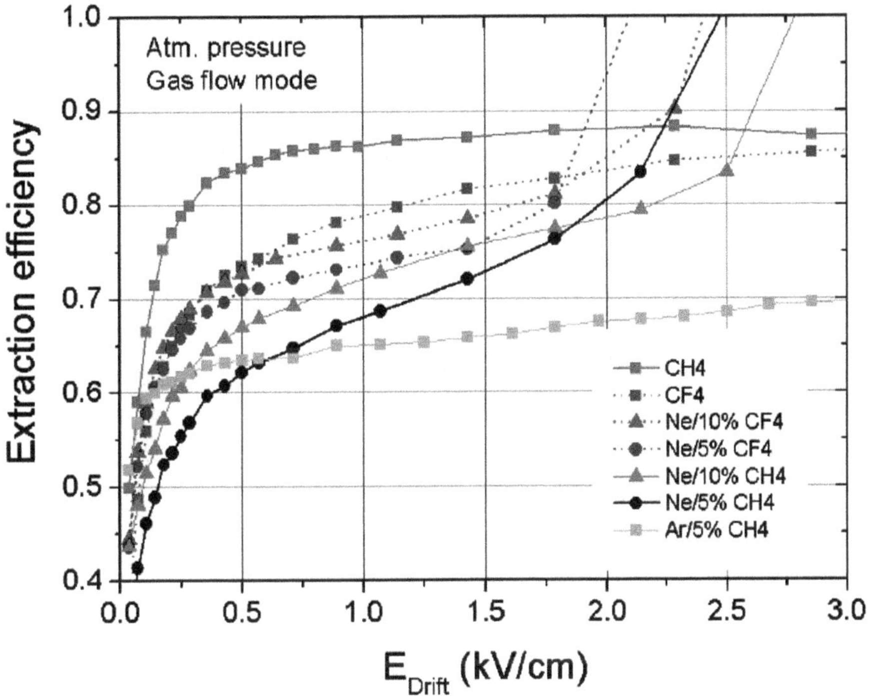

$$QE_{gas} = QE_{vac}(E_\nu) f\{(E/p),E_\nu\}, \quad (9)$$

where QE_{vac} is the quantum efficiency of the cathode measured in vacuum. The function $f\{(E/p),E_\nu\}$, sometimes called extraction efficiency, reflects the probability that an electron is extracted into a gas from a surface after correction for the so called photoelectron back diffusion. Photoelectron back diffusion are photoelectrons which when they leave the photocathode may after a few collisions with surrounding atoms and molecules return to the cathode and are hence not emitted to the surrounding gas (Bradbury, 1932; Theobald, 1953).

The extraction efficiency for some gases often used in photodetectors is presented in Figure 2.

2.1.3 Ionization by Collisions with Surrounding Atoms and Molecules

If an excited atom or a molecule has energy higher than the ionization potential of some of the surrounding atoms or molecules, it can ionize them via the process:

$$M_1^* + M_2 = M_1 + M_2^+ + e^-, \quad (10)$$

At atmospheric pressure and in the gas mixtures used in photodetectors this process is usually quite fast (typically less than 10^{-8} seconds). Hence, most of the electrons created via this process will be inside the avalanche volume. These electrons contribute to the total number of electrons produced in the avalanche process.

In old books, one can often read about the creation of feedback electrons appearing in the detector volume with a delay of between 1 ms and 1 s after the avalanche formed. They are attributed to metastable states of excited atoms (see for example Funfer, 1965). This is a feature of the low gas pressure used where atoms and molecules can stay excited for a long time without being de-excited by collisions with the surrounding atoms or molecules. At atmospheric pressure such large delays are not observed.

Hence, one can conclude that at relatively low gas gain the process of ionization between excited atoms and molecules with the surrounding neutral atoms and molecules (III-10) contributes to the total number of electrons created during the avalanche development. They do not trigger feedback avalanches located far away from the primary avalanche.

2.2 Creation of Secondary Electrons due to Ion Recombination on the Cathode

As was mentioned earlier, the avalanche electrons drift fast and are in a short time collected on the detector anode. The positive ions, which move much slower than the electrons, will at this moment still remain in the avalanche gap. They continue their slow movement towards the cathode which they finally reach after a time t_+. When the ions approach the metallic surface of the cathode, an electron will tunnel from metal surface to the ion and recombine it resulting in a neutral atom or molecule again.

If

$$E_{ion} > 2\varphi, \quad (11)$$

the process of ion recombination can then be accompanied by the emission of an electron from the cathode. In vacuum the probability of this single electron emission is, in first approximation:

$$\gamma_v \approx k_+ (E_{ion} - 2\varphi), \quad (12)$$

where k_+ is a coefficient (see for example Nappi, 2013 and references therein).

As was mentioned earlier, in the gas these electrons appear as back diffusion, so the probability a new free electron will appear in the gas is:

$$\gamma_+ = f\{(E/p), (E_i - 2\varphi)\}\, k_+(E_i - 2\varphi), \quad (13)$$

where the function $f\{(E/p),(E_i - 2\varphi)\}$ is close to the function $\{(E/p),E\nu\}$, shown in Figure 2.

This free electron emitted from the cathode is able to trigger a new avalanche which will be delayed with respect to the primary avalanche by a time $t_+ + t_e$, where t_e is the time necessary for the electron to drift from the cathode to the anode. Normally, t_e is much shorter than t_+.

In some gas mixtures(for example in photodetectors filled with argon with high ionization potential as the main gas, mixed with vapors having small ionization potential), ions with high ionization potential may ionize the molecules with small ionization potentials via the charge exchange process:

$$M_1^+ + M_2 = M_1 + M_2^+, \tag{14}$$

In this case, as follows from formula (13), γ_+ will be reduced, so the probability of the creation of secondary avalanches due to the recombination of ions at the cathode will be reduced. The feedback process in photodetectors will be described in more detail later, where also some practical designs of photodetectors filled with photosensitive gases or combined with high efficient solid photocathodes will be described.

3. THE MAIN DESIGNS OF AVALANCHE GASEOUS DETECTORS AND THE PRINCIPLE OF THEIR OPERATION

Nowadays four types of multiplication structures are commonly used in photodetectors:

1. Multiwire proportional chambers,
2. Parallel plate avalanche counters,
3. Resistive plate chambers,
4. Hole type gas electron multipliers.

Below a brief description of each of these detectors is given.

3.1 Multiwire Proportional Chambers

The multiwire proportional chamber (MWPC) was invented in 1968 by Georges Charpak (Charpak, 1968). This detector is an electronic imaging device for both charged particles and photons. It has successfully been used in many high-energy physics experiments at CERN and elsewhere and enabled many important discoveries. For this great invention Charpak was awarded the Nobel Prize in Physics in 1992.

An illustration of the MWPC is shown in Figure 3. It is basically two metallic plates separated by a gap typically of 6-8mm. In the middle of the gap is an array of parallel metallic wires. The wires typically have a diameter around 20-30 μm, and the pitch is about 3mm.

When a positive voltage is applied to all wires (or negative voltage is applied to the metallic plates), an electric field as shown in Figure 4 will be formed. As can be seen, it forms periodically repeating independent cells, each of them having the same electric field line distributions. Each cell is electrically separated from one another without any mechanical walls. Hence, a charge formed in one cell cannot move to another cell, except negligible leakage through diffusion. The electric field strength is uniform far away from the wires. Close the wire the radial geometry rapidly increases the electric field strength.

The cells operate independently from one another. A negative charge, an electron, appearing in the MWPC far away from a wire (i.e. much further away than the wire radius) will drift towards the wire at constant velocity in the parallel, uniform electric field. When the electron approaches the wire the electric field strength increases rapidly and significantly. The electron gains velocity and energy the closer to the wire it is. If this electron is now only collected on the wire, the detected signal will be $V = e/C$, which in a typical system with capacitance $C = 10$ pF gives $V = 0.016$ μV (e is the electric charge of the electron $1{,}6 \cdot 10^{-19}$ C). This is far below any possibility of detection.

Figure 3. Illustration of the design of a multiwire proportional chamber; it consists of two parallel cathodes and a plane of parallel wires placed in the middle of the gap. When a charged particle traverses the MWPC it creates a track of primary electrons along its path. These electrons are attracted towards the anode wires by an electric field. During their drift they induce a signal in the anode wires. The collection of the electron charges on the wires normally contributes to less than 1+% of the total signal. In the drawing the signals are shown with inversed polarity as measured on the output of the charge sensitive amplifiers. The signals in the wires are negative. From: http://en.wikipedia.org/wiki/wire_chamber.

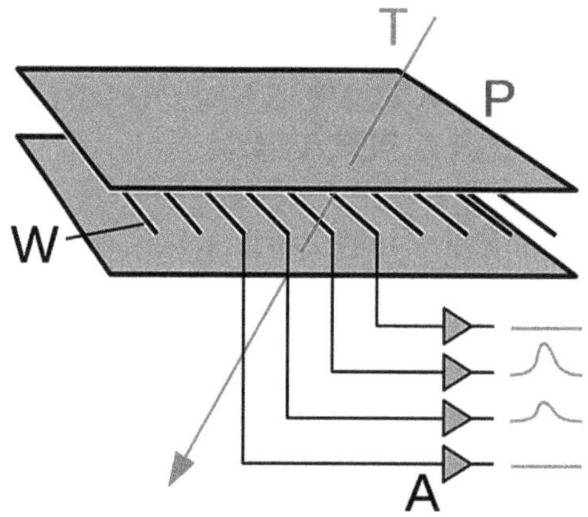

If the electric field strength is strong enough, in the region closest to the anode wire (starting a few wire radii out from the wire), the electron gets high enough energy to trigger an electron avalanche.

The variation of the electric field around a wire in a MWPC is shown in Figure 5.

Far away from the wire in the Y-direction (i.e. towards the cathode) the electric field strength is low and constant. Approximately 100 wire radii away from the wire (~1 mm) the electric field strength starts to grow exponentially. 2-3 wire radii out from the wire the electric field strength is strong enough to trigger electron avalanches. In the X-direction (i.e. towards the neighboring wire), the electric field strength decreases exponentially away from the wire and reaches 0 at the point between two wires. This is where the electric field lines bend up towards the cathode in Figure 4. The small region between two wires with weak electric field is normally a dead region in a MWPC. Any photoelectron produced here will drift so slowly in the weak electric field that it will often recombine before it is accelerated towards the wire and can trigger an electron avalanche.

The electron avalanches are only formed in a small space close to the anode wires. As the wire radius is of the order of 10 μm, this volume is less than $1/1000^{th}$ of a mm^3. The electrons formed in the avalanche only have a short distance to drift at high velocity in the strong electric field. Hence, the electron avalanche only last a short time, typically less than one ns. The heavy positive ions formed in the avalanche move much slower away from the wire in the decreasing electric field. They can remain in the avalanche region around the wire for quite some time, typically of the order of μs. During this time their charge in movement distorts and weakens the electric field around the wire making this part of the wire insensitive to form a new avalanche as long as the ions remain in the volume. This limits the maximum counting rate of MWPCs.

The time development of an avalanche is shown in Figure 6. Because of lateral diffusion through elastic collisions between the electrons and the gas atoms and molecules, and the small radius of the

Figure 4. The electric field line distribution in a multiwire proportional chamber (MWPC); an independent electrical cell is formed around each anode wire. The electric field strength is uniform far away from the wires. Close the wire the radial geometry rapidly increases the electric field strength. An electron appearing in the MWPC far away from the wire (i.e. much further away than the wire radius) will drift towards the wire at constant velocity in the parallel, uniform electric field. When the electron approaches the wire the field strength increases, and the electron gains velocity and energy the closer to the wire it is. In this region it gets high enough energy to form an electron avalanche. Hence, the electron avalanches are only formed in the small space close to the anode wires. The electrons formed in the avalanche only have a short distance to drift at high velocity in a strong electric field. The electron avalanche only last a short time (typically in the order of ns).

From http://www.scholarpedia.org/article/File:Electric_field_lines_in_a_typical_MWPC.jpg

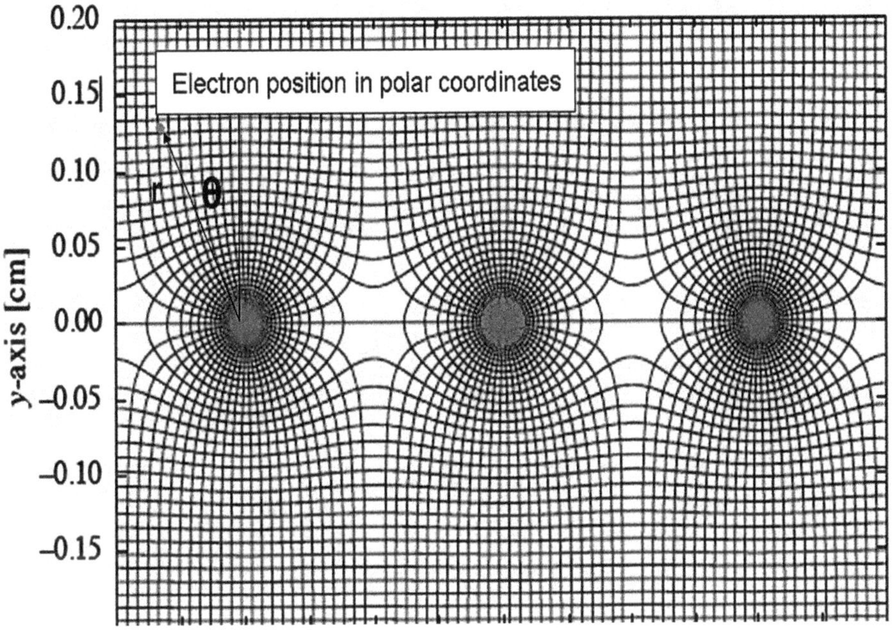

anode wire, the avalanches surround the wire asymmetrically. The electrons are rapidly collected on the wire and the positive ions begin to drift slowly towards the cathode.

A typical drift velocity for the electrons in the avalanche region is 5 cm/μs. The avalanche begins a few wire radii from the anode wire, which typically is 20 μm in diameter. This means that the whole amplification process takes less than 1 ns. At that instant, the electrons have been collected on the anode wire and the positive ion sheath will drift towards the cathode at decreasing velocity. The detected signal, negative on the anode wire and positive on the cathode, is a consequence of the change in energy of the system due to the movement of charges. The electrons in the avalanche are produced close to the anode, and therefore their contribution to the total signal will be small (<1%). Positive ions, instead, drift across the counter, often several cm, and generate most of the signal.

The ions' drift velocity is much lower than the drift velocity of the electrons, and the total drift time of the ions will be quite long, typically some 100 μs. But, and this is important, half the signal is developed after 1/1000th of the total time, some 100 ns. It is therefore normal practice to terminate the

Figure 5. The variation of the electric field in an MWPC; (X) is along an axis perpendicular to the wire plane and centered on the wire in an MWPC. (Y) is along a direction parallel to the wire plane pointing towards the next wire and centered on the wire in an MWPC. Far away from the wire in the X-direction (i.e. towards the cathode) the electric field strength is low and constant. ~100 wire radii away from the wire (~1 mm) the electric field strength starts to grow exponentially. 2-3 wire radii out the electric field strength is strong enough to trigger electron avalanches. In the Y-direction, the electric field strength decreases exponentially away from the wire towards 0 approximately 100 wire radii away from the wire. This is where the electric field bends up towards the cathode in Figure 4.
From Sauli 1977.

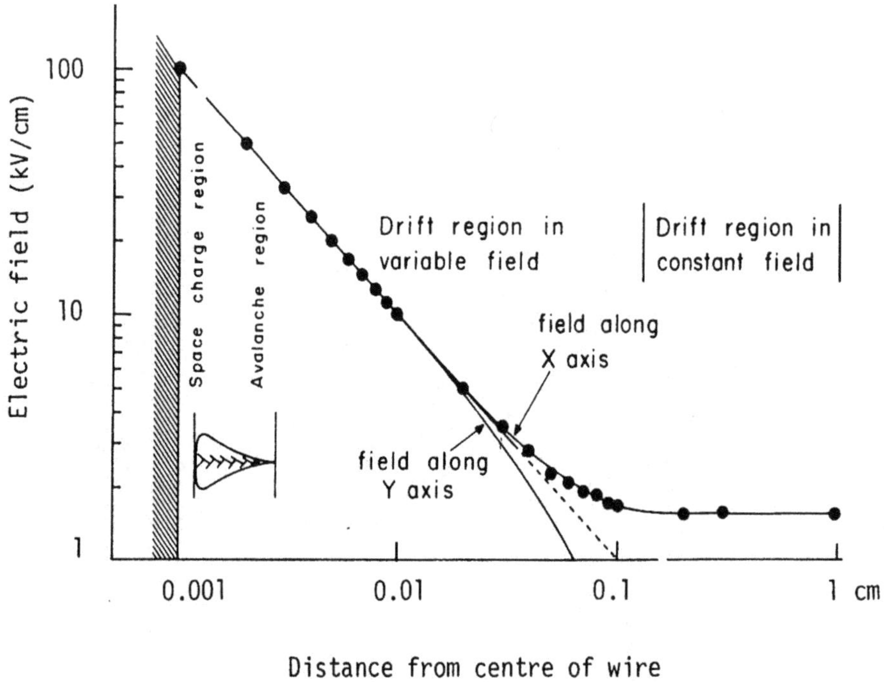

counter with a resistor R, such that the signal is differentiated with a time constant $\tau = RC$. Short pulses can be obtained using low impedance terminations, with the aim of increasing the rate capability of the counter. At the limit as $R \rightarrow 0$, one speaks rather of a current signal than a voltage.

The gas gain can be calculated using a formula similar to (6):

$$\ln A = \int_{a}^{b(r,\theta)} \alpha(r, \theta)\, dr, \quad (15),$$

where $b(r,\theta)$ is the position of the primary electron in polar coordinates, as shown in Figure 4.

Modern simulation tools allow us to accurately calculate the avalanche development in any gas and for various detector geometries. As an example, Figure 7 shows a simulated electron avalanche cloud development near an anode wire. The simulation not only take into account the effect of the electric field on the electron, but also elastic collisions between the electrons and ions in the avalanche with the gas atoms and molecules; ionization, excitation and de-excitation processes in the avalanche; as well as electric repulsion by the charges in the avalanche.

Figure 6. The time development of an electron avalanche: a) an electron is formed through ionization or photoionization far away from the wire (i.e. many wire radii away from the wire). b) The electron triggers and electron avalanche at some distance from the anode wire, typically 2-3 wire radii away. c) The electron avalanche grows rapidly and exponentially in charge. On this time scale the heavy ions remain where they are liberated. d) The avalanche surrounds the wire as the electrons scatter at wide angles through elastic collisions with the gas atoms and molecules on their way towards the wire. e) The electrons are collected on the wire and the positive ions start to drift slowly away from the wire, following the electric field lines towards the cathode.
From Sauli, 1977.

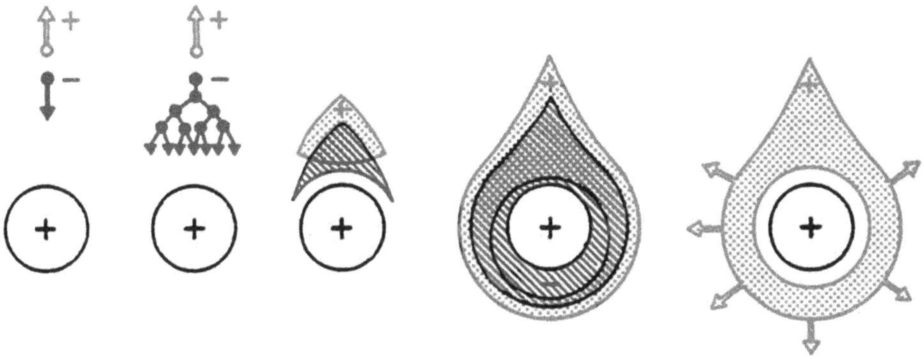

The simulation illustrates the drift of the electron towards the anode wire. During the drift it scatters in wide angles by elastic collisions with the much heavier gas atoms and molecules. At a few wire radii away from the wire it triggers an electron avalanche. The avalanche formation is shown as formation of new branches of tracks, where each branch is a new charged particle. The majority of the charges in the avalanche are formed close to the wire (< 1 wire radius away) in the exponentially growing avalanche. The whole process is over in less than 100 ps. The positive ions left behind slowly drift away from the wire, following the electric field lines towards the cathode. As they are much heavier than the electrons they scatter significantly less during their drift.

Typically photosensitive gaseous detectors operate at gas gains of 10^4-10^5. At such high gains the pulse height distribution of the signals from each electron avalanche follow an exponential distribution. More specifically, the distribution is shown to follow a special case of the negative binomial distribution, normally called the Pólya distribution, describing the signal strength from a single photoelectron avalanche (Byrne 1962; Alkazov 1970):

$f_Z(z) = m^m / \Gamma(m)\ z^{m-1}\ e^{-mz}$, (16),

Here, z = N/A where N is the number of electrons in the avalanche and A is the average number of electrons in a large number of avalanches, i.e. the average gas gain. m is a parameter in the Pólya distribution that is shown to be of the order 1.5 for normal wire detectors (Lanciart and Morucci 1962). At low gas gains the parameter m approaches 1 and the distribution becomes an exponential distribution.

The agreement of the Pólya distribution with the signal distribution from MWPCs has been verified several times (Neumann 1980; Francke 1991). Since the size of the avalanche is quite small, <<1/1000th of a mm^3, each electron avalanche can be treated independently even for quite high counting rates. Since the electric field around each anode wire forms an individual cell, the signal amplitude from a cylindrical proportional counter and from an MWPC are the same up to quite high counting rates.

The single photoelectron detection efficiency ε_{phe} depends on the gas gain A and on the front-end electronics threshold q_{thr}:

$$\varepsilon_{phe} \sim \exp(-q_{thr}/A), \quad (17)$$

To detect single photoelectrons with efficiency close to 100% with modern charge sensitive amplifiers a gas gain of more than 10^4 is required.

If all primary electrons are created in one cell, the avalanche will induce a negative signal on the anode wire and a positive signal on the cathode plane. The signal is mainly induced by the movement of the positive ions as they drift away from the anode wire to the cathode plane. The electrons in the avalanche are produced close to the anode wire (half of them less than one wire radius, ~10 µm, away from the wire). Therefore, their drift length is short, and their contribution to the total signal will be small (as mentioned earlier < 1%).

The latter is often used for more precise determination of the avalanche position along the wire. One of the cathodes of the MWPC is usually segmented in isolated parts like wires, strips or pads (see Figures 8 and 9). In the case of wires or strips, they are normally positioned perpendicular to the anode wires. As the avalanche always occurs close to a wire, the position of the primary charge cannot be determined with a better precision, σ_y, than the wire spacing, or more correctly:

$$\sigma_y = d / \sqrt{12}, \quad (18)$$

since the position of each wire follows a square distribution function. d is the wire spacing.

Each one of the segments is connected to its own amplifier. The size of the segmented parts is normally chosen such that the induced signal from each avalanche is spread over several of them. The largest part of the signal originates from the drift of the ions while they are still quite close to the anode wire, and hence quite far from the cathode segments. The result is that the distribution of the induced signal amplitudes on the segmented cathode electrodes is a two-dimensional Gaussian distribution in space (Mathieson, 1984). The position resolution measured in the cathode plane can be significantly better than the size of the cathode segments by interpolating the signals recorded on adjacent elements. There are several methods of weighting these signals in the calculation of the most probable position of the avalanche. The most common method is the center of gravity method (Endo, 1981). In this method, the signal amplitude in each segment is used as a weight when calculating the position of the avalanche. Another method is the Gaussian potential method, where each segment is given a probability that the signal belongs to the avalanche due to the distance between the segment and resulting avalanche position (Ullrich, 1996). The most accurate position determination is obtained if the true distribution of the induced signal on the cathode looks. This can be determined e.g. by simulations, or by measurements with a well collimated beam of light or charged particles.

Figure 10 shows the result from simulated and measured induced signals on the cathode strips of the MWPC shown in Figure 9.

Figure 8. Photograph of the cathode segmented into pads used in the ALICE Ring Imaging Cherenkov detector (RICH); the pads are 8x8 mm². The total areas the cathode is 20x40 cm². Every photoelectron detected in the RICH detector will induce a signal in several pads. The position of the photoelectron is then determined with higher accuracy than the pad size by fitting the measured distribution of the signals on the pads with the expected signal distribution.
From http://cds.cern.ch/record/628218/files/hmpid-2003-001.jpg?version=1.

Figure 9. Schematic drawing of an MWPC where the cathode is segmented into strips; the cathode strips are perpendicular to the anode wires to determine the position of the avalanche, and hence the original photoelectron, with the highest possible resolution along the wire.
From Endo, 1981.

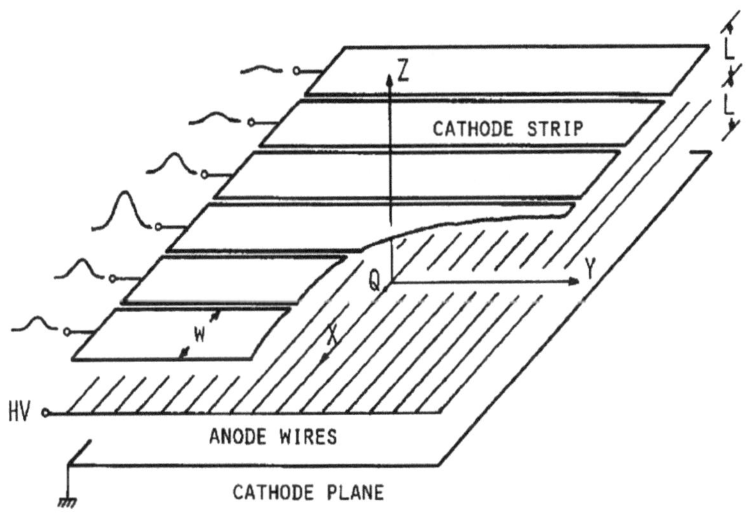

Figure 10. Induced signals on 9 mm wide cathode strips in a MWPC from a single photoelectron; the initial electron is produced near the center of the 5th strip. The left figure shows the expected signal distribution from simulated signals. The right figure shows the measured signal distribution. The signals from a single photoelectron are distributed over several strips. Please note the good agreement indicating a good understanding of the processes. Since the signal is spread over several strips, the position of the avalanche can be determined with significantly higher precision than the strip width by fitting the measured distribution with the expected one (where one knows the position with infinite precision). *From Endo, 1981.*

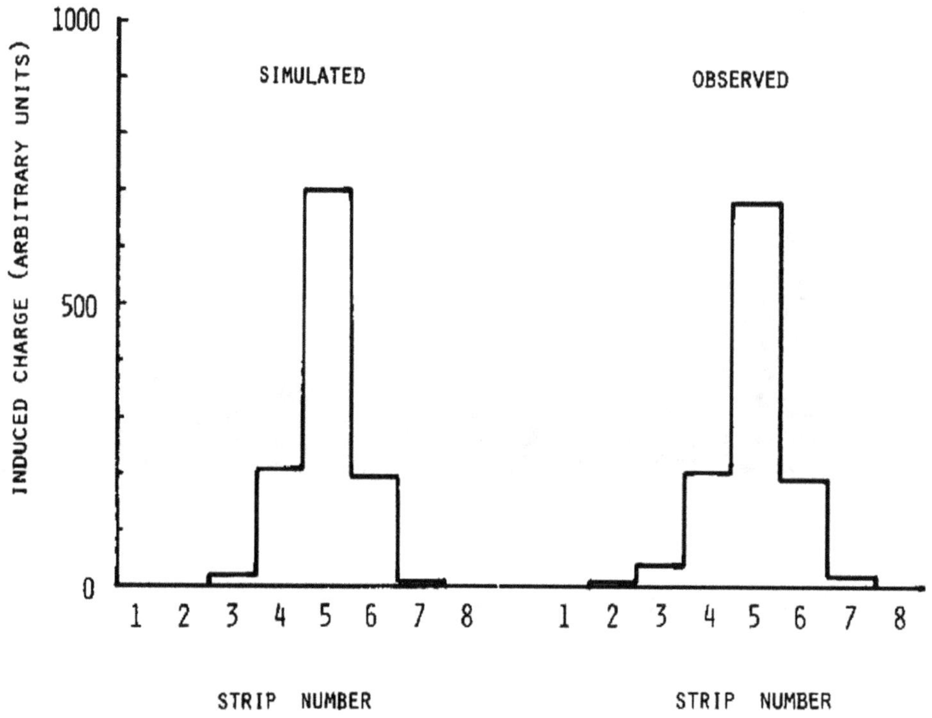

3.2 Parallel Plate Avalanche Counters

A parallel plate avalanche counter (PPAC) is another detector for photons and charged particles. It consists of two parallel metallic plates separated by a gap of 2 – 5 mm. A high voltage is applied between the electrodes. If a primary electron is created near the cathode, it starts drifting towards the anode. If the electric field in the gap in strong enough, it can trigger a Townsend electron avalanche in the gap. The average number of electrons produced in the avalanche in the gap will be:

$$A = \exp(\alpha\, d_{gap}), \tag{19}$$

where d_{gap} is the gap width.

Historically parallel plate avalanche chambers were mainly used to study avalanche and feedback processes (Raether, 1964). In photon detectors, parallel plate avalanche chambers were initially used as a preamplification structure for MWPCs (Charpak, 1978, Charpak, 1979).The early designs of photo-

Figure 11. A cascaded photosensitive detector; it consists of a parallel plate avalanche preamplification structure above a MWPC. Ultraviolet photons enter the detector through a UV-transparent window. They are converted into photoelectrons in a conversion gap above the parallel plate avalanche chamber. Each photoelectron creates a first electron avalanche in the parallel plate chamber. The electrons from the avalanche are transferred into the MWPC where each electron produces a new electron avalanche. The resulting signal is large enough to be detected with good efficiency.

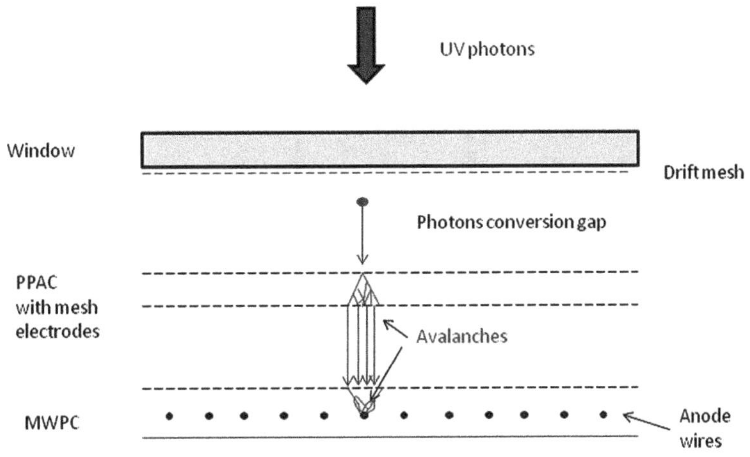

sensitive MWPCs did not have enough gas gains to be able to detect single photoelectrons. They needed a preamplification structure to produce larger signals for each single photoelectron. The total gain of the parallel plate chamber and the MWPC was high enough to efficiently detect single photoelectrons. Such a cascaded photosensitive detector is shown in Figure 11.

The electrodes of the drift mesh and the PPAC are made of metallic meshes (typically having 80% of optical transparency). The gap between the mesh electrodes is between 2 and 5 mm. The top electrode of the MWPC also is also made of a mesh or parallel wires.

Ultraviolet photons enter the detector volume through a UV-transparent window. They interact through photoelectric effect with a photosensitive compound in the photo-conversion gap between a drift mesh and the parallel plate avalanche chamber and produce photoelectrons. Each primary photoelectron drifts under the influence of an electric field towards the parallel plate avalanche chamber, and enters it through the mesh. In the parallel plate avalanche chamber it triggers an electron avalanche close to the cathode. As all avalanches are triggered approximately at the same distance from the cathode of the parallel plate avalanche chamber, each photoelectron will give similar signal strengths (see Equation 19).

As some of the electrons in the avalanche diffuse by elastic scattering with the gas atoms, they can escape out of the parallel plate avalanche chamber into the drift volume between the parallel plate avalanche chamber and the MWPC. This corresponds to the electrons that by diffusion surround the wire in Figure 6d). These escaping electrons are attracted by a stronger electric field towards the MWPC where most of them enter through the cathode mesh rather than being collected on the mesh. The fraction of electrons that manages to pass through the MWPC cathode mesh into the MWPC depends on the electric field ratio above and below the top MWPC electrode. Inside the MWPC each electron drift towards the anode wires where they trigger an electron avalanche. The signal from all these electron avalanches, originating from a single photoelectron before the parallel plate avalanche chamber, is then

Figure 12. A BaF_2 calorimeter using a photosensitive parallel plate avalanche chamber; scintillation light is produced in a crystal of BaF_2 by X-rays, gamma rays or charged particles, and detected by the photosensitive parallel plate avalanche chamber.
From Charpak, 1993.

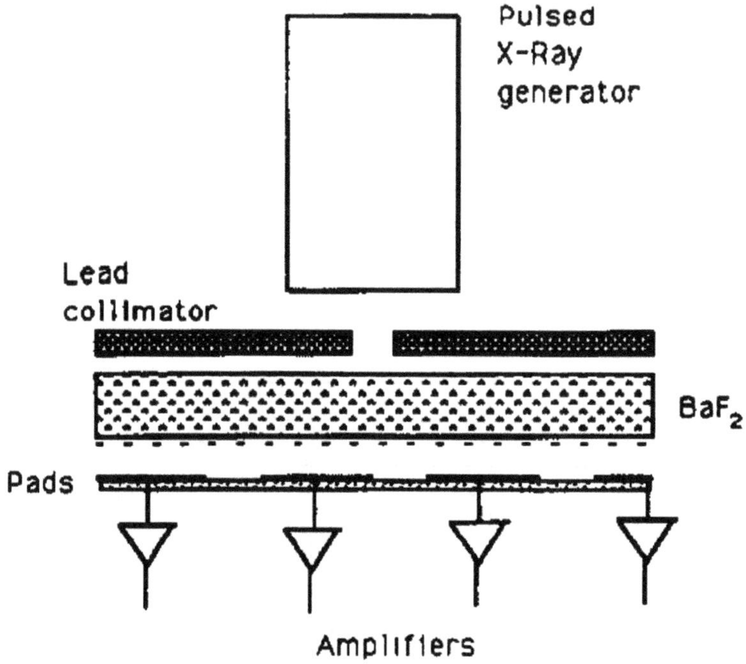

detected. Besides the MWPC, parallel plate avalanche chambers were also tested in the final stage of cascade multiplication (see for example (Charpak, 1982)).

Later it was demonstrated that a parallel plate avalanche chamber alone can also detect single photoelectrons from single UV-photons even without the MWPC (Charpak, 1993, Fonte, 1999). A prototype of such a detector with an anode segmented into pads for position-sensitive measurements was developed as a photosensitive detector in a BaF_2 calorimeter, see Figure 12 (Charpak, 1993).

BaF_2 emits scintillation light over a wide wavelength region, 135 – 450 nm, where the longer wavelengths have a long decay time from metastable states in the crystal. A fast light pulse is emitted in the UV region 135-300 nm, which is well suited for photosensitive gaseous detectors. The ultraviolet photons refract out of the crystal and enter into a photosensitive parallel plate avalanche chamber through the upper cathode mesh. In the PPAC the photons are via photoelectric effect in a photosensitive gas converted into photoelectrons, which in turn trigger an electron avalanche. The position of the light is determined by the signal amplitude in the pads. The sum of all signals in all pads is an accurate measure of the amount of scintillation light produced, which is proportional to the energy deposited in the crystal by the X-ray, gamma ray or charged particle.

3.3 MICROMEGAS

A variation of the parallel plate avalanche chamber is the MICROMEGAS detector. It was developed by Charpak in 1994 (Charpak, 1995; Giomataris 1996). It is a parallel plate avalanche chamber with a

Figure 13. The MICROMEGAS detector is a parallel plate avalanche chamber with a narrow gap between the electrodes, only 50 – 100 μm. The mesh is supported by columns built up lithographically on the anode. Photons enter the detector from above and interact with a photosensitive compound in the drift gap above the MICROMEGAS producing photoelectrons. They follow the electric field lines towards the amplification gap between the electrodes where they are multiplied in an electron avalanche. The signals from the avalanche are read out by the readout electrodes on the anode.
From Barouch, 1999.

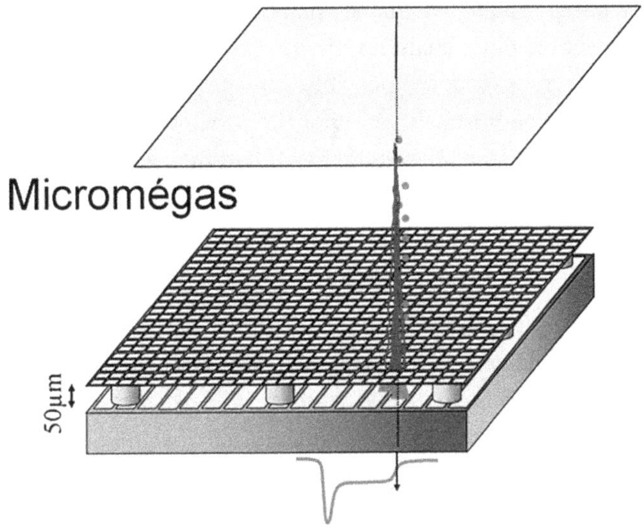

small gap between the electrodes, typically 50 – 100 μm (see Figure 13). The cathode is made of a thin metallic mesh, only a few micrometers thick. The openings in the mesh are as small as 15-20 μm, and the distances between the openings is around 25 μm. The anode is a printed circuit board where the readout electrodes are etched into the desired pattern, strips or pads. Above the MICROMEGAS is a second mesh defining a drift and conversion volume.

To keep the mesh at a constant distance from the anode, small columns or pillars are built on the anode. In early designs the distances were made by fishing lines wrapped around the anode. In more modern designs the pillars are built up lithographically and have well defined height.

Due to the small dimensions, a MICROMEGAS can operate with quite low voltages, 400 – 800 V.

Light enters the detector from above through a window. In the drift volume the photons interact with a photosensitive compound liberating photoelectrons. These photoelectrons follow the electric field lines in the drift space towards the cathode mesh of the MICROMEGAS. Due to the much stronger electric field in the gap of the MICROMEGAS, most of the photoelectrons enter through the mesh into the amplification region. Inside the MICROMEGAS they trigger Townsend avalanches. The induced signals from the avalanches are read out by the electrodes on the anode.

The gas gain of a MICROMEGAS can be high, a few times 10^4, but not as high as a normal parallel plate avalanche chamber which can reach gas gains above 10^5 (Charpak, 2002). The position resolution is good, down to 25 μm (Papaevangelou, 2011).

Today, MICROMEGAS is one of the most popular designs of a gaseous detector and it is used widely in high energy physics (see for example Reserva, 2007).

3.4 Resistive-Plate Chambers

In some applications of photosensitive parallel plate avalanche chambers the electrodes are made of high resistivity materials rather than metals. Such materials can be GaAs with a resistivity of 10^8 Ω cm, Pestov glass with a resistivity 10^{10} Ω cm, or doped silicon where the surface resistivity can be chosen between 10^{-3} to 10^4 Ω cm by varying the degree of doping (see for example Francke, 2004). Such a detector is called a resistive plate chamber or RPC. The design and the principle of operation of the resistive plate chamber are similar to the parallel plate avalanche chamber. However, it is a much more robust device. If a discharge occurs in a resistive plate chamber, the resistive electrodes restrict the discharge current and protect the front end electronics from the possible damage. Figure 14 shows a photosensitive resistive plate chamber filled with a photosensitive vapor, i.e. a vapor with low ionization potential. The detector consists of two parallel resistive electrodes separated by a gap of 0.4 mm. The inner part of the anode electrode is covered with metallic readout strips with a pitch of 30 μm. Each strip is connected to a charge sensitive amplifier.

Figure 14. A photosensitive resistive plate chamber; the upper cathode is made of GaAs and the lower cathode is made of Pestov glass. Metallic strips are evaporated on the inside of the Pestov glass anode plate. They allow the position of each electron avalanche to be determined with high accuracy in the direction perpendicular to the anode strips. Along the anode strips one cannot tell where the avalanche occurred. The UV photons enter the detector parallel and close to the cathode. They interact with the photosensitive gas and liberate photoelectrons. Each photoelectron triggers an electron avalanche which quickly develops towards the anode in the strong electric field between the plates, inducing a fast signal in the anode strips.
Francke, 2004.

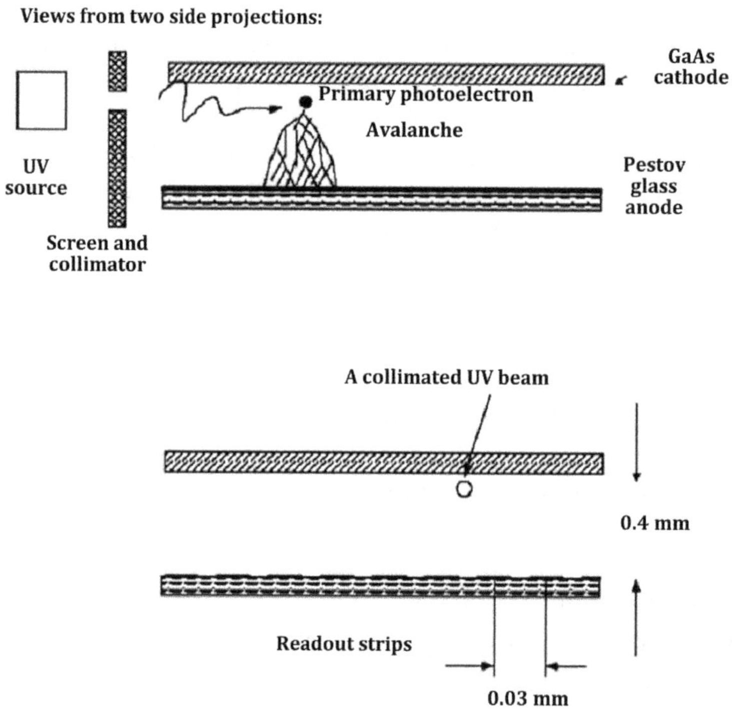

A collimated beam of UV light enters parallel to the electrodes and close to the cathode where it photoionize the gas and liberates photoelectrons. Each photoelectron triggers the formation of a Townsend avalanche as described by formula 19. The avalanche induces signals on the anode strips. If the strips are narrow enough the signal is spread out over several strips enabling the avalanche position to be determined in one dimension (perpendicular to the anode strips) with a high accuracy by interpolating the signal in each strip. In some gas mixtures the precision can be as good as ~30 µm.

Should a discharge occur in the gap between the electrodes, only a fraction of the charges on the cathode are fed to the anode (a discharge can be thought of as a short circuit between the cathode and the anode since the plasma in the discharge has quite low resistivity). Charges further away from the discharge on the cathode will have to travel through the resistive material in the cathode for some distance creating a voltage drop which limits the current and slows down the discharge.

3.5 Hole-Type Electron Multipliers

Photosensitive gaseous detectors based on hole-type electron multipliers are becoming increasingly interesting. This type of detector was developed quite long ago for detecting charged particles (Del Guerra, 1987; Sakurai, 1966). Their complex manufacturing limited their popularity until their advantages as photosensitive detectors were realized. The most important advantages are:

1. Efficient suppression of ion and photo feedback from the cathode. This allows a significantly higher gas gain compared to photosensitive MWPC or PPAC,
2. The possibility to build these detectors from materials (glass, ceramics, etc.) with comparably low levels of contaminants as in sealed gaseous photodetectors.

Many different hole-type electron multipliers are used in photodetectors:

- Glass and ceramic capillary plates,
- Gas electron multipliers,
- Thick gas electron multipliers,
- Resistive electrode gas electron multipliers,
- Microhole-microstrip structures.

3.5.1 Capillary Plate Detectors

Glass capillary plates are used in some photosensitive detector designs. At the first glance they seem similar to the well-known microchannel plates (MCPs). However, there are fundamental defenses. Glass capillary plates are thicker, the holes are larger in diameter and, what is most important, the physical processes are different. Microchannel plates operate in a high vacuum and secondary electron emission from the inner capillary walls are used for electron multiplication. Glass capillary plates operate in gas at atmospheric pressure. Electron multiplication occurs when primary electrons liberated in the gas trigger the formation of Townsend electron avalanches in a strong electric field created inside the holes.

Figure 15 shows a glass capillary plate. It is made off a glass or ceramic plate 1 mm thick. The plate is perforated by narrow holes with a diameter of 10 to 1000 µm. The plate is metalized on both surfaces but not on the walls in the holes.

The metalized surfaces serve as electrodes. When a voltage V_0 is applied between the electrodes, the field lines experience a focusing effect inside the holes as shown in Figure 16. At high enough voltage, primary electrons produced above the plate can initiate Townsend avalanches inside the holes.

In some designs the capillary plate is combined with a drift space, as shown in Figure 17. In this case a weak electric field is formed between the cathode and the top capillary electrode. Primary electrons created through photoionization in the drift space drift towards the holes and trigger electron avalanches inside the holes. A readout plate with strip or pad electrodes is normally located underneath the capillary plate. As in the case of the mesh parallel plate avalanche chambers described above, some of the avalanche electrons will be extracted from the holes and collected on the readout plate.

From the signals induced on the segmented readout plate, one can determine the position the hole in which each avalanche occurred.

3.5.2 Gas Electron Multiplier

A gas electron multiplier (GEM) is another hole-type detector (Sauli, 2006). It is made of a 50 μm thin Kapton foil which is metalized on both sides, as shown in Figure 18. The Kapton foil is perforated with narrow holes, 70 μm in diameter and 140 μm apart.

To make the holes accurate and smooth enough for stable electron avalanche formation they are made by photolithographic methods and later etched in a process similar to high precision printed circuit board manufacturing. An electron microscope image of a cross section through a GEM is shown in Figure 19.

A voltage is applies between the metalized surfaces of the GEM. This creates a focused electric field inside the holes as shown in Figure 20; a dipole-like electric field in each hole. The electric field lines are strongly focused in the holes, creating an electric field which is strong enough to develop electron avalanches. The electric field lines in the center of the hole propagate out of the hole towards the cathode and anode above and below the GEM. The field lines attract any photoelectrons produced in the drift field above the GEM into the holes where they trigger electron avalanches. Similar, a fraction of the electrons formed in the avalanches in the holes manage to escape the holes along the field lines towards a lower anode where they are collected and induce signals.

As in the case of the capillary plates, GEMs are usually combined with an anode readout plate placed 2-3 mm below the GEM, as shown in Figure 21.

For position resolution measurements the readout plate may consist of an array of isolated metallic strips or pads. Two layers of strips, isolated from each other, can be arranged in perpendicular directions allowing one to perform 2D measurements. These are often called "x and y strips". The number of readout channels in such a configuration is lower than when using pad readout. However, if the counting rate is high it may cause ambiguities. In the event of two simultaneous avalanches it is not possible to determine which avalanche belongs to which x and y pair of signals.

The operation principle of a GEM is the same as for a capillary plate. Primary electrons are created in the drift volume through photoionization of the gas by external radiation. Under the influence of the drift electric field they move towards the nearest holes. If the electric field inside the hole is strong enough, the primary electrons trigger Townsend avalanches in the holes. The multiplication factor, or a gas gain, A, can be as high as 10^4. If another electric field (often called an "induction field") is applied between the GEM anode and the readout plate, part of the avalanche electrons is captured by this field, escape out of the holes and drift to the readout plate. During the fast drift they induce signals in the

Figure 15. A schematic view of a glass capillary plate on the left; to the right is shown a photograph of such a plate. It is made off a glass or ceramic plate 1 mm thick. The plate is perforated by narrow holes with a diameter of 10 to 1000 μm. The plate is metalized on both surfaces, but not on the walls in the holes.
From http://www.hp.phys.titech.ac.jp/msgc/prepri/spie44_nishi.pdf.

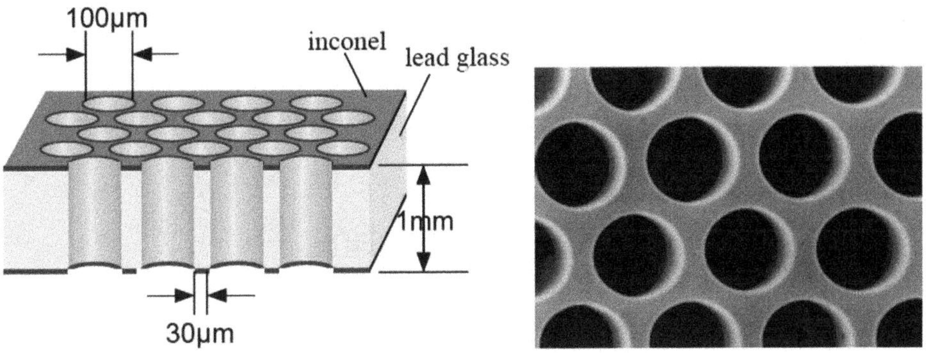

Figure 16. The cross section of a glass capillary plate detector; when a voltage V_0 is applied between the metalized surfaces, the field lines are compressed/focused inside the holes. If the electric field in the holes is strong enough electron avalanches can be triggered by photoelectrons produced above the cathode and which follows the electric field lines into the holes.
From Del Guerra, 1987.

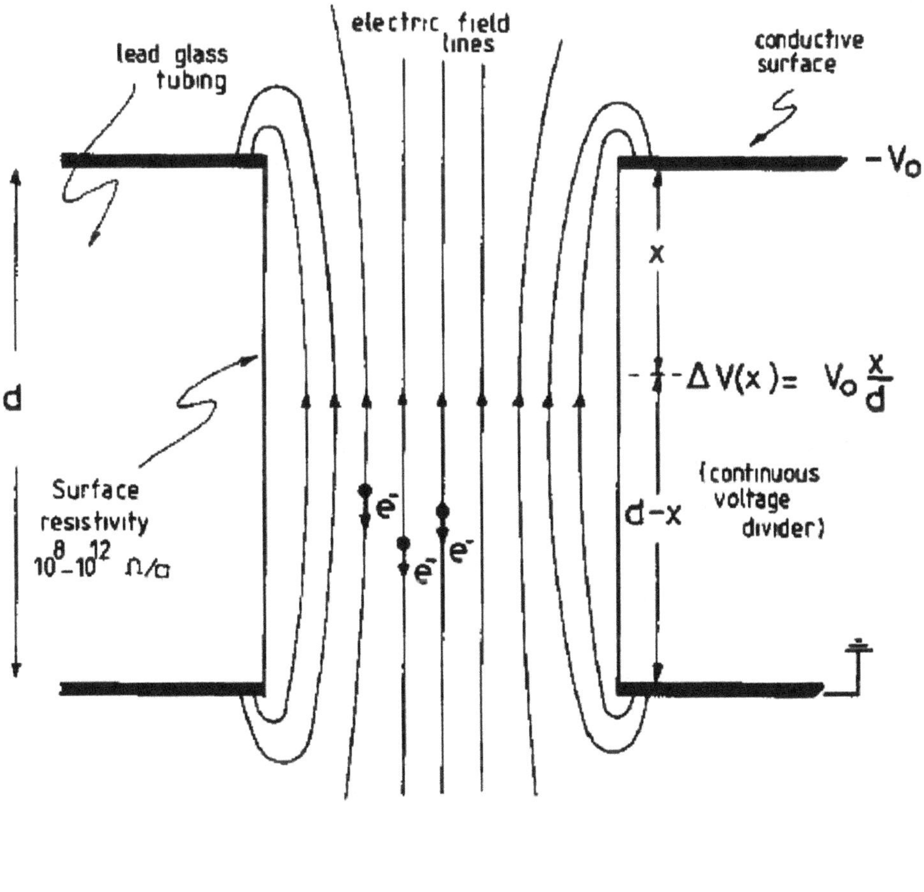

Figure 17. Schematic drawing of a capillary plate combined with a drift region and a readout plate segmented in pads or strips allowing one to determine the position of the hole in which the avalanche occurred; photoelectrons are formed in the drift volume. They drift in a weak electric field towards the nearest hole in which they trigger electron avalanches. Some of the electrons from the avalanche are extracted out of the capillary plate and towards the readout plate where they induce signals which makes it possible to determine the position of each avalanche with high precision.
From http://www.hp.phys.titech.ac.jp/msgc/prepri/spie44_nishi.pdf.

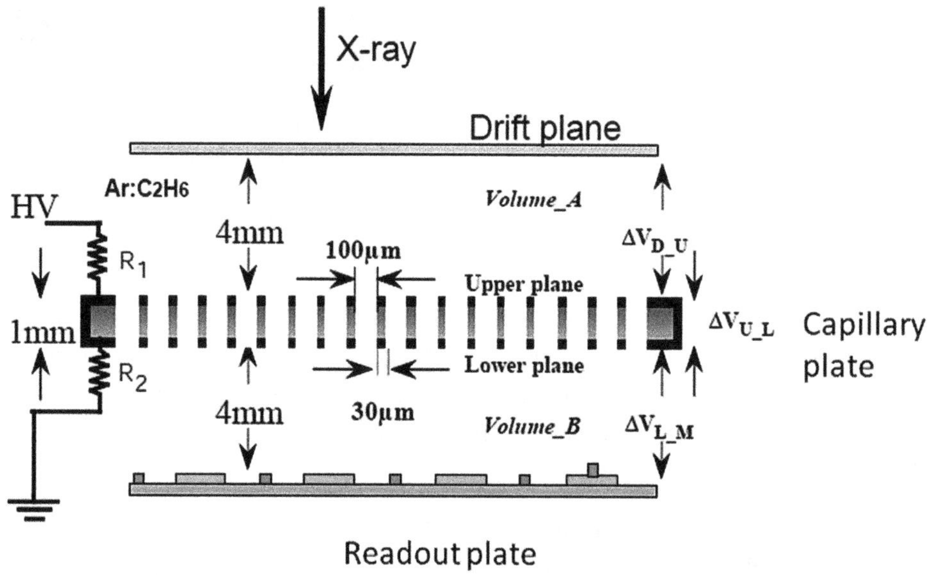

Figure 18. A magnified photograph of a gas electron multiplier (GEM); the GEM is made of a thin foil of Kapton. The Kapton is metallized on both sides. In a lithographic etching process narrow holes perforate the foil. Each hole acts as an independent electron avalanche structure.
From http://gdd.web.cern.ch/GDD.

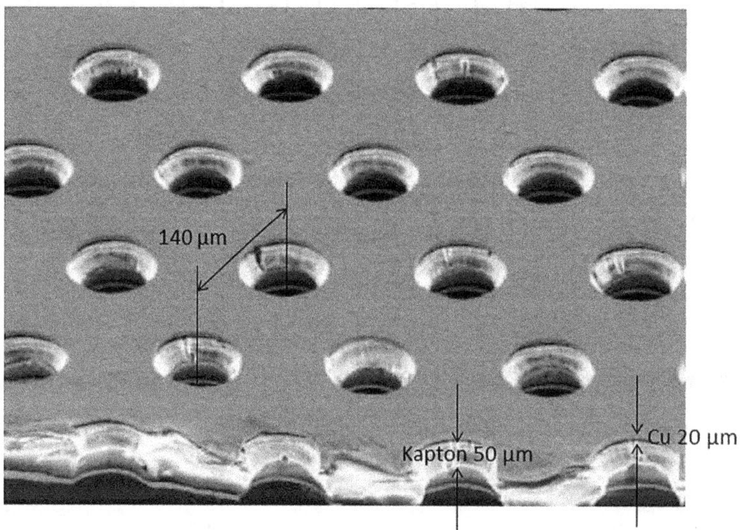

Figure 19. Electron microscope image of a cross section through a GEM; the etching process is tunes to shape the whole into a narrow structure forcing a maximum focusing of the electric field lines in the hole.
From Sauli, 2006.

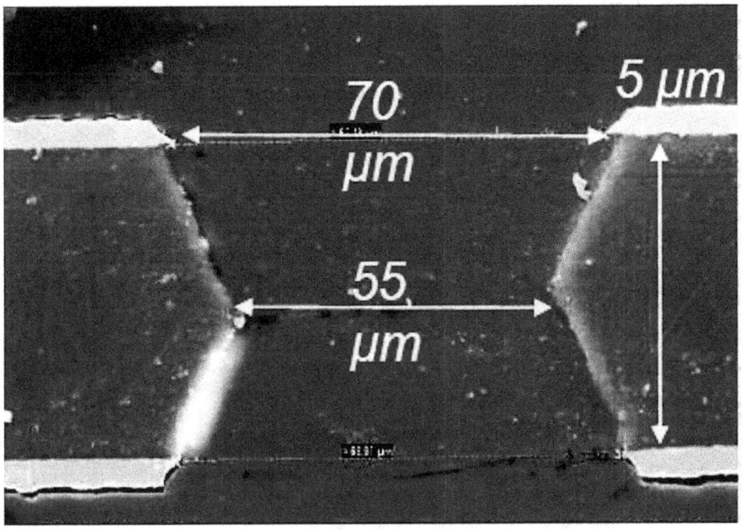

Figure 20. Schematic drawing of the electric field lines and equipotential lines in a GEM detector; a voltage applied between the top and bottom side of the Kapton foil creates a dipole like electric field in the holes. The electric field lines are strongly focused inside the holes creating an electric field strong enough to form electron avalanches.
From http://cerncourier.com/cws/article/cern/27921.

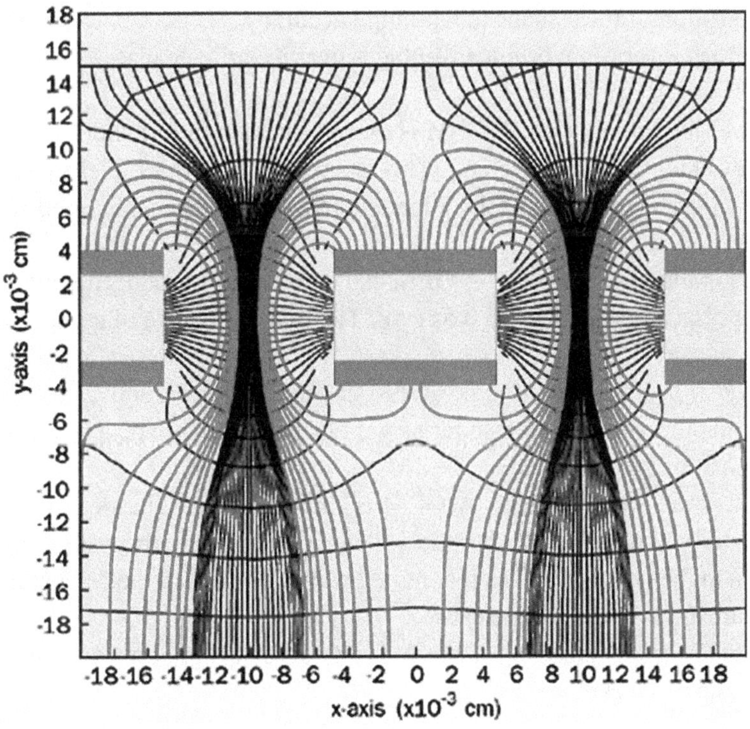

Figure 21. A GEM detector layout; usually the GEM is combined with a drift cathode above the GEM and a readout anode plate below. The latter is segmented on metallic strips or pads. Each of them is connected to its own amplifier. Two orthogonal layers of strips, isolated from each other, can be used instead of pads for 2D localization of the avalanches. This reduces the number of readout channels. *From http://gdd.web.cern.ch/GDD/.*

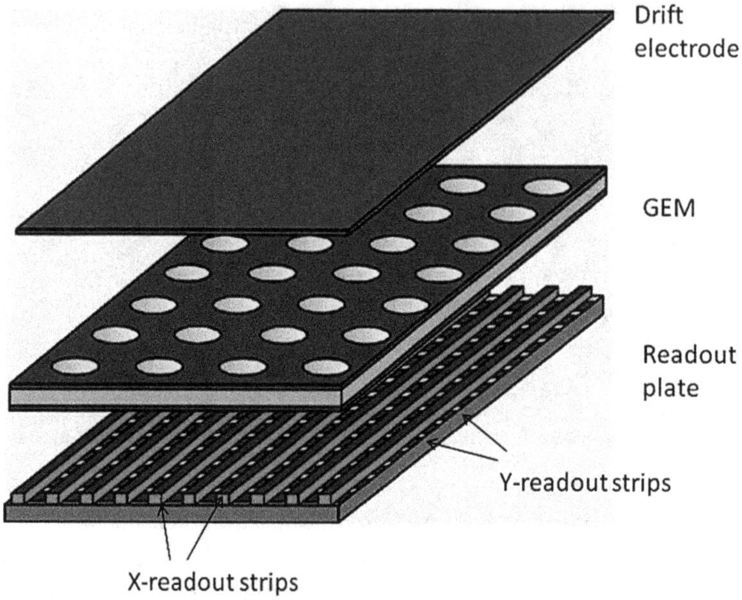

anode strips or pads. By measuring the distribution of the induced signals on the strips (or pads) one can determine the position of the avalanche with high accuracy.

A GEM detector has several important advantages over a capillary plate:

1. *A GEM is easier to manufacture.* A GEM is made of a Kapton foil and in which a hole pattern is etched by photolithography and etching. This is a standard process in electronics manufacturing. Glass capillary plates require special equipment and are only manufactured by a few specialized manufacturers.
2. *A GEM can be made in large areas.* A GEM can be routinely produced with a size of 30x30 cm^2 and in some special cases even up to 0.5x1 m^2. The maximum size of a glass capillary plate is 10 x 10 cm^2.
3. *A GEM has higher position resolution.* A GEM has smaller holes, and a smaller pitch between the holes than the capillary plates used so far as gas electron multipliers. The closer the holes are the higher is the position resolution.
4. *A GEM operates at lower voltages.* A GEM is thinner than a glass capillary plate and can be operated at lower voltages. The higher the voltage, the higher is the risk of electric breakdowns. Furthermore, a higher voltage means that more energy is stored in the capacitance system, and the larger is the damage a discharge can cause.
5. *A GEM can handle higher counting rates.* A GEM can detect photons or charged particles at high counting rates, up to 10^5Hz/cm^2.

Figure 22. Photograph of an early TGEM made from a normal glass fiber printed circuit board; 0.4 mm holes are drilled as dense as possible in a honeycomb pattern.

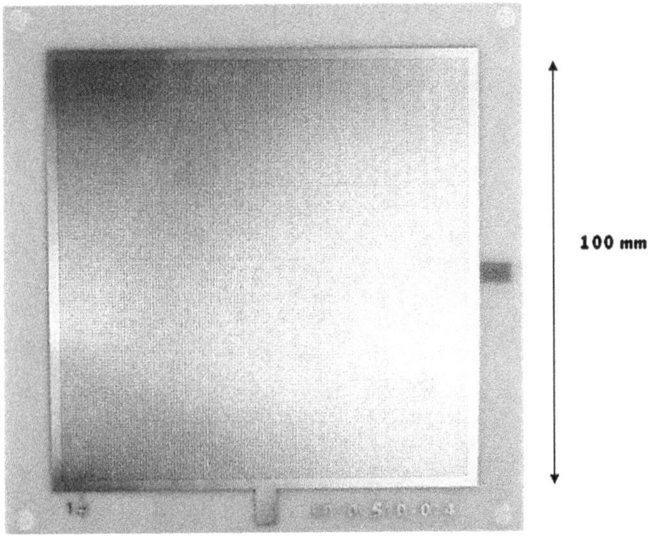

For these reasons, GEM is today the most popular among hole-type gaseous detectors.

3.5.1 Thick GEM

A Thick GEM (TGEM) is a GEM- like detector, but significantly thicker than an ordinary GEM. It can have thicknesses up to 2-3 mm and also larger holes with diameters of 0.3 – 0.8 mm spaced at similar distances (Periale, 2001; Periale 2002). From a geometrical point of view the TGEM more resembles a capillary plate rather than a GEM. However, the technology used for TGEM manufacturing is different from both GEM and capillary plate manufacturing. A TGEM is made from a printed circuit board 0.4- 3 mm thick. The board is metalized on both sides. The holes are simply drilled with a computerized numerical control (CNC) drilling machine. This means that TGEMs can be made in any well-equipped mechanical workshop, and it can be made in large areas sometimes up to a square meter. In mass production, the cost of TGEM per unit of area is significantly lower than for capillary plates or GEMs.

A photograph of one of the first successful TGEM used as a photosensitive detector is shown in Figure 22. It is made from a standard glass fiber printed circuit board with cupper on both sides. 0.4 mm holes are drilled with 0.8 mm spacing in a honeycomb pattern.

In some modern designs of TGEMs dielectric rims are manufactured after the drilling, as shown in Figure 23. The main purpose of this modification is to avoid sharp metallic edges from the drilling process. As smooth edge decreases the electric field around the edges and in this way reduces the risk of sparks. The rims are often made of photoresist in a photolithographic process.

As capillary plates, TGEMs operate at higher voltages than GEMs. On the other hand, their maximum achievable gains are typically 10 times higher than the best GEM designs can offer, as shown in Figure 24. This is of great importance for photosensitive detectors. For 100% detection efficiency of single photoelectron gas gains around 10^5 or higher are needed.

Figure 23. Magnified photograph of a TGEM with dielectric rims around the edges of the holes; the smooth dielectric rims reduce the risk of spark development from sharp metallic edges at the border of the whole.
From Shalem, 2006.

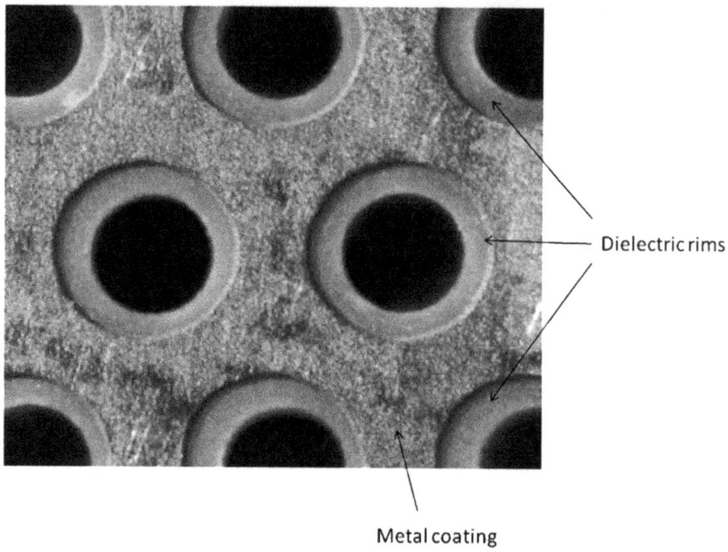

The TGEM is a robust detector. The large hole diameter makes them less sensitive for dust and other micro-particles in the holes. As a consequence, a TGEM in contrast to a GEM does not require extremely clean conditions (like a clean room) during manufacturing or during the installation inside the gas chamber. The more robust design allows TGEM to withstand occasional sparks and it usually remains undamaged even after intense sparking.

A TGEM can beneficially be used in applications which do not require high position resolutions such as in Ring imaging Cherenkov detectors (RICH).

3.5.4 Resistive Electrode TGEM

A gas gains larger than 10^5 is required for 100% efficient single photoelectron detection. This is on the limit of what hole-type detectors can achieve, and they continuously risk sparking. There are several causes of sparking in hole type detectors.

In applications where hole-type gaseous multipliers are used for the detection of charged particles discharges may be caused at high gains by:

1. Defects on the metal coating, especially near the edges of holes
2. Micro-particles and dust presented inside and outside the holes. These particles can also form regions of high conductivity in the holes which leads to sparks.

High density ionization produced by charged particles, especially by heavily ionizing particles and recoils. When the total charge in the avalanche reaches the so-called Raether limit a discharge is formed. The Raether limit is given by:

Figure 24. Gain curves measured with three different hole-type detectors: GEM (left curve), capillary plate (middle curve) and TGEM (right, called G10). The detectors are irradiated with alpha particles from an ^{241}Am source (α), X-rays from a 50 kV X-ray tube and gamma rays (γ). The TGEM requires higher voltages than GEM and capillary plates to form electron avalanches. On the other hand the maximum achievable gain is significantly higher for the TGEM. This is important to be able to detect single photoelectrons which is required in a good photosensitive detector.
From Östling, 2003.

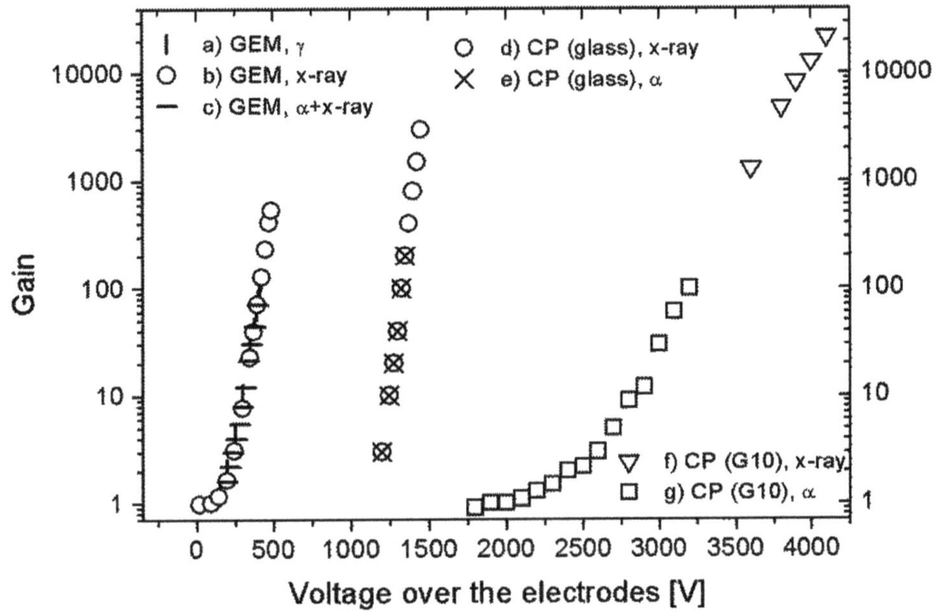

$$A_{max}n_0=10^6\text{-}10^7 \text{ electrons,} \tag{20}$$

where A_{max} is the maximum achievable gain before breakdowns appears and n_0 is the number of primary electrons produce by the charged particle (Fonte,1999, Peskov, 2001, Nappi, 2013).

4. High flux of charged particles which may lead to avalanches overlapping in time and space.
5. Electron emission from the cathode induced by the flux of positive ions (Fonte, 1999, Peskov, 2009, Nappi, 2013).

The situation is quite different when a hole-type detector is used for detecting visible or UV photons. In this case, the main mechanism of discharges is a feedback loop. The photon interacts with a solid, liquid or gaseous photocathode emitting a photoelectron. The primary photoelectron triggers the formation of a Townsend electron avalanche in the holes of the detector. This avalanche can produce secondary electrons outside of the avalanche, which in turn triggers new avalanches. The process is fast and can lead to a breakdown. The secondary electrons can be caused by many effects. It can be photons emitted in the initial avalanche process which irradiates the photocathode producing new photoelectrons. It can be ions drifting back to the top of the hole structure emitting electrons into the drift volume when they recombine.

The discharge appears when

$$A\,\gamma = 1, \tag{21}$$

where γ is the probability that the avalanche creates a secondary electron. Hence, caution shall be taken in the design of the detector so the probability of creating secondary electrons is minimized.

Even if capillary plates and TGEMs, in contrast to GEMs, can withstand sparks the energy released in the discharge can be high enough to damage the front end electronics. Hence, high current sparks shall be avoided by all means.

There several methods of minimizing or even avoiding damages caused by sparks. The front-end electronics can be protected by a chain of input diodes which drains the current from a spark to ground rather than feeding it into the amplifier. The capacitance of the detector can be reduced by segmenting the detector electrodes into smaller parts. This reduces the energy in a possible spark etc.

An alternative approach is to use resistive electrodes instead of metallic (Bidault, 2007; Oliveira, 2007). When a high voltage is applied to the resistive electrodes they charge up to an equipotential surface creating the same electric field structure as in a normal GEM. However, in case of a discharge, the spark is fed with charges from the whole electrode. The current in the spark is restricted be the resistance in the electrodes. Such a detector is a hybrid between a resistive plate chamber (RPC) and a GEM. It is often called a resistive electrode GEM or RETGEM.

In advanced designs the RETGEM the electrodes are divided into resistive strips as shown in Figure 25. Metallic readout strips located under the resistive strips and are used as antennas to pick up the induced electric signals (Di Mauro, 2009). A thin dielectric layer is placed between the resistive and the metallic strip. The dielectric is only 50-200 μm thick. The dielectric layer fully decouples the resistive strips from the metallic ones, and thus blocks any discharge current which wants to propagate to the input of the front-end electronics.

3.5.5 Cascaded Hole-Type Detectors

Gaseous photodetectors need to operate at gas gains close or above 10^5 to generate a large enough signal from a single photoelectron to detect it with 100% efficiency. The overall gas gain can be increased by cascading several hole-type detectors. One possible such arrangement is shown in Figure 26. The function of a cascaded hole-type detector is similar to what was earlier described to the combination of a MWPC in combination with a mesh parallel plate avalanche chamber.

The incoming photons interact with a photosensitive compound in the drift region creating photoelectrons. The photoelectrons follow the electric field lines into the holes of the top GEM where they initiate an electron avalanche. The gas gain of the top GEM is kept low, typically $10^3 - 10^4$ to minimize the risk of electrical breakdown. Most of the electrons in the first avalanche are collected on the lower electrode of the first GEM, but a significant fraction follow the electric field lines towards the holes of the second GEM. As they drift towards the second GEM they diffuse laterally which decreases the charge density of the electron cloud. When they reach the holes of the second GEM they initiate new electron avalanches. Also this GEM is operated at quite low gas gain, typically 10^3, to minimize the risk of breakdown. The induced signals from the second avalanche are detected by the pads in the readout plane.

The main advantages of cascaded hole-type detectors are:

Figure 25. Magnified photograph of a RETGEM; the strip electrodes shown in the figure consists of three layers. On top is a resistive layer which forms the electric field for the electron avalanche development. In the bottom is a metallic layer which acts as readout electrodes that pick up the induced signals from the avalanche. Between the two layers is a thin dielectric layer which electrically separates the two conductive layers. If a spark is formed on the resistive layer its resistance limits the current in the spark. The dielectric layer prevents the spark current from reaching the metallic layer and the readout electronics connected to it.
Di Mauro, 2009.

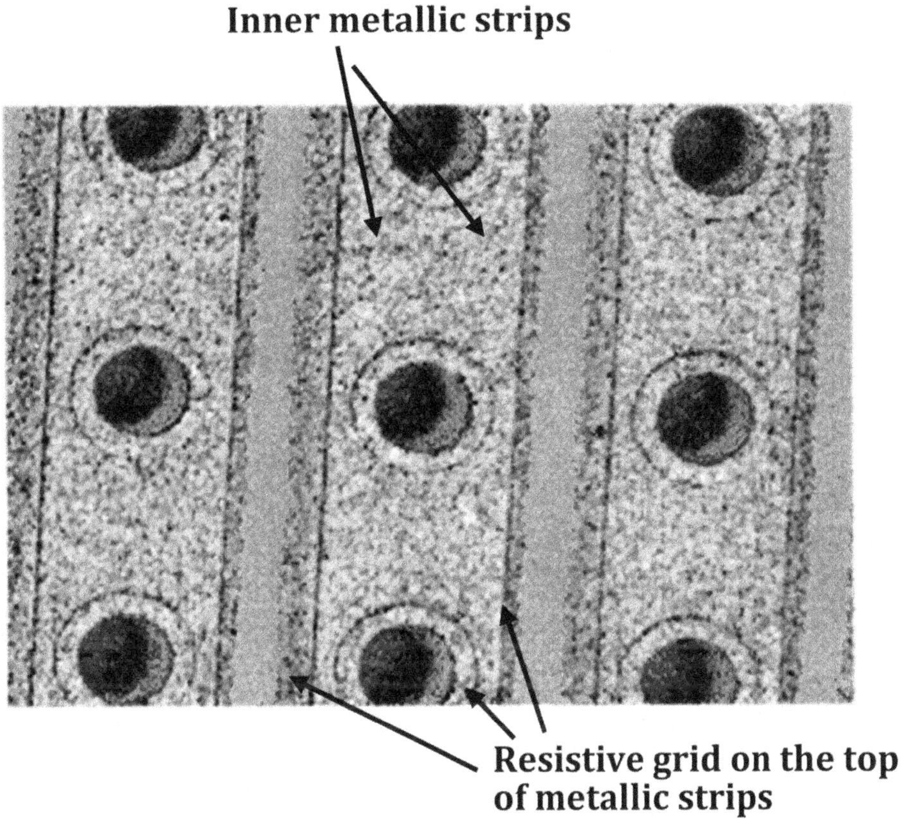

1. The total gain of the combined detectors can be higher than for a single detector as the avalanche is fed from one detector to the next.
2. Each electron amplifier in the cascaded detector operates at lower voltages and gains than for a single detector. This minimizes, or even eliminates, the occurrence of sparks.
3. Since the electrons that exit the holes of each electron multiplier diffuse laterally before they reach the next electron amplifier, hence decreasing the charge density, the value of the Raether limit for the entire cascaded detector increases (Fonte, 1999a; Peskov 2001). As a result, the probability of discharges greatly reduced at a maintained gas gain of 10^5 as desired for photosensitive detectors.

Two to four hole type detectors can be cascaded in this way. One of the hole-type detectors can also be replaced by a parallel plate avalanche chamber (Fonte, 1998; Fonte 1999a).

Figure 26. A cascaded hole-type detector consisting of two GEMs operating in tandem mode; the photoelectrons created in the drift region from the incoming photons are first amplified in an electron avalanche in the top GEM. A fraction of the electrons from the first GEM propagates to the second GEM where they are further amplified in a second electron avalanche. The induced signals from the second avalanche are read out by the pads in the readout plane.
From Park, 2007.

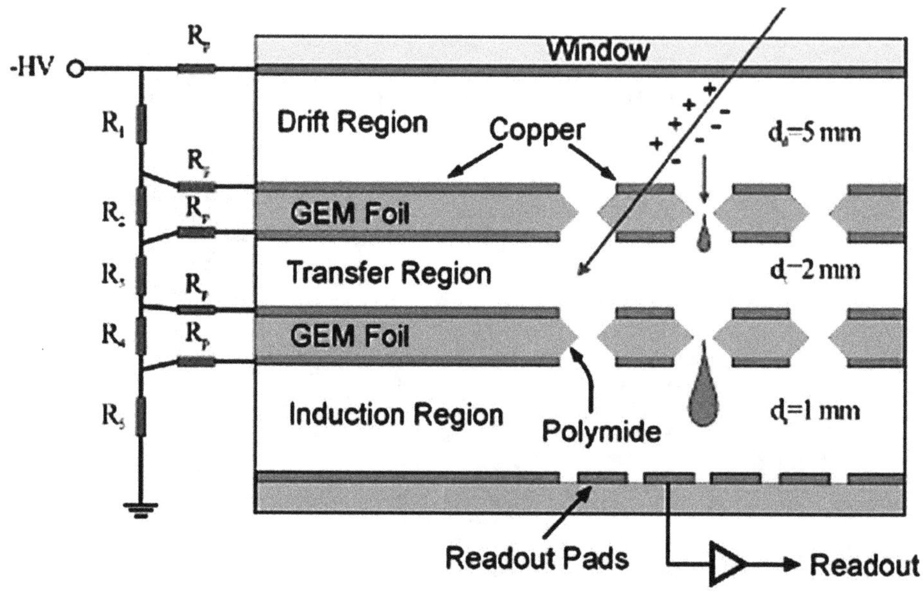

CONCLUSION AND REMARKS

In this chapter the main designs of gaseous photomultipliers has been reviewed: MWPC, PPAC, capillary plates, GEM and GEM-like detectors. More information about various gaseous detectors can be found in recent paper and books (Hilke, 2011, Titov, 2012, Nappi, 2013; Francke, 2014). Since decades MWPC or MWPC combined with PPAC are widely used in various high-energy physics experiments e.g. as Ring Imaging Cherenkov light detectors. UV sensitive MWPC are used in plasma diagnostics to visualize plasma motion and instabilities. In the last years hole-type detectors, and especial cascaded GEM or TGEM/RETGEM, are being used for photon detection in high energy physics experiments.

These types of detectors represent the ultimate photoelectron detectors. They have nearly 100% detection efficiency for single photoelectrons. A single photoelectron is the smallest charge there is. They can be manufactured at reasonable cost in large areas and in large volumes. They are robust and have a long life time. They are easy to operate.

They rely on one important factor to also be efficient photon detectors. They must be combined with an efficient photosensitive material which with high efficiency can convert a single photon into a photoelectron. The coming chapters are devoted to this task.

REFERENCES

Alkazov, G. D. (1970). Statistics of electron avalanches and ultimate resolution of proportional chambers. *Nucl. Instr. and Meth. in Phys. Res.*, *A89*, 155.

Aoyama, T. (1985). Generalized gas gain formula for proportional counters. *Nuclear Instruments & Methods in Physics Research. Section A, Accelerators, Spectrometers, Detectors and Associated Equipment*, *234*(1), 125–131. doi:10.1016/0168-9002(85)90817-4

Azevero, C. D. R., Cortesi, M., & Lyashenko, A. V. (2010). Towards THGEM UV-photons detectors for RICH: On single-photon detection efficiency in Ne/CH_4 and Ne/CF_4. *JINST*, *5*(P01002), 1–12.

Barouch, G. (1999). Development of a fast gaseous detector: Micromegas. *Nucl. Instr. and Meth.in Phys. Res.*, *A432*, 32–48.

Bellazzini, R., & Spandre, G. (2003). MicroPattern Gas Detectors with pixel read-out. *Nuclear Instruments & Methods in Physics Research. Section A, Accelerators, Spectrometers, Detectors and Associated Equipment*, *513*(1-2), 231–238. doi:10.1016/j.nima.2003.08.038

Bellazzini, R., Spandre, G., Brez, A., Minuti, M., Baldini, L., Latronico, L., & Braem, A. et al. (2007). Imaging with the invisible light. *Nuclear Instruments & Methods in Physics Research. Section A, Accelerators, Spectrometers, Detectors and Associated Equipment*, *581*(1-2), 246–253. doi:10.1016/j.nima.2007.07.098

Bidault, J.-M., Crotty, I., Di Mauro, A., Martinengo, P., Fonte, P., Galy, F., & Zanette, O. et al. (2007). The first applications of newly developed gaseous detectors with resistive electrodes for UV imaging in daylight conditions. *Nuclear Instruments & Methods in Physics Research. Section A, Accelerators, Spectrometers, Detectors and Associated Equipment*, *580*(2), 1036–1041. doi:10.1016/j.nima.2007.06.061

Black, J. K., Deines-Jones, P., & Readys, S. E. (2003). X-ray polarimetry with an active-matrix pixel proportional counter. *Nuclear Instruments & Methods in Physics Research. Section A, Accelerators, Spectrometers, Detectors and Associated Equipment*, *513*(3), 639–643. doi:10.1016/j.nima.2003.07.015

Bradbury, N. E. (1932). Photoelectric current in gases between parallel plates as a function of the potential difference. *Physical Review*, *40*(6), 980–987. doi:10.1103/PhysRev.40.980

Byrne, J. (1962). Proc. Roy. Soc. Edinburgh. *Sect.*, *A66*, 33.

Charpak, G. (1982). Some recent trends in the evolution of gaseous detectors. *Nuclear Instruments and Methods*, *196*(1), 1–9. doi:10.1016/0029-554X(82)90608-5

Charpak, G. (1999). *High resolution position detector for high-flux ionizing particles*. Patent WO-1996FR01576.

Charpak, G. (2002). *Micromegas, a multipurpose gaseous detector*. Nucl. Instr. and Meth.in Phys. *Res.*, *A478*, 26–36.

Charpak, G., Bouclier, R., Bressani, T., Favier, J., & Zupančič, Č. (1968). The use of multiwire proportional counters to select and localize charged particles. *Nuclear Instruments and Methods in Physics Research*, *62*(3), 262–268. doi:10.1016/0029-554X(68)90371-6

Charpak, G., Gauden, I., & Giomataris, Y. (1993). Some studies of the applications of CsI photocathodes in gaseous detectors. *Nuclear Instruments & Methods in Physics Research. Section A, Accelerators, Spectrometers, Detectors and Associated Equipment, 333*(2-3), 391–398. doi:10.1016/0168-9002(93)91181-L

Charpak, G., Majewski, S., Melchart, G., Sauli, F., & Ypsilantis, T. (1979). Detection of far-ultraviolet photons with the multistep avalanche chamber. Application to Cherenkov light imaging and to some problems in high-energy physics. *Nuclear Instruments and Methods in Physics Research, 164*(3), 419–433. doi:10.1016/0029-554X(79)90074-0

Charpak, G., Melchart, G., Petersen, G., & Sauli, F. (1979). High- accuracy localization of minimum ionizing particles using the cathode –induced charge centre-of-gravity read-out. *Nuclear Instruments and Methods, 167*(3), 455–464. doi:10.1016/0029-554X(79)90227-1

Charpak, G., & Sauli, F. (1978). The multistep avalanche chamber: Anew high-rate, high accuracy gaseous detectors. *Physics Letters, 78B*(4), 523–528. doi:10.1016/0370-2693(78)90502-6

Del Guerra, A., Conti, M., Gorini, G., Fujieda, I., Mulera, T. A., & Perez-Mendez, V. (1987). High resistance lead glass tubing for RICH counters and electromagnetic calorimeters. *Nuclear Instruments & Methods in Physics Research. Section A, Accelerators, Spectrometers, Detectors and Associated Equipment, 257*(3), 609–613. doi:10.1016/0168-9002(87)90971-5

Di Mauro, A., Lund-Jensen, B., Martinengo, P., Nappi, E., Oliveira, R., Peskov, V., & Santiard, J. C. et al. (2007). Development of innovative micro-pattern gaseous detectors with resistive electrodes and first results of their applications. *Nuclear Instruments & Methods in Physics Research. Section A, Accelerators, Spectrometers, Detectors and Associated Equipment, 581*(1-2), 225–231. doi:10.1016/j.nima.2007.07.083

Di Mauro, A., Martinenego, P., & Nappi, E. (2009). GEM with double layered micropattern electrodes and their applications. *IEEE Transactions on Nuclear Science, 56*(3), 1550–1557. doi:10.1109/TNS.2009.2020864

Endo, I., Kawamoto, T., & Mizuno, Y. (1981). Systematic shifts of evaluated charge centroid for the cathode readout multiwire proportional chamber. *Nuclear Instruments & Methods in Physics Research. Section A, Accelerators, Spectrometers, Detectors and Associated Equipment, 188*, 51–58.

Fonte, P., Ferreira-Marques, R., Peskov, V., & Policarpo, A. (1999). Single-electron pulse-height spectra in thin-gap parallel plate chambers. *Nuclear Instruments & Methods in Physics Research. Section A, Accelerators, Spectrometers, Detectors and Associated Equipment, 433*(1-2), 513–517. doi:10.1016/S0168-9002(99)00309-5

Fonte, P., Martinengo, P., Nappi, E., Oliveira, R., & Peskov, V. (2009). Progress in developing hybrid RPC:GEM-like detectors with resistive electrodes. *Nuclear Instruments & Methods in Physics Research. Section A, Accelerators, Spectrometers, Detectors and Associated Equipment, 602*(3), 850–853. doi:10.1016/j.nima.2008.12.154

Fonte, P., Peskov, V., & Ramsey, B. (1998). Rate and gain limitations of MSGC and MGC combined with GEM and other preamplification structures. *Nucl. Instr. and Meth.in Phys. Res., A419*, 405–409.

Fonte, P., Peskov, V., & Ramsey, B. (1999a). A study of breakdown limits in microstrip gas counters with preamplification structure. *Nucl. Instr. and Meth.in Phys. Res., A416*, 23–31.

Fonte, P., Peskov, V., & Ramsey, B. (1999b). The fundamental limitations of high-rate gaseous detectors. *IEEE Transactions on Nuclear Science, 46*(3), 321–325. doi:10.1109/23.775537

Fonte, P., Peskov, V., & Sauli, F. (1991). *VUV emission and breakdown in parallel plate chambers*. Nucl. Instr. and Meth.in Phys. *Res., A310*, 140–145.

Fonte, P. T., Francke, T., Pavlopoulos, N., Peskov, V., & Rodionov, I. (2005). Novel single photon detectors for UV imaging. *Nuclear Instruments & Methods in Physics Research. Section A, Accelerators, Spectrometers, Detectors and Associated Equipment, 553*(1-2), 30–34. doi:10.1016/j.nima.2005.08.002

Francke, T. (1991). *A Ring Imaging Cherenkov Counter using a Solid Radiator.* (Ph. D. Thesis). Royal Institute of Technology, Stockholm, Sweden.

Francke, T., & Peskov, V. (2014). *Innovative applications and developments of micro-pattern gaseous detectors*. IGI Global. doi:10.4018/978-1-4666-6014-4

Francke, T., Peskov, V., Rodionov, I., & Fonte, P. (2004). High rate (up to 10^5 Hz/cm^2), high position resolution (30 μm) photosensitive RPCs. *Nuclear Instruments & Methods in Physics Research. Section A, Accelerators, Spectrometers, Detectors and Associated Equipment, 533*(1-2), 163–168. doi:10.1016/j.nima.2004.07.021

Francke, T. V., Peskov, V., Rodionov, I., & Fonte, P. (2004). High rate (up to 10^5 Hz/cm^2), high position resolution (30 μm) photosensitive RPCs. *Nuclear Instruments & Methods in Physics Research. Section A, Accelerators, Spectrometers, Detectors and Associated Equipment, 533*(1-2), 163–168. doi:10.1016/j.nima.2004.07.021

Funfer, E., & Neiert, H. (1965). *Detectors of radiation*. Moscow: Atomizdat.

Giomataris, Y., Rebourgeard, P., Robert, J. P., & Charpak, G. (1996). MICROMEGAS: A high-granularity position-sensitive gaseous detector for high-flux environment. *Nuclear Instruments & Methods in Physics Research. Section A, Accelerators, Spectrometers, Detectors and Associated Equipment, 376*(1), 29–35. doi:10.1016/0168-9002(96)00175-1

Hilke, H.G. (2011). *Gaseous Detectors for Particles and Radiation. Part 1: Principles and Methods.* Landolt-Börnstein - Group I Elementary Particles, Nuclei and Atoms Volume 21B1, 72-106.

Lanciart, A., & Morucci, J. P. (1962). Article. *Journal de Physique et le Radium, 23*(suppl. 6), 102A. doi:10.1051/jphysap:01962002306010200

Llopart, X., Ballabriga, R., Campbell, M., Tlustos, L., & Wong, W. (2007). Tmepix, a 65k programmable pixel readout chip for arrival time, energy and/or photon counting measurements. *Nuclear Instruments & Methods in Physics Research. Section A, Accelerators, Spectrometers, Detectors and Associated Equipment, 581*(1-2), 485–494. doi:10.1016/j.nima.2007.08.079

Mathieson, E., & Gordon, J. S. (1984). Cathode charge distribution in multiwire chambers, II. Approximate and empirical formulae. *Nuclear Instruments & Methods in Physics Research. Section A, Accelerators, Spectrometers, Detectors and Associated Equipment, 227*, 277–282. doi:10.1016/0168-9002(84)90133-5

Meinschad, T., Ropelewski, L., & Sauli, F. (2004). GEM-based photon detector for RICH applications. *Nuclear Instruments & Methods in Physics Research. Section A, Accelerators, Spectrometers, Detectors and Associated Equipment*, *535*(1-2), 324–329. doi:10.1016/S0168-9002(04)01670-5

Melai, J., Breskin, A., Cortesi, M., Bilevych, Y., Fransen, M., van der Graaf, H., & Schmitz, J. et al. (2011). A UV sensitive integrated Micromegas with Timepix readout. *Nuclear Instruments & Methods in Physics Research. Section A, Accelerators, Spectrometers, Detectors and Associated Equipment*, *628*(1), 133–137. doi:10.1016/j.nima.2010.06.301

Nappi, E., & Peskov, V. (2013). Imaging gaseous detectors and they applications. Wiley-VCH.

Neumann, W. (1980). *The ultimate energy resolution of the proportional counter and the analysis of unresolved low energy spectra*. 7th Symp. on Microdosimetry, Oxford, UK.

Oliveira, R., Peskov, V., Pietropaolo, F., & Picchi, P. (2007). First tests of thick GEMs with electrodes made of a resistive Kapton. *Nuclear Instruments & Methods in Physics Research. Section A, Accelerators, Spectrometers, Detectors and Associated Equipment*, *576*(2-3), 362–366. doi:10.1016/j.nima.2007.03.010

Östling, J., Brahme, A., Danielsson, M., Francke, T., Iacobaeus, C., & Peskov, V. (2003). Study of hole-types gas multiplication structures for portal imaging and other high count rate applications. *IEEE Transactions on Nuclear Science*, *50*(4), 809–819. doi:10.1109/TNS.2003.814562

Papaevangelou, T. (2011). Micromegas technology. *Proc. 6th DITANET Topical Workshop on Detection Technique*.

Park, S., Hie-Hanng, C., & Ha, Y. (2007). Development of digital imaging device based on the gas electron multiplier. *Radiation Measurements*, *42*(10), 1661–1665. doi:10.1016/j.radmeas.2007.09.009

Periale, L., Peskov, V., & Carlson, P. (2001). *Detection of the primary scintillation light from dense Ar, Kr and Xe with novel photosensitive gaseous detectors*. ArXive Physics/0106048

Periale, L., Peskov, V., Carlson, P., Francke, T., Pavlopoulos, P., Picchi, P., & Pietropaolo, F. (2002). Detection of the primary scintillation light from dense Ar, Kr and Xe with novel photosensitive gaseous detectors. *Nuclear Instruments & Methods in Physics Research. Section A, Accelerators, Spectrometers, Detectors and Associated Equipment*, *478*(1-2), 377–383. doi:10.1016/S0168-9002(01)01779-X

Peskov, V., & Fonte, P. (2009). *Research on discharges in micropattern and small gap gaseous detectors*. ArXiv:0911.0463

Peskov, V., Fonte, P., Danielsson, M., Iacobaeus, C., Ostling, J., & Wallmark, M. (2001). The study and optimization of new micropattern gaseous detectors for high-rate applications. *IEEE Transactions on Nuclear Science*, *45*(4), 1070–1074. doi:10.1109/23.958725

Raether, H. (1964). *Electron avalanches and breakdown in gases*. London: Butterworths.

Reserva, R., Arogancia, D., Bacala, A., Boudjemline, K., Burke, D., Colas, P., & Zerguerras, T. et al. (2007). Studies on the drift properties and spatial resolution using a microMEGAS-equipped time projection chamber. *Pramana Jorn. of Phys.*, *69*(6), 989–994. doi:10.1007/s12043-007-0217-z

Rutherford, E., & Geiger, H. (1908). An electrical method of counting the number of alpha particles from radio-active substances. *Proc. R. Soc. Lond., Ser. A, 81*, 141–161. doi:10.1098/rspa.1908.0065

Sakurai, H., Tamura, T., Gunji, S., & Noma, M. (1966). A new type of proportional counter using a capillary plate. *Nuclear Instruments & Methods in Physics Research. Section A, Accelerators, Spectrometers, Detectors and Associated Equipment, 374*(3), 341–344. doi:10.1016/0168-9002(96)00226-4

Sauli, F., (1977). *Principles of operation of multiwire proportional and drift chambers*. CERN Yellow report 77-09.

Sauli, F. (1997). GEM: A new concept for electron amplification in gas detectors. *Nuclear Instruments & Methods in Physics Research. Section A, Accelerators, Spectrometers, Detectors and Associated Equipment, 386*(2-3), 531–534. doi:10.1016/S0168-9002(96)01172-2

Sauli, F. (2006). *Gas electron multiplier detectors: performance and applications*. Presentation at the MPGD Workshop, CERN, January 2006. Retrieved from http://gdd.web.cern.ch/GDD/

Shalem, C., Chechik, R., Breskin, A., & Michaeli, K. (2006). Advances in Thick GEM-like gaseous electron multiplier-Part-1: Atmospheric pressure operation. *Nuclear Instruments & Methods in Physics Research. Section A, Accelerators, Spectrometers, Detectors and Associated Equipment, 558*(2), 475–489. doi:10.1016/j.nima.2005.12.241

Theobald, J. K. (1953). Investigation of back diffusion of photoelectrons in various standard gases as it affects secondary electron emission coefficients. *Journal of Applied Physics, 24*(2), 123–128. doi:10.1063/1.1721223

Titov, M. (2012). Gaseous Detectors. In Handbook of Particle Detection and Imaging, (pp. 239-264). doi:10.1007/978-3-642-13271-1_11

Townsend, J. S. E. (1903a). The conductivity produced in gases by the aid of ultraviolet light. *Philosophical Magazine, 5*(28), 389–398. doi:10.1080/14786440309462937

Townsend, J. S. E. (1903 b). The genesis of ions by motion of positive ions in gas and a theory of the sparking potential. *Philosophical Magazine, 6*(35), 598–618. doi:10.1080/14786440309463061

Townsend, J. S. E., & Kirkby, M. A. (1901). Conductivity produced in Hydrogen and carbonic acid by motion of negatively charged ions. *Philosophical Magazine, 1*(6), 630–642. doi:10.1080/14786440109462656

Ullrich, T. (1996). Cherenkov ring fitting techniques for the CERES RICH detectors. *Nucl. Instr. and Meth.in Phys. Res., A371*, 243.

KEY TERMS AND DEFINITIONS

Calorimeter: A detector used in high energy and astrophysics experiments to measure the energy deposited inside it by electro-magnetic and hadronic showers. Some calorimeters are capable to determine also the energy of absorbed gamma photons and the energy deposited by passing charged particles.

Cascaded Detectors: Detectors consisting from several multiplication structures so that the primary electrons experience several consequent multiplication processes. Cascaded detector are used to suppress feedback processes and to achieve high overall gains.

Dipole Electric Field: An electric field formed by two opposite but equal charges separated by a certain distance.

Electron Microscope: A microscope using accelerated electrons instead of light as a source of illumination. Electron microscopes have higher resolving power compare to light microscopes, because electrons wavelengths on orders of magnitude are shorter than visible light.

Electron Multiplier: A device in which a single primary electron via various processes can create up to 10^6 or even more secondary electrons. This allows electronically detect single electrons. Electron multiplier can be vacuum, gaseous or solid-state types.

Gaseous Photomultiplier: A gas–filled device converting UV or visible photons to primary photoelectrons which in turn trigger Townsend avalanches, allowing electronically detect primary electrons.

Quenching Gases: A mixture of gases (typically noble gases with some molecular gases) in which secondary processes are strongly suppressed. This allows reaching high gas gains.

UV Photons: Photons with wavelength roughly between 30 and 400 nm.

VUV Photons: Photons emitting in spectral interval between 30 and 185 nm.

APPENDIX: SIGNAL DEVELOPMENT, READOUT TECHNIQUES, AND HOW POSITION MEASUREMENTS ARE PERFORMED

In this chapter the main designs of gaseous electron multipliers have been reviewed, e.g. multiwire proportional chambers, parallel-plate avalanche counters, resistive plate chambers, and hole–type gas electron multipliers. One of the main features of gaseous photomultipliers is their position resolution, so it is important to understand in more detail how the position of avalanches is determined.

The signal induced by an avalanche in the gaseous detector electrodes consists from two components, a fast and a slow as shown in Figure 25. The fast component is related to the avalanche electron collection, while the slow component is induced by slow ion movement towards the cathode (see paragraph 1 and Figure 1). After reaching the maximum the signal starts dropping due to the finite RC time constant (R is the resistivity and C is the capacitance) of the input electronic circuit. Slow ion movement makes the signal too long, so in practice a shaping amplifier is used to restrict the contribution of the slow component in the signal formation.

Let's consider concrete detectors.

MWPC

An avalanche moving towards the anode wire induces a negative signal on this wire and positive signals on surrounding electrodes, i.e. the neighboring anode wires and the cathode. The negative signal indicates at which wire the avalanched is developed. This simple approach was applied in photosensitive MWPCs used in plasma diagnostics for visualization of its instabilities in VUV region as well as for spectroscopic measurements.

In more sophisticated designs of photosensitive MWPCs their cathodes are segmented in isolated parts, either wires or strips, oriented perpendicular (or at some angle) to the anode wire, or in pads. The coordinate of the avalanche along the wire is usually determined by one of two readout methods:

1. By a charge division method (for this resistive anode wires are used), or
2. From the signals induced on the cathode strips/pads

The first method relies on the fact that the charge collected at either end of a resistive anode wire is divided in proportion to the length of wire from the point at which the charge is injected (see Figure 26).

If Q_A and Q_B are the charges collected then the coordinate along the wire is then the coordinate along the wire is:

$$y = L\, Q_A/(Q_A+Q_B), \tag{20}$$

where L is the length of the wire.

This readout method was used in SLD RICH (SLD stands for SLAC-Stanford linear accelerator center- Large Detector.

The second method is based on determining the center of distribution of induced signals on segmented cathode (see Figures 7 and 8), each connected to its own amplifier. The induced signals are largest on the strip closest to the avalanche and diminish proportionately with distance from the avalanche point.

Figure 7. A detailed simulation of the development of a Townsend electron avalanche near an anode wire in an MWPC; the simulation illustrates the drift of the electron towards the wire, and how it scatters by collisions with the gas atoms. At a few wire radii away from the wire it triggers an electron avalanche, visible as the formation of new branches of tracks. The majority of the charges are formed close to the wire (< 1 wire radius away) in the exponentially growing avalanche. The whole process is over in less than 100 ps.
From http://www.science.mcmaster.ca/medphys/images/files/courses/4R06/4R6Notes3_GasFilled_Detectors.pdf.

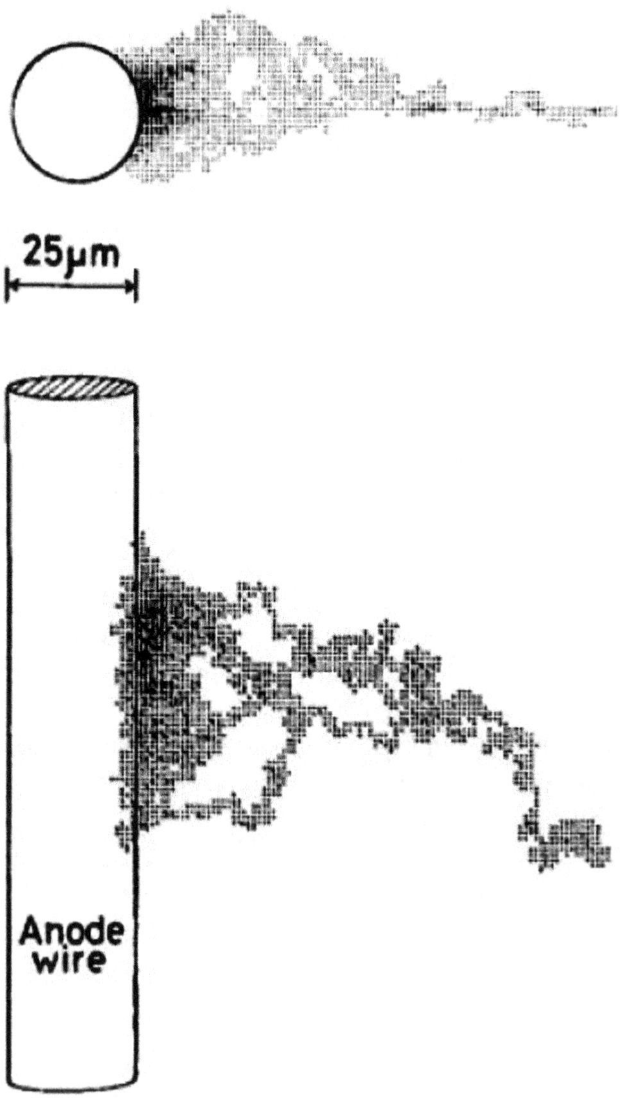

If y_i is the coordinate of the ith strip and Q_i is the measured charge on that strip, then the avalanche point can be estimated by calculating the center of gravity,

$$y = \Sigma(Q_i - b)y_i \,/\, \Sigma(Q_i - b), \tag{21}$$

where b is a small bias which is subtracted from each Q_i in order to correct for the dispersive effects of noise, see for details (Charpak, 1979)

Center of gravity method was widely used in MWPCs flushed with TEA or TMAE vapors. For example in DELPHI the coordinates of the Cherenkov ring were measured through the induced pulse in the cathode strips perpendicular to the wires.

In the latest generation of RICH detectors, in which MWPC are combined with CsI photocathodes, pad cathodes are used (see Figure 8) with individual analog readout. Such detectors are used in several currently running high energy physic experiments.

2. PPAC

CsI- PPAC used for the readout of UV scintillators had pad readout, see reference (Charpak, 1979) for more details.

Beside the traditional PPAC, some groups try to adopt for UV imaging new innovative designs of PPACs, having a very small gap (~100 μm) between the cathode micro mesh and the anode, called MICROMEGAS. Figure 22 in Chapter 15 shows the full detector assembly, the MICROMEGAS structure (the micro mesh supported by micro pillars defining the amplification gap), the CsI photocathode deposited on the micromesh, or sometimes on the additional mesh, and anode plate. In the latest designs, the Timepix chip (Llopart, 2007) containing a matrix of 256 x256 square pixels with a size of 55 μm is used for position measurements. An additional mesh cathode often was placed 3-5 mm above the micromesh. In some measurements the cathode was kept at the same potential as the micromesh to ensure that all field lines extend through the holes into the amplification region below the grid. The field distribution served two purposes; to extract the primary photoelectrons from the photocathode and to guide them towards the holes. In some cases this region is used for the preamplification. The primary electrons are collected into the multiplication region where they experience a multiplication by the high field strength below the grid (e.g. 40–60 kV/cm). The resulting avalanche charge cloud was collected at the pixelated anode allowing to electronically visualize UV radiation.

Each of the pixels has a metal input pad connected to an input capacitor. The collected avalanche charge is read out with a charge sensitive pre-amplifier and is compared with a threshold value (which can be adjusted per pixel). There are different readout modes. In Medipix2 mode, the chip can be used to count the number of events above a selected threshold during a defined shutter period. In Timepix-mode the arrival time of the first hit above threshold is recorded per pixel. In Time Over Threshold -mode the14 bit counter is incremented as long as the signal remains above threshold. This mode can be used to measure the input charge.

3. PHOTOSENSITIVE RPC

In existing photosensitive RPCs strip readout is used and in most of applications they were exploited in digital mode, when a counting rate from the strips was measures for signals above the given threshold (Francke, 2004; Fonte, 2005).

4. HOLE–TYPE GAS ELECTRON MULTIPLIERS

Usually hole–type electron amplifies are combined with readout plates (see Figure 21). Avalanche electrons extracted from the holes are drifting towards to the readout plate induced signals there. In first prototypes the readout plate was segmented into to pick up strips each connected in its own amplifier (Meinschad, 2004). The coordinate of the photons was determined by the center of gravity method.

Later in UV imaging applications digital readout from strips was implemented as more adequate for these measurements, see for example (Bidault, 2007; Di Mauro, 2007).

Pixelated readout introduced earlier in pioneering works (Black, 2003; Bellazzini, 2003) also was used in some prototypes. In work (Bellazzini, 2007) a GEM combined with a semitransparent CsI photocathode was coupled to a custom CMOS VLSI pixel array (CMOS stands for complementary-symmetry metal–oxide–semiconductor and VLCI stands for Very-large-scale integration; it is the process of creating an integrated circuit by combining thousands of micro-transistors into a single chip) which served, at the same time, the charge collecting electrode and the analog, self-triggering, readout electronics. The fine pitch of both the read-out system and of the amplification structure (50 μm) and the possibility to work in single photon detection allows reaching avalanche reconstruction accuracy of 4 μm.

In TGEM and photodetectors used for Cherenkov light detection, when spatial resolutions of about few mm are sufficient pad readout is used. The pad could be as large as 8x8 mm^2.

As can be seen from the short description of the readout methods used in the case of position-sensitive gaseous photomultipliers, most of them are based on simultaneous recording of signals from segmented cathodes. The charge stored in the input capacitors is readout after receiving a trigger

This inevitably leads to systems equipped with thousands of read-out elements (strips, pixels) and consequently the same number of electronic channels. In some cases, it increases the total cost of a project. In general, the size of the electronics can be reduced considerably by connecting several read-out elements to a single channel through an appropriate multiplexing pattern: the charges stored in the input capacitors are readout in an optimized way after receiving a trigger.

Chapter 3
Position-Sensitive Gaseous Photomultipliers Filled with Photosensitive Vapours

ABSTRACT

The ultimate goal of all development of photosensitive detectors is to find a detector capable of detecting single photons with high efficiency. Furthermore, the photon shall not only be detected as a photon somewhere. We want to know where it was with high precision in space, often down to a few micrometers. We want to know when it was there, preferably with a precision of less than a nanosecond. We want to know where and when for each individual photon in a high flux of photons. Sometimes we even want to know the polarization of each photon. Position-sensitive gaseous photomultipliers filled with photosensitive vapours are capable of all of this. It is a challenging task. A single photon is the weakest light there is. For UV and visible light the energy in the photon is so low that it can barely emit a single electron through photoelectric effect with a gas. This photoelectron has practically no kinetic energy when it is released. A single electron at rest is the weakest possible electrical signal there is, so the detector must be able to amplify this extremely weak signal without any noise. We will here describe the history of photosensitive gaseous detectors, their applications and what the state of the art technology is today.

1. INTRODUCTION

A breakthrough in the development of instruments for detecting individual photons was the invention of the first position sensitive gaseous photomultiplier. As often happens in science, the first prototypes of these detectors were developed by two independent groups, one at CERN (Séguinot, 1977) and the other one in P.L. Kapitza laboratory in Moscow (Bogomolov, 1978). The two historical papers describing these first instruments were submitted in 1976 only two months apart. Both teams used the newly invented multiwire proportional chamber (MWPC) as amplification structure (Charpak, 1968). The breakthrough was to fill the MWPC with vapours with low ionization potential, which made the detector photosensitive. A photosensitive gas had been used long time ago in a single-wire counter (Chubb, 1955). However,

DOI: 10.4018/978-1-5225-0242-5.ch003

this detector had no position resolution. It could just tell whether there was light or not. The big step was the capability of the MWPC to determine a two dimensional position of each avalanche and hence the position of each absorbed photon. These photomultipliers became the first UV imaging detectors.

The photosensitive MWPC has several other important features. For example, one can build a photosensitive MWPC detector with a very large sensitive area (~m^2). As it is operated at atmospheric pressure there are no mechanical constraints on the size of the window. The detector can be used with a continuous flow of gas, which makes it less sensitive to outgassing, oxygen leaking in and other impurities in the gas.

There were some important differences between the MWPC designs used by the two groups (Séguinot, 1977; Bogomolov, 1978). In the first experiments Séguinot et al. used a small standard MWPC with an anode wire diameter of 20 μm. The detector was either filled with a mixture of argon (Ar) with 11 Torr of benzene vapour or argon +99Torr of CO_2+11Torr of benzene vapour. In both cases the total pressure was 1 atmosphere. The ionization threshold of benzene (E_i) is 9.15 eV, meaning that UV photons with a wavelength shorter than 137 nm will be able to ionize the gas and liberate a photoelectron, which in turn can trigger an electron avalanche in the MWPC.

The gas gains in these particular gas mixtures were limited by photon feedback. There are many more UV photons produced by de-excitation of atoms in the electron avalanche than secondary electrons. Hence, if the photosensitive vapour is sensitive to the wavelength of these secondary photons and the absorption length is long enough, the secondary UV photons will trigger secondary avalanches. The maximum achievable gas gain with the two gas mixtures was 10^4 and 10^5, respectively.

All early MWPC designs had this drawback. At high gas gains the photon feedback frequently caused discharges to appear at the dielectric surfaces supporting the anode wires and the cathode mesh. Later, to avoid feedback and discharge problems the CERN group introduced a multistep photosensitive chamber. It was a MWPC combined with one or two parallel plate avalanche chambers made of meshes (Charpak, 1979). This device was successfully used for detection of Cherenkov photons and was the foundation for the first RICH detectors. Eventually, the MWPC was modified and made capable to operate at high gas gains without photon feedback and without a preamplification stage.

The photosensitive MWPC of Bogomolov et al. was flushed with a mixture of argon and 20 Torr of toluene. The ionization threshold of toluene (E_i) is 8.8 eV, meaning that UV photons with a wavelength shorter than 155 nm will be able to ionize the gas, liberate photoelectrons, and trigger electron avalanches in the MWPC. In this mixture the maximum achievable gas gain they obtained was close to 10^6. This exceptionally high gas gain was achieved due to two important features of their MWPC:

1. They used much thicker anode wires (in diameter) than in the case of the detector of Séguinot et al. As will be shown later, the spectrum of the secondary UV photons emitted in the electron avalanche not only depends on the gas mixture, but also on the strength of the electric field, and hence on the anode wire diameter. This has important effects on the amount of photon feedback.
2. Their MWPC had a specially designed dielectric interface between the anode and the cathode electrodes. Rectangular grooves were cut out making the distance from the conductive part of the cathode to the conductive part of the anode significantly longer, as shown in Figure 1. This prevents discharges from forming along the dielectric surface.

Figure 1. Drawing of the first photosensitive MWPC developed for visualization of UV photons by Bogomolov et al. Groves were cut out in the dielectric frame supporting and separating the cathode from the anode (hashed region). This makes the distance from the conductive part of the cathode (2) to the anode wires (5) significantly longer. In this way the formation of discharges along this dielectric surface is eliminated, or greatly reduced.

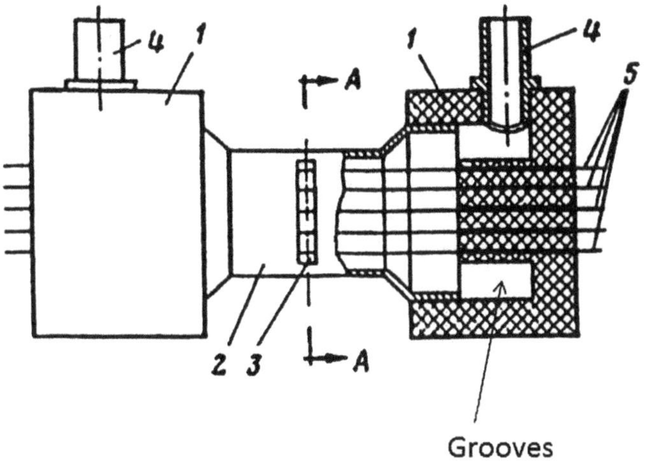

This detector was later extensively used for plasma diagnostics and in particular for visualization of plasma instability by imaging the UV photons emitted by the plasma.

Whatever the similarities and differences between Séguinot's and Bogomolov's MWPCs, these two studies initiated a new direction in the development of photodetectors. The high energy physics community rapidly picked up on these results and developed them further.

These further developments were performed in two main directions:

1. The search for photosensitive vapours with lower ionization potential than benzene or toluene to be able to detect photons with wavelengths longer than 155 nm.
2. The development of large area Ring Imaging Cherenkov (RICH) detectors.

The search for gases with low ionization potential will be described below. The development of RICH detectors will be described later in the application chapters.

2. SEARCH FOR VAPOURS WITH LOW IONIZATION POTENTIAL

Inspired by pioneering developments, several high-energy physic and plasma physics groups started to search for other gases and vapours than benzene or toluene with low ionization potential. It turned out not to be an easy task.

There are several problems associated with many vapours with low ionization which limit their use in a practical detector. The typical ones are:

1. Some substances have too low vapour pressure at room temperature to be an efficient convertor of UV photons to photoelectrons.
   Quantitatively this can be described as follows. Let's for simplicity consider a mixture of a noble gas and a photosensitive gas. The mean free path of the photoelectrons in the photosensitive gas is:

   $$l_{ph} = 1/(n_{vap}\, \sigma_{abs}(\lambda)), \tag{1}$$

   where n_{vap} and $\sigma_{abs}(\lambda)$ are the density and the ionization cross section of the photosensitive gas, respectively. λ is the photon wavelength.
   The number of photoelectrons created from light with a spectrum $I(\lambda)$ and assuming full absorption is:

   $$N_{pe}(\lambda) = k_{pe} \int_{\lambda w}^{\lambda i} I(\lambda)\, Q(\lambda)\, d\lambda, \tag{2}$$

   where k_{pe} is a coefficient and $Q(\lambda)$ is the quantum efficiency of the photosensitive gas. The integration is performed from the window's transmission threshold, λ_w, to the wavelength λ_i corresponding to the ionization potential E_i, where $\lambda_i = hc/E_i$. Here h is Plank's constant and c is the speed of light. If the light passes through a gas layer with the thickness of L where L is not long enough to give full absorption, then the number of created photoelectrons will be:

   $$N_{pe}(\lambda) = k_{pe} \int_{\lambda w}^{\lambda i} I(\lambda)\, Q(\lambda)\, (1 - \exp(-L/l_{ph}))\, d\lambda, \tag{3}$$

   In the case where $L << l_{ph}$:

   $$N_{pe}(\lambda) = k_{pe}\, (L/l_{ph}) \int_{\lambda w}^{\lambda i} I(\lambda)\, Q(\lambda)\, d\lambda, \tag{4}$$

   As can be seen, in this case the efficiency of the detector is inversely proportional to the absorption length of the photons in the gas, l_{ph}, and therefore can be very low.
2. Some photosensitive gases or vapours are chemically aggressive. They can react with the construction materials, or the other gases used for avalanche multiplication and destroy them.
3. Some photosensitive gases react with air, which makes them difficult to handle and to transfer into the gas chamber.
4. Some photosensitive gases are poisonous or hazardous in other ways.

In the end, few gases are suited to be used in photosensitive detectors. As a result these gases are widely used.

The most common photosensitive gases and vapours used in photosensitive MWPCs are:

- Trimethylamine (TMA),
- Triethylamine (TEA),
- Ethylferrocene (EF),
- Tetrakis(dimethylamino)ethylene(TMAE), usually pronounced as "tammy".

Each of these gases will be described in more detail below.

2.1 Trimethylamine (TMA)

The ionization potential of trimethylamine, ($(CH_3)_3N$) is 7.8 eV making it useful for detecting UV light with a wavelength shorter than 160 nm.

TMA is actually a gas at room temperature, but it is often used in water solution to make it a liquid. It smells like rotten fish in low concentrations and ammonia in high concentrations. The high vapour pressure (see Figure 2) makes it difficult to adjust the TMA concentration in a gas mixture. If it is not actively adjusted by e.g. measuring the flow of TMA and the other gases, the TMA concentration easily gets too high and the amount of amplification gas too low for high enough gas gain in the detector. TMA is slightly reactive with metals like copper and brass.

In the 60-ies and 70-ies TMA was used as a photo-converter in commercial stainless steel single-wire photosensitive counters called CFM-3. The detectors were sealed and the TMA was used both as photo-converter and amplification gas. They were produced by the Institute for vacuum technique in Moscow (Diatroptov, 1972). The CFM-3 was widely used in various spectroscopic measurements, e.g. in focal planes of vacuum monochromators.

The quantum efficiency of a CFM-3 detector is presented in Figure 3.

Later the quantum efficiency of TMA vapour in full absorption was measured by Séguinot (Séguinot, 1989). He used a gaseous avalanche chamber where the photo-conversion gap was thick enough for all photons to convert into photoelectrons. His results are shown in Figure 4. As can be seen, the quantum efficiency of TMA is as high as the best vacuum photomultipliers, offering a competitive alternative to the expensive photomultipliers in some applications.

Figure 2. The vapour pressure as a function of temperature of TMA and TEA TMA is actually a gas at room temperature. The high vapour pressure makes it difficult to mix it with other amplification gases in a correct concentration. TEA on the other hand has a low vapour pressure and can be used in saturated mode.

Swift, 1945; Dutt, 1982, Anderson, 1988.

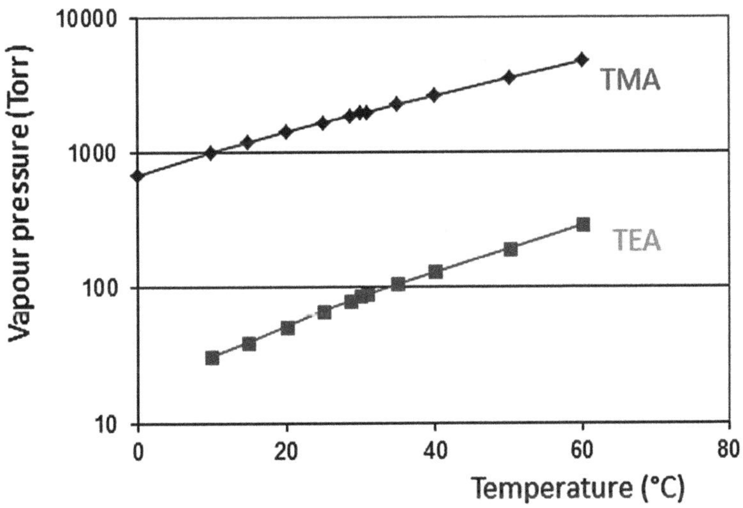

Figure 3. The sensitivity of the commercial single-wire detector CFM-3, which used TMA as a photo-converter; it was often used to calibrate the wavelength of vacuum monochromators. The sensitivity is quite low as the detector was not thick enough to absorb all photons.
From Diatroptov, 1972.

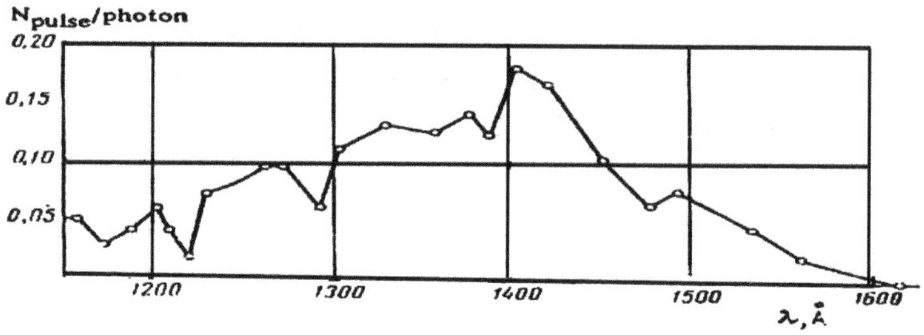

Figure 4. The quantum efficiency of TMA vapour measured at full absorption using a gaseous avalanche chamber with methane, CH4, and helium as an amplification gas, respectively. TMA is sensitive for photons with a wavelength shorter than 160 nm. The cut-off at 135 nm in the curve where CH_4 is used as amplification gas is due to absorption of the photons by the CH_4. Helium is transparent to much shorter wavelengths.
From Séguinot, 1989.

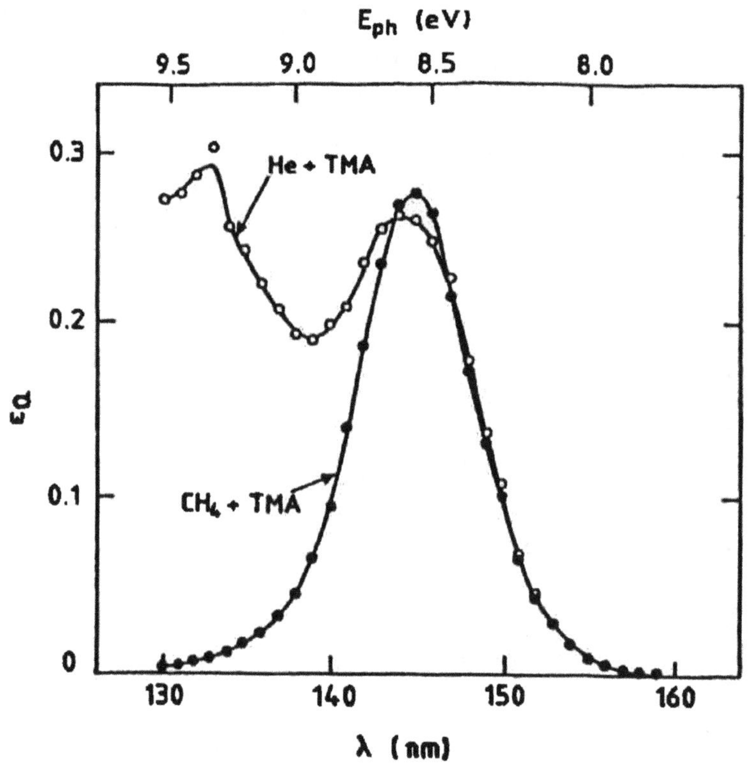

A continuous flow of TMA was used for the first time in a photodetector to detect secondary UV light from a gas scintillation counter (Peskov, 1980).Other applications of TMA as a photo-converter are described in (Cebrian, 2013: Alavarez, 2014).

2.2 Triethylamine (TEA)

Between 1978 and approximately 1983 triethylamine ($N(CH_2CH_3)_3$), TEA, was the most popular photosensitive compound. The main application was the detection of the Cherenkov photons. TEA has similar characteristics as TMA. It smells of rotten fish in low concentrations, and of ammonia in higher concentration.

TEA has two advantages over TMA:

1. A slightly lower ionization potential (7.5 eV), which makes it sensitive to photons with wavelength shorter than 165 nm.
2. A lower, but still sufficiently high, vapour pressure. The vapour pressure is 50 Torr at 20°C, see Figure 2. This makes it easier to handle and control its concentration in gas mixtures. The vapour pressure is sufficient for full absorption of UV photons at relatively short distances. A 2 mm absorption gap is sufficient at 20°C.

The quantum efficiency of the TEA is quite similar to that of TMA (compare Figure 4 and Figure 5).

The first tests of a photosensitive MWPC filled with a mixture of TEA + Ar showed that the gas gain could not be high enough for efficient single photoelectron detection. At the high gas gains necessary for

Figure 5. The quantum efficiency of TEA vapour at full absorption; the quantum efficiency is similar to TMA. The lower vapour pressure quickly made TEA more popular than TMA.
From Séguinot, 1989.

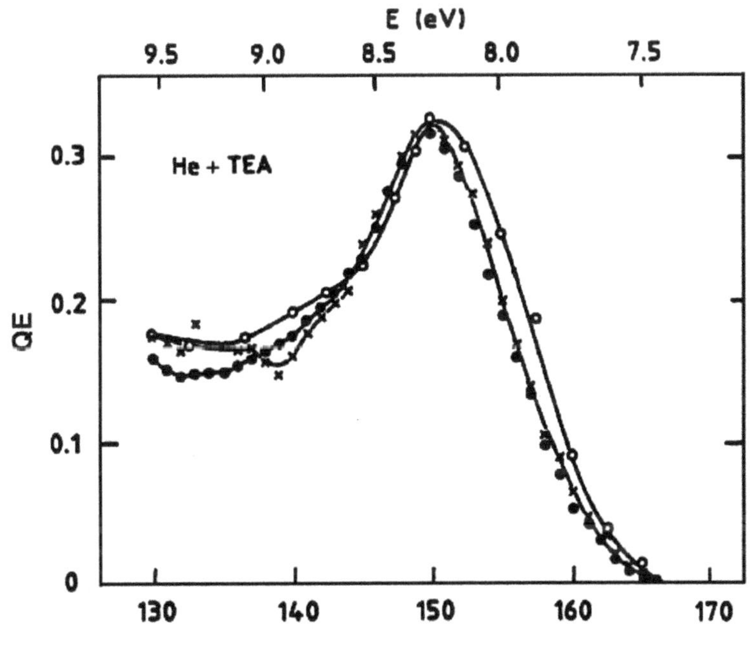

single photoelectron detection, ~10^5, photon feedback occurs. Photons emitted by the avalanche ionize the TEA vapour and creates secondary electrons at various distances from the primary avalanche. These then drift towards the anode wire and trigger secondary avalanches.

Higher gains without photon feedback problems were possible to achieve only with a multistep detector. Often the MWPC was combined with one or two parallel plate avalanche chambers used as a preamplification structures. If the gap between the parallel plate avalanche chamber and the MWPC and is more than 2 l_{ph}, (where l_{ph} is the absorption length of the feedback photons in TEA), then most of the feedback photons produced in the avalanches of the MWPC do not reach the conversion space between the parallel plate avalanche chamber and the entrance window. In this way, any photoelectrons produced from feedback photons created in the MWPC will not be multiplied by the parallel plate avalanche structure and hence give too small signals to be detected. On the other hand, Cherenkov photons entering the detector volume via the window and produce photoelectrons in the conversion gap, will obtain full multiplication, both in the parallel plate avalanche chamber and in the MWPC. The signals produced from the Cherenkov photons are at least 10 times higher than the signals from the secondary electrons produced by the avalanche feedback photons.

The first Ring Imaging Cherenkov (RICH) detectors had this multistep design. Later, and after many tests, it was discovered that gains close to 10^6 can be achieved without feedback even in a single-stage MWPC if the argon amplification gas is replaced my methane, CH_4 (Séguinot, 1980; Ypsilantis, 1981). The reason is that the feedback photons emitted in electron avalanches in methane have longer wavelength than in argon. The TEA is less sensitive to these photons and the photoionization by avalanche feedback photons becomes less efficient.

After this important discovery most of the TEA-based RICH detectors were dramatically simplified exploiting a single MWPC filled with a mixture of TEA and CH_4 or TEA and ethane, C_2H_6, which gives an even more stable operation than methane (Artuso, 2003).

More detailed description of various detectors filled with TEA vapours will be given in the chapter on applications.

2.3 Ethylferrocene (EF)

Ethylferrocene, $C_{12}H_{14}Fe$, is an organometallic compound. It was originally used in military or space applications as an additive in rocket propellant, to promote the burning rate. Gaseous detectors (single-wire counters as well as MWPCs) filled with gas mixtures containing EF vapours were successfully used in 1969-1986 for the detection of UV light from plasma and visualization of plasma instabilities.

The main attraction of EF is that the ionization potential (E_i=6.08 eV) is less than for TMA and TEA. It is sensitive to photons with wavelengths up to 204 nm. The quantum efficiency is slightly lower than for TMA and TEA, as shown in Figure 6.

EF is chemically inert and can be exposed to air without spoiling its properties. However, it has a big disadvantage, its low vapour pressure (see Figure 7).

To be able to use EF as an efficient photo-convertor one either have to use a long drift space (to ensure a sufficient absorption of the photons), or heat the EF to 60-80ºC.

In plasma studies where one wants to study the emission of UV photons from plasma, the EF-based photodetectors are heated. In other applications, for example for the detection of soft X-rays from the plasma, room temperature EF can be used since the iron atom in the molecule has a good absorption efficiency for soft X-rays (see the chapter on applications).

Figure 6. The quantum efficiency of ethylferrocene, EF; it has a lower ionization potential than TMA and TEA, meaning that it is sensitive to photons of longer wavelength. The quantum efficiency is lower than for TMA and TEA, why it did not reach as much popularity as them.
From Charpak, 1989.

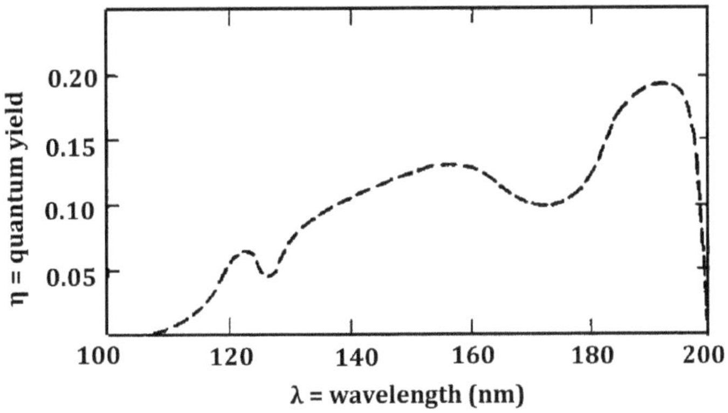

Figure 7. The vapour pressure of EF as a function of temperature; the vapour pressure is far too low at room temperature to be able to have an efficient absorption of photons in a reasonable short distance. For EF to be efficiently used is has to be heated to 60°C or more, which complicates the design of the detector.
Open circles (from Melin, 1974), Diamond and crosses-data from (from Peskov, 1989).

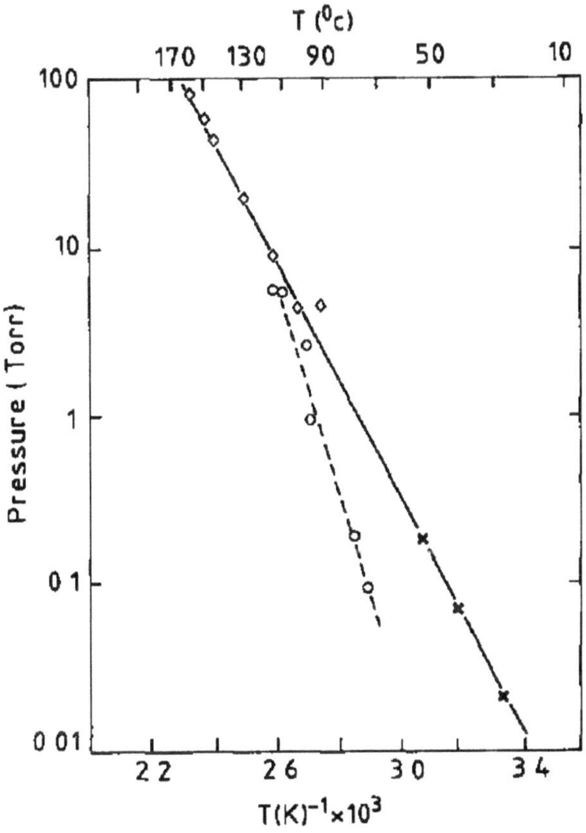

In 1986 EF was tested in the Charpak group at CERN where it was used for the detection of scintillation light from various solid scintillators. This is how Charpak describes this fascinating time in his book "Research on particle imaging detectors" (Charpak, 1995):

We were enriched by the coming of a Russian physicist Vladimir Peskov. [...] He found new photosensitive vapours compatible with chambers. But it is clear from our work, the spreading of new techniques in detector physics is strongly coupled to the environment. [...] Coming to CERN, Peskov found an opportunity to make important new contributions by bringing his experience to several groups. He came with his own bottle of ethylferrocene, a liquid made of a substance with a low ionization potential.

Below is a true and fun story describing some tests of EF in the Charpak group reprinted from Sauli's book "Seminal paper by Georges Charpak 1924-2010 (Sauli, 2011):

Magic Ethylferrocene

Developing new photo detectors served as the Charpak group's main focus at CERN when I (Vladimir Peskov) joined George's team in 1986...

From Moscow I (Vladimir Peskov) brought with me a small bottle filled with ethylferrocene (EF), another photosensitive liquid, which we previously used in the P.L. Kapitza laboratory for the visualization of UV radiation from plasma. EF had slightly higher ionization potential than TMAE (6.08 eV) and lower quantum efficiency; however, it was not reactive at all with air and commonly used materials. Georges immediately apprehended this attractive property of EF; subsequently, my task involved building a small prototype of a BaF_2 *calorimeter readout by a gaseous detector filled with EF. Georges frequently entered the room while I laborated to gather the latest results. I know Georges valued my work, and this recognition makes me proud and motivated. One of Georges' worries was that EF would slowly polymerize in air, so each morning he examined the vessel with EF to inspect if any changes occurred.*

One day Georges arrived at the laboratory dressed well in a nice grey suit, white shirt and red neck tie. As you know, physicists usually do not pay much attention to their wardrobes. He certainly did not feel comfortable in this formal attire, and I could perceive that it restricted his natural movements. He then announced, "Today we will have important visitors, members of the Nobel physics committee. Maybe they will come to see your work, but you work as usual, there is no need to prepare special demonstrations". Then, as was his custom, he took the bottle of EF and shook it, a bit clumsily this time, to determine if any polymerization occurred. Abruptly, half the bottle of EF sprayed all over his nice clothes. His white shirt was covered with unsightly brown spots. He looked terrible; and, of course, there was no time for him to go home and change. This was the first time I had ever seen him upset! Some of his co-workers generously offered him their clothes, but as a big, brawny man he couldn't fit into any of them. He panicked and put on a technician's white dressing-gown, but he looked ludicrous in it.

Fortunately, I knew that by applying heat, EF stains can be removed. Irritated and skeptical, he watched as I brought into the laboratory a hot air blower and proceeded to direct the air flow toward the EF spots. "You will polymerize EF, and it will stay forever," George shouted. However, the EF spots slowly shrunk and then vanished. George smiled, his characteristic optimism restored! Soon after, a few modest-

looking people walked around the laboratory. They spent nearly an hour in Georges' office and then made a return to briefly visit to the laboratory. Approaching me Georges gave the following introduction: "This is my favorite project! All future photodetectors will be filled with EF. It is not corrosive, not reactive. It is magic!". He asked me to put a drop of EF on his sleeve in order to show the visitors how the brown spot disappeared when heated. The visitors exclaimed, "Ia ooh," and George joked, "if you want to try yourself, Vladimir can spill EF on your clothes."The doubtful and amused visitors politely declined his offer and rushed to leave the laboratory. Elated by the day's events, Georges invited me to dinner to celebrate.

So many years passed since this eventful day. I almost forgot about it since Georges interacted with numerous interesting visitors on a daily basis! But I recalled this wonderful day when in 1992 Georges was awarded the Nobel Prize in Physics. Perhaps, the fact that he met the Nobel committee members without ugly spots on his jacked played a favorable role! Maybe, they were impressed by my demonstration? Or, maybe his sheer genius led to an indelible contribution to the development of new detectors? Whatever was the real reason culminating in his Nobel honor, George remained dedicated to advancing EF research throughout rest of his life. Our last project with him also involved EF, the development of gaseous detectors for early forest fire detection. Indeed, EF may be magic!

2.4 TMAE

For many years, photosensitive gaseous detectors used the photosensitive vapours described above: benzene, toluene, TMA, TEA or EF. Other vapours were also tested, but with less favorable results, e.g. cis-2-butene (E_i= 9.35 eV), acetone (E_i= 9.65 eV), ammonia (E_i= 10.3 eV), isobutane (E_i= 10.6 eV), DMA (E_i=8.3 eV) etc.

The real boom in this field happened when David Anderson introduced tetrakis (dimethylamino) ethylene, a vapour more often called TMAE and pronounced "Tammy" (Anderson, 1980). TMAE has a photoionization potential as low as 5.36 eV, and it is photosensitive to photons with a wavelength of up to 231 nm. It is an oily fluid with a melting point of -4°C. It was originally used and manufactured by US Naval Ammunition Department (Nakato, 1972). TMAE oxidizes rapidly in air and produces a strong fluorescence light. In large quantities it can light up the whole sky, which is probably why it was of interest for military use.

During oxidation in air TMAE also produces a dense white smoke. Actually, a bottle of TMAE closed down the international airport of El Paso, Texas, for 7 hours in 1993. One of the authors (T. Francke) had a bottle of TMAE shipped from CERN to the US to be used in a space experiment. The transportation firm dropped the package at El Paso airport and the bottle of TMAE broke. Unfortunately, it broke just next to an air inlet for the ventilation system and the thick white smoke quickly filled the whole airport. No one was injured. This put TMAE on the front page of El Paso Times.

G. Charpak describes his first encounter with TMAE in his book "Research on particle imaging detectors" (Charpak, 1995):

He (D. Anderson) had found a prolific stock in the US Army of tetrakis (dimethylamino)-ethylene (TMAE), which has the lowest ionization potential, 5.36 eV, of all known vapours. In the liquid form it was easy to obtain in large quantities. The vapour pressure was easy to control at moderate temperatures and it had considerable quantum efficiency in the VUV range. I invited him to join our group at CERN. He came with his own counter and found the support of a group which had some expertise in gaseous detectors. He discovered with us that in high-energy physics the small overlap between the emission spectrum of BaF_2 *and the range of sensitivity of TMAE vapour was sufficient for the conception of a new style of fast electromagnetic calorimeters.*

However, TMAE found its most important use in RICH detectors rather than in calorimeters. The low ionization threshold of TMAE allowed a wide choice of solid, liquid and gaseous Cherenkov radiators combined with cheap, large area fused quartz windows. One of the most impressive implementations of a TMAE based detector was the DELPHI RICH detector built at CERN. This detector as well as some other Cherenkov detectors which used TMAE will be described in great detail in the application chapters.

The TMAE vapour pressure at room temperature is not as high as for TMA and TEA, see Figure 8, but it is much higher than for EF. Nevertheless, for efficient recording of UV light a detector filled with TMAE should be heated, typically to 40-60 °C.

Figure 8. The vapour pressure of TMAE as a function of temperature (in °C and in the reciprocal of the absolute temperature T); the vapour pressure is low at room temperature, and the TMAE as well as the detector needs to be heated to 40-60 °C to have an efficient absorption of single photons.
From Anderson, 1988.

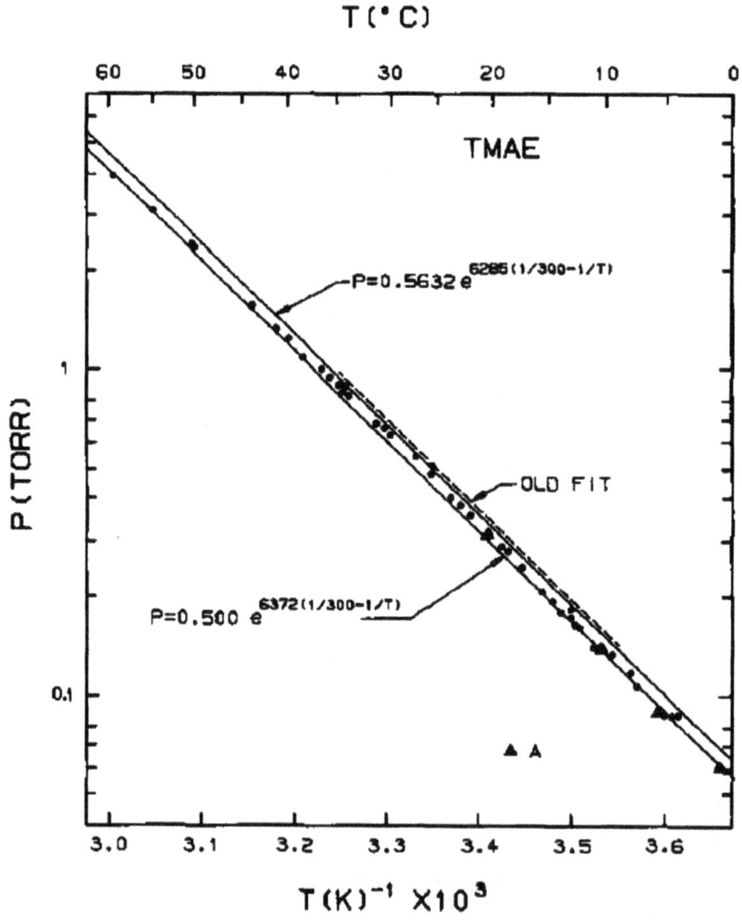

The quantum efficiency of TMAE is as high as 30-60% in the wavelength interval of 130-190 nm, as shown in Figure 9. This makes it the most efficient gaseous photocathode.

To use TMAE in a MWPC is not a simple task. The main difficulties are:

1. TMAE reacts with air (oxygen). Special measures should be done when handling the TMAE.
2. TMAE reacts with many materials changing their properties and causing leakage current along the surface many dielectrics.
3. TMAE molecules also penetrate inside some materials, for example aluminium, and remain there for a long time (months!) even after removing the TMAE from the gas. This cause a slow outgassing of TMAE vapours and makes it difficult to open and work on the chamber after exposure to TMAE.
4. Necessity to clean commercially available TMAE since the purity is not sufficient for use in photodetectors.

Anderson noticed several of these unpleasant features of TMAE. It reacts with many organic materials such as rubber O-rings and vacuum grease. Viton is found to be the least reactive of all O-ring materials tested and ought to be used without grease. He also found that TMAE which had been exposed to air can be purified by vacuum distillation. Evidently, some impurities with a vapour pressure higher than that of TMAE is being removed in this process.

Figure 9. The quantum efficiency of TMAE; TMAE is a good photo-converter for photons with wavelengths up to 210 nm. At wavelengths shorter than 190 nm it has a quantum efficiency of 30-60% at full absorption. This is the highest quantum efficiency of all gaseous photocathodes.
From Séguinot, 1989.

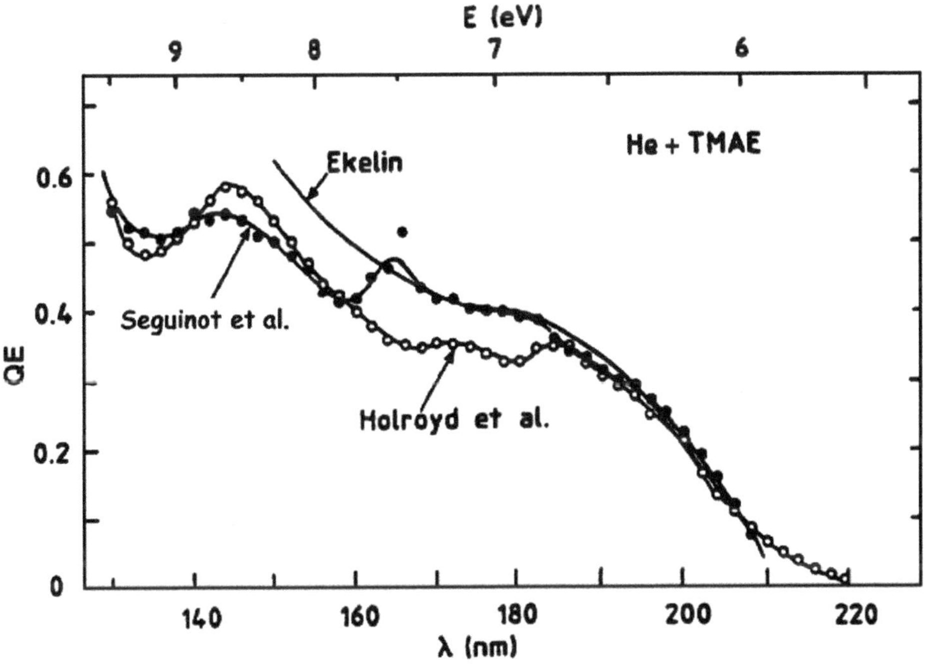

Heller improved the purification process (Heller, 1980). In a nitrogen atmosphere the TMAE is shaken with a small amount of water and separated with a separating funnel. This is repeated until the water is clear. This treatment removes the oxidation products of trimethyl oxamide (TMO), tetramethyl urea (TMU) and dimethyl amine. Then the TMAE is dried with a molecular sieve and must be stored in a tightly closed bottle. It should only be opened under nitrogen. He also notes that distillation under vacuum removes TMO and maybe TMU but not the amine unless the volatile first fraction is discarded.

The oxidation of TMAE is catalyzed by materials containing OH groups including ethanol and paper tissues. A drop of TMAE placed on a tissue will, if oxygen is present, turn slightly brown and glow green (visible in the dark to dark adapted eyes). This catalyzed reaction can be used to test the oxygen levels in a nitrogen filled glove box when TMAE is handled.

The high quantum efficiency of TMAE, and its sensitivity to photons of long wavelengths, made it worth the effort to try to find practical solutions to overcome the challenges with TMAE.

The first problem is to prevent the TMAE from oxidizing in air. It is normally solved by handling the TMAE in a nitrogen filled glove box. The gloves fixed in these cupboards make it possible to clean the TMAE, transfer it from a bottle to an airtight container with valves and tube connections etc. It is a complication, but it works quite well after some practice. There are even flexible glove boxes made like a plastic bag with gloves on. The equipment is placed in the bag, all air is sucked out and it is filled to a desired volume with nitrogen. Then the TMAE and the equipment are handled through the thin gloves in the bag.

The second problem is to use only materials in the detector and gas system which withstands TMAE. Gas chambers using TMAE are often made of stainless steel, the dielectric materials are made of ceramics, printed circuit boards are coated with a special protective layer, all O-rings are made of Viton, all tubes are made of cupper or stainless steel etc.

The major challenge is how to clean the TMAE, and to keep it clean during operation. In detectors having a small drift region (2- 3 mm) the TMAE cleanliness may be not a big issue. However in many applications, requiring photoelectrons drift on longer distances, TMAE vapours should be clean enough to minimize or fully avoid electron capture by electronegative molecules. In detectors with a low gas flow, the oxygen and water vapour levels have to be kept at a minimum to prevent the TMAE from oxidizing and loosing quantum efficiency. This is normally done by circulation or flushing the gas through oxygen and water absorbers.

There are two essential steps in TMAE purification:

1. Washing it in water (it is not soluble in water, while its oxidation products are soluble), and
2. Distillation/drying.

Since TMAE reacts with air, the water used must be free of oxygen. In practice, the water is dried by bubbling argon or nitrogen through it for a long period of time (many hours). Then in the same gas atmosphere the liquid TMAE is shaken with a small portion of water. The contamination products are separated from the TMAE using a separating funnel. This process is repeated many times until the water remains clear. The cleaned TMAE is then stored in a bottle containing a molecular sieve which removes any remaining water.

Further water removal is usually performed by distillation. Since TMAE has a relatively low vapour pressure (0.4 Torr at 20 °C), the distillation should be carried out at temperatures around 50-60°C. A simple method often used in practice is to pump TMAE vapour under vacuum from a bottle to another vessel which is cooled by liquid nitrogen where the TMAE freezes on the walls. The distilled TMAE is heated up again and pumped to another container which is cooled. Such a procedure should be repeated a few times. Normally one discards the first centiliters distilled away from the original TMAE vessel.

The practical problems of using TMAE do not stop here. When the TMAE is used in the photodetector new problems arises such as photon feedback, aging, leakage currents etc. These problems will be addressed in the next paragraphs.

3. EXPERIENCES LEARNED WHEN OPERATING GASEOUS DETECTORS FILLED WITH PHOTOSENSITIVE VAPOURS

3.1 Photon Feedback

We have already learned that in MWPCs filled with photosensitive vapours, in particular with TEA or TMAE, undesirable photon feedback may appear at high gas gains (~10^5). Let us discuss this effect in more detail.

For simplicity let us consider a binary gas mixture, for example argon +TEA or argon + TMAE at atmospheric pressure. Let us look at two extreme cases, relatively weak electric fields, i.e. a few kV/cm as in the case of a parallel plate avalanche chamber, and strong electric fields, i.e. 10^6V/cm which can be created in the vicinity of a thin wire.

The probability of creating a secondary photoelectron in the gas volume from photons emitted in the electron avalanche process can be expressed by an expression similar to Equation (3):

$$\gamma_{ph} \sim {}_{\lambda w}\int^{\lambda i} Q(\lambda) \{1\text{-}\exp(-L/l_{ph)}\} I_{av}(E,\lambda)\, d\lambda, \qquad (5)$$

where $I_{av}(E,\lambda)$ is the emission spectrum of the photons in the electron avalanche for this gas mixture. This spectrum depends on the electric field strength, E, or in a more general case of E/p, where p is the gas pressure. The emission spectrum is unique for each gas mixture.

Usually photodetectors operate with full absorption of the light ($L >> l_{ph}$), so that

$$\gamma_{ph} \sim {}_{\lambda w}\int^{\lambda i} Q(\lambda)\, I_{av}(E,\lambda)\, d\lambda, \qquad (6)$$

In the case of low an electric field (i.e. a few kV/cm), the electrons slowly gain kinetic energy during their drift in the gas. After a number of elastic and inelastic collisions they reach a kinetic energy sufficient to excite or ionize the photosensitive vapour. Figure 10 illustrates schematically the energy levels of a hypothetical photosensitive molecule.

The energy levels of the photosensitive molecule consist of a number of energy bands, the ground state and one or several excited bands. Inside each energy band there are a number of vibration and rotation energy levels. In reality, the energy band structure is much more complicated. After the loss of kinetic energy during an inelastic collision (excitation or ionization of the photosensitive vapour), the electron again starts gaining energy during its drift in the electric field and the process repeats, another inelastic collision, new excitation or ionization, further gain of kinetic energy between the collisions etc.

The ionization potential and the first excitation potential of argon are 15.7 eV and 11.5 eV, respectively. These are considerably higher than the ionization potential of TEA or TMAE. Hence, any electron

Figure 10. Schematic representation of the energy bands (ground states and excited states) of a hypothetical diatomic molecule as a function of the distance between atoms; the blue arrow indicates the excitation of the molecule by an electron. The green arrow indicates the de-excitation of the molecule where a photon or even a photoelectron (in some special cases) may be emitted.

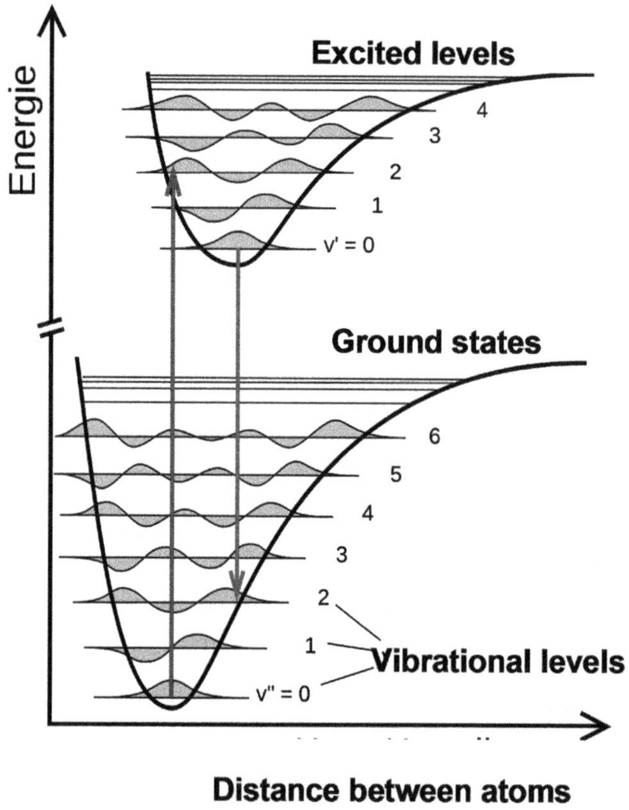

Figure 11. The emission spectrum of light produced in electron avalanches in a parallel plate avalanche chamber filled with argon +2%CH_4 +0.04% of TMAE; none of the characteristic argon lines are observed. From Charpak, 1988.

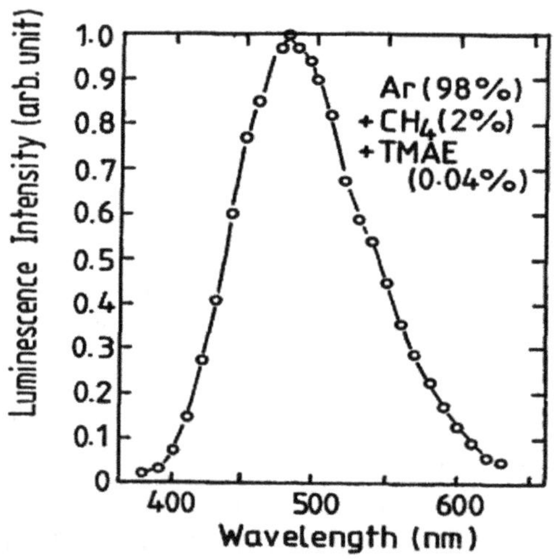

gaining energy in an argon + TEA or TMAE mixture will primarily lose its energy by excitation and ionization of the TEA or TMAE. A negligible fraction of the drifting electrons will ever reach enough kinetic energy to excite an argon atom.

Experimental evidence of this can be seen in a parallel plate avalanche chamber operating with an electric field of few keV/cm. By measuring the spectrum of the emitted light in the avalanche process, only the emissions spectra of TEA or TMAE are observed, not the characteristic lines of a monoatomic gas such as argon. This is regardless of if the amplification gas is pure argon or a mixture of argon with another molecular gas if the latter has an ionization potential sufficiently higher than the ionization potential of the photosensitive vapours. This is illustrated in Figure 11 (Charpak, 1988).

The process is quite different in a strong electric field ($E \sim 10^6$V/cm), e.g. around the anode wire of a MWPC. Here the electrons gain sufficient energy between each collision not only to ionize the photosensitive gas, but even to excite and ionize the argon atoms. The emission spectrum $I_{av}(E, \lambda)$ dramatically changes. It contains now components in the far UV region (usually from excited argon atoms as well as the emission spectrum of the photosensitive gas and its fragments). The emission spectra from electron avalanches produced in single wire counters filled with mixtures of argon have been studied in detail (see for example Bogomolov, 1978b; Peskov 1979a; Peskov 1979b).

The UV photons emitted by argon and other gases are capable to ionize the TEA or TMAE and create secondary photoelectrons (Fonte, 1991) which is the cause of photon feedback.

Let us now consider another binary gas mixture, where the main gas component is a molecular gas, for example CH_4+TEA or CH_4+TMAE. MWPCs operating in this gas mixture can develop gas gains close 10^6 with much lower probability of feedback photons appearing. The probability to create a secondary photoelectron in such a gas is:

Figure 12. The transmission spectra of some gases and windows often used in photosensitive gaseous detectors; for convenience, the TMAE quantum efficiency is shown in the same graph (dashed line). Some emission lines of carbon (which are frequent fragments of molecular gases appearing in electron avalanches) are also shown.
From Séguinot 1989.

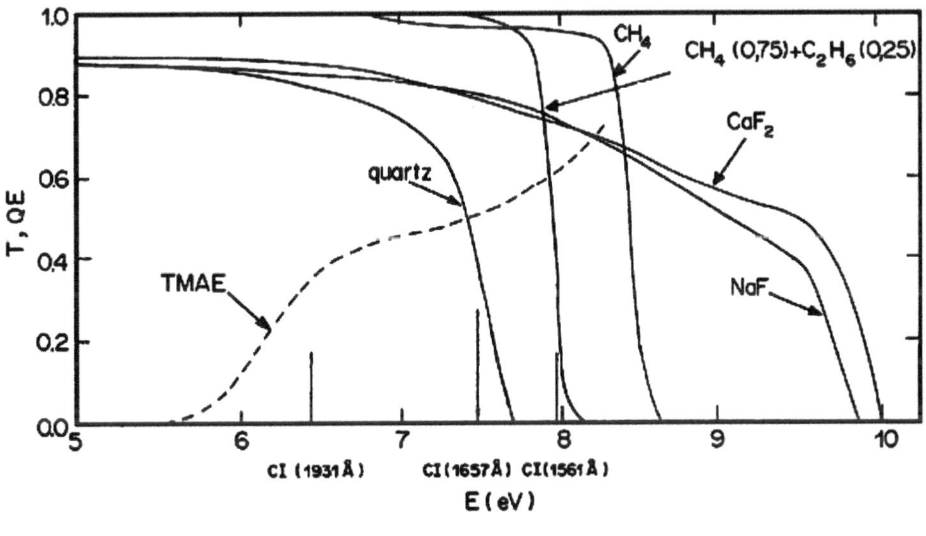

$$\gamma_{ph} \sim {}_{\lambda w}\int^{\lambda i} Q(\lambda) \{1\text{-}\exp(-(L/l_{ph} + L/l_{abs}))\} I_{av}(E,\lambda)\, d\lambda, \quad (7)$$

where $\exp(-L/l_{ph})$ and $\exp(-L/l_{abs)}$ describes the absorption of the photons produced in the electron avalanche process by the photosensitive vapours and the amplification gas (CH_4 in this case), respectively. This absorption leads to photoionization or excitation of the gas. If the photosensitive detector operates with full light absorption (which is usually a case) then,

$$\gamma_{ph} \sim {}_{\lambda w}\int^{\lambda i} Q(\lambda) \{1\text{-}\exp(-L/l_{abs})\} I_{av}(E,\lambda)\, d\lambda, \quad (8)$$

The transmission (equal to 1- close to zero absorption probability) of UV photons is shown for some molecular gases in Figure 12.

As can be seen, at atmospheric pressure the CH_4 transmission for UV photons has a sharp cutoff at 8.3 eV (~140nm). This simplifies Equation (8) to be:

$$\gamma_{ph} \sim {}_{\lambda th}\int^{\lambda i} Q(\lambda)\, I_{av}(E,\lambda)\, d\lambda, \quad (8)$$

where λ_{th} is the transmission threshold of the amplification gas (CH_4 in this case). It may also be the transmission threshold of the detector window if it has a lower threshold than the amplification gas. Often windows with good UV transmission are used to maximize the detection efficiency of the detector, e.g. CaF_2 or MgF_2.

Let us again consider the two cases but for methane this time; weak electric fields, i.e. a few kV/cm, and strong electric fields, i.e. 10^6V/cm. In the first case, the electrons mainly lose their kinetic energy by excitation and ionization of the photosensitive vapour, and partially on excitation of the lowest vibrational and rotational levels of the methane amplification gas. For a molecular gas like methane, these levels are more numerous than for a monoatomic gas like argon. As before, in a strong electric field the electrons will loss of their energy not only through excitation and ionization of the photosensitive vapour, but also through excitation of the higher energy levels of the methane molecules, and even by ionizing the methane molecules.

The different physical processes of the electron energy dissipation is clearly observed in the avalanche emission spectra $I_{av}(E,\lambda)$. Some molecular gases, such as H_2 or N_2, have quite strong emission in the UV region of the spectra. Figure 13 shows the hydrogen energy levels.

In contrast to the monoatomic gases which emit light at distinct lines from pure excitation levels, molecular gases have, besides the excitation levels, many rotational and vibrational levels which emit light at many more wavelengths than the monoatomic gases.

There is a wide energy gap between the ground level and the excited bands in the molecular hydrogen spectrum. The UV emission spectrum with its strong peak around 160 nm is created by transitions of the electrons from the excited band to the ground band. In the emission spectra from an electron avalanche one can also observe more exotic emission lines. The high temperature in the avalanche may cause the hydrogen molecule to temporarily dissociate into semi-individual atoms before recombining to a molecule again. This is represented by the dashed curve. Many hydrogen molecules are also ionized in the avalanche and again produce new emission bands. The transitions from excited levels to the repulsive state associated with the dissociation of a H_2 molecule have a maximum emission at 220 nm. Excited hydrogen atoms, which may appear after the dissociation of the hydrogen molecule, emit the classical

Figure 13. Energy level diagram for H_2 as a function of the distance between the hydrogen nuclei; the solid curve is for neutral hydrogen with its well-known energy levels. The high temperature in the avalanche may cause the hydrogen molecule to temporarily dissociate into semi-individual atoms before recombining to a molecule again. This is represented by the dashed curve. The dot-dash curve represents the energy levels for an ionized hydrogen molecule.
From Senftleben, 2010.

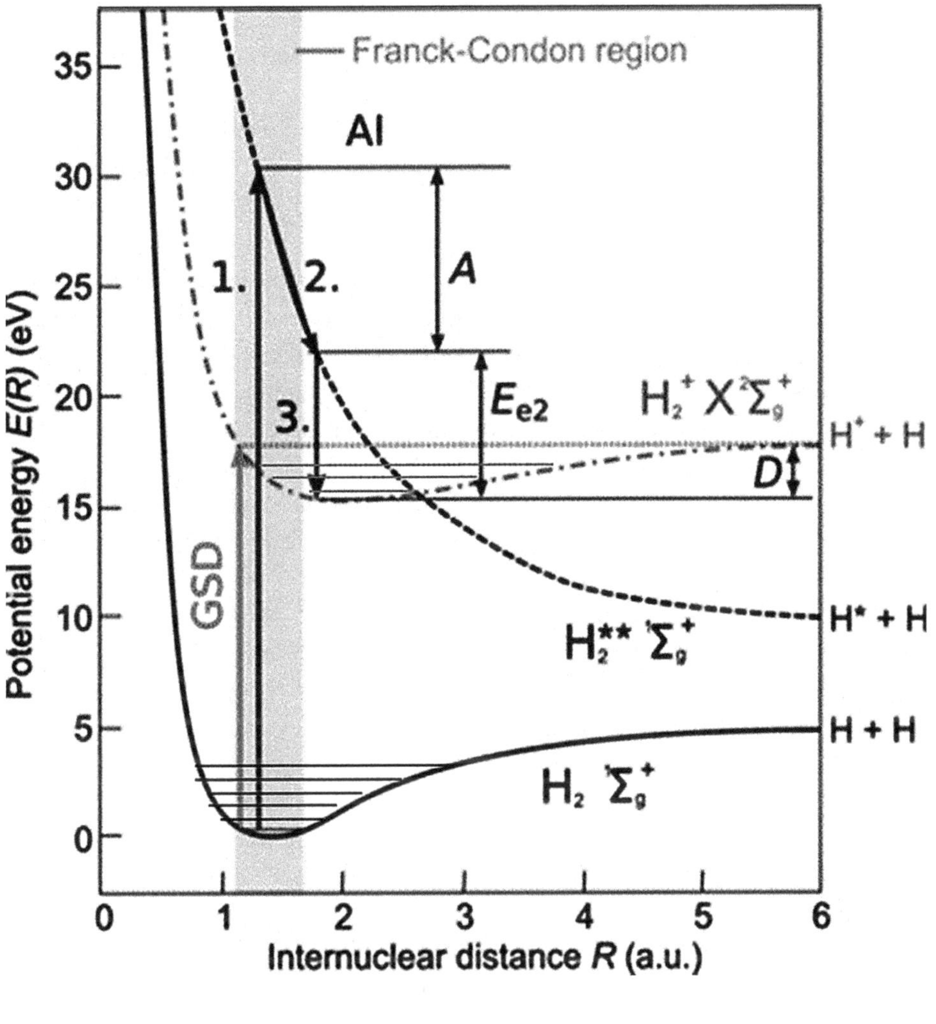

L_α line (121.6 nm). Due to the wide emission spectra, with these types of gases photodetectors one can only operate without photon feedback at relatively low gas gains, usually below 10^3. This is not sufficient for the detection of single photoelectrons with a high efficiency, so nitrogen and hydrogen are not well suited as amplification gases in photosensitive detectors.

Other gases like CH_4, isobutane, etc. do not have a strong emission in the UV region and do not by themselves produce photon feedback which disturbs the operation of the photodetector. However, in the avalanche process the CH_4 may fragment into small amounts of carbon and hydrogen, which do emit photons in the UV region and can ionize photosensitive vapours to some extent. Examples of carbon emission lines are shown in Figure 12.

However, this emission is much weaker for methane and isobutane than for e.g. argon which explains why photosensitive detectors that operated with CH_4 or isobutane can reach much higher gas gains than detectors operating with argon.

To conclude, three main factors contribute to the creation of the photon feedback in molecular gases:

1. The emission spectra of the molecules .
2. The emission from fragments of the molecules produced in the avalanche.
3. The value of the threshold absorption wavelength, λ_{th}, in the detector.

All these three factors are reflected in Equation 8. The first two are describe by the emission spectra $I_{av}(E,\lambda)$, the last one is included through the integration interval.

The photon feedback problem has been addressed and reduced in a number of different and quite ingenious ways in practical applications. The first Cherenkov ring imaging detectors used a multistep chamber to suppress the effect of photon feedback. Recall that it consisted of one or two parallel plate avalanche chambers working as a preamplification structure above an MWPC. The Cherenkov photons were absorbed in the conversion gap above the top parallel plate avalanche chamber. The photoelectrons produced were multiplied first in the parallel plate avalanche chamber, and then in the MWPC. This results in a large signal for each photoelectron. Any UV photons produced in the electron avalanche in the MWPC were absorbed in the thin gap between the MWPC and the lower parallel plate avalanche chamber. These secondary photoelectrons were amplified only in the MWPC, and produced significantly weaker signals. So weak that they are not detected. Hence, amplitude discrimination can efficiently suppress photon feedback. Note that the first multistep photon detectors operated in a mixture of argon and a photosensitive vapour, for example TEA (Séguinot, 1980).

Later, when the origin and mechanisms behind photon feedback was better understood, much simpler devices could be used. By using CH_4, iso-C_4H_{10} or any mixture of them as amplification gas, the gas gain can be high enough with a single-step MWPC to efficiently detect single photoelectrons. Another important feature of $CH_4$is that the pulse-height distribution shows a clear maximum (Polya type) rather than a pure exponential distribution (Furry type) typical for gas mixtures with argon. This makes it possible to discriminate the signals just below the peak without losing to much efficiency. The pulse height spectra will be discussed in more detail in the next chapter.

When CH_4 oriso-$C_4H_{10}$is used as the main gas in a photosensitive MWPC photon feedback only appears at gas gains exceeding 10^5. So a single photoelectron will not produce any noticeable photon feedback. However, when a charged particle traverses the MWPC conversion gap it can generate about 100 electrons through ionization of the gas. These 100 electrons can trigger a common avalanche in which the photon feedback can be significant. Secondary avalanches from these photons may complicate the recognition of Cherenkov rings.

To suppress photon feedback in photosensitive detectors filled with TEA or TMAE, special mechanical designs can restrict the solid angle at which the feedback UV light may reach the conversion volumes. In this way the feedback photons are absorbed by the structure before they can convert to photoelectrons and initiate secondary avalanches. Examples of such blind designs are shown in Figures 14, 15, and 16.

Finally, it was discovered that a MWPC filled with ethane and TMAE vapour at 40°C can operate at gas gains ~10^5 without any signs of photon feedback. Such a photosensitive MWPC was successfully used in the CAPRICE experiments in space (Carlson, 1994).

Figure 14. One design of a MWPC with a special blind structure of electrodes used to restrict the solid angle at which UV photons from the avalanche may propagate to the photosensitive volume of the photodetector
From Anassontzis, 1992.

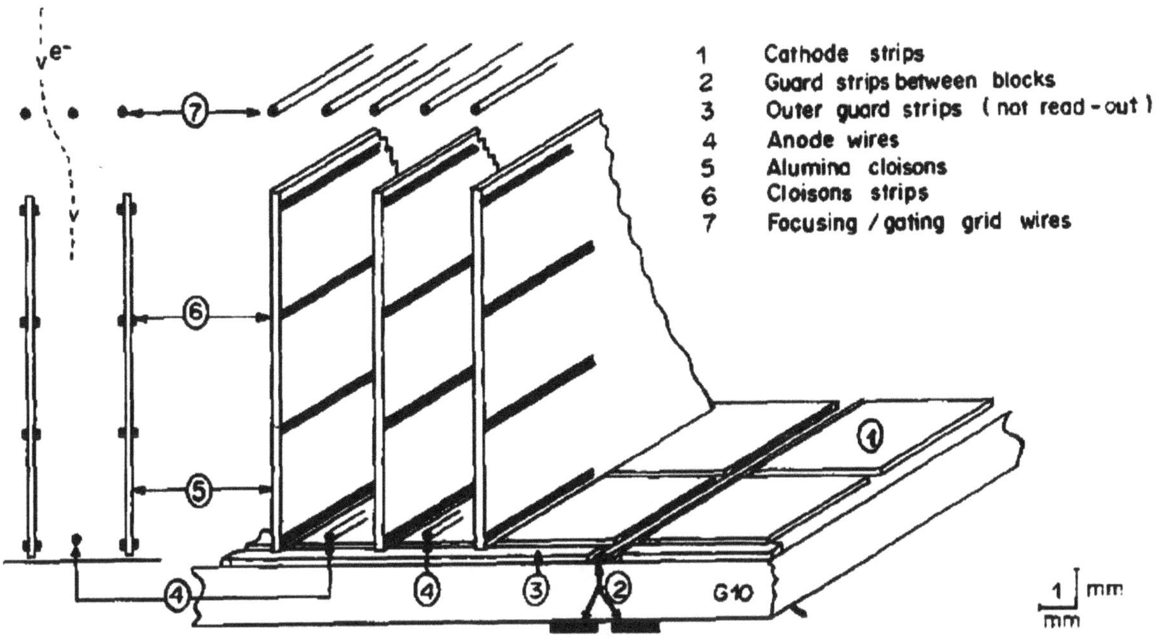

Figure 15. An alternative design of light shielding plates preventing UV photons from the avalanche to propagate back to the photosensitive volume

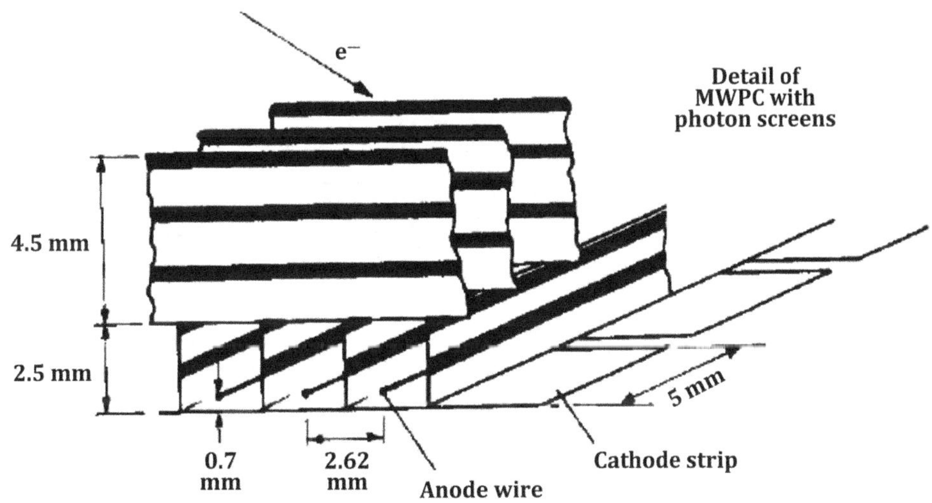

Figure 16. A third way of preventing feedback photons from the avalanche to propagate back into the photo-conversion region is to use an array of single wire type counters; the slit in each cathode cylinder is wide enough to let the photoelectrons enter into the amplification region around the wire, but small enough to shield a major fraction of the UV photons produced in the avalanche from leaking back into the photo-conversion region.
From Arnold, 1988.

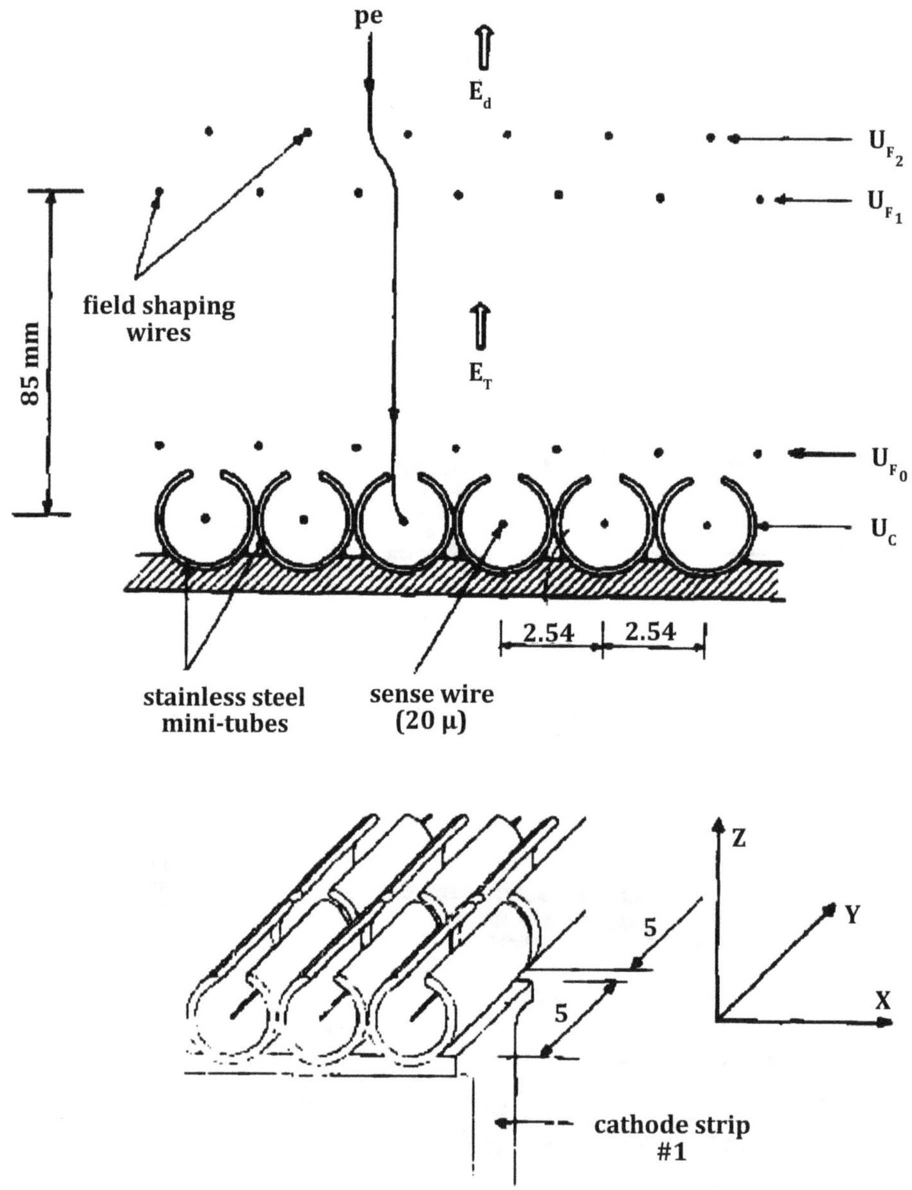

3.2 Ion Feedback

Ion feedback is caused by ions produced in the electron avalanche and slowly drifting away from the avalanche region. When they recombine at the cathode they may emit secondary electrons which may trigger secondary avalanches. Ion feedback is usually not an issue in photosensitive detectors.

There are two main reasons for this:

1. In gaseous detectors filled with photosensitive vapours the probability that the ions reaches the cathode and extract secondary electrons from it is small. Most ions recombine on the way back to the cathode.
2. In several applications, for example Cherenkov detectors, the MWPC operate in triggered mode. They record only the photoelectrons arriving at the anode wires with a given delay with respect to a trigger signal produced by the passing charged particle. This makes it possible to discriminate photoelectrons from secondary electrons created by ions which occur much later than the signals from the initial electron avalanche.

Let us again consider the now familiar two cases; weak electric fields (a few kV/cm) and strong electric fields (10^6 V/cm).

In the first case, the electrons in the avalanche will mainly have enough energy to ionize the photosensitive gas. Hence, most of the ions will be ions of the photosensitive compound. The probability to extract electrons from the cathode is described by Equation (13) in Chapter 2. This probability is larger than zero only if the ionization potential of the ion is larger than two times the value of the cathode work function. The work function for the cathodes used in MWPCs are around 3.5-5 eV. This means that only TEA ions can extract electrons from the MWPC cathode and the probability of this process is very low according to Equation (13) in Chapter 2. TMAE ions are not able to extract any secondary electrons at all from the cathode, sometimes not even to recombine. The TMAE ions accumulate on the cathode, which leads to other interesting effects which will be described later.

The situation will not change at strong electric fields even though ions from the amplification gas are now present. Via a charge exchange process, these ions will be almost immediately be neutralized by ionization of the photosensitive vapour. It is again only ions of the photosensitive compound which will drift towards the cathode and interact with it as in the case of weak electric fields.

4. THE SINGLE PHOTOELECTRON PULSE HEIGHT SPECTRUM

Besides the quantum efficiency, i.e. the probability that a photon will create a photoelectron, the second most important characteristic of any photosensitive detector is its single photoelectron detection efficiency. The single photoelectron detection efficiency is the probability that a single photoelectron will produce a large enough a signal to be detected.

At small and medium gas gains (typically $\leq 10^5$), the pulse height distribution, S_a, measured at the input of the charge sensitive amplifier has an exponential shape (a Furry distribution):

$$S_a = (1/q_{amn})\{\exp(-q_a/q_{amn})\} = (1/A)\,\{\exp(-q_a/A)\}, \qquad (9)$$

where q_{amn} is the average charge in the produced in the avalanche, which is the same as the gas gain A.

Due to the threshold of the electronics, defined by its noise level, the single photoelectron detection efficiency is:

$$\eta_{se} = ({}_{qth}\int^{\infty} S_a\, dq) \,/\, ({}_{0}\int^{\infty} S_a\, dq) \sim \exp(-q_{th}/A), \tag{10}$$

A typical pulse height distribution is illustrated in Figure 17.

Only signals above a certain threshold in the electronics are detected, which is illustrated by the green area in the figure. The signals below this threshold go undetected as indicated by the black area. The fraction of detected signals is the single photoelectron detection efficiency, η_{se}.

The spectrum from single photoelectrons shows an exponential Furry distribution. If the number of primary electrons is larger than 1, the spectrum has clear a maximum as shown in Figure 18.

As seen in the figure, the signals originating from a single photoelectron are particularly difficult to detect with good efficiency due to the shape of the distribution. Even by using electronics with low noise levels, and putting a low threshold in the distribution, a significant number of photoelectrons are undetected and the single photoelectron detection efficiency is reduced. For electron avalanches induced by two electrons or more, the implication of the threshold is much less severe. Even a quite high threshold will give high detection efficiency. This is why efficient single photoelectron detectors have been so difficult to develop.

By increasing the gas gain the distribution starts to deviate from pure exponential shape. It becomes more and more rounded, and the relative width decreases. The maximum is shifted towards higher and higher values, as shown in Figure 19.

Typically, photosensitive gaseous detectors operate at gas gains of 10^4-10^5. At such high gains the pulse height distribution of the signals from each electron avalanche deviate from a pure exponential distribution. More specifically, the distribution is shown to follow a special case of the negative bino-

Figure 17. The pulse height distribution of a typical photosensitive detector operating at gains around 10^4; only signals above a certain threshold are detected (green area) while the signals below this threshold goes undetected (black area). The single photoelectron detection efficiency, η_{se}, is the fraction of all signals which are detected (green).
From Riegler, 2008.

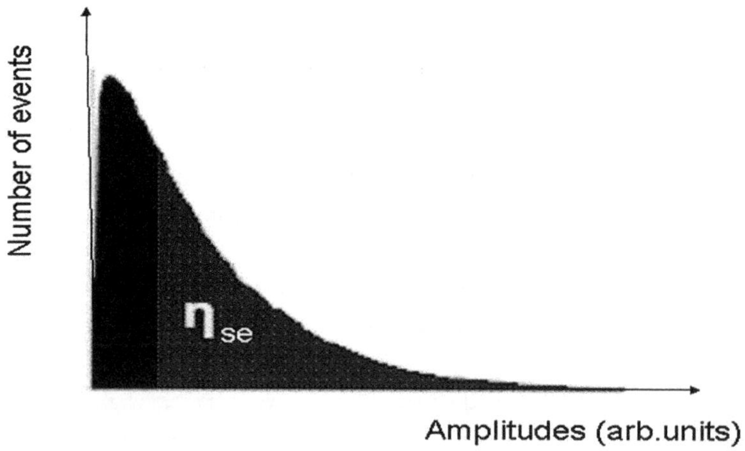

Figure 18. The pulse height spectra for electron avalanches initiated by n=1, 2, 5 and 10 primary electrons
From Sauli, 2009.

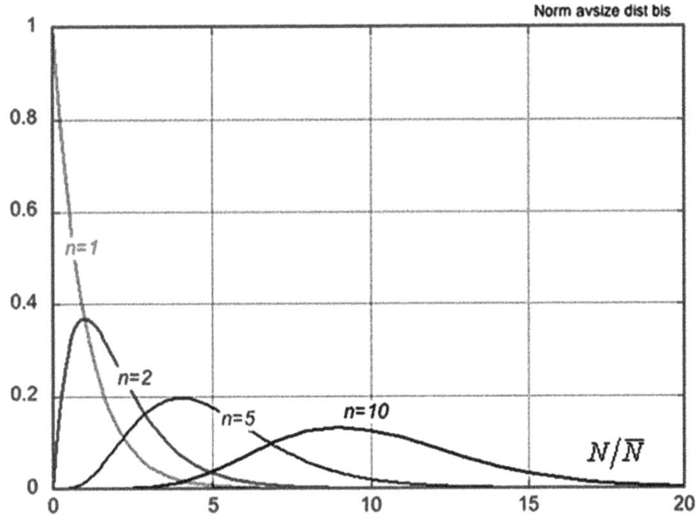

Figure 19. The evolution of the pulse height spectrum for increasing gas gains; at low gains the distribution is a pure exponential distribution. As the gas gain increases the distribution approaches a Pólya distribution.
From Schlumbohm, 1959.

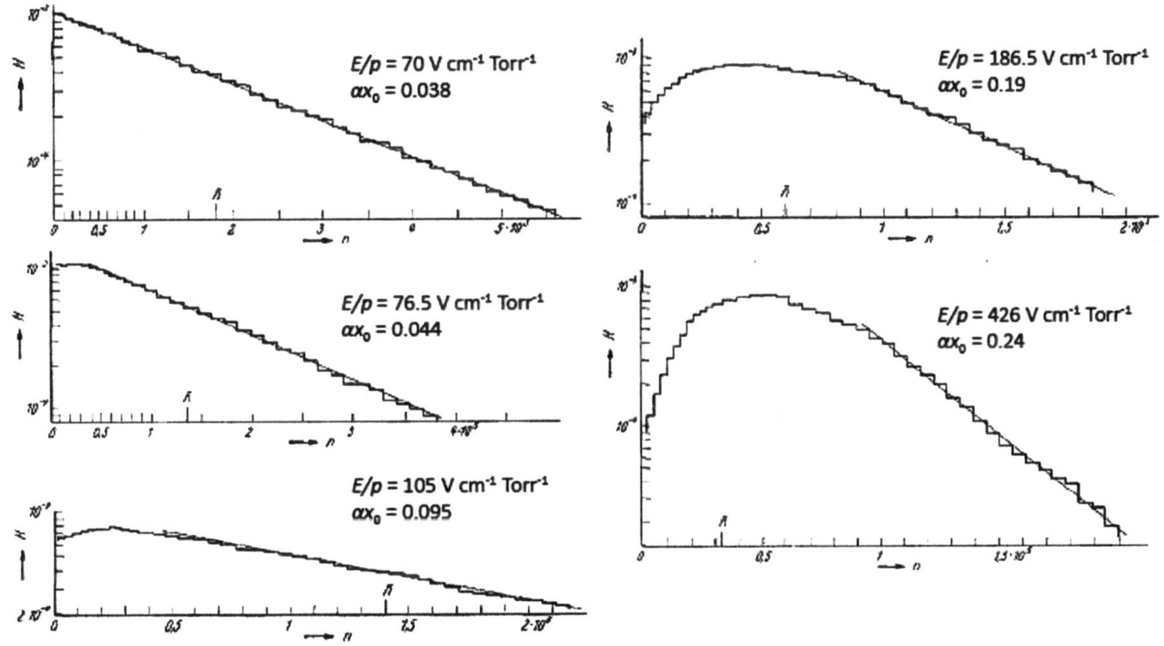

mial distribution, normally called the Pólya distribution (equation 16, Chapter2). Recall for clarity that it describes the signal strength from a single photoelectron avalanche:

$f_Z(z) = m^m / \Gamma(m)\, z^{m-1}\, e^{-mz}$,

Let us repeat that here z = N/A where N is the number of electrons in the avalanche and A is the average number of electrons in a large number of avalanches, i.e. the average gas gain. m is a parameter in the Polya distribution that is related to the first Townsend coefficient and depends on the gas and the gain. m is shown to be of the order 1.5 for normal wire detectors. At low gas gains the parameter m approaches 1 and the distribution becomes an exponential distribution.

The shape of the Pólya distribution illustrates well why photosensitive detectors that are capable to operate at high gas gains ($10^4 - 10^5$) can have high detection efficiency even for single photoelectrons. As long as the threshold of the electronics is below the peak of the distribution the detection efficiency is high, as illustrated in Figure 20.

Photosensitive gaseous detectors using ethane, isobutane or ethane as amplification gas show a Pólya type spectrum at gas gains $>10^5$. They are the best gases known to be used for single photoelectron detection.

Also parallel plate avalanche chambers show the same evolution of the single photoelectron spectrum from an exponential distribution to a Pólya type distribution with increasing gas gains (Fonte, 1999).

Figure 20. The signal distribution from single photoelectrons initiating electron avalanches in a gaseous detector operating at high enough gas gain to follow a Pólya distribution; by applying a threshold in the electronics, signals higher than the threshold value will be detected (green area) while lower signals go undetected (Black area). As long as the threshold of the electronics is below the peak of the distribution, the detector has high detection efficiency for single photoelectrons.
From Schindler, 2008.

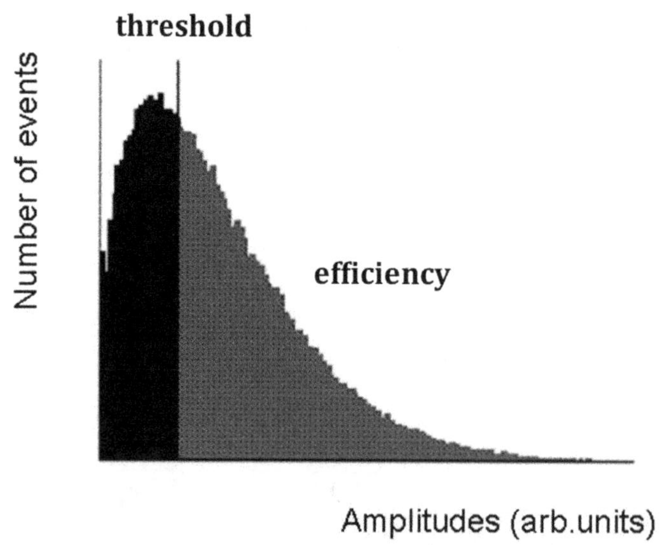

5. AGING OF PHOTOSENSITIVE GASEOUS DETECTORS

All gaseous detectors suffer from an aging effect where the performance of the detector degrades with time, and it can even lead to discharges, eventually. Photosensitive gaseous detectors are no exception. Aging is a slow degradation of the detector performance under irradiation. The aging phenomenon is very complex and poorly understood, and it includes many effects (see for example Kadyk, 1991; Va'vra, 1992). One observable effect is accumulation of polymers on the anode wires, as shown in Figure 21. The polymer formation is created by chemical reactions occurring in avalanche plasma near anode of the MWPC.

The temperature in the electron avalanche is very high, many thousands of degrees C. It also contains a high concentration of free charges. This is often called hot plasma. In such plasma molecules break up in collisions with electrons, de-excitation of atoms, and UV absorption processes. Exotic new molecules are produced, and materials that are normally inert become quite reactive at these conditions. Whereas ionization processes often require electron energies in the excess of 10 eV, the breaking of covalent bonds and formation of free radicals requires only 3-4 eV. This leads to a higher concentration of free radicals than ions in the electron avalanches. Consequently, free radical polymerization is regarded as the dominating mechanism of wire chamber aging. The free radicals either recombine to form the original molecules, or they break up into smaller and often volatile molecules, or they form new cross-linked molecular structures of increasing molecular weight. The latter leads to the formation of conductive or insulating deposits on the electrode surfaces. The processes are illustrated in Figure 22.

The result of the deposition of polymers on the electrodes, especially on the anode wires, is a decrease of the gas gain due to the modification of the electric field, excessive dark current, sparking and even

Figure 21. Maginified photographs of anode wires with various amount of polymer debris created by chemical processes in the plasma produced by the electron avalanche
Kadyk, 1991.

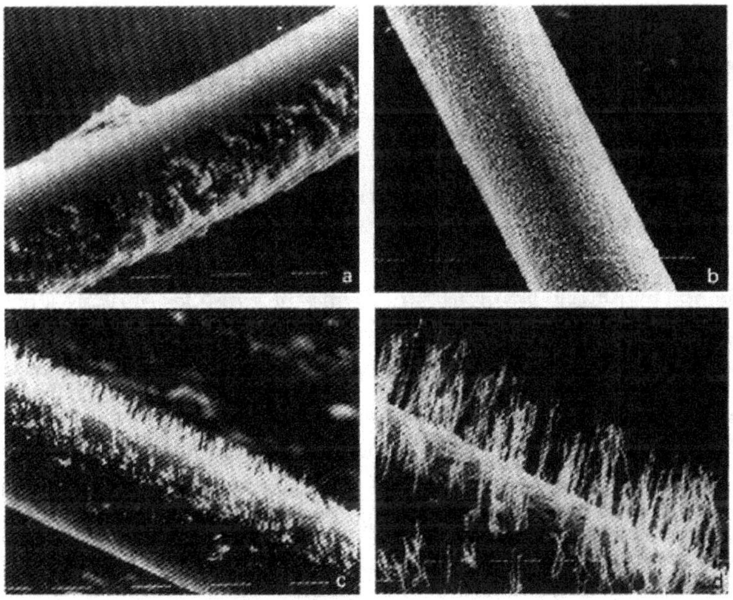

Figure 22. Schematic representation of various chemical processes occurring in the plasma close to the detector electrodes; the processes are triggered by molecular fragments produced in the avalanche. Titov, 2003.

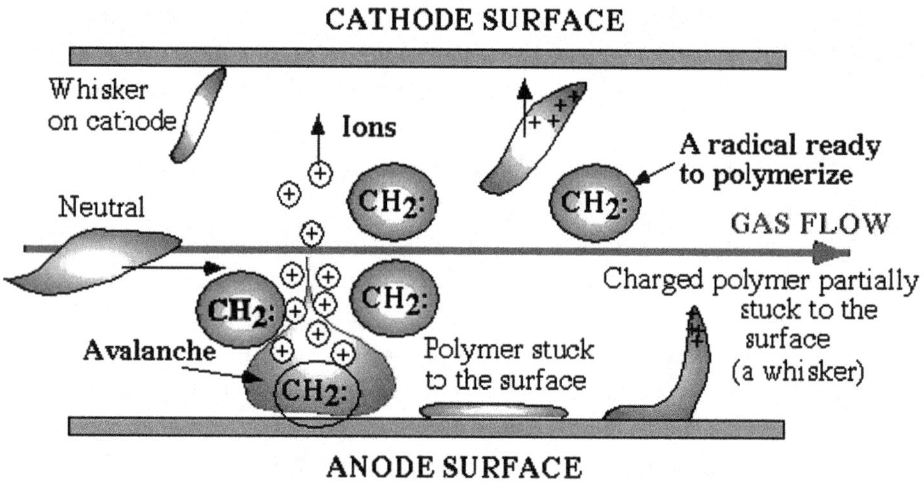

self-sustained discharges. The rate of polymer formation depends on many parameters which make the process complicated to predict and prevent. For example, the radiation induced degradation observed in wire chambers depends on the choice of gases, the purity of the gases, materials used in contact with the gas, materials of electrodes, strength of the electric field etc.

Aging of TEA and TMAE filled detectors have been investigated by several groups (see for example Woody, 1988; Va'vra, 1991; Va'vra, 1995; Va'vra, 1997; Pyrlik, 1998; Korpar, 1999; Lau, 2003). In all studies it was found that the gas gain drops proportional to the accumulated charge per unit length of the anode wire although the exact behavior can be quite different in each case. As an example, Figure 23 shows the gas gain degradations as a function of the charge accumulated on the anode wire for detectors filled with a mixture of ethane and TEA or TMAE.

Both TEA and TMAE show a significant decrease in gas gain, proportional to the absorbed charge. The decrease is faster in TMAE than in TEA. On the other hand seems the degradation to flatten out for TMAE at higher absorbed charges.

It is also observed that the gain drop, or in other words a wire aging rate, is inversely proportional to the anode wire diameter and hence to the strength of the electric field at the vicinity of the wire.

Another interesting observation is that the anode wire surface may be regenerated or partially cleaned after heating the anode wire by passing a small current through it.

An analysis of the polymer products on the anodes and the cathodes of photosensitive wire chambers have only been done for TMAE (Va'vra, 1987). It was found that on the anode wires the main depositions are not oxidized TMAE, but rather a tertiary amide. It remains unclear what fragments of TMAE can react with oxygen to form tertiary amide. There have also been findings of some unidentified heavy organic compounds with atomic masses of 202, 110 and 100. On the cathode wires was found carbon in the form of graphite, small fractions of amides (N-C=O), silicone $[R2SiO]_n$ and amines NR3 (R=C,H) (the silicon probably comes from the oil bubbler). The elements ratio was 83% C, 8.2% O, 5.9% N, 2.7% Si.

Figure 23. Gain degradation vs. accumulated charge per unit length of wire in detectors filled with C_2H_6+TEA and C_2H_6+TMAE gas mixtures; both TEA and TMAE show degradation in gas gain for increasing absorbed charge. The degradation is significantly faster in TMAE than in TEA. The degradation in TMAE seems to flatten out for high absorbed charges.
From Va'vra, 1996.

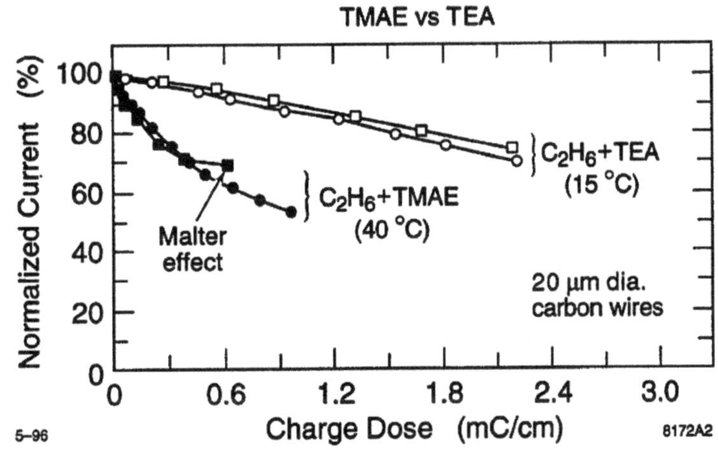

Thus many types of organic species coated the cathode. From the first glance it looks as if at least part of the problem comes from contaminations (for example Si and O). It is also obvious that the avalanche breaks up the TMAE in its smallest constituents.

Aging studies of TMA and EF have been performed in ultraclean conditions (Francke, 2003). One should note that it is difficult to ensure an extremely clean gas system when it is continuously flushed with the gas. The main problem comes from the continuous outgassing of the materials used for in the gas system, test chamber and the detector itself, as well as contaminants from the oil bubblers, tubes, leaks and the gas itself. It is also difficult and takes a long time to outgas an experimental set up operating in a gas-flushing mode. It is much more efficient to outgas the set up by vacuum pumping on it while heating it to a reasonably high temperature, and then use the detector in a sealed mode (Francke, 2003).

The set up for ultraclean studies of aging of photosensitive detectors using different photosensitive vapours is presented in Figure 24. It is basically a station for production of sealed single-wire counters filled with photosensitive vapours. It has been used for manufacturing of sealed detectors for plasma diagnostics. The station includes a single wire counter, a heating and pumping system and a gas system. The cylindrical cathode of the single wire counter is made of stainless steel and has a diameter of 15 mm. The anode wire can be chosen from two different diameters, 0.05 mm and 0.2 mm, and are made of molybdenum. The dielectric interface between the ends of the anode wires and the cathode are made of glass. The detector has a LiF entrance window with a diameter of 2 mm.

Initially the single wire counter is cleaned by pumping it down to a vacuum of ~10^{-6} Torr and heating it to 150°C for several days. The gas system contains two small glass bottles, one empty and another one filled with the photosensitive liquid one wishes to study (EF or TMA). The system is also connected to a glass cylinder with spectroscopic clean amplification gas. In the tests was used argon, krypton or xenon with 99.999% cleanness level. The EF and TMA were chemically cleaned at the Chemical Lab of the Moscow State University (D. Lemenovsky lab). Before introducing them into the single wire counter,

Figure 24. A station for manufacturing of sealed single-wire counters; the system can also be used to study aging of gaseous detectors with photosensitive compounds, under ultraclean conditions. All parts of the system can be individually cleaned by heating and vacuum pumping on them.

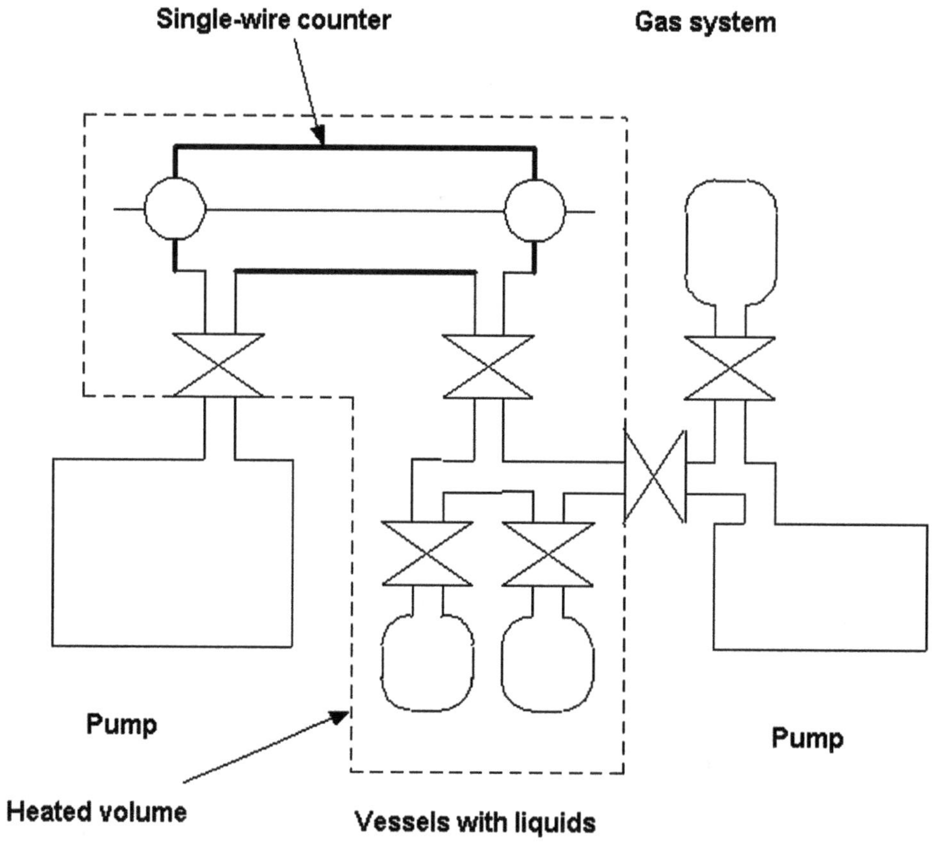

the liquids were distilled 6-10 times by cryogenic and vacuum transfer from one glass bottle to another, in a process described above.

The results of the aging tests obtained with TMA and EF are presented in Figure 25 (Francke, 2003). For comparison, in the same figure is also presented some published data for TEA and TMAE vapours (Va'vra, 1995; Va'vra, 1997). One can see that a photosensitive detector using ultra clean EF and TMA suffer less from aging than a detector using TEA or TMAE under normal conditions. Since TMA and TEA are chemically similar, one can speculate that the difference in aging properties is due to the amount of contamination is the system.

The results strongly indicate that cleanness has a major impact on aging. The cleaner the gases and the system are the slower the chamber ages. It is also shown that it is rather a strong electric field that causes aging effects than high gas gains. The safest way to avoid aging as much as possible is to use as clean a system as possible, assure a good cleaning system for the gases and try to use as weak an electric field as possible in the amplification region.

TMAE has shown to cause the fastest and worst aging effects. The effects can be strongly suppressed by:

Figure 25. Results of aging tests using several photosensitive vapours: TMAE (open diamonds) in a not very clean gas chamber (Va'vra, 1995); TMAE (open squares and triangles) in a clean gas chamber [(Va'vra, 1995); TEA (open circles) (Va'vra, 1997). The measurements using the specially cleaned and sealed single wire counter with an anode wire diameter of 50 μm and operated at a gas gain of 10^3 is for shown for EF (filled squares) and TMA (filled circles).The measurements in the single wire counter with an anode radius of 200 μm and operating at a gain of 10^8 in streamer mode is similar for both EF and TMA (filled triangles). In this last case no aging is observed even for quite high absorbed levels of charge. Francke, 2003.

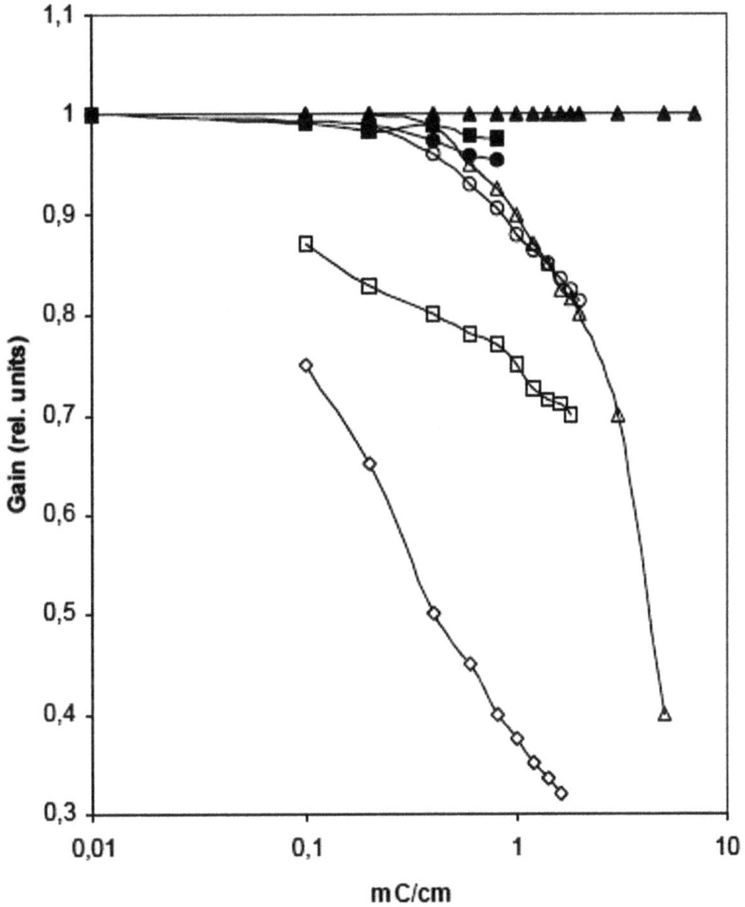

- Using materials in the chamber which can easily be made extremely clean, preferably by heating and pumping, or flushing with dry and clean nitrogen. Avoid porous materials, plastics, organic materials, grease, glue etc. Preferably only use glass and stainless steel.
- Carefully clean and dry the chamber before it is closed.
- Only use ultra clean amplification gases, and preferably clean them further with a good cleaning system before letting them into the gas system.
- Use as high gas flow as possible.

- Be careful in the choice of oil in the gas locks, i.e. the bubblers on the gas inlet and even outlet. Only use an oil with extremely low vapour pressure and make sure it is clean and of highest quality.
- Make sure the system is absolutely gas tight not to let in any air or humidity.
- Carefully clean the TMAE (or any other photosensitive vapour used). Assure it cannot come in contact with air, humidity or any organic material, especially not substances containing OH-groups.
- Clean the system as often as possible.
- Regenerate the wires by putting as strong a current as possible through them to heat them up and burn away any depositions.
- Use as thick amplification wires as possible.

6. CATHODE EXCITATION

It has been mentioned that electron feedback from positive ions do not play an important role in the operation of most gaseous detectors filled will with photosensitive vapours.

This is normally correct, but only for detectors which do not have any problems with respect to aging. In detectors affected by aging, where a thin polymer layer is formed on the cathode, the situation can be dramatically different. Here the positive ions flowing back onto this polymer layer can cause several undesirable macroscopic effects such as the appearance of sudden instabilities or even discharges. These instabilities are attributed to the so called Malter effect (Malter, 1936). The positive ions will accumulate on the thin dielectric polymer layer. As the ion charge increases it creates a very strong electric field across the layer. This triggers a field emission from the cathode, not only in the form of single electron emission as was originally suggested by Malter, but mainly in the form of burst or jets of electrons. The effect is called explosive field emission and has been described in great detail by Nappi (Nappi, 2013). In the strong electric field over the polymer layer, electrons from the metallic cathode can penetrate into the dielectric polymer layer, accumulate there and then sporadically be released to the gas volume in the form of electron jets. Such an emission usually happens from well localized spots on the cathode which could be either be permanent or randomly appearing and disappearing with time.

Malter effects and explosive field emission effects are clearly observed in TMAE filled detectors. The TMAE molecules have a low ionization potential and are the weakest one (compared to TEA, TMA or EF) with respect to breaking into fragments which can cause polymer creation.

Figure 26 shows an example of explosive field emission observed by the CRID TPC team at Stanford Linear Accelerator Center (Va'vra, 1995). Figure 26a shows an active spot on the aged cathode from which sporadic bursts of electrons are observed. Figure 26b shows the frequency of these bursts in time. They appear at random intervals and may appear both before and after the particle beam crossing in the detector volume.

Another type of cathode excitation is lowering of the work function of the cathode. Some metals often used in photosensitive gaseous detectors have a high work function, φ, e.g. iron, nickel, chromium, copper, aluminium etc. If the work function is higher than the energy required to neutralize the ionized photosensitive gas, the ions of the photosensitive gas produced in the avalanche and drifting back slowly to the cathode will not be neutralized when they reach the cathode. Instead, they will accumulate on the metallic cathode surface. This is mainly observed for TMAE which has the lowest recombination energy. Hence, in some cases charged TMAE ions will accumulate on the cathode and create a charged

Figure 26. Malter-type bursts of electrons from the cathode of an aged MWPC-based TPC filled with TMAE vapour: a) an image of a spot on the cathode from which bursts of secondary electrons occurs, b) the random distribution in time when bursts of secondary electrons(small peaks) occurs with respect to the particle beam crossing time (the large peak in the middle of the figure)
From Va'vra, 1995.

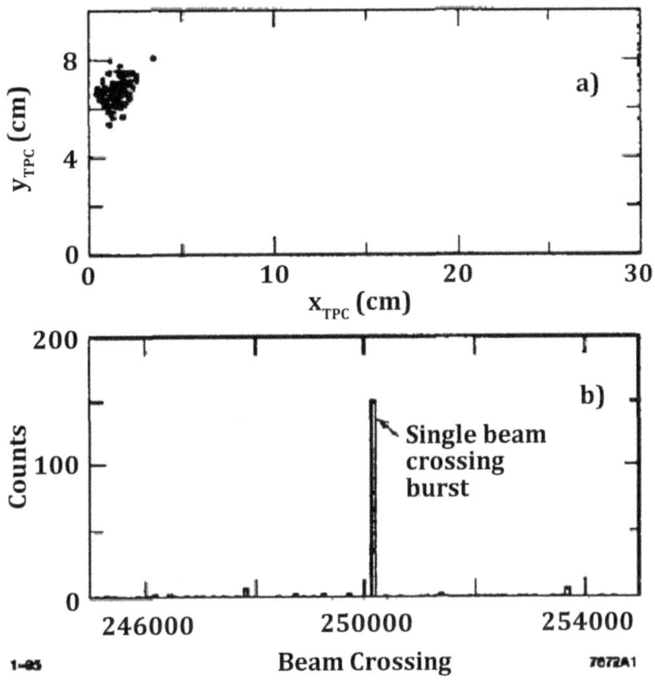

layer with a strong internal electric field, even without the presence of a polymeric film. The TMAE ions will actually form a thin liquid and polarized layer which due to the high electric field across it will lower the work function, of the metal. As a consequence, the metal surface coated with adsorbed TMAE exhibit an increased sensitivity to light. We have obtained a photosensitive surface.

Although this effect is strongest when using TMAE, the lowering of the work function and the temporal increased sensitivity to light has been observed with other gases, even not photosensitive, when the metallic cathode is under intense ion bombardment (Karabadzhak, 1983). The lowering of the work function and other effects causing an increased emission of secondary electrons from the cathode is commonly called the cathode excitation effect (Nappi, 2013). As will be shown later in this book, the cathode excitation effect can sometimes be beneficial and is used in gaseous detectors from time to time, e.g. in combination with solid photocathodes.

To summarize, cathode excitation effect, i.e. sporadic secondary emission of electrons from the cathode, may appear when:

1. A dielectric layer is formed on the cathode which charges up if the ion deposition rate is higher than the recombination rate. The strong electric field produced will extract electrons or even bursts of electrons from the metal underneath.

2. The photosensitive ions are not neutralized on the surface due to a high work function. The ions accumulate and create a charged layer on the cathode. This charged surface is photosensitive and will have a lower work function than the metallic surface.
3. Special ignition mechanisms which are strong enough to trigger electron emission from the cathode.

Examples of a dielectric formation could be polymer films formed due to the aging, insulating deposits left from discharges, resistive oxides, poor plating etc.

Examples of ignition mechanisms which facilitate the cathode excitation effect are e.g. highly ionizing particles creating dense avalanches, creation of sharp spikes on the cathode, sparks, surface erosion produced by sparks, etc.

The cathode excitation effect is often a disturbing effect and will be considered again in several parts of this book.

7. LEAKAGE CURRENT IN DETECTORS FILLED WITH TMAE VAPOURS

TMAE molecules and ions can be absorbed on dielectric surfaces. This will make the dielectric surfaces slightly conductive and may cause an undesired leakage current, especially if the surface connects the high voltage anode to ground. This may affect the detector performance.

An example is GEM detectors. Many observations have shown that standard GEMs or RETGEMs exposed to TMAE vapours become noisy and even unstable due to the leakage current through the holes. Only in gas mixtures with EF vapours GEMs and RETGEMs show a stable performance. However, EF is seldom a realistic choice as photo-converter since it requires heating the whole detector to 70-80 °C to achieve high photon detection efficiency.

Figure 27. A wall-less GEM made of two resistive plates (cathode and anode) with matching holes, separated by special spacers; in this way TMAE and polymers may deposit on the electrodes without causing a resistive path between the anode and the cathode which may lead to a leakage current.

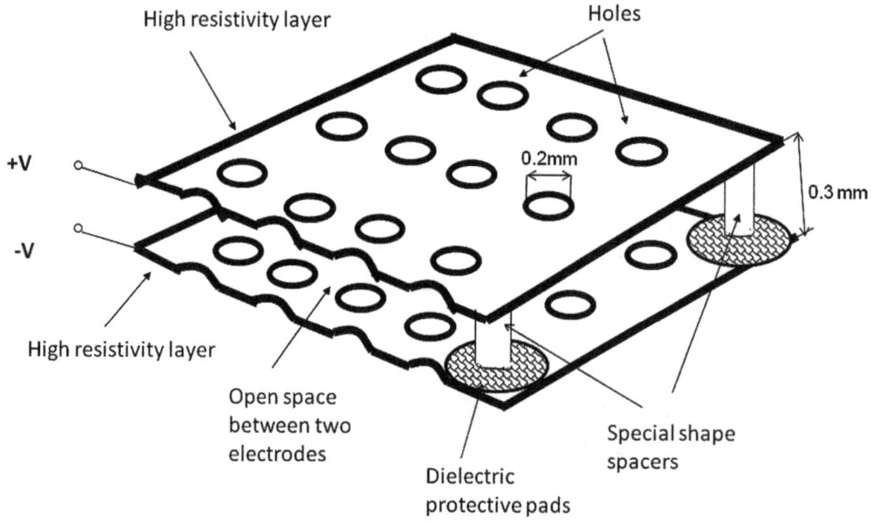

Figure 28. The shape of the spacers used in the wall less GEM shown in Figure 27; the spacers are placed far away from the matching hole pairs where the reactive TMAE debris is formed, and where most polymerization occurs. They have a shape common in high voltage support structures and are placed on large isolating pads to maximize the path on the dielectric surface from the anode to the cathode.

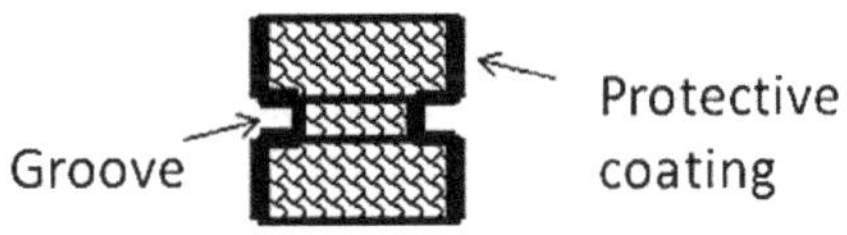

Luckily, one can modify a GEM detector to make it work with TMAE vapours with practically no leakage current (Breuil, 2013).Such a detector consists of two parallel resistive electrodes (the top electrode is the cathode, the bottom electrode is the anode) with matching holes. The two electrodes are separated by specially shaped spacers located far away from the holes. It is called a "wall-less GEM" and is shown in Figures 27 and 28.

The spacers are placed far away from the matching hole pairs, where most the reactive TMAE debris is formed and where most polymerization occurs. The spacers have a shape common in high voltage support structures to maximize the distance for possible leakage currents, and are placed on large isolating pads to further maximize the path along a dielectric surface from the anode to the cathode. The holes on the top and bottom electrodes are carefully aligned with respect each other allowing the formation of an electric field very similar to the one in a standard GEM. Such a structure can operate stable and without spurious pulses at gas gains of 10^5 in a gas mixture with TMAE.

In a more sophisticated version of this detector the electrodes are made of resistive strips separated by dielectric strips, as shown in Figure 29.

Figure 29. A more sophisticated design of a wall-less GEM; both electrodes are segmented into strips, separated by narrow insulating dielectric strips. The array of resistive strips on the anode plate is oriented perpendicular to similar resistive strips placed on the cathode plate.

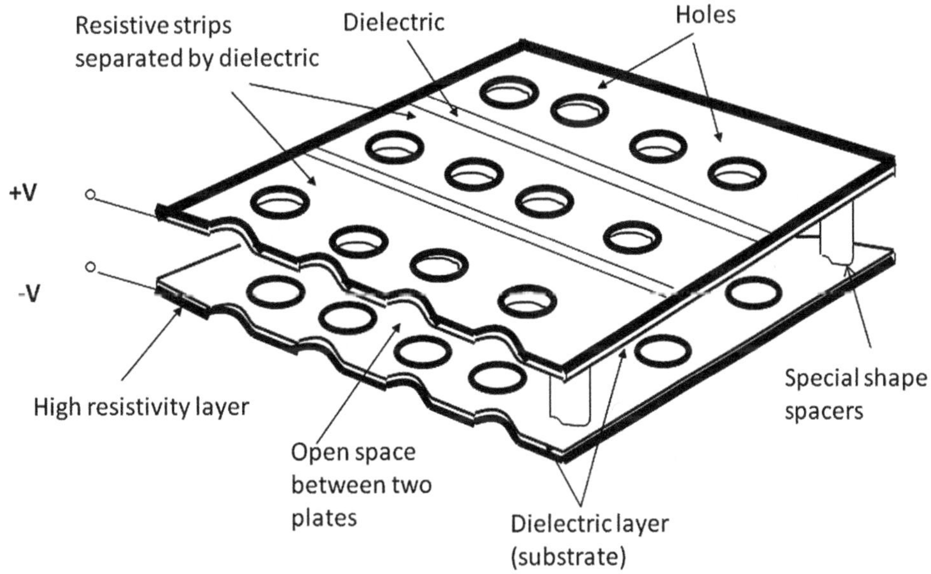

The strip design better protects the detector if a spark should occur. In a possible spark only the charge stored in the capacitance of one strip in the anode and an orthogonal strip in the cathode will add energy to the discharge. Since the strips are resistive as well, a possible spark is further reduced by the voltage drop caused during the discharge by the resistivity in the strips. The segmentation into strips also enables a rough position measurement at low counting rates without using a separate readout plate. For example, if the strips on one electrode are perpendicular to the strips on the other electrode, one can detect a 2D coordinate of the avalanche position by recording the induced signals in them in coincidence mode. This is only possible at low counting rates where the avalanches are well separated in time. If avalanches start to overlap in time, but not in position, ambiguities arise in the reconstruction.

Nore that the RPC can also operate with TMAE vapour without any special problems since it has spacers located far away fro the region of high electric field (Francke, 2004).

8. CONCLUSIVE REMARKS

MWPCs filled with gases with low ionization potential (e.g. benzene, toluene, TMA, EF, TEA, TMAE etc.) were the first successful position sensitive photomultipliers. When they were developed in the 1980ies they revolutionized the photon detection technique. They are fast, accurate and efficient. Furthermore, these detectors can be built in large areas, which made them attractive for e.g. Cherenkov photon detection. TMAE filled multiwire proportional chambers were successfully used in various Ring Imaging detectors. They were used in relatively small size experiments such as CERES (Baur, 1994) and CAPRICE (Francke, 1999), as well as in gigantic experiments such as DELPHI (Albrecht, 1999) and SLD (Va'vra, 1995).

Position sensitive gaseous photomultipliers are also used in plasma diagnostics, including the visualization of UV light emitted from plasma instabilities (Bogomolov, 1987). Recently, they have been used for more practical applications such as remote detection of various flames, including the forest fires, in broad daylight (Charpak, 2009).

There are also applications using photosensitive vapours beyond the detection of UV photons. One interesting example was the optical chamber which imaged the light produced along the tracks of charged particle as they traverse TEA or TMAE (Charpak, 1988).

The initial position sensitive photodetectors using gaseous photocathodes revolutionized many physics experiments, and was the foundation for many important scientific discoveries. Today, 35 years later, gaseous photosensitive vapours are still widely used, mainly due to the simplicity to make large area photosensitive detectors. In many cases the gaseous photocathodes are being replaced by solid photocathodes, still operating in gaseous avalanche detectors. Solid state photosensitive detectors are gaining as well as position sensitive photomultiplier tubes, but they are still difficult to produce for large area applications.

The next chapters will describe the interesting development of liquid solid photocathodes.

REFERENCES

Adam, W., Albrecht, E., Ambec, I., Augustinus, A., Barnoux, C., Boštjančič, B., & Zevgolatakos, E. et al. (1994). The forward ring imaging Cherenkov detector of DEPHI. *Nuclear Instruments & Methods in Physics Research. Section A, Accelerators, Spectrometers, Detectors and Associated Equipment*, *338*(2-3), 284–309. doi:10.1016/0168-9002(94)91314-5

Alavarez, V., Aznar, F, & Borges, F.I.G.M. (2014). Description and commissioning of NEXT-MM prototype: first results from operation in a xenon-trimethylamine gas mixture. *JINST,* 9P030101, 1-20.

Albrecht, E., van Apeldoorn, G., Augustinus, A., Baillon, P., Battaglia, M., Bloch, D., & van Vulpen, I. et al. (1999). Operation, optimization and performance of the DELPHI RICH detectors. *Nuclear Instruments & Methods in Physics Research. Section A, Accelerators, Spectrometers, Detectors and Associated Equipment*, *433*(1-2), 47–58. doi:10.1016/S0168-9002(99)00320-4

Anassontzis, E. G., van Apeldoorn, G., & Aria, E. (1992). The barrel ring imaging Cherenkov counter of DELPHI. *Nuclear Instruments & Methods in Physics Research. Section A, Accelerators, Spectrometers, Detectors and Associated Equipment*, *323*, 351–362.

Anderson, D. F. (1988). Measurement of TMAE and TEA vapour pressure. *Nuclear Instruments & Methods in Physics Research. Section A, Accelerators, Spectrometers, Detectors and Associated Equipment*, *270*(2-3), 416–418. doi:10.1016/0168-9002(88)90709-7

Arnold, R., Baillon, P., Besch, H. J., Bosteels, M., Christophel, E., Dracos, M., & Ypsilantis, T. et al. (1988). A ring imaging Cherenkov detector, the DELPHI barrel RICH prototype. Part A: Experimental studies of the detection efficiency and the spatial resolution. *Nuclear Instruments & Methods in Physics Research. Section A, Accelerators, Spectrometers, Detectors and Associated Equipment*, *270*(2-3), 255–288. doi:10.1016/0168-9002(88)90695-X

Artuso, M., Ayad, R., Bukin, K., Efimov, A., Boulahouache, C., Dambasuren, E., & Smith, A. et al. (2003). Construction, pattern recognition and performance of the CLEOIII LiF-TEA RICH detector. *Nuclear Instruments & Methods in Physics Research. Section A, Accelerators, Spectrometers, Detectors and Associated Equipment*, *502*(1), 91–100. doi:10.1016/S0168-9002(02)02162-9

Baur, R., Breskin, A., Chechik, R., Drees, A., Faschingbauer, U., Fischer, P., & Wurm, J. P. et al. (1994). The CERES RICH detector system. *Nuclear Instruments & Methods in Physics Research. Section A, Accelerators, Spectrometers, Detectors and Associated Equipment*, *343*(1), 87–98. doi:10.1016/0168-9002(94)90537-1

Bogomolov, G. D., Dubrovski, I. Yu., & Letunov, A. A. (1987). Experimental investigation of non-stationary streamer formation in a microwave plasma column at high pressure. *Soviet Physics, JETP*, *66*, 295–302.

Bogomolov, G.D., Dubrovski Yu., Peskov, V.D. (1978a). Multiwire Gas Counter for Coordinate Measurements in the VUV Region. *Instr. Exp. Tech, 21*, 778-782.

Bogomolov, G. D., Dubrovski, Yu., & Peskov, V. D. (1978b). Photoelectrical Readout for Multiwire Counters. *Instruments and Experimental Techniques*, *21*, 639–642.

Breuil, P., Fonte, P., & Nappi, E. (2013). Advances in development of micropattern gaseous detectors with resistive electrodes. *Eur. Phys. J. Plus*, *160*, 1–16.

Carlson, P., Francke, T., Löfgren, S., Weber, N., & Suffert, M. (1994). Results from beam tests of the CAPRICE RICH detector. *Nuclear Instruments & Methods in Physics Research. Section A, Accelerators, Spectrometers, Detectors and Associated Equipment*, *349*(2-3), 577–591. doi:10.1016/0168-9002(94)91229-7

Cebrian, S., Dafni, T., Ferrer-Ribas, E. (2013). Micromegas-TPC operation at high pressure in xenon-trimethylamine mixtures. *JINST,* 8P01012, 1-16.

Charpak, G. (1995). *Research on particle imaging detectors*. Singapore: World scientific.

Charpak, G., Benaben, P., Breuil, P., Martinengo, P., Nappi, E., & Peskov, V. (2009). Progress in development of a S-RETGEM-based detector for an early forest fire warning system. *JINST*, *4*(12P12007), 1–16. doi:10.1088/1748-0221/4/12/P12007

Charpak, G., Bouclier, R., Bressani, T., Favier, J., & Zupančič, Č. (1968). The use of multiwire proportional counters to select and localize charged particles. *Nuclear Instruments and Methods in Physics Research*, *62*(3), 262–268. doi:10.1016/0029-554X(68)90371-6

Charpak, G., Domini, W., & Fabre, J. P. (1988). Some applications of the imaging proportional chamber. *IEEE Nucl.Sci*, *35*(1), 483–486. doi:10.1109/23.12770

Charpak, G., Majewski, S., Melchart, G., Sauli, F., & Ypsilantis, T. (1979). Detection of far- ultraviolet photons with the multistep avalanche chamber. Application to Cherenkov light imaging and to some problems in high-energy physics. *Nuclear Instruments and Methods in Physics Research*, *164*(3), 419–433. doi:10.1016/0029-554X(79)90074-0

Charpak, G., Peskov, V., Sauli, F., & Scigocki, D. (1989). Ethylferrocene in gas, condensed, or adsorbed phases: Three types of photosensitive elements for use in gaseous detectors. *Nuclear Instruments & Methods in Physics Research. Section A, Accelerators, Spectrometers, Detectors and Associated Equipment*, *277*(2-3), 537–546. doi:10.1016/0168-9002(89)90787-0

Chubb, T. A., & Friedman, H. (1955). Photon counters for far ultraviolet. *The Review of Scientific Instruments*, *26*(5), 493–498. doi:10.1063/1.1771334

Diatroptov, D. B., & Peskov, V. D. (1972). Vacuum ultraviolet spectra of a free plasma filament in a microwave high-pressure discharge. *Soviet Phys. JETF*, *34*, 554–558.

Dutt, N. V. K., Kahol, A. P., & Jyothlraju, P. (1982). Vapour pressure of trimethyl phosphate and trimethylamine. *Journal of Chemical & Engineering Data*, *27*(4), 369–370. doi:10.1021/je00030a001

Ekelöf, T., Séguinot, J., Tocqueville, J., & Ypsilantis, T. (1981). The Cherenkov ring-imaging detectors: Recent progress and future development. *Physica Scripta*, *23*(4B), 718–724. doi:10.1088/0031-8949/23/4B/023

Fonte, P., Ferreira-Marques, R., Peskov, V., & Policarpo, A. (1999). Single-electron pulse-height spectra in thin-gap parallel-plate chambers. *Nuclear Instruments & Methods in Physics Research. Section A, Accelerators, Spectrometers, Detectors and Associated Equipment*, *433*(1-2), 513–517. doi:10.1016/S0168-9002(99)00309-5

Fonte, P., Peskov, V., & Sauli, F. (1991). UV emission and breakdown in parallel-plate chambers. *Nuclear Instruments & Methods in Physics Research. Section A, Accelerators, Spectrometers, Detectors and Associated Equipment*, *310*(1-2), 140–145. doi:10.1016/0168-9002(91)91013-L

Francke, T., Bergstrom, D., Boezio, M., Carlson, P., & Suffert, M. (1999). A gas- RICH detector for space. *Nuclear Instruments & Methods in Physics Research. Section A, Accelerators, Spectrometers, Detectors and Associated Equipment*, *433*(1-2), 87–91. doi:10.1016/S0168-9002(99)00311-3

Francke, T., Peskov, V., Rodionov, I., & Fonte, P. (2004). High rate (up to 10^5 Hz/cm^2), high position resolution (30 μm) photosensitive RPCs. *Nuclear Instruments & Methods in Physics Research. Section A, Accelerators, Spectrometers, Detectors and Associated Equipment*, *533*(1-2), 163–168. doi:10.1016/j.nima.2004.07.021

Heller, C., (1980). *Tetrakis(dimethylamino)ethylene (TMAE) purification report.* CERN.

Kadyk, J. (1991). Wire chamber aging. *Nuclear Instruments & Methods in Physics Research. Section A, Accelerators, Spectrometers, Detectors and Associated Equipment*, *300*(3), 436–479. doi:10.1016/0168-9002(91)90381-Y

Karabadzhak, G. F., Peskov, V. D., & Podolyak, E. R. (1983). Coordinate gas counters with super high energy resolution for plasma diagnostics. *Nuclear Instruments and Methods*, *A217*(1-2), 56–64. doi:10.1016/0167-5087(83)90109-6

Korpar, S., Krizan, P., Stanovnik, A., Staric, M., & Skrk, D. (1999). Aging and rejuvenation of a TMAE + methane multiwire photon detector. *IEEE Transactions on Nuclear Science*, *46*(3), 317–320. doi:10.1109/23.775536

Lau, K., Atiya, M., Broemmelsiek, D., Hamacher, T., Ispiryan, M., Korpar, S., & Staric, M. et al. (2003). Test-beam aging studies of a TMAE prototype for the HERA-B RICH. *Nuclear Instruments & Methods in Physics Research. Section A, Accelerators, Spectrometers, Detectors and Associated Equipment*, *515*(1-2), 313–318. doi:10.1016/j.nima.2003.09.016

Malter, L. (1936). Thin film field emission. *Physical Review*, *50*(1), 48–58. doi:10.1103/PhysRev.50.48

Melin, G. (1974). Handbuch der Anorganischen Chemie. Springer-Verlag.

Nakato, Y., Ozaki, M., & Tsobomura, H. (1972). Photoionization and Rydberg States of Tetraaminoethylenes in Organic Solutions. *Journal of Physical Chemistry*, *76*(15), 2105–2112. doi:10.1021/j100659a008

Nappi, E., & Peskov, V. (2013). Imaging gaseous detectors and their applications. Wiley-VCH. doi:10.1002/9783527640294

Peskov, V. (1980). Coordinate gas scintillation x-ray and UV photon counters. *Instruments and Experimental Techniques*, *23*, 507–510.

Peskov, V., Charpak, G., Sauli, F., Scigocki, D., Diep, V., & Janjic, D. (1989). Organometallic photocathodes for parallel-plate and wire chambers. *Nuclear Instruments & Methods in Physics Research. Section A, Accelerators, Spectrometers, Detectors and Associated Equipment*, *283*(3), 786–791. doi:10.1016/0168-9002(89)91458-7

Peskov, V. D. (1979a). VUV Spectrum of Townsend Avalanches. *Zh. Prikl. Spectr*, *30*, 860–866.

Peskov, V.D. (1979b). On the VUV Radiation of Townsend Avalanches. *J. Phys. (France)*, *40*, C7-331-C7333.

Pyrlik, J., Atiya, M., Broemmelsiek, D., Hamacher, T., Ispiryan, M., Korpar, S., & Starič, M. et al. (1998). Aging measurements of a TMAE based photon detector for the HERA-B RICH. *Nuclear Instruments & Methods in Physics Research. Section A, Accelerators, Spectrometers, Detectors and Associated Equipment*, *414*(2-3), 170–181. doi:10.1016/S0168-9002(98)00645-7

Riegler, W., Schindler, H., & Veenhof, R. (2008). *Avalanche statistics*. Report at the RD-51 collaboration meeting. Retrieved from http://indico.cern.ch/event/35172/other-view?view=standard

Sauli, F. (2009). *Principles of gas detectors*. Talk given at KEK. Retrieved from http://view.officeapps.live.com/op/view.aspx?src=http%3A%2F%2Fcompassweb.ts.infn.it%2Frich1%2Fjarda%2FPresentations%2FTIPP09_japan%2FSAULI_PRINCIPLES.ppt

Sauli, F. (2011). Seminal Papers by Georges Charpak. In *Nuclear Instruments and Methods in Physics Research Section A: Accelerators, Spectrometers, Detectors and Associated Equipment*. Elsevier.

Schindler, H., & Veenhof, R. (2008). *Avalanche statistics*. Retrieved from http://webcache.googleusercontent.com/search?q=cache:Oml9YMQffgAJ:https://indico.cern.ch/event/58953/session/0/contribution/1/material/slides/+&cd=1&hl=en&ct=clnk&gl=ch

Schlumbohm, H. (1958). Zur statistic der electronenlawinen im ebenen feld. *Zeitschrift fur Physik*, *151*(5), 563–572. doi:10.1007/BF01338427

Séguinot, J (1989). *Lescompteurs Cherenkov: applications et limites pour l'identification des particules, developments et persepectives*. Preprint CERN-EP/89-92, LPC/89-25, 1-77

Séguinot, J., Tocqueville, J., & Ypsilantis, T. (1980). Imaging Cherenkov detectors: Photoionization of triethylamine. *Nuclear Instruments and Methods*, *173*, 283–298.

Séguinot, J., & Ypsilantis, T. (1977). Photo-ionization and Cherenkov ring imaging. *Nuclear Instruments and Methods*, *142*(3), 377–391. doi:10.1016/0029-554X(77)90671-1

Senftleben, A., Pflüger, T., Ren, X., Al-Hagan, O., Najjari, B., Madison, D., & Ullrich, J. et al. (2010). Search for interference effects in electron impact ionization of aligned hydrogen molecules. *Journal of Physics. B, Atomic, Molecular, and Optical Physics*, *43*(8), 081002–081007. doi:10.1088/0953-4075/43/8/081002

Swift, E. Jr, & Hochanadel, H. P. (1945). The vapour pressure of trimethylamine from 0 to 40°. *Journal of the American Chemical Society*, *67*(5), 880–881. doi:10.1021/ja01221a508

Titov, M. (2003). *Radiation damage and long-term aging in gas detectors.* A talk given at the INFN Eloisotron project's 42nd Workshop on Innovative detectors for supercolliders, Erice, Italy.

Va'vra, J. (1987). Wire aging of hydrocarbons with TMAE additions. *IEEE Transactions on Nuclear Science, NS-34*(1), 486–490. doi:10.1109/TNS.1987.4337390

Va'vra, J. (1992). Wire chamber gases. *Nuclear Instruments & Methods in Physics Research. Section A, Accelerators, Spectrometers, Detectors and Associated Equipment, 323*(1-2), 34–47. doi:10.1016/0168-9002(92)90266-7

Va'vra, J. (1995). Can TMAE photocathode be used for high rate applications? *Nuclear Instruments & Methods in Physics Research. Section A, Accelerators, Spectrometers, Detectors and Associated Equipment, 367*(1-3), 353–357. doi:10.1016/0168-9002(95)00536-6

Va'vra, J. (1997). Wire aging with TEA photocathode. *Nuclear Instruments & Methods in Physics Research. Section A, Accelerators, Spectrometers, Detectors and Associated Equipment, 387*(1-2), 183–185. doi:10.1016/S0168-9002(96)00986-2

Va'vra, J., Kady, J., & Wise, J. (1996). Study of photosensitive mixtures of TMAE and helium, hydrocarbons or CF4-based carrier gases. *Nuclear Instruments & Methods in Physics Research. Section A, Accelerators, Spectrometers, Detectors and Associated Equipment, 370*(2-3), 352–366. doi:10.1016/0168-9002(95)00842-X

Woody, C. L. (1988). Aging effects in low pressure wire chambers containing TMAE. *IEEE Transactions on Nuclear Science, 35*(1), 498–497. doi:10.1109/23.12772

KEY TERMS AND DEFINITIONS

Aging Effect: Degradation of the detector performance with time under irradiation. One of the reasons is accumulation polymers on the detector electrodes, created during avalanche process.

Cathode Excitation Effect: Spontaneous emission of single electrons or burst of electrons from the cathode bombarded by positive ions. This emission may continue for some time even after the end of ion bombardment.

Gas Gain: A number of secondary electrons created by one primary electron due to the avalanche multiplication.

Ion Feedback: A feedback loop caused by secondary electrons created by positive ions. This mainly happened during ion recombination on the cathode surface.

Multiwire Proportional Counter (MWPC): A gaseous detector consisting from two parallel cathode planes with an array of parallel anode wires located in between the cathode planes.

Photon Feedback: A feedback loop caused by electrons created due to the photoionization of the gas inside the detector or extracted by a photoelectric effect from its cathode.

Photosensitive Vapors: Vapors with small ionization potentials, which can be ionized by VUV photons. Such vapors are added to the gaseous photodetectors serving as convertor of VUV photons to primary electrons.

Pulse-Height Spectrum: A histogram in which the horizontal axis represents the pulse amplitude and the vertical axis shows the number of events recorded for each amplitude.

Quantum Efficiency (QE): Usually meaning the ratio of created electrons per incident photon. However in the case of photosensitive vapors QE is often expressed in terms of electrons per absorbed photons.

RICH Detector: A position sensitive detector capable recording the Cherenkov ring and determining the angle of the Cherenkov light cone.

Ring Imaging Cherenkov (RICH): A technique allowing to identify charged elementary particles through the detection of the angle of the Cherenkov light which these particles produce (under certain conditions) traversing a medium with reflective index > 1.

Chapter 4
Liquid Photocathodes

ABSTRACT

Liquid photocathodes were studied intensely for a number of years around 1988. Initially it was observed that the gaseous compounds used as photocathodes were absorbed on surfaces, making them slightly photosensitive. This opened up the dream of a well-defined photosensitive layer. In a photosensitive detector with a gaseous photoelectric converter it is difficult to know where each individual photoelectron is actually emitted. The conversion volume has to be made thick enough to allow efficient conversion of the incoming photons. This smears the position resolution, and reduces the time resolution. A thin layer photocathode would eliminate this smearing in space and time. Furthermore, the gas system might be simplified, or even removed with such a liquid cathode. We summarize the results of these studies, which led to the important development of solid photocathodes in gaseous detectors.

1. INTRODUCTION: EARLY OBSERVATIONS

In the previous chapter we described highly efficient UV sensitive gaseous detectors filled with photosensitive gases. These detectors ionize vapours with low ionization potential, so they are sensitive to wavelengths:

$$\lambda < \lambda_{ion}, \qquad (1)$$

where λ_{ion} is the ionization threshold of the vapour.

Experiments have shown that these detectors may also have some very small photosensitivity (usually several orders of magnitude less) also for wavelengths longer than λ_{ion}, if the light hits the cathode of the detector. This is mainly due to the classical photoelectric effect on the metallic surface.

The photoelectric effect in a metallic surface happens in there steps:

1. Absorption of the light in the thin surface layer of the cathode where the photon transfers its energy to one of the electrons in the conduction band,
2. Ballistic transport of the excited/hot electron to the surface,

DOI: 10.4018/978-1-5225-0242-5.ch004

3. The electron escapes from the metal surface into the gas.

The maximum possible kinetic energy of the electron after it has escaped from the metallic surface is described by the famous Einstein formula mentioned earlier:

$$E_k = h\nu - \varphi, \tag{2}$$

where h is the Plank constant and ν is the frequency of the incident light.

Metals have high reflectivity in the visible and the near ultraviolet region of the spectrum, so the majority of the photons are reflected on the surfaces and to not cause any photoelectric effect. The good reflectivity is one of the explanations why metals have low quantum efficiency to light of these wavelengths. The quantum efficiency is usually expressed in terms of electrons per incident photon.

Metals have a high abundance of free electrons why the exited hot electron will experience many collisions where it gradually loses its excess kinetic energy. It will be "thermolized" within a short distance. As a result only electrons created very close to the surface will have some chance to escape if, of course, its kinetic energy is still above the metal's work function. The combination of good reflectivity and energy losses within the metal before the electron has a chance to escape from the surface leads to very low quantum efficiency for metals to visible and UV light.

The quantum efficiency also depends on which type of gas the metallic surface is in contact with (Mc. Daniel, 1964; Funfer, 1965).

There are two reasons for this:

1. Some atoms and molecules change the surface barrier/work function of the metal.
2. Some gases can form a dielectric layer on the metallic surface. Electrons escaping from the metal into the gas must penetrate through this layer, where the electrons can be absorbed before escaping into the gas.

Photosensitive gaseous detectors practically never use a metallic cathode as a photo-converter due to the unacceptably low quantum efficiency.

A breakthrough idea came from David Anderson (Anderson, 1982). He suggested using a photosensitive metallic cathode covered with a thick layer of absorbed TMAE. He called it a "liquid photocathode". He used the fact that the threshold energy for photoionization in liquids is generally lower than in gases owing to the polarization energy P_+ of the positive ion and the lowering of the electron conduction band energy, V_0.The relation between the gas phase ionization energy E_i and the liquid threshold E_{liq}, as was mentioned earlier, is given by the relation (Schmidt, 1997):

$$E_{lid} = E_i + P_+ + V_0 + E_{val}, \tag{3}$$

Sometimes E_{val} can be neglected, the polarization energy is always negative, whereas V_0 may be of either sign. In TMAE, V_0=-0.26 eV which extends the photosensitivity of its liquid phase to 350 nm, whereas gaseous TMAE is only photosensitive up to wavelengths of 230 nm (Anderson, 1982).

One way to form a liquid photocathode is to cool a metallic plate and expose it to TMAE vapour.

Further studies performed by Anderson and his colleagues revealed that a liquid photocathode can be used in two basic modes:

1. A thick enough layer where the UV light is fully absorbed.
2. A very thin layer, where it remains almost transparent to the UV light, but it changes the work function, and therefore the quantum efficiency, of the metal.

Anderson and his colleagues proposed to use a liquid TMAE layer as the photocathode in a MWPC (Anderson, 1992; Anderson 1993). With a liquid photocathode they expected to build UV detectors with good sensitivity in the wavelength range up to 300 nm, which may open prospects for new applications in high energy physics and other areas of science. These perspectives attracted a lot of researchers to actively study the liquid photocathode option. Subsequent investigations have shown, however, conflicting results. For example, (Schotanus, 1987) confirmed the liquid TMAE sensitivity using spectral measurements, whereas some other investigators have failed to confirm Anderson's results (see for example Michau, 1986; Suffert, 1986). To clarify this controversy the Charpak and Ypsilantis groups at CERN launched a systematic study in 1986 of TMAE and other liquid photocathodes. The results of these exciting studies are described in the next paragraph.

2. SYSTEMATIC STUDIES OF TMAE BASED PHOTOCATHODES

2.1 TMAE Based Photocathodes

The results of the systematic studies of various liquid photocathodes performed at CERN are summarized in several publications (see for example Peskov, 1988; Charpak, 1989; Séguinot, 1990; Charpak, 1991).

Let us start with the TMAE based photocathodes. The key measurements were performed using the apparatus shown in Figure 1. It consists of a UV lamp, a monochromator and a single-wire counter. The photon beam from the monochromator is split into a reflective beam, which was monitored by a vacuum photomultiplier and a transmitted beam which was parallel-focused to the cathode of the single-wire counter. The counter has a special cathode allowing discs of different designs and manufactured from various materials to be fixed on its surface. The cathode could be cooled or heated separately from the rest of the single-wire counter volume.

The measurements done with TMAE vapour can be divided in three main categories:

1. An absorbed layer of TMAE on the cathode at room temperature.
2. A thick absorbed layer of TMAE on a cooled metallic cathode.
3. A very thick layer of TMAE (1-3 mm), i.e. a real liquid photocathode.

2.1.1 A Thin Absorbed Layer of TMAE

The quantum efficiency of Cu and Al cathodes as a function of wavelength and measured at room temperature in Ar+10%CH_4 at atmospheric pressure is shown in Figure 2.

The admittance of even an extremely low pressure of TMAE ($p<0.1$mbar) over a period of 20 hours is sufficient to form a photosensitive layer with a quantum efficiency increasing by more than a factor of 10 both at short and long wavelengths. The quantum efficiency of this type of photosensitive layer does not depend on the pressure or composition of amplification gas mixtures used in the studies.

Figure 1. Apparatus for studies of TMAE-based photocathodes; the components are as follows: 1-D_2 arc lamp, 2-UV monochromator, 3-CaF_2 beam splitter, 4-filter station, 5-transparent metallic meshes for compensation the electric field distortion by the CaF_2 window, 6- single-wire counter, 7-copper cathode which can be cooled or heated, 8-cryogenic bath, 9-TMS, TMP or neopentane liquid container, 10-TMAE or TEA container, 11-oxysorb cartridge, 12-carrier gas bottle supply, 13-wavlength shifter deposited on the photomultiplier window, 14-photomultiplier, 15 -separating window.

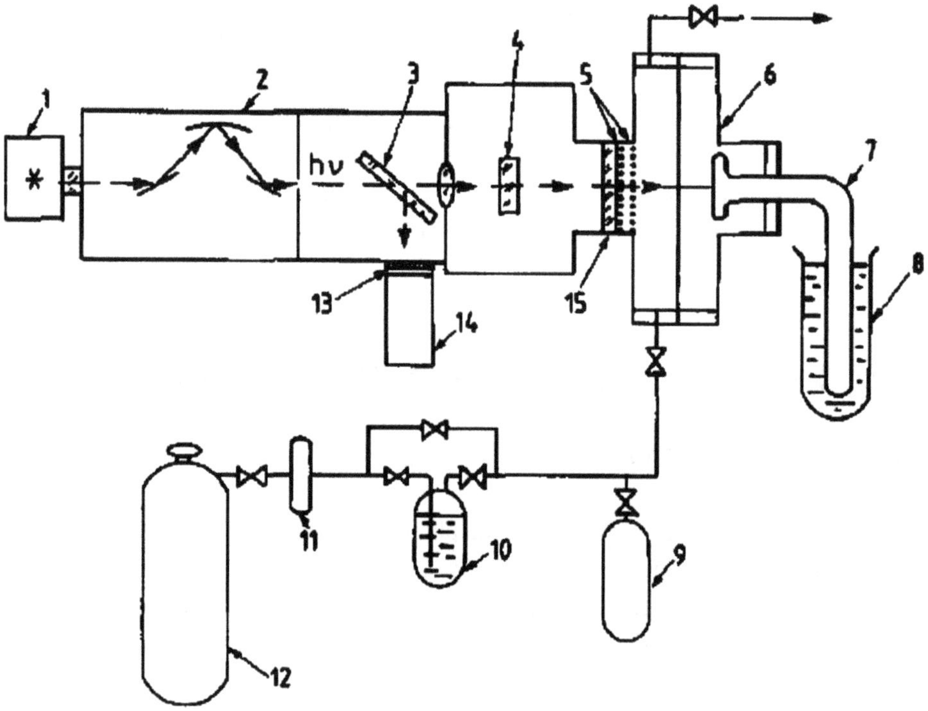

This photosensitive layer is extremely tenacious, as can be concluded from Figure 3. The whole single wire counter and the associated gas system had to be heated to about 60°C for three days and flushed with a clean gas for nine days in order to clean away the TMAE and recover the initial quantum efficiency of the metallic cathode.

The TMAE that remains in the system is not only difficult to clean away from the cathode. It will actually outgas from every surface and coat any new clean surfaces added to the system.

The formation time of the TMAE photocathode at 20°C is shown in Figure 4. The previous TMAE thin-film cathode (triangles) was replaced at time zero with a clean copper cathode and then flushed with a clean gas (which did not contain TMAE vapour). However, a new TMAE film was automatically formed on the metallic cathode due to the TMAE trapped in the gas system (outgassing from the gas system's inner surfaces). The chamber reached its previous quantum efficiency in about 20 hours.

This process can be repeated many times without adding any new TMAE to the system. Hence, the TMAE layer only needs to be a few nm thick to have full quantum efficiency.

Figure 2. The quantum efficiency vs. wavelength of copper and aluminium cathodes without and with exposure to TMAE vapour (p<0.1mbar) for a period of 20 hours; the absorbed layer of TMAE increases the quantum efficiency one order of magnitude for copper, and two orders of magnitude for aluminium.

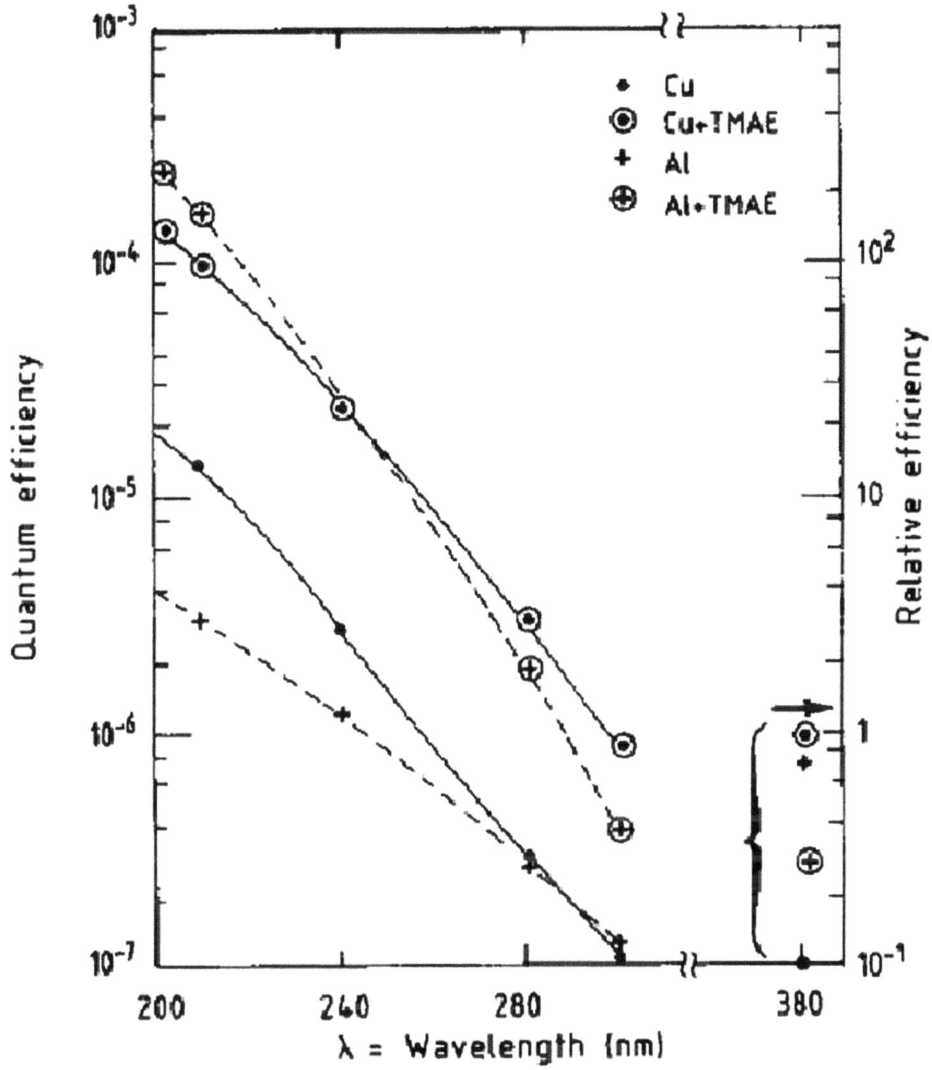

2.1.2 A Thick Absorbed Layer of TMAE

If the copper cathode is cooled to T_{cath}=0°C in gas with some TMAE vapour (even as low as 0.1 mbar of TMAE), intensive TMAE condensation begins. One can see by bare eyes that the cathode surface changes colour and becomes wet. The QE grows by factors between 10 and 100 at the wavelength region λ<250 nm. At the same time the sensitivity to light between 290 and 400 nm is reduced by two orders of magnitude.

Qualitatively the same results are obtained at elevated TMAE pressures, p > 0.1 mbar, as shown in Figure 5.

Figure 3. Measurements illustrating the evaporation of a thin TMAE layer on a copper cathode as a function of time; after formation of the TMAE layer at time zero, the system was flushed for nine days with Ar + 50% CH_4 (no TMAE). The gas flow rate was 3 l/h (the single-wire counter volume was 0.25 l). For the first three days the entire single wire counter system was heated to 50-70°C. The measurements were performed with light with wavelength λ=380 nm.

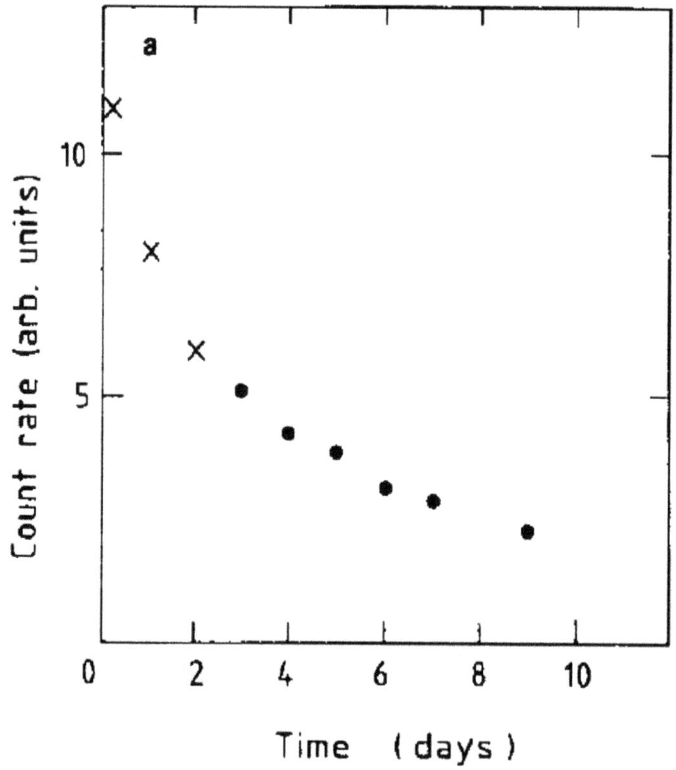

Figure 4. The formation of a photosensitive TMAE layer on a copper cathode kept at 20°C; the TMAE being deposit was coming from TMAE impurities in the gas system (surface outgassing). After heating the entire system for 10hours, a new clean cathode was installed at time zero with the same result. The gas flow was 3l/h. As in the previous case, the measurements were done with light with a wavelength λ=380 nm.

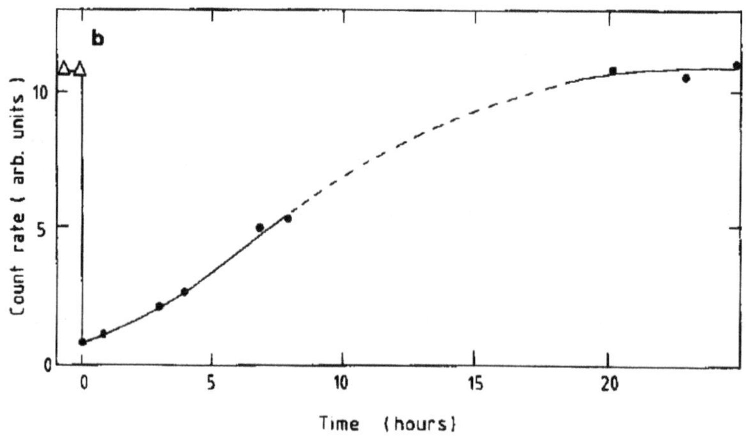

Figure 5. Results of quantum efficiency measurements vs. photon wavelength for various liquid and solid photocathodes deposited on a copper surface at a total pressure of 1bar unless otherwise noted: 1and 1'- are the curves for bare copper and copper + TMAE, respectively, reproduced from Figure 2, 2 - CH_4+TMAE, T_{cath} =0°C, 3-the same as 2, but the cathode is heated for several minutes to 40-50 °C, 4 - CH_4+TMAE+ TMP, T_{cath}= -15°C, 5 - CH_4+TMAE+TMS, p =1.5 bar, T_{cath}= - 5°C, 6 - CH_4+TMAE+ NP, T_{cath}= 0 °C, 7- CH_4+TMAE+ NP, T_{cath}= -20 °C (see the next chapter), 8 - CH_4+ diethyferrocene, T_{cath}= -40 °C.

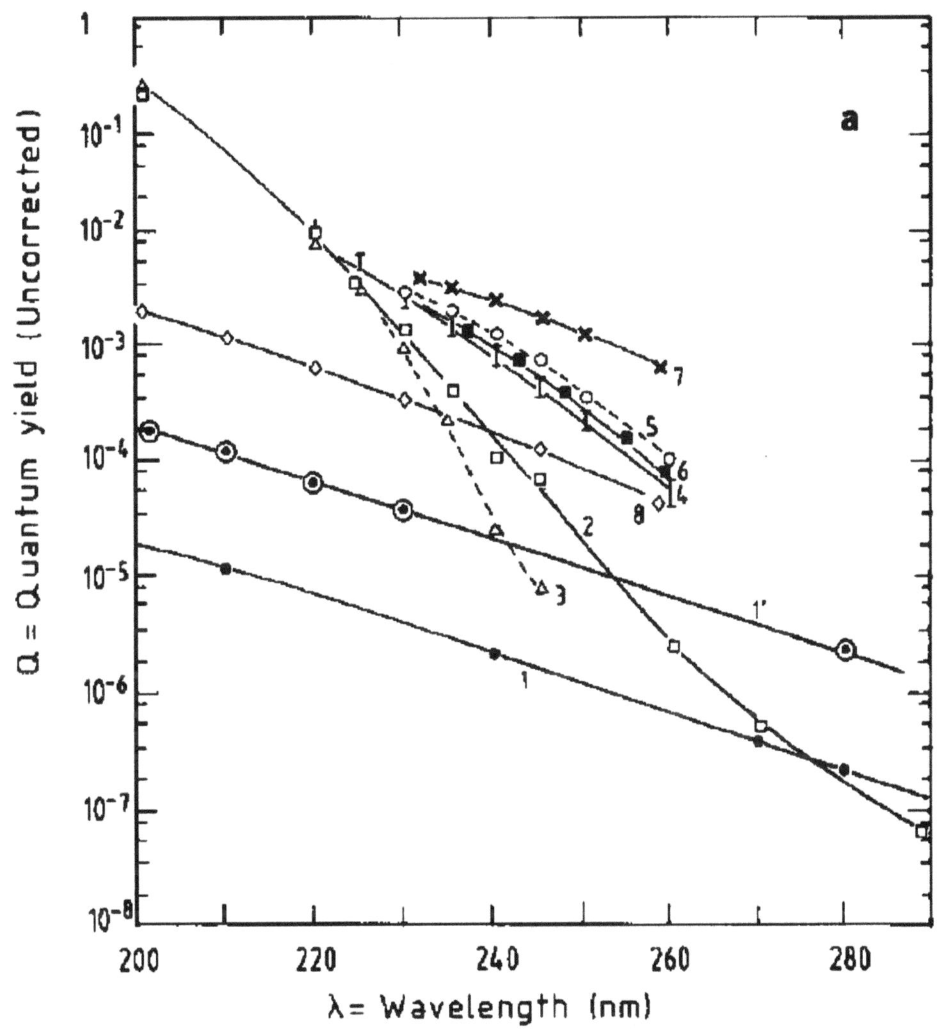

The difference from the quantum efficiency of a thin layer of TMAE on the copper cathode at low pressure of TMAE (curve 1') and higher pressure of TMAE (>.0.1 mBar, curve 2) is that one has to take into account the absorption in the TMAE gas and also its photoionization which gives the steep rise of the quantum efficiency for λ< 220 nm.

The net results for liquid photocathodes in the region where there is no interference from the photo-ionization in the gaseous TMAE (λ >230 nm) and corrected for the light absorption in the single-wire counter are presented in Figure 6.

Figure 6. The same results as presented in Figure 5, but corrected for light absorption in the detector volume; 2 - CH_4+TMAE, T_{cath} =0°C, 4 - CH_4+TMAE+ TMP, T_{cath}= -15°C, 5 - CH_4+TMAE+TMS, p =1.5 bar, T_{cath}= - 5°C, 6 - CH_4+TMAE+ NP, T_{cath}= 0 °C, 7- CH_4+TMAE+ NP, T_{cath}= -20 °C. Clearly a thick layer of TMAE on the cathode dramatically increases the quantum efficiency compared to a thin layer.

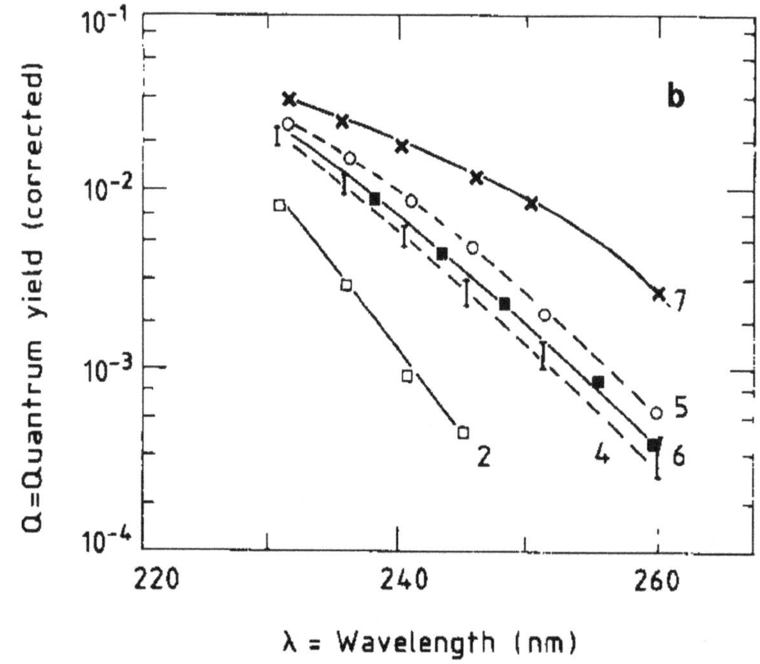

As can clearly be seen, a thick absorbed layer of TMAE is sensitive to light above the ionization threshold of gaseous TMAE, and in the wavelength interval 230-260 nm it is considerably higher the quantum efficiency of a thin TMAE layer. When the TMAE photocathode is heated to T_{cath}=40-50°C for several minutes to boil off the TMAE from the cathode, the sensitivity for λ >230 nm decreases significantly indicating that the enhanced quantum efficiency observed is indeed due to TMAE condensation on the cathode.

With the cathode is kept at room temperature the observed quantum efficiency is intermediate between those of the thin layer and the thick and cool layer. This implicates that the thicker the TMAE layer the higher is the quantum efficiency.

In all the measurements of Figure 5 and 6 the methane was continuously flushed through TMAE (and the other liquids: TMS, TMP, NP) and it was the main constituent of the amplification gas in the single wire counter. If, however, isobutane was used for the same purpose, the quantum efficiency of the liquid photocathode was reduced by a factor of 2 to 3. This is presumably due to the formation of an absorbed layer of isobutane on the cathode, creating an obstacle for photoelectrons.

2.1.3 A Very Thick Layer of TMAE (1–3 mm), a Real Liquid TMAE Photocathode

It is technically difficult make a real liquid TMAE photocathode (1-3 mm thick) in the setup described above. This particular photocathode was studied in another experiment, which followed closely the original setup of Anderson. It used an assembly shown schematically in Figure 7.

Figure 7. Schematic view of the apparatus used to study a very thick liquid TMAE photocathode; it includes a BaF_2 crystal scintillator, a wire plane (a) touching the crystal, cathode wire planes (b, d), anode wire plane (c) and a bath filled with liquid TMAE (e). The BaF_2 crystal is irradiated with gamma rays to make it emit UV light over a wide wavelength range.

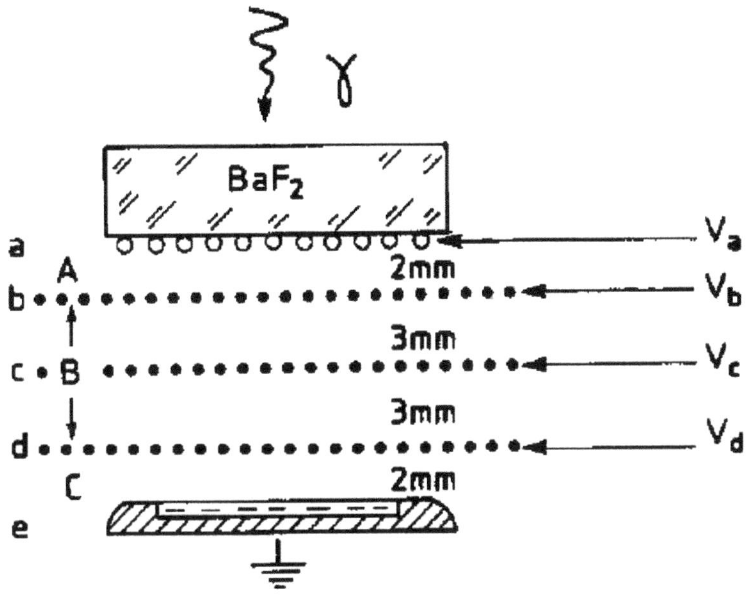

The setup consists of a low-pressure (≤ 3 mbar) MWPC located inside a thermostatic box, a gas-flow system, a liquid cathode receptacle, a pump and detection electronics. A BaF_2 crystal is used to emit UV light over a wide wavelength range when irradiated by gamma rays. The aluminum cathode had a groove forming a cathode bath which held the liquid TMAE on the cathode. It was filled either by vapour condensation, as the bath was cooled or from a special liquid reservoir.

The thermostatic box (and the gas and liquid fill lines) could be operated between -20 °C and + 70 °C. The BaF_2 scintillator was excited either with γ-rays from a^{60}Co source (1.17 and 1.33 MeV) or by X-rays from an ^{241}Am source (59 keV).

The emission spectrum of the BaF_2 scintillator and how it overlaps with the quantum efficiency of TMAE and EF vapours are shown in Figure 8.

Detailed studies performed with this setup fully confirmed Anderson's observations of the high quantum efficiency of liquid TMAE. It was shown that two contributions give rise to the increased quantum efficiency of the chamber: the formation of a thin absorbed film of TMAE on the BaF_2 surface and extraction of photoelectrons from the bath containing liquid TMAE (Peskov, 1988; Séguinot, 1990). For example, by introducing liquid TMAE into the cathode bath (in the temperature interval of 0-4°C in order to minimize the contribution from gaseous TMAE in the chamber) increased the sensitivity of the MWPC to the UV light emitted by the BaF_2crystal by a factor of more than 3, indicating that the liquid provides a contribution that is at least three times larger than that of the film absorbed on the BaF_2 crystal.

There was also an attempt to investigate a solid TMAE photocathode by cooling the gas chamber below the freezing point of TMAE, -4°C. In this case, however, the solid layer of TMAE was also formed on the BaF_2 crystal and this attenuated the intensity of the UV light reaching the solid TMAE

Figure 8. The emission spectrum of BaF_2 (1) and the quantum efficiency of EF (2) and TMAE (3) at full absorption; EF is only sensitive to the fast emission component of BaF_2 peaked around 190 nm. TMAE is also sensitive to part of the component peaked at 220 nm.

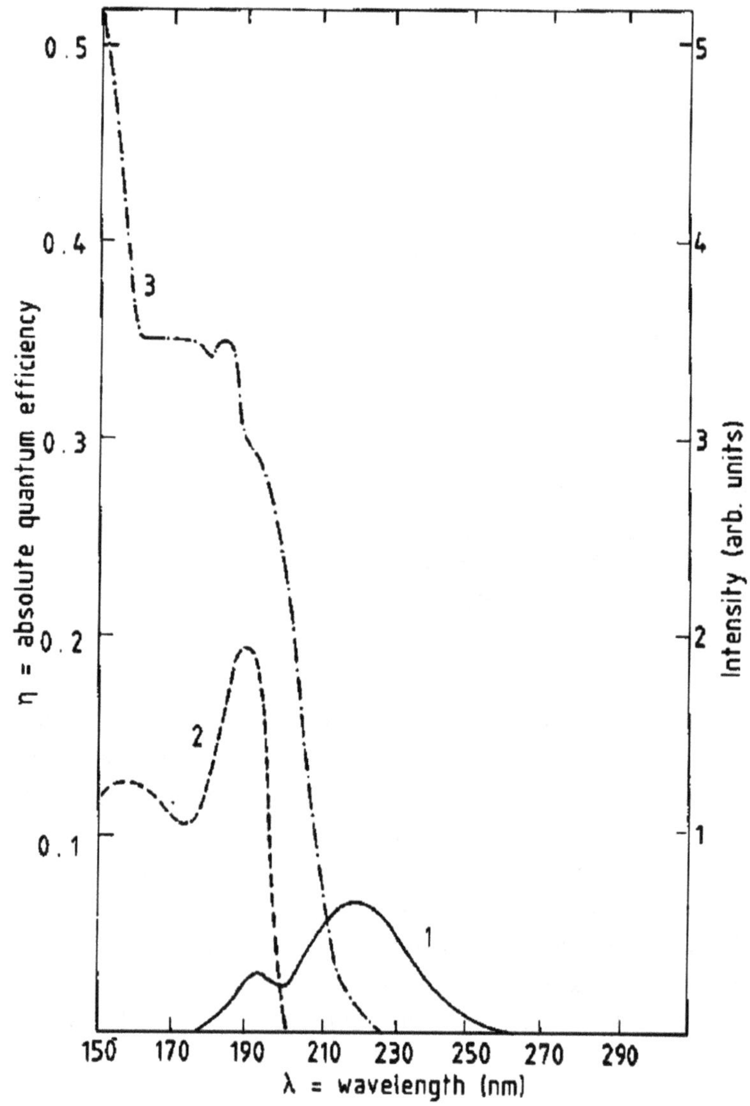

on the photocathode in an unpredictable way. So the quantum efficiency of the solid TMAE layer on the cathode was difficult to estimate reliably. However, the measurements show that the efficiency of the solid layer formed on the BaF_2 crystal was almost the same as the liquid one.

2.2 The Physics behind a TMAE Photocathode

The results of the extensive tests of liquid photocathodes in Charpak's and Ypsilantis' groups have been summarized by Charpak (Charpak, 1988). We will summarize the explanations of the results below.

2.2.1 An Absorbed Layer of TMAE

When a metallic cathode is brought into contact with TMAE vapour, or with any other gas mixture, some of the gas molecules are attached to the surface by molecular forces and produce a very thin photocathode (one or a few monolayers thick, i.e. less than 1 nm). The increased quantum efficiency observed in the metallic cathode has a simple explanation. The molecules attached to the surface change the work function, φ, of the metal. In the particular case of a thin film of TMAE, the metal work function is lowered mainly due to the predominant orientation of the absorbed TMAE molecules. The TMAE molecule forms a dipole on the surface of the metal. The dipole creates an electric field which helps to extract electrons from the metal which are now partly part of the wave function of the TMAE molecule. In such a way that the π-electrons enter the metal and electrons from the metal enters the TMAE molecule and form an electric dipole layer. These electrons have a lower work function than the original metallic surface (see Figure 9).

This explains why a metallic cathode covered with a thin layer of TMAE has a quantum efficiency that is an order of magnitude higher than the bare metallic surface, observed up to 400 nm, which is below the ionization threshold of liquid.

This also explains why it is so difficult to remove a TMAE layer from a metallic surface. It is attached to the surface through a molecular dipole bond. It is only by heating for a long period of time (a few days) and flushing with a clean gas or vacuum pumping that the surface can be completely cleaned.

2.2.2 A Thin Condensed Layer of TMAE

The effect of a thin molecular film on the quantum efficiency of the cathode may be explained as follows. In a classical case, in order to escape from the surface of the metal, the photoelectron must overcome a barrier defined by the work function, φ. However, when there is a contact between the metallic surface

Figure 9. Schematics of the TMAE dipole formation on the metallic surface, b) changes in the efficient work function of the metal due to the double electric filed formed on the metallic surface by the absorbed TMAE molecules

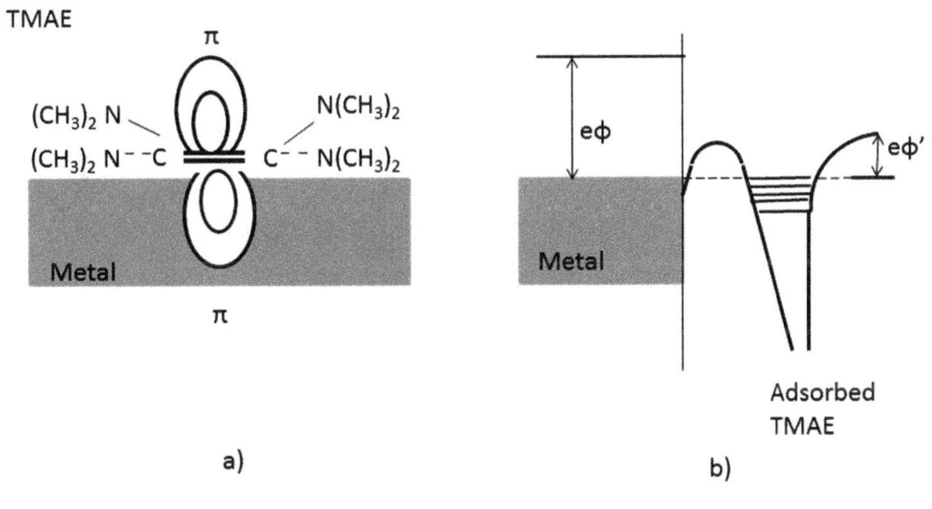

and a liquid film possessing a conduction band energy, V_0, the work function of the metal decreases if $V_0<0$. In this way, electrons find it easier to overcome the metal-liquid barrier than the metal-gas barrier. However, to escape then from the liquid film the electrons must overcomeV_0, so their kinetic energy E_k should be greater than V_0. Depending on the relationship between these three factors φ, V_0 and E_k, the liquid TMAE film can either increase or reduce the quantum efficiency of the cathode. Of course, some photoelectrons are created by UV light in the liquid layer itself. In this case, the whole physical process would be identical to the case considered in the next paragraph, describing a thick layer of TMAE.

2.2.3 A Thick Layer of TMAE

If the liquid layer is thick, the light does not reach the metal because it is absorbed in the liquid, and the photoelectrons are produced mainly through photoionization of the liquid's molecules. The electrons liberated after the photoionization at a depth x in the liquid layer of thickness d_{liq} ($x<<d_{liq}$) have a kinetic energy, as shown in Figure 10.

$$E_k = h\nu - E_{liq} \tag{4}$$

After some elastic and inelastic collisions the electron will lose its excess energy during a time t_{therm} and will be thermalized. In hydrocarbons t_{therm} is short (~ps) owing to large inelastic losses in every collision. In liquid noble gases it is longer because there it is mainly elastic collisions between the electron and the atoms, where less energy is lost in each collision. In a thermolization process the electron will

Figure 10. Schematic illustration of the photoionization process in a liquid (a) and the energy diagram of an electron escaping from the surface of the liquid due to the excess of its kinetic energy. The kinetic energy required to escape from the liquid could be lower than inside the metal.

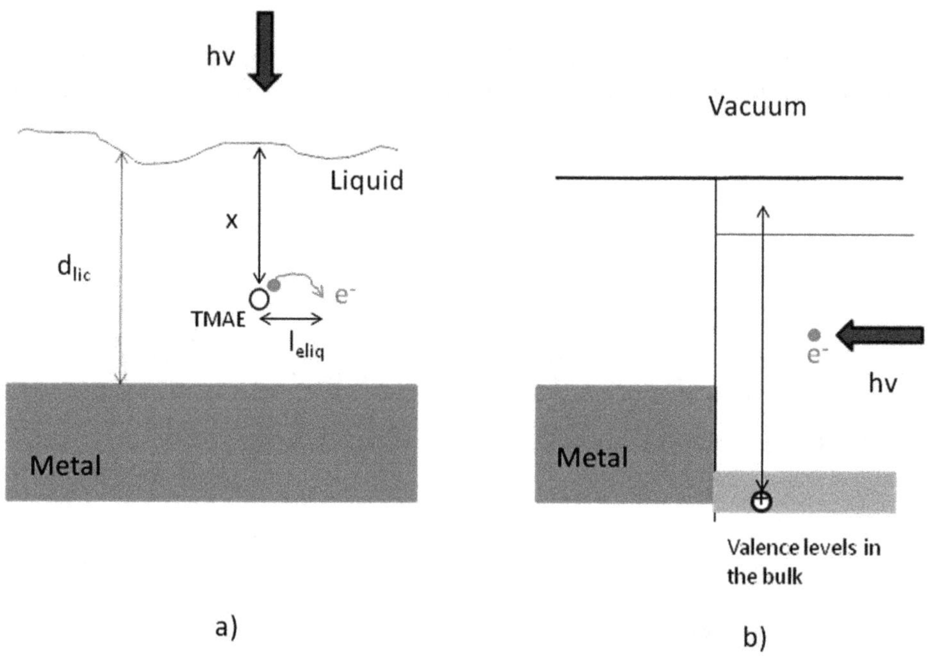

be separated from its positive ion by a distance of l_{eliq} (see Figure 10a). This distance is about 2 nm for polar liquids, between 5 nm and 20 nm for hydrocarbons and much longer for noble liquids (Holroyd, 1988; Yakovlev, 1985). The probability that this electron will escape recombination (without an external electric field) is:

$$P_{esc} = \exp(-r_c/l_{eliq}), \tag{5}$$

with

$$r_c = e^2/\varepsilon kT, \tag{6}$$

where e is the electron charge, ε is the dielectric constant of the liquid, k is the Boltzmann constant and T is the temperature (Holroyd, 1985a). The probability decreases when l_{eliq} increases. In the presence of a strong external electric field, P_{esc} would decrease (Holroyd, 1985a: Holroyd, 1984b). Electrons that escape recombination start to move in the electric field and acquire kinetic energy during its random motion (Schmidt, 1984; Schmidt 1997):

$$T_e = D_{liq}/\mu_{liq} > V_{dliq}, \tag{7}$$

where D_{liq} is the diffusion coefficient, μ_{liq} is the mobility and V_{dliq} is the electron drift velocity.

When $T_e > V_0$, the electron has enough energy to cross the liquid-gas barrier and escape into the gas phase, if of course, it is not trapped by impurities. The impurity concentration has to be low enough so that the lifetime of the electrons in the liquid before they escape from the surface is much longer than their drift time across the liquid layer. The most likely impurities remaining in the liquid after purification are oxygen and carbon dioxide. In liquids the rate of reaction K_{react} of electrons with these impurities has been measured and values in the order of 10^{11} and 10^{12} l/mol·s has been reported (Schmidt, 1974).

The electron life time τ_{tr} before trapping is given by:

$$\tau_{tr} = 1/K_{react}\, n_{at}, \tag{8}$$

where n_{at} denotes the concentration of the electron attaching impurity in mol/l. The drift time t_{dliq} of an electron with mobility μ_{elliq} in an electric field over a liquid layer thickness d_{liq} is given by:

$$t_{dliq} = d_{liq}/\mu_{eliq}\, E, \tag{9}$$

In order to extract most of the electrons produced in the liquid photocathode, the lifetime of the electrons should be at least 10 times of the drift time. Combining equations (8) and (9) efficiencies an expression for the maximum impurity concentration:

$$n_{max} = \mu_{eliq}\, E/10K_{react}\, d_{liq}, \tag{10}$$

With μ=50 cm^2/Vs, E=1kV/cm, $K_{react}=10^{11}$ l/mol·s and d_{liq}=0.1 mm, a maximum tolerable concentration of 50x10^{-8} mol/l is obtained. This example is meant to illustrate the great importance of the purification procedure.

One way of increasing the quantum efficiency is to apply a strong as possible electric field across the liquid layer. This will increase the probability of the electron overcoming the potential barrier and reduce the likelihood of geminate recombination and trapping by impurities. This effect has been clearly observed experimentally (Peskov, 1988).

For the small fraction of the electrons created near the surface of the liquid ($x<l_{eliq}<<d_{liq}$), their escape is easier owing to their excess of kinetic energy $E_k=hv-E_{lid}$ (see Figure 10b).

In some wavelength intervals and for liquids having a small absorption cross section, this effect is negligible in thick layers, but in a thin layer it can play an important role. In TMAE, the thickness of the layer where it is easier for the electron to escape is about 100 nm in the short wavelength region (Schmidt, 1997). This explains the good quantum efficiency of thin layers of TMAE for short wavelengths.

The quantum efficiency of photoionization Q_{liq} can be rather high in some liquids. For example, for TMAE it reaches 1.5% at $\lambda=150$ nm. Obviously, this medium is more efficient than a bare metal. Of course, with the extraction of the photoelectrons in the gas phase, the real efficiency would depend on the value of V_0 and the applied electric field, E. Experimentally it has been found that for liquid photocathodes with $V_0\approx$ - 0.2eV and E≈ 1 kV/cm the quantum efficiency is close to the value Q_{liq} obtained with a liquid ionization chamber (Peskov, 1988; Holroyd, 1988). For liquids with $V_0\approx$ -0.5 eV and E≈ 1kV/cm the efficiency is (0.2-0.3) Q_{liq}.

Together, this explains why a thick layer of liquid TMAE is more efficient than a thin layer or an absorbed layer.

It is known that in liquid noble gases where a strong electric field is applied across the "liquid- gas" border electrons can be extracted from the liquid due to their elevated temperature $T_e>V_0$.This is, however, very difficult to achieved in molecular liquids due the many inelastic collisions of the electrons with the molecules. In liquid noble gases the collisions are mainly elastic with small or negligible energy transfer. There have been some short reports indicating that field extraction is possible to achieve also in warm liquids (Aprile, 2006). However, more systematic studies should be done to confirm this effect.

2.3 Why Did Some Authors Not Reproduce Anderson's Results?

The results and their explanations presented above make it possible to link up all disparate findings of earlier published literature which were mentioned in the introduction to this chapter.

These measurements were done in two distinct ways:

1. Measurements without spectral resolution, and
2. Measurements with spectral resolution.

The experiments without spectral resolution focused its efforts on observing the part played by a thin absorbed TMAE film in a wire chamber. This was observed by measuring the time structure of the sequences of pulses coming from the different electrodes and the different gas gaps (Michau, 1986; Suffert, 1986). Based on the results of the systematic studies presented above the proportion of the signal due to the surface photo-effect can be estimated. To do this it is enough to calculate the convolution of the BaF_2 emission spectrum in the 180-300 nm range with the quantum efficiency of TMAE vapour (see Figure 8) and of the surface photoelectric effect. Our calculations show that for a wire chamber with a1 cm gap and with methane flushing, the surface contribution would be only 6%. For isobutane the effect would be half as much, i.e. it would correspond to ~3%. For gaps larger than 1 cm the contribution from the surface relative to the gas is negligible and can only be observed under conditions of a very good statistics.

Let us now turn to the experiments carried out with spectral resolution. In the introduction it was pointed out that when a monochromator is used, the photoelectric effect from TMAE condensed on the surface was clearly observed (Schotanus, 1987). In the work of Holroyd et al. (Holroyd, 1986) the surface photoelectric effect was investigated on a dielectric window, so to compare these results with one presented in (Peskov, 1988) would be incorrect. Note, however, that in (Holroyd, 1986) an effect of unexpected low threshold of TMEA was observed, which could be manifestation of a liquid photocathode. In (Michau, 1986) the spectral measurements were carried out by means of filters. The authors measured that the ratio of signal at 230 nm to the signal at 210 nm is $7x10^{-2}$, which actually coincide closely with the result presented in Figure 5.

As can be seen from this cursory analysis the results presented in (Peskov, 1988) do not conflict with other publications.

2.4 Photocathodes with TMAE in Solution

2.4.1 Experimental Results

Liquid photocathodes made by condensing solutions of TMAE with other liquids, e.g. tetramethyl silane (TMS), tetramethyl pentane (TMP) and neopentane (NP) were also tested and shown in Figure 1 (Peskov, 1988). These gases were added from the cylinder (9) to the main gas supply (12) when it passed through a flow meter (11) and the TMAE bubbler. All these liquids were specially purified.

The electrons in these liquids have the highest drift mobility and the longest thermolization length of any known organic compounds (Yakovlev, 1985, Holroyd, 1988). Hence, the maximum quantum efficiency can be expected from them. Before the measurements it was not evident which of the three liquids would be the best, since one has maximum mobility, the second has maximum thermolization length and the third has maximum V_0 (see Table 1).

Attempts were made to form photocathodes from these mixtures when the copper cathode was at room temperature. Although liquid condensation did not occur, a thin absorbed film was formed which had some sensitivity to UV light. However, the liquid states were roughly three times better. Further studies where done by cooling the cathode down to T_{cath}<0 °C, when the vapours of TMAE (with TMS, or TMP, or NP) were flushed through the counter. It is difficult to calculate the TMAE concentration accurately in this kind of liquid photocathode, so it was assumed it to be at least equal to the ratio of partial pressures. The TMAE concentration could be slightly altered by changing the velocities at which a gas was flushed through the system. The maximum efficiencies attained are given in Figures 5 and

Table 1. Electron mobilities, band energies at 295 K, and thermolization length at $h\nu - E_{liq} \geq 1$ eV

	Mobility (cm^2/Vs)	V_0 (eV)	Thermolization Length (nm)
TMS	100	-0.55	~10
NP	70	-0.43	~20
TMP	24	-0.33	~10

Yakovlev, 1985, Holroyd, 1988.

Figure 11. The counting rate vs. cathode voltage (or cathode electric field) for two different photon wavelengths, λ= 200 and 240 nm, respectively; the gas flow was methane + TMS + TMAE at 20 °C and 1 bar, with the copper cathode at T_{cath}=0 °C. At λ= 200 nm the count rate is mostly due to gas phase photoionization of the TMAE, but at λ= 240 nm the signal mainly originates from the liquid photocathode.

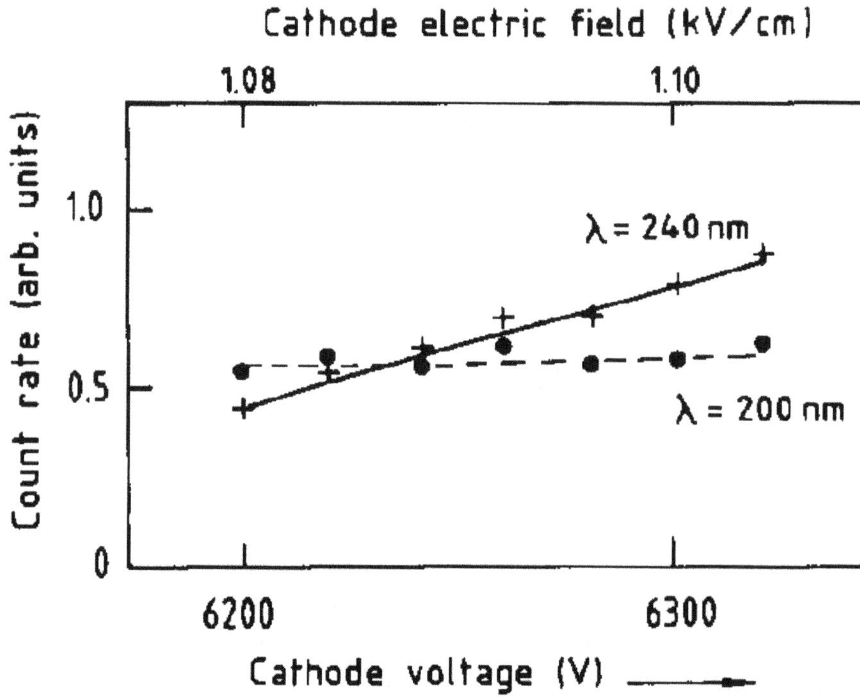

6. Curves 5 and 6 show the best liquid cathode efficiencies obtained. As can be seen, the sensitivity of TMAE in NP and TMAE in TMS gave a quantum efficiency of ~2% at λ=235nm. The TMAE+TMP cathode was always a factor of 1.5 to 2 times less efficient than the TMAE+TMS cathode.

The quantum efficiency of the liquid photocathodes increased with increasing electric field across the cathode surface, as shown in Figure 11.

The electric field strength could not be varied very much under the experimental conditions of the work (Peskov, 1988) because of a concomitant increase of the electric field at the wire of the counter, causing breakdown or instability. Even though the electric field strength at the cathode surface increased from 1.06 to only 1.1kV/cm, the count rate doubled, indicating a very strong field dependence at λ= 240 nm. No electric field effect was seen at λ= 200 nm, where the majority of the photons convert in the gaseous TMAE everywhere. A striking confirmation of this dependence was provided by replacing the solid cooper cathode by a cathode of parallel 100 μm wires with 1 mm spacing. At the wire surface the electric field was calculated to be a factor of 5 higher than at the flat copper cathode. The measured quantum efficiency at λ= 235 nm for a TMAE+TMS cathode was ≈6% at room temperature, which should be compared with 1% for the flat cathode also at room temperature. When correcting for the active area of the wires for the incoming photon flux compared to the flat copper cathode (10% coverage), even higher quantum efficiency was obtained, close to 60%. This indicates that it is possible to have highly efficient photocathodes at room temperature if a strong extraction field is applied across the liquid photocathode.

Figure 12. The count rate vs. time when the chamber was flushed with pure methane at an accelerated gas rate of 2 l/min; the curve labelled with triangles is for λ= 200 nm and corresponds to an initial TMS (or TMP) + TMAE photocathode, the dots represent measurements at λ = 245 nm and for an initial TMS + TMAE photocathode, and crosses correspond to the results obtained for λ = 245 nm and an initial solid TMP + TMAE photocathode (see the next chapter).

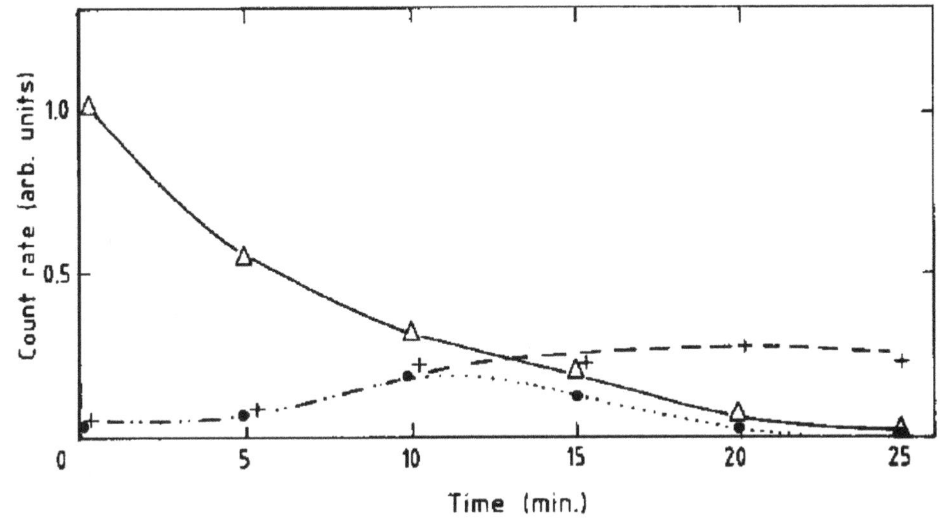

The stability of the liquid photocathode mixtures were tested as follows. After forming the mixed cathodes, pure methane was flushed through the single wire counter by bypassing the TMAE and the TMS (or TMP) reservoirs (see Figure 1) and by increasing the flow velocity by a factor of about 40, to several liters per minute. The results are shown in Figure 12.

At λ = 200 nm the signal falls quickly because it is predominately due to gas phase photoionization, whereas at λ= 245 nm the signal initially increased in both TMS and TMP cathodes because of increased gas transparency. The intensity of the avalanche pulses from the solid TMP photocathode continued to increase and stabilized after 20 min to a value about five times the initial value (this is the correction for photon absorption), whereas the TMS photocathode reaches a peak after 10 min and then decreases to a negligible value after 25 min. This proves that the liquid photocathode (TMS+TMAE) has evaporated.

2.4.2 Interpretation of the Results

The efficiencies observed with liquid photocathodes made from small concentration of TMAE in solvents of TMS, TMP or NP are always higher than those observed with pure TMAE. In these types of liquid photocathodes the photoionization process does not take place in the solvent, but in the TMAE due to it slower photoionization threshold. After the photoelectrons are produced in the liquid photocathode with a kinetic energy, E_k, they lose their kinetic energy in elastic and inelastic collisions, a process called thermolization. In the solvents the value of the l_{eliq} is larger than for pure TMAE. Therefore, the recombination decreases and the quantum efficiency increase (Holroyd, 1985b).

In the case of liquid photocathodes in gaseous detectors, the photoelectrons have to be extracted from the layer into the gas phase. The values of V_0 and T_e that induced the extraction efficiency are different for different solvent media. For TMA, TMP and NP the value of V_0 is slightly higher and T_e is considerably larger than for TMAE. This increases the quantum efficiency compared with that from pure TMAE photocathodes. At strong electric fields, i.e. >1 kV/cm, all electrons will be extracted so that the efficiency will basically be determined by the value of l_{eliq} (Schmidt, 1984; Aprile, 2006). From this follows that photocathodes made from TMAE in solvents will have higher efficiencies than are achieved with pure liquid TMAE photocathodes. This is also confirmed by all measurements.

3. OTHER LIQUID PHOTOCATHODES

TMAE based photocathodes, especially mixed with high mobility solvents has high quantum efficiency and are interesting to use in photosensitive detectors. Unfortunately, it is not easy to handle these photocathodes. They interact strongly with air, are corrosive for many materials and also may have bad ageing effects when use in gaseous wire chambers. Thus, a complicated technology is required in order for these substances to be deposited as photocathodes in UV sensitive gaseous photodetectors. It is necessary to work under very clean conditions and to be able to cool the surface where the photocathode will be condensed to the liquid phase. For applications in high energy physics, such as Cherenkov detectors or UV scintillator based calorimetry, where huge detectors may be built, it will be more practical to use a simpler technology.

For these reasons, new photosensitive liquid substances, which do not interact with air and ideally can be used at room temperature using a simple technology of deposit them on surfaces as photocathodes, have also been intensively investigated.

Good candidates are TEA and some organometallic compounds, for example diethylferrocene and ethylferrocene. The latter is relatively easy to produce, they are not toxic, do not interact chemically with oxygen or with gases commonly used in gaseous detectors and can therefore be manipulated in air.

3.1 TEA Photocathode

As was mentioned earlier, some spectroscopic studies of TEA deposited on a dielectric layer indicated the characteristics of a liquid photocathode; its ionization threshold shifted from 165 nm to 240 nm, Figure 13. The observation, which varied in magnitude from measurement to measurement, was attributed to the photoelectric effect from a liquid film of TEA on the windows of the test chamber (Holroyd, 1986).

Holroyd's et al. observations were confirmed by Charpak's team (Charpak, 1989a). They used a monochromator system to measure the quantum efficiency of liquid TEA (see Figure 1) as well as a BaF_2 crystal coupled to a wire chamber (similar to one shown in Figure 5, but without a cooling bath) to investigate such a photocathode in a real experimental situation. The technique for obtaining liquid photocathodes was very simple. Vapours of TEA were introduced into the MWPC at room temperature and absorbed layers were formed everywhere in the detector. The spectroscopic measurements indeed revealed the shoulder in the sensitivity curve similar to one shown in Figure 13.

To exclude the possibility that the observed efficiency in the wavelength interval 160 -200 nm is not partially due to scattered light in the monochromator, scrupulous measurements were performed with a BaF_2 scintillator, which emission spectrum is below the ionization of TEA vapours. Test with cosmic muons depositing about 35 MeV in the crystal showed a typical exponential single-electron spectrum identical to one obtained with 59.5 keV x-rays from ^{241}Am. This allowed them to estimate the number

Figure 13. The photoionization quantum efficiency of TEA vs. wavelength performed at a pressure of 1.05 Torr

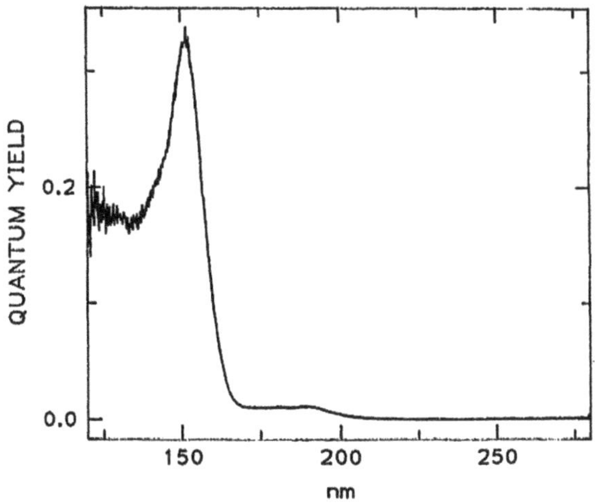

Figure 14. The quantum efficiency, expressed in photoelectrons per MeV deposited in the BaF_2 crystal, as a function of the electric field applied on the photocathode for TEA absorbed layers at different TEA vapour pressures (75 Torr, 48 Torr and 39 Torr) and for a TEA condensed layer deposited on the BaF_2 surface (open circles)
From Charpak, 1989a.

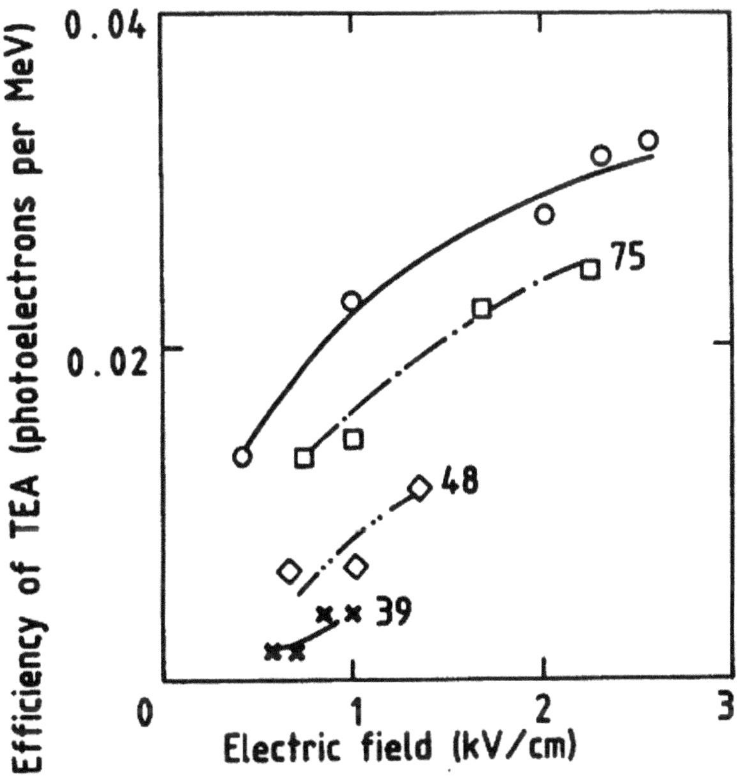

of photoelectrons created from the liquid TEA layer per MeV deposited into the BaF_2 crystal, as shown in Figure 14.

As can be seen, the TEA photocathode is sensitive to the fast emission of BaF_2, but has poor quantum efficiency. Even when the BaF_2 crystal is slightly cooled to provoke condensation of a thicker liquid layer of TEA on the crystal, the maximum efficiency was about 0.3 photoelectrons per MeV deposit energy.

3.2 EF Photocathode

Ethylferroscene is an easy to handle. It does not react with oxygen, nor any other material commonly used in gaseous detectors. Some results of spectral measurements of the quantum efficiency of an EF absorbed layer, performed with the setup shown in Figure 1, are depicted in Figure 8. As can be seen it is about 1% at λ=193 nm (the first peak of the fast BaF_2 emission spectrum, (Charpak, 1989a). The quantum efficiency depends on the materials used as the support for the EF photocathode and also on the quality of the EF (Figure 15).

Figure 15. The quantum efficiency of several photocathodes as a function of photon wavelength: 1-Clean Cu cathode, 2-an EF absorbed layer deposited on a Cu cathode at room temperature (the EF was produced in the USSR). 3-7) various EF photocathodes formed from EF synthetized at the University of Geneva):3-deposited on a Cu cathode, 4-deposited on an indium cathode,5-deposited on Al cathode, 6- when the metallic substrates (Cu, In and Al) were kept at 80°C, 7- a 1 mm thick EF liquid layer.
From Charpak, 1989a.

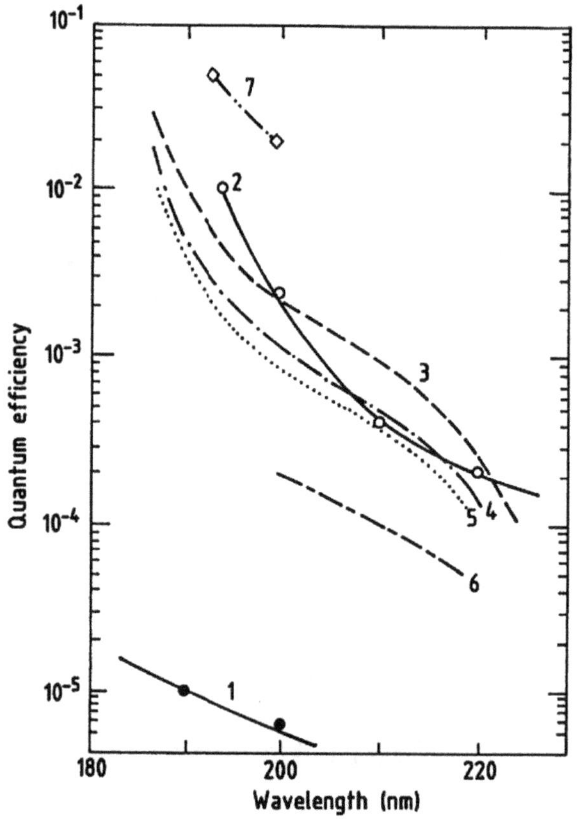

With a thick liquid EF layer the QE was ~7% at λ=193 nm. When the metallic substrate was heated, to boil off the EF and reduce the effect of the absorbed layer the quantum efficiency was correspondingly decreased. However, it still remained higher than in the case of the pure Cu metal indicating that some absorbed layer of EF was sticking to the surface.

The efficiency achieved with EF layers is higher than with TEA layers making it more attractive in applications. This triggered systematic studies of EF photocathodes at experimental conditions and setups oriented towards real applications, e.g.as a photosensitive element for the readout of UV crystal scintillators like BaF_2 and others.

One of the aims of these studies was to understand the contribution of the photosensitive layer formed on the detector window or on the surface of the BaF_2 scintillator in creating photoelectrons as was earlier observed in the case of TEA. Sometimes this absorption of photosensitive gases on windows has practical applications as will be discussed further in the chapters on applications.

Figure 16. Experimental set-up for the investigation of the EF photocathode under various experimental conditions. 1- a H_2 lamp (or a ^{22}Na source for coincidence measurements with the PMT), 2- is the position of the PMT for the measurements the internal radioactivity of BaF_2.
From Charpak, 1989b.

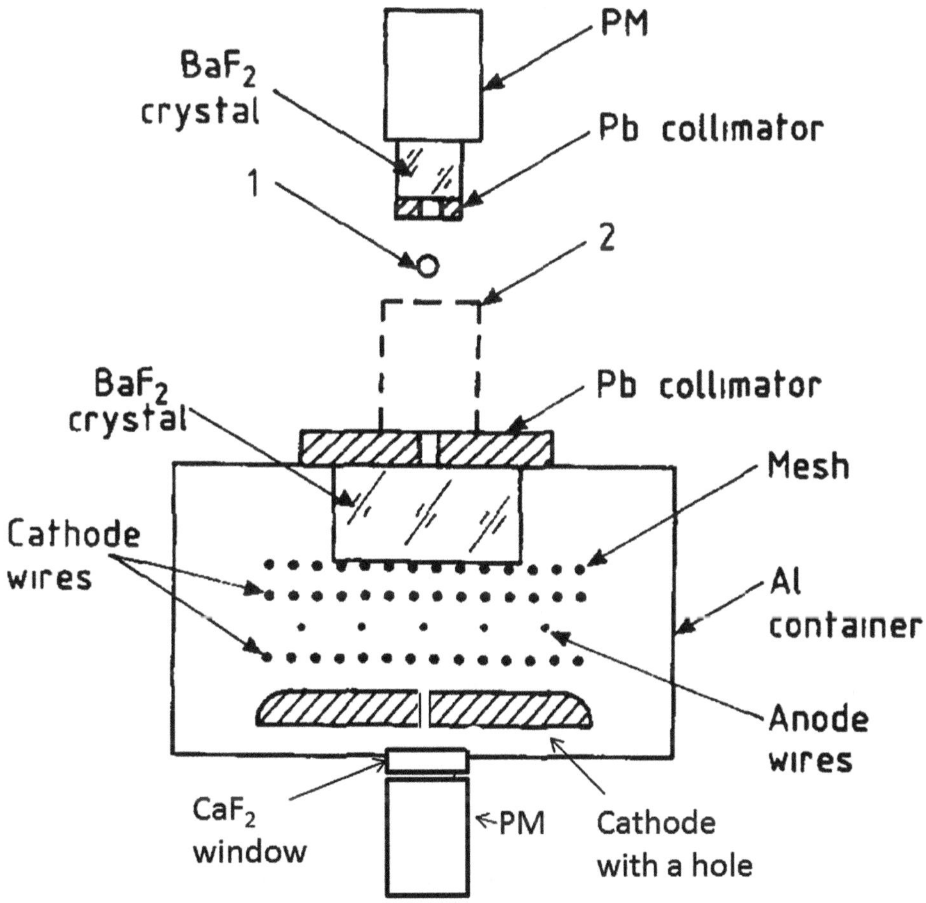

3.2.1 Detailed investigations of wire and parallel plate chambers with an EF photocathode

3.2.1.1 Experiments with a Pulsed UV Lamp

One of the experimental set ups for the studies EF photocathodes is shown in Figure 16.

It was very similar to the prototype of the BaF_2 calorimeter combined with TMAE vapours used earlier by Anderson (Anderson, 1983) and shown in Figure 7. Additional light sources were added: a pulsed UV lamp, radioactive sources and charged particles (Charpak, 1989b). One of the features of this setup was the photomultiplier tube (PMT) allowing monitoring the scintillation light emitting from the BaF_2 scintillator and assessing in this way the energy deposited by the radioactive sources or by charged particles. There was also a possibility to irradiate the wire chamber through the BaF_2 crystal by a hydrogen lamp (having light pulse duration of 25 ns, FWHM). The output window of the lamp was covered by an interference filter with a maximum transmission at 193 nm and a FWHM of 18 nm.

Because of the short pulse duration of both the UV lamp and the fast emission of the BaF_2 crystal around 195 nm, measuring the drift time enabled the authors to recognize in which part of the wire chamber the photoelectrons were created. By applying a positive voltage on the Al plate or on the mesh in contact with the BaF_2 crystal, it was possible to exclude the contributions from the adjoining drift regions and extract the contribution from one electrode at a time. The conclusion from these studies was that absorbed layers on both the Al electrode and the BaF_2 surface gave the main contribution to the photoelectron creation (see Charpak, 1989b for details). At elevated temperatures (about 70 °C) the contribution from the EF gas becomes significant and it was used to estimate the quantum efficiency of the chamber to the radiation with λ=193 nm.

From the known values of this quantum efficiency and by measuring the signals at room temperature (where the gaseous EF has negligible contribution to the quantum efficiency) the quantum efficiency of the absorbed layer was estimated to be between 0.7% and 1.5%, increasing with the electric filed. This is close to the value obtained from spectroscopic measurements (as shown in Figure 15). In the case of a thicker condensed EF layer (the entire sealed chamber was cooled, so the vapours inside were condensed on all surfaces) the efficiency was between 1.5 and 2%, which is about 7% of that of TMAE vapours under full absorption. It should be noted that the efficiency of the EF layer increased with the electric field at the surface and the concentration of the quencher gas. For example, in pure CH_4 at p=0.5 atm and E$\approx$ 6 kV/cm, the quantum efficiency of the EF layer for λ= 193 nm was about 6.5% which is $\approx$ 22% of TMAE.

3.2.1.2 Measurements of the Absolute Efficiency of the Wire Chamber with a ^{22}Na Source

In these measurements, the PMT with the other BaF_2 scintillator was placed 10 cm away from the $BaF_2$of the wire chamber (see Figure 16). The BaF_2 crystal in the wire chamber and the one on the PM were irradiated by a ^{22}Na source, placed between them on the common axis. They were thus irradiated simultaneously by the two 511 keV gamma rays emitted back to back in the positron annihilation in the ^{22}Na source. The gamma rays propagate in opposite directions and produce scintillation pulses in both crystals. The pulse amplitude spectrum of BaF_2VUV emission consists in the peak produced by the 511 keV gamma rays, the background being 1 MeV gamma rays emitted by the^{22} Na and by the internal radioactivity of the BaF_2. After subtraction of this background, the number of coincidences, N_{coin}, (within a 200 ns window) measured at different electric fields E were the following. At E = 200 V/cm, N_{coin} =

Figure 17. The EF quantum efficiency in photoelectrons per MeV deposited in the BaF_2 crystal as a function of the light absorption at 193 nm in the EF semitransparent photocathodes formed on the surface of the BaF_2 crystal; the arrow indicates the EF layer conditions of the beam test.
From Charpak, 1989a.

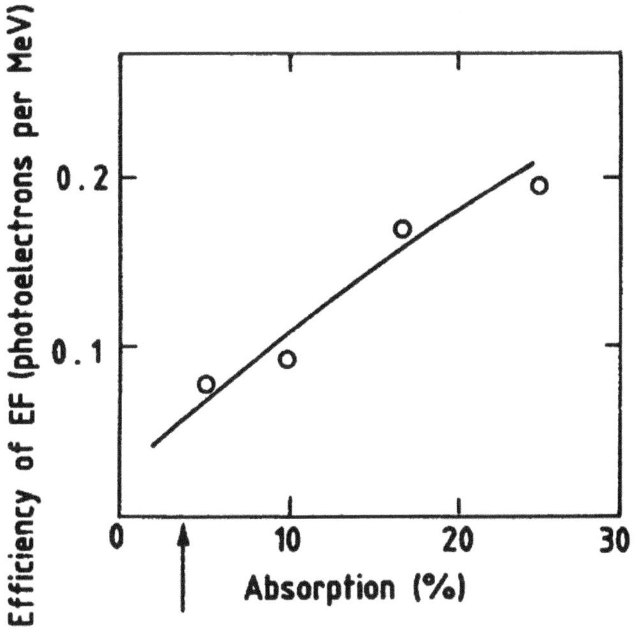

1/200; at E = 400 V/cm, N_{coin} = 1/130; at E = 2kV/cm, N_{coin} = 1/100. Since the PM records the bursts of scintillation light from the BaF_2 with 100% efficiency, the value of the absolute efficiency of the EF layer, for λ= 193 nm, can be estimated to be between 0.5% and1%, depending on the electric field.

It was also clearly observed that the efficiency of the EF layer increases with its thickness, as shown in Figure 17. Note that in this figure the EF layer thickness is presented in the units of the light absorption at λ=193 nm in the condensed layer, which, at small values of attenuation is proportional to the thickness.

3.2.1.3 Measurements in a Test Beam

Tests also were made with charged particles at the CERN PS accelerator. In order to investigate the EF photocathode under "clean" experimental conditions, i.e. without any contribution from EF vapour, a parallel-plate avalanche chamber (PPAC) containing a BaF_2 crystal, 2.5 x 2.5 x 7 cm^3 was constructed (see Figure 18).

As in the case of the wire chamber, the EF vapour formed a photosensitive layer, which was deposited everywhere, including on the surface of the BaF_2 crystal and of the Cu cathode made of separated pads. The distance between the PPAC electrodes was 4 mm. The photoelectrons extracted from the BaF_2 and the cathode mesh surfaces produced avalanches, so that the number of electrons reaching the anode was larger by a factor A, corresponding to the gas gain. In principle, the photons emitted by the BaF_2crystal can also ionize the EF vapour; but the feature of this device was that photoelectrons that are created at a distance from the cathode larger than the electron mean free path before ionization λ_{ie}, will produce avalanches with a much smaller gas gain. For example, for A = 10^5, λ_{ie} ~ 0 .3 mm; therefore only 1/10 of

Figure 18. Construction of the PPAC used for beam test; the thick arrow shows the direction of the beam.

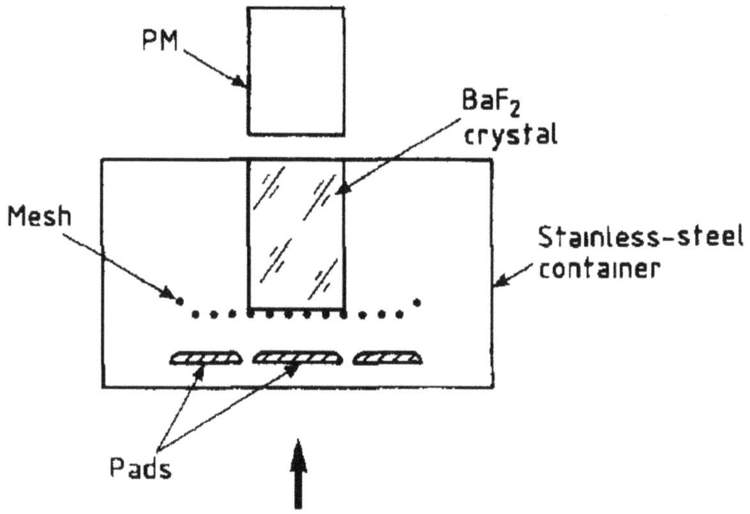

the gap can in principle contribute to the creation of photoelectrons due to photoionization. Under these conditions and at room temperature, one can really neglect the contribution of the gas photoionization.

Tests were done with unseparated charged particle of about 7 GeV/c energy. The beam also included electrons which are not minimum ionizing in the crystal. By inserting a thick absorber, only muons were filtered out. The energy deposited by the muons in the BaF_2 crystal was approximately 49 MeV, and the maximum energy deposited by the electrons was about 700 MeV.

This BaF_2 crystal was without Al reflector usually used to achieve a better light collection. Calculations and measurements show that this reduces the intensity of the BaF_2 light entering the PPAC by ~ 30%. With the muon beam, the efficiency of the PPAC in coincidence with the PM was about 100%. The amplitude of the signal from the PM for the fast component was ~50 photoelectrons (the absolute efficiency of the PM to this light was about 25%). This means that in the case of the PPAC, with an efficiency of 0.7% for the EF layer absorbed on Cu, one should expect about 2 photoelectrons for 49 MeV deposited energy by the muons, and about 50 photoelectrons for 700 MeV from the electrons. As expected, in the case of the muon beam, the pulse-height distribution had a broad maximum. For the 700 MeV electron signal (the trigger was a signal from the PM, corresponding to the maximum energy loss in the BaF_2) the energy resolution was ~20%. From these data it could be concluded that the efficiency of an EF absorbed layer was really about 1%. This means that in the case of a high quality crystal with 90% transparency and with an Al reflector one can expect an energy resolution at 1 GeV to be about 10%. One should note that in the case of TMAE under the condition of full absorption of the scintillation light from the BaF_2crystal (with a transparency of the crystal of 75% and with an Al reflector), the energy resolution for 1 GeV was 6%, improving to 3.9%/ $\sqrt{E_p}$ (where E_p is the particles energy) at high energies (Bouclier,1988) . With the EF such an improvement with increasing energy can also be expected, which would make the calorimetry with an absorbed layer attractive, even at the present level of sensitivity.

We can conclude that all the experiments described above show that the absolute efficiency of an absorbed layer of EF is about1% at λ = 193 nm. When the EF layer is condensed, this efficiency increases by a factor of 3 to 4, which makes such a photosensitive element useful for some applications (as described in the later chapter on applications).

4. CONCLUSIVE REMARKS

Results obtained with TMAE, TEA and EF thin liquid photocathodes are summarized in Figure 19. The quantum efficiency of these substances at λ=193 nm is expressed as a function of the thickness of the layer, expressed in units of light absorption in liquid layer.

The wavelength of 193 nm was chosen for comparisons because in concise with the fast emission peak of the BaF_2 scintillator in the far UV, and one of the possible applications of a liquid photocathodes is to readout a BaF_2 scintillator calorimeter.

As can be seen from the curves presented in Figure 19, TMAE, certainly, is the most efficient photocathode, but let us remind that TMAE is difficult to handle. It is reactive with air and also with some material which complicates its application. Although EF has a rather modest efficiency it is quite easy to handle making it attractive in practice.

Liquid photocathodes have a lot of potentials in applications. One of them, is the readout of crystal scintillators emitting UV light. Another example is to use TMAE and EF photocathodes for detecting and visualization flames in air in broad daylight. It is based on the fact that flames emit UV photons capable to photoionize these substances. This approach, offers much greater sensitivity to flames than any commercial sensors. Liquid photocathodes allow this detector to operate in wide range of temperatures (Bidault, 2007).

Figure 19. The quantum efficiency of semitransparent photocathodes deposited on the surface of a BaF_2 crystal as a function of its thickness, as measured by the light absorption in the layer (at a wavelength λ=193 nm); the labels for EF and TEA are exchanged in the figure.
From Charpak, 1991.

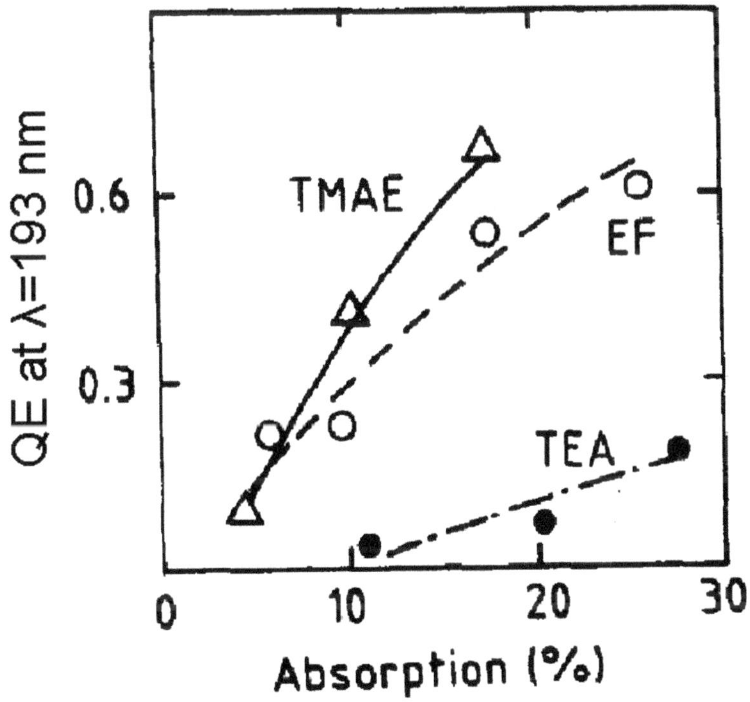

In spite if a lot of advantages, liquid photocathodes have not been used yet in a wide scale. The main reason is that shortly after the discovery of the liquid photocathodes, another efficient photocathode was developed. It is based on solid substances. The fast and impressive progress in this direction will be discussed in the next two chapters.

However, the study of liquid photocathodes allowed to deeply understanding the role of absorbed gases in the operation of solid photocathodes and this experience was used later for the optimization of gas mixtures used in detectors combined with photocathodes. For example, that gas mixtures containing isobutane reduces the quantum efficiency of the photocathode due to the absorbed layer of isobutane. In contrast, CH_4 does not create any problems in electron liberation from the solid photocathodes and for this reason it is the most common gas used in practical devices. In some works TMAE and EF are used to enhance the quantum efficiency of solid photocathodes. It is not excluded therefore that in the future new interest in liquid photocathode may appear.

REFERENCES

Anderson, D. F. (1982). Extraction of electrons from a liquid photocathode into a low pressure wire chamber. *Physics Letters*, *118B*(1-3), 230–232. doi:10.1016/0370-2693(82)90634-7

Anderson, D. F., Bouclier, R., & Charpak, G. (1983). Coupling of a BaF_2 scintillator to a TMAE photocathode and a low-pressure wire chamber. *Nuclear Instruments & Methods in Physics Research. Section A, Accelerators, Spectrometers, Detectors and Associated Equipment*, *217*, 217–223.

Aprile, E., Bolotnikov, A. E., & Bolozdynya, A. I. (2006). Noble gas detectors. Wiley-VCH. doi:10.1002/9783527610020

Bidault, J.-M., Crotty, I., Di Mauro, A., Martinengo, P., Fonte, P., Galy, F., & Zanette, O. et al. (2007). The first applications of newly developed gaseous detectors with resistive electrodes for UV imaging in daylight conditions. *Nuclear Instruments & Methods in Physics Research. Section A, Accelerators, Spectrometers, Detectors and Associated Equipment*, *580*(2), 1036–1041. doi:10.1016/j.nima.2007.06.061

Boucher, R. G., Charpak, G. W., & Gao, W. (1988). Test of an electromagnetic calorimeter using BaF_2 scintillators and photosensitive wire chamber between 1 and 9 GeV. *Nuclear Instruments & Methods in Physics Research. Section A, Accelerators, Spectrometers, Detectors and Associated Equipment*, *267*(1), 69–86. doi:10.1016/0168-9002(88)90631-6

Charpak, G., Lamb, D., & Peskov, V. (1989a). New developments in calorimetry based on VUV scintillators coupled to photosensitive gaseous detectors. *Proceedings of the ECFA Study Week on Instrumentation Technology for High-luminosity Hadron Colliders*. Doi:10.5170/CERN-1989-010-V-2.593

Charpak, G., Peskov, V., & Sauli, F. (1989b). Organometallic photocathode for parallel-plate and wire chamber. *Nuclear Instruments & Methods in Physics Research. Section A, Accelerators, Spectrometers, Detectors and Associated Equipment*, *283*(3), 786–791. doi:10.1016/0168-9002(89)91458-7

Charpak, G., Peskov, V., Sauli, F., & Scigocki, D. (1989). Ethylferrocene in gases, condensed or absorbed phases: Three types of photosensitive elements for use in gaseous detectors. *Nuclear Instruments & Methods in Physics Research. Section A, Accelerators, Spectrometers, Detectors and Associated Equipment, 277*(2-3), 537–546. doi:10.1016/0168-9002(89)90787-0

Charpak, G., Peskov, V., & Schmidt, W. F. (1988). Liquid and solid photocathodes based on TMAE or TMPD: physics and application.*Proceedings of the Internat. Workshop on Liquid-state electronics.*

Charpak, G., Peskov, V., Scigocki, D., & Lemenovski, D. (1991). Liquid and solid photocathodes and their applications for fast gaseous particle detectors. *IEEE Transactions on Electrical Insulation, 26*(4), 623–630. doi:10.1109/14.83681

Funfer, E., & Neiert, H. (1965). *Detectors of radiation, Atomisdat.* Moscow. (in Russian)

Holroyd, R. A., & Anderson, D. F. (1985a). The physics and chemistry of room –temperature liquid-filled ionization chambers. *Nuclear Instruments & Methods in Physics Research. Section A, Accelerators, Spectrometers, Detectors and Associated Equipment, 236*(2), 294–299. doi:10.1016/0168-9002(85)90164-0

Holroyd, R. A., Ehrenson, S., & Preses, J. M. (1985b). Electron mobility, ion efficiencies, and photoconductivity in liquid tetrakis (dimethylamino) ethylene. *Journal of Physical Chemistry, 89*(20), 4244–4249. doi:10.1021/j100266a019

Holroyd, R. A., Preses, J. M., & Woody, C. L. (1986). *Measurements of the quantum efficiency of TMAE and TEA from threshold to 120 nm.* Report at the, XXIII International Conference on High Energy Physics, Berkeley, CA. Retrieved from http://www.iaea.org/inis/collection/NCLCollectionStore/_Public/18/052/18052811.pdf

McDaniel, E.W. (1964). *Collision phenomena in ionized gases.* Wiley.

Michau, J. C., Pichard, B., Soirat, J. P., Spiro, M., & Vignaud, D. (1986). The performance of a UV sensitive multiwire proportional chamber filled with TMAE. *Nuclear Instruments & Methods in Physics Research. Section A, Accelerators, Spectrometers, Detectors and Associated Equipment, 244*(3), 565–570. doi:10.1016/0168-9002(86)91085-5

Peskov, V., Charpak, G., Mine, P., Sauli, F., Scigocki, D., Séguinot, J., & Ypsilantis, T. et al. (1988). Liquid and solid photocathodes. *Nuclear Instruments & Methods in Physics Research. Section A, Accelerators, Spectrometers, Detectors and Associated Equipment, 269*(1), 149–160. doi:10.1016/0168-9002(88)90872-8

Peskov, V., Charpak, G., Sauli, F., Scigocki, D., Diep, V., & Janjic, D. (1989). Organometallic photocathodes for parallel-plate and wire chamber. *Nuclear Instruments & Methods in Physics Research. Section A, Accelerators, Spectrometers, Detectors and Associated Equipment, 283*(3), 786–791. doi:10.1016/0168-9002(89)91458-7

Schmidt, W. (1997). Liquid-state electronics of insulating liquids. SRS Press.

Schmidt, W. F. (1974). Electron migration in liquids and glasses. In *Electron solvent and anion solvent interaction.* Amsterdam: Elsevier.

Schmidt, W.F. (1984). Electronic conduction processes in dielectric liquids. *IEEE Trans. Electric. Insul., El-19*, 389-418.

Schotanus, P., Van Eijk, C. W. E., Hollander, R. W., & Pijpelink, J. (1987). Development study of a new gamma camera for positron emission tomography. *Nuclear Instruments & Methods in Physics Research. Section A, Accelerators, Spectrometers, Detectors and Associated Equipment*, *252*(2-3), 255–259. doi:10.1016/0168-9002(86)91186-1

Séguinot, J., Charpak, G., Giomataris, Y., Peskov, V., Tischhauser, J., & Ypsilantis, T. (1990). Reflective UV photocathodes with gas-phase electron extraction: Solid, liquid and absorbed thin films. *Nuclear Instruments & Methods in Physics Research. Section A, Accelerators, Spectrometers, Detectors and Associated Equipment*, *297*(1-2), 133–147. doi:10.1016/0168-9002(90)91359-J

Suffert, M., Charpak, G. (1986). *Some studies of a BaF_2 scintillator coupled to a photosensitive wire chamber: the influence of the temperature on the photocathode efficiency and the existence of a possible absorbed layer of TMAE.* Preprint CERN-EP-INT-86-03, 1-13.

Yakovlev, B. S., & Lukin, L. V. (1985). Photoionization in Non-Polar Liquids. *Advances in Chemical Physics*, *60*, 99–160. doi:10.1002/9780470142844.ch3

KEY TERMS AND DEFINITIONS

Electron Temperature: A mean kinetic energy of electrons in media (gas, liquid, solid) which in electric field can be much higher than the temperate of the media.

Organometallic Compounds: A class of substances containing at least one metal to carbon bond in which the carbon is a part of an organic group.

Outgassing: Release of the gas adsorbed on surfaces or trapped inside some materials.

Parallel-Plate Avalanche Counter (PPAC): A gaseous detector in which avalanche multiplication is happening between two parallel metallic electrodes.

Thermostatic Box: A vessel in which the temperature is maintained constant and if necessary can be changed in a controllable way.

Wire Chamber: (Sometimes is called a multiwire proportional counter or MWPC) a detector consisting from two parallel cathode plane with an array of parallel anode wires located between the cathode planes.

Chapter 5
Early Work on UV Sensitive Solid Photocathodes for Gaseous Detectors

ABSTRACT

Ultimately, a solid photocathode is advantageous over gaseous and liquid photocathodes. Gaseous detectors with solid photocathodes are simpler to manufacture and have better time resolution than detectors filled with photosensitive gases. They should be able to operate at high gas gains and detect single photoelectrons. Finally, compared to vacuum photomultipliers they have little sensitivity to magnetic fields and can measure the coordinate of photon conversion and can be made with large areas and arbitrary shapes. All this proves that gaseous photomultipliers with solid photocathodes open a new page in photosensitive detection technique. We will here review the early development of solid photocathodes, and the road towards the modern photocathodes which have revolutionized the photosensitive detector technology.

1. INTRODUCTION

The threshold of sensitivity of photosensitive detectors is determined by the ionization potential of the gases, E_i, or the ionization threshold, E_{liq}, in the case of liquids. E_{liq} is related to E_i as described in Formula (3) in Chapter 1. The gases with the lowest ionization potential, which are used in physics experiments, are TMA, TEA, EF and TMAE. They were discovered a long time ago and afterwards there has actually been no real progress in finding suitable photosensitive substances, i.e. materials with lower ionization potentials. This is mainly because the substances with low ionization potential are usually chemically aggressive and/or unstable in air. This is why there has always been an interest to combine gaseous detectors with solid photocathodes.

The potential advantages of such an approach are:

DOI: 10.4018/978-1-5225-0242-5.ch005

1. Very low thresholds of spectral sensitivity, i.e. they are sensitive to light of long wavelengths. This is given by the work function of the material.
2. A high quantum efficiency.
3. A high time resolution (better than 1 ns) since there are no space jitters in creating photoelectrons (they all come from the well-defined cathode).
4. They can be manufactured with a large sensitive area since there is mechanical constrains on the window size.

What solid photocathodes can be use and what has been tested?

We already discussed earlier that metals, even those with a low work function, φ, are not efficient photocathodes for two main reasons. They have high reflectivity for UV and visible light, and the electron quickly loses their kinetic energy in collisions with other electrons inside the metal. As a result, only electrons created very close to the surface (typically a few atomic layers) can escape from the metal. Therefore, other types of photocathodes should be considered.

One should note that for vacuum photomultipliers there have been a lot of developments of various solid photocathodes. The main focus of these developments was on photocathodes sensitive to visible light (Sommer, 1968). It was shown, both experimentally and theoretically, that among various photocathode materials the most efficient are the semiconductors. In simple words this is because in a semiconductive material the electrons interact with phonons and the corresponding energy losses are small. Therefore, photoelectrons from deeper regions can reach the surface with energies above the mean thermal energy. Moreover, most semiconductors have a much greater absorption coefficient than metals.

There were early attempts to use some of the photocathodes developed for vacuum photomultipliers in gas filled photodiodes operating at small gas gains: 10-100 (see for example Sauli, 1982; Charpak, 1983 and references therein). Such devices were not only studied in the laboratories, but were even commercially available from various photonic companies. Examples are RCA Phototubes and photocells, EMI Industrial Electronics, ITL Instrument technology, Hamamatsu etc., (see for example RCA Technical manual PT-60, 1963). They are now replaced by solid-state detectors, and they have virtually disappeared from the market.

Studies show that the maximum achievable gains of these photodiodes are limited. It depends on gas type, and is mainly limited either by ion feedback or both by ion and photon feedback. High gains sufficient to detect single photoelectrons, are achieved only in quenched gases and with photocathodes having modest sensitivity in the UV region (although superior to liquid TMAE photocathodes) (Peskov, 1988). However, high stability with time is achieved only with very clean gases and this is why these photodiodes can be used only in sealed mode. Attempts to use them in gas-flush mode have been unsuccessful (Charpak, 1992).

The challenge therefore is to develop gaseous detectors with solid photocathodes with high quantum efficiency. They shall be capable to operate at gas gain $>10^4$, sufficient to detect single photoelectrons. Preferably they shall be able to operate in gas flow mode, which tremendously simplifies the design especially in the case of the position-sensitive photomultipliers which requires a lot of feedthroughs for determination of the avalanche positions.

The first success comes from the detectors sensitive to UV light. Such photocathodes have relatively high work functions (often called "affinity" in the case of semiconductive photocathodes). This reduces the probability of ion feedback. In quenched gases the probability of photon feedback is also small. As a result, with such detectors one can reach rather high gas gains.

Figure 1. A MWPC combined with a solid photocathode

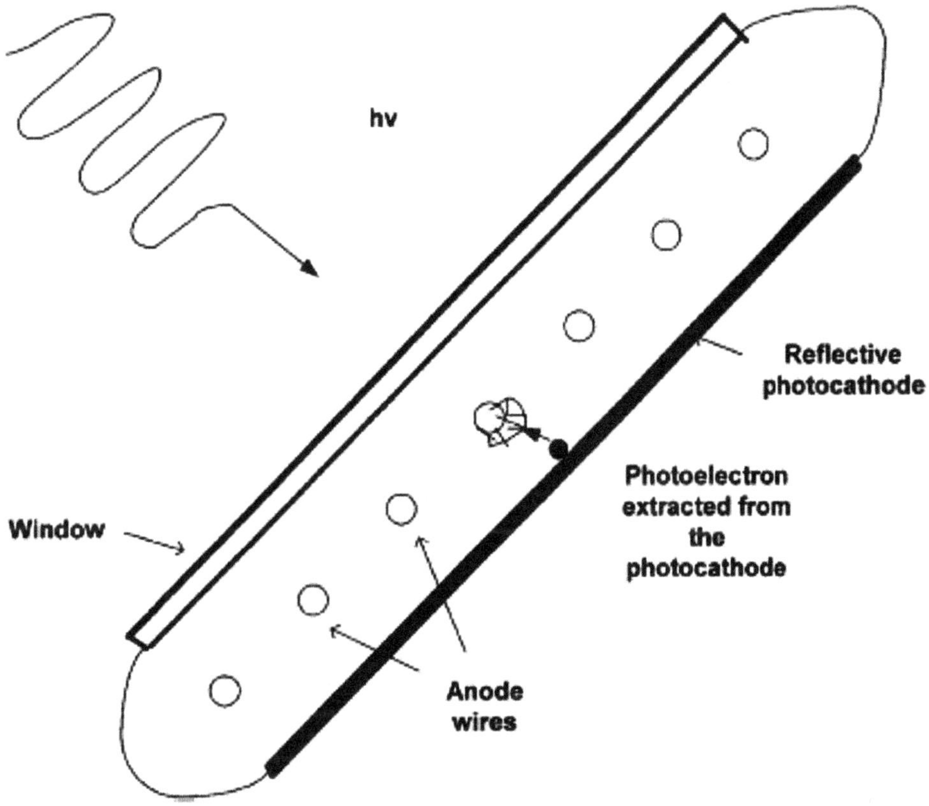

Later, after solving many serious problems gaseous detectors sensitive to visible light were developed as well.

In this and the next chapter we will focus on gaseous detectors combined with UV sensitive solid cathodes. The overview of these developments will be given in their historical sequence.

2. A MWPC COMBINED WITH A CUI PHOTOCATHODE

The first MWPC combined with a UV sensitive solid photocathode, CuI, was develop and successfully used for plasma diagnostics (Peskov, 1980). The CuI photocathode was made by treating a clean Cu surface with a solution of I_2 dissolved in dry ethanol. Because such a photocathode interacts slightly with water vapour, it was installed in the detector in a dry-air atmosphere. Note that the reaction of the CuI with water is not irreversible, and in principle a cathode damaged by H_2O could be "repaired" by heating it (Peskov, 1980, Charpak, 1991).

The design of a MWPC combined with a CuI photocathode is schematically presented in Figure 1.

UV photons penetrating through the window cause photoeffect from the CuI cathode (no photoionization of the gas). The photoelectrons extracted from the cathode drift towards the anode wire where they trigger Townsend avalanches.

Figure 2. The quantum efficiency of a MWPC combined with a CuI photocathode and operating in with CH_4 as amplification gas
Peskov, 1980, Charpak, 1991.

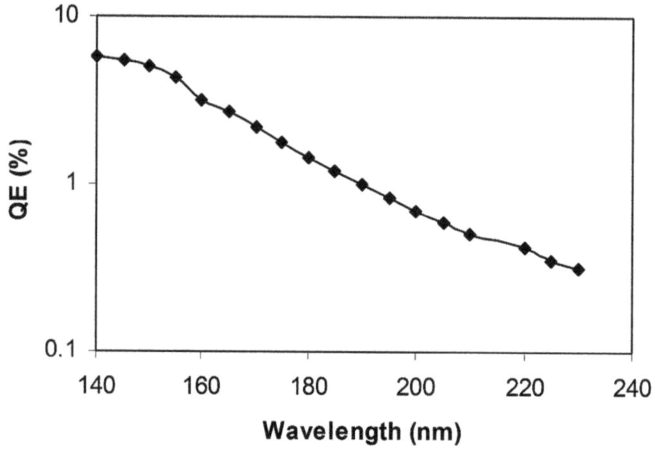

The quantum efficiency of the CuI photocathode is presented in Figure 2 (Peskov, 1980).

Comparing the curves presented in Figure 5 and Figure 6 in Chapter 4 one can conclude that at least at $\lambda > 230$ nm the quantum efficiency of a CuI photocathode is lower than Cu covered with an absorbed layer of TMAE. However, CuI is much easier to handle as it can some in contact with air as long as humidity is avoided.

Note also the value of the CuI quantum efficiency depends on many factors. It varies with gas composition, temperature treatment, the photocathode history (for how long it was kept in air or humidity, was it heat treated etc.), the strength of the electric field near the cathode. Therefore, the values of the quantum efficiency measured by different authors vary in the literature.

Let us briefly discuss the most important of these factors. An absorbed layer of some gases, like isobutane, may reduce the quantum efficiency of the CuI photocathode by a factor of 2- 3 compared to "good" gases, for example CH_4. Adding TMAE vapour to practically any gas mixture commonly used in gaseous detectors, *increase* the quantum efficiency at least on a factor of 2-3, even if the detector is flushed with a gas not containing TMAE after an initial contact with TMAE (Charpak, 1991).

The quantum efficiency of CuI is enhanced by heating it. This was first used by Peskov (Peskov, 1980; Charpak, 1991), and later later reproduced and further studied by (Buzulutskov, 1995). The results are shown in Figure 3.

In this work the quantum efficiency was measured in vacuum immediately after preparation, after 15h of heating at 50 °C and after further heating at 165 °C. In the latter case the photocathode was kept at a temperature of 165°C for half an hour and then gradually cooled down. As in the work by Peskov it was observed that heat treatment in vacuum is quite stable in time: once the quantum efficiency is stabilized at the elevated temperatures it does not change anymore after cooling down to room temperature.

One should stress the difference of the quantum efficiency when measured in vacuum and in gas. In a gas atmosphere the motion of photoelectrons liberated from the photocathode is affected by diffusion. Due to this some photoelectrons, after several collisions with the gas atoms and molecules, return back to the cathode. This effect, called electron back diffusion was studied a long time ago (see for example

Figure 3. The quantum efficiency of a CuI photocathode vs. wavelength, measured in vacuum, after heating for different times; data from Charpak. 1991 are also shown for comparison.
Buzulutskov, 1995.

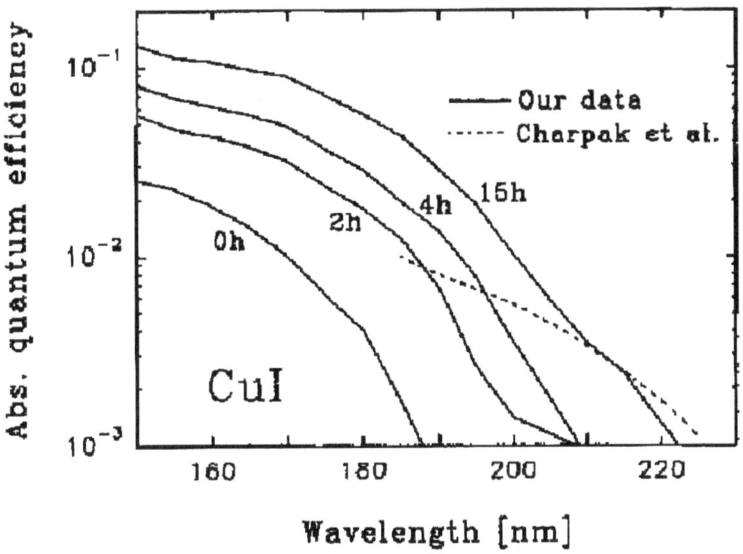

Figure 4. The dependence of k(E) vs. the strength of the electric field for several gases and gas mixtures
From Breskin, 2002.

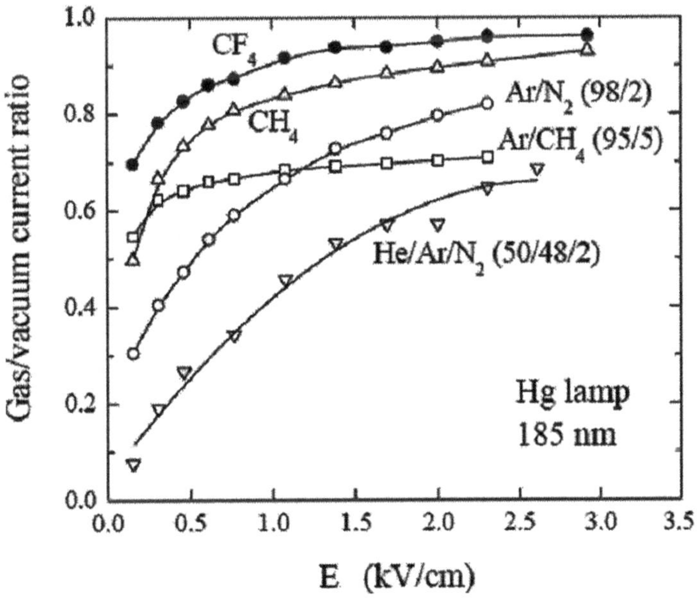

Bradbury, 1932, Theobald, 1953). Therefore the quantum efficiencies measured in the gas Q_{gas} and in the vacuum Q_{vac} are different:

$$Q_{gas} = k(E)\, Q_{vac}, \tag{1}$$

where k(E) is a coefficient depending on the electric field on cathode, E (k(E) < 1). Usually, k(E) monotonically increases with E and in quenched gases reaches a plateau value. At this plateau Q_{gas} reaches a maximum, as shown in Figure 4.

As can be seen, k(E) has the highest values (~0.9) in CF_4 and CH_4. This is one reason why the MWPC with the CuI photocathode is best operated in CH_4. In this gas the maximum gain above which photon feedback appears is about 10^5, sufficient for single photoelectron detection.

For a long time, CuI was the only photocathode used in practice in MWPCs flushed with CH_4. The search for new solid photocathodes obtained new momentum after the discovery and studies of the liquid photocathodes.

3. OTHER SOLID PHOTOCATHODES

3.1 TMPD

Some solid photocathodes, for example based on solid mixtures of TMAE and neopentane, with a quantum efficiency much higher than Cu (about 4% at λ=235 nm), were developed (Peskov, 1988a). The results are shown in Figure 6 (Chapter 4). The general observation made in work was that solid photocathodes have a sensitivity that is equal to or even better than that of the corresponding liquids. This fact has a natural explanation because in the solid state V_0 as well as μ_{eliq} is usually higher than in liquids. These two factors contribute to a higher sensitivity for VUV. In the special case of neopentane, V_0 becomes positive in the solid state (Grand, 1977) so that photoelectrons near the surface are "pushed" out from the photocathode into the vacuum or to the gas. Indeed, this cathode exhibits a spontaneous electron emission of about 10 to 30 photoelectrons per second even without illumination.

The problem was that such a photocathodes are quite complicated to implement in practice since the cathode must be cooled to temperatures of ~-20 °C and maintained there.

This is why other photocathodes were considered as well, even if some of them had lower quantum efficiency. A good example can be tetramethyl-p-phenylenediamine (TMPD). The molecular structure of TMPD is similar to that of TMAE, but it is solid at room temperature and is not reactive with oxygen in air (for short exposures). A TMPD cathode of about 1 μm thickness can easily be made by vacuum evaporation on a Cu cathode. The TMPD is dissolved in ethyl alcohol (100%), poured on a warm copper cathode substrate and then the alcohol is distilled away under vacuum. Typical results obtained with solid a TMPD photocathode are shown in Figure 5 (triangles).

One can see that the quantum efficiency of TMPD in the wavelength interval 180-240 nm is rather close to the CuI photocathode used in the MWPC (Figure 2). For completeness, in the same figure are also shown the quantum efficiency of TMPD photocathodes in gas mixtures with TMS or TMP or NP. These mixed cathodes were made by flushing the cooled copper cathode covered by TMPD (from T_{cath} = -5 to -40 °C) with methane plus the vapours of TMS (or TMP, or NP), a solution was formed by

Figure 5. The quantum efficiency of various TMPD photocathodes vs. photon wavelength; triangles - solid TMPD (room temperature), crosses - TMPD cooled to -5 °C and flushed with a mixture of methane with TMP, squares - TMPD cathode at T_{cath} = -40 °C and flushed with methane + neopentane; curves 1-3 are data obtained with a liquid ionization chamber with TMPD dissolved in TMS, TMP and NP, respectively.

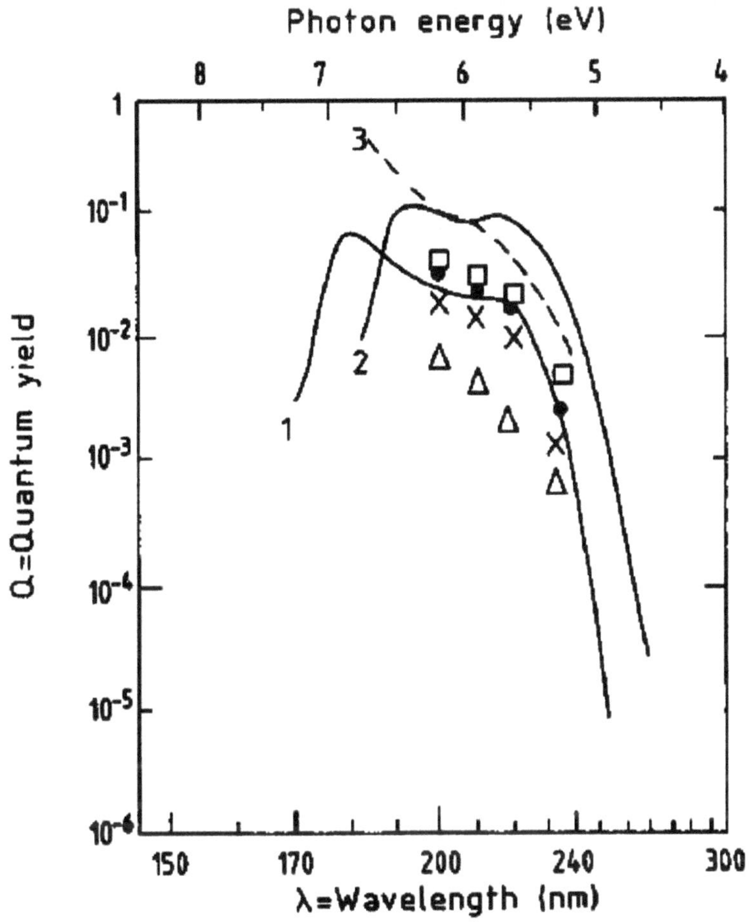

absorption of these vapours in TMPD. The curves labeled 1, 2 and 3 in Figure 5 are the results of measurements with liquid ionization chamber (no photoelectron extraction) (Choi, 1982; Bottcher, 1984).

The measured points are somewhat lower than the liquid ionization measurements. The difference is probably due to differences in the electric field strength.

Another solid photocathode tested in is diethyl ferrocene whose gas-phase photoionization threshold is λ = 184 nm (Peskov, 1988a). The results are shown as curve 8 in Figure 5, Chapter 4. It shows a lower quantum efficiency than liquid TMAE, but it is still a factor 80 higher than for a bare copper cathode up to λ = 260 nm.

It can be conclude that although some solid photocathodes, like TMAE+NP, may have quite high quantum efficiency, most of them, except TMPD are quite difficult to implement in a practical detector. This why the search for efficient photocathodes, allowing to be easily combined with MWPC or other type of position-sensitive gaseous, detectors was continued.

The main requirement on such a photocathode is that it can be used in detectors operating in flushed-gas mode. This is the simplest approach especially in applications requiring large sensitive areas. It is also important that photocathodes are capable to operate stably not only in ultraclean gases, but in gas mixtures containing some impurities, which is usually the case in flushed detectors made of cheap and light materials like stezalit, printed circuit boards etc.

The further studies followed two directions; the search of air stable solid metalloorganic photocathodes, and tests of known photocathodes, used in vacuum tubes, to measure their stability in air.

Unfortunately, the first brief tests of some photocathodes used in vacuum photomultipliers, e.g. CsI, (Peskov, 1988a), CsTe (Charpak, 1992) etc. were unsuccessful. They showed low quantum efficiency after contact with air.

On the other hands, the molecular structure of organometallic compounds enables them to combine the desirable properties of low ionization potential and little chemical activity (Astruc, 1986). For this reason these compounds seem to be very attractive for use as photocathodes in gaseous detectors. For several years from 1987 till 1990 the main focus of some research groups at CERN and elsewhere was on metalloorganic photocathodes. Below a few examples will be given.

3.2 CPIHMB

CPIHMB stands for cyclopentadienyl-iron-hexamethyl-benzene, chemical formula: $Fe(\eta\text{-}C_5H_5)\eta\text{-}(C_6Me_6)$, where “Me” is methyl groups. It is a brown powder with a vapour pressure of ~10^{-6} Torr at room temperature having a photoionization potential in the gas phase of $E_i = 4.5$ eV (275 nm). This is considerably below TMAE, which makes this substance very attractive.

One of the ways to cover a metallic plate with a few monolayers of CPIHMB is by cryogenic pumping. A vessel containing CPIHMB and the entire gas system is heated to 70 °C whereas the metallic cathode, located in a gas chamber is kept at a few degrees C. The gas flow carries CPIHMB vapours from the container to the gas chamber housing the cathode and deposits it on its cool surface. After forming this layer, all measurements are made at room temperature.

The quantum efficiency of the CPIHMB photocathode measured in the range 200 to 300 nm is shown in Figure 6.

In this figure, curve 1 is the sensitivity of a clean copper cathode, curve 2 is the efficiency of the Cu cathode with a layer of CPIHMP, curve 3 gives the sensitivity obtained with the layer of CPIHMB when TMS vapours was presented in the gas mixture. It can be seen that after deposition of the CPIHMB layer, the efficiency of the metallic cathode increase by 1 to 2 orders of magnitude in the wavelength region $\lambda < 260$ nm. As expected, the measured photoionozation threshold of CPIHMB in the condensed form ($E_{isolid} = 4.2$ eV or 295 nm) is smaller than in the gas phase.

The time profile of the photocathode formation and its stability are shown in Figure 7.

In these particular measurements the single-wire counter (see Chapter 4, Figure 1) was flushed with CH_4 + CPIHMB vapour at room temperature. Cooling of the metallic cathode was started at time t=0 and stopped at t_{stop}. It was then kept at room temperature, as were the counter and the whole gas system. The photocathode remained stable for days, even after the cooling was stopped, a property that is favorable for practical applications.

Although the efficiency of this cathode is lower than CuI, one cannot exclude that after some optimization of the photocathode deposition technique (for example after introduction of the post heating treatment) its quantum efficiency can be considerably increased.

Figure 6. The quantum efficiency as a function of wavelength for different photocathodes; 1 -clean Cu, 2 – Cu + condensed CPIHMB, 3 – Cu + condensed CPIHMB + TMS; for comparison the quantum efficiency of Cu with condensed EF is shown (curve 4).
From Peskov, 1989.

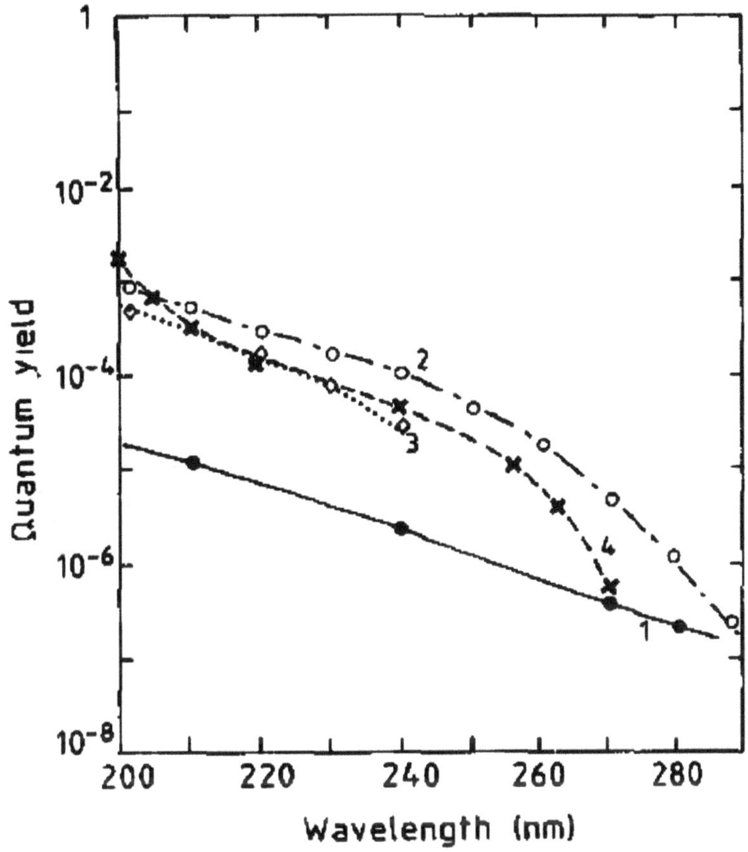

Figure 7. The quantum efficiency of a CPIHMB photocathode as a function of the deposition time (up to t_{stop}) and the stability of the sensitivity (as from t_{stop})
From Peskov, 1989.

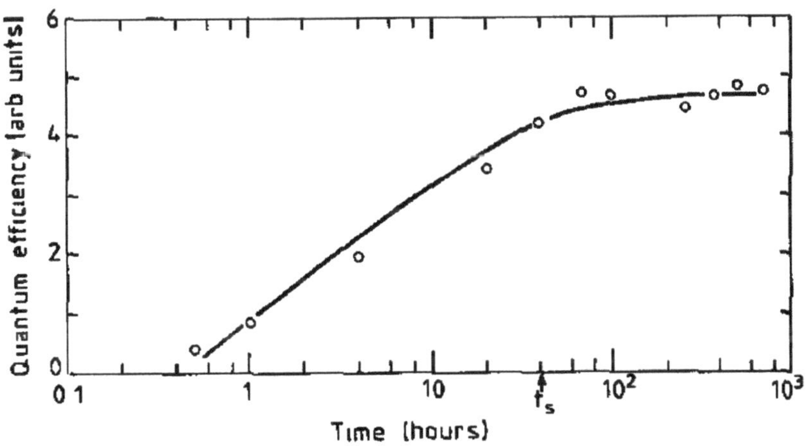

Figure 8. An electron microscope photograph of a polycrystalline structure obtained after evaporation of a solution containing diferrocenyl mercury; the white lines are 1 μm in length.

3.3 Diferrocenyl Mercury Photocathodes

Diferrocenyl mercury (chemical formula: $C_2H_{18}Fe_2Hg$) is a stable substance that is solid at room temperature and is not reactive with air, water or polar solvents. A few cubic millimeters of diferrocenyl mercury are introduced into 30 cm^3 of xylene (Charpak, 1991). The mixture is then heated to the boiling point (~120 °C). Under these conditions the diferrocenyl mercury dissolved. One drop of the solution is placed on the surface of the copper substrate which is then evaporated.

In another method a few mg of diferrocenyl mercury was dissolved in 50 ml of toluene (Imrie, 1992). A freshly prepared, vacuum deposited copper substrate was warmed to 60-80 °C and covered with a ~1 mm layer of the solution, which was allowed to evaporate.

In both cases, after evaporation the substrate was covered with a layer of crystals between 1 and 10 μm in size, as shown in Figure 8.

The results of the quantum efficiency measurements of such a layer are presented in Figure 9.

As can be seen, the quantum efficiency in the wavelength interval $\lambda > 190$ nm is considerably lower than CuI. However, the polycrystalline structure of the layer makes it attractive as a secondary electron emitter, allowing improving the efficiency and timing properties of a PPAC.

3.4 DiferroTriphenylphosphinegold Ferrocene and d, d-Dimethyl-Trimethylene-Ferrocenephane Photocathodes

Other examples of air-stable metalorganic solid photocathodes are diferrotriphenylphosphinegold ferrocene and d, d-dimethyl-trimethylene-ferrocenephane. The first substance is prepared by the same

Figure 9. The quantum efficiency of pure Cu (curve 1), a Cu substrate covered by diferrocene mercury (2) and diferrocene mercury + EF (3)

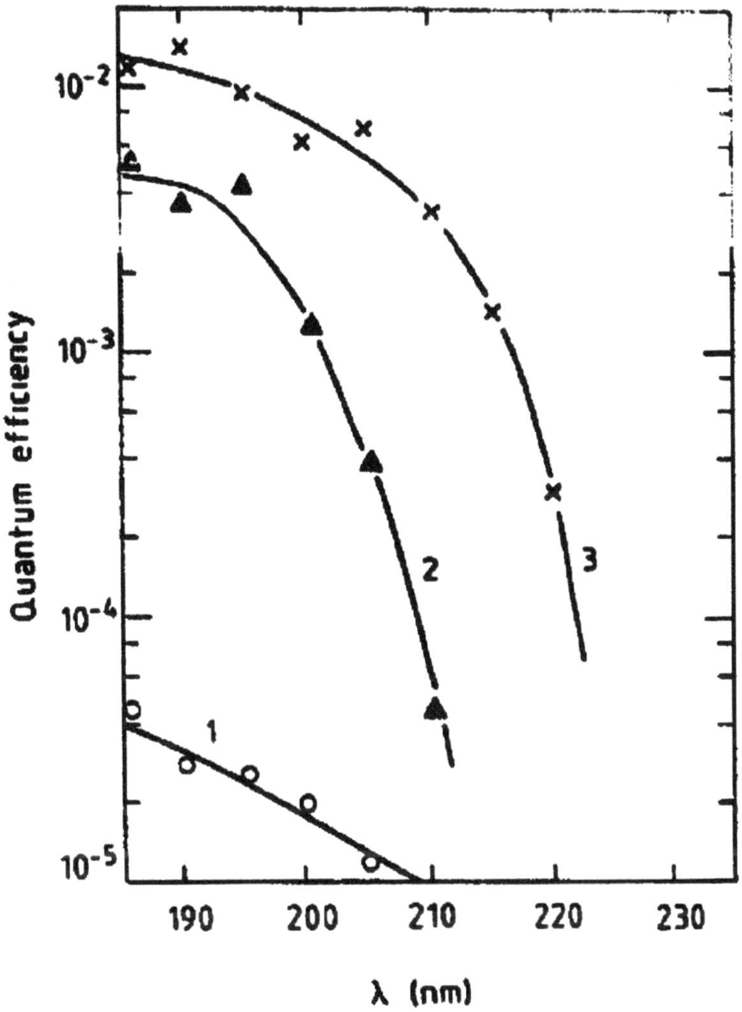

procedure as the diferrocenyl mercury (see previous paragraph). The second is deposited by sublimation in an argon atmosphere at a temperature of 80-140 °C (for more details see Imrie, 1990).

The quantum efficiency of these substances is shown in Figure 10.

As can be seen the quantum efficiency of d, d-dimethyl-trimethylene-ferrocenephane is about 1% at 200 nm and about 0.1% at 220 nm which is lower than CuI and liquid TMAE photocathodes.

3.5 Quantum Efficiency Enhancement of Metalloorganic Photocathodes after a Short Contact with TMAE or EF Vapours

The quantum efficiencies of all tested metalloorganic compounds, similar to bare metallic cathodes, are increased by a factor 2-3 after a short contact with TMAE or EF vapours. The effect remains for a long time even if the detector is flushed with gases not containing these vapours. This is shown in Figures

Figure 10. The quantum efficiency vs. wavelength for bare diferrotriphenylphosphinegold ferrocene (a) and d, d-dimethyl-trimethylene-ferrocenephane (b) photocathodes and these photocathodes covered with absorbed layers of EF and TMAE
From Imrie, 1992.

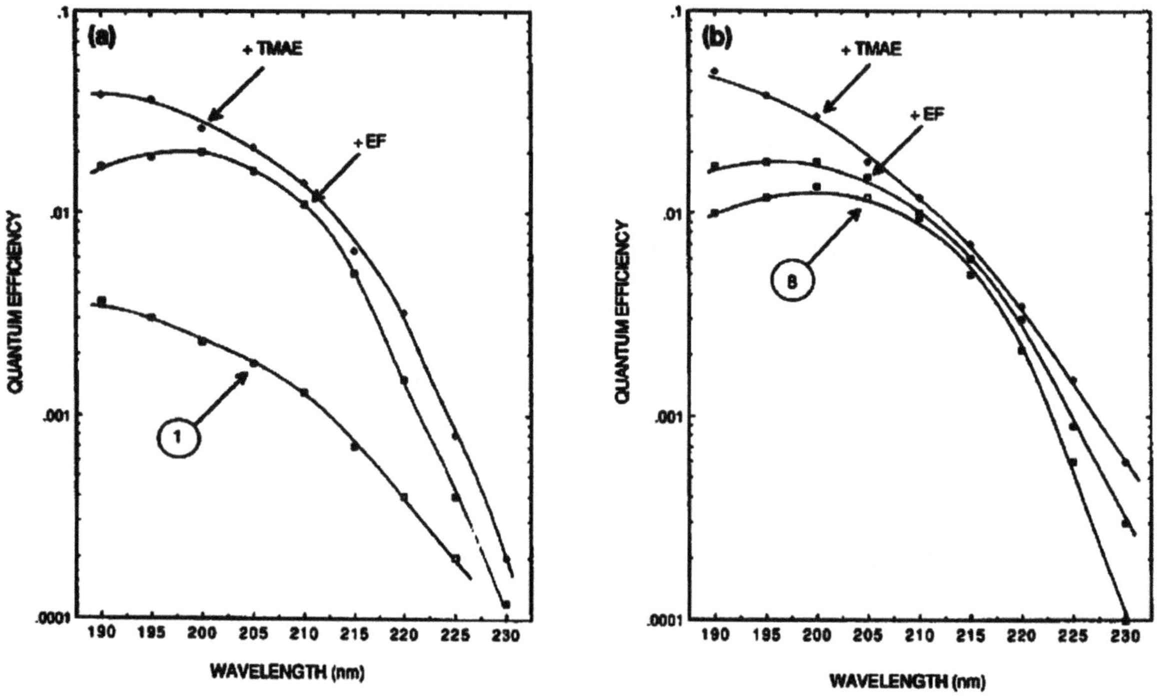

9 and 10. This offers a simple method of the quantum efficiency enhancement, which can be used in practice. The most attractive, of course is the enhancement caused by EF vapours since it is air-stable allowing it to be used in gas-flushed MWPCs.

4. CONCLUSIVE REMARKS

Various solid UV sensitive photocathodes have been described, CuI as well as some metal organic compounds.

The main goal of these studies is to find an alternative to TMAE vapour, which is chemically aggressive and also requires detector heating in order to reach high quantum efficiency. Air-stable solid photocathodes could be a convenient alternative to TMAE simplifying the detector design. Moreover, the time jitter in creating primary electrons is much smaller for solid photocathodes than for photosensitive gases. A detector with solid photocathodes will have excellent timing properties.

However, the studies were not very successful since all solid photocathodes tested had quantum efficiency lower than TMAE vapour, although the sensitivity of some of them in the wavelength interval 220-240 nm is comparable to a liquid or absorbed layer of TMAE.

So, what is the practical value of the listed photocathodes?

Photodetectors having narrow sensitivity in the far UV region are attractive in some applications. They allow spectroscopic measurements without using filters. An example could be visualization of plasma in the UV region in a strong background of ultraviolet and visible light, see the chapter application for more details. Some experimental results obtained with a MWPC combined with CuI photocathode will be presented there as well.

Another application is the readout of VUV crystal scintillators such as BaF_2, LiF_2, $KLuF_4$, K_2YF_5 of $KMgF_3$ (Peskov, 1991; Charpak, 1989; Charpak, 1989a; Buzulutskov, 1990). These scintillators are fast (fluorescent decay time <1 ns) and emit at short wavelength ($\lambda < 200$ nm).

From Figure 10 it is clear that a suitably chosen photocathode treated with TMAE or EF can have a sufficiently good quantum efficiency in this region with the results that the convolution of the fluorescent spectrum of the scintillator with the quantum efficiency of the photocathode is expected to be acceptably high. This was clearly demonstrated in the work (Imrie, 1992) in which a PPAC having and a d, d-dimethyl-trimethylene-ferrocenephane photocathodes covered with an absorbed layer of EF, combined with one of the following scintillators: $KLuF_4$, K_2YF_5 or $KMgF_3$, 3 or10 mm thick. The results of their investigations indicate that 1.2 - 2 photoelectrons per MeV of the deposit energy in the crystal were obtained from $KLuF_4$ and K_2YF_5 and about 5 photoelectrons per MeV from $KMgF_3$. The spatial resolution of the centre of gravity method was 2-3 mm. $KLuF_4$, which has a high density (about 5 g/cm^3 and a radiation length of about 2 cm, its emission peaking at $\lambda \approx 160$ nm) could be used as a high-energy photon detector. One of its applications considered in work (Peskov, 1991) was a pre-shower detector. The pre-shower detectors in front of a calorimeter provide an improved e/π rejection (Massam, 1963). The main advantages of this type of pre-shower detector are the following: it is radiation hard, it can measure the position of the shower and the deposit energy, and it makes all calorimetry system more homogeneous which allows reaching better energy resolution.

Another interesting application is a pre-preshower detector. A thin (2-3 mm) layer of crystal scintillator with small Z, placed before the pre-shower detector or calorimeter may help to improve e/γ rejection (Charpak, 1991a; Peskov, 1991). Indeed, this thin layer practically will not be sensitive to high-energy photons, but electrons and other charged particles will interact as minimum ionizing particles and will produce scintillation light. The scintillation light recorded by the gaseous detector with solid UV sensitive photocathode will give the information about the interaction point of the charged particles with the crystal. The feasibility of this approach was demonstrated experimentally with a 5 mm thick $KMgF_3$ scintillator, coupled to a PPAC with a solid photocathode mentioned above. With this detector about 10 photoelectrons was detected per MeV of the deposit energy. This efficiency is enough to record minimum-ionizing particles. The spatial resolution achieved in this test was about 4 mm.

The results described in this section demonstrated that gaseous detectors with solid photocathodes, beside the plasma diagnostics and spectroscopic measurements could also be used for the readout of the VUV scintillators. They are simpler and have better time resolution than detectors filled with photosensitive gases. Compared to photodiodes mentioned at the beginning of this chapter, they can operate at high gas gains and detect single photoelectrons. Finally, compared to vacuum photomultipliers they have little sensitivity to magnetic fields and can measure the coordinate of photon conversion and can be made with large areas and arbitrary shapes. All this proves that gaseous photomultipliers with solid photocathodes open a new page in photodetection technique. The next breakthrough happened after a technology of manufacturing CsI photocathodes which can combine with gaseous detector was mastered. This exciting developments, which open large scale applications of gaseous detectors with solid photocathodes in high energy physics and beyond will be described in the next chapter.

REFERENCES

Astruc, D. (1986). Organoiron electron-reservoir complexes. *Accounts of Chemical Research, 19*(12), 377–383. doi:10.1021/ar00132a001

Bottcher, E. H. (1984). *Experimentelle Untersuchung der photoelektrischen Leitung reiner und mit aromatischen moluculen dotierler organisher.* (PhD Thesis). Free University Berlin.

Bradbury, N. E. (1932). Photoelectric Currents in Gases between Parallel Plates as a Function of the Potential Difference. *Physical Review, 40*(6), 980–987. doi:10.1103/PhysRev.40.980

Breskin, A., Buzulutskov, A., & Chechik, R. (2002). GEM photomultiplier operation in CF_4. *Nuclear Instruments & Methods in Physics Research. Section A, Accelerators, Spectrometers, Detectors and Associated Equipment, 483*(3), 670–675. doi:10.1016/S0168-9002(01)01940-4

Busulutskov, A. F., Turchanovich, L. K., Vasilchenko, Y. L., Jansosn, Y. L., & Rachko, Z. A. (1990). Investigation of the crystal scintillators in the VUV region. *Nuclear Instruments & Methods in Physics Research. Section A, Accelerators, Spectrometers, Detectors and Associated Equipment, 288*(2-3), 659–661. doi:10.1016/0168-9002(90)90165-3

Buzulutskov, A., Breskin, A., & Chechik, R. (1995). Heat enhancement of the photoefficiency from CsI, NaI and CuI photocathodes. *Nuclear Instruments & Methods in Physics Research. Section A, Accelerators, Spectrometers, Detectors and Associated Equipment, 366*(2-3), 410–412. doi:10.1016/0168-9002(95)00863-2

Charpak, G., Dominik, W., Sauli, F., & Majewski, S. (1983). The gas photodiode as a possible large-area detector. *IEEE Transactions on Nuclear Science, NS-30*(1), 134–137. doi:10.1109/TNS.1983.4332236

Charpak, G., Lamb, D., Peskov, V., & Valbis, J. (1989). New developments in calorimetry based on VUV scintillators coupled to photosensitive gaseous detectors.*Proceedings of the ECFA Study Week on Instrumentation Technology for High-luminosity Hadron Colliders.*

Charpak, G., Lemenovski, D., Peskov, V., & Scigocki, D. (1991). New photocathodes for fast gaseous detectors. *Nuclear Instruments & Methods in Physics Research. Section A, Accelerators, Spectrometers, Detectors and Associated Equipment, 310*(1-2), 128–132. doi:10.1016/0168-9002(91)91010-S

Charpak, G., Peskov, V., & Sauli, F. (1992). Preliminary results of the study of gaseous detectors with solid photocathodes sensitive in the spectral region from 105 to 300 nm. *Nuclear Instruments & Methods in Physics Research. Section A, Accelerators, Spectrometers, Detectors and Associated Equipment, 323*(1-2), 445–451. doi:10.1016/0168-9002(92)90330-7

Charpak, G., Peskov, V., & Scigocki, D. (1989b). *Study of BaF_2 calorimeters in future colliders.* Report at the 9th INFN Eloisotron project workshop: perspectives for new detectors in future supercolliders, Erice, Italy.

Charpak, G., Peskov, V., & Scigocki, D. (1991a) New readout for BaF2 calorimeters: gaseous detectors with solid photocathodes. *Proc.Large Hadron Collider Workshop.*

Charpak, G., Peskov, V., Scigoscki, D., & Valbis, J. (1989a). New scintillators for photosensitive gaseous detectors. *Proceedings of the Sympos. on particle identification at high-luminosity hadron colliders.* doi:10.1142/9789812795878_0057

Choi, H. T., Sethi, D. S., & Braun, C. L. (1982). Geminate charge recombination in the photoionization of *N,N,N',N'*-tetramethyl-*p*-phenylenediamine (TMPD) in various solvents. *The Journal of Chemical Physics*, *77*(12), 6027–6039. doi:10.1063/1.443846

Grand, D., & Bemas, A. (1977). The ionization potential of a solute and the ground state energy of the excess electron. *Journal of Physical Chemistry*, *81*(12), 1209–1211. doi:10.1021/j100527a018

Imrie, D. C., Moghul, A. K., & Wells, K. (1990). Decamethylferrocene-a new photocathode for the detection of BaF_2 scintillation light. *Nuclear Instruments & Methods in Physics Research. Section A, Accelerators, Spectrometers, Detectors and Associated Equipment*, *297*(1-2), 179–182. doi:10.1016/0168-9002(90)91364-H

Imrie, D. C., Peskov, V., & Tamadoni-Jahromi, R. (1992). Air-stable solid photocathodes with good ageing properties for fast gaseous detectors. *Nuclear Instruments & Methods in Physics Research. Section A, Accelerators, Spectrometers, Detectors and Associated Equipment*, *317*(1-2), 92–96. doi:10.1016/0168-9002(92)90595-U

Malamud, G., Mine, P., Vartsky, D., Equer, B., Breskin, A., & Chechik, R. (1994). The photo-efficiency of CsI, amorphous silicon and organometallic reflective photocathodes. *Nuclear Instruments & Methods in Physics Research. Section A, Accelerators, Spectrometers, Detectors and Associated Equipment*, *343*, 121–128. doi:10.1016/0168-9002(94)90541-X

Massam, T., Muller, T., & Zichichi, A. (1963). A telescope to identify electrons in the presence of pion background. Preprint CERN/63-25, Nucl. Phys Div., 1-10.

Mine, Ph., Malamud, G., Vartsky, D., & Vasiliais, G. (1997). Organic UV photocathodes for gaseous particle detectors. *Nuclear Instruments & Methods in Physics Research. Section A, Accelerators, Spectrometers, Detectors and Associated Equipment*, *387*(1-2), 171–175. doi:10.1016/S0168-9002(96)00983-7

Moghal, A. (1991). *Ultraviolet photocathodes for use with solid scintillator proportional counter*. Brunel University.

Peskov, V. (1980). *Complex/multiparamers plasma diagnostics in UV and X-ray regions.* (Doctor of Science Thesis). Kapitza Institute for Physical Problems, USSR Academy of Sciences.

Peskov, V., Charpak, G., Mine, P., Sauli, F., Scigoscki, D., Seguinot, J., & Ypsilantis, T. et al. (1988a). Liquid and solid photocathodes. *Nuclear Instruments & Methods in Physics Research. Section A, Accelerators, Spectrometers, Detectors and Associated Equipment*, *269*(1), 149–160. doi:10.1016/0168-9002(88)90872-8

Peskov, V., Charpak, G., Sauli, F., Scigoscki, D., Diep, V., & Janjic, D. (1989). Organometallic photocathodes for parallel-plate and wire chambers. *Nuclear Instruments & Methods in Physics Research. Section A, Accelerators, Spectrometers, Detectors and Associated Equipment*, *283*(3), 786–791. doi:10.1016/0168-9002(89)91458-7

Peskov, V., & Zichichi, A. (1991). A new scintillator, $LiBaF_3$, and its basic properties. In *New technologies for supercolliders* (pp. 49–55). New York: Prenum Press. doi:10.1007/978-1-4684-1360-1_4

Peskov, V., & Zichichi, A. (1992). Fast (< 1 ns) gaseous detectors with solid photocathodes for frontier physics. *Nuclear Instruments & Methods in Physics Research. Section A, Accelerators, Spectrometers, Detectors and Associated Equipment*, *315*(1-3), 77–81. doi:10.1016/0168-9002(92)90684-V

Peskov, V. D. (1988). Gas detectors and sources of UV radiation based on Cs cathodes. *Journal of Applied Spectroscopy*, *48*, 316–318.

RCA Technical manual PT-60. (1963). Radio Corporation of America.

Sauli, F. (1982). Rediscovering the gas-filled photodiode. *Nuclear Instruments and Methods*, *203*(1-3), 601–604. doi:10.1016/0167-5087(82)90681-0

Sommer, A. H. (1968). *Photoemissive materials: preparation, properties and uses*. New York: Wiley.

Theobals, J. K. (1953). Investigation of Back Diffusion of Photoelectrons in Various Standard Gases as It Affects Secondary Electron Emission Coefficients. *Journal of Applied Physics*, *24*(2), 123–127. doi:10.1063/1.1721223

Vasileiadis, G., Malamud, G., Miné, P., & Vartsky, D. (1996). Solid and vapour phase UV photocathodes for gaseous detectors. *Nuclear Instruments & Methods in Physics Research. Section A, Accelerators, Spectrometers, Detectors and Associated Equipment*, *372*(1-2), 31–34. doi:10.1016/0168-9002(95)01294-X

KEY TERMS AND DEFINITIONS

Back Diffusion of Electrons: An effect of flow of some photoelectrons emitted from the cathode back to the cathode due to the diffusion in the gas. This effect leads to the reduction of the efficient quantum efficiency of solid photocathode in the gas compared to one measured in vacuum.

Plasma Diagnostics: Experimental technique used to determine the key plasma parameters such as electron and ion temperatures, density, spatial profiles, dynamic, etc.

Pre-Shower Detector: A position-sensitive detector combined with an absorber having a relatively small radiation length. Pre-shower detector is a part of the followed after him calorimeters: electromagnetic and hadronic. Its good spatial resolution is essential to identify the source of the deposited energy, e.g, to distinguish between single photons and multiple particles.

Photoelectron: An electron created in the detector volume due to the photoeffect. In the case of gaseous detectors is appear either via a photoelectric effect from the cathode or due to the photoionization of the gas.

Shower: A cascade production of secondary particles when a high-energy particle interacts with dense matter. Shower can be electromagnetic and hadronic. The first one is triggered by photons, electrons or positrons entering a thick layer of material and producing bremsstrahlung radiation and electron-positron pairs. The hadronic shower is caused by hadrons via strong nuclear forces, it includes hadron production, nuclear de-excitation, pion and muon decays.

Solid-State Detectors: Detectors of charged particles or photons made of semiconductive materials. These detectors have many designs and can operate either in charge collection mode or in avalanche mode.

APPENDIX

In this Chapter only some promising researches on solid photocathodes for gaseous detectors were mentioned. These pioneering works woke a great interest to this new approach and triggered further studies performed by various groups. Although these complimentary investigations did not lead to any breakthrough in this direction, for completeness we will mention some of these works. Few new solid metalloorganic photocathodes were tested in the work (Peskov, 1992). The highest quantum efficiency, about 1% at 200 nm was achieved with d,d-dimethil-trimethilen-ferrocenophane.

In the work (Malamud, 1994), besides several other photosensitive materials (see next Chapter), were also studied two solid metalloorganic compounds: bis(cyclopentadienyl)-ruthenium (ruthenocene)- {$(C_5H_5)_2$ Ru} and decamethyl-ferrocene (DMFe) {$(C_5(CH_3)_5)_2$ Fe}, which was studied earlier by (Imrie, 1990). The quantum efficiencies of these substances was in the same range as the metalloorganic photocathodes described earlier in this chapter.

Figure 11. The quantum efficiency of decamethylferrocene in methane and under different vacuum and temperature conditions (from Mine, 1997); for comparison data obtained in work (Imrie, 1990) are presented (triangles).

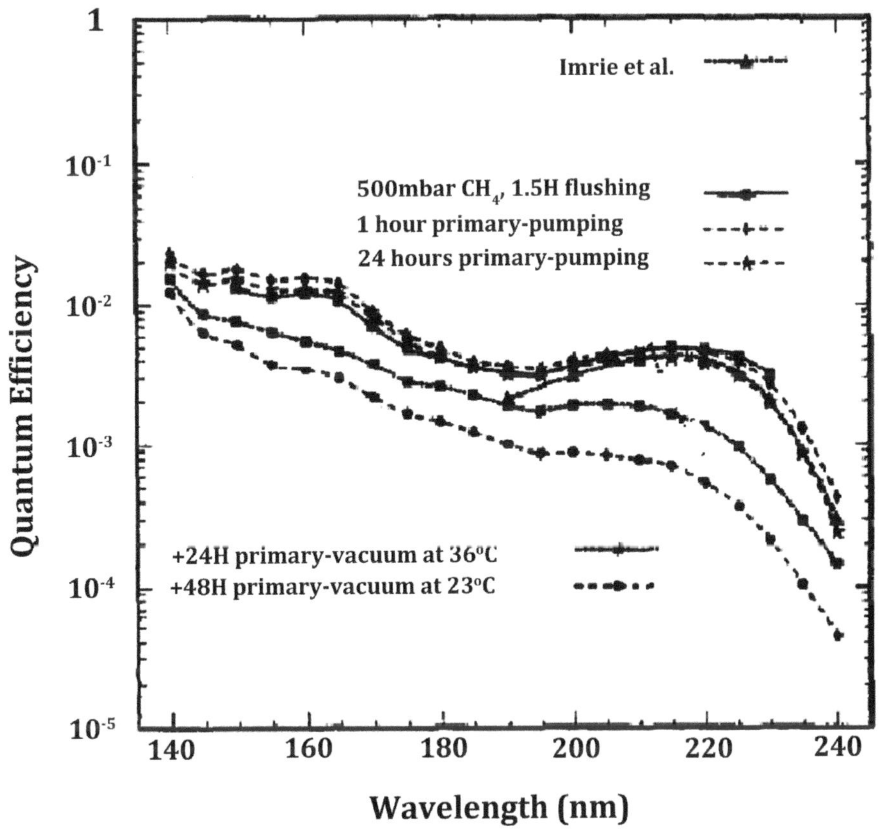

In the work (Mine, 1997) DMFe was studied systematically and under different conditions. However, the highest quantum efficiency achieved was quite close to that measured in (Imrie, 1990), as shown in Figure 11.

Tetrathiafulvalene and bis(cyclopentadieny1) magnesium $(C_5H_5)_2Mg$, solid photocathodes were studied in the work (Vasileiadis,1996).

Since both these compounds are in powder form, the samples were prepared by vacuum deposition of a thin layer on a copper disk. The samples were stored under argon atmosphere. A small exposure to air, of about 5 min, was necessary during the transport to the measuring apparatus. In spectral measuremenst the detected signal was weak in both cases, resulting in relatively low quantum efficiency throughout the whole spectral range.

Results of detailed studies of various solid photocathodes can be found in (Moghal, 1991).The following conclusion was drown by this author among organometallic photocathode, decamethyl ferrocene gave one of the best overall performance in terms of efficiency and ease of preparation.

Chapter 6
CsI and Cs_2Te Photocathodes

ABSTRACT

The development of CsI and Cs_2Te solid photocathodes presented the breakthrough in photosensitive gaseous photodetectors we all had been waiting for, and put so much effort in finding. It was actually going back to basic. CsI had for a long time been used in vacuum photomultipliers. It proved to also work well in gas floes, and can even be exposed to air for shorter times. We will describe the development work that led to the current state of the art in solid photocathodes.

1. INTRODUCTION

The tests of photocathode materials used since a long time in vacuum photomultipliers, to see if they can also operate in gas flows and even be air-proof gave positive results eventually. A first successful test of a CsI photocathode combined with a parallel plate avalanche chamber and flushed with CH_4 was developed by Charpak (Charpak, 1989). This photocathode, manufactured on a stainless steel cathode by a vacuum deposition technique, showed excellent stability and the highest quantum efficiency in the UV region among all solid photocathode tested earlier in gaseous detectors. Surprisingly enough, the quantum efficiency of this photocathode remained high after transfer from the evaporation system into gaseous detector in air without any special precaution.

CsI had been tested before, but not with successful results (Peskov, 1988). The problem turned out to be that the CsI was evaporated on a copper cathode and then during storage and transportation in air to the test facility it was exposed to sunlight. Both these factors (Cu cathode and sunlight) damaged the CsI coating.

The high and stable quantum efficiency of CsI opened for a wide range of new applications for photosensitive gaseous detectors. The possible applications ranged from readout of crystal scintillators to detection of Cherenkov light. This is why the report by Charpak made at a conference at Fermilab (Fermi National accelerator laboratory) in 1989 triggered a great interest in the high energy physics community to this approach (Charpak, 1989). It was also at this conference that the two authors (T. Francke and V. Peskov) met and formed a lifelong friendship and close collaboration.

DOI: 10.4018/978-1-5225-0242-5.ch006

Shortly after that a wire chamber with a CsI photocathode was developed (Dangendorf, 1990), and systematic studies were performed of CsI photocathodes and ways of enhancing their quantum efficiency further (Séguinot, 1990). These three publications initiated a new direction in the development of photosensitive gaseous detectors. Inspired by these initial results other groups started to actively investigate CsI photocathodes. At CERN the RD-26 collaboration was established with focus on production and characterization of large area photocathodes, susceptible to equip fast UV-photon imaging devices (Nappi, 1992; Paic, 1996).

A true fun story, originally written by one of us (V. Peskov) and recited in Sauli´s book (Sauli, 2011), reflects well the enthusiasm the spirit of these times:

Surely, the development of photosensitive gaseous detectors with CsI photocathodes was a pivotal achievement of George's group around 1990. We built the first prototype, a parallel plate chamber with a reflective CsI photocathode mounted inside the chamber in air. Then the chamber was flushed with CH_4 at atmospheric pressure. Using this simple device, we performed initial measurements of the quantum efficiency of the CsI photocathode operating in various gases, and it turned out to be higher than for ethylferrocene vapours. This detector operated at high gas gain, and some gas mixtures reached gains up to 10^5. George, elated by this development, encouraged me to make a presentation featuring these promising results at the Fermi Lab symposium on Particle Identification at High Luminosity Colliders (Fermi Lab, 1989). Many prominent detector physicists including D. Anderson, A. Breskin, J. Séguinot, T. Ypsilantis, L. Lederman attended this symposium. Subsequently, in a few months they launched new studies of photo-detectors based on CsI photocathodes.

Highly competitive as always, Georges drove us to rapidly conduct new photosensitive detector experiments at CERN. I split my time between Georges' laboratory focusing on the application of CsI-based detectors for the readout of the scintillation light from various crystals and in the Séguinot-Ypsilantis Lab working on new RICH detectors. To bolster our group's productivity, Georges hired Ian, a shy, amiable technician who was somewhat in awe of Georges. He proved instrumental and helped us build a first prototype of a BaF_2 calorimeter combined with CsI-based gaseous detectors. Georges, however, thought that we worked too slow and put a great deal of pressure on us, so much so, he even requested a beam test of this prototype although it was not fully tested in the laboratory. Working well past ten o'clock the night before the beam test, we finished assembling the prototype. Ian and I were both very tired and during the final operation of the BaF_2 crystal installation the O-ring which surrounded the BaF_2 crystal was over pressed resulting in a slight crack in the crystal. We didn't worry too much about this small crack since we didn't think it would make a significant difference nor effect the light collection nor create any gas leak in the parallel-plate chamber. We simply continued to assemble it.

Georges, in good spirits and well rested unlike us, arrived at the laboratory early the next morning, and told us to move our device to the CERN PS beam test area. Our set up was fully assembled and the BaF_2 crystal with the CsI coated parallel-plate avalanche chamber were inside the aluminum shielding box with a window on the top. We planned to mount a photomultiplier tube to this window for independent detection of the BaF_2 scintillation light. Quite pleased, Georges walked around the setup. Then, just before we started to mount the photomultiplier, he decided to look through the window inside the assembled detector. I do not know how he was able to notice the tiny crack, but he did notice it! He asked the technician, "Don't you think there is a crack in the crystal? Look inside!" Ian flushed and looking *inside the shielding box said, "I do not see any crack." Georges looked again through the window, and insisted, "But there is a crack!" Again, Ian looked through the window, and repeated, "I do not see anything". Georges knew something was wrong. He turned to me and asked, "Vladimir, do you see the crack?" I did not want to contradict the new technician, and tarnish his credibility, so I looked inside and said, "No, I do not see anything." I really couldn't see inside the dark shielding box the tiny crack. Georges again peered inside and shouted, "But there is a crack!" Now it was Ian's turn to have another look inside, and this time he affirmed, "There is something, but it does not look like a crack." Once more Georges peeked inside, "Certainly, there is a crack!" Again, Ian peered in and said, "Yes, there is something!" "Vladimir," Georges called, "can you look inside?" I looked inside, "Yes, there is something that looks like a small crack." "How did this happen? This crystal is very expensive!" yelled George. "I don't know" said the technician. And in an attempt to help him, I said, "Maybe the O-ring made too much pressure and the crystal spontaneously cracked at night?" "You want to say that if we build a full calorimeter one day it may fall to pieces?" skeptically replied George. We were racing against the clock, so we started moving to the beam test area without clarification about what really happened. It took another few days of hard work before we achieved accurate results. But despite this tiny crack, we succeeded in detecting the scintillation light from the BaF_2 crystal with our CsI-based detector. The results were so promising that we submitted a paper to the journal Nuclear Instruments and Methods in Physics Research.*

A few months later, seating with Georges in the CERN cafeteria, we told him the real story. Ian was quite embarrassed all through the meal as Georges and all the people around laughed loudly. As a memento, Ian still keeps the crystal in his cabinet at CERN!

So why was the high–energy physics community so excited by the results obtained with CsI photo-convertor if it was known for decades and widely used in vacuum photodetectors? What new was done by Charpak?

The breakthrough was in developing a simple technology to deposit a CsI layer operating in reflection mode with high quantum efficiency. The quantum efficiency is close to one of TMAE vapour. It therefore represents a serious alternative to TMAE in many applications.

Moreover it was proven that it can work under the conditions gaseous detectors offer where, in contrast to vacuum devices, the photocathodes are under intense bombardment by ions coming from the avalanches. It was proven also that after a short contact with air the CsI photocathode can be "regenerated" by a long flushing with a dry gas, for example argon. This effect was later confirmed by Séguinot and Ypsilantis (Séguinot, 1990). In the most common gas mixtures used in MWPCs the CsI quantum efficiency remains very stable and is not affect by outgassing of the gas chamber or by other small impurities in the gas. This strongly simplifies its use compared to TMAE, which reacts with many materials.

One should remember that in the early works on vacuum based devices it was observed that CsI is somehow damaged after exposure to air and its quantum efficiency becomes significantly lower than before the exposure to air (see for example Heroux, 1966; Simons, 1987).

In most of the ancient work on vacuum based photomultipliers it was concluded that the quantum efficiency of CsI at 179 nm is about 10%, but in these vacuum detectors mainly used semitransparent CsI photocathodes where the photo-conversion is not 100%.

There were also early suggestions to replace TEA vapours by a CsI photocathode in a RICH detector based on a needle amplification structure (Comby, 1990). However, the group measured very low quantum efficiency values (0.5%) in the wavelength range 165-200 nm using CH_4 at atmospheric pressure as amplification gas, which made them abandon the idea.

Some attempts to adopt a CsI photocathode to gaseous detectors were done in the Ypsilantis group, but again the quantum efficiency was measured to be very low and this discouraged them to continue the efforts (Ypsilantis, 1986).

Therefore, answering the question posed above it can be concluded that the roles of Charpak, Dangendorf and Séguinot were that for the first time it was demonstrated that CsI photocathodes can be used in avalanche gaseous detectors capable of operating stable at high gas gains (Charpak, 1989; Dangendorf, 1990; Séguinot, 1990).

After the success with the CsI photoconvertor investigations the groups went on to study if other alkali halide photocathodes can be combined with gaseous detectors as well (see the following Chapter and Buzulutskov, 2008). They showed that a Cs_2Te photocathode has a high sensitivity up to 300 nm. Unfortunately such a photocathode is quite difficult to manufacture and it can operate stable only in ultraclean sealed gaseous detectors. However, such detectors may have many applications as will be described in the chapter on applications.

2. CsI PHOTOCATHODES

2.1 CsI Photocathode Manufacturing Methods

2.1.1 Conventional (with Exposure to Air)

Today gaseous detectors with CsI photocathodes have replaced almost every gaseous photosensitive vapour in most of applications, including RICH detectors (see for example Nappi, 2001). However, it took some time and hard work to reach this state of art and fully understand the small details that are so important for getting reproducible results. The main problem comes from the fact that the surface photoelectric effect is very sensitive to the surface condition, i.e. any layer of absorbed substances on top of the CsI (Mc Daniel, 1964). Being exposed to air after vacuum deposition, the CsI photocathode can adsorb water and other gases which change the quantum efficiency. In additional, after water absorption the uniformity of the CsI can be affected. Often the layer is cracked and may consist of separate micro-islands of CsI. In order to obtain some reproducibility one should follow a procedure to outgas the photocathode after exposure to air. The details of this procedure were gradually developed by experiments of various groups. It can be summarized as follows:

1. It is very important to use pure CsI for the vacuum deposition (Anderson, 1994). The best results are obtained with cleaved and undoped CsI crystals, as shown in Figure 1.
2. The evaporating system should be also as clean as possible and it should not have been previously used for evaporation of other, especially organic, materials.
3. The CsI used for the coating should be melted and some initial fraction of it evaporated on something else before the true evaporation.
4. The substrates shall be very clean, e.g. stainless steel covered by an Al layer (Séguinot, 1990). One should avoid copper substrates, since, as was mentioned earlier, Cu reacts chemically with CsI when exposed to light.

Figure 1. The measured quantum efficiency vs. the photon wavelength, λ, of a 500 nm thick reflective CsI photocathode after 1, 17 and 41 hours of methane gas flow through the gas chamber; the solid line without data points is the quantum efficiency of TMAE gas. The dashed line is results obtained with another 500 nm thick CsI photocathode showing the good reproducibility of the method described above. The dash-dotted line is one of the best quantum efficiency measurements for a vacuum based detector with a reflective CsI photocathode.
Carruthers, 1969 from Séguinot, 1990.

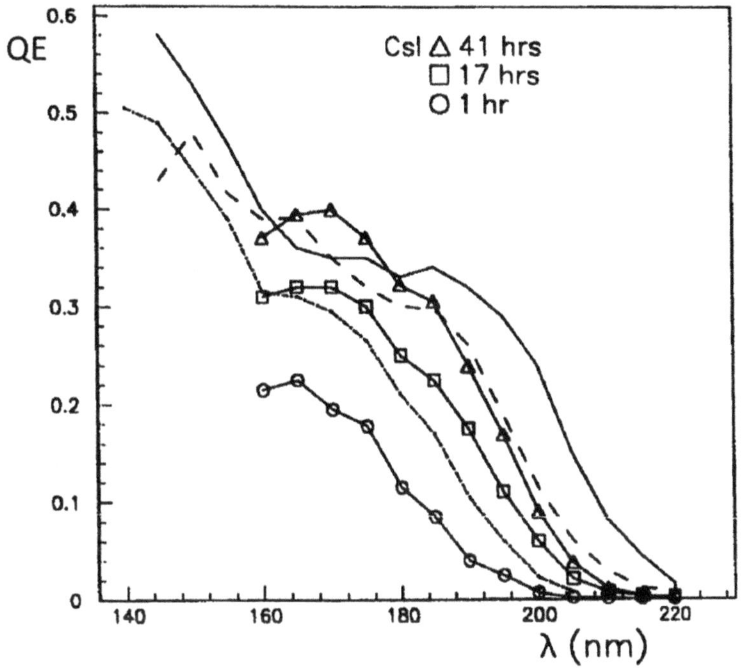

5. The substrate shall be heated to 60-70°C during the evaporation in order to keep it dry.
6. Minimize the time the CsI is in contact with air when the photocathode is transferred to the gas detector.
7. Heat and outgas the CsI photocathode, but not the entire gas chamber, inside the detector at 60-70°C after the installation (Anderson, 1991).
8. Use proper gases in the detector (see below).

If one follows these procedures it is possible to obtain quantum efficiency as high as Séguinot obtained in his original work (Séguinot, 1990). Séguinot also found that CsI quantum efficiency is improving with time after flushing the gas chamber and the photocathode with a dry gas, e.g. argon, as shown in Figure 1. This indicated a slow removal of moisture from the photocathode. After 41 hours of gas flushing the quantum efficiency of the CsI photocathode has almost reached the quantum efficiency of TMAE vapours. Note that this quantum efficiency is even higher than the best CsI quantum efficiency measured in vacuum based detectors (Carruthers, 1969).

Using a stainless steel substrate covered with aluminium has shown to give the best results. It is also important to carefully optimize the photocathode thickness, as shown in Figure 2. It turned out that the

Figure 2. The quantum efficiency vs. the photon wavelength for CsI photocathodes of different thicknesses; for comparison, the quantum efficiency of gaseous TMAE is also presented as the solid line without data points.
From Séguinot, 1990.

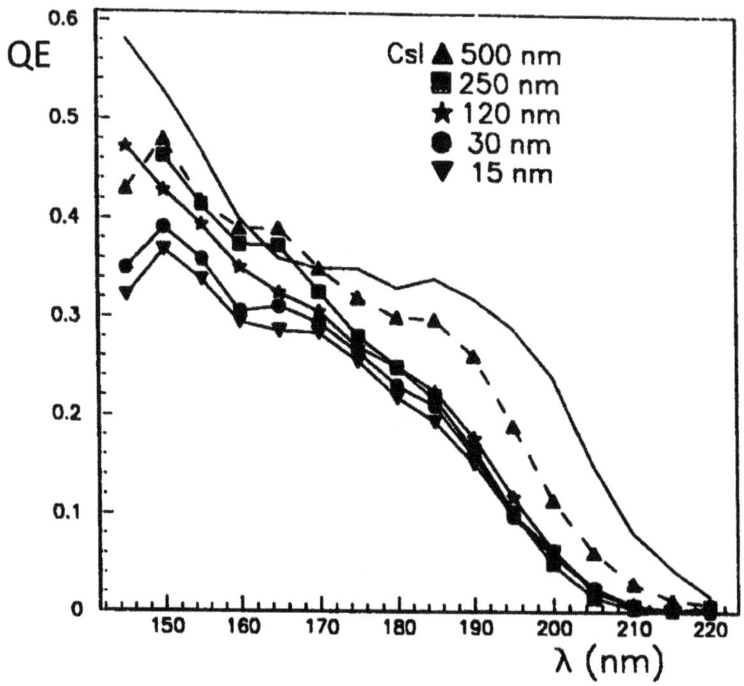

best results are obtained with a 500 nm thick cathode. If it is thinner, the quantum efficiency drops. If it is thicker, ions from the electron avalanche accumulate on the CsI and charges up the cathode. This lowers the quantum efficiency as the ions shield the CsI. It can also cause sporadic electron emission from the cathode.

Further studies revealed that one of the key procedures, in addition to the gas flushing, is heating the photocathode above 60°C, as shown in Figure 3. The heating procedure efficiently outgases water and other gases from the CsI.

As can be seen from Figure 3, after flushing a freshly prepared CsI with CH_4 its quantum efficiency increases during the first 5 hours and then reaches a plateau value. However, after the photocathode is heated to 97 °C a further enhancement of the CsI quantum efficiency is observed (Anderson, 1991). Also Séguinot heated the CsI photocathode to 30 °C in his initial results. Heating was soon implemented in the deposition process by most groups.

Another remarkable contribution from Anderson's group was the observation that highest CsI quantum efficiency was achieved when the evaporation process used freshly cleaved CsI crystals (undoped) rather than CsI powder which had been grinded long time ago (Anderson, 1994). This can be clearly seen in Figure 4.

In the beginning some groups had difficulties to reproduce the high quantum efficiency reported by Séguinot (Séguinot, 1990). The measured quantum efficiencies were often only half of Séguinot's results, (see for example Malamud, 1994; Lu, 1994; Breskin, 1994; Braem, 1994; Aleksan, 1994).

Figure 3. The quantum efficiency as a function of time of a freshly deposited 500 nm thick CsI photocathode; the photocathode is flushed with a 40 l/min flow of methane, after 20 hours the photocathode is heated to 97 °C. Both the gas flushing and the heating increase the quantum efficiency of the CsI photocathode dramatically.
From Anderson, 1991.

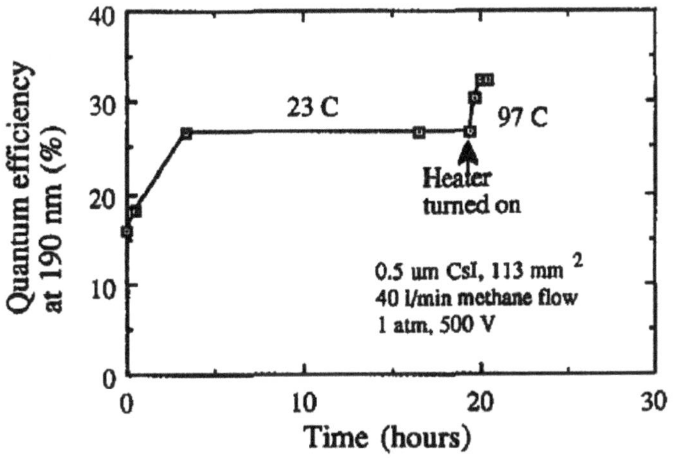

Figure 4. The quantum efficiency as a function of wavelength measured in methane for a CsI photocathode produced from a CsI scintillator (Sc) or CsI powder (P); V means it is vacuum deposited on the plate, and Sp means that CsI is dissolved in alcohol and sprayed on the plate.
From Anderson, 1994.

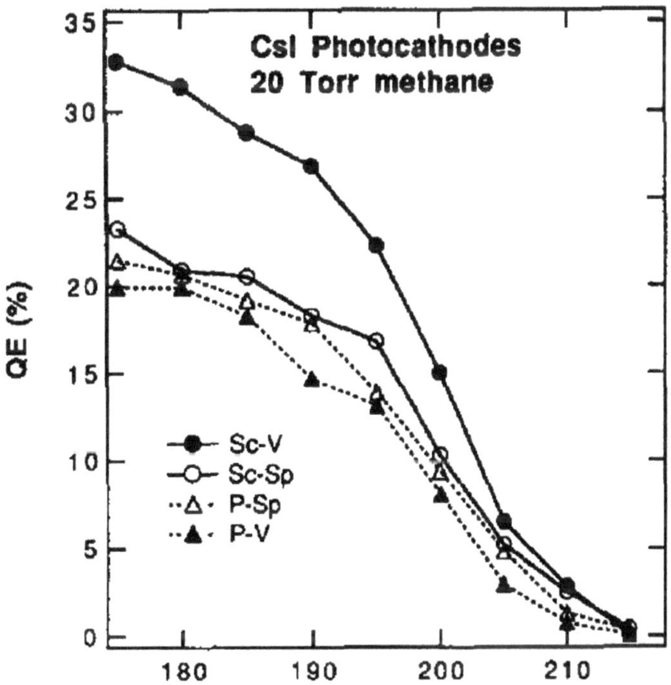

Figure 5. The quantum efficiency of CsI photocathodes deposited on printed circuit boards with various coatings: Cu/Au, Ni/Au and stainless steel coated with Al layer
The data ware extracted after analysis of beam test performed with RICH detectors (Piuz, 1996; Almedia 1995). Data marked "RD-26 reference" were obtained by the CERN RD-26 project for small samples. The plots are from (Almedia, 1995).

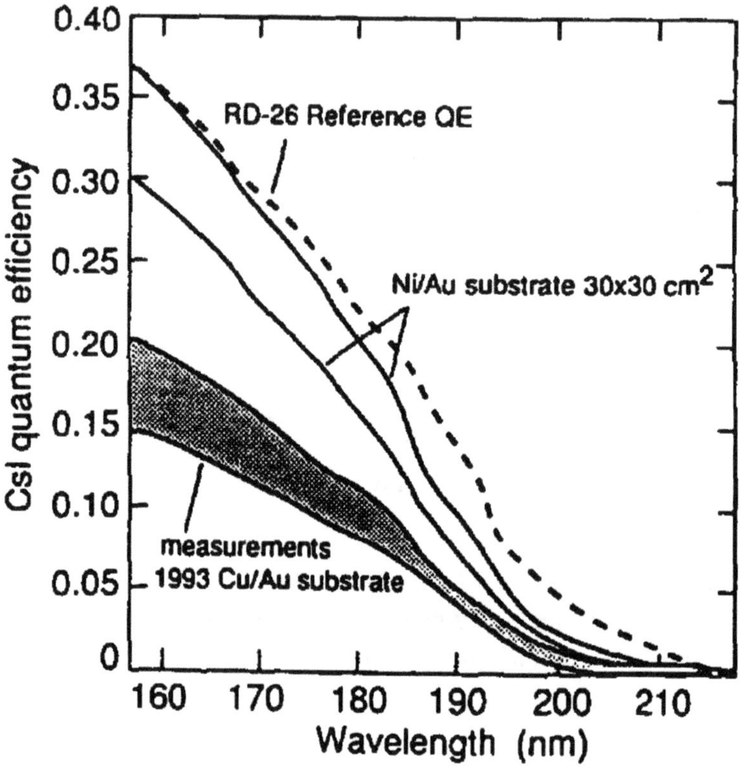

Consequent studies showed that there were two main reasons for this. First, the calibration procedure was different between groups, second some groups deviated from the 8 golden rules listed above. The differences in calibration was connected to the fact that some groups (e.g. Breskin) used as a quantum efficiency reference a vacuum photosensitive detector calibrated by Hamamatsu Inc., whereas most other groups used as a reference either the quantum efficiency of TMAE vapours or some data obtained from recording Cherenkov UV photons. The last two methods were in a rather good agreement, and Hamamatsu later admitted a mistake in their calibration.

Experience gained by many groups after 1994-1995 show that any small deviation from the 8 golden rules indeed prevents achieving a high CsI quantum efficiency. For example, some authors heated the whole gas chamber whereas the correct way is to heat only the substrate.

The importance of using the correct substrate for the CsI photocathode is illustrated in Figure 5. It confirms that the best results are obtained with a stainless steel substrate coated with a thin Al film.

2.1.2 The Spray Technique

CsI photocathodes are normally manufactured by a vacuum evaporation technique. More detailed information about this technology will be given later in this chapter.

In some brief laboratory studies, when one do not aim to reach the highest possible quantum efficiency, another very simple technique can be used. It is often referred to as the "spray-on technique" (Periale, 2002). Here the CsI is dissolved in a solvent and spayed in air on a heated substrate (Anderson, 1992; Anderson, 1993). By this technique large area photocathode can be produced without access to a vacuum deposition system. The process is quite simple. Pieces of a CsI scintillator are dissolved either in pure ethanol or in distilled water. In the latter case the optimum concentration of the CsI is 10% by weight. This solution is then diluted to about four times its original volume by the addition of more ethanol. This solution is then sprayed with a simple air brush on a heated substrate until a desired thickness is achieved. To control the thickness of the layer a standard piezoelectric thickness monitor is placed next to the photocathode and also heated to evaporate the solvent. Initially such photocathodes are saturated with water. By heating them this water slowly evaporates. The quantum efficiency of sprayed photocathodes as function of wavelength is shown in Figure 4. As can be seen, such photocathodes have about 2/3 the quantum efficiency of the photocathodes made by scintillator vacuum deposition and following the 8 golden rules.

2.1.3 CsI Photocathodes Not Exposed to Air

Among various systematic studies of CsI photocathodes some tests were done with samples from the evaporation system transferred into the gaseous detectors without exposure to air (Peskov, 1995). The first attempts were done with the single wire setup described above (Peskov, 1994). The preliminary conclusion is that such photocathodes have quantum efficiency only slightly higher in the long wavelength region. Practically no improvements are observed for short wavelengths. However, this technique allows a better reproducibility. For example, with 15 photocathodes transferred into the detector in this way the maximum deviation in the quantum efficiency was around 20%.

Approximately at the same time the RD51 collaboration at CERN started developing a dedicated evaporation system allowing CsI photocathode transfer to the gas chambers without exposure to air (Braem, 1995). It later became an evaporation facility at CERN which everyone could use to produce CsI photocathodes even for large area MWPCs used in various high-energy physics experiments. It is schematically shown in Figure 6a). A photograph of the system is shown in Figure 6b).

The system consists of two main parts united to a single system. One half is a vacuum evaporation chamber, and the second half is the photocathodes quality control chamber. CsI photocathodes as large as 60x60 cm^2 can be manufactured in this evaporation unit. After evaporation the photocathode is moved to the second chamber where its quantum efficiency is measured in vacuum. The photocathode is then moved back to the evaporation chamber and by mechanical manipulators packed in the sealed gas chamber used for the transportation of the ready photocathode to the MWPC.

Modern MWPC-based photodetectors often have the CsI photocathode coated on the pads of the readout cathode. This readout cathode is often made of a matrix of metallic pads (typically 8x8 mm^2) separated from each other by a thin isolating wall. Each pad is connected to its own charge sensitive amplifier. The induced signals on these pads give the position of the electron avalanche.

The cheapest option for such a pad plane is to use a printed circuit board (PCB). In standard PCBs copper is used. As we have already mentioned, copper is not well suited for CsI photocathodes. Therefore, the copper layer is often coated with Ni and Au between the copper and the CsI. The nickel layer protects the CsI from reactions with the copper to form CuI, and the gold ensures an inert substrate. After being glued onto an aluminium frame the PCB substrates undergo an elaborate sequence of cleaning and testing

Figure 6. a) Layout of the CsI production facility developed at CERN; left: the evaporation chamber, right: the quantum efficiency measurement side; b) a photograph of the CsI evaporation facility at CERN
a) From Hoedlmoser, 2006. b) From http://alice-hmpid.web.cern.ch/alice-hmpid/.

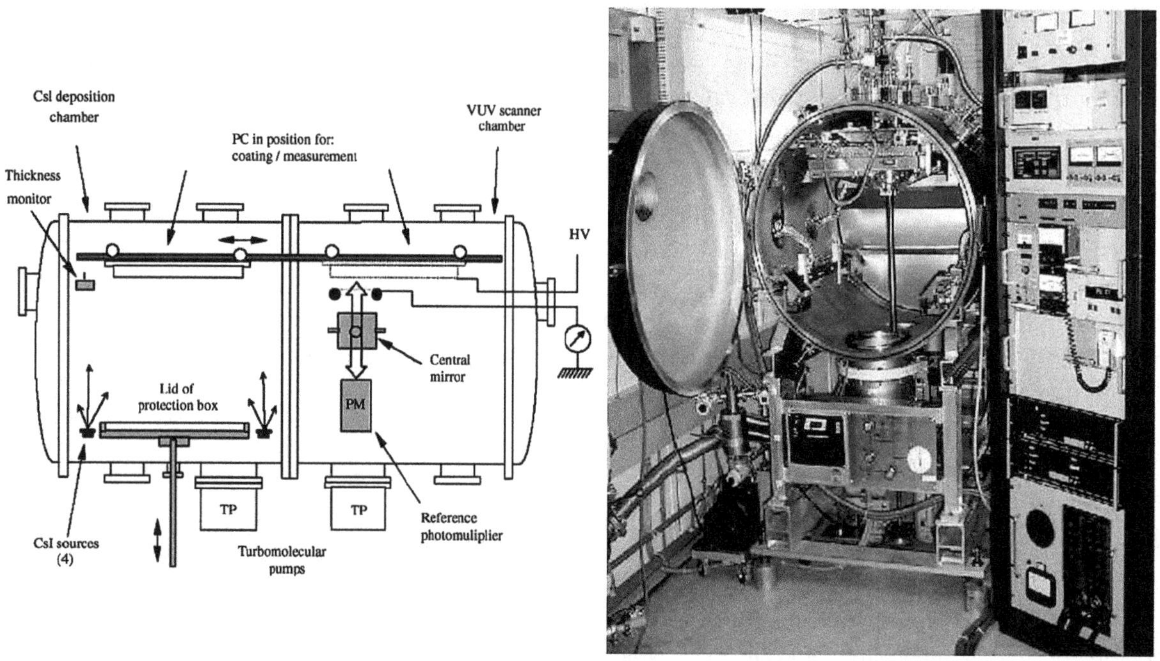

procedures (cleaning with strong detergents, ultrasonic baths, demineralized water, pure alcohol, tests of gas-tightness and high voltage capability within the MWPC). Polluted substrates and outgassing may give rise to operational difficulties during or after the CsI coating process, e.g. long vacuum pumping to reach the required vacuum, pollution of the CsI evaporator or a low photocathode quantum efficiency. Subsequently, the substrates arrives at the CsI deposition workshop, where they are cleaned again with demineralized water, alcohol and nitrogen jet, followed by one day in a drying oven at 60°C before they are finally installed in the CsI evaporation facility. The PCB substrate is mounted horizontally above four containers with pre-melted cleaved CsI crystals. The CsI sources are located at a fixed distance from the corners which gives a uniform CsI film thickness. The thickness precision is normally ~5%. After that the vacuum chamber is pumped for a minimum of two days at 60°C. 60°C is the highest temperature the PCB can stand during the vacuum bake out. The final pressure achieved in the chamber before CsI deposition is approximately 4x 10^{-7} Torr. Residual gas analyses prior to the evaporation normally show high concentrations of water and hydrocarbons, as shown in Figure 7.

The pumping prior to the evaporations should be done for up to three days to remove as much water as possible from the PCB and the chamber walls. During installation (and extraction) the chamber is kept open to air for as short time as possible, and it is being kept at a temperature higher than the room (typically 30 °C) at all stages to minimize absorption of humidity in the hygroscopic CsI layers present in the chamber from previous evaporations. Periodically the evaporation chamber is cleaned to avoid accumulation of too much CsI on the walls. However, good results have been obtained after up to 15 evaporation cycles without cleaning. The CsI deposition is carried out with the full system at 60 °C.

Figure 7. Typical composition of the residual gas in the evaporation chamber routinely measured prior to CsI evaporation
From Hoedlmoser, 2006.

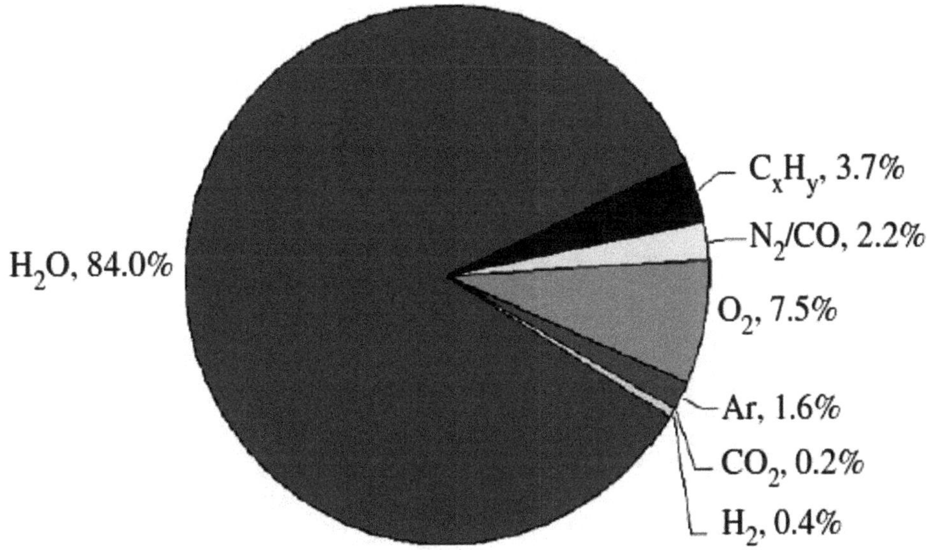

The evaporation is performed by means of resistive heating of four molybdenum boats each filled with 0.75 g pre-melted CsI and symmetrically arranged below the substrate. The heating current running through the sources is increased slowly (within 15–30 min) until the evaporation starts. The thin CsI film is a superposition from four the CsI sources. The CsI sources are carefully prepared in a dedicated vacuum set-up, pre-melting the CsI during a few minutes at 600 °C to provoke outgassing of eventual impurities and to obtain an optimal thermal contact of the CsI with the molybdenum walls of the sources.

During the evaporation the deposition rate is kept at ~1 nm/s. The thickness and the deposition rate is measure with the help of two quartz oscillators. Given the precise amount of pre-melted source material a final layer thickness of 300–500 nm is achieved when all CsI has evaporated.

After the CsI deposition, the photocathode is transferred to the quantum efficiency measurement side. It measures the photo-current produced by a H_2 lamp which is scanned across the whole photocathode surface. In such a way a 2D map of the quantum efficiency is obtained (see Figure 8).

The quality of the CsI photocathode depends on the heat treatment after the coating. Initially, the quantum efficiency is rather low but it increases if the photocathode is heated to 60 °C for several hours after deposition, especially in the long wavelength range (≥190 nm). Normally the photocurrent increases by 20–50% within the first 2–10 h after coating. Before opening the deposition chamber, the CsI photocathode is encapsulated in vacuum in an air tight transfer box. After the extraction of the box from the evaporation system it is filled with Ar and can be stored without any quantum efficiency degradation for years.

When it is necessary, the container with the photocathode can be opened in a special glove box (see Figure 9) flushed with Ar, e.g. to install it in the gaseous detector.

The technique described above for photocathode production and installation in the gaseous detectors is now applied in most of the laboratories manufacturing CsI photocathodes for gaseous detectors. It offers good photocathode quality and reproducibility and quantum efficiency close to 30%. As an example

Figure 8. 2D image of the photocathode quantum efficiency (in arbitrary units) measured in vacuum through the photo-current generated by a H_2 lamp scanned across the whole photocathode
From Hoedlmoser, 2006.

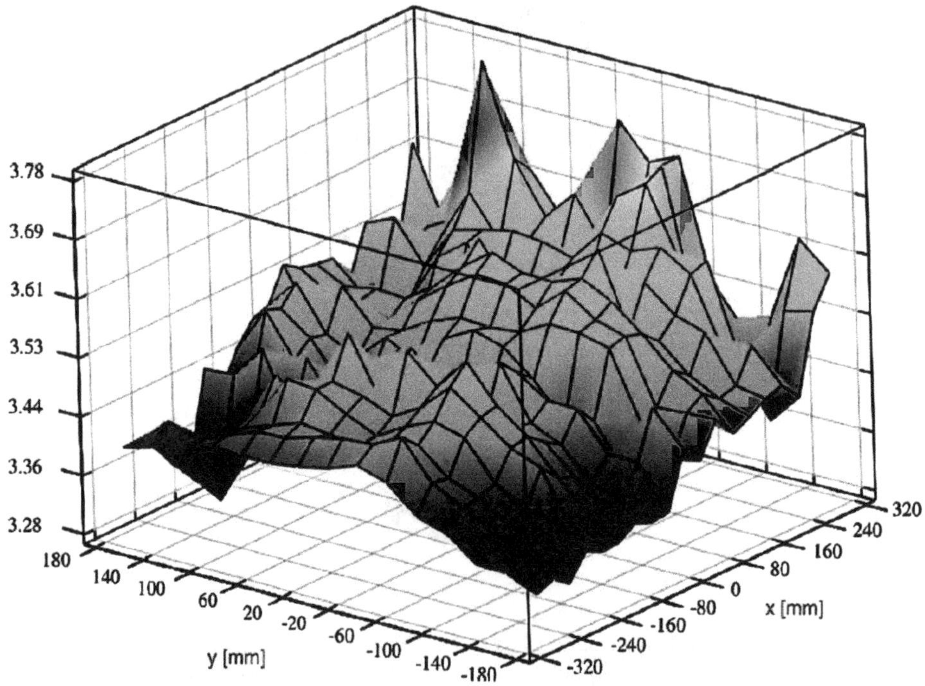

Figure 9. Photograph showing the transfer chamber with the CsI photocathode installed inside a glow box flushed with Ar
The operation is performed by J. Vanbeelen (CERN).

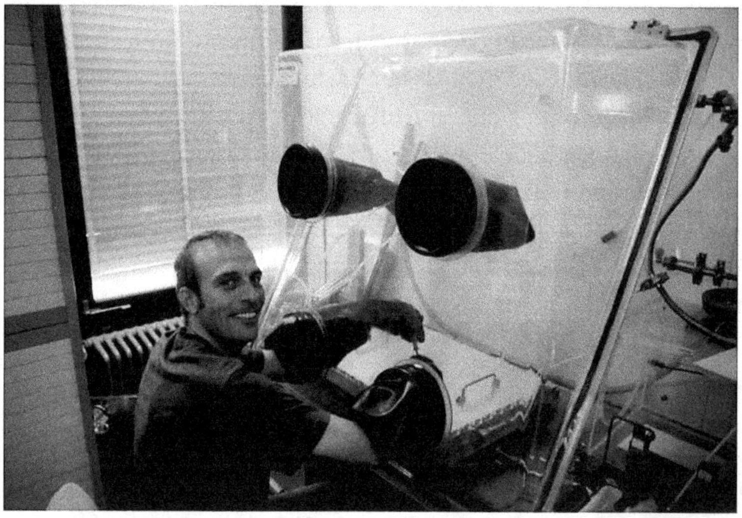

Figure 10. The average normalized photocurrent measured in vacuum for 45 CsI photocathodes produced at CERN for the ALICE RICH detector; rhomb - before the post heat enhancement procedure, squares – after the heat enhancement, triangles –after the photocathodes were stored in the sealed box and flushed with Ar for weeks/months after production, circles - later rescans of some photocathodes after they were stored in a box flushed with Ar for several months; the green line defines the minimum requirement to accept a photocathode for use in the ALICE detector. The red dash line shows the best quantum efficiency obtained by the CERN RD 26 collaboration with small CsI samples.
From Hoedlmoser, 2006.

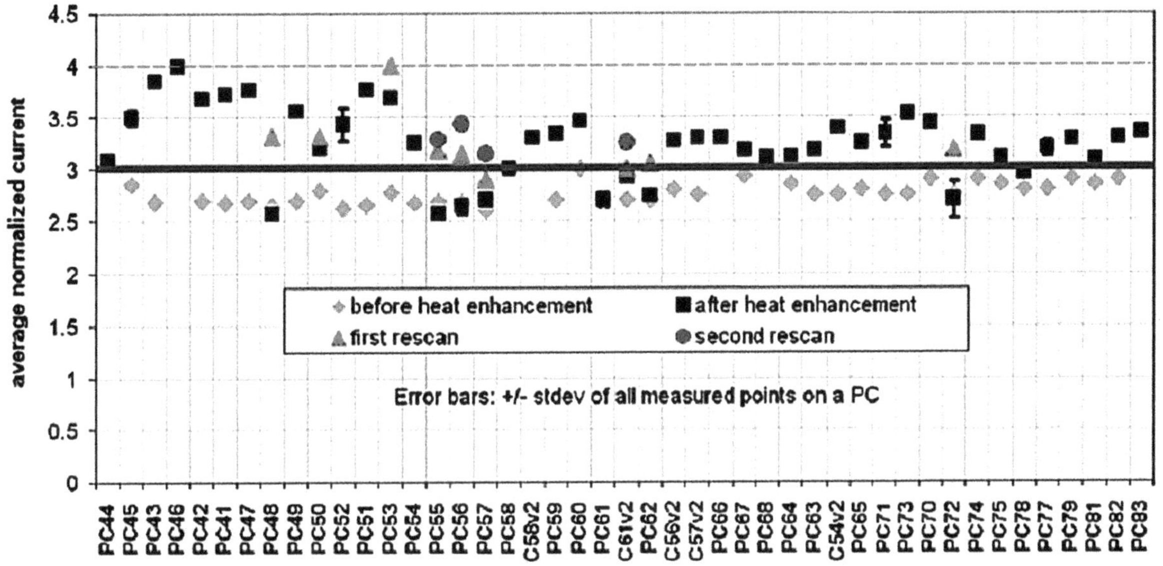

Figure 10 shows the quality of 45 photocathodes produced at CERN for the ALICE RICH detector. The average photocurrent current measured inside the vacuum chamber (see Figure 6) shows a variation from 2.5 to 4 with a mean value of 3.42.

As can be seen in the figure, the photocurrent is normally increased by 30-50% by the heat treatment. In some cases the post heat enhancement effect is not observed. This phenomena, named a "fatigue" effect, has been investigated by the evaporation team (Braem, 2007). It is repeatedly observed that after a successful production of about 7–10 photocathodes the subsequent ones are unsuccessful, delivering a photocurrent below the acceptance level. It was found empirically that in order to restore the photocathodes production to the normal level, maintenance is needed. The evaporation unit is then carefully cleaned to remove the CsI deposited on its inner walls followed by pumping on it at elevated temperatures (~70 °C) until the vacuum reaches a value of 10^{-7} Torr. After this, it is possible to resume the production of photocathodes with the normal quantum efficiency.

From the analysis of test beam results of several photocathodes the absolute quantum efficiency shows good agreement with Monte Carlo simulations (Di Mauro, 1999). Figure 11 shows the quantum efficiency for two of the photocathodes in Figure 10, # 44 and # 46, in comparison with the reference quantum efficiency obtained by the CERN RD26 project for small samples.

In conclusion, all measures described above (the substrate preparation, pumping to high vacuum, heat treatment, periodical evaporation system cleaning, and photocathode transfer to the detector without

Figure 11. Examples of absolute quantum efficiency for two photocathodes (# 44 and # 46) derived from a Monte Carlo analysis of test beam results in comparison to the reference quantum efficiency obtained by the CERN-RD26 project for small samples
From Hoedlmoser, 2006.

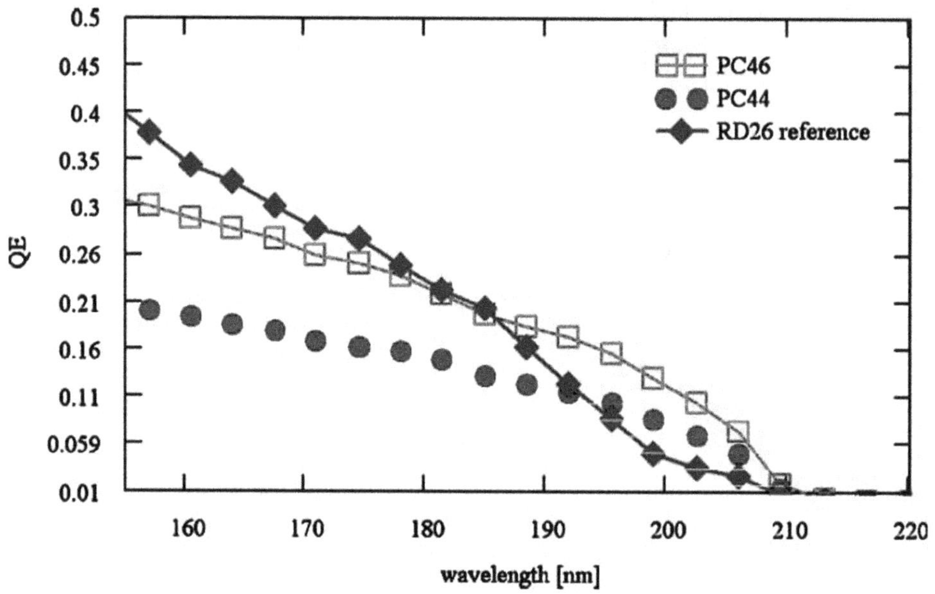

exposure to air etc.) are important to reliably produce large area CsI photocathodes. These photocathodes are then successfully used for a long time (years) in high-energy physics experiments such as ALICE, COMPASS, etc. The same set up is also used for CsI coatings on TGEMs and RETGEMs which are modern promising photodetectors that we will return to later.

2.2 The Physics of CsI Photocathode Production

We will try to summarize the current understanding of the physical explanation behind the processes involved in the CsI deposition, heat enhancement and the fatigue effect as derived from the experience gained by many groups.

2.2.1 The Fatigue Effect

In order to understand the fatigue effect authors one group tried to correlate the photocurrent with various parameters measured before and during the evaporation; vacuum level, chemical composition of the residual gas, temperature, evaporation rate etc. They could not find any clear correlation (Hoedlmoser, 2006). The conclusion from these studies was that the migration of residual water inside the evaporation plant is the main factor influencing the quality of CsI photocathodes.

The effect of water on the quality of the CsI photocathode is known and has been studied for a long time. If the vacuum is not ultra-high, e.g. in systems like the CERN CsI deposition facility, the residual gas is dominated by water (see Figure 7). Water plays a special role in vacuum technology due to its strong absorption properties on metallic walls and other surfaces. CsI is a highly hygroscopic material.

The presence of CsI on the chamber walls strongly affects the vacuum conditions and water migration since the CsI layers act like a getter pump. During the production of CsI photocathodes a rather thick and visible layer of CsI is formed on the inner walls of the evaporation chamber. Hereafter it will be called "the parasitic layer".

After each evaporation, the set up is cooled down to room temperature and then exposed to air in order to extract the photocathode box. Every time the chamber is opened, the parasitic layer absorbs water from the ambient air. During the subsequent evaporation, the accumulated water can migrate to the photocathode changing its properties. For example, CsI molecules in the top layer of the photocathode can dissociate into Cs^+ and I^- ions, which increases the work function and makes it more difficult for the photoelectrons to pass through this layer into the gas.

After understanding the role of the parasitic layer, besides the regular cleaning of the evaporation system mentioned above, additional measures were also implemented to minimize other sources of water:

1. During the photocurrent measurement, the right half of the vessel is kept at 60 °C while the left half, with the parasitic layer, is cooled down to room temperature.
2. During the time between two evaporations the door of the evaporation vessel is kept closed and the vessel itself is kept at 30 °C, i.e. the heating is never switched off completely.

This gives noticeable results. The fatigue effect is no longer observed and the reproducibility of the CsI quantum efficiency is increased (Braem, 2007).

2.2.2 The Structure of the CsI Layer

Studies show that the heat enhanced CsI film transits from a uniform structure to a columnar structure, where each grain is typically 100 nm on the side and 300 – 400 nm tall. To answer the question how this structure arises and how it is influenced by the film deposition conditions, it can be very instructive to refer to the "structural zones" model which is described in many books on thin films (see for example Ohring, 1992). In this model the film structure is mainly defined by the rate of arriving molecules (absorption rate) and by their transversal diffusion capabilities on the surface of the film. The diffusion rate depends on the ratio T_{sub}/T_{mat} (where T_{sub} is the substrate temperature, T_{mat} is the melting point of the film material) and the film structure is predicted to change as a function of this ratio.

At $T_{sub}/T_{mat} \leq 0.3$, when the surface diffusion is negligible. The CsI solidifies immediately it reaches the substrate. The molecules arriving from the evaporator cannot move far along the substrate before they are covered by the newly arriving ones which leads to a highly disordered porous columnar structure with small diameters ~10 nm (structure Z1 in Figure 12). In this case the substrate surface roughness has a significant influence on the final film structure, e.g. shadowing effects, voids etc.

At $0.3 \leq T_{sub}/T_{mat} \leq 0.5$ a structure like type Z2 in Figure 12 is formed. In this case the transversal surface diffusion becomes increasingly important and the deposited molecules have enough time to form larger micro islands which will gradually grow into hexagonal crystal columns with facetted tops.

Finally, at $T_{sub}/T_{mat} \geq 0.5$, when not only surface but also bulk diffusion becomes relevant during the film growth process, a Z3 type of structure is formed (see Figure 12). Re-crystallization happens during the evaporation leading to larger equiaxed grains, with grain sizes up to film thickness. The transition between the different structures usually is quite sharp (Ohring, 1992).

Figure 12. Schematic representation of the crustal structure of the CsI film formed during evaporation under vacuum as a function of T_{sub}/T_{mat}
From Ohring, 1992.

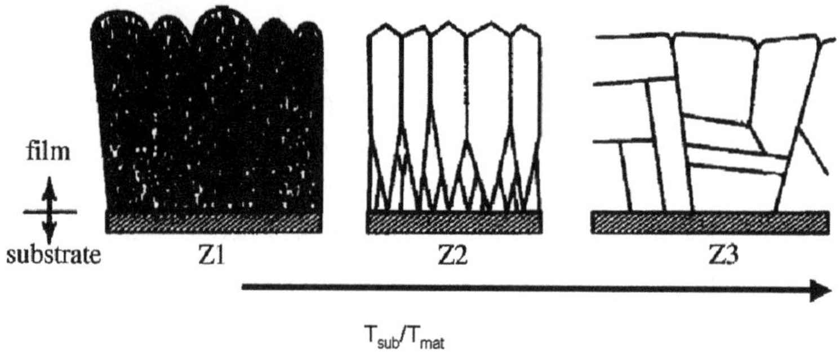

According to the morphological classification described above, surface diffusion is an important physical process in the formation of a CsI photocathode. In addition, one should seriously consider the effect of water on the CsI crystal structure. Due to the high impingement rate of water molecules during the coating process (~ 0.1–1 monolayer per sec in vacuum higher than 10^{-7} Torr) and taking into account the hygroscopic features of CsI, a significant amount of water might be incorporated into the film which grows with a rate of ~1 nm (~2.2 monolayer) per second.

At this evaporation rate the water conditions in the chamber will affect the CsI structure by influencing the surface diffusion rate during the film formation. Of course, during the evaporation not only CsI molecules hit the photocathode substrate, but also other molecules as well which also can affect the grain growth and their structure.

The differences in grain morphology, and the size of the grains, are likely to influence the quantum efficiency of the photocathode. The form of the grain defines the surface which is available for photoelectron extraction, i.e. a very rough surface of the CsI layer surface should increase the quantum efficiency.

Another important factor is that the evaporated CsI does not form an ideal crystalline structure. It contains a number of defects: cracks, vacancies, insertions, dislocations etc., as well as chemical impurities. Lattice defects change the electronic levels in the vicinity of the defects and they create additional energy levels within the band gap. This in turn changes the light absorption characteristics and may also create traps for holes and electrons (so called V and F centers). Together this will influence the quantum efficiency as a function of wavelength as well as the absolute value of the quantum efficiency (see for example Dekker, 1958). It is known that also the heat treatment changes the population and dynamics of the additional energy levels and thus causes changes in quantum efficiency.

2.2.3 A Semitransparent CsI Photocathode

Above we have described reflective CsI photocathodes, which are quite robust and offer high quantum efficiency in the VUV region. On the other hand, for some applications a semitransparent CsI photocathode may have advantages. For example, in the case of scintillating crystal readout it can be expected that more scintillation light will penetrate a semi-transparent CsI photocathode because the difference in refractive index is smaller between the crystal and the CsI, than between CsI and gas. Hence, less photons will be reflected off the surface of the CsI photocathode and not emit a photoelectron.

Semitransparent photocathodes has been studied and used in vacuum photodetectors since a long time (see for example Carruthers, 1969). It was early found that the thickness of any semitransparent photosensitive film, including the CsI, has a critical optimum value. If the thickness exceeds the escape depth of the photoelectrons, the sensitivity is reduced because the electrons produced by the light absorbed far away from the escape zone cannot be emitted into the vacuum. On the other hand, if the thickness is considerably smaller than the escape depth, the sensitivity is reduced since a fraction of the incident light is transmitted through the CsI rather than absorbed.

Note that in semiconductors the absorption coefficient and the initial energy of the photoelectrons are both wavelength dependent and they increase with photon energy. Therefore, photons with larger energy produce photoelectrons which have a greater escape length. This means that the quantum efficiency of the semitransparent photocathode can be optimized by finding the optimum thickness for the given spectral region in which the maximum sensitivity is required.

Probably the first attempt to combine a high gas gain photodetector with a semitransparent CsI photocathode was done by Charpak in 1993 (Charpak, 1993). The aim was to verify if such a thin CsI layer remained stable after a short exposure to air and under avalanche ion bombardment. The quantum efficiency measured for this semi-transparent photocathode is presented in Figure 13 and compared to the quantum efficiency of a reflective CsI photocathode. It can be concluded from this figure that the

Figure 13. The quantum efficiency of a reflective and a semi-transparent CsI photocathode (50 nm thick); none of the cathodes are exposed to air.
From Charpak, 1993.

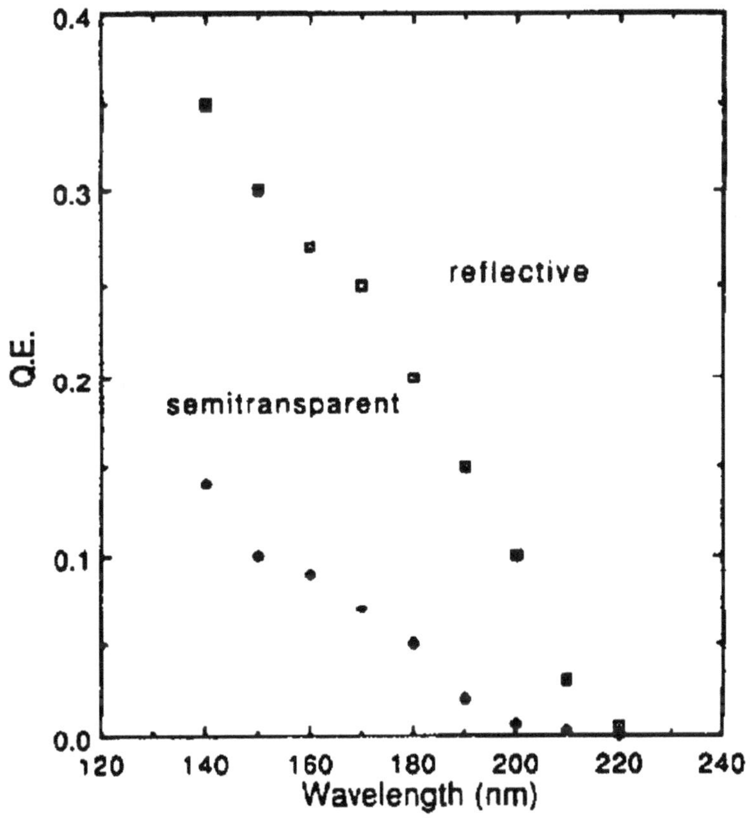

Figure 14. The quantum efficiency of a semi-transparent CsI photocathode as a function of its thickness for several wavelengths
From Lu, 1994.

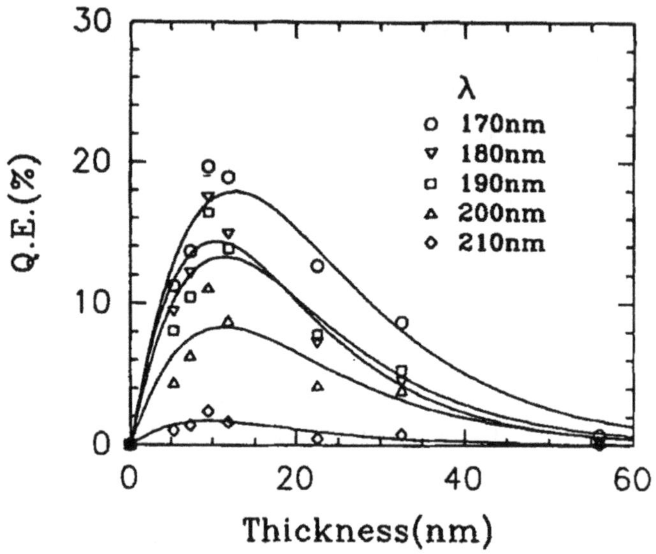

quantum efficiency of the semi-transparent CsI photocathode is only half of the reflective one. The preliminary conclusion from these tests was that although the reflective CsI photocathode has higher quantum efficiency and is simpler to fabricate because its thickness is not critical, the semi-transparent CsI photocathode can also be used in avalanche gaseous detectors.

Later, semitransparent CsI photocathodes has been systematically studied. For example, measurements were done aiming to find an optimum thickness of the semi-transparent CsI photocathode as shown in Figure 14 (Lu, 1994). It was observed that 10-15 nm is the optimal thickness of a semi-transparent CsI photocathode. The quantum efficiency is then about 80% of a reflective CsI photocathode.

Akkerman calculated the escape depths of photoelectrons for several alkali halides as shown in Figure 15) (Akkerman, 1994). Comparing Figure 15 with Figure 14, it can be concluded that the measurements and the calculation are in a reasonably good agreement. The large electron escape length of CsI, of the order of 16 nm for 1 eV electrons, is due to an efficient electron transport mechanism dominated by electron-phonon interactions, characterized by very small energy losses between collisions.

Note that at thicknesses of only a few tens of nm, the surface tension of CsI will pull the CsI together in islands. The islands have high quantum efficiency, but the areas between the islands have no quantum efficiency. Therefore, in some cases the optimum thickness can actually be larger than 10-20 nm to reduce the effect of island formation. The island formation also depends on the surface structure of the substrate.

2.2.4 CsI Quantum Efficiency Enhancement by Contact with TMAE

An absorbed layer of TMAE increases the quantum efficiency not only of metal cathodes, but also of CsI photocathodes (Séguinot, 1990). Figure 16 shows the results obtained with a 500 nm thick CsI photocathode covered with various TMAE layers: solid, liquid, and absorbed. They were formed by flowing methane, mixed with 0.04 Torr of TMAE through a gas chamber housing the plate with the CsI

Figure 15. Calculated escape probability for electrons with an initial kinetic energy of 1 eV as a function of their creation depth in CsI, NaCl and KCl; the calculations are based on a microscopic model of electron transport in alkali halides.
From Akkerman, 1994.

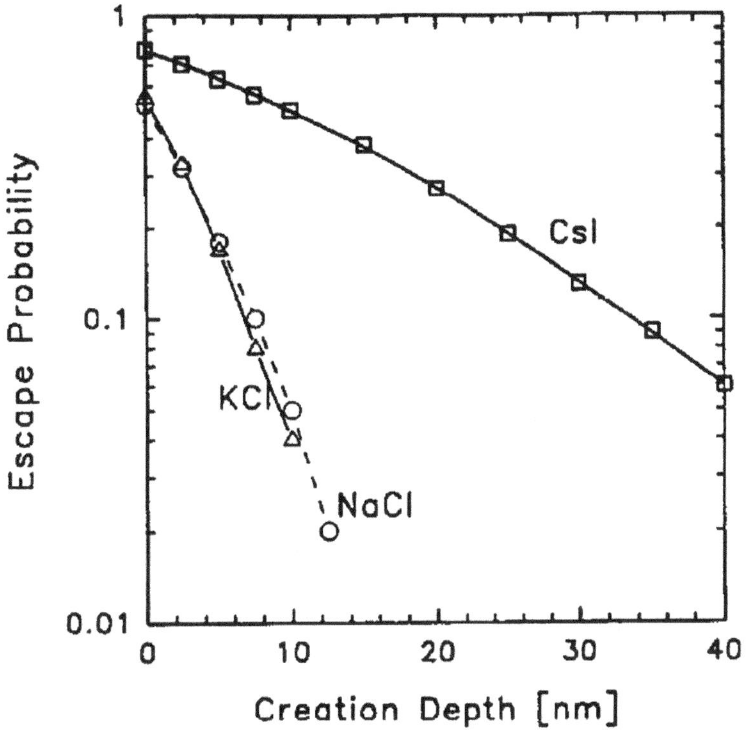

photocathode. The temperature of the plate was 248, 274, and 303 K, to form solid, liquid and absorbed TMAE, respectively. In the case of the solid and liquid layers, the quantum efficiency was lower than for a pure CsI photocathode. However, for the absorbed layer it was significantly higher, even higher than the quantum efficiency of gaseous TMAE. Even more remarkable is the fact that the quantum efficiency of the absorbed TMAE layer stayed unchanged when the TMAE partial pressure was reduced to zero after some time, as shown by the curve labeled 303 (0), and it remained stable even after the CsI photocathode had been flushed for three weeks with pure methane. This photocathode was later exposed to air by opening the gas chamber for one hour. Its quantum efficiency returned to the previous level after several hours of vigorous methane gas-flow (80 1/h).

These interesting results were fully reproduced by Anderson as shown in Figure 17.

Note that the high values of quantum efficiency were obtained after flushing the CsI photocathodes with methane while maintaining a temperature of ~40°C (Séguinot, 1992). Both factors are now believed to be crucial for obtaining a good result.

We have already discussed which effect an absorbed layer of the carrier gases on the photocathode has on the quantum efficiency. A CsI photocathode operating in an isobutane atmosphere will always have lower quantum efficiency than when operated in methane or ethane. This is attributed to the formation of a liquid layer of isobutane on the CsI surface making it difficult for some photoelectrons (for example those having low kinetic energy) to penetrate through this layer. Heating the CsI photocathode

Figure 16. The measured quantum efficiency versus the photon wavelength for a 500 nm thick CsI photocathode onto which a solid, a liquid, or an absorbed film of TMAE has been deposited; the labelling gives the photocathode temperature (K), followed by the TMAE partial pressure (Torr) in brackets. The solid line without data points is the TMAE gas-phase quantum efficiency and the dashed line is the pure CsI efficiency.
From Séguinot, 1990.

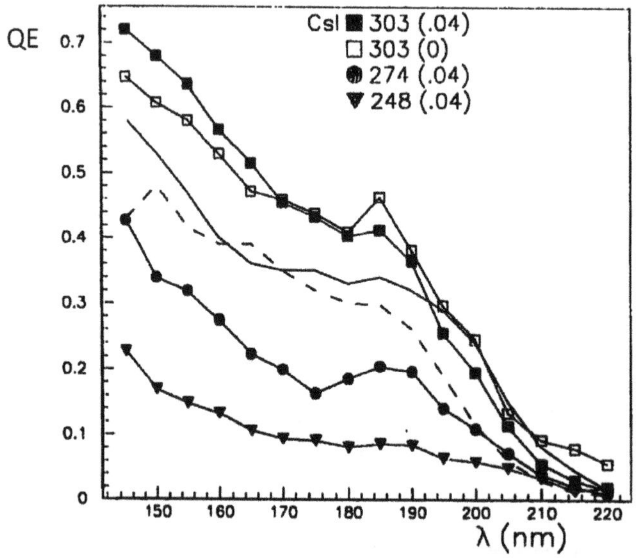

Figure 17. The quantum efficiency as a function of wavelength for CsI and CsI-TMAE photocathodes at 25°C in isobutane and methane; the quantum efficiency of pure CsI at 65°C in isobutane and for optically dense TMAE gas is presented as well.
From Anderson, 1992.

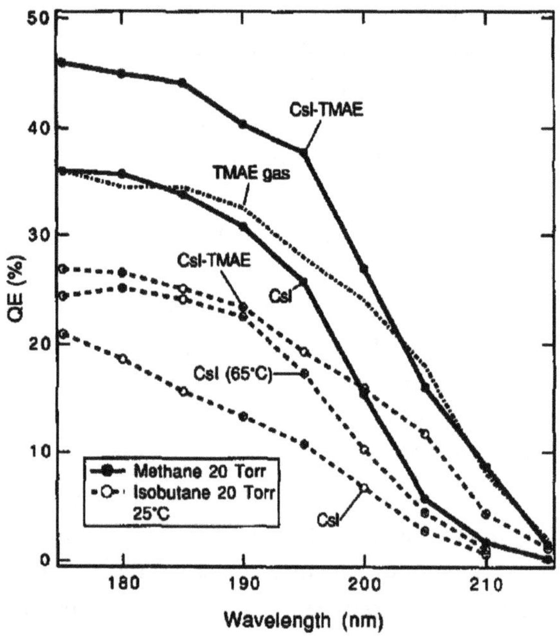

in isobutane to 65°C increases the quantum efficiency at longer wavelengths by about 50%, as shown in Figure 17 (Anderson, 1992). This is probably due to thinning of the absorbed layer by heating. The effect is most pronounced at longer wavelengths where the energy of the created photoelectron is lowest. The quantum efficiency enhancement by an absorbed layer of TMAE is also strongest in this region, since it reduces the work function of the surface layer.

A CsI photocathode with an absorbed TMAE layer has several advantages in RICH detectors as well as in other applications. It has a high quantum efficiency, it is fast (very small time jitter in creation of primary photoelectrons) and it is solar blind, which is attractive in outdoor applications such as flame detectors.

2.3 Operation of Gaseous Detectors with CsI Photocathodes

2.3.1 Operation of a Parallel Plate Avalanche Chamber with a CsI Photocathode

The first operation of a CsI photocathode combined with a gaseous detector was done with a parallel plate avalanche chamber (Charpak, 1989; Charpak, 1991).

There are several phenomena which make the operation of gaseous detectors with CsI photocathodes different from vacuum photodetectors. This includes photoelectron back diffusion, photon and ion feedback, the cathode excitation effect, CsI aging due to the ion bombardment etc.

Figure 18. Schematic view of the experimental setup used for studying the operation of a parallel plate avalanche chamber combined with a CsI photocathode
From Charpak, 1991.

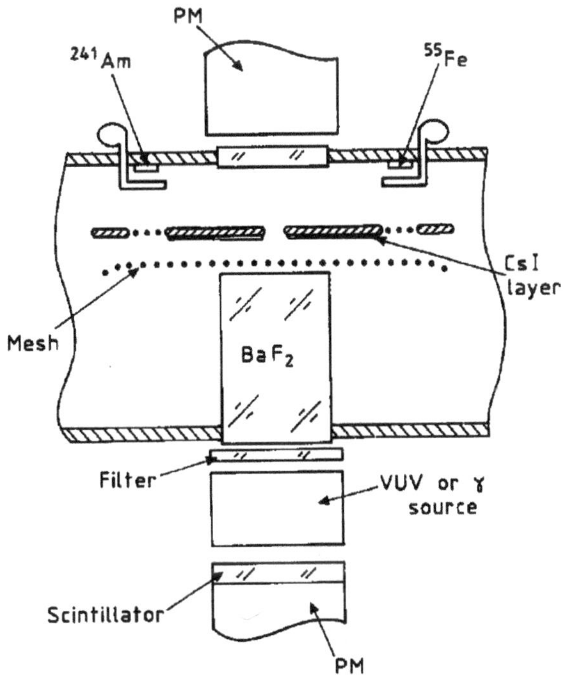

For electric fields at the cathode of less than 3 kV/cm the best photoelectron extraction efficiency from the CsI photocathode, close to 90%, is achieved in CH_4. To achieve high gas gains with a parallel plate avalanche chamber filled with CH_4 at atmospheric pressure is quite difficult since rather high voltages are required. Gas gains up to 10^6 are much easier to reach in gas mixtures of He and Ar with hydrocarbons.

Charpak's setup used is shown in Figure 18. It contains a gas chamber, a pulsed VUV source (or radioactive sources ^{241}Am or ^{55} Fe) and a photomultiplier (PM) to monitor the VUV source. The gas chamber contains a BaF_2 crystal and a parallel plate avalanche chamber with a 4 mm gap between electrodes. The anode mesh of the parallel plate avalanche chamber is 1 mm from the BaF_2 scintillator. The photocathode is made of from a stainless steel plate, 15 cm in diameter covered with a 5 nm thick CsI layer. The cathode has two holes, each covered by a mesh, through which the primary electrons created by the radioactive sources may be introduced into the avalanche gap of the parallel plate avalanche chamber. The space between the sources and the cathode plate acts as a conversion region when the radioactive source is activated by opening a lead shutter.

The VUV source is a compact pulsed hydrogen lamp. It produces periodical UV pulses with duration of 15 ns FWHM which create photoelectrons from the CsI photocathode. In most of measurements the VUV light is filtered by a narrow band filter with a transmission centered at 193 $\pm$10 nm. This filter simulates the fast scintillation emission of BaF_2 to which the CsI is most sensitive. Part of this light also reaches a PM through a small hole made in the photocathode surface and a CaF_2 window.

This setup, equipped with various primary electron sources, allows detailed studies of the parallel plate avalanche chamber operation: its gas gain, feedback, time resolution etc.

2.3.1.1 Feedback Investigation

Since the CsI photocathode has a quite high efficiency in the UV region one can expect some photon feedback problems. The parallel plate avalanche chamber signal as a function of the applied voltage is shown in Figure 19.

The secondary avalanches appearing at gas gains higher than 10^5 originate from photon feedback from the avalanche, reaching the cathode and emitting new photoelectrons. The relations between the primary charge, the initial avalanche and the first secondary avalanche are:

$$N_{av1} = A\, n_0, \tag{1}$$

$$N_{av2}/N_{av1} = A\, \gamma_{ph}, \tag{2}$$

where N_{av1} is the charge in the primary avalanche and N_{av2} is the charge in the first secondary avalanche (symbols A and γ_{ph} were defined in Chapter 3).

To estimate the feedback efficiency γ_{ph} as a function of the gas constituents the total charge in the avalanche was measured at a fixed feedback level $A\gamma_{ph}$ of 25%. Since the primary charge is known (~3×10^3 electrons), the gas gain can be calculated. The raw data showing the charge at 25% photon feedback for different gas mixtures are shown in Figure 20.

These data were taken with gas gains ranging from 2×10^5 to 4×10^6. As can be concluded, the feedback is essentially independent of the concentration of hydrocarbons at the partial pressures from 6 to 60 Torr.

No trace of ion feedback was observed up to the breakdown limit.

Figure 19. Typical signals from a parallel plate avalanche chamber combined with a CsI photocathode, recorded at various gas gains: a) top - UV signal from the pulsed lamp detected with a PM, bottom – avalanche signal from the parallel plate avalanche chamber at a gas gain of 5x10^4 delayed on the avalanche development time (~100 ns); b) pulses from the parallel plate avalanche chamber at a gas gain of 5x10^5, c) the same at gas gain of 10^6, d) discharges appearing at a gas gain of > 4x10^6
From Charpak, 1991.

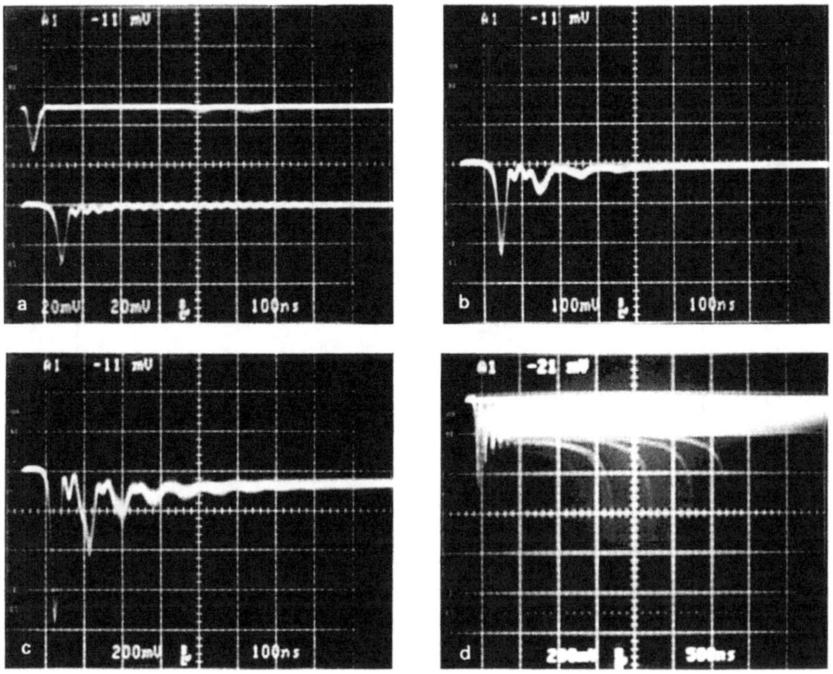

2.3.1.2 Avalanche Light Emission in Parallel Plate Avalanche Chamber

In some gas mixtures, for example He + C_2H_6, the photon feedback is only 25% at gas gains as high as ~10^6. This makes CsI-PPAC very attractive, since already at 2-3 times lower gas gains, single photoelectrons are detected with 100% probability and without photon feedback.

The origin of this feedback is well understood. Photons emitted from the avalanches reach the CsI photocathode and can create photoelectrons there which then initiate secondary avalanches. Mathematically this can be described by the formula, similar to 8 in chapter 3:

$$\gamma_{ph} \sim \int_{\lambda th}^{\infty} Q(\lambda)\, I_{av}(E,\lambda)\, d\lambda, \quad (3)$$

At a feedback level of 25%:

$$0.25 \sim \int_{\lambda th}^{\infty} Q(\lambda)\, I_{av}(E,\lambda)\, d\lambda, \quad (4)$$

In gas mixtures of noble gases with quenchers having an ionization potential lower than the noble gas and in weak electric fields (~few kV/cm) electrons mainly excite and ionize the quencher gas, and

Figure 20. The charge in the avalanche at which 25% feedback occurs as a function of the partial pressure of quencher gases (various hydrocarbons and hydrocarbons with EF additives); for readability an error bar is shown only for one point, the other errors are similar.
From Charpak, 1991.

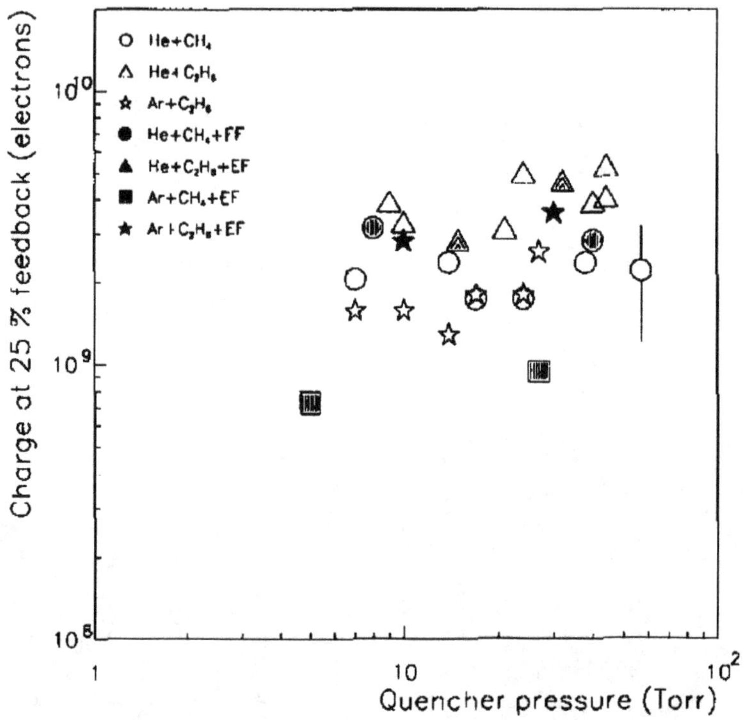

not the noble gas. Only the UV photons emitted from the quencher molecules, which has a wavelength overlapping with the CsI quantum efficiency, may efficiently create photoelectrons.

The VUV emission from the avalanches (in the range 120-170 nm) for the same mixtures, as indicated in Figure 20 was previously measured in work (Fonte, 1991), using a parallel plate avalanche chamber with a metallic cathode (not coated with a CsI layer). The UV light emission decreases rapidly, by several orders of magnitude, with increasing levels of hydrocarbons in the gas (6 – 60 Torr). A similar behavior can be expected for the photon feedback if the VUV light is responsible for it. However, such a behavior is not observed with a CsI photocathode. Since the quantum efficiency for UV light of the CsI photocathode is several orders of magnitude higher than for stainless meshes used in studies of (Fonte,1991), it can be concluded that in the case of the CsI-PPAC, photon emission in the avalanches at longer wavelengths (longer than 175 nm) is responsible for the photon feedback.

To verify this measurement, the light emission from electron avalanches in the wavelength interval 200-500 nm was performed with the help of the PM that was used for monitoring the VUV lamp emission in Figure 18. This is possible because the photons originating from the electron avalanches are delayed with respect to the emission signal from the lamp by the time it takes for the electrons to drift from the photocathode to the amplification volume in the parallel plate avalanche chamber, around 100 ns.

The results (the ratio of the light signal measured by the PM to the charge signal measured in the parallel plate avalanche chamber) are presented in Figure 21.

Figure 21. Light emission per electron in electron avalanches vs. the quencher partial pressure; error bars are shown for only one point.
From Charpak, 1991.

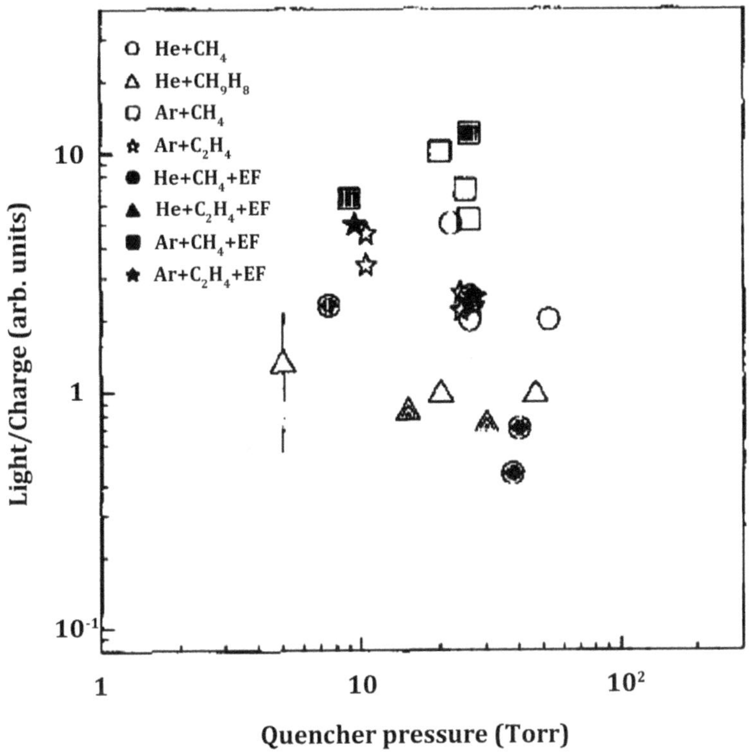

Since the measurements were done at different gas gains and the ratio of the light to the charge signal strongly depend on gas gain (see Peskov, 1998), the dispersion of the experimental points is large. However, the same qualitative features as observed for the photon feedback efficiency can be recognized: only a weak dependency on quencher concentration. This behavior, which contrasts with the behavior observed with VUV emission, is similar to what was observed earlier for the molecular emission in gases containing TEA, TMAE or H_2O vapours (Peskov, 1998).

This allows us to conclude that the avalanche photon emission in the wavelength region λ>200 nm is not a fragment emission, but a molecular emission (Fonte, 1991).

The fact that the molecular emission is much stronger than the fragment emission is natural since generally the excitation thresholds are lower in a molecule than in individual atoms, and the cross sections are higher for the molecular emission.

2.3.1.3 Energy and Time Resolutions

Other important characteristics of any detector for photons and charged particles are its energy and time resolutions. In the case of the CsI-PPAC both these characteristics were carefully measured and evaluated.

Alkhazov had shown that the energy resolution of avalanche gaseous detectors, if one neglects the contribution from electronic noise, can be described by (Alkhazov, 1970):

$$(\delta S_A/S_A)^2 = \{F + (\delta A/A)^2\}/n_0, \quad (5)$$

where S_A is the signal amplitude. F is the Fano factor, a constant related to the statistics when creating primary electrons in the ionization processes (for most of gases $F \leq 0.2$) (Fano, 1947).

Often the energy resolution is expressed in the unit full width at half maximum, FWHM, of the amplitude distribution, where FWHM = 2.35 δS_A. Typically the energy resolution of gaseous detectors is in the range of FWHM = 18-22% for 6 keV X-rays, creating around 200 primary electrons for each X-ray.

In the case of the CsI-PPAC, the energy resolution was extracted from measurements with the pulsed UV lamp. To correct for fluctuations in the UV lamp emission, the lamp signal (both the current and the light emission) were continuously monitored. A signal proportional to the lamp discharge current was fed to a standard CAMAC ADC along with the fast signal of the CsI-PPAC (initial avalanche). The signals were divided with each other after the acquisition. The results are presented in Figure 22.

The inserts on the left side of the figure present the raw data-the pulse-height spectra of the parallel plate avalanche chamber and the lamp. The main graph shows the normalized CsI-PPAC energy resolution after correcting the signal with the amplitude of the lamp signal. The typical energy resolution is 8-10% FWHM, depending on the gas mixture and the chamber gain (which in these measurements was below 2×10^5).

Figure 22. The results of the CsI-PPAC energy resolution measurements obtained with a pulsed VUV lamp; the inserts show the raw chamber and lamp signal distributions, whereas the main plot shows the normalized chamber distribution corrected for lamp fluctuations.
From Charpak, 1991.

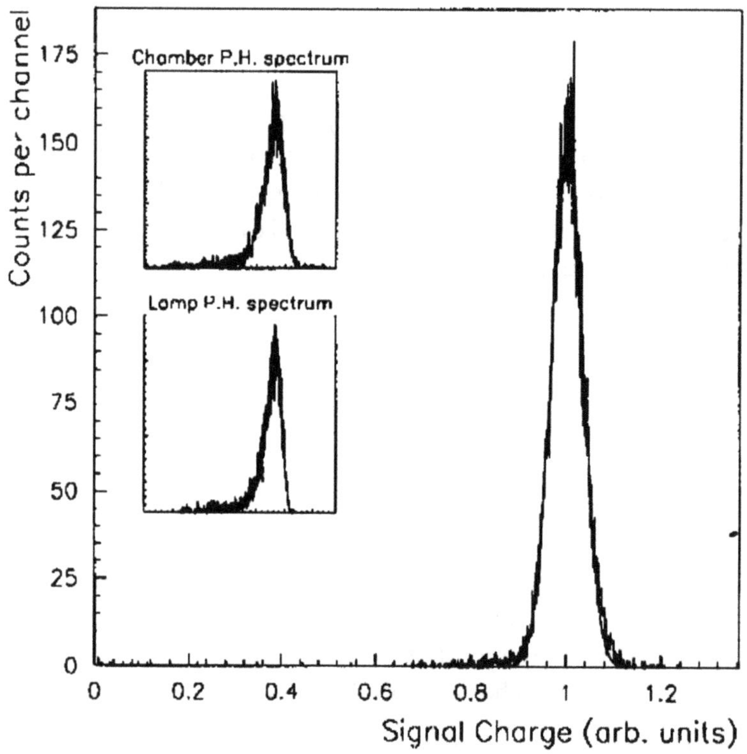

Of these 8%, the statistical fluctuations of the primary charge (~3x10^3 electrons) created by the light from the VUV lamp of is around 5% FWHM (FWHM = {(√3000)3000)}*2.35) for a Gaussian distribution). There are also gain fluctuations which also spoil the energy resolution by a factor of two, especially at high gains. Using formula 5 one can recalculate the energy resolution to 200 primary electrons (which as was mentioned above are often used as a reference for the detector energy resolution evaluation as emitted from 6 keV X-rays). The resulting energy resolution of 20% FWHM is close to what is typically measured with gaseous detectors, so one can conclude that the CsI photocathode does not affect the energy resolution of the parallel plate avalanche chamber at gas gains <10^5. At higher gas gains, depending on how the signals from the parallel plate avalanche chamber are treated (for example if the integration time of the amplifier is long enough to include signals from secondary avalanches) the energy resolution could be also affected by the photon feedback.

As was already mentioned few times earlier, Gaseous detectors filled with photosensitive vapours, in particular MWPCs, do not exhibit good time resolution. This is because the external VUV photons, entering the detector volume through the window, create photoelectrons along their pass. These photoelectron therefore arrive to the anode wires at different times (spreading from zero to a maximum drift time defined by the drift distance, the gas and the strength of the electric field in the drift volume). This creates a time jitter in the order of μs.

In contrast, in a solid photocathodes all photoelectrons are liberated from the photocathode, i.e. at the same distance from the amplification structure, and therefore the jitter in their arriving time is small. Hence, one can expect that such detectors will have considerably higher time resolutions. This is also observed experimentally. Among the various designs of gaseous detectors with solid photocathodes, a special place belongs to the parallel plate avalanche chamber. In this geometry, having a well-defined distance between the photocathode and the meshes of the parallel plate structure the time jitter in the creation of the primary photoelectron is minimum (in the ideal case it is close to zero ns).

The response time distribution of a CsI-PPAC was measured by the time difference between the leading edge of the VUV lamp monitor signal, detected by fixed-threshold discrimination, and the peak of the initial avalanche signal from the chamber (Charpak, 1991). The peak detection was implemented by differentiating the chamber signal with a 5 ns time constant and determining the zero-crossing with a fixed-threshold discriminator. A typical time distribution is presented in Figure 23.

The time resolution is ~500 ps FWHM or better for all gas mixtures measured. Anderson measured a similar time resolution even with single photoelectrons using a CsI-PPAC operated at a gas gain of 10^7 and at low gas pressures (Anderson, 1992).

Later in this chapter the time resolutions of other amplification structures combined with CsI photocathodes will be presented and discussed as well.

2.3.1.4 Breakdown

As in any gaseous detector, at high enough gas gains discharges appear. The origin of these breakdowns is a feedback loop of secondary electrons from the avalanche generating new photoelectrons in a non-quenching way, which becomes critical at some gain A_{max}:

$$A_{max}\,\gamma_{ph} = 1, \qquad (6)$$

The gain at which the breakdown occurs is related to the photon feedback efficiency.

Figure 23. Time distribution of a CsI-PPAC; the time resolution (FWHM) is 500 ps.
From Charpak, 1991.

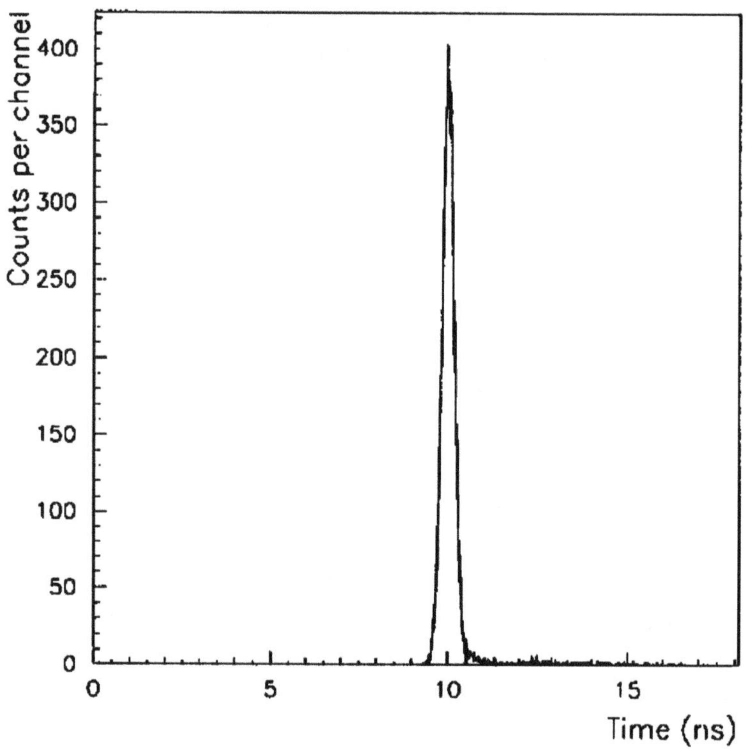

2.3.1.5 The Cathode Excitation Effect

The cathode excitation effect was observed for a long time (Karabadzhak, 1983 Nappi, 2013).

Under the intense ion bombardment, metallic cathodes or cathodes covered with a photosensitive layer exhibit a temporary increase its quantum efficiency to visible and ultraviolet light, and consequently an increase of γ_{ph} and γ_+. After the ion bombardment is stopped the quantum efficiency slowly (in a few minutes) returns to its original value. The explanation, as in all effects mentioned above, is a formation of a layer of positive ion on the on the cathode surface. These ions create a strong local electric field which reduces the work function and also may cause electrons jets to be emitted, also known as explosive field emission (see Fonte, 1999).

This phenomenon is directly related to the Malter effect (Malter, 1936). Strictly speaking the Malter effect is usually used to explain after-emission of single electrons from thin dielectric films bombarded by positive ions. Cathode excitation is beyond the "classical" Malter effect; the electron emission from the cathode is rather in the form of electron jets rather than single electrons.

As a consequence, after intense ion bombardment after-pulses may appear. The effect slowly decays with time. After a possible breakdown one cannot immediately apply the same voltage as it was before the breakdown because the breakdown produces a large ion bombardment of the cathode. If one tries to turn up the high voltage after a breakdown, new breakdowns appear at considerably lower voltages due to cathode excitation. One has to wait for some minutes for the cathode to discharge before the nominal voltage can be applied. This is sometimes quite inconvenient as the detector is not operable for several minutes.

2.3.1.6 Aging

Degradation of the photocathode performance with time is usually named with the general term "aging". Among other effects, aging includes quantum efficiency degradation with time under intense UV radiation, or a combination of UV radiation and positive ions. Aging of gaseous detectors combined with CsI photocathode has been investigated in many studies. However, the data obtained by different groups often contradict each other and are often not reproducible. It turns out that CsI photocathode degradation under irradiation of light and ions depends on many factors, which are sometimes difficult to control. It depends in the structure of the CsI layer, its history (was it exposed to air or not, was it heated or not etc.), impurities in the system, intensity of the VUV photon flux, gas gain etc.

For these reasons, despite of long history of studies, the aging effects on CsI are not still fully understood as it is usually a combination of several effects. We will try to summarize the main results obtained with a CsI-PPAC. Later, aging studies for other types of gaseous detectors will be presented as well.

To sort out various effects contributing to the CsI aging, the tests with a CsI-PPAC were performed at different conditions. For example, one system was used to study photocathode aging without exposure to air. This was achieved by special manipulators, which allowed a transfer of the photocathodes from the evaporation system to the test chamber without exposing the photocathode to any air (Peskov, 1995). The test chamber was then flushed with gas at atmospheric pressure. Such a procedure shows very small aging effects, as shown in Figure 24.

Systematic studies of CsI photocathode aging in a parallel plate avalanche chamber operating at low pressure were performed by Anderson's group. Low pressure was chosen because it allows operating

Figure 24. Relative drop of the photocurrent I/I_0 (or quantum efficiency) as a function of the collected charge on the cathode, for a CsI photocathode that was aged in vacuum (curves 1,2) or in an Ar + 10% CH_4 mixture at 1 atmosphere (3,4). In cases1 and 3, the CsI photocathodes were previously exposed to air. Curve 1 had no gas amplification, while curve 3 was obtained at a gas gain of 10. In cases 2 and 4, the photocathode was not exposed to air. Curve 2 was obtained without gas amplification and curve 4 was obtained at a gas gain of 100.

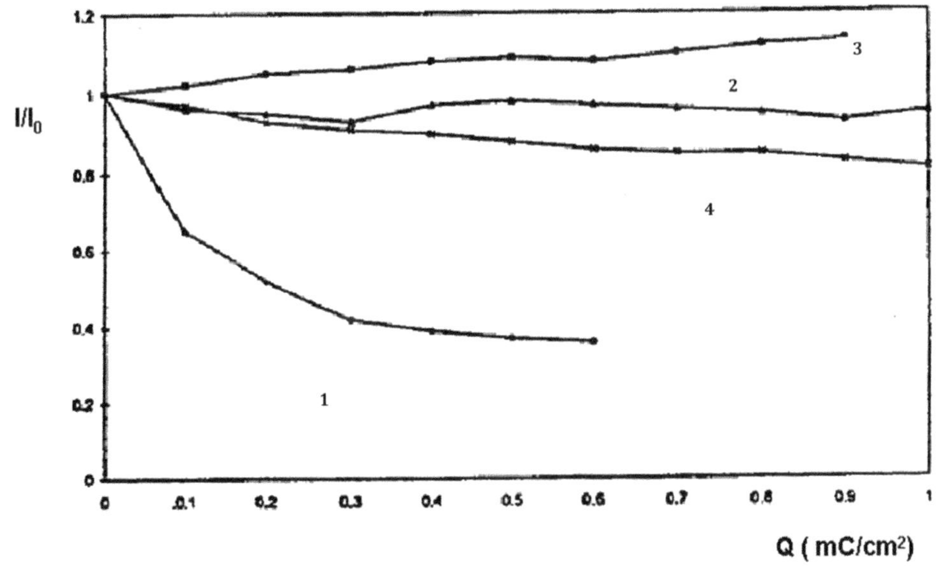

Figure 25. The photocurrent as a function of time for a CsI photocathode (previously exposed to air), measured in vacuum and illuminated with UV light
From Anderson, 1991.

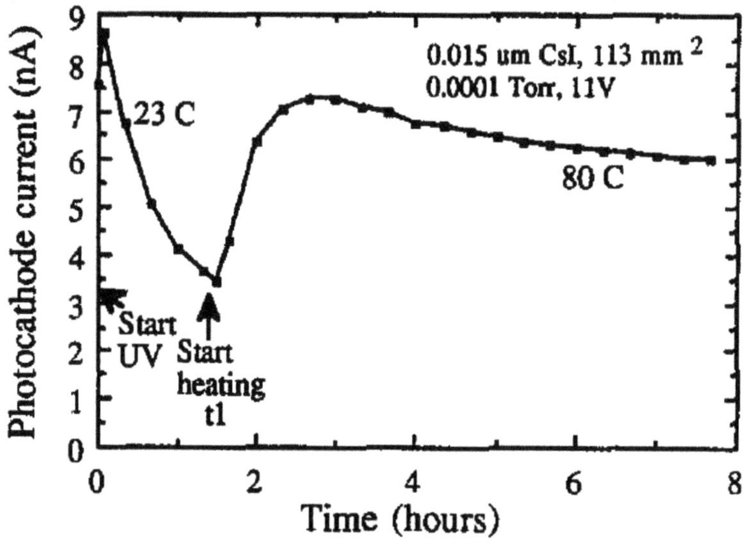

the CsI-PPAC in various molecular gases, whereas at atmospheric pressure, due to the practical limits on the applied high voltages, only mixtures of noble gases with quenchers can be used. The results obtained by this group reveal many interesting details. As an example, Figure 25 shows the quantum efficiency degradation of the CsI photocathode measured in vacuum in a continuously pumped parallel plate avalanche chamber under continuous illumination of UV-light. As can be seen, initially, during the first 10 minutes, the quantum efficiency increases (accumulated charge of about 6 $\mu C/cm^2$). Then it starts to decrease. However, if the photocathode is heated, the quantum efficiency is quickly restored, then the rate of the degradation is considerably reduced and eventually it flattens out at an accumulated charge of 0.2 mC/cm^2.

To illustrate the difficulties of these studies, the data presented in Figure 25 (Anderson, 1991) are in contradiction to the results obtained in later studies (Peskov, 1995), curve 1 in Figure 24. This can be due to that Anderson used a thinner CsI photocathode (0.1 μm thick) while Peskov used a 0.5 μm thick layer of CsI. Of course, the gas chamber used in the two studies may also have had different levels of cleanness, which certainly affects the aging rate.

Anderson also found great improvement in the aging characteristics of the CsI photocathode (previously exposed to air) when it was operated in the presence of a gas, as shown in Figure 26.

These important results (presented in figures 24 and 26) were also verified for a ultra-clean and sealed detector containing CsI photocathodes not exposed to air (Francke, 2003). The photocurrent was measured with Hamamatsu photodiodes R1187, used in vacuum and in argon at atmospheric pressure using a technique described in (Charpak, 1992). The results are shown in Figure 27. The same tendency was observed as by Anderson. The aging rate in the gas atmosphere is considerably less than in vacuum. The accumulated charge for the gas filled photodiode was 30 mC/cm^2.

Figure 26. Normalized photocurrent as a function of the collected charge for CsI photocathodes aged in vacuum and in five different gases at 20 Torr, and illuminated with UV light. No gas amplification was used (gas gain = 1).
Anderson, 1992.

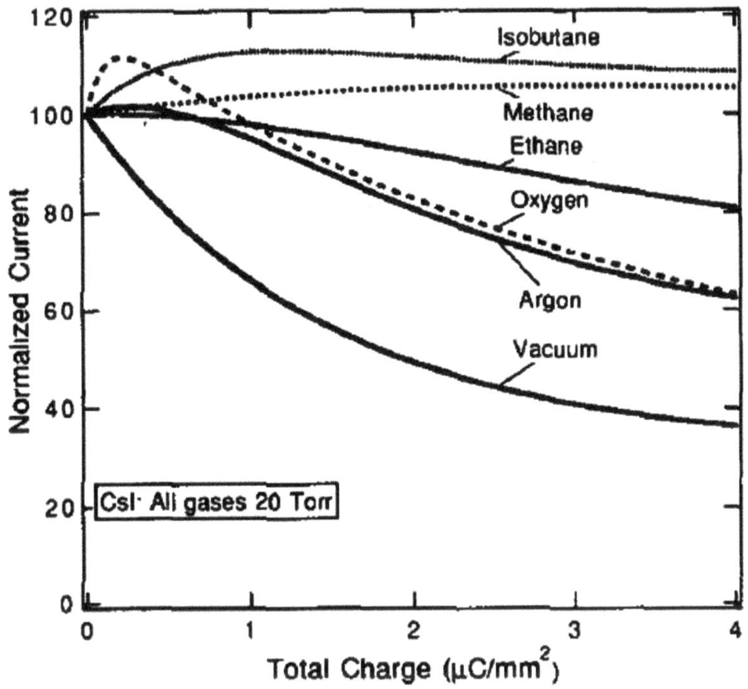

For a CsI photocathode exposed to air, only a weak correlation between aging rate and gas gain was observed, at least at low accumulated charges < 4 μC/mm², as shown in Figure 28. The results obtained with a CsI + TMAE photocathode, also presented in this figure, will be commented later in this paragraph.

The results show that at a gas gain of 10^5 at least $2x10^{10}$ photons/cm² can be detected without major problems with aging which is sufficient for most of applications.

The data on aging of CsI + TMAE photocathodes are available only for a low pressure parallel plate avalanche chamber (see Figures 28 and 29). The data that exist indicate that the aging of a CsI + TMAE photocathode is faster than measured with a pure CsI photocathode.

Due to the lack of experimental data and the complexity of the interaction between CsI and TMAE and its dissociation products (as well as other possible impurities in the chamber) it is too early to make any attempts to interpret the results.

2.3.2 Resistive Plate Chambers Combined with CsI Photocathodes

The main requirements on a CsI-PPAC are sufficient gas gains to detect single photoelectrons, high quantum efficiency, high position resolution, and high reliability. The last is absolutely necessary for many applications. However, at high gains discharges can be unavoidable. They can be caused by heavily ionizing particles, air showers, the cathode excitation effect etc. Hence, most detectors must be spark protected.

Figure 27. The results of aging studies for CsI photocathodes (photocurrent vs. time) using Hamamatsu photodiodes R1187 operated in vacuum (open diamonds), and operated in argon (filled squares); note that there were breaks in the measurements between days 5 and 10, 15 and 20, 22 and 25, 26 and 27. One can see that during the breaks the quantum efficiency was partially restored.
From Francke, 2003.

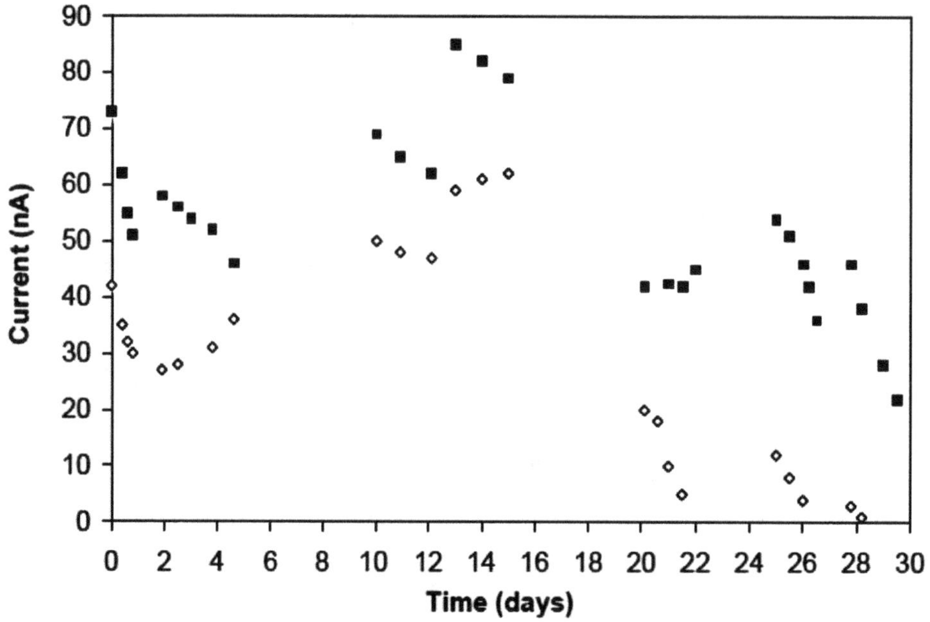

Figure 28. The normalized photocurrent as a function of the collected charge for Csl and CsI-TMAE photocathodes aged in 20 Torr of isobutane at a variety of charge gains

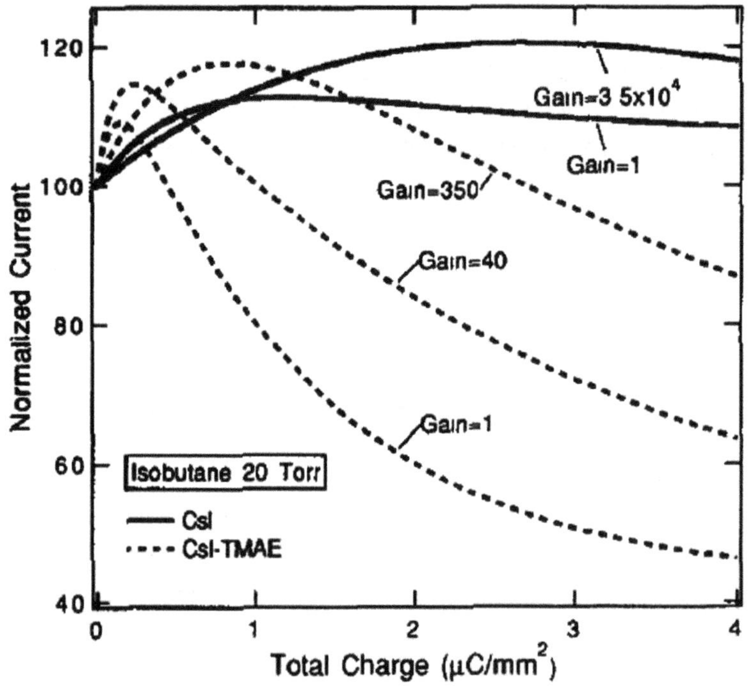

Figure 29. The photocurrent of a CsI-TMAE photocathode as a function of time; The gas gain was 260 in methane gas at a pressure of 20 Torr. The UV light illumination was interrupted at time t_1 and reestablished for brief moments at times t_2-t_5.
From Anderson, 1991.

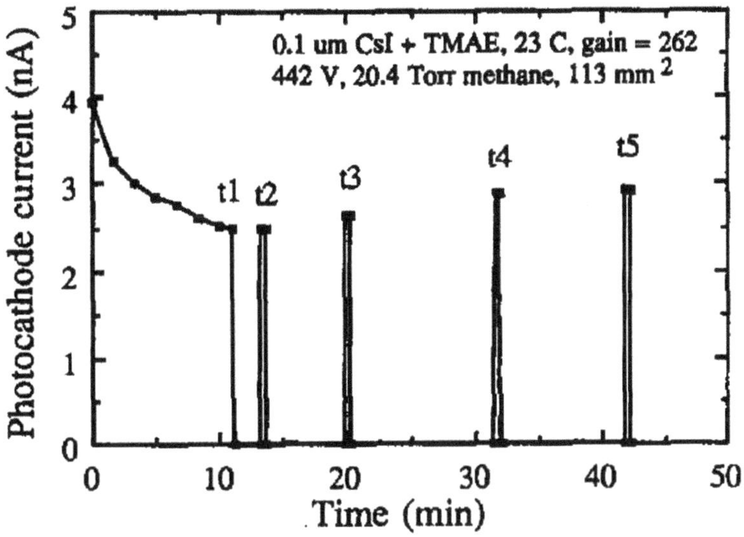

In Chapter 2 we described parallel plate avalanche chambers with resistive electrodes, called resistive plate chambers. They are often interesting alternatives to parallel plate avalanche chambers at low counting rate applications.

Several designs of resistive plate detectors with semitransparent CsI photocathodes has been developed, tested and used in various applications. One of them is shown in Figure 30.

This detector has a cathode made of a metallized MgF_2 window coated with a 20 nm thick CsI layer and the anode made of Pestov glass with metallic strips at 30 μm pitch, 1x1 mm^2 pads. The light beam is introduced perpendicular to the electrode surfaces. Such a CsI resistive plate chamber was systematically studied (Fonte, 2005). As expected, the quantum efficiency of this detector is around half of that of a reflective CsI photocathode due to that not all photons do convert in the photocathode.

In principle, a resistive plate chamber is inherently spark-protected. It can operate at high gas gains, up to the appearance of 'streamer' discharges, without any risk of damage. However, the presence of a photosensitive photocathode will introduce photon feedback above a certain gas gain level. Photon feedback spoils the position resolution for gains above 10^5 in the He + 0.8%CH_4 + EF mixture. In a gas mixture of $C_2H_2F_4$ +10% SF_6 + 5% isobutane the photon feedback effect is significantly weaker. It can reach a counting plateau, which is important for single photon counting, at gains above 10^5, as shown in Figure 31). Unfortunately, this gas mixture causes a small degradation of the CsI quantum efficiency with time, calling for further investigations.

One should note that the addition of EF to the He + 0.8% CH_4 mixture is crucial to suppress photon feedback and to achieve a good performance. This is likely connected to the EF low ionization potential, efficiently quenching any light emission from CH_4 in the avalanche multiplication, and EF has a weak own light emission.

Figure 30. Schematic drawing of a resistive plate chamber with a semitransparent CsI photocathode
From Francke, 2004.

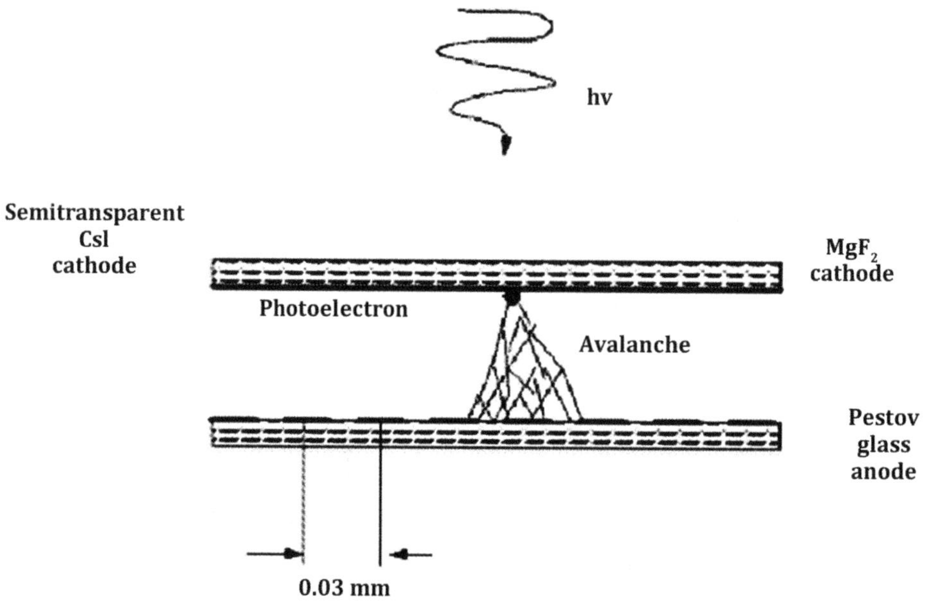

Figure 31. The counting rate vs. gas gain in a $C_2H_2F_4$ +10% SF_6 + 5% isobutane gas mixture; the data indicate a hint of a counting plateau above gas gains of 10^5.

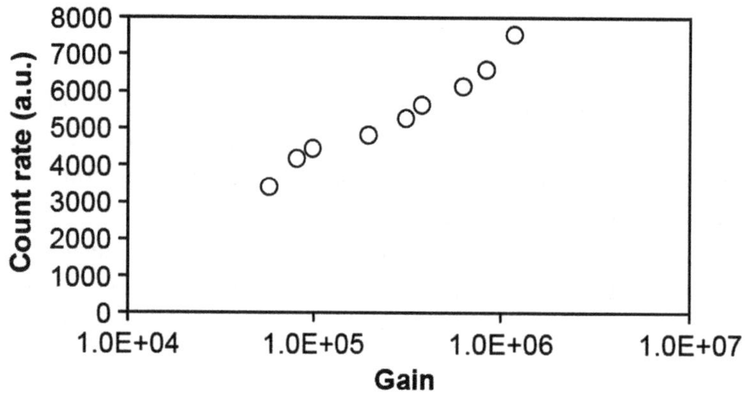

The signal amplitude as a function of the counting rate at gas gains of 1.5 x10^6 and 5 x10^6 in the $C_2H_2F_4$ +10%SF_6 + 5% isobutane mixture is shown in Figure 32. The presented data suggest that counting rates up to a few hundred Hz/cm^2 may be sustained without prohibitive amplitude loss.

The corresponding rate of spurious pulses is rather low, close to 10 Hz from the whole active area. Images obtained with pixel readout show that such noise pulses originated mainly from fixed spots on the cathode surface and is probably due to explosive electron emission from the photocathode (Fonte, 2005a). After strong irradiation of the resistive plate chamber the noise pulse rate could increase more than 10 times, but returns to the normal rate in 0.1–0.5 s after the irradiation stops (as the results of discharging the cathode excitation effect).

Figure 32. The signal charge vs. counting rate for gas gains of 5 x 10^6 (upper curve) and 1.5 x10^6 (lower curve) for a resistive plate chamber with a transparent CsI photocathode in a $C_2H_2F_4$ +10%SF_6 + 5% isobutane mixture.

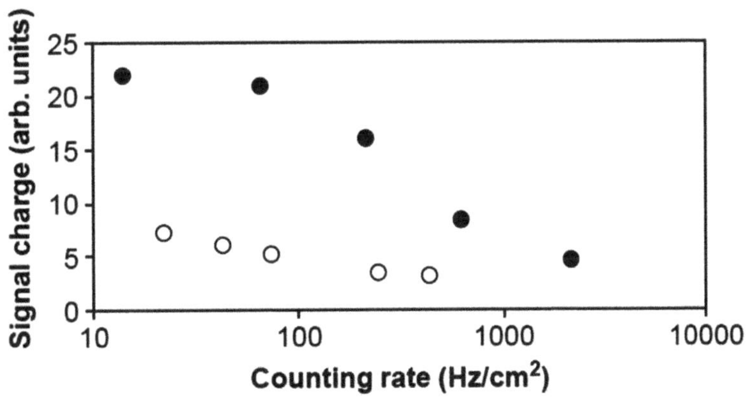

Figure 33. The counting rate vs. strip number measured with a collimated UV beam of 30 µm diameter for different gains and gas mixtures: He + 0.8% CH_4 + EF and "triple mix"

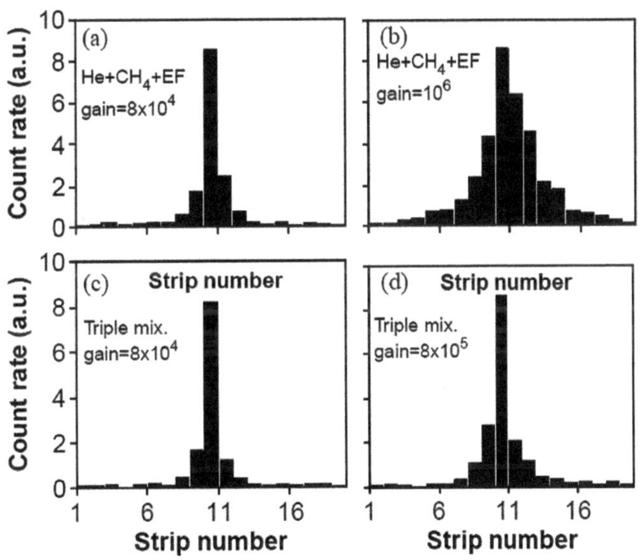

The position resolution of the chamber was measured by illuminating a single anode strip with either a cylindrical 30 µm diameter VUV beam or a rectangular 5 x 0.03 mm^2 beam. The anode strips are connected individually to an ASIC readout chip that for each event localizes the strip nearest to the maximum of the induced charge profile. The charge is distributed over several strips, so by discriminating all strips but the one closest to the maximum significantly increases the position resolution. The number of such hits was then counted for each strip (digital readout mode). This readout method has been successfully used for X-ray imaging resistive plate chambers (Francke, 2001).

In the He + 0.8% CH_4 + EF gas mixture and at gas gains close to 10^5 the position resolution is comparable to the strip pitch of 30 µm (Figure 33a). For larger gas gains the resolution is spoiled by photon feedback (Figure 33b). The $C_2H_2F_4$ +10% SF_6 + 5% isobutane gas mixture is essentially free from this photon feedback effect (Figures 33c and d), but suffers from degradation of the CsI quantum efficiency with time.

$CH_2H_2F_4$ + 10% SF_6 + 5% isobutane; the position resolution is comparable to the strip pitch of 30 μm, except for the He + 0.8% CH_4 + EF mixture at gains above 10^5 (b), where the resolution is spoiled by photon feedback.

Such a CsI resistive plate chamber has been used in several applications, e.g. for the UV visualization in day light conditions, hyperspectroscopy, etc.

2.3.3 A MWPCs with CsI Photocathode

Nowadays, among all photosensitive gaseous detectors the most attention is given to MWPCs combined with CsI photocathodes. This is because their good ability to detect single photoelectrons by operating at high enough gas gains, close to 10^5. For such high gas gains, a parallel plate avalanche chamber operated at atmospheric pressure requires a quite high voltage to be applied across the avalanche gap. In high energy physics environment, where heavily ionizing particles are abundantly present, the parallel plate avalanche chamber may suffer from sparking at these voltages. The latest studies of this phenomenon were done by Fonte (Fonte, 1991a). The breakdown limit (as opposed to the photon feedback) is found to be quite independent of the choice of gas. It depends mainly on the total charge in the avalanche, the so called Raether limit (Raether, 1964). Strictly speaking this unstable operation is observed with an ordinary parallel plate avalanche chamber (no CsI photocathode) combined with a drift space. In the presence of a CsI photocathode, the drift region is not necessary which reduces the spark probability. Moreover, in the presence of a CsI photocathode the origin of the breakdown is different. It now happens via a photon-photoelectron feedback loop, resulting in that this detector is more spark resistant than ordinary parallel plate avalanche chambers.

In practice the operational voltage is an important parameter which is desirable to keep as low as possible. This is why most applications that are currently in operation uses a MWPC combined with a CsI photocathode.

2.3.3.1 Main Designs

The study of CsI-MWPC based photomultipliers was initiated by Dangendorf et al. (Dangendorf, 1990). Their photodetector consisted of a MWPC preceded by a parallel plate avalanche chamber made of mesh electrodes to form a pre- amplification gap and operating without a transfer region, as shown in Figure 34. The UV-light enters the detector through a CaF_2 window. It traverses the MWPC and interacts with the CsI photocathode placed at the other end of the chamber. The photoelectrons emitted are first pre-amplified in a parallel plate avalanche chamber before they enter into the MWPC. Only the anode wires are read out in this detector.

Later it was realized that in CH_4 at atmospheric pressure, a high enough gas gain is possible to achieve even with a single step MWPC combined with a CsI photocathode eliminating the need for the parallel plate avalanche chamber pre-amplification stage. In analogy with a CsI-PPAC we will call it a CsI-MWPC. An optimized design of such a detector for RICH applications is shown in Figure 35.

It consists of a Cherenkov radiator (for example liquid C_6F_{14}) and a CsI-MWPC placed at some distance from it to expand the cones of Cherenkov light into Cherenkov rings of desirable size (see the

Figure 34. A schematic view of a double step chamber consisting of a parallel plate avalanche chamber and a MWPC; the chamber was used to measure the quantum efficiency and the stability of the CsI photocathode.
From Dangendorf, 1990.

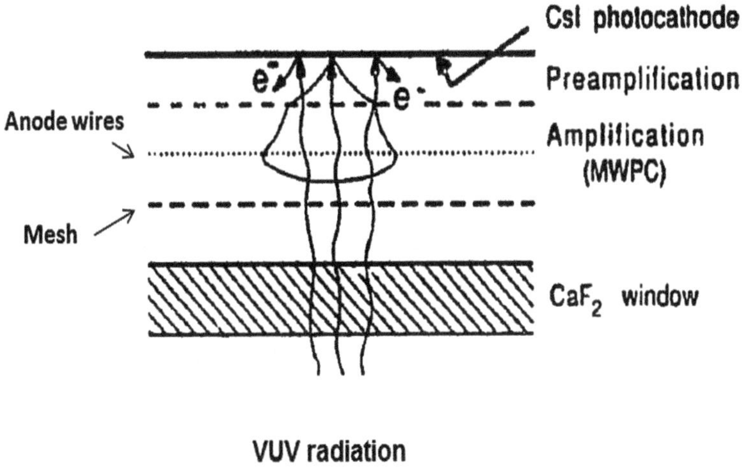

Figure 35. A cross-section of the CsI-RICH layout; the ionization created by the charged particles in the gap between the Cherenkov radiator and the CsI-MWPC is drained by the collection electrode located close to the quartz window. Cherenkov photons emitted from the radiator emit photoelectrons from the CsI photocathode and trigger avalanches in the MWPC. The position of the avalanches is determined from the signals induced on the pad cathode.

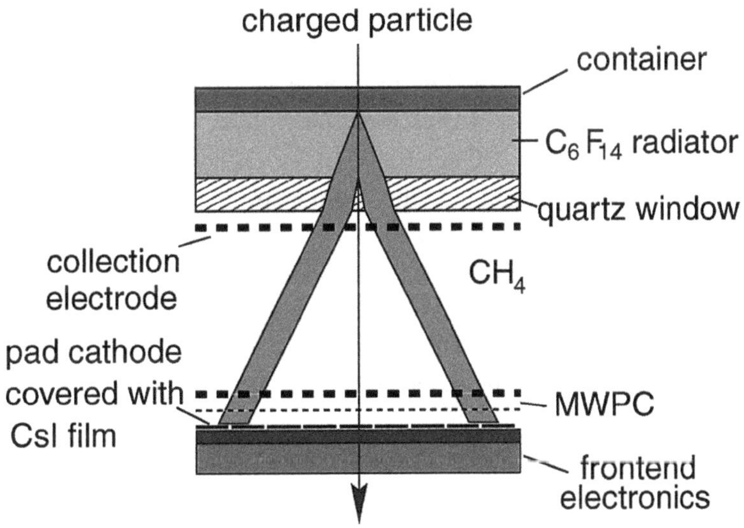

chapter on applications for more details). The primary ionization created in the expansion by the passing charged particles are drained by the positively voltage on the collection electrode located close to the quartz window. Cathode and collection electrode planes are made out of 100 μm diameter gold-plated Cu/Be wires, stretched at 200 and 50 g/wire, respectively, by using crimping pins.

The MWPC anode plane is made of gold-plated tungsten-rhenium wires of 20 μm diameter, with 4 mm pitch, soldered on a G-10 printed board with a precision of 0.1 mm and a tension of 47 g. On both edges of the anode plane, thicker guard wires are installed to resist the boundary discontinuity of the electrostatic field. At half-length of the module a Macor insulating support structure is placed between the pad cathode and the anode plane in order to minimize the sag of the sensing wires due to the electrostatic force.

A positive voltage of 2050 V is applied to the anode wires (which is only half of what is required in a CsI-PPAC). The cathodes are connected to ground. The Cherenkov photons interact with the CsI photocathode and liberate photoelectrons. These photoelectrons drift towards the anode wire and trigger Townsend avalanches there.

The CsI cathode is segmented into pads of 8 × 8.4 mm^2. Each pad is connected to its own preamplifier. The pad cathode is built by gluing together two printed circuit boards to achieve a stiff and flat plane made of light materials. The Cu-clad printed circuit board is electrolytically coated with a 7 μm thick Ni layer followed by a 0.5 μm thick Au layer and accurately polished by mechanical and chemical treatments. The photosensitive layer is finally obtained by depositing 300 nm of CsI onto the cathode substrate.

The photocathodes are prepared at CERN in the large evaporation station described above. The active area of the individual modules of the CsI-MWPC could be as large as 64 × 40 cm^2. Similar designs of CsI-MWPC based RICH detectors are developed for several large-scale high-energy physics experiments, including ALICE, COMPASS etc. They will be described more in detail in the chapter on applications.

2.3.3.2 The CsI Quantum Efficiency and Long Term Stability

In most of studies, the quantum efficiency of the CsI-MWPCs for RICH applications, are reconstructed from the analysis of data obtained during the detection of Cherenkov photons produced by charged particles. This method is based on the fact that the number of Cherenkov photons emitted by the radiator can be accurately calculated. The quantum efficiency is then calculated from the relation between the emitted and the detected photons, correcting for geometry, window transmission, gas transmission, MWPC single photoelectron detection efficiency etc.

Manufacturing a large area CsI photocathode (64 × 40 cm^2) is not a trivial task. Only after respecting all rules and steps formulated in paragraph 2.1 one could achieve quantum efficiency close to the best values first obtained by Séguinot (Séguinot, 1990).

From the experiments with charged particles it is also possible to determine that ~20% photon feedback typically appears at a gas gain of $4x10^4$. When detecting Cherenkov light the photon feedback on this level is not disturbing since the feedback avalanches are grouped close to the primary Cherenkov photon due to the small distance between the avalanche and the CsI photocathode. They even help to better to identify the ring.

Besides the quantum efficiency and the feedback threshold, another important characteristic of the CsI–MWPC is its long-term stability. As was shown in this and previous chapters, the quantum efficiency of any photocathode, including the CsI, is strongly dependent on the surface conditions and in particular on any absorbed layer. In the beginning, the use of CsI as a photosensitive element for gaseous detectors created a lot of skepticism among scientists. Many researches believed that it would be impossible to achieve any long term stability of a CsI photocathode combined with a gaseous detector. However, further studies revealed that if the CsI photocathode is operating in a continuous flow of gas its surface changes and the CsI is protected by a thin absorbed layer.

Figure 36. The long term stability of a CsI photocathode, measured at different wavelength in methane at 1 atmosphere
From Breskin, 1996.

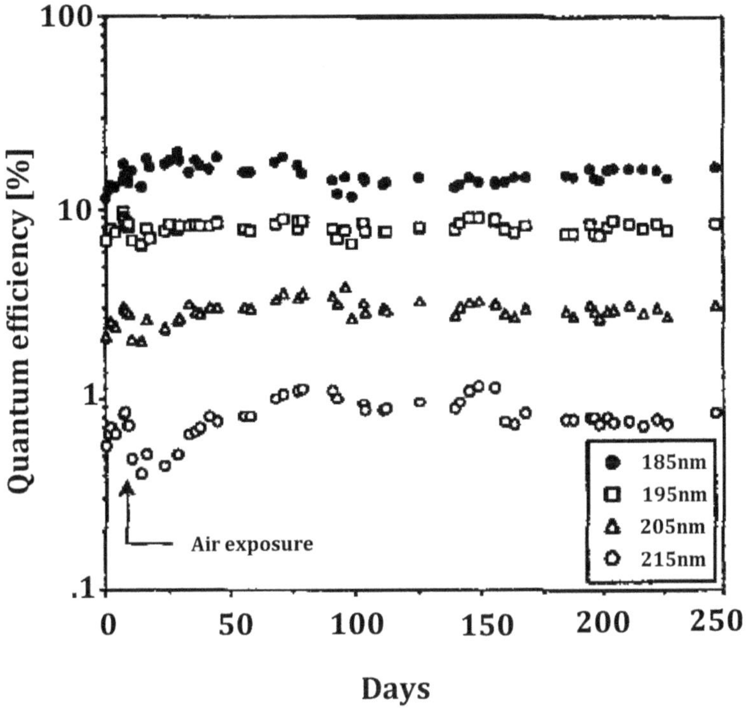

An example of how stable a CsI photocathode is when continuously flushed with methane is shown in Figure 36.

Changes of the quantum efficiency with time are observed only under strong UV illumination and correspondingly strong ion feedback bombardment. It is believed that such a strong ion bombardment disturbs the absorbed layer on the CsI, and also creates a charged layer on top of the CsI. Please also note from Figure 36 that even after temporary exposure to air, the continuous flow of methane restores the original quantum efficiency within a few days.

The longest tests of the CsI stability known to us was performed by the ALICE RICH group, who monitored the quantum efficiency of the CsI photocathodes for several years. Practically no changes in the quantum efficiency were observed during this remarkable period of time (Piuz, 2007).

In addition, we have now new data from the CsI-MWPCs in the ALICE experiment during various LHC injection tests in 2008 and 2009 and the LHC run till its temporary shutdown 2013.

2.3.3.3 Cathode Excitation Effects

Based on the experience from many groups and experiments such as ALICE, STAR, etc., it is well established now that CsI-MWPCs are in general operating stable, especially at low particle fluxes.

One important exception is when CsI-MWPCs are exposed to a high flux of charged particles and/or photons. A relevant example is the COMPASS experiment. All its MWPCs with CsI photocathodes, operating in a methane atmosphere, have been extensively tested in the laboratory to provide gas gains between10^5 and 10^6 with operation voltages between 2300 and 2400 V, respectively. But in an intense particle beam (for example minimum ionizing particle fluxes up to 1 kHz/cm^2 and photon fluxes up to 50 kHz/cm^2 they can be stably operated only at gains not exceeding $3x10^4$ (2000 V). In the COMPASS experimental, operating them at higher voltages occasionally causes electrical instabilities. For larger gas gains discharges may appear which require a long (~1 day) recovery time. The effect is clearly related to the combination of the applied voltage and the ionizing particle flux. The same behavior is also observed in a series of optimization studies and validation tests performed at the CERN Gamma Irradiation Facility.

Such instabilities at high particle fluxes requiring a long recovery time is clearly related to the cathode excitation effect. Some additional studies of this effect in relation to CsI photocathodes will be presented in the following paragraphs.

2.3.3.4 Aging

As in the case of the CsI-PPAC, a lot of studies on CsI photocathode aging have been done for CsI-MWPCs. The results were often contradictory, which is not astonishing since, as was mentioned earlier,

Figure 37. The quantum efficiency vs. collected charge for several photocathodes measured under corona discharge (gain ~10^5) at one atmosphere; 1 - CsI, 2 – CsI + EF, 3 – Cu + I_2 + TMAE, 4 - diferrocenyl mercury + EF, 5 - a CsI photocathode at low pressure
Charpak, 1991.

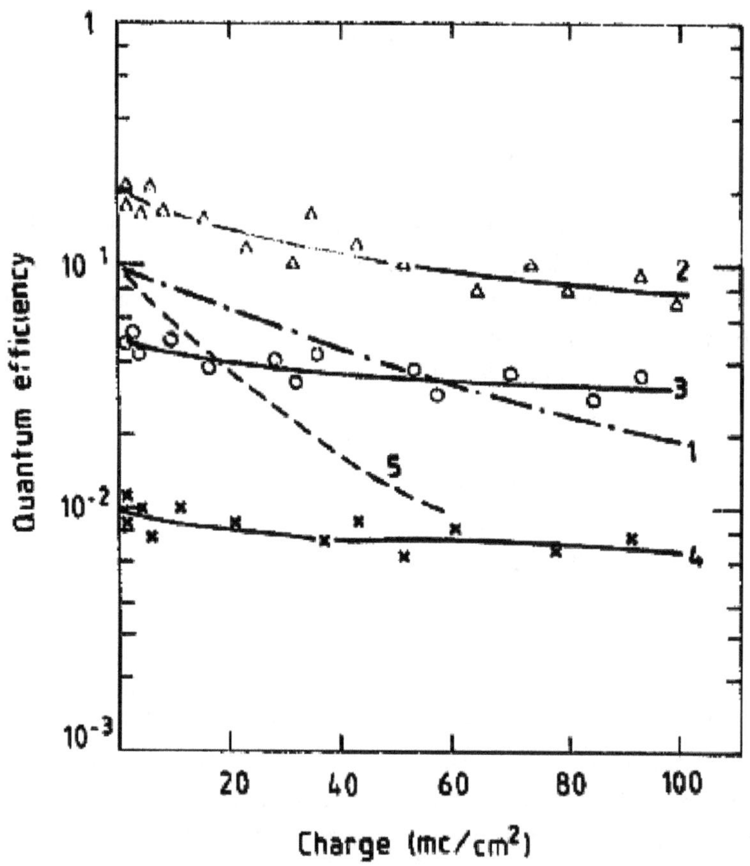

Figure 38. The relative decay of the photocurrent from a CsI photocathode in a MWPC at 1 atmosphere of CH_4 with a gas gain of 10^5, as a function of the accumulated charge; the photocathode is 900 nm thick CsI on Sn/Pb-coated printed circuit board substrate.
From Krizan, 1995.

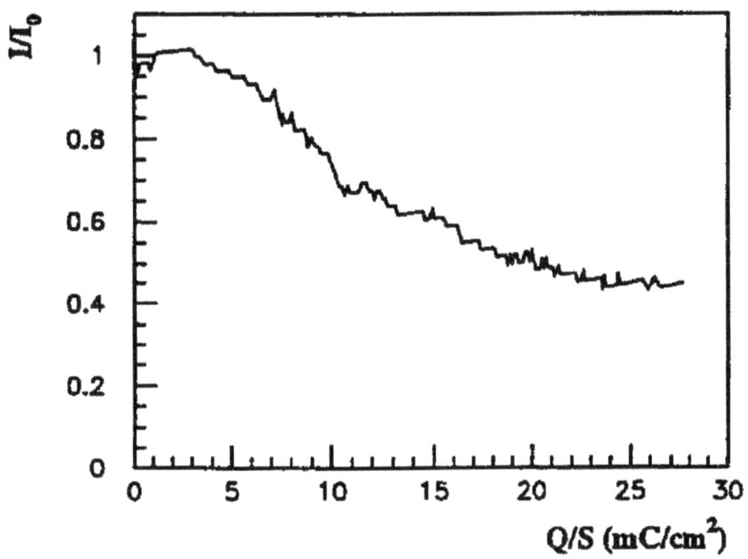

the aging process depends on many factors. It depends e.g. on the photocathode history, gas pressure, gas cleanness, gas gain, charging up effect etc. Below a few examples are given.

In early work of Charpak, aging tests of various photocathodes, including CsI, were done with a single–wire counter operating at high gas gain in a corona discharge (Charpak, 1991a). The results obtained in these studies are shown in Figure 37.

As can be seen, CsI quantum efficiency degradation is observed only after a collected charge of 20-30 mC/cm^2, which is much higher than for a low pressure parallel plate avalanche chamber.

Rather similar results were obtained by Krizan as shown in Figure 38 (Krizan, 1995).

These are the best results of CsI aging obtained so far. Other groups have observed quantum efficiency degradation at much lower, sometimes a factor of hundred lower, accumulated charges (Breskin, 1996). Part of the discrepancy may come from the difference in measurement procedure. Charpak, for example, before each measurement the VUV irradiation was stopped for 30 min to allow the CsI surface to get rid of the surface charge so that only a net effect of aging is observed. This procedure was also used later in some works (Anderson, 1991). These measurements show a slower aging effect than measurements done with a continuous illumination.

Note, that all the CsI-MWPC aging measurements described above were performed with photocathodes exposed to air prior to the installation in the gas chamber. Therefore, it will be especially interesting to examine the latest result obtained with CsI photocathodes not exposed to air. Such studies have been carried out, for example, by the ALICE RICH group (Hoedlmoser, 2006).

Their setup is shown in Figure 39.

The aging tests without exposing the CsI photocathode to air was performed using one of the ALICE RICH prototypes. The Cherenkov radiator was replaced by an array of three collimated ^{90}Sr sources. As shown in Figure 39, the axis of each radioactive source intersect the cathode plane within the Cheren-

Figure 39. Drawing of the ALICE RICH prototype used to measure aging without exposing the CsI photocathode to air: (a) the CsI-MWPC with a pad cathode coated with a CsI layer; the locations of the three irradiation spots (^{90}Sr) are indicated. The anode wires are horizontal (X-coordinate). The Cherenkov annular zone overlaps with the three spots. (b) Cross-section of the detector along a plane containing the centre of the radiator and ^{90}Sr spots 1 and 3; the MWPC has a 2 mm half-gap, 4.2 mm anode pitch. The pad size is 8.0 x8.4 mm², all coated with CsI. The chamber gas is CH_4 at 1atmosphere.
Hoedlmoser, 2009.

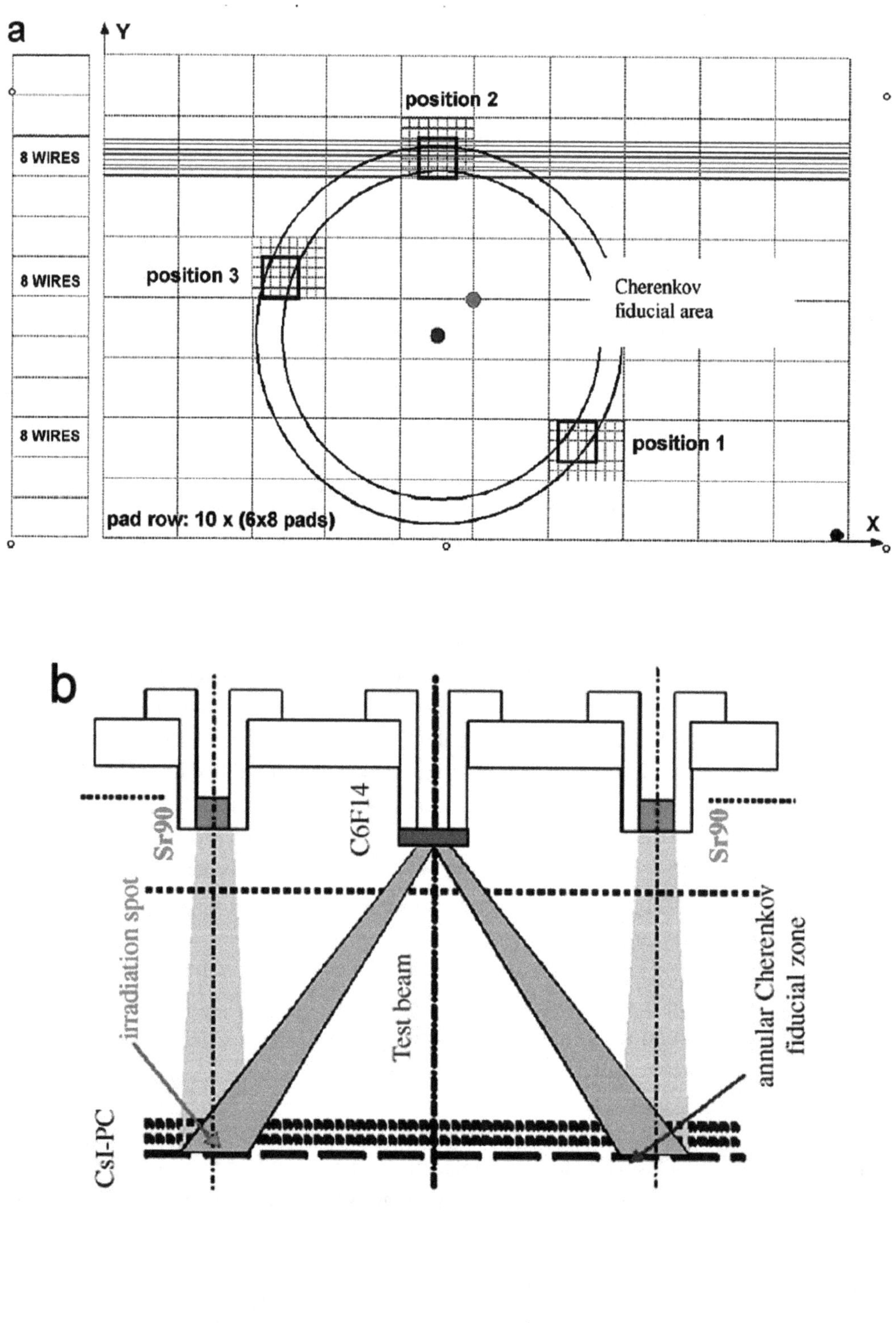

Figure 40. The normalized photocurrent (proportional to the quantum efficiency) of one of the CsI photocathodes exposed to radioactive sources; the accumulated charges are: in position #1 -6.31 mC/Cm², in position #2 - 6.8 mC/cm², in position #3 - 1.54 mC/cm² and in position #4 ~1mC/cm². The measurements were performed several months after the irradiation tests.
From Hoedlmoser, 2007.

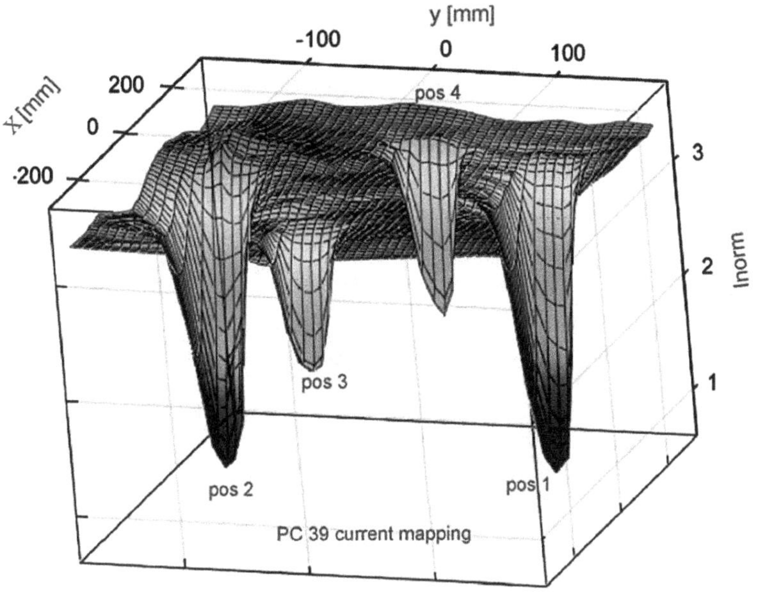

kov ring area (see Figure 39a). The position of each radioactive source is chosen so that the irradiated zones do not overlap the same anode wires in order to facilitate the anode and pad current read-out. The collimator is designed such that each ^{90}Sr source irradiates and area equivalent to an array of 4 x4 pads. The impurity levels in this detector were under continuous monitoring. The levels of oxygen and water impurities were always lower than 10 and 5 ppm, respectively. The levels of hydrocarbons in the methane during operation was <200 ppm C_2H_6 and <50 ppm for other hydrocarbons C_nHm. To avoid any contact with air, the CsI photocathode transfers to and from the detector were done in the glove box.

Several CsI photocathodes were studied a different source intensities ranging from 3 to 260 MBq. The gas gain was about 3x10^4 resulting in accumulated charges on the cathode between 0.2 and 7 mC/cm².

The results are presented in Figures 40 and 41. Figure 40 shows the normalized photocurrent measured for one of the CsI photocathodes exposed to strong radioactive sources. A clear degradation of the photocurrent in the four irradiated areas is visible after an accumulated charge of ≥1 mC/cm².

In Figure 41 results obtained at 1 mC/cm² and 0.2 mC/cm² are plotted. In this case no degradation of the quantum efficiency was observed after an accumulated charge of 0.2 mC/cm².

One should note that the photocathodes used in ALICE are of high quality. They have a quantum efficiency at least a factor two higher than the one used in the earlier aging tests (Charpak, 1991a) (Krizan, 1995). Therefore one can speculate if high quality photocathodes are more sensitive to ion bombardment.

If one assumes that the safe value of the accumulate charge is ≤ 0.2 mC/cm², is it sufficient for the CsI photocathode to survive in the ALICE radiation environment? The ALICE experiment has a low event rate, in the range of a few 10^4 and 10^5 Hz in Pb–Pb and p–p collisions, respectively. However, high multiplicities are expected, reaching for central Pb–Pb collisions up to ~100 particles/m² in the RICH

Figure 41. Scan of another CsI photocathode exposed to the radioactive sources of different activities; the irradiated positions are marked with blue crosses above the plot. The positions irradiated with 0.2 mC/cm² exhibit no measurable quantum efficiency drop, not immediately after irradiation, nor up to 250 days after irradiation.
From Hoedlmoser 2007.

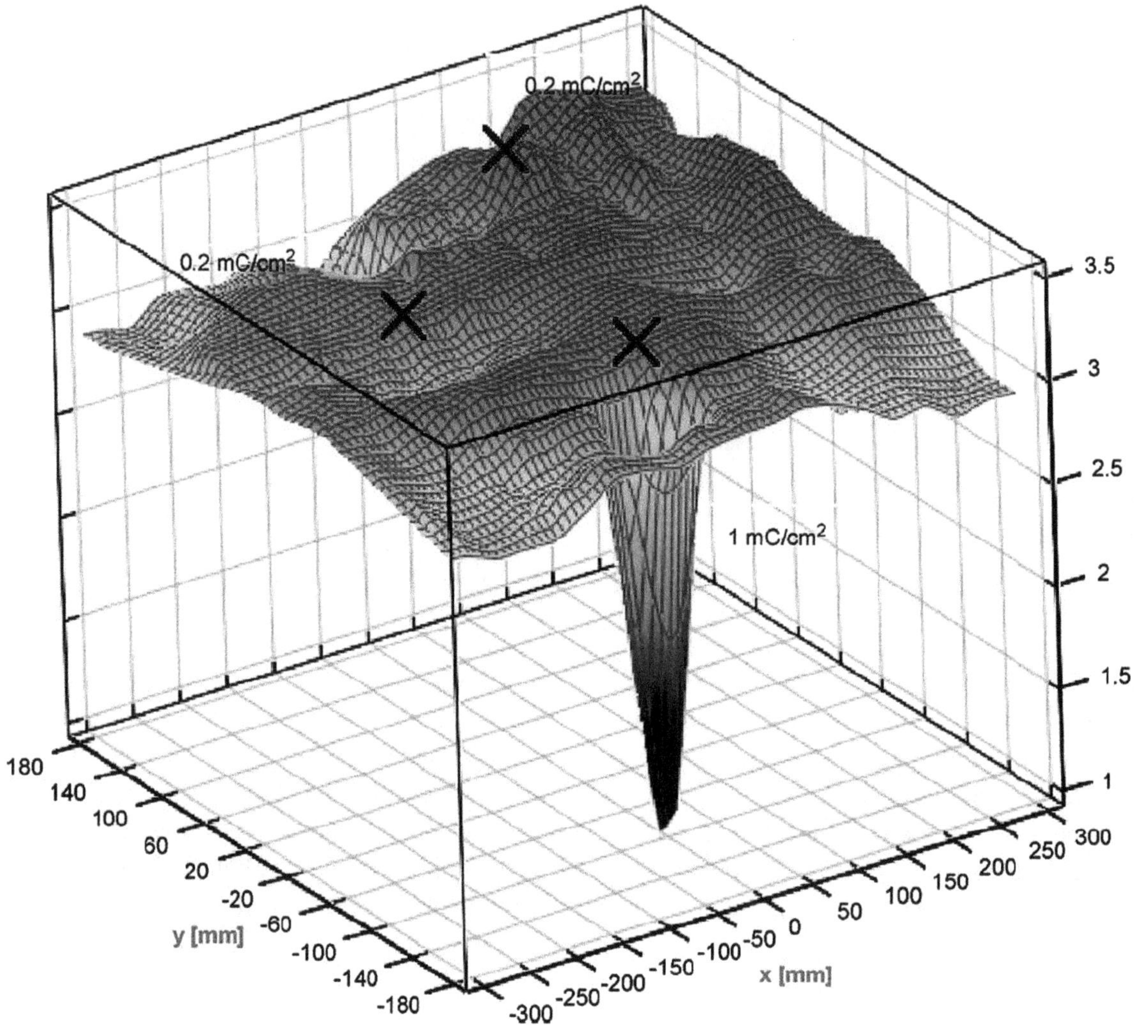

modules located at 5 m from the vertex. The expected charge dose for 10 years of operation is estimated to be 0.52 mC/cm², shared as 0.45 and 0.07 mC/cm2 between p–p and Pb–Pb events, respectively (Hoedlmoser, 2006). One can conclude therefore that performed test indicate that CsI- MWPCs can operate safely at least for several years. Probably much longer since it is known that usually accelerated aging tests give the worse results than test performed with a weaker source, but for longer period of time.

2.3.4 Sealed Single-Wire Counters with CsI Photocathodes

CsI-PPAC and CsI-MWPC operating in with a continuous gas flow is a simple and economical way to build large area detectors. For some applications, it is more convenient to use sealed detectors.

Sealed single–wire counters filled with photosensitive vapours have been used for a long time and are successfully used in various applications. Attempts to manufacture compact sealed gaseous with CsI or other solid photocathodes, capable of detecting single photons, started around 1999 by several groups.

The first such detectors were manufactured based on commercially available single-wire counters, obtained from the Institute for Vacuum Technique, Moscow (Carlson, 2003). They are made of stainless-steel tubes with diameters of 15 and 20 mm, 5 cm in length having a 0.1-mm anode wire in the centre. The wire is supported by a glass structure at both ends between. The detectors have LiF_2 windows. The photocathode is deposited directly inside the detector by opening the window of the glass end plugs, respecting the golden rules of CsI photocathode manufacturing. The detector is then pumped to a vacuum of 10^{-6} Torr, heated to 50°C, and finally filled with Ar + 10% CH_4 + ethylferrocene vapour at atmospheric pressure and then sealed. The quantum efficiency of such a detector is around 20% at 180 nm, which is close to the best values obtained with CsI photocathode manufactured at CERN.

Later, a collaboration between the CERN ALICE RICH group and Oxford Instruments Inc. developed sealed photosensitive detectors using solid photocathodes (Periale, 2007). These detectors are stainless steel cylindrical single wire counters (most of the prototypes were 110 mm long with a photocathode of 20 mm diameter) and sealed by quartz or MgF_2 windows. Such a detector is shown in Figure 42.

The gold plated tungsten anode wire is quite thick, with a diameter of 0.1 or 0.2 mm. The anode is stretched between ceramic supporting structures at each end of the cylinder. Prior assembly, all parts are carefully cleaned with various solvents in an ultrasonic bath.

Figure 42. Photograph of a sealed gaseous detector with a solid CsI photocathode developed in collaboration between CERN and Oxford Instruments Inc.
From Periale, 2007.

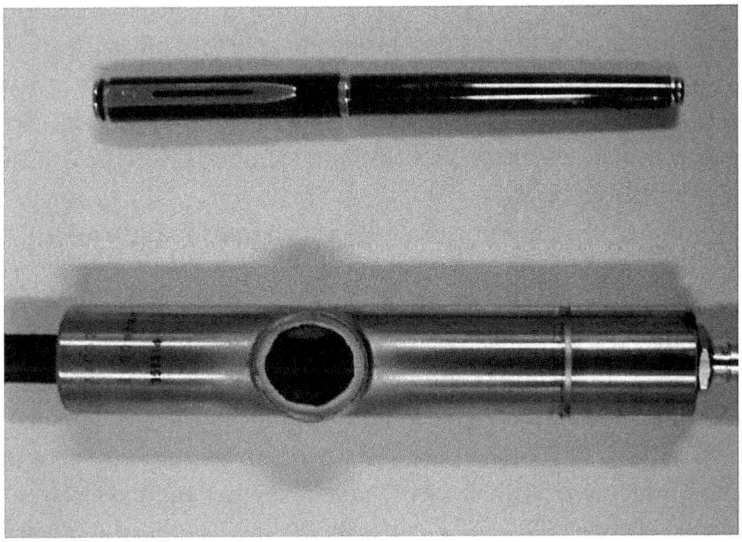

Before the window was mounted, a CsI layer is deposited on the inner part of the cylindrical cathode opposite to the window at a small evaporation plant at CERN. The cylinder is then extracted from the evaporation plant and exposed to air during the transport from CERN to Oxford Instruments' lab in Helsinki. At Oxford Instruments' lab the window is attached to the cathode cylinder and sealed. The detector is then heated to 65 °C and pumped to a vacuum of better than 10^{-6} Torr for several days. It is then cooled to room temperature and filled with Ar + 10% CO_2 or Ar + 10% CH_4 at atmospheric pressure. The quantum efficiency of these detectors are good, close to 20% at 180nm which is quite surprising keeping in mind the CsI is exposed to air for few days.

Stability tests of sealed detectors from both types of production have been performed for 8 years now, and no degradation has been observed proving the reliability of this technique. These detectors can be used e.g. for the detection of flames and dangerous gases which will be discussed later.

2.4 Hole Type Gaseous Detectors Combined with CsI Photocathodes

So far, we have mainly spoken about CsI photocathodes combined with classical detectors like parallel plate avalanche chamber and MWPC. In the parallel plate avalanche chamber all avalanche ions are collected on the CsI cathode. In an MWPC at least 50% of them are collected on the CsI cathode. The other 50% drift to the opposite metallic cathode. Hence, in these detectors the bombardment of the cathode by positive ions is severe, and may cause CsI aging and cathode excitation effects.

In hole-type electron multipliers, described in Chapter 2, significantly less positive ions drift to and bombard the CsI photocathode. Also, in this geometry only a small fraction of the photons emitted in the avalanche reaches the cathode reducing the risk of photon feedback.

The first attempt to combine hole-type detector with solid photocathodes and operating them at high enough gas gains to detect single photoelectrons were reported in 1998 at the RICH Conference in Ein-Gedi, Israel (Peskov, 1999; Garty, 1999). Peskov used a capillary plate operating atmospheric pressure, whereas Garty used low pressure GEMs.

2.4.1 GEM with CsI Photocathodes

GEMs can be combined either with semi-transparent or reflective photocathodes. Reflective photocathodes generally offer twice as high quantum efficiency compared to semi-transparent photocathodes. In the case of GEMs, a layer of CsI is deposited on the top electrode to serve as a reflective photocathode. Such coatings similar to those previously and successfully have been used in vacuum microchannel plates (Fraser, 2009). A semi-transparent CsI photocathode can alternatively be deposited on the entrance window of the detector above the GEM.

A cascaded GEM with reflective CsI photocathode is presented in Figure 43.

It consists of four GEMs operating in tandem mode allowing them to boost the overall gas gain up to 10^6. It has full single-photoelectron detection efficiency if the voltage settings are correctly chosen. The electric drift field above the GEM, labelled E_{drift} in Figure 44, should be weak, to allow a good focusing of the photoelectrons into the GEM holes by the GEM electric field. The weak reversed drift field makes the detector practically insensitive to ionizing particles, permitting a stable operation in a high radiation background. The transfer field, E_{trans} below each GEM, should be carefully optimized to guarantee a high electron extraction from the GEM holes, but without suppressing the electron collection into the subsequent multiplication element. The GEM potential, ΔV_{GEM}, is crucial. It creates the electric field

Figure 43. Schematic view of a quadruple GEM with reflective photocathode on top
From Morman, 2004.

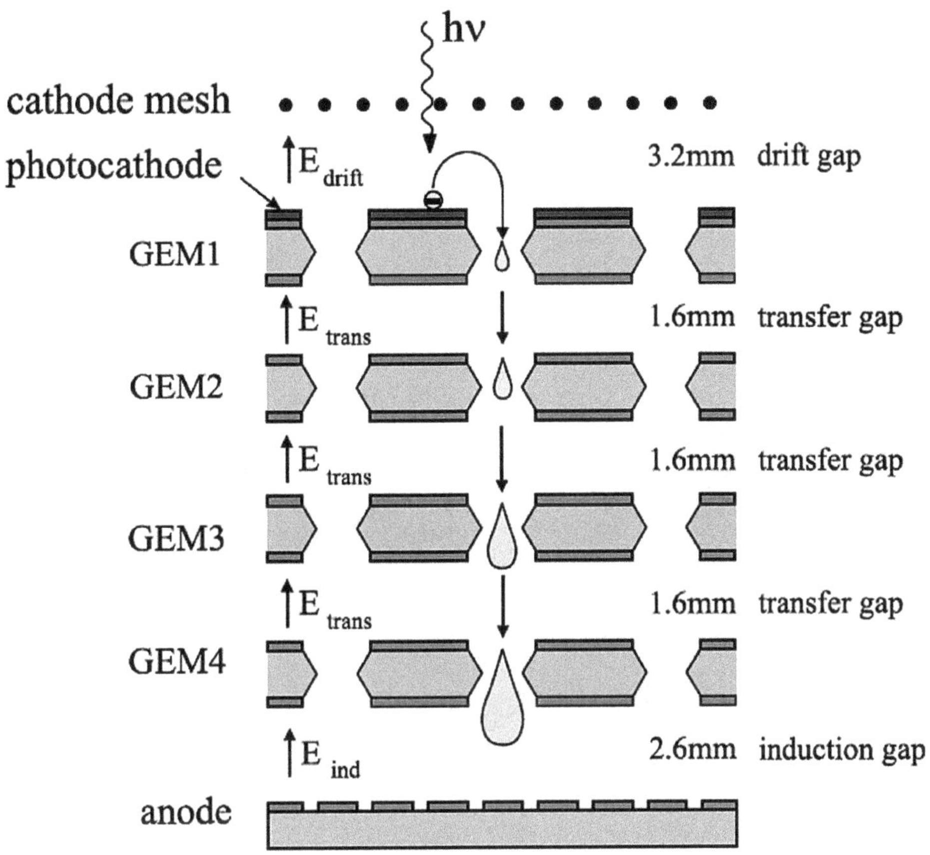

that extracts the photoelectrons from the photocathode surface and guides them into the GEM holes. Furthermore, the amplification in the GEM holes is important for minimizing electron losses to the bottom-GEM electrode at the hole exit.

One of the problems associate with the use of GEMs is charging up of inner dielectric holes by avalanche ions. This leads to gain variations with time, which may reach up to 20-30% (Azmount, 2006; Simon, 2007). This effect can be minimized if the HV is constantly applied across the GEMs. In this case stability around 1% can be achieved (Peskov, 2013).

As an example, the PHENIX Cherenkov detector uses a large area (~1m^2) triple GEM with a reflective CsI photocathode (subdivided into 12 detector modules, which each is 23 x 27 cm^2).

There have also been several attempts to develop small size (active are of few cm^2) sealed detectors, consisting of three or four cascaded GEMs coupled to a semitransparent CsI photocathodes. These detectors operate with Ar/CH_4 (95/5) at atmospheric pressure. Such a detector is shown in Figure 45.

Due to the semi-transparent photocathode, the quantum efficiency of this detector is low. Its aging characteristics as well as a long-term stability has shown to be worse than for a sealed single-wire counter with a reflective CsI photocathode, as shown in Figure 46.

The main reason that the sealed GEM show worse long term stability is outgassing from the Kapton and the printed circuit boards, whereas in a single-wire counter only materials with low outgassing prop-

Figure 44. The principle of a GEM with reflective photocathode; photoelectrons are emitted from a reflective photocathode on top of a GEM electrode. They are extracted from the photocathode (a) with an efficiency ε_{extr} and guided into the GEM apertures with an efficiency ε_{hole} (b) where they multiplied (c). A fraction of the avalanche electrons is extracted and transferred with an efficiency ε_{trans} into a following GEM; another fraction is lost to the bottom GEM electrode.
From Morman, 2004.

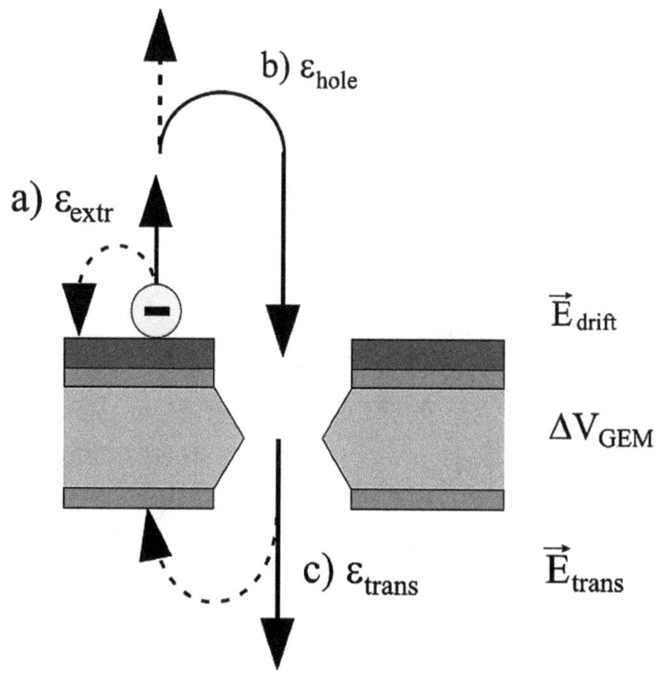

erties used, such as glass, ceramics and stainless steel. To this one should add that a semi-transparent CsI photocathodes is more sensitive to aging effects being so thin.

Sealed multistep GEM detectors were also developed by Bellazzini group for X-ray imaging and VUV photon detection (Bellazzini, 2007).

2.4.2 Capillary Plates with CsI Photocathodes

Capillary plates (CP) are of particular interest because they are made of glass which is well suited for the high level of cleanliness necessary for the operation of sealed detectors (Peskov, 1999; Iacobaeus, 2004). The capillary plates can be of several geometries: traditional single–step or multistep capillary amplification mode, where the main multiplication occurs within the capillaries. It can also be a double capillary plate combined with CsI-coated GEM, as shown in Figure 47, operating in “parallel-plate amplification” mode, where the main amplification occurs between the capillary plate and the readout plate. In the latter, the capillaries are used as a feedback ‘suppressor, as illustrated in Figure 47c. Both reflective and semitransparent CsI photocathodes can be used. The readout plates comprise of isolating metallic strips enabling position-resolution measurements.

The gain characteristics of these detectors are presented in Figure 48.

The position resolutions of some of the capillary plate geometries are shown in Figure 49.

Figure 45. Photograph one of a sealed cascaded GEM with a semitransparent CsI photocathode deposited on the window
From Breskin, 2002.

Figure 46. Results of aging studies and a long-term stability of two sealed detectors with semitransparent CsI photocathodes containing either a triple or a quadruple GEM; the graph shows the evolution of the anode current density with the accumulated anode charge, in three ageing runs with different gas gains (A), voltage drops across a single GEM (ΔVGEM) and operation modes. The avalanche ion fraction reaching the CsI photocathode (IBF) is indicated as well.

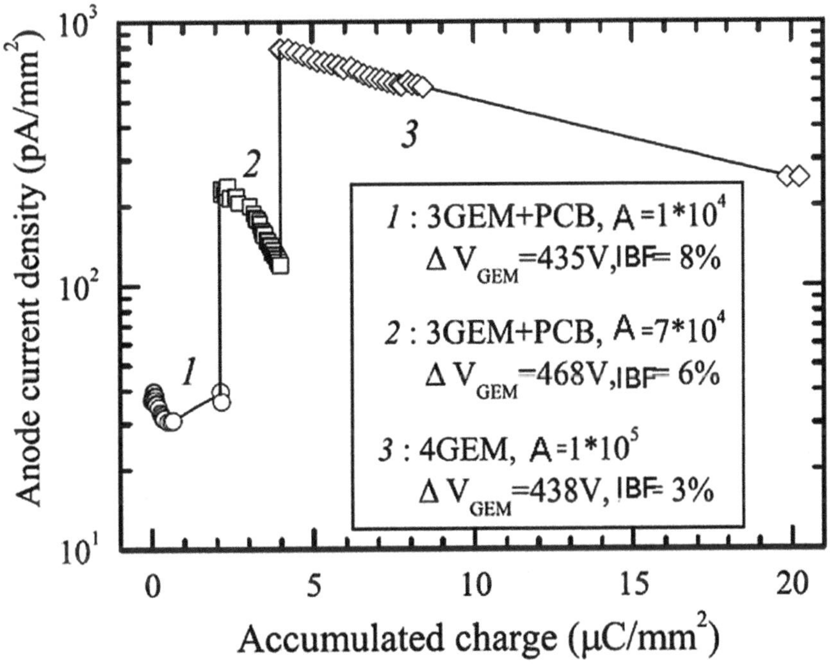

Figure 47. Various geometrical arrangements of single- and double-capillary plates: (a) double capillary plates combined with a CsI-coated GEM, (b) triple capillary plates; in (b) and (c) the photoelectrons are emitted from a semitransparent CsI photocathode. (c) A capillary plate operating in a parallel-plate avalanche mode
From Iacobaeus, 2004.

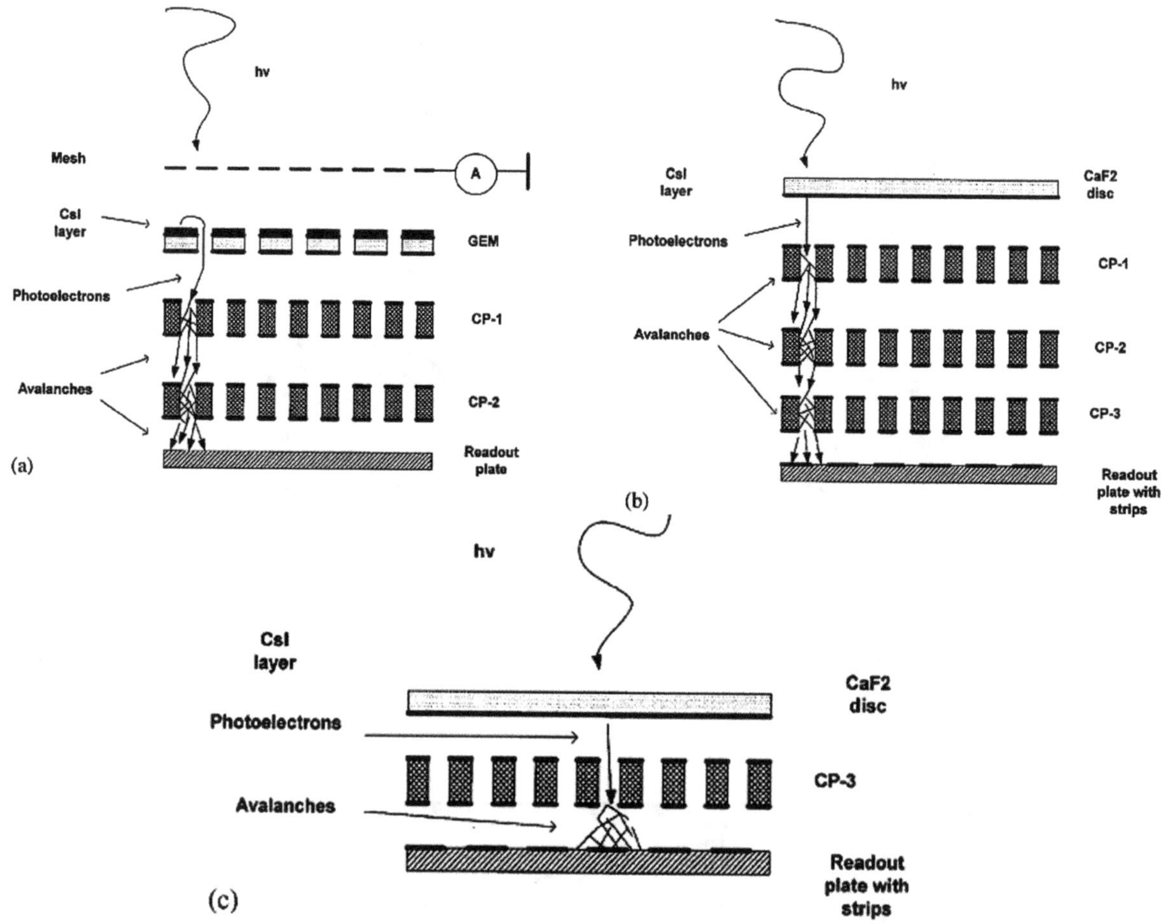

The results clearly demonstrate that among various capillary detectors with CsI photocathodes high gains can be reached either with triple CPs or in a single CP operating in parallel-plate avalanche mode. In the latter, the capillary plate has the role of transporting (not amplifying) electrons, stopping avalanche-induced photon-feedback and reducing ion-feedback effects. The avalanche amplification takes place in the parallel electric field between the bottom of the capillary plate and the readout anode plate.

The capillary plate operating as a parallel plate detector also provides rather narrow charge distributions on the readout circuits, resulting in a good position resolution.

The largest glass capillary plates which can be manufactured today are about 10x10 cm^2, which limit their applications.

At low counting rate, <100 Hz/cm^2, the gas gain of a capillary plate is stable with time. At high counting rates, >1000 Hz/cm^2, gain variations arises. If the capillary plate is hydrogen treated, where the resistivity of the glass surface is reduced, it can operate stably at counting rates up 10^5 Hz/cm^2 (Silin, 1999, Ochi, 2002).

Figure 48. Gas gain vs. the voltage applied across the CsI-capillary plates, measured with UV-photons at atmospheric pressure in Ar + 5% CH_4; filled diamonds: the gas gain of a single CP; filled square: the gas gain of double capillary plate at equal voltages over CP-1 and CP-2; filled triangles: the overall gain of a double capillaries at a potential over CP-2 kept 50 V higher than on CP-1; crosses: the overall gain of a triple capillary plate at potential over CP-3 kept 50 V higher than on CP-2 and that on CP-2 kept 50 V higher than on CP-1; stars: the gain of a single CP detector (see Figure 47) operating in a parallel plate avalanche mode

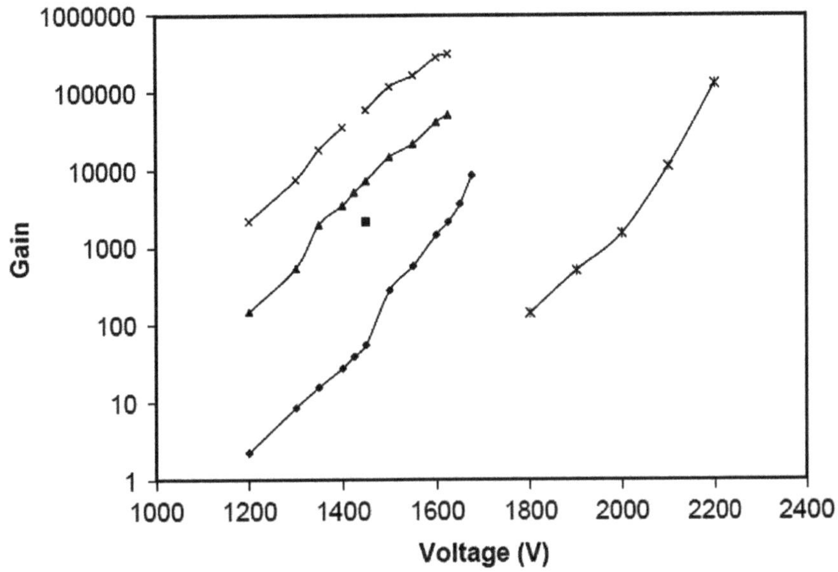

Figure 49. The induced charge profile measured with a UV-source illuminating the detector through a narrow slit collimator; open diamonds: triple CPs operating in a capillary amplification mode, squares: a single CP operating in a parallel-plate avalanche mode. The gas mixture is Xe (40%) + Kr (49%) + CO2 (20%) at a pressure of 2 atmospheres.

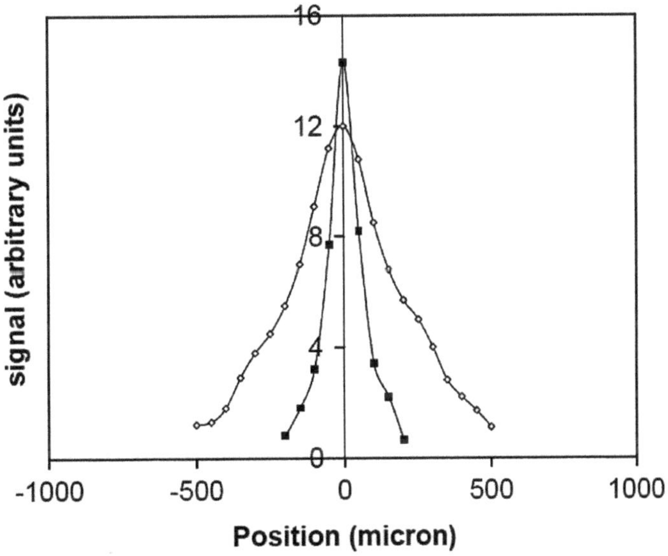

2.4.3 CsI Coated Thick GEMs

In some applications a high position resolution, is not necessary. For example, in RICH detectors a position resolution of about 1-2 mm is often sufficient. Then it is beneficial to use thick GEMs (TGEMs). These detectors can be commercially produced at low price (1 USD for 10^3 holes) and with large sensitive area (~0.5 m^2) (Periale, 2001). Does CsI- TGEM has any advantages over CsI -MWPC?

2.4.3.1 Comparative Studies of CsI-TGEM and CsI-MWPC

Figure 50 shows a setup to compare CsI-TGEM and CsI-MWPC.

It consists of a gas chamber housing a triple CsI-TGEM detector, a CsI-MWPC included for comparative purposes, a gas system allowing gas circulation with various gas mixtures, a UV lamp, a lamp emitting only visible light, an X-ray tube and several radioactive sources. The chamber has appropriate windows on both sides, to allow illumination of each detector separately with any of the above radiation sources.

Some results of these comparative studies are shown in Figures 51 and 52.

Figure 51 shows gain curves, measured in current mode with UV photons, in single-, double- and triple-TGEMs, irradiated simultaneously with the UV lamp and with either ^{55}Fe or ^{106}Ru.

The highest point on each gain curve corresponds to the onset of discharges (~1 per 10 min).

These measurements are relevant for photon detectors operating under a background of highly ionizing radiation. Similar results (slope, maximum gain) were reached for a bare TGEM and with a CsI-coated TGEM. This indicates that the CsI coating does not induce noticeable photon feedback. Thus, the maximum achievable gain is determined only by the Raether limit, as in the bare TGEM. With a

Figure 50. An experimental setup for comparative investigation of CsI-TGEM and CsI-MWPC with single photons, x-rays and b-electrons

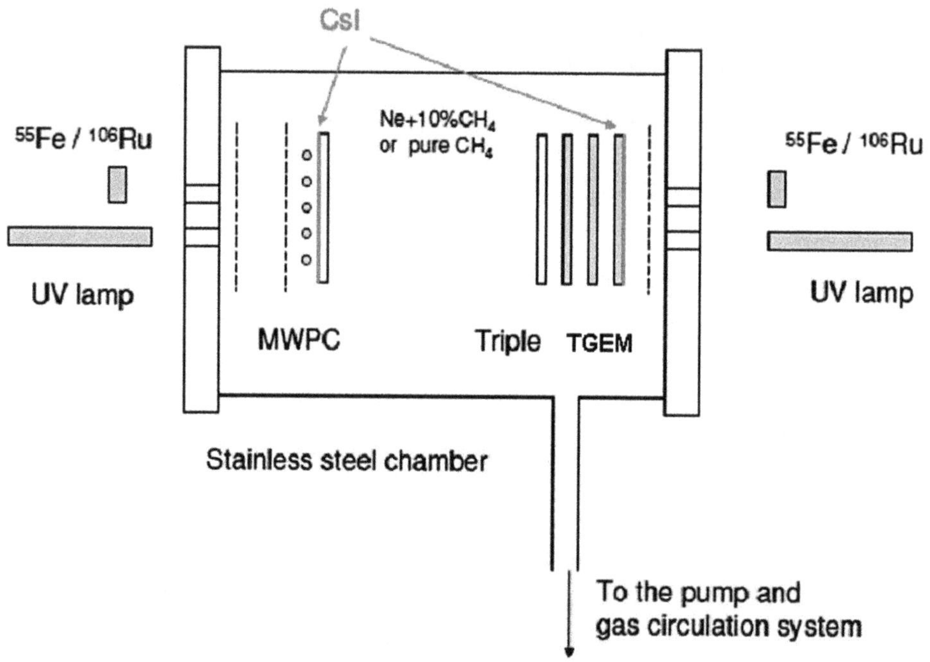

Figure 51. Gain curves measured with single-, double- and triple-TGEMs with hole diameters of 0.4 mm, in Ne + 10% CH_4; the detectors were irradiated simultaneously with UV light and with a radioactive source, either ^{55}Fe or ^{106}Ru. Results are shown for detectors with a CsI photocathode (open symbols) and without CsI (filled symbols).
From Peskov 2010.

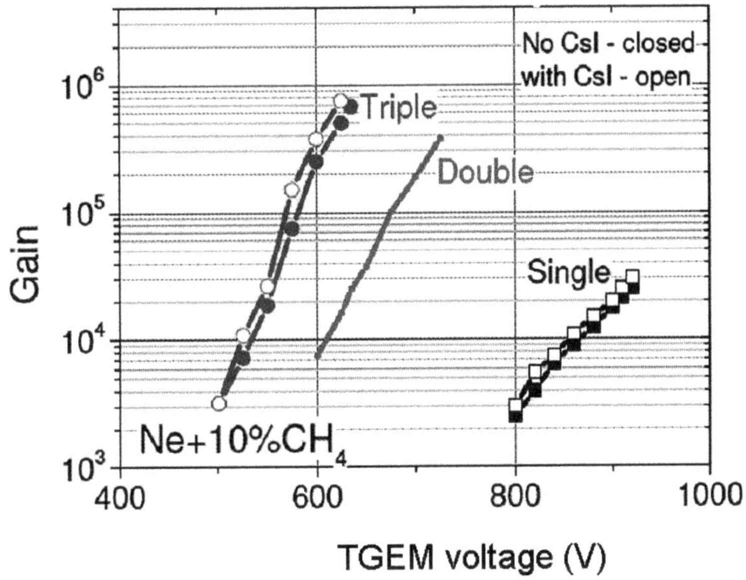

Figure 52. Gain curves of the bare MWPC and the CsI-MWPC, measured with UV light (for gains below 10^4) and with ^{55}Fe x-rays (for gains above 10^3); the arrows indicate the gains at which feedback pulses appeared in the CsI-MWPC and their amplitude ratio. The amplification gas was CH_4.
From Peskov 2010.

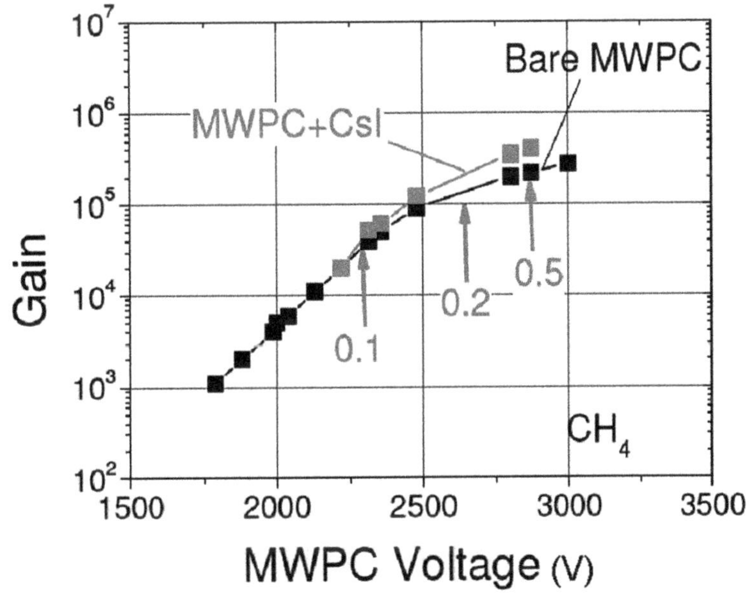

reversed drift-field (as used in the case of CsI-GEM), the CsI-TGEM can operate at high gains even in the presence of ionizing particles.

Figure 52 shows the gain curves of the bare MWPC and the CsI-MWPCs, measured in CH_4 with an ^{55}Fe source. As seen in Figure 52, for voltages below 2250 V (gain < $5x10^4$), the gas gain increased exponentially with the voltage. Gain saturation due to space charges, which are typical to wire chambers, starts at voltages above 2500V (gain > 10^5). The feedback is on the level of 10% already at a gas gain of $3x10^4$, which coincides with described earlier results obtained with ALICE RICH modules. As a consequence, at low counting rates the CsI-MWPC must be operated at a gain almost ten times lower than the CsI-TGEM. Corona discharge in the MWPC appears at a certain maximal gain $A_m=1/\gamma$, practically independent of the nature of the ionizing radiation or the counting rate.

The long term gain stability of TGEM's has been studied under various experimental conditions (see for example Chechik, 2006; Alexeev, 2009). In TGEMs having dielectric material exposed to avalanche electrons and ions, e.g. the rims around the holes and hole walls, the gas gain varies (increases or decreases) over time periods of up to several hours after the voltage is applied. The gain variation is typically of a factor of two prior to stabilization. One efficient way to stabilize TGEMs with rims is to keep the voltage across TGEMs permanently on (Peskov, 2012). Under these conditions the gain variations are only a few %.

TGEMs without rims around the holes, originally developed be Periale et al. (Periale, 2001; Alexeev, 2009), reaches typically 10 times lower gains compared to those with rims. However, they exhibited excellent stability with time.

The data suggest that the long term gain stability, also observed in GEMs (Azmount, 2006), is related to charges slowly accumulating at the rim of the hole, and on the dielectric walls in the amplification holes. The time required for reaching equilibrium depends on the type of gas (positive ions), the electric field strength, the avalanche charge and the counting rate, and on other parameters such as the amount of the exposed dielectric, its structure, the degree of absorbed moisture (affecting surface and volume resistivity) etc. Note that the latter depend on the detector operation conditions, e.g. detector pumping prior to gas introduction, time of gas flushing etc.

It can be conclude that the TGEM gain stability is not a serious issue. One can either use TGEMs without rims or minimize the charging up effect by keeping the voltage across TGEMs for a long time.

Another potential source of the instability is the CsI excitation effect. Intense ion bombardment of the CsI photocathode has shown to cause temporal increase of its sensitivity to UV and visible light. The CsI quantum efficiency resumes to its original values only after 20–30 minutes after the illumination has stopped. During this time an increased rate of spurious pulses is observed. This phenomenon was named the *cathode excitation effect*.

A frequent consequence of cathode excitation in CsI-MWPC detectors is that, following breakdown, the nominal operation voltage cannot be resumed for a substantial period of time. To avoid these phenomena some of the COMPASS RICH CsI-MWPCs had to be operated at lower voltages, which naturally affected the photon detection efficiency.

The "cathode excitation effect" in a CsI-triple-TGEM is observed after intense UV irradiation. However, RICH detector usually operates at much lower UV flux than what is required for triggering a noticeable effect.

The "cathode excitation effect" has been studied by intensely irradiating a CsI-TGEM with x–rays (Peskov, 2010). In a first series of measurements, the photon detector, operating at a gain of 10^4, was irradiated by a collimated X-ray beam, delivering ~$3x10^4$ Hz/mm^2 x-rays to the irradiated spot. In these

Figure 53. Counting rate of spurious pulses and visible-light induced pulses vs. time from the cathode excitation effect in a CsI-coated triple TGEM; the gas mixture was Ne + 10% CH_4. The gas gain was ~10^4. From Peskov 2010.

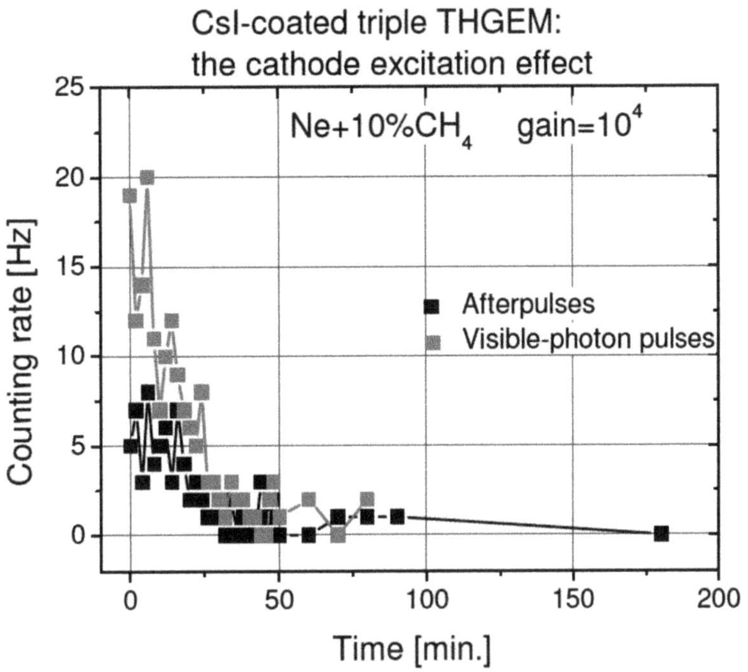

conditions the breakdown rate was ~1 per 10 minutes. After 5-10 minutes of irradiation, the X-ray tube was switched off and the rate of after-pulses, and pulses induced by visible light were measured (recall that normally a CsI photocathode is "solar blind" with a cut-off at ~ 220 nm).

The results are shown in Figure 53, showing a measurable rate of spurious pulses and visible-light induced pulses, lasting for about 30 minutes and fully subsiding after about 100 minutes. These results were reproducible, provided the detector was not irradiated for a couple of hours.

In another set of measurements, the "cathode excitation" was induced in the same way but was followed by an immediate increase of the triple-TGEM voltages to a gain of ~ 10^5, ensuring sensitivity to single electrons. The results are presented in Figure 54. Both the spurious-pulse rate and the visible photon sensitivity increased by about a factor 10, but subsided after a period of time comparable to that of the lower-gain (10^4) operation.

The onset of discharges in these experiments always occurred in the last TGEM, at total gain of (2–3) x10^5. The recovery time of the triple-TGEM, for resuming its voltages, was of about

30–60 seconds. The data indicate that "cathode excitation" could be initiated in the CsI-triple-TGEM, with its characteristic increased spurious-pulses rate and sensitivity to visible light, but with no significant increase of the recovery time.

For completeness in the same setup was used to study the "cathode excitation effect" in the CsI-MWPC, induced by intense irradiation with an X-ray gun. The x-ray intensity was kept similar to that used with the triple-TGEMs. Figure 55 shows the counting rate vs. time of spurious after-pulses and pulses induced by visible light, following the exciting x-ray irradiation.

Figure 54. Counting rate of spurious pulses and visible-light pulses vs. time after cathode excitation induction at a gain 10^4, followed by a 10-fold gain increase for a CsI-coated triple TGEM operated with Ne + 10% CH_4 at a gas gain ~10^5
From Peskov 2010.

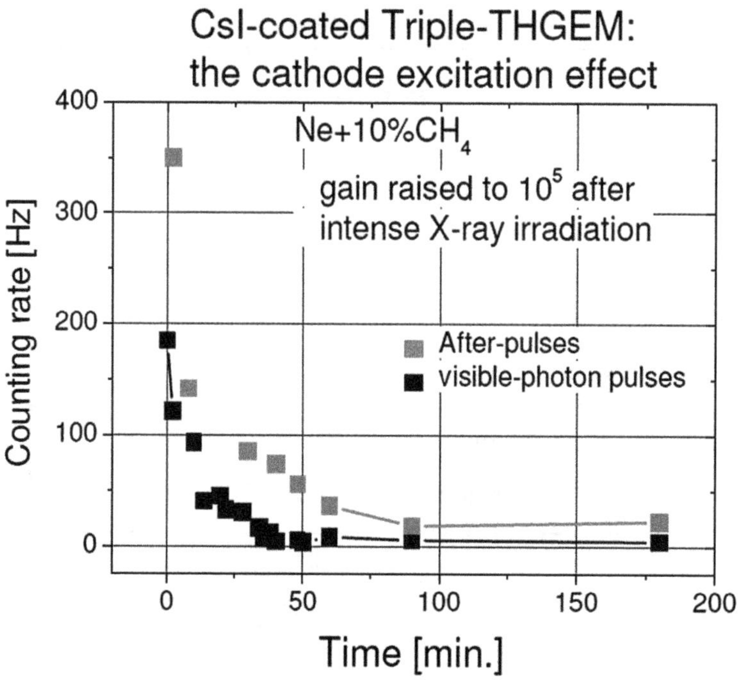

Figure 55. Counting rate vs. time of spurious pulses and pulses induced by visible photons, measured after the CsI-MWPC irradiation with an intense x-ray beam; the X-ray radiation was stopped at time zero. Gas: CH_4; gain: $5x10^4$.
From Peskov, 2010.

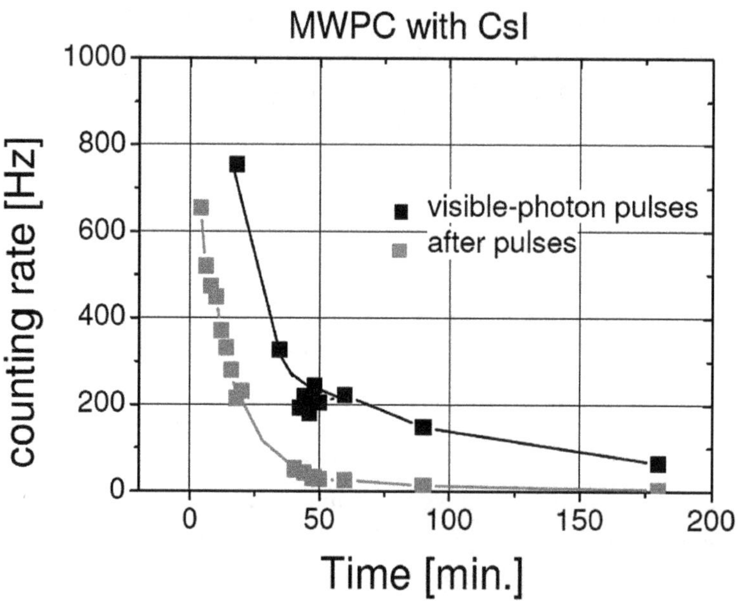

An elevated rate of both types of pulses was observed for more than three hours. After 4-5 hours it resumed its original level of ~2 Hz and ~20 Hz for spurious pulses and visible-photon pulses, respectively. The cathode excitation effect at a gain of $5x10^4$ was by far more intense compared to the triple-TGEM (investigated at gains of 10^4 and 10^5), manifesting both higher rate of spurious pulses and elevated sensitivity to visible light for longer periods of time. This may be due to the "open geometry" of the MWPC in which the CsI photocathode is receptive to the full avalanche-ion and -photon flux.

In contrast, at the same gain, only a small fraction of the avalanche ions will reach the photocathode in the cascaded-TGEM. The ion backflow in cascaded-TGEM can be further reduced, e.g. with ion-trapping strips patterned on the hole-electrodes, as will be described in greater details in the next chapter.

An elevated voltage was applied on the CsI-MWPC, sufficient to trigger a corona discharge. Following this discharge, and even though it lasted for only a few seconds, we could no longer re-apply the voltage for almost one hour. Recall that similar long recovery times of CsI-MWPC were observed in the COMPASS RICH. This is different from similar experiments with CsI-TGEM, in which high voltage could be restored within ~30 sec following a discharge. This difference may be attributed to the fact that TGEMs screen the majority of discharge-ions and photons.

Discharges in the CsI-MWPC appeared only at some critical gain. At lower gains, no discharges were triggered by low- or medium-intensity sources. However, it was observed that at high X-ray intensities the value of the critical gain was reduced by about a factor of two.

The main conclusion drawn from these studies is that the cathode excitation effect is more intense in CsI-MWPC than in CsI-triple-TGEM. It should be seriously taken into consideration in the choice of high-rate UV detectors for RICH applications.

2.4.3.2 CsI –TGEM Efficiency

The effective photon detection efficiency of the photosensitive TGEMs is:

$$\varepsilon_{eff}(\lambda) = QE(\lambda)\, S_{eff}\, \varepsilon_{extr}(\lambda, V_{TG}, E_D)\, \varepsilon_{hole}(V_{TG}), \tag{7}$$

where $QE(\lambda)$ is the CsI vacuum quantum efficiency as a function of the wavelength λ, S_{eff} is the CsI-coated fraction of the TGEM surface, $\varepsilon_{extr}(\lambda, E_D)$ is the photoelectron extraction efficiency in gas as a function of the applied voltage across the TGEM, V_{TG}, and the drift field E_D. $\varepsilon_{hole}(V_{TG})$ is the efficiency of the photoelectron collection into the TGEM holes.

In the case of the parallel plate avalanche chamber the photoelectron extraction efficiency from the solid photocathode, (governed by the back diffusion) depends on the gas and the electric field. The extraction efficiency $\varepsilon_{extr}(\lambda, V_{TG}, E_D)$ from the CsI photocathode in the uniform electric field was measured in work (Azevedo, 2010).

Calculations show that the electric field on the top surface of a TGEM (and similarly on a GEM) is not uniform. It has high values in the narrow regions close to the hole-edges and a kind "plateau" in between the holes (see Figure 56).

As a rough estimation of the mean electric field on the TGEM surface one can use the value of the plateau since E_D is applied for the most of the surface. For example, if the voltage applied across the TGEM is 900 V the mean value of E_D can be considered as 2 kV/cm (see Figure 56). Then using plots presented in Figure 3 in Chapter 5 or new data (Azevedo, 2010) from one can quite precisely evaluate $\varepsilon_{extr}(\lambda, V_{TG}, E_D)$.

Figure 56. The electric field strength on TGEM top surface as calculated by MAXWELL along the line interconnecting two hole centres
From Azevedo, 2010.

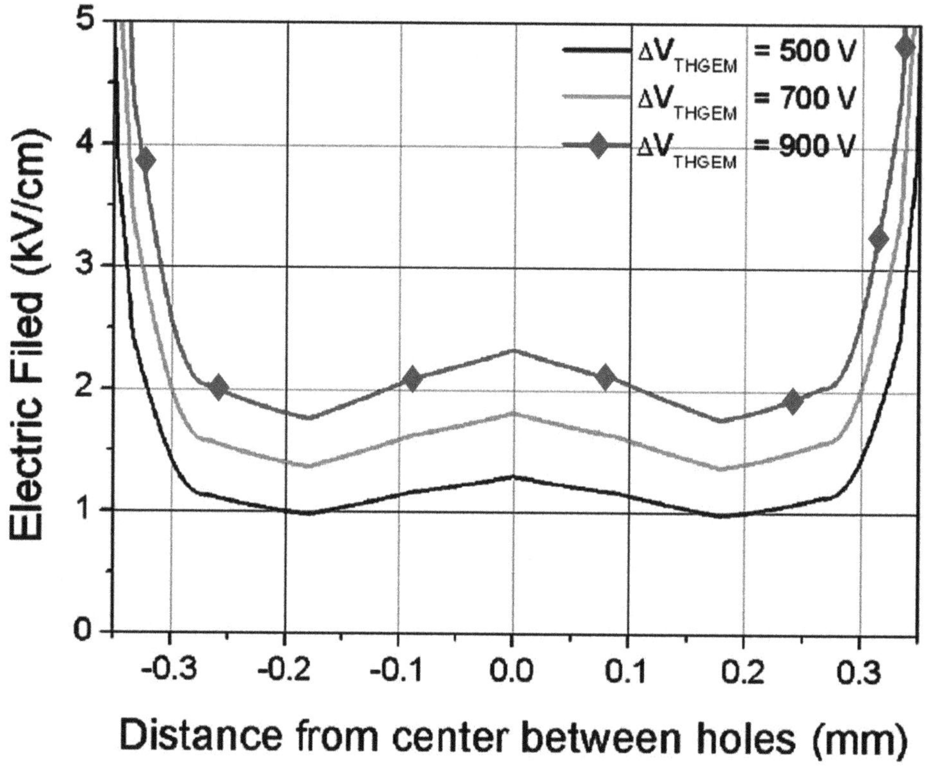

The photoelectron collection efficiency ε_{hole} (V_{TG}) was also measured by Azevedo (Azevedo, 2010). It increases with the potential across the holes, reaching unity at TGEM gains of 25–1000, depending on the type of gas (Figure 57).

It can be concluded that full collection efficiency is achieved for realistic working conditions. Therefore two features determine the efficiency of the CsI-TGEM: it's active surface and the electric field above CsI layer. Bothe these parameters are less than in the case of CsI-PPAC and CsI-MWPC. It is obvious that S_{eff} is geometrically smaller due to the light "loses" in the hole openings.

The qualitative comparison between CsI-TGEM and CsI-MWPC was performed with charged particles producing Cherenkov light in a liquid radiator. Both detectors have a similar pad readout plane and the signals from a single pad or several adjusted pads (clusters) were measured. With CsI –TGEM ~9 raw clusters per event (minimum ionizing article or MIP) was measured, whereas in the case of the CsI-MWPC operating in CH_4 it was ~15 per MIP (Peskov, 2012).

Therefore, the CsI-TGEM efficiency is about 60% of CsI-TGEM. This is mainly due to the geometrical factor by light losses in the TGEM holes. Of course, some geometrical TGEM optimizations are possible which may increase the sensitivity. There were suggestions, for example, to coat the first two TGEMs in the quadruple TGEM configuration with CsI and operate them at gains close to unity, which may allow to partially compensating for the light losses (Peskov, 2012). In spite of lower efficiency, the CsI-TGEM seems to be an interesting alternative to the CsI-MWPC technology especially since it allows the use of

Figure 57. Single-photoelectron collection efficiency in Ne/CH4 and Ne/CF4 mixtures, measured in pulse counting mode, versus the voltage across the TGEM; the threshold gain values for reaching full collection efficiency are indicated for each mixture.
From Azevedo, 2010.

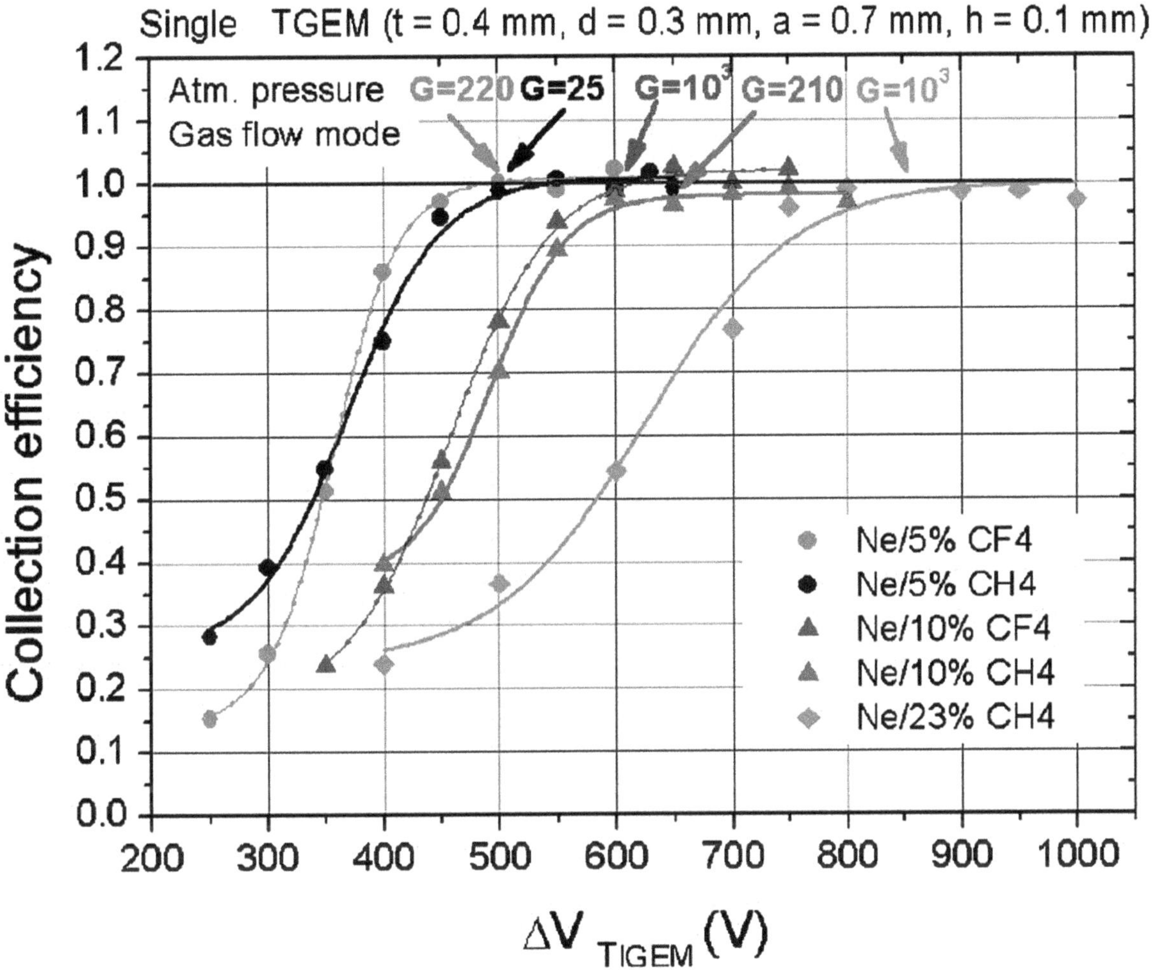

nonflammable gases and thus resolve a safety issue which can be serious in many large–scale experiments, e.g. the upgraded ALICE detector. The CsI-TGEM approach, due to its capability to suppress ion back flow and thus reduce the cathode excitation effect, has also advantages in applications dealing with high flux of charged particles or photons.

Inspired by the ALICE group success in detection of Cherenkov light with TGEMs (Martinengo, 2011; Peskov 2012), several other groups launched R&D programs aiming to further explore this promising technique. For example, a systematic study of the TGEM detector response as a function of its geometrical parameters and electrodes applied voltage was performed by a COMPASS team (Alexeev, 2012; Alexeev, 2012a: Alexeev, 2013). There are also efforts to reduce even more the ion back flow in cascaded TGEMs detectors with staggered hole alignment (Sauli, 2006).

Following the same idea, the COMPASS group tested a triple TGEM where TGEM1 and TGEM3 have aligned holes while there is a complete misalignment of the holes of TGEM2 with respect to TGEM1 and TGEM3. With this arrangement it was possible to reduce the ion back flow to values below 3%, with a measured effective gain of $2x10^5$.

2.4.4 CsI Coated Resistive Thick GEM

In Chapter 2 we described a resistive TGEM (RETGEM), having electrodes made not from metals as in ordinary TGEM, but from resistive materials, most often resistive Kapton 100XC10E5 or a resistive paste Encre MINICO used for transistors in printed circuits. These resistive electrodes protect the detector and the front-end electronics in case of occasional discharges and thus make the detector more robust and reliable in operation.

CsI photocathodes have high quantum efficiency and are stable in time only if they are deposited on specially prepared substrates, for example stainless steel coated with Al layer. During studies of RETGEMs it was discovered that the CsI photocathode remained stable even if it was evaporated directly on one of the resistive materials (Martinenego, 2011).

The RETGEM top electrodes were coated with CsI by following the standard procedure used earlier for the ALICE RICH photocathodes production (Hoedlmoser, 2006). After the evaporation of a 0.3 nm thick CsI layer, the response of the RETGEM was evaluated by the VUV scanning system described in earlier in this Chapter (see (Hoedlmoser,2006, for more details). Figure 58 shows the mean ratio of the RETGEM photo-currents to the reference PM as a function of the time after the evaporation.

The response enhancement a few hours after the coating is similar to that obtained with ALICE –RICH CsI photocathodes. After the installation into the gas chamber, the quantum efficiency of the CsI-RETGEM was estimated using a Hg lamp combined with a narrow band filter (185 nm). The induced photocurrent was measured between the top electrode of the CsI-RETGEM and the drift mesh as a function of the voltage applied across this gap. The saturation value of this current was then compared with the saturation value of the current produced by the same light beam in a TMAE filled single–wire counter. From these measurements the quantum efficiency of the CsI photocathode was evaluated to be ~17 ±2% at185 nm. Similar results (14–16%) were obtained in pulse counting mode (Agocs, 2008).

The long-term stability of CsI photocathode evaporated on the top of resistive layers has been studied by several groups. The typical results are presented in Figures 59 and 60 proves that the photocathodes remained stable at least for 100-130 days.

Of course, the rate characteristics of RETGEMs are worse than for TGEMs, but due to their efficient spark-protection RETGEMs are attractive in some applications, e.g. in RICH detectors, in visualization of UV sources in day light conditions etc.

In recent years CsI-coated REGEMs has passed through several modifications (Peskov, 2009; Peskov, 2013). One of them was the development a double-layered photosensitive RETGEM having electrodes made of resistive strips with metallic strips under them and used to detect the induced signals (Peskov, 2009; Di Mauro, 2009). This type of detector was named S-RETGEM. The results from such a detector are shown in Figure 61.

An S-RETGEM is manufactured in the following way. First, on the side of the bare G-10 plate (Figure 62a) Cu strips with circular dielectric opening (Figure 62b) were created by a photolithographic technology. The width of the strips, depending on the particular design was 0.7–1 mm, their pitch was1–1.4 mm and the circle diameter was 0.5–0.8 mm, respectively. Subsequently, the surface of each metallic strip

Figure 58. Ratio of the photocurrent values measures for different CsI-coated GEM, TGEM and RET-GEM immediately after the evaporation and 15 h latter
From Martinenego, 2011.

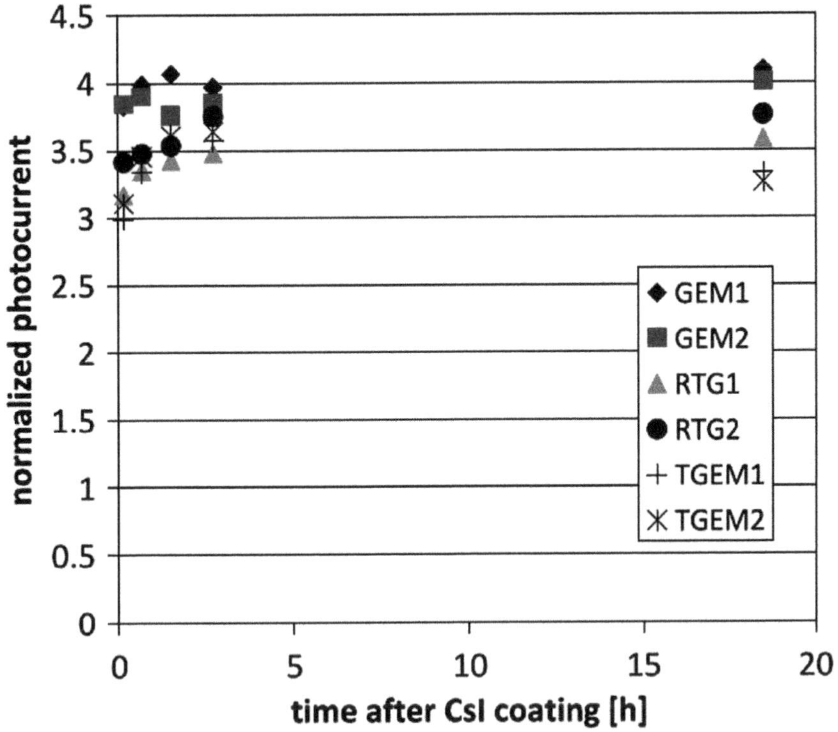

Figure 59. Results of long-term stability of a double CsI-RETGEM measured in Ar + 10% CO_2 at 1 atmosphere with two different CsI photocathodes evaporated on its top electrode

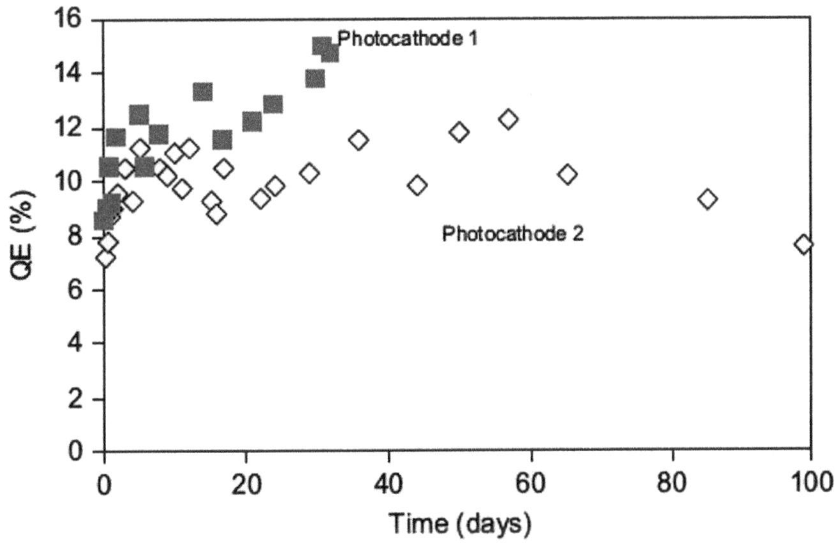

Figure 60. The quantum efficiency vs. time measured in Ar + 5% CO_2 (rhombs) and Ne (squares) at1 atmosphere; the Ar(Hg) lamp was kept on and the HV was applied to the detector only during the measurements.
From Peskov, 2009.

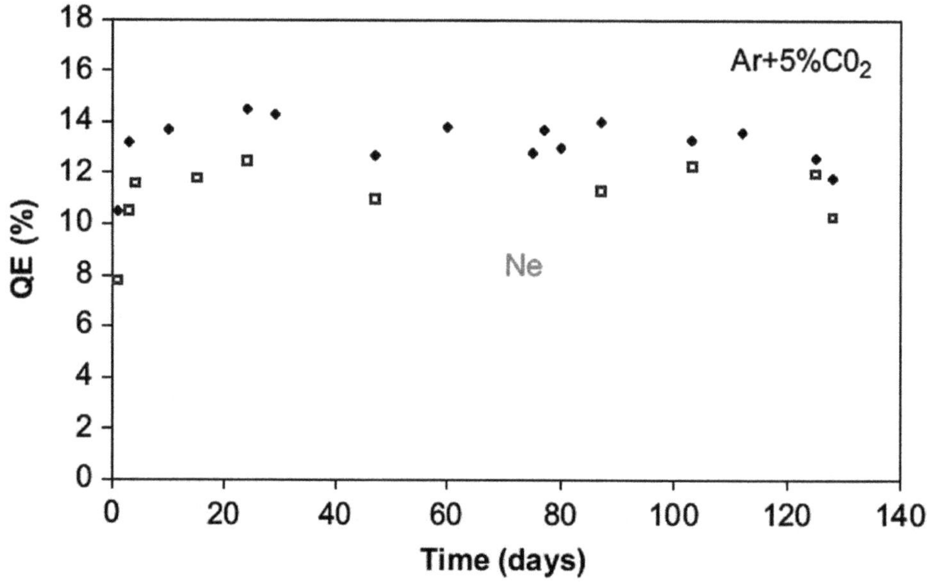

Figure 61. Gain versus rate characteristics measured for the RETGEM with CrO electrodes (open symbols), and for RETGEM with resistive kapton coated with a CsI layer (filled symbols); in the latter case, the measurements were performed with the help of an Hg lamp. Stars indicate the maximum reachable rate (for the given signal amplitude), following which sparks appear.
From Di Mauro, 2007.

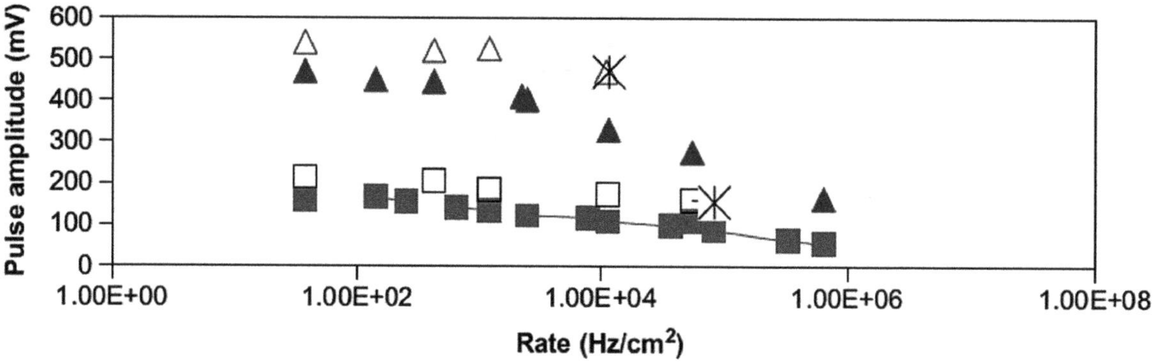

was coated by a screen-printing technology with a15 μm thick resistive layer, forming the resistive strips (Figure 62c). The gap-width between neighboring resistive strips was 0.1 mm. These double-layered plates were then cured in an oven at temperatures of 200°C followed by drilling the holes with a CNC machine in the middle of each circle opening. Their diameter was in the range of 0.2–0.5 mm. Note that it was essential to keep the diameter of the drilled holes smaller than the diameter of the uncoated by Cu circles/opening on the metallic strips.

Figure 62. Steps in the photosensitive S-RETGEM manufacturing process: (a) bare G-10 plate, (b) metallic strips with circle opening are manufactured on the G-10 plate by a photolithographic technology, (c) resistive strips are manufactured on the top of metallic strips by a screen printing technique and holes are drilled across and (d) the top surface of the S-RETGEM is coated with a CsI layer

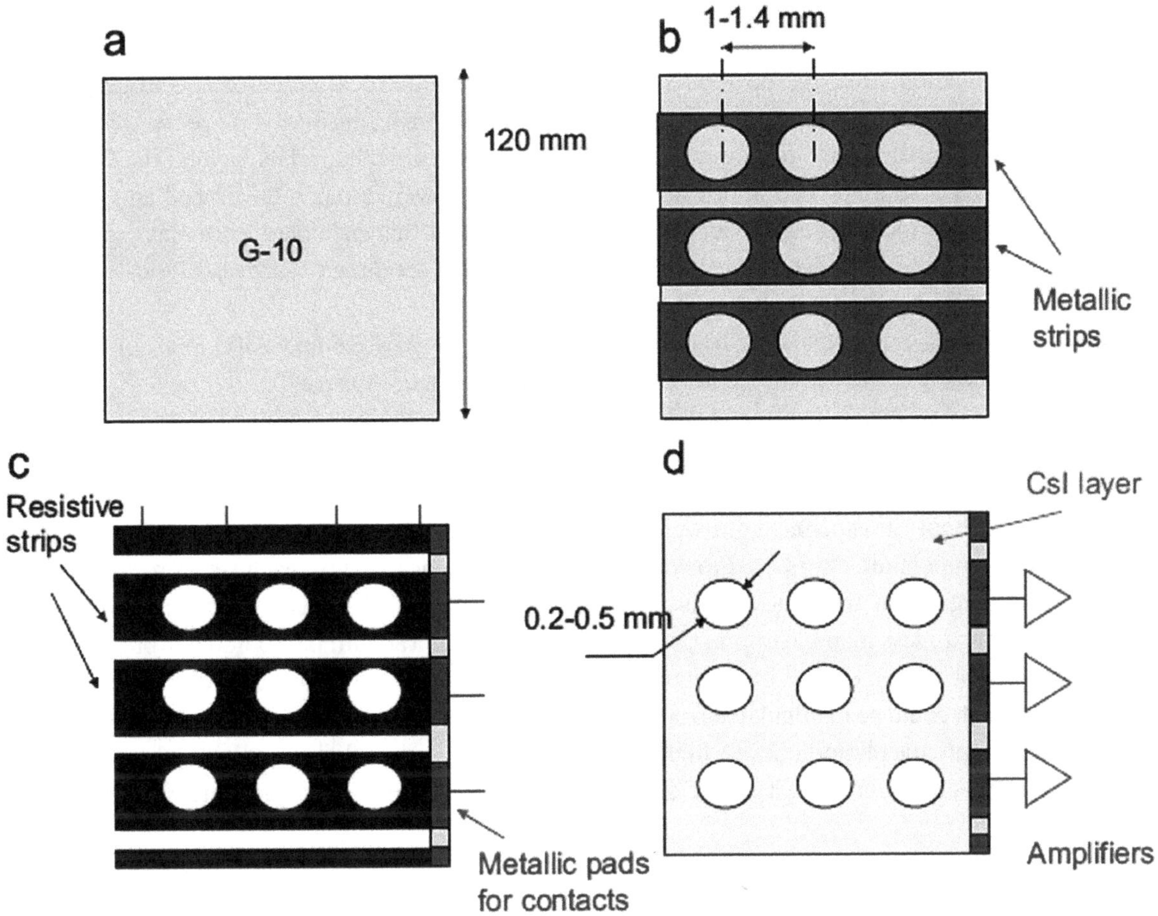

This feature provided an efficient spark-protection. S-RETGEMs with active areas of 10x10 and10x20 cm^2 have been manufactured following this technique. Finally, the cathode of each S-RETGEM was coated with a 0.35 mm-thick CsI layer (Figure 62d). The inner metallic strips have been used to collect the charge produced by avalanches in holes situated inside each strip. The resistive grid made the detector intrinsically spark protected: in the case of sparks, the resistive layers strongly restricted the current and thus the destructive power.

This approach allows building large-area detectors. Additionally, the strip geometry offers a safe operation (small individual strip's capacity and the resistive coatings makes discharges mild) and the possibility to perform 2-D position measurements. Moreover, in the case of persisting discharges in some particular holes, the strip on which these holes are located could be electrically "disabled" allowing the remaining detector surface to operate at high gas gains.

Designs with additional dielectric layers between the resistive and metallic strips were developed and tested as well.

3. A Cs_2Te PHOTOCATHODE

In some measurements, for example plasma visualization, detectors sensitive to narrow spectral intervals are often desired. Such detectors can be gaseous photomultipliers filled with vapours sensitive in narrow band interval or with gaseous detectors combined with appropriate solid photocathodes, for example various other alkali halide: CsI, KCl etc. (Carruthers, 1969).

In other applications, like the position-sensitive detector in the focal plane of spectrographs, the requirement can be exactly opposite; detectors with wide wavelength sensitivity are preferable.

This motivated researches to test other low threshold UV sensitive materials beyond the CsI. Solid photocathodes sensitive up to visible or even in infrared region were already developed and were and are widely used in various vacuum devices, but it was not obvious that such photocathodes can be combined with high-gain gaseous detectors although some of them were already successfully used in some commercial gas-filled photodiodes.

One of interesting candidate is Cs_2Te photocathode which is sensitive up to 300 nm. For the study of its stability in the gas flushed mode the set up shown in Figure 63 was used.

It consists of a glass tube with a quartz window and two metallic flanges, equipped with stainless steel tubes for the gas flow, and a special device, allowing one to make some mechanical manipulations inside the tube. This test tube was flushed with gas through an Oxysorb gadget. The gases tested were CH_4, Ar or He with or without EF vapours, e.g. specially cleaned EF. In the case of He a zeolite trap, cooled to liquid nitrogen temperatures, was used to additionally clean the gas. Before each set of measurements the system was cleaned by warming and flushing the chamber for 5-7 days. This procedure reduced the level of impurities to less than 0.01 ppm. Commercially available vacuum photodiodes were used. The photocathode to be tested, e.g. Cs_2Te, was placed inside the test tube. The test chamber was equipped with a knife, which could be manipulated from outside to break the glass envelop of the photodiodes. The gas then entered into the photodiodes. A high voltage up to a few kV could be applied to the anode. The cathode signal was read out using a charge-sensitive amplifier or a picoammeter. Mercury or hydrogen

Figure 63. Schematic view of the glass test chamber used to measure the quantum efficiency and evaluate the stability of commercial photocathodes sensitive in wide spectral interval and operating in gas atmosphere
From Charpak, 1992.

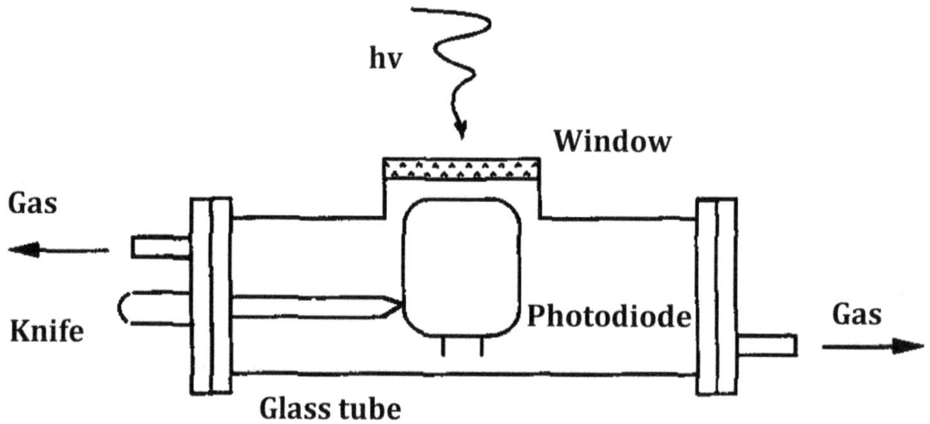

Figure 64. The quantum efficiency for a Cs_2Te photocathode as a function of time; the photocathode is exposed to different gases.
From Charpak, 1992.

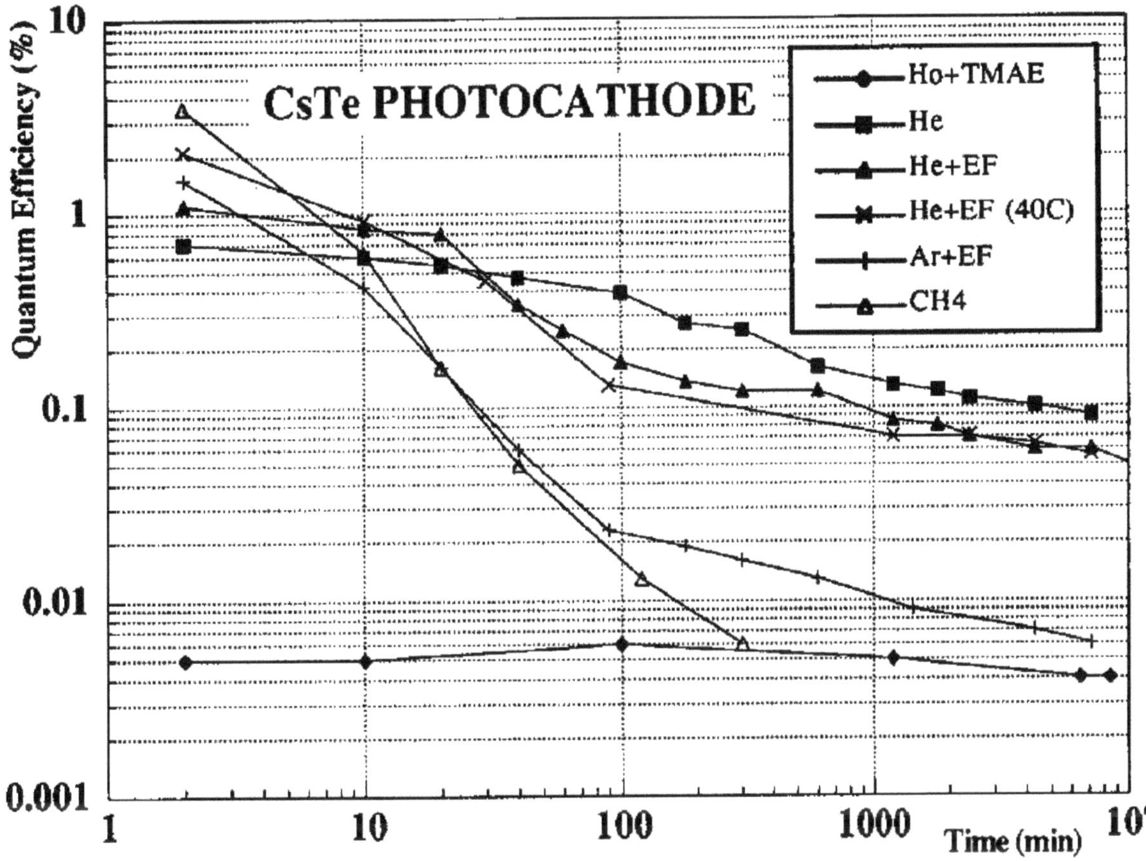

lamps with narrow-band filters, or a light beam from a monochromator, were used to illuminate both the photodiode. The changes with time of the quantum efficiency of the photocathodes were then measured.

Figure 64 shows the measured quantum efficiency as a function of time for Cs_2Te photocathodes exposed to various gases at atmospheric pressure.

Time zero corresponds to the moment when the glass envelope of the vacuum photodiodes was broken. It can be seen that the quantum efficiency degrades rather quickly in all cases, except for the cleanest gas used, He. This is not astonishing, since photodiodes filled with ultraclean gases are known to be stable. The important fact is that one can reach some stability of operation by flushing with a clean He gas under rather simple laboratory conditions. At short wavelengths, $\lambda < 200$ nm, the degradation was 3-5 times less than at longer wavelengths.

Although small impurities in the gas cause degradation of the Cs_2Te quantum efficiency, one can considerably reduce this effect by condensing a layer of clean EF on the photocathode surface. The rate of diffusion of gaseous impurities through this layer appears to be rather slow, and this allows one to expose the photocathode even to air for a short time, a few seconds. It was found that by afterwards flushing with clean He and warming the photocathode it is possible to evaporate the liquid EF layer and in this way restoring the quantum efficiency at 225 nm to about 10% of the initial value. Even better results, close to 50% of the initial value, were obtained when an organo-metallic compound was condensed on the photocathode surface.

Tests with a well outgases chamber showed that if it was just filled with an ultraclean gas without a continuous flow of gas, the quantum efficiency of the CsTe photocathode initially degrades by a factor of two during the first hour, but then remains quite stable. This indicates that the photocathode acts like a getter adsorbing and chemically reacting with the impurities in the system, such as oxygen or water, until they are fully removed. For this reason the main focus in further studies was done in sealed detectors.

Enomoto showed that the quantum efficiency of a Cs_2Te photocathode is reduced only by 5% after a 1 hour exposure to clean nitrogen (Enomoto, 1994). A short exposure to air is dramatic. The quantum efficiency dropped within seconds to less than 1%.

3.1 Cs_2Te Photocathode Manufacturing in the Laboratory

Although Cs_2Te photocathodes are routinely used in many commercial devises, e.g. multichannel plates, it was not easy to obtain information about how to best manufacture them since photonic companies usually do not make the most important details public. Some general information can be found in the excellent book of Sommer (Sommer, 1968). However, to implement the described procedures each research laboratory has to gain its own experience, which was not an easy task.

The first Cs_2Te photocathodes for gaseous detectors were manufactured using an experimental facility for manufacturing prototypes of vacuum MCPs, shown schematically in Figure 65 (Francke, 2002; Carlson, 2003; Fonte, 2005).

The system is the same that is used for the CsI photocathode production. It consisted of two connected vacuum chambers. A latch separated the evaporation chamber from the detector assembly chamber when needed. The evaporation chamber was used for the photocathode manufacturing. The evaporated photocathode was then moved with the manipulator to the adjacent assembly chamber where it was integrated in a gaseous detector or sealed and stored in a special gas tight container.

To achieve a vacuum better than 10^{-6} Torr and low levels of outgassing, the setup was pumped for a few days at elevated temperature (100-150 °C) before the evaporation process started.

First a telluride layer was evaporated on a metallic substrate or on a window surface. In the latter case, the window surface had already been coated with a thin tungsten film with a transmission for UV light of about 90%. An indium telluride compound (Te generator) was heated to 450°C where Te vapour was released and deposited on the photocathode substrate. The Te film was then exposed to Cs vapours, released from a commercially available Cs generator. At this stage the entire vacuum chamber is heated to 120° C to avoid Cs condensation on the cool parts of the setup. During the Cs evaporation, the photocathode was continuously illuminated by a UV lamp, and the photoelectric emission from the cathode was continuously monitored. The Cs generator was closed when the photocurrent reached a maximum.

An anode ring was placed in front of the photocathode substrate to collect photoelectrons emitted from the photocathode when the UV was activated. The potential of the anode ring was 150 V.

After evaporation, the setup was cooled to room temperature. The substrate with the Cs_2Te photocathode was then moved to the assembly chamber.

The evaporation chamber can also measure the quantum efficiency of the photocathode either in vacuum or by filling the chamber with different gases.

Figure 65. The setup used for manufacturing of Cs_2Te photocathodes

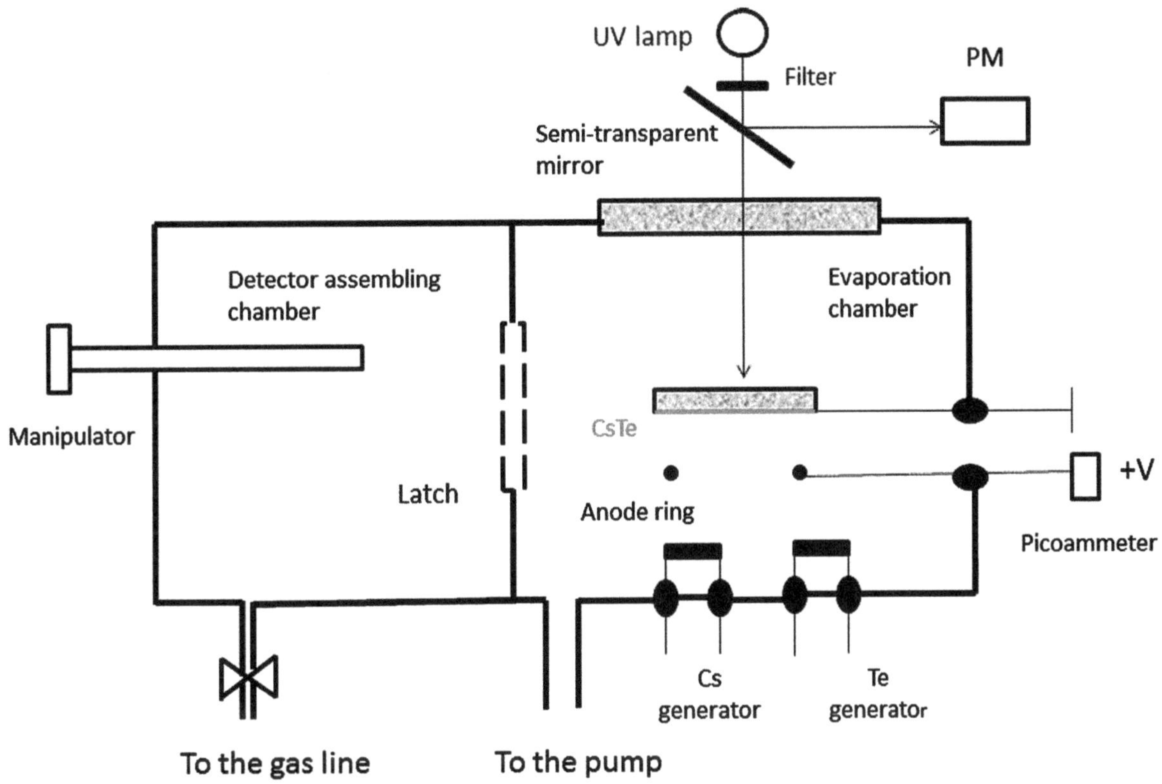

3.3 Laboratory Tests

Measurements confirmed that the quantum efficiency of Cs_2Te photocathodes exposed to gas is always a factor 2-3 less than in vacuum (Francke, 2002; Carlson, 2003; Fonte, 2005). This is due two main reasons:

1. Photoelectron back diffusion, which depending on which type of gas that is used, reduces the quantum efficiency to 10-50% compared to the quantum efficiency in vacuum.
2. Interaction with residual impurities in the gas may give another 30-50% reduction.

Figure 66 shows the quantum efficiency of one Cs_2Te photocathode produced in the laboratory and measured in a gas atmosphere at voltages ensuring a photocurrent plateau.

For comparison, Figure 67shows the Cs_2Te quantum efficiency measured in vacuum (Enomoto, 1994).

The first tests of a Cs_2Te photocathode operating in gas atmosphere in avalanche mode were done in a parallel plate geometry. Typically, gas gains of about 10^3 were achieved (see Figure 68), which is 100-1000 times lower than for CsI photocathodes.

A Cs_2Te photocathode has high quantum efficiency in much wider spectral range than a CsI photocathode. Therefore, it is more sensitive to photon feedback. Also due its lower affinity, ion feedback is not negligible. This limits the maximum gas gain.

Figure 66. The quantum efficiency of a Cs_2Te photocathode measured in He + 0.8% CH_4 + EF mixture at 1 atmosphere
From Francke, 2002.

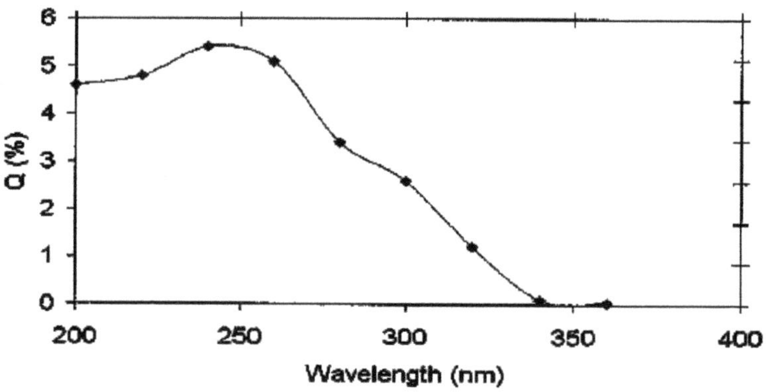

Figure 67. Quantum efficiencies measured in vacuum: (a) CsI with a MgF_2 window; (b) CsI with a quartz window, and (c) Cs_2Te with a quartz window

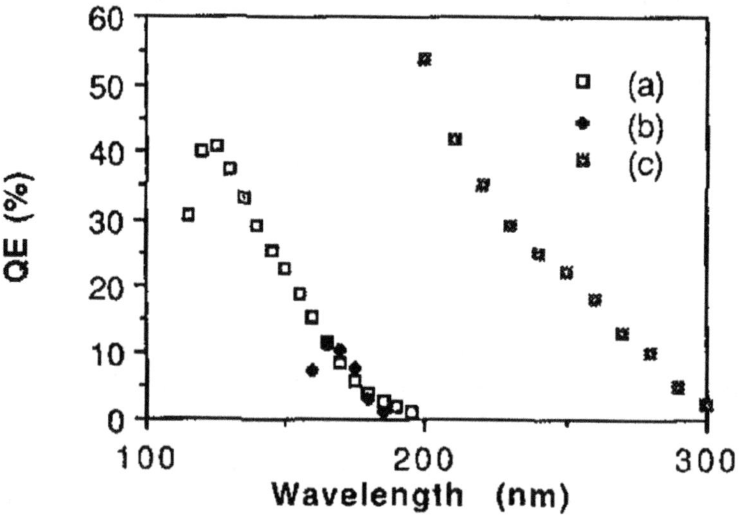

Almost a 10 times higher gas gain was achieved with a mesh parallel plate avalanche chamber combined with a capillary plate, as shown in Figure 69. In this device the voltage applied to the capillary plate is high enough to ensure high photoelectron transmission through it, but not sufficiently high for electron multiplication. The role of the capillary plate, which had inclined holes, was to attenuate the avalanche light from reaching the photocathode and to collect part of the avalanche positive ions. The multiplication process takes place in the parallel plate avalanche chamber only.

Some results from Cs_2Te aging measurements are presented in Figures 70 and 71.

As can be seen, Cs_2Te photocathodes are quite stable in Ar+ CH4 gas mixtures. However, they degrade rather fast in an Ar + isobutane gas mixture. Please recall that isobutane also reduced the quantum efficiency of CsI photocathodes.

Figure 68. Gain vs. applied voltage for a gas filled photodiode with Cs_2Te photocathode in He + 0.8% CH_4 + EF mixture
From Fonte, 2005.

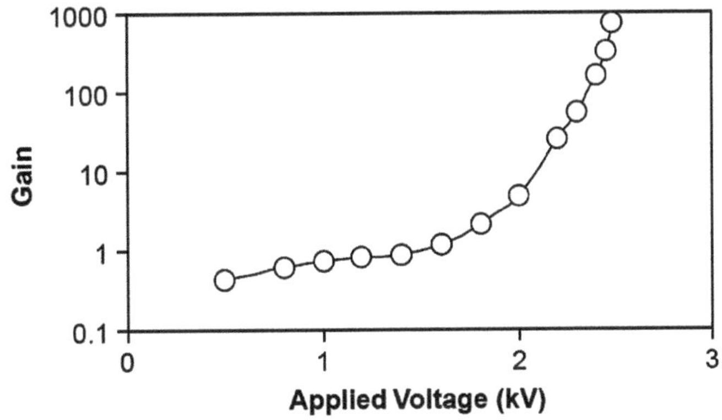

Figure 69. A schematic drawing of the gaseous detector with the Cs_2Te photocathode and the capillary light attenuator
From Francke, 2002

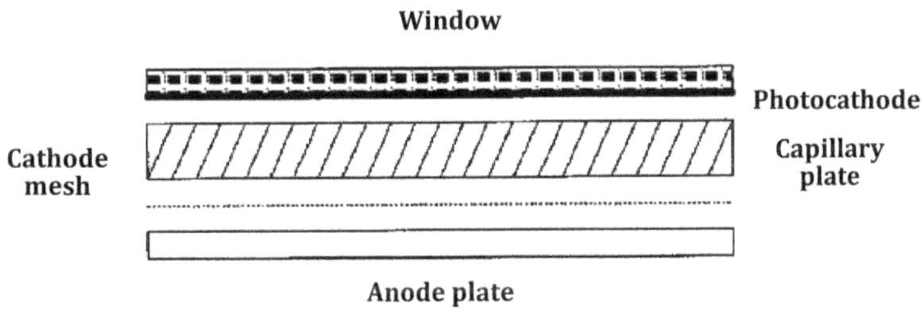

The attempts to manufacture Cs_2Te photocathode in the laboratory were not always successful. Due poorly controlled process parameters some photocathodes had much lower quantum efficiency compared to commercially produced ones. For this reason the alternative approach used by various researches was to purchase ready photocathodes delivered in sealed containers or sealed gas-filled detectors from photonic companies.

4. CONCLUSIVE REMARKS

Figure 72 summarizes the quantum efficiencies of the most important photosensitive materials used in gaseous photomultipliers.

As can be seen, the highest quantum efficiency is offered by TMAE vapours. However, it is chemically aggressive which makes it quite difficult to handle. In addition, TMAE causes strong anode wire ageing. Its low vapour pressure requires the detector to be heated. Sometimes, special blind mechanical structures are required to prevent spurious avalanches from photon feedback.

Figure 70. Aging measurements for vacuum (at applied voltages of 100 and 1000V), and gas filled photodiodes (Ar + CH_4) with Cs_2Te photocathodes at a gain of 10 (upper curve)
From Fonte 2001a.

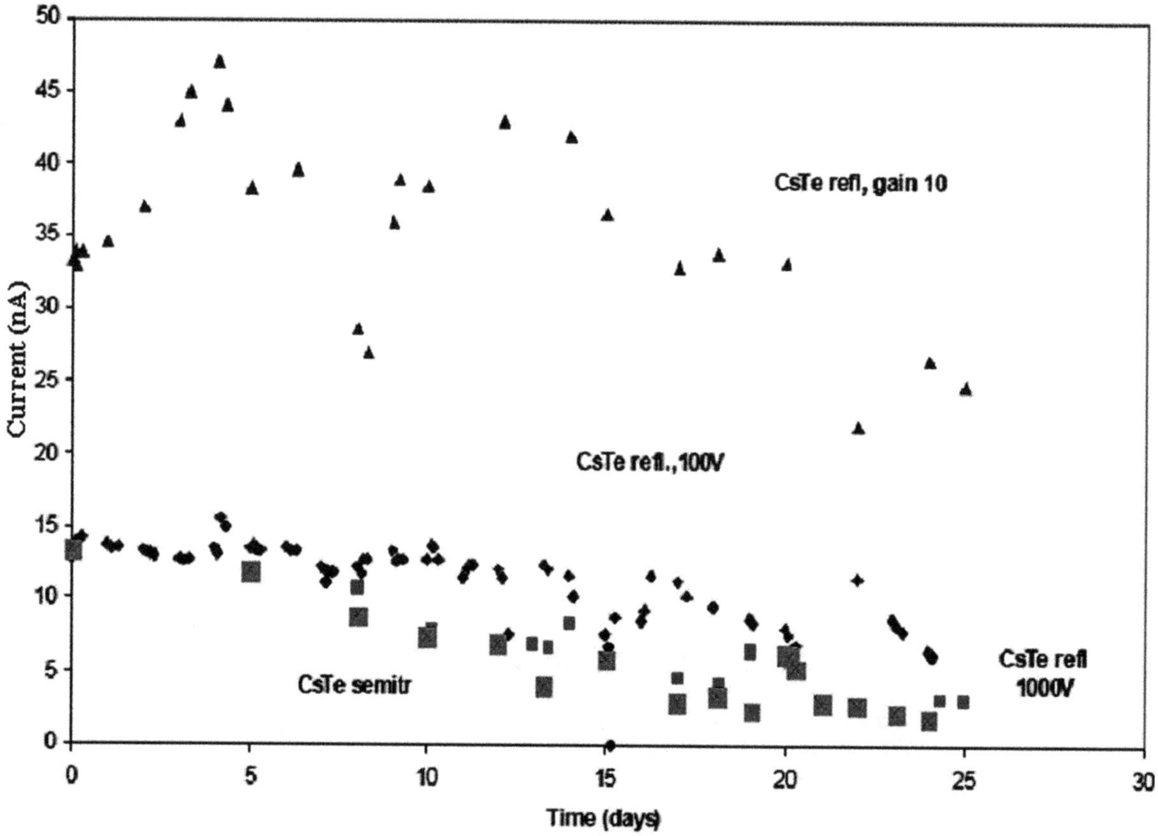

Solid CsI photocathodes offer a good alternative with only slightly lower quantum efficiency over a similar wavelength interval.

Detectors with solid CsI photocathodes have several important advantages:

1. CsI is not chemically aggressive.
2. CsI can be exposed to air for a short time without degradation of its quantum efficiency (if a proper recovering procedure is applied). Hence, it is rather easy to handle compared to other photocathodes.
3. CsI photocathodes can be easily mounted inside avalanche gaseous detectors in a glove box, and even in air.
4. CsI photocathodes are stable in in a continuous flow of gas, and do not exhibit any strong aging effect under avalanche ion bombardment.
5. In contrast to vacuum photomultipliers, CsI photodetectors can be used as large area position–sensitive detectors (tens of m^2).

Figure 71. Results of aging measurements for photodiodes with Cs_2Te photocathodes in an Ar + isobutane gas mixture
From Fonte, 2005a.

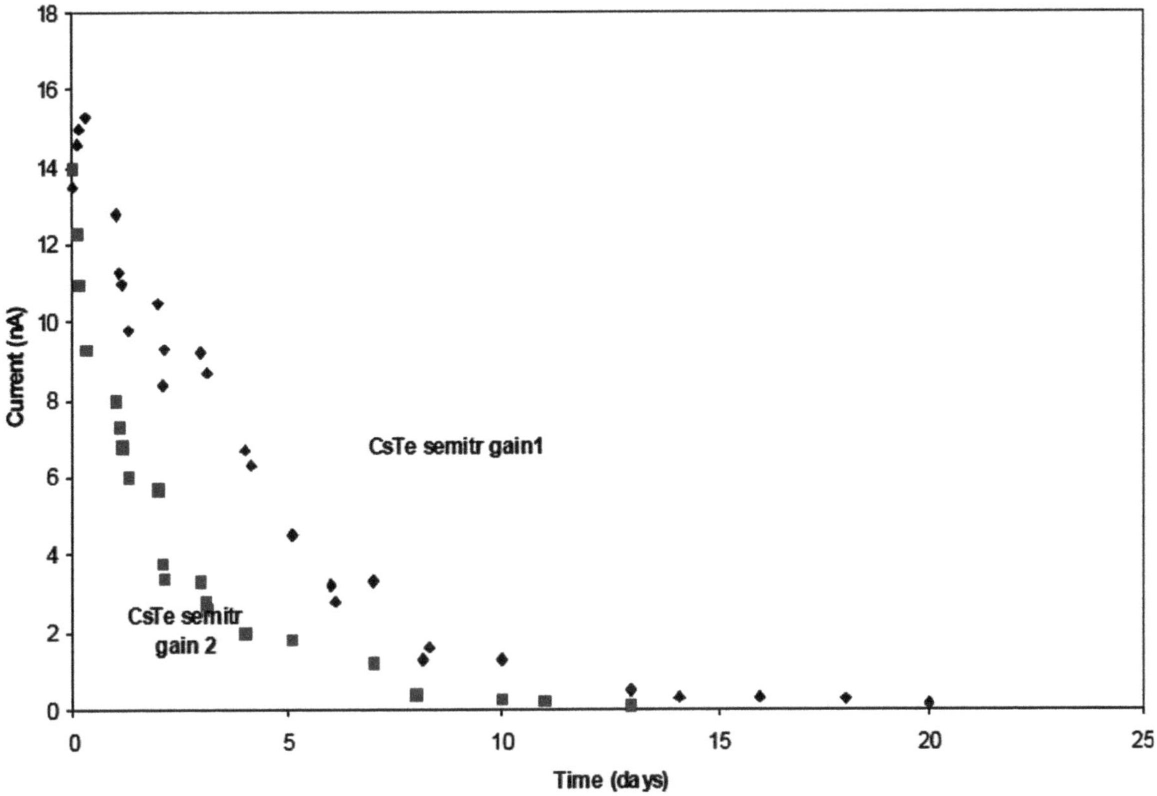

Figure 72. Quantum efficiency of the most common photosensitive materials used in gaseous detectors: filled triangles: TMAE vapours, filled squares: EF vapours, filled diamonds: TEA vapours, open diamonds and open triangles: solid CuI and CsI solid photocathodes, respectively
From Francke, 2004a.

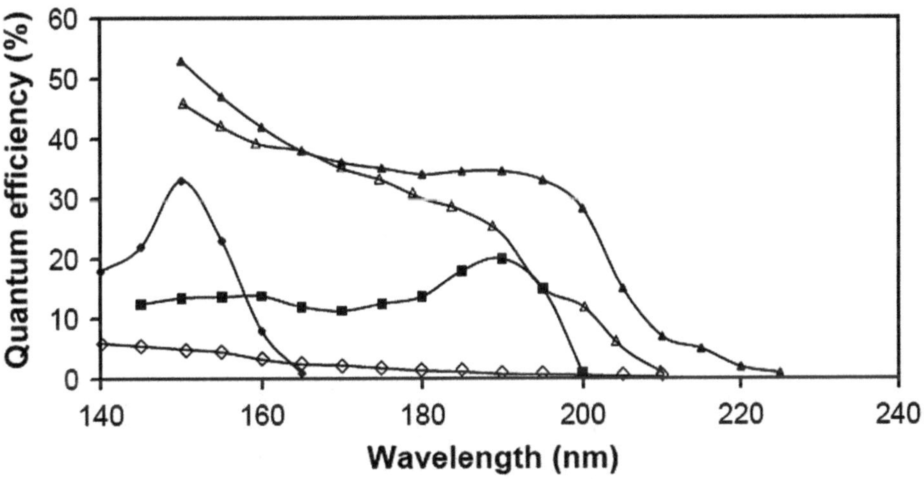

Due to these advantages, the CsI-based photon detectors represent the most effective solution concerning cost and performance for large detector area applications in relatively low rate (or low occupancy) experiments.

The main application of CsI-based photodetectors is as Cherenkov ring imaging (RICH) detectors. For example, large area CsI-MWPCs are used in the ALICE ($11m^2$) and COMPASS ($5.8m^2$) experiments at CERN. There are other applications as well. Readout of crystal scintillators, x-ray imaging, and detection of UV light under daylight conditions, etc.

There are attempts to develop gaseous photomultipliers sensitive up to 300 nm, for example with Cs_2Te photocathodes. Some other photocathode materials are also potentially interesting (see for example Guzhov, 1969; Neeraj Tripathi, 2011).

In the next Chapter, more exotic gaseous detectors having high sensitive for visible light will be described. The solid photocathodes used there (SbCs, bi alkali) are today stable only in very clean gases. Efforts were done to make them more robust by coating their surface with thin CsI layers, preventing direct contact with gases which ease the requirements on the gas cleanness.

REFERENCES

Agocs, A. G., Di Mauro, A., Ben David, A., Clark, B., Martinengo, P., Nappi, E., & Peskov, V. (2008). Study of GEM-like detectors with resistive electrodes for RICH applications. *Nuclear Instruments & Methods in Physics Research. Section A, Accelerators, Spectrometers, Detectors and Associated Equipment*, *595*(1), 128–135. doi:10.1016/j.nima.2008.07.031

Akkerman, A., Boutboul, T., Breskin, A., Chechik, R., & Gibrekhterman, A. (1994). Low-energy electron transport in alkali halides. *Journal of Applied Physics*, *76*(8), 4656–4662. doi:10.1063/1.357303

Akkerman, A., Boutboul, T., Breskin, A., Chechik, R., & Gibrekhterman, A. (1994). Low-energy electron transport in alkali halides. *Journal of Applied Physics*, *76*(8), 4656–4662. doi:10.1063/1.357303

Aleksan, R., Besson, P., Bourgeois, Ph., Garganne, P., Hamel, G., & Monchenault, P. (1994). Measurements of CsI photocathode quantum efficiency in methane. *Nuclear Instruments & Methods in Physics Research. Section A, Accelerators, Spectrometers, Detectors and Associated Equipment*, *343*(1), 173–177. doi:10.1016/0168-9002(94)90547-9

Alexeev, M., Barbosa, F., Birsa, R., Bradamante, F., Bressan, A., Chiosso, M., & Wollny, H. et al. (2012). Detection of single photons with Thick GEM-based counters. *Nuclear Instruments & Methods in Physics Research. Section A, Accelerators, Spectrometers, Detectors and Associated Equipment*, *695*, 159–162. doi:10.1016/j.nima.2011.11.079

Alexeev, M., & Birsa, , R., Bradamante, F., Bressan, A., Chiosso, Ciliberti, P., … Veloso, J.F.C.A. (2012a). Development of TGEM-based Photon Detectors for COMPASS RICH-1. *Physics Procedia*, *37*, 781–788. doi:10.1016/j.phpro.2012.02.422

Alexeev, M., Birsa, R., Bradamante, F., Bressan, A., Büchele, M., Chiosso, M., & Veloso, J. F. C. A. et al. (2014). Status and progress of the novel photon detectors based on TGEM and hybrid MPGD architectures. *Nuclear Instruments & Methods in Physics Research. Section A, Accelerators, Spectrometers, Detectors and Associated Equipment*, *766*, 133–137. doi:10.1016/j.nima.2014.07.030

Alexeev, M., Birsa, R., Bradamante, F., Bressan, A., Chiosso, M., Ciliberti, P., & Wollny, H. et al. (2009). The quest for a third generation of gaseous photon detectors for Cherenkov. *Nuclear Instruments & Methods in Physics Research. Section A, Accelerators, Spectrometers, Detectors and Associated Equipment, 610*(1), 174–177. doi:10.1016/j.nima.2009.05.069

Alkhazov, G. D. (1970). Statistics of electron avalanches and ultimate resolution of proportional counters. *Nuclear Instruments and Methods, 89*, 155–165. doi:10.1016/0029-554X(70)90818-9

Almedia, J., Amadon, A., Besson, P., Bourgeous, P., Braem, A., Breskin, A., & Williams, T. D. et al. (1995). Review of the development of cesium iodine photocathodes for applications to large RICH detectors. *Nuclear Instruments & Methods in Physics Research. Section A, Accelerators, Spectrometers, Detectors and Associated Equipment, 367*(1-3), 332–336. doi:10.1016/0168-9002(95)00571-4

Anderson, D., Kwan, S., Honeisen, B., & Peskov, V. (1991). *Technical memo on new results on CsI photocathodes: enhancement and aging.* Fermilab technical Memo *TM-1753.* Retrieved from: http://lss.fnal.gov/archive/test-tm/1000/fermilab-tm-1753.pdf

Anderson, D., Kwan, S., & Peskov, V. (1994). CsI and some new photocathodes. *Nuclear Instruments & Methods in Physics Research. Section A, Accelerators, Spectrometers, Detectors and Associated Equipment, 343*(1), 109–116. doi:10.1016/0168-9002(94)90539-8

Anderson, D. F., Kwan, S., & Peskov, V. (1993). CsI photocathode quantum efficiency and a simple production technique. *Nuclear Instruments & Methods in Physics Research. Section A, Accelerators, Spectrometers, Detectors and Associated Equipment, 326*(3), 611–612. doi:10.1016/0168-9002(93)90866-G

Anderson, D. F., Kwan, S., Peskov, V., & Hoeneisen, B. (1992). Properties of CsI and CsI-TMAE photocathodes. *Nuclear Instruments & Methods in Physics Research. Section A, Accelerators, Spectrometers, Detectors and Associated Equipment, 323*(3), 626–634. doi:10.1016/0168-9002(92)90007-Q

Azevedo, C. D. R., Cortesi, M., Lyashenko, A. V., Breskin, A., Chechik, R., Miyamoto, J., & dos Santos, J. M. F. et al. (2010). Towards TGEM UV-photon detectors for RICH: On single-photon detection efficiency in Ne/CH4 and Ne/CF4. *JINST, 5*(01P01002), 1–12. doi:10.1088/1748-0221/5/01/P01002

Azmount, B., Anderson, W., Crary, D., Durham, J., Hemmik, T., Kamin, J…Woody, C (2006). A study of gain stability and charging effects in GEM foils. *IEEE Nucl. Sci. Conf. Record, MP3-4*, 3847-3854.

Bellazzini, R., Spandre, G., Minuti, M., Baldini, L., Latronico, L., Omodei, N., & Lempinen, E. et al. (2007). A sealed gas pixel detector for X-ray imaging. *Nuclear Instruments & Methods in Physics Research. Section A, Accelerators, Spectrometers, Detectors and Associated Equipment, 579*(2), 853–858. doi:10.1016/j.nima.2007.05.304

Braem, A. (1995). Private communication

Braem, A., De Cataldo, G., Davenport, M., Di Mauro, A., Franco, A., Gallas, A., & Peskov, V. et al. (2007). The current progress of the ALICE Ring Imaging Cherenkov Detector. *Nuclear Instruments & Methods in Physics Research. Section A, Accelerators, Spectrometers, Detectors and Associated Equipment, 580*(2), 1144–1147. doi:10.1016/j.nima.2007.06.050

Braem, A., DiMauro, A., Nappi, E., Ljubicic, A. Jr, Paic, G., Piuz, F., & Williams, T. D. et al. (1994). Fast RICH detector with a cesium iodide photocathode at atmospheric pressure. *Nuclear Instruments & Methods in Physics Research. Section A, Accelerators, Spectrometers, Detectors and Associated Equipment*, *343*(1), 163–172. doi:10.1016/0168-9002(94)90546-0

Breskin, A. (1996). CsI UV photocathodes: History ad mystery. *Nuclear Instruments & Methods in Physics Research. Section A, Accelerators, Spectrometers, Detectors and Associated Equipment*, *371*(1-2), 116–136. doi:10.1016/0168-9002(95)01145-5

Breskin, A., Buzulutskov, A., Chechik, R., Sinh, B. K., Bondar, A., & Shekhtman, L. (2002). Sealed GEM photomutiplier with a CsI photocathode: Ion feedback and aging. *Nuclear Instruments & Methods in Physics Research. Section A, Accelerators, Spectrometers, Detectors and Associated Equipment*, *478*(1-2), 225–229. doi:10.1016/S0168-9002(01)01752-1

Breskin, A., Chechik, R., Vartsky, D., Malamud, G., & Mine, P. (1994). A correction of the quantum efficiency of CsI and other photocathodes due to the recalibration of the reference photomultipliers. *Nuclear Instruments & Methods in Physics Research. Section A, Accelerators, Spectrometers, Detectors and Associated Equipment*, *343*(1), 159–162. doi:10.1016/0168-9002(94)90545-2

Buzulutskov, A. F. (2008). Gaseous photodetectors with solid photocathodes. *Physics of Particles and Nuclei*, *39*(3), 424–453. doi:10.1134/S1063779608030052

Carlson, P., Francke, T., Lund-Jensen, B., & Peskov, V. (2003). Gaseous photomultipliers with solid photocathodes for the detection of sparks, flames and dangerous gases. *Nuclear Instruments & Methods in Physics Research. Section A, Accelerators, Spectrometers, Detectors and Associated Equipment*, *505*(1-2), 207–210. doi:10.1016/S0168-9002(03)01053-2

Carruthers, G. R. (1969). Magnetically focused electronographic image convertors for space astronomy applications. *Applied Optics*, *8*(3), 633–638. doi:10.1364/AO.8.000633 PMID:20072272

Charpak, G., Benaben, P., Breuil, P., Di Mauro, A., Martinenego, P., & Peskov, V. (2009). Development of new hole-type avalanche detectors and the first results of their applications. *IEEE Transactions on Nuclear Science*, *55*(3), 1657–1663. doi:10.1109/TNS.2008.924084

Charpak, G., Fonte, P., Peskov, V., Sauli, F., Scigocki, D., & Stuart, D. (1991). Investigation of operation of a parallel-plate avalanche chamber with a CsI photocathode under high gain conditions. *Nuclear Instruments & Methods in Physics Research. Section A, Accelerators, Spectrometers, Detectors and Associated Equipment*, *307*(1), 63–68. doi:10.1016/0168-9002(91)90131-9

Charpak, G., Gaudean, I., Giomataris, Y., Peskov, V., Scigodki, D., Sauli, F., & Stuart, D. (1993). Some studies of the applications of CsI photocathodes in gaseous detectors. *Nuclear Instruments & Methods in Physics Research. Section A, Accelerators, Spectrometers, Detectors and Associated Equipment*, *333*(2-3), 391–398. doi:10.1016/0168-9002(93)91181-L

Charpak, G., Lemenovski, D., Peskov, V., & Scigocki, D. (1991a). New photocathodes for fast gaseous detectors. *Nuclear Instruments & Methods in Physics Research. Section A, Accelerators, Spectrometers, Detectors and Associated Equipment*, *310*(1-2), 128–132. doi:10.1016/0168-9002(91)91010-S

Charpak, G., Peskov, V., & Sauli, F. (1992). Preliminary results of the study of gaseous detectors with solid photocathodes sensitive in the spectral region from 105 to 300 nm. *Nuclear Instruments & Methods in Physics Research. Section A, Accelerators, Spectrometers, Detectors and Associated Equipment, 323*(1-2), 445–451. doi:10.1016/0168-9002(92)90330-7

Charpak, G., Peskov, V., Scigoscki, D., & Vlalbis, J. (1989). Mew scintillators for photosensitive gaseous detectors. *Proceedings of the Sympos. on particle identification at high-luminosity hadron colliders.* Retrieved from: http://lss.fnal.gov/conf/C890405/

Chechik, R., Cortesi, M., Breskin, A., Vartsky, D., Bar, D., & Dangendorf, V. (2006). *Thick GEM-like (TGEM) detectors and their possible applications.* Retrieved from: http://arxiv.org/ftp/physics/papers/0606/0606162.pdf

Comby, G., Mangeot, Ph., Prugne, P., Chalot, J. F., Coulareau, E., Quidort, J., & Tichit, J. et al. (1980). Multi-needle detector with cathode focusing. *Nuclear Instruments and Methods, 176*(1-2), 313–321. doi:10.1016/0029-554X(80)90722-3

Dangendorf, V., Breskin, A., Chechik, R., & Schmidt-Boecking, H. (1990). A gas-filled UV –photon detector with CsI photocathode for the detection of Xe light. *Nuclear Instruments & Methods in Physics Research. Section A, Accelerators, Spectrometers, Detectors and Associated Equipment, 289*(1-2), 322–324. doi:10.1016/0168-9002(90)90276-C

Dangendorf, V., Breskin, A., Chechik, R., & Schmidt-Boecking, H. (1990a). An X-ray imaging gas scintillation detectors with CsI-wire chamber UV photon readout. *Proceedings SPIE 1235, Instrumentation in Astronomy VII.*

Dekker, A. J. (1962). *Solid State Physics.* Englewood Cliffs, NJ: Prentice-Hall.

Dekker, A. L. (1958). *Solid State Physics.* London: MacMillan.

Di Mauro, A., Cozza, D., Davenport, M., Di Bari, D., Elia, D., Martinengo, P., & Williams, T. D. et al. (1999). Performance of large area CsI-RICH prototypes for ALICE at LHC: For the ALICE collaboration. *Nuclear Instruments & Methods in Physics Research. Section A, Accelerators, Spectrometers, Detectors and Associated Equipment, 433*(1-2), 190–200. doi:10.1016/S0168-9002(99)00300-9

Di Mauro, A., Lund-Jensen, B., Martinengo, P., Nappi, E., Oliveira, R., Peskov, V., & Santiard, J. C. et al. (2007). Development of innovative micro-pattern gaseous detectors with resistive electrodes and first results of their applications. *Nuclear Instruments & Methods in Physics Research. Section A, Accelerators, Spectrometers, Detectors and Associated Equipment, 581*(1-2), 225–231. doi:10.1016/j.nima.2007.07.083

Di Mauro, A., Martinenego, P., Nappi, E., Oliveira, R., Peskov, V., Pietropaolo, F., & Picchi, P. (2009). GEM with double layered micropattern electrodes and their applications. *IEEE Transactions on Nuclear Science, 56*(3), 1550–1557. doi:10.1109/TNS.2009.2020864

Enomoto, R., Sumiyoshi, T., Fujita, Y, (1994). Test of various photocathodes. *Nucl. Instrum. and Methods in Phys. Res., A343*, 117-120.

Fonte, P., Francke, T., Pavlopoulos, N., Peskov, V., & Rodionov, I. (2005a). *Novel single photon detectors for UV imaging*. Retrieved from http://www.arxiv.org/abs/physics/0502027

Fonte, P., Peskov, V., & Ramsey, B. (1999). The fundamental limitations of high-rate gaseous detectors. *IEEE Transactions on Nuclear Science*, *46*(3), 321–325. doi:10.1109/23.775537

Fonte, P., Peskov, V., & Sauli, F. (1991). VUV emission and breakdowns in parallel-plate chambers. *Nuclear Instruments & Methods in Physics Research. Section A, Accelerators, Spectrometers, Detectors and Associated Equipment*, *310*(1-2), 140–145. doi:10.1016/0168-9002(91)91013-L

Fonte, P., Peskov, V., & Sauli, F. (1991a). Feedback and breakdown in parallel-plate chambers. *Nuclear Instruments & Methods in Physics Research. Section A, Accelerators, Spectrometers, Detectors and Associated Equipment*, *305*(1), 91–110. doi:10.1016/0168-9002(91)90522-R

Fonte, P., Francke, T., Pavlopoulos, N., Peskov, V., & Rodionov, I. (2005). Novel single photon detectors for UV imaging. *Nucl. Instrum. and Methods in Phys. Res.*, *A553*, 30-34.

Francke, T., Eklund, M., Ericsson, L., Kristofferson, T., Peskov, V., Rantanene, J., & Ullberg, C. et al. (2001). Dose reduction in medical X-ray imaging using noise free photon counting. *Nuclear Instruments & Methods in Physics Research. Section A, Accelerators, Spectrometers, Detectors and Associated Equipment*, *471*(1-2), 85–87. doi:10.1016/S0168-9002(01)00920-2

Francke, T., & Peskov, V. (2003). Aging in gaseous photodetectors. *Nuclear Instruments & Methods in Physics Research. Section A, Accelerators, Spectrometers, Detectors and Associated Equipment*, *515*(1-2), 292–301. doi:10.1016/j.nima.2003.09.013

Francke, T., & Peskov, V. (2003). Aging in gaseous photodetectors. *Nuclear Instruments & Methods in Physics Research. Section A, Accelerators, Spectrometers, Detectors and Associated Equipment*, *515*(1-2), 292–301. doi:10.1016/j.nima.2003.09.013

Francke, T., & Peskov, V. (2004a). Photosensitive gaseous detectors and their applications. *Nuclear Instruments & Methods in Physics Research. Section A, Accelerators, Spectrometers, Detectors and Associated Equipment*, *525*(1-2), 1–5. doi:10.1016/j.nima.2004.03.017

Francke, T., Peskov, V., Rodionov, I., & Fonte, P. (2004). High rate (up to 10^5 Hz/cm^2), high position resolution (30 μm) photosensitive resistive plate chambers. *Nuclear Instruments & Methods in Physics Research. Section A, Accelerators, Spectrometers, Detectors and Associated Equipment*, *533*, 163–168. doi:10.1016/j.nima.2004.07.021

Francke, T., Peskov, V., Rodionov, I., & Sokolova, T. (2002). Novel position-sensitive gaseous detectors with solid photocathodes. *IEEE Transactions on Nuclear Science*, *49*(3), 977–983. doi:10.1109/TNS.2002.1039600

Fraser, G. W. (2009). X-ray detector in astronomy. Cambridge University Press.

Garty, G., Breskin, A., Chechik, R., & Shefer, E. (1999). Single photoelectron detection with a low-pressure gas electron multiplier coupled to a CsI photocathode. *Instrum. and Methods in Phys. Res.*, *A433*, 476–481.

Guzhov, A. A., & Shuba, Yu. A. (1969). Measurement of photoelectron energy spectra with a secondary electron multiplier. *Journal of Applied Spectroscopy*, *10*(6), 692–693. doi:10.1007/BF00614424

Heroux, L., McMahon, W. J., & Hinteregger, H. E. (1966). The influence of cathode thickness and aging on the photoelectric efficiencies of LiF and CsI in XVUV. *Applied Optics*, *5*(8), 1338–1339. doi:10.1364/AO.5.001338 PMID:20057543

Hoedlmoser, H., Braem, A., De Cataldo, G., Davenport, M., Di Mauro, A., Franco, A., & Peskov, V. (2006). Production technique and quality evaluation of CsI photocathode for the ALICE/HMPID detector. *Nuclear Instruments & Methods in Physics Research. Section A, Accelerators, Spectrometers, Detectors and Associated Equipment*, *566*(2), 338–350. doi:10.1016/j.nima.2006.07.047

Hoedlmoser, H., Braem, A., De Cataldo, G., Davenport, M., Di Mauro, A., Franco, A., & Schyns, E. et al. (2007). Long term performance and ageing of CsI photocathodes for the ALICE/HMPID detector. *Nuclear Instruments & Methods in Physics Research. Section A, Accelerators, Spectrometers, Detectors and Associated Equipment*, *574*(1), 28–38. doi:10.1016/j.nima.2007.01.101

Iacobaeus, C. A., Breskin, A., Danielsson, M., Francke, T., Mormann, D., Ostling, J., & Peskov, V. (2004). Advances in capillary-based gaseous UV imaging detectors. *Nuclear Instruments & Methods in Physics Research. Section A, Accelerators, Spectrometers, Detectors and Associated Equipment*, *525*(1-2), 42–48. doi:10.1016/j.nima.2004.03.022

Karabadzhak, G. F., Peskov, V. D., & Podolyak, E. R. (1983). Coordinate gas counters with supehigh energy resolution for plasma diagnostics. *Nuclear Instruments & Methods in Physics Research. Section A, Accelerators, Spectrometers, Detectors and Associated Equipment*, *217*, 55–64.

Krizan, P., Cindro, M., Cindro, V., Korpar, S., Omahen, G., Stanovnik, A., & Starič, M. (1995). Measurements of the quantum efficiency of CsI photocathodes in a multiwire proportional chamber. *Nuclear Instruments & Methods in Physics Research. Section A, Accelerators, Spectrometers, Detectors and Associated Equipment*, *364*(2), 243–252. doi:10.1016/0168-9002(95)00471-8

Lu, C., & McDonald, K. T. (1994). Properties of reflective and semitransparent CsI photocathodes. *Nuclear Instruments & Methods in Physics Research. Section A, Accelerators, Spectrometers, Detectors and Associated Equipment*, *343*(1), 135–151. doi:10.1016/0168-9002(94)90543-6

Maier-Komor, P., Bauer, B. B., Friese, J., Gernhäuser, R., Kienle, P., Körner, H. J., & Zeitelhack, K. et al. (1995). Preparation of reflective CsI photocathodes with reproducibly high quantum efficiency. *Nuclear Instruments & Methods in Physics Research. Section A, Accelerators, Spectrometers, Detectors and Associated Equipment*, *362*(1), 183–188. doi:10.1016/0168-9002(95)00314-2

Malamud, G. P., Mine, P., Vartsky, D., Equer, B., Breskin, A., & Chechik, R. (1994). The photo-efficiency of CsI, amorphous silicon and organometallic reflective photocathode materials. *Nuclear Instruments & Methods in Physics Research. Section A, Accelerators, Spectrometers, Detectors and Associated Equipment*, *343*(1), 121–128. doi:10.1016/0168-9002(94)90541-X

Malter, L. (1936). Thin film field emission. *Physical Review*, *50*(1), 48–58. doi:10.1103/PhysRev.50.48

Martinengo, P., Peskov, V., Breskin, A., Di Mauro, A., Mayani Paras, D., Molnar, L., & van Hoorne, J. et al. (2011). R&D results on a CsI-coated triple thick GEM-based photodetector. *Nuclear Instruments & Methods in Physics Research. Section A, Accelerators, Spectrometers, Detectors and Associated Equipment*, *639*(1), 126–129. doi:10.1016/j.nima.2010.10.033

McDaniel, E.W. (1964). *Collision phenomena in ionized gases*. Wiley.

Mormann, D., Breskin, A., Chechik, R., & Shalem, C. (2004). Operation principles and properties of the multi-GEM gaseous photomultiplier with reflective photocathode. *Nuclear Instruments & Methods in Physics Research. Section A, Accelerators, Spectrometers, Detectors and Associated Equipment*, *530*(3), 258–274. doi:10.1016/j.nima.2004.03.212

Nappi, E. (2001). CsI RICH detectors in high-energy physics experiments. *Nuclear Instruments & Methods in Physics Research. Section A, Accelerators, Spectrometers, Detectors and Associated Equipment*, *471*(1-2), 18–24. doi:10.1016/S0168-9002(01)00910-X

Nappi, E., & Peskov, V. (2013). Imaging gaseous detectors and their applications. Wiley-VCH. doi:10.1002/9783527640294

Nappi, E., Posa, F., Satalino, G., Scognetti, T., Tomasicchino, G., Braem, A., & Vranic, D. (1992). Development of a large area advanced fast RICH detectors for particle identification at the large hadron collider operated with heavy ions. Detector R&D proposal. CERN/DRDC 92-3. *DRDC*, *P35*, 1–19.

Ochi, A., Tanimori, T., Nishi, Yu., Nishi, Ya., Nagayoshi, T., Ohashi, Y., & Toyokawa, H. et al. (2002). Use of a Micro-Strip Gas Chamber conductive capillary plate for time-resolved X-ray area detection. *Nuclear Instruments & Methods in Physics Research. Section A, Accelerators, Spectrometers, Detectors and Associated Equipment*, *477*(1-3), 48–54. doi:10.1016/S0168-9002(01)01907-6

Ohring, M. (1992). *The Materials Science of Thin Films*. New York: Academic Press. doi:10.1016/B978-0-08-051118-4.50007-4

Paic, G., Piuz, F, Nappi, E., Manzari, V., Posa, F., Tomasicchio, G., ...Tustonic, T. (1996). *Status report of the CsI-RICH Collaboration 1995: development of a large area advanced fast RICH detector for particle identification at the LHC operated with heavy ions*. LDRB- Status report-RD-26., CERN-LHCC-96-020, 1-25

Periale, L., Peskov, V., Braem, A., Di Mauro, A., Martinenego, P., Picchi, P., & Sipila, H. et al. (2007). Development of new sealed UV sensitive gaseous detectors and their applications. *Nuclear Instruments & Methods in Physics Research. Section A, Accelerators, Spectrometers, Detectors and Associated Equipment*, *572*(1), 189–192. doi:10.1016/j.nima.2006.10.272

Periale, L., Peskov, V., Carlson, P., Francke, T., Pavlopoulos, P., Picchi, P., & Pietropaolo, F. (2001). *Detection of the primary scintillation light from dense Ar, Kr and Xe with novel photosensitive gaseous detectors*. ArXive; Physics/0106048

Periale, L., Peskov, V., Carlson, P., Francke, T., Pavlopoulos, P., Picchi, P., & Pietropaolo, F. (2002). Detection of primary scintillation light from dense Ar, Kr and Xe with novel photosensitive gaseous detectors. *Nuclear Instruments & Methods in Physics Research. Section A, Accelerators, Spectrometers, Detectors and Associated Equipment*, *478*(1-2), 377–383. doi:10.1016/S0168-9002(01)01779-X

Peskov, V. (1995). Gaseous detectors with solid photocathodes. *Proceedings of the first International four seas Conference*. Trieste, Italy: CERN. Retrieved from: http://cds.cern.ch/record/282759/files/CERN-97-06.pdf

Peskov, V. (2013). *Stability tests of GEMs with cylindrical holes fo the ALICE TPC upgrade*. Retrieved from https://indico.cern.ch/getFile.py/access?contribId=1&sessionId=1&resId=1&materialId=slides&confId=245535

Peskov, V., Bencze, G., Di Mauro, A., Martinenego, P., Mayani, D., Molnar, L., & Shukla, I. et al. (2012). First observation of Cherenkov rings with a large area CsI-TGEM-based RICH prototype. *Nuclear Instruments & Methods in Physics Research. Section A, Accelerators, Spectrometers, Detectors and Associated Equipment*, *695*, 154–158. doi:10.1016/j.nima.2011.11.045

Peskov, V., Borovik-Romanov, A., Sokolova, T., & Silin, E. (1994). Gaseous detectors of ultraviolet and visible photons. *Nuclear Instruments & Methods in Physics Research. Section A, Accelerators, Spectrometers, Detectors and Associated Equipment*, *353*(1-3), 184–188. doi:10.1016/0168-9002(94)91634-9

Peskov, V., Charpak, G., Dominik, W., & Sauli, F. (1989). Investigation of light emission from a parallel-plate avalanche chamber filled with noble gases and with TEA, TMAE and H_2O vapours at atmospheric pressure. *Nuclear Instruments & Methods in Physics Research. Section A, Accelerators, Spectrometers, Detectors and Associated Equipment*, *277*(2-3), 547–556. doi:10.1016/0168-9002(89)90788-2

Peskov, V., Charpak, G., Mine, P., Sauli, F., Scigocki, D., Seguinot, J., & Ypsilantis, T. et al. (1988). Liquid and solid photocathodes. *Nuclear Instruments & Methods in Physics Research. Section A, Accelerators, Spectrometers, Detectors and Associated Equipment*, *269*(1), 149–160. doi:10.1016/0168-9002(88)90872-8

Peskov, V., Cortesi, M., Chechik, R., & Breskin, A. (2010). Further evaluation of a TGEM UV-photon detector for RICH – comparison with MWPC. *JINST*, *5*(11P11004), 1–25. doi:10.1088/1748-0221/5/11/P11004

Peskov, V., Martinengo, P., Nappi, E., Oliveira, R., Paic, G., Pietropaolo, F., & Picchi, P. (2009). Progress in the development of photosensitive GEMs with resistive electrodes manufactured by a screen printing technology. *Nuclear Instruments & Methods in Physics Research. Section A, Accelerators, Spectrometers, Detectors and Associated Equipment*, *610*(1), 169–173. doi:10.1016/j.nima.2009.05.070

Peskov, V., Silin, E., Sokolova, T., & Rodionov, I. (1999). First attempts to combine capillary tubes with photocathodes. *Nuclear Instruments & Methods in Physics Research. Section A, Accelerators, Spectrometers, Detectors and Associated Equipment*, *433*(1-2), 492–501. doi:10.1016/S0168-9002(99)00346-0

Piuz, F. (1996). CsI-photocathode and RICH detectors. *Nuclear Instruments & Methods in Physics Research. Section A, Accelerators, Spectrometers, Detectors and Associated Equipment*, *371*(1-2), 96–115. doi:10.1016/0168-9002(95)01442-X

Raether, H. (1964). *Electron avalanches and breakdown in gases*. London: Butterworth.

Rodionov, I., Bidault, J.-M., Crotty, I., Fonte, P., Galy, F., Peskov, V., & Zanette, O. (2005). *Advanced gaseous photodetectors for hyperspectroscopy and other applications*. 2005 IEEE Nuclear Science Symposium Conference Record.

Sauli, F. (1982). Rediscovering the gas-filled photodiode. *Nuclear Instruments and Methods in Physics Research, 203*(1-3), 601–604. doi:10.1016/0167-5087(82)90681-0

Sauli, F. (2011). Seminal Papers by Georges Charpak. Elsevier, Nuclear Instruments and Methods. *Physics Research Section A: Accelerators, Spectrometers, Detectors and Associated Equipment.*

Sauli, F., Ropelewski, L., & Everaerts, P. (2006). Ion feedback suppression in time projection chambers. *Nuclear Instruments and Methods in Physics Research, 560*(2), 269–277. doi:10.1016/j.nima.2005.12.239

Sauli, F., Ropelewski, L., & Everaerts, P. (2006). Ion feedback suppression in time projection chambers. *Nuclear Instruments and Methods in Physics Research Section A, 560*(2), 269–277. doi:10.1016/j.nima.2005.12.239

Séguinot, J., Charpak, G., Giomataris, Y., Peskov, V., Tischhauser, J., & Ypsilantis, T. (1990). Reflective UV photocathodes with gas-phase extraction: Solid, liquid and absorbed thin films. *Nuclear Instruments & Methods in Physics Research. Section A, Accelerators, Spectrometers, Detectors and Associated Equipment, 297*(1-2), 133–147. doi:10.1016/0168-9002(90)91359-J

Simon, F., Azmount, B., Becker, U., Burns, L., Crary, D., Kearney, K., & Woody, C. et al. (2007). Development of tracking detectors with industrially produced GEM foils. *IEEE Nucl. Sci, 54*(6), 2646–2652. doi:10.1109/TNS.2007.909912

Simons, D. G., Fraser, G. W., De Korte, P. A. J., Pearson, J. F., & De Jong, L. (1987). UV and XUV quantum detection efficiencies of CsI-coated microchannel plates. *Nuclear Instruments & Methods in Physics Research. Section A, Accelerators, Spectrometers, Detectors and Associated Equipment, 261*(3), 579–586. doi:10.1016/0168-9002(87)90371-8

Sommer, A. H. (1968). *Photo-emissive materials: preparation, properties, and uses.* New York, NY: Wiley.

Tripathi, N., Bell, L. D., Nikzad, S. H. O. U. L. E. H., Tungare, M., Suvarna, P. H., & Sandvik, F. S. (2011). Novel Cs-Free GaN Photocathodes. *Journal of Electronic Materials, 40*(4), 382–387. doi:10.1007/s11664-010-1507-7

KEY TERMS AND DEFINITIONS

Afterpulses: A series of pulses appearing at the detector output at some conditions, e.g.feedback, after some delay with respect to the primary pulse produced by the external radiation.

Bq-Becquerel: Units of radioactivity expressed in number of nuclear decays per second.

CsI-MWPC: A multiwire proportional counter combined with a CsI photocathode.

CsI-PPAC: A parallel-plate avalanche chamber combined with a CsI photocathode.

Energy Resolution: One of the characteristics of photons and charged particle detectors: its capability to resolve two close lying energies. If monochromatic energy photons are entering the detector, the pulse-height spectrum of the output electronic pulses will be not a delta function, but rather a line having some width coming from fluctuations in the number of ionization and excitation process as well as due to the electronics. The energy resolution is often expresses in terms of full width of half maximum (FWHM) of the output signal or as a ratio of FWHM to the mean value of the pulse amplitude.

Evaporation Chamber: A high vacuum chamber in which CsI photocathode is evaporated on the substrate surface.

Evaporation Plant: A compete apparatus which includes the evaporation chamber in which photocathodes are manufactured, a chamber where their quantum efficiency are measuring and a device for transporting the photocathode to a special sealed gas chamber.

Fatigue Effect: In the context of this book: degradation of the quality photocathode as function of the number of produced photocathodes even if all of them are produced by the same technology and in the same evaporation chamber.

Glove Box: A sealed container with manipulators (such as special gloves) allowing to perform mechanical actions with objects when contact with atmosphere should be avoided, for example for handling substances which chemically reacting with air. Glow box is usually flushed withy argon or nitrogen.

MAXWELL: A software package allowing to calculate electromagnetic fields for 3-D and 3-D electrode arrangements. MAXWELL is used by researches and engineers for designing and analyzing various electromagnetic and electromechanical devices.

Printed Circuit Boar (PCB): A single layer or a multilayer board with conductive tracks/electrical connectors (pads, strips, etc) etched from cupper layers laminated into dielectric substrate. In the case of the multilayer PCB conductors on different layers are connected with plated through holes/ vias.

Quantum Efficiency: In the case of solid photocathodes is the ration of created photoelectrons to the number of incident photons.

RETGEM: Resistive Electrode Gas Electron Multiplier.

S-RETGEM: Thick gas electron multiplier with electrode made of resistive strip s.

TEA: Triethylamine.

TGEM: Thick gas electron multiplier.

Time Resolution: The minimum time difference where two evens can still be separated.

TMAE: Tetrakis(dimethylamino)ethylene.

UV: Ultraviolet light. In the context of photocathodes UV light often refers to light in the wavelength interval 185 – 400 nm. However, since upper layer a of the atmosphere, due to the ozone layer, absorbs radiation till 260 nm, some authors define UV region between 300 and 400 nm.

VUV: Very far ultraviolet light. In the context of photocathodes it often refers to light with wavelength shorter than 185 nm.

APPENDIX

Experimental data show that aging of CsI photocathodes is not a big problem for most applications. However, it is still interesting to understand the physics behind this phenomenon as it can have implications on other photocathode materials. According to a review paper by Breskin, the CsI photocathode aging under intense photon flux is due to a photolysis process, leading the dissociation of CsI molecules (Breskin, 1996):

$$hv + CsI \rightarrow Cs^{+} + I + e^{-}, \tag{8}$$

The iodine atoms evaporate and the excess of cesium, with a higher electron affinity, leads to a reduction of the quantum efficiency. Another possible effect of photon induced aging, especially for photocathodes exposed to air, is the formation of radicals following the dissociation of H_20 contaminants and their reaction with CsI (Maier-Komor, 1995). During avalanche amplification in gases, the CsI photocathode is subject to positive ion (A^+) bombardment. These ions damage the photocathode surface by ion sputtering and by ion-induced dissociation of CsI (Breskin, 1996):

$$A^{+} + CsI \rightarrow A + CsI^{+} \rightarrow Cs^{+} + I, \tag{9}$$

Also here, the evaporation of iodine causes an enhancement of the electron affinity of the surface resulting in a loss in quantum efficiency. In most gaseous detectors operating at atmospheric pressure, the positive ions do not even have enough energy to dissociate the CsI crystals. Hence, ion dissociation of CsI crystals is only possible at low gas pressures where the ion kinetic energy is higher.
In the case of an ideal CsI crystal, the absorption of UV photons should lead to the creation of exitons and free electrons in the conduction band, as illustrated in Figure 73. They would not cause any dissociation (Dekker, 1962).

Figure 73. The energy bands for an ideal CsI crystal (a) and for CsI containing various defects (b)
From Francke, 2003.

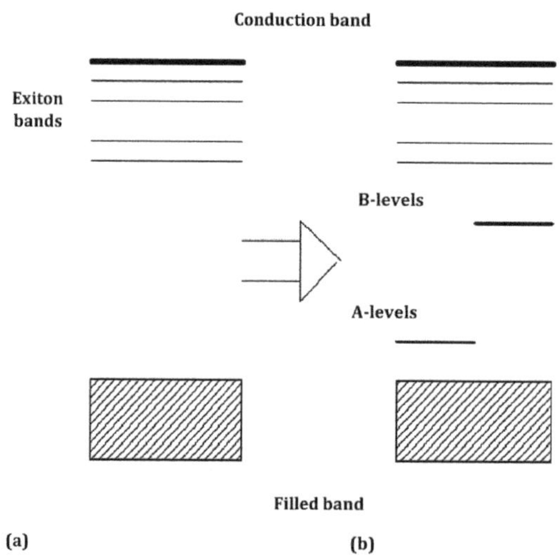

In reality CsI is not an ideal crystal. It contains cracks, vacancies, interstitials, dislocations, impurities, insertions, and so on. This creates extra energy levels indicated as "A-levels" and "B-levels" in Figure 62b. In this model, A-levels are occupied states above the filled band, and B-levels are unoccupied states below the conduction band. As a result electron traps are formed and new levels appear, from which the photoemission may occur. Note that a CsI photocathode may also have stoichiometric deviations, i.e. contain excess of Cs or I due to, for example, exposure to light during the evaporation or dissociation in water. This also creates extra levels in the energy bands. Dramatic and irreversible changes occur when the CsI photocathode is exposed to air. Its initial poly-crystal structure cracks into granules (Breskin, 1996).

Absorbed water further damages the photocathode structure and creates a "CsI-water" solution in which Cs and Iodine may dissociate. If an aging test is done with non-monochromatic light, some contribution to the photoemission come from the newly created energy levels and this will enhance the photocurrent. On the other hand, the new energy levels can also trap free electrons inside the CsI structure. This may cause a charge-up effect. Thus there are several competing processes, which contribute to the measured photocurrent. If the aging test is interrupted (for example, the light is blocked), some relaxation processes may occur (Dekker, 1962), where the degraded quantum efficiency can be partially restored. The same restoration effect can be expected with photocathode heating. Note that both these predictions are based on solid-state theory (Dekker, 1962) have been confirmed experimentally (Anderson 1991).

Let us now discuss the possibility of CsI dissociation by ion impact. Although the positive ions from the avalanche do not have enough kinetic energy to dissociate the CsI, they can create insertions and disrupt the "ideal" crystalline structure. As a result, the photocathode aging characteristics, especially for photocathodes exposed to air, may depend quite extensively on the photocathode history: cleanliness, duration of exposure to air, and so on. To this a charging up effect by positive ions should be added, which may also introduce some additional variations of the quantum efficiency with time.

This may explain the great range of the experimental data mentioned above. One can therefore speculate if aging is a result of a gradual destruction of the ideal crystalline structure through the creation of defects, extra levels, insertions, and dissociation in water layers with a possible further dissociation of Cs and I atoms (Francke, 2003). Due to the defects, the electron transfer properties will be diminished and the photocathode will, layer by layer, be destroyed. As a result the quantum efficiency will drop. This model also explains the heating effect as not only causing removal of water, but also partial restoration of the quantum efficiency value due to relaxation effects (Dekker, 1962). In this case, a photocathode will have the best aging properties if the photocathode is never exposed to air. The experimental data tends to support this prediction.

Chapter 7
Gaseous Detectors Sensitive to Visible Light

ABSTRACT

In this Chapter the latest generation of gaseous photomultipliers, sensitive up to visible region of spectra are described. So far we have described the successful development of gaseous detectors with high sensitive to ultraviolet light, capable of detecting single photoelectrons with 100% efficiency and which can be operated stable for many years in high flux environments. The next step is to develop such detectors sensitive to visible light.

1. INTRODUCTION

The currently most interesting developments in the field of gaseous photodetectors are electron multipliers with a high sensitivity to the visible light. In order to emphasize the great importance of these efforts, it will be useful to start with a review of other types position-sensitive photomultipliers, vacuum and solid-state, which appeared in the market in the 1990s.

Non position-sensitive vacuum photomultipliers which are sensitive visible light has existed since the 1930ies. In the 1950ies came the first position-sensitive vacuum devices such as TV camera/ tubes and image intensifiers. Due to high levels of electronic noise they were not sensitive to single photons.

Vacuum photomultipliers use various compounds as highly efficient reflective and semitransparent photocathodes, e.g. SbCs, bialkali and many others. The spectral sensitivities as a function of wavelength for some of them are shown in Figure 1.

Note that the sensitivity of any photodetector can be expressed either in units of the quantum efficiency as was used in the previous chapters, or in the units of the radiant sensitivity as used in Figure 1. The latter is defined as the photocurrent from the photocathode (in mA) per radiated power (in W) in a certain wavelength interval.

As can be seen, photocathodes optimized for the visible region of spectra have quantum efficiency between 25 and 50%. The active areas of these devices are typically between 3 and 15 cm^2. Some special designs with a few cm thick entrance windows to withstand the atmospheric pressure may have active

DOI: 10.4018/978-1-5225-0242-5.ch007

Figure 1. Radiant sensitivity of some photocathodes used in vacuum photomultipliers; the dashed lines indicate levels of equal quantum efficiency.
From Gys, 2005.

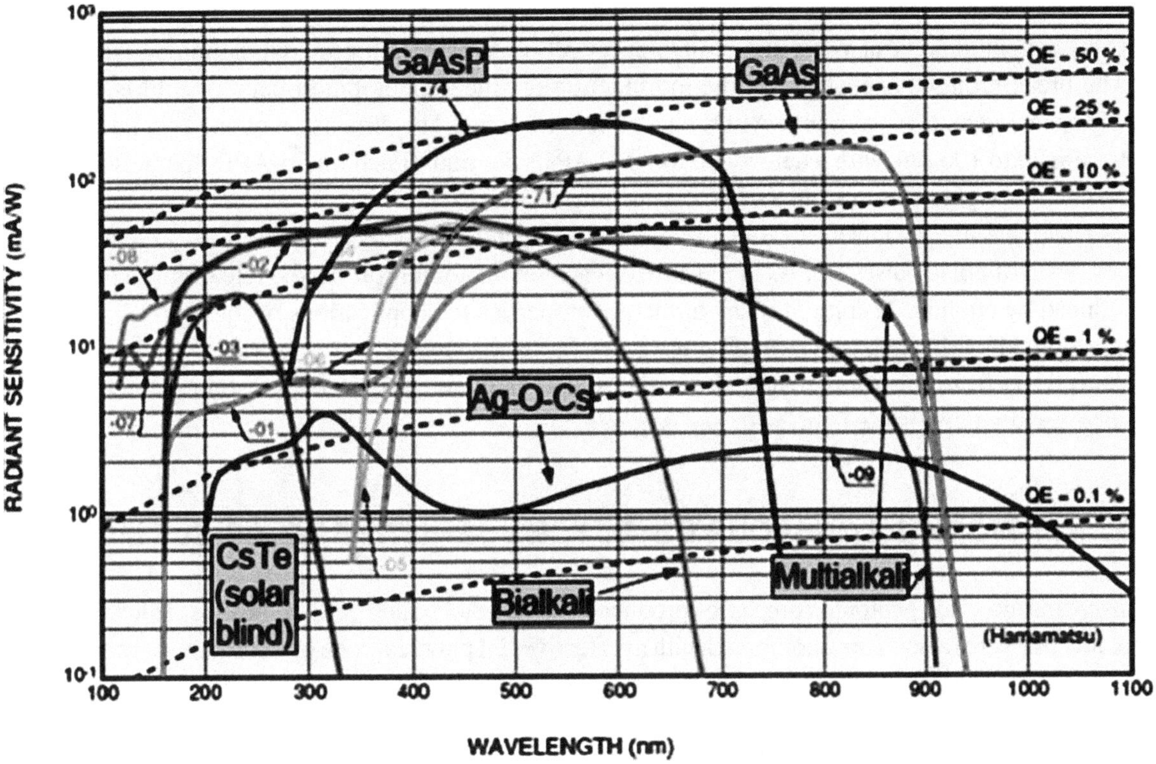

areas up to1200-1500 cm^2. These gigantic photomultipliers were manufactured in the past in Russia and also by Hamamatsu photonics in Japan. To produce photomultipliers with larger active surfaces is difficult due to the serious mechanical constrains on the window size.

In the last decades position-sensitive PMs were introduced into the market having grid type dynodes combined with a multi-anode readout structure. The typical spatial resolution of these detectors is about 1 mm and their maximum effective area is about 30 cm^2.

Another type of position-sensitive photomultipliers is vacuum microchannel plates (MCPs). These devices are usually combined with semi-transparent photocathodes emitting photoelectrons which are multiplied inside the capillary holes via a secondary-electron emission process. Some of these devices offer a position resolution about10 μm. However, the maximum sensitive area of this type of detector is around 100 cm^2.

Other types of position sensitive photodetectors, which recently has started to compete with vacuum PMs, are the solid-state devices, e.g. avalanche photodiodes (APD) and silicon photomultipliers (SiPM).

APD is a highly sensitive electronic device which converts light to primary electrons via photoelectric effect. Its unique feature is that at high reverse bias voltage (typically 100-200 V), an avalanche multiplication process starts inside the semiconductor material and a current gain around 100 can be reached. Some advanced silicon APDs with alternative doping allows an even greater bias voltage (> 1500 V) without breakdown resulting in gains of >1000.

Much higher gains, up to 10^5-10^6, can be reached in certain APDs operating in Geiger mode. In these devices the reverse bias voltage is above the APD's breakdown limit. As in the case of gaseous Geiger counters, such APD needs to have means for quenching the breakdown rapidly. The Geiger mode is particularly useful for single photon detection provided that the dark current event rate is sufficiently low. An APD is not positions sensitive. A matrix of Geiger APDs can be used for rough position measurements.

The breakthrough happened after the introduction of silicon photomultipliers, or SiPMs, which is simply speaking an array of micro-APDs on a common carrier. The dimension of each single APD can vary from 20 to 100 μm, with a density up to 1000 APDs per mm^2. Each micro-APD operates in Geiger mode and is coupled with the others via a quenching resistor.

The applicability and usefulness of these devices depends on many parameters, e.g. quantum efficiency, level of noise pulses etc. As in the case of vacuum photomultipliers, the rate of noise pulses can be reduced by cooling. In spite of many attractive properties the applications of these novel devices is mainly limited by the high cost per unit area.

As follows from this short review, it would be extremely attractive to develop large area gaseous detector sensitive to visible light, with a reasonable cost.

2. EARLY WORK ON GASEOUS DETECTORS SENSITIVE TO VISIBLE LIGHT

Many years ago some photonic companies produced commercial sealed gas photodiodes filled with pure Ar at low pressure (a few Torr), combined with a reflective CsI photocathode or with some photocathodes sensitive to visible light. Such detectors had coaxial geometry and could operate stable only at gas gains below 10-20. They were widely used in applications where the need for moderate gains and limited bandwidths did not demand more sophisticated devices, such as in decoding the film sound track in movies.

Probably the first developments of sealed gas photodiodes filled with quenched gases (methane or isobutane) and having sensitivity from 330 to 630 nm was done by Charpak et al. (Charpak, 1982).These prototypes had transparent planar window coated on the inside with a light-transmitting conducting layer overlaid with a thin micromesh to improve charge collection and constituting the anode facing a thin metal plate coated with the photosensitive layer (often a bialkali).The active photocathode diameter ranged from 20 to 80 mm. The quantum efficiencies of the methane filled photodiode and some other photodetectors are shown in Figure 2.

The quantum efficiency of the RCA 8850 phototube (upper curve in the figure) at the peak is 30%. Therefore the quantum efficiency of the gas-filled photodiode is about 7% in the wavelength interval between 400 and 550 nm. Note that this is only half of the vacuum photodiode. There are two reasons for the low quantum efficiency in the gas atmosphere: photoelectrons back diffusion (see Chapter 2, Figure 2) and photocathode degradation when it adsorbs residual impurities in the gas. The residual impurities cause an initial degradation of the photocathode (the first 20-30 min) after which the quantum efficiency becomes stable almost for unlimited time.

Figure 3 shows the photocurrent as a function of applied voltage for methane filled photodiode compared with a vacuum photodiode.

These detectors were only able to reach gas gains of about 60-70 at low pressure of methane. Higher pressures required higher voltages which the photodiodes were not able to withstand.

Later, attempts were made to combine a Cs-based photocathode with sensitive up to 450 nm with a single–wire counter filled with a mixture of a noble gas and hydrogen or methane (Peskov, 1988). The

Figure 2. Relative quantum efficiency vs. wavelength of several devices having bialkali photocathodes: phototube RCA8850 operating at a gain of one, commercial vacuum photodiode and methane filled photodiode
From Charpak, 1982.

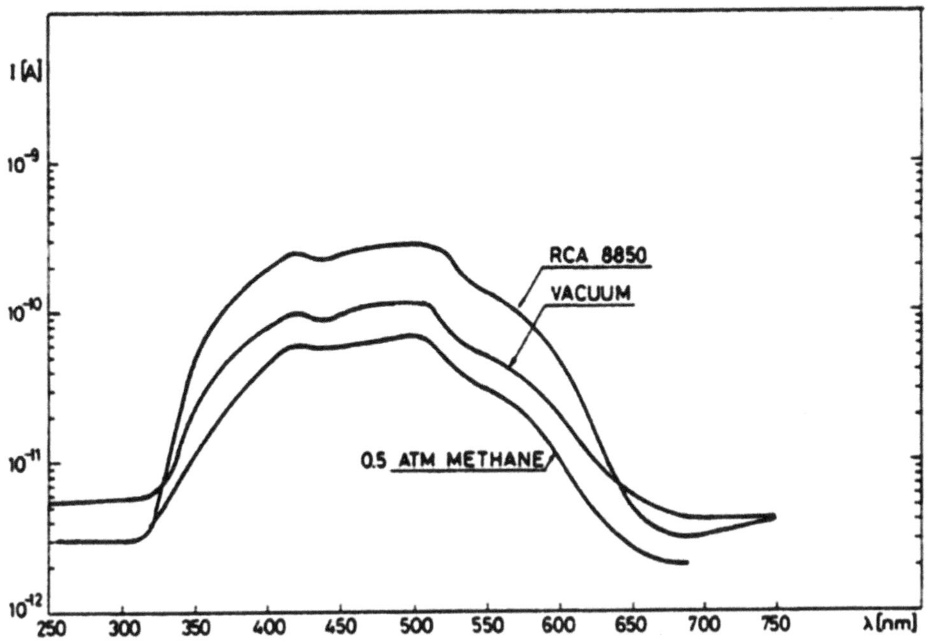

Figure 3. Typical photocurrent vs. voltage characteristics of a methane filled photodiode compared to a reference vacuum photodiode in the voltage range below electron multiplication (collection region); the gap between the anode and the cathode was 5.5 mm.
From Charpak, 1982.

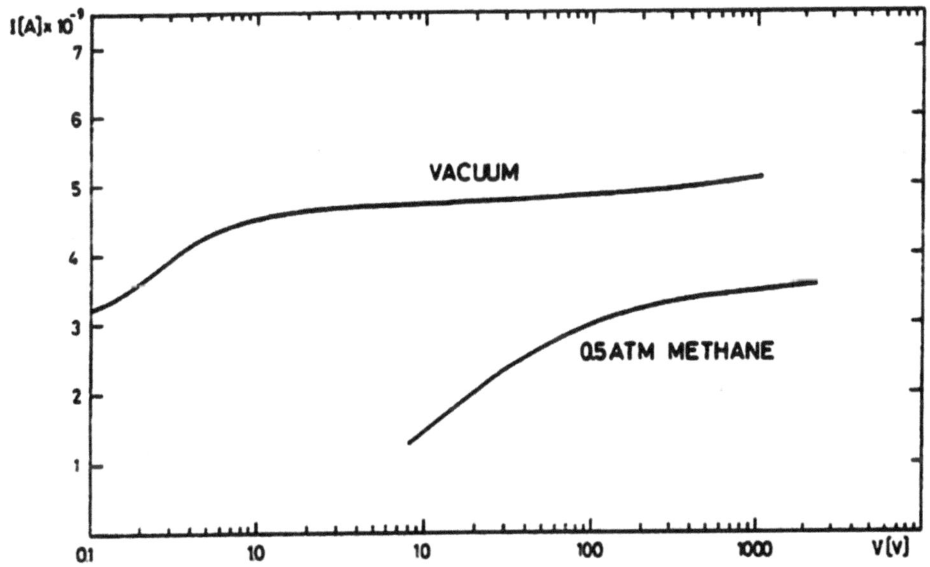

Figure 4. Schematic drawing of the setup used for manufacturing Cs-based photocathodes inside a single wire counter: 1 - a glass–based single-wire counter, 2 - a glass vessel containing a Cs generator 7, 3 - electrodes used for the ohmic heating of the Cs generator, 4-valves, 5 - glass cylinder containing ultraclean gas mixture, 6 - pressure gauge, 7 - Cs generator

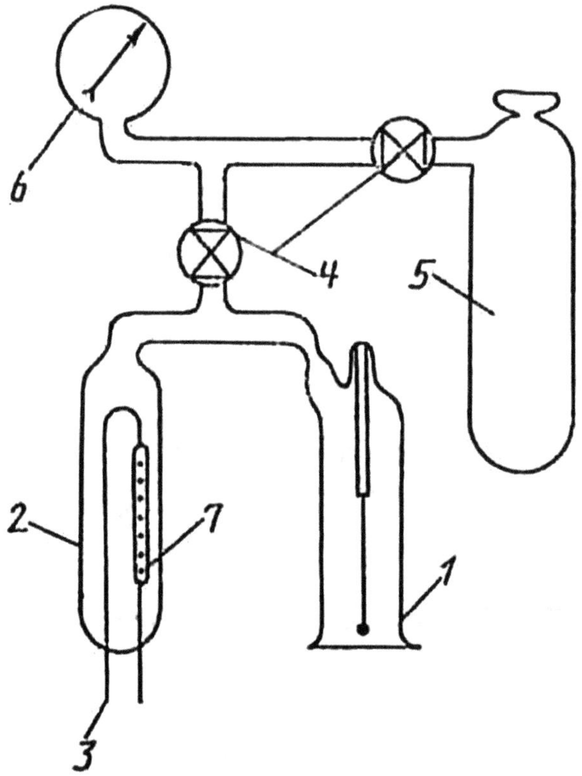

quantum efficiency of this first laboratory prototype was not high. However, the detector could operate at a gas gain of close to 10^5 which is sufficient for the single photoelectron detection.

It is difficult to manufacture photocathodes of these materials, as they are easily destroyed by air and humidity. Step by step the research community gained experience in manufacturing efficient photocathodes compatible with avalanche gaseous detectors. Besides the book of Sommer there was no access to the photocathode technologies of the photonic companies (Sommer, 1968).

The first experimental apparatus used by Peskov was simple and it is shown in Figure 4 (Peskov, 1988). It consists of a glass–based single wire counter coupled to a glass tube containing a commercial Cs generator and a gas system allowing filling the detector with various ultraclean gases.

The single-wire counter was made of glass with a copper cathode evaporated on its inner wall. The anode wire was made of tungsten with 0.1 mm diameter. To avoid discharges, a glass ball with 0.5 mm diameter was manufacture at the end of the anode wire (see Figure 4).

The Cs generator was a tube, 0.3 mm in diameter, made of nickel and having several holes; it was filed with an alloy consisting of 3% Al, 17% Zr and 80% $CsCrO_4$. The gas system was cleaned by heating it to 400°C and pumping it to a vacuum of about 10^{-6} Torr during 10 hours. After that a current of 5 A was passed through the generator to heat it to a temperature high enough to release caesium vapour.

Figure 5. The quantum efficiency of a Cs-coated copper photocathode (upper curve) and Cu treated with TMAE vapour
From Peskov, 1988.

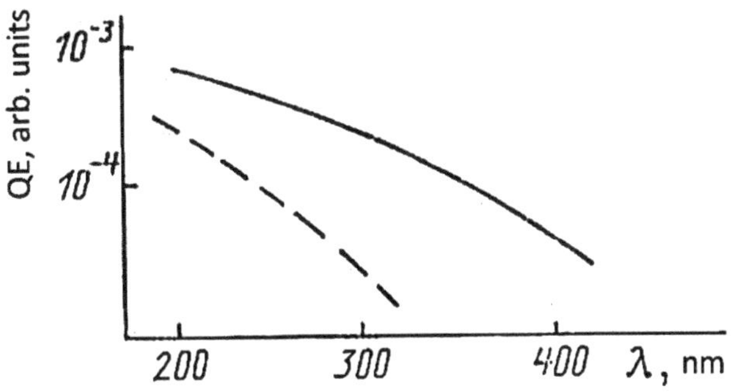

The vapour was deposited on the inner wall of the glass vessel. After that the glass vessel was heated to about 300°C and the Cs vapour entered into the single-wire counter. The counter volume was kept at room temperature which made the caesium condensate on its walls. Then the single wire counter was finally filled with the gas to be tested to atmospheric pressure.

The quantum efficiency of the Cs photocathode (with respect to a TMAE absorbed layer) is shown in Figure 5.

Nowadays it is know that the quantum efficiency of the TMAE absorbed layer depends on its thickness, governed by the residual TMAE vapour pressure in the detector (see Chapter 4). Unfortunately, the precise conditions were not well controlled in this study, since it this times (the paper was submitted to the publisher in 1986) there was little understanding of the TMAE layer properties (Peskov, 1988). Therefore, from Figure 5 only a rough conclusion can be drawn. From later studies it is known that the quantum efficiency of Cu coated with a TMAE absorbed layer is, depending on the thickness of the TMAE layer, between ~$2x10^{-4}$and 10^{-1} (Peskov, 1988a; Séguinot, 1990). From these values, looking at the curves presented in Figure 5 it can be concluded that the Cs photocathode efficiency at 400 nm is also between 10^{-4} and 10^{-1} whereas the efficiency of an untreated Cu cathode is 10^{-7}. Hence, the Cs coating increased the quantum efficiency of the cathode at least on three orders of magnitude.

The next important contribution in the development of sealed gaseous photomultipliers came from Edmends et al. (Edmends, 1988). This team succeeded to combine various designs of parallel plate avalanche chambers with home-made semitransparent SbCs photocathodes. Their simplest apparatus, called phototetrode, is shown in Figure 6. It had a glass window of 40 mm diameter, with the caesium-antimonide photocathode deposited on the inside.

Before assembling this device an antimony film was vacuum-deposited on the inside face of the cathode window. After assembling the stack of discs and rings for the parallel plate avalanche chamber meshes, the detector was connected to a vacuum pump. Inside the chamber a Cs generator, similar to the one described above and a getter was also installed. The device and the gas filling system were baked for several hours at 250°C while pumping, and both the caesium source and the getter were degassed by passing a carefully adjusted current through them. The temperature was then reduced to 150°C and the getter was activated by passing a nominal current through it. After the caesium source was fired, the photoemission current was monitored continuously during the evaporation using a blue light source. This procedure turned out to be extremely important and was missing during the Cs cathode manufacturing in

Figure 6. Cross section through the phototetrode with its envelope, Cs generator and the getter
From Edmends, 1988.

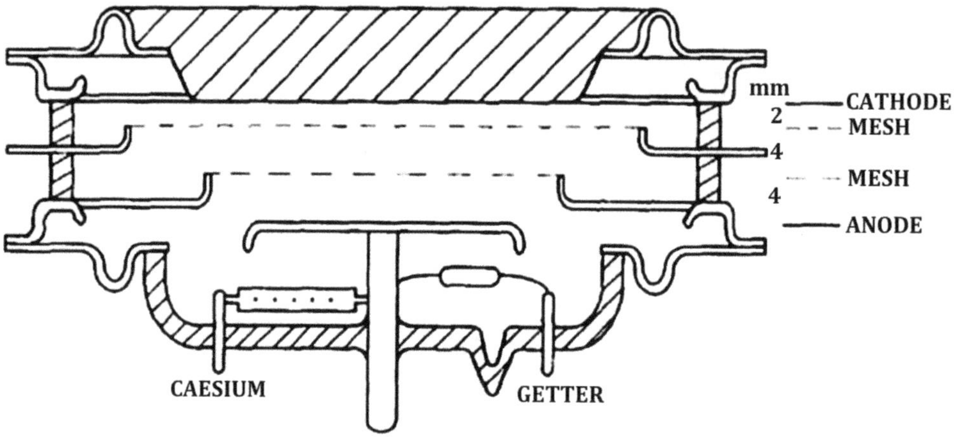

previous studies (Peskov, 1988). Initially the photocurrent rose rapidly as the caesium reacted with the antimony cathode. Caesium condensation on the glass walls and stainless steel structure was prevented by keeping the chamber at a high temperature. Excess caesium was removed from the detector by continuous pumping. At some point the photocurrent reached a plateau and the caesium evaporation was halted.

To preserve the photocathode, which is reactive with oxygen and water vapour, extreme cleanliness was maintained in the gas-filling system. In addition, active cleaning was also performed with two getters: one installed in the filling line and the other one inside the detector. For better outgassing, the surfaces of the inner walls of the system was alternately filled and pumped down before finally being sealed off with 20 Torr of methane.

The prototype shown in Figure 6 did not reach gas gains sufficient for single photoelectron detection and was mainly used for studies the photocathode quantum efficiency and its life time.

Its quantum efficiency as a function of wavelength measured immediately after the gas filling is show in Figure 7 (the curve labeled α). As can be seen, its reaches 3.7% at 480 nm.

After the test the photocathode was bombarded with ions which caused the quantum efficiency to drop to 1.5% (curve β in Figure 7). After the ion current stopped, the photocathode remained stable for a quite long time (curve γ). The author's draw the conclusion that the photocathode was permanently destroyed by the ions. Note that a similar effect was not observed with commercial visible-light sensitive photodiodes filled with noble gases. Therefore, the question of what was the cause of the photocathode degradation remained open. Another possible reason of the quantum efficiency decay could be absorption of some residual impurities in the detector volume.

To achieve higher gas gains, a multistep parallel plate avalanche chamberwith alternating low field and high field gaps was used. The authorscalled it a photohexode and it is shown in Figure 8. With this multiplication structure, operating in 20 Torr of CH_4, gas gains of $8x10^3$were obtained.

The research started by Edmends et al. was later continued by several groups. For example, Borovik-Romanov successfully reproduced Edmends results and with the help of the Cs generators other photocathodes were also manufactured, in particular GaAS(Cs) and AuCs, and their operation was investigated in various quenching gases (Borovik-Romanov, 1994). The setup used in this work was essentially the same as in the work of Edmends, as shown in Figure 9.

Figure 7. Quantum efficiency of a home-made SbCs photocathode measured immediately after the manufacturing (α), after intense ion bombardment (β) and a few days later when there was no exposure (γ)
From Edmends, 1988.

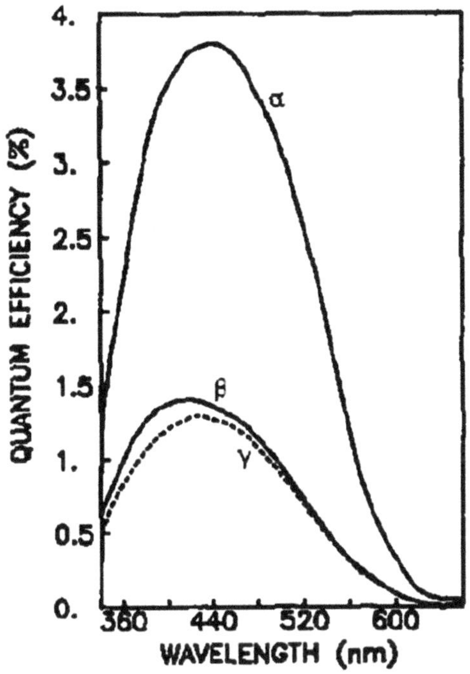

Figure 8. Cross section of a multistep parallel plate avalanche chamber photodetector
From Edmends, 1988.

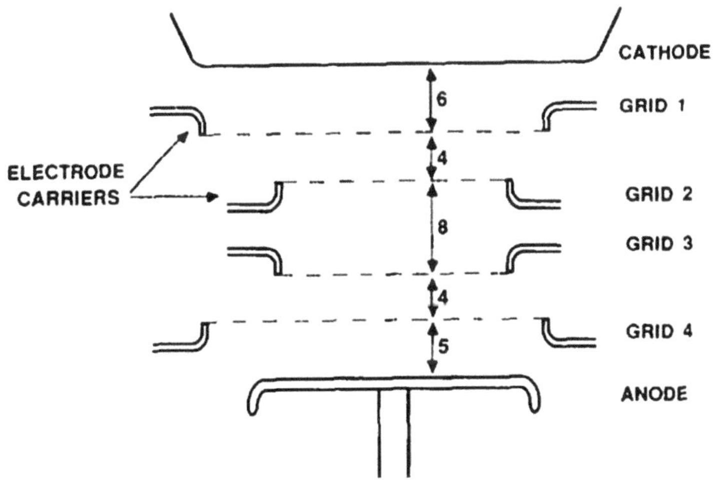

Figure 9. Setup for the manufacturing and testing various Cs-based photocathodes
From Borovik-Romanov, 1994.

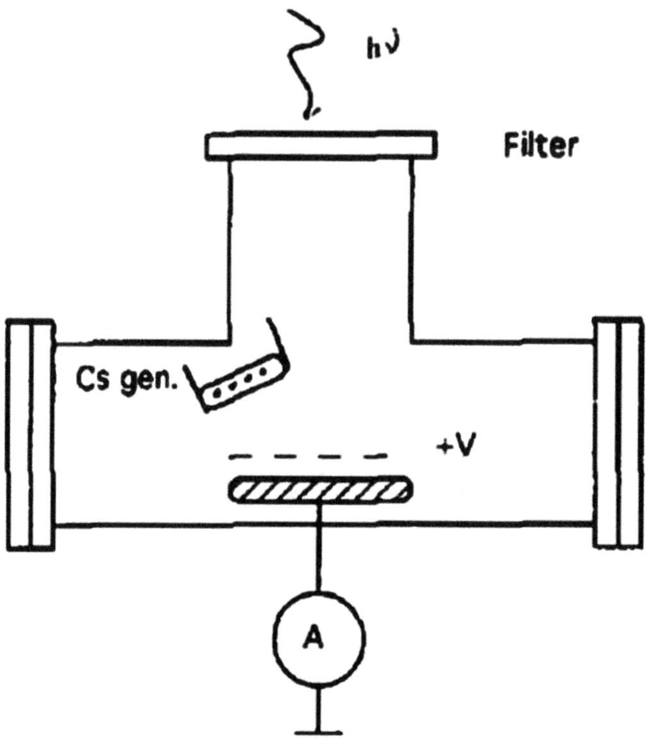

A small parallel plate avalanche chamber with a Sb, GaAs or Au cathode was mounted inside a T-shaped glass tube with a quartz window. An anode electrode mesh plane was placed 2 mm above the cathode. The Cs generator was placed 1-3 cm in front of the cathode. Before evaporation of Cs from the generator, the chamber was heated to 100 °C and pumped to 10^{-6} Torr for 96 hours. A mercury lamp with a narrow band filter was used as a light source to monitor the photocurrent. The quantum efficiency of the photocathode was measured with respect to the quantum efficiency of TMAE and several phototubes calibrated by Hamamatsu.

During the evaporation, the photocurrent produced by the Hg lamp was measured holding the mesh at a fixed voltage of 10V. When the current passed through the maximum the evaporation was terminated. After the Cs-based photocathode was formatted, the photocurrent was measured as a function of the voltage on the mesh, first in vacuum and then in gases (Ar, CH_4 or isobutane with and without TMAE vapour) at pressures ranging from 20 Torr to 1 atmosphere.

The results of the quantum efficiency measurements of various photocathodes are presented in Figure 10.

In methane a quantum efficiency of a few to 10% was achieved in the spectral interval 180-450 nm.

The detector with Cs_3Sb photocathode could operate at gas gain of 10^3 without strong feedback problems. Discharges appear at gas gains of (3-5) x 10^3.

Borovik-Romanov also repeated the aging studies of the Cs_3Sb photocathode in gas atmosphere first performed by Edmends et al. (Edmends, 1988). Similar results were obtained: In 20 Torr of CH_4 and a gas gain of 10 the quantum efficiency dropped a factor of two at a total collected charge of ~0.2 μC/cm^2, see Figure 11.

Figure 10. Typical quantum efficiency obtained in CH_4 (20 Torr) with GaAs and Cs_3Sb photocathodes; for comparison the data from (Edmends, 1988) are also drawn.
From Borovik-Romanov, 1994.

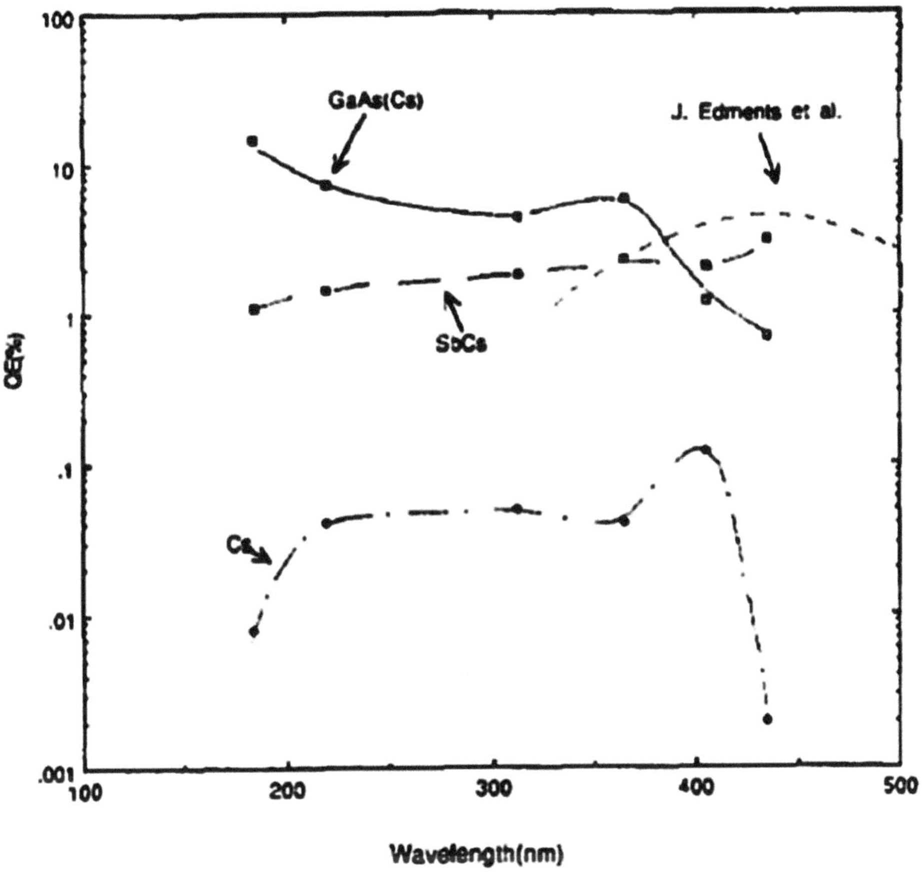

The Cs-based photocathodes described above have remarkable quantum efficiency in the visible range of the spectra, but they are extremely sensitive to O_2, H_20 and other impurities. As a result, they can only be used in a clean gaseous atmosphere which complicates the detector design and the gas filling system.

There were attempts to develop more robust photocathodes, which can be used in more practical conditions including micro-leaks, medium quality gases etc. (Borovik-Romanov, 1994; Enomoto, 1994). Borovik-Romanov tested two types of photocathodes with coating to protect them while exposed for a short time to air or not extremely clean gases. They used TMAE or CsI as a protective layer. The additional thin layer on the top of the photocathode prevent or slow down the diffusion of the gas from the chamber whereas photoelectrons created from the photocathode have sufficient kinetic energy to penetrate through this layer and be recorded in gas detector (see Figure 12).

Although the preliminary results obtained with a TMAE protecting layer were encouraging, the further focus of research was on CsI protective layers. This was because handling TMAE is not a simple task. On the other hand, it is known that a thin layer of CsI is transparent for photons and for photoelectrons as is used in semitransparent photocathodes.

Figure 11. Results of aging tests of Cs_3Sb photocathodes in methane at various conditions
From Borovik-Romanov, 1994.

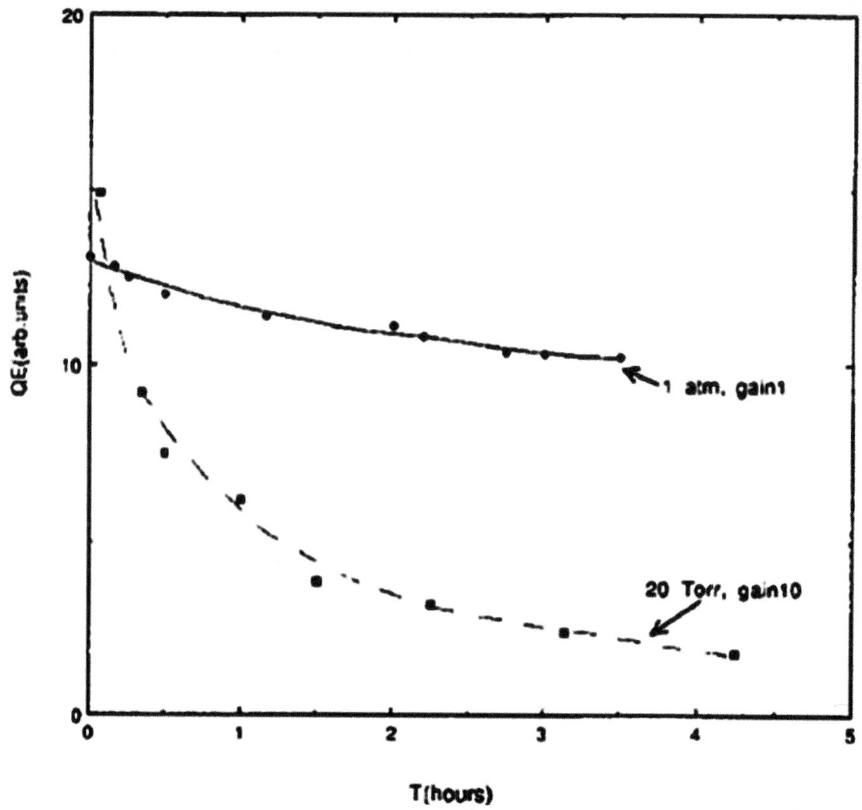

Figure 12. Schematic drawing illustrating the principle of the photocathode protection; the first studies of the photocathodes protection in gaseous detectors were done by Charpak and Anderson (Charpak, 1992; Anderson, 1994). It was found that air sensitive photocathodes can be temporally protected by covering them with a thin layer of d,d-dimethyl-trimethylene-ferrocene-phane or ethylferrocene, or TMAE. They studied the time stability of SbCs and SbCs+ TMAE photocathodes taken from commercial phototubes. The latter photocathode had some air durability due to the reaction of oxygen with TMAE, which removes oxygen from the sealed gas volume and work like some kind of getter.

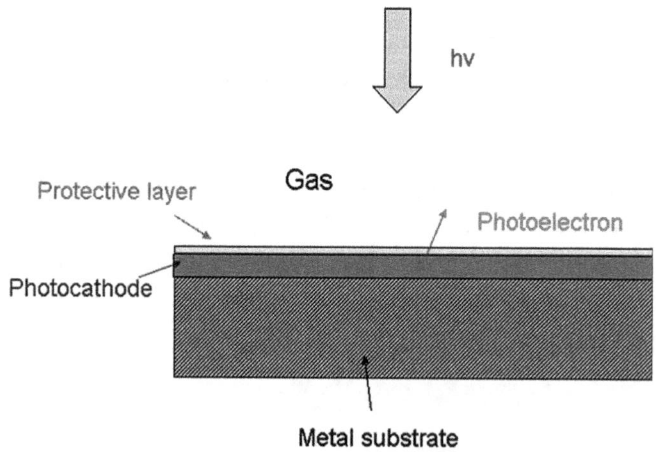

Figure 13. An evaporation system used for the production various Cs-based photocathode coated with a CsI protective layer

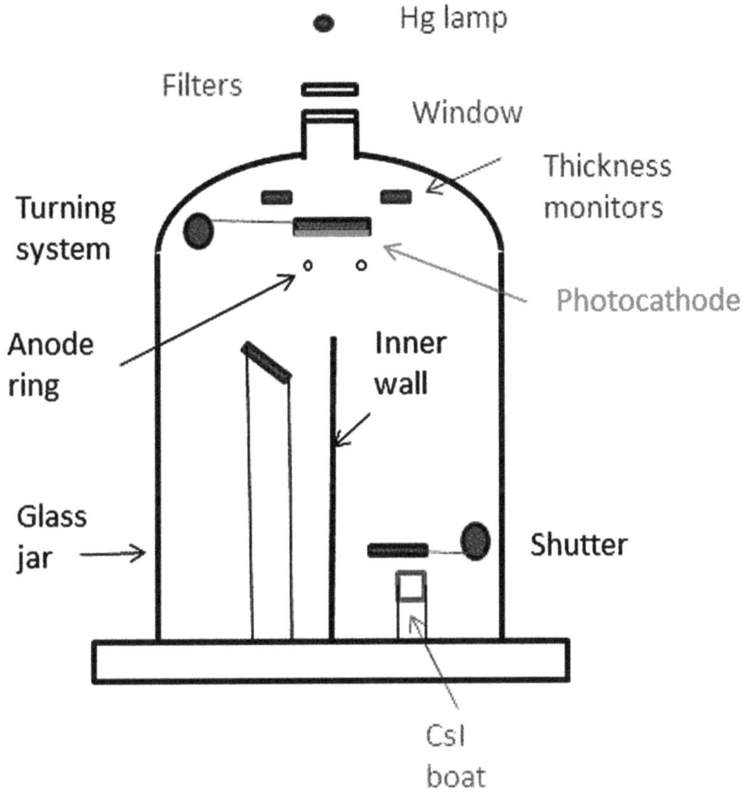

An evaporation system for manufacturing CsI coated photocathodes is shown in Figure 13. It consisted of a standard glass vacuum jar inside which a photocathode disc, a Cs generator, a boat with CsI crystals, and quartz thickness monitors was installed.

After the Cs-based photocathode, e.g. Cs_3Sb, is manufactured it is flipped by the magnetic turning system. The quantum efficiency is then measured with filtered UV-light from a mercury lamp. The quantum efficiencies measured are presented in Figure 14. As can be seen the CsI coating reduces the quantum efficiency for long wavelengths.

In the first test, the CsI protected Cs_3Sb photocathode was exposed to air when mounted in the detector. The exposure to air caused cracks in the CsI. The gas then leaked through these cracks causing a slow degradation of the photocathode. In later studies the photocathode was transferred to the detector without exposure to air.

The conclusion from these studies is that indeed the protective layer helps to preserve the photocathode when exposed to gases with ordinary quality, like in a glove box flushed with argon. The price to pay is a significant loss of quantum efficiency for wavelengths longer than 250 nm.

Similar attempt to protect a Cs_3Sb photocathode with a thin layer of CsI was done by Enomoto et al. (Enomoto, 1994). In their studies the photocathode coating with CsI was carried out in steps, each step only a few nm thick. After each step, the quantum efficiency was observed to be reduced by 10-20%.

Figure 14. The quantum efficiency of Cs_3Sb photocathodes vs. the wavelength under different conditions: room temperature 20°C and 60°C, with and without a CsI protective layer one can see that heating to 60°C helps to increase the quantum efficiency.
From Peskov, 1994.

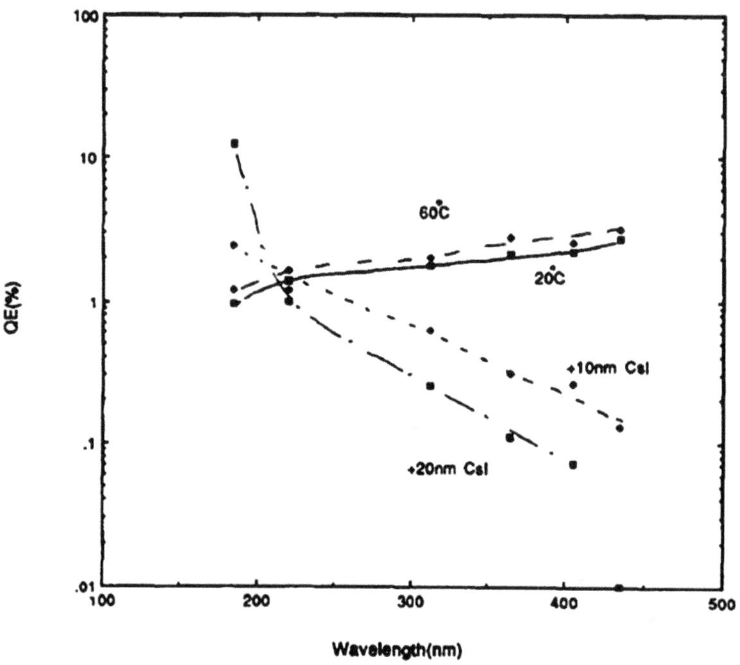

After 6 steps they exposed the photocathode to air and then measured its quantum efficiency. At 250 nm it was reduced by a factor 10.

3. MASTERING PHOTOCATHODE MANUFACTURING

After the first encouraging results obtained with gaseous detectors combined with home-made photocathodes, more serious efforts were invested in developing highly efficient photocathodes.

Maximum precautions were taken to ensure that the system is clean enough and made from appropriate materials, which must have low outgassing properties (Peskov, 1995). The highest quantum efficiencies (30 to 50%) and the most stable photocathode operations were obtained with glass chambers. Before manufacturing the photocathode the whole system was heated up to 250-300°C and pumped to a vacuum better than 10^{-6} Torr for a week or two. In order to do an additional cleaning of the surface to be coated (Au, Sb, GaAs, Si) a glow discharge in noble gas atmosphere at a few Torr was performed inside the glass chamber. Then the pumping continued for a few days.

During the evaporation the temperature was reduced to 150°C and the Cs generator was activated bypassing a 5-8 A current through it. Cs vapour filled the entire volume of the test chamber. After a few more hours of pumping the Cs vapour was practically completely pumped out. The only remaining Cs is what had chemically reacted with the surfaces. The Cs reacted with Sb or Au layers or with a Si or GaAs wafer (Peskov, 1995). The cooling of the chamber and the whole system down to room temperature should be done over a day or more. Devices in vacuum are well thermally isolated and retain their temperature a long time. Any temperature gradient will cause re-evaporation of residual Cs from the surfaces and cause instability of the photocathodes or even spoil the quantum efficiency.

It was found that if everything was done according to this procedure it was not even necessary to monitor the quantum efficiency during the manufacturing. It was observed that if the system is well outgassed, the photocathodes will be stable even without a getter. In some experiments, attempts were made to enhance the quantum efficiency by introducing oxygen. In this case the quantum efficiency was monitored as a function of time and the oxidation was stopped when the quantum efficiency reached a maximum.

It should be noted that in commercial phototubes the absolute value of the quantum efficiency and the shape of the spectral response varies by 30-100% from photocathode to photocathode. The same is true for the photocathodes described here. In Figure 15 only "typical" values are presented (Peskov, 1995).

Exposing SbCs photocathodes to O_2 may increase the quantum efficiency by a factor of 2 (to ~20%), but the deviation between different photocathodes is large reflecting how sensitive this procedure is.

No ageing was observed for these photocathodes in vacuum and at low fluxes (photocurrent<10^{-14} A). These measurements were performed up to a total collected charge of10 mC/cm^2. At high fluxes (photocurrents in the order of pA) some variation in quantum efficiency with time was observed. This effect was more pronounced when the intensity of the source was varied with time, for example when it was blocked for some time. It was attributed to a "fatigue" effect (Sommer, 1968). Heating the photocathodes to 80-100°C increased their quantum efficiency. The same behavior is observed for CsI photocathodes.

Figure 15. Typical quantum efficiency of Sb_3Cs and GaAs-Cs photocathodes fabricated in a glass chamber and measured in vacuum

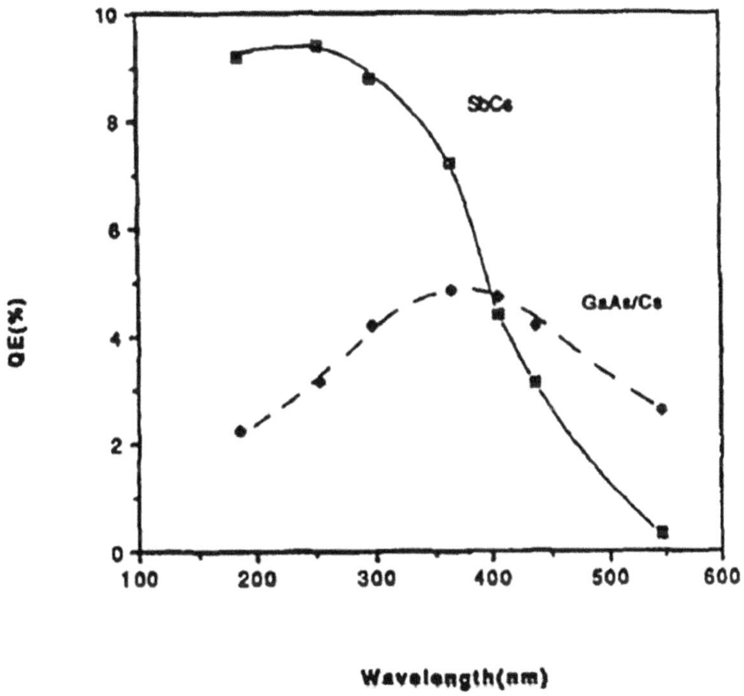

Figure 16. Photocurrent in relative units versus voltage for the parallel plate avalanche chamber combined with a Sb_3Cs photocathode measured in vacuum (1) and in CH_4at 80 Torr (2)
From Peskov, 1995.

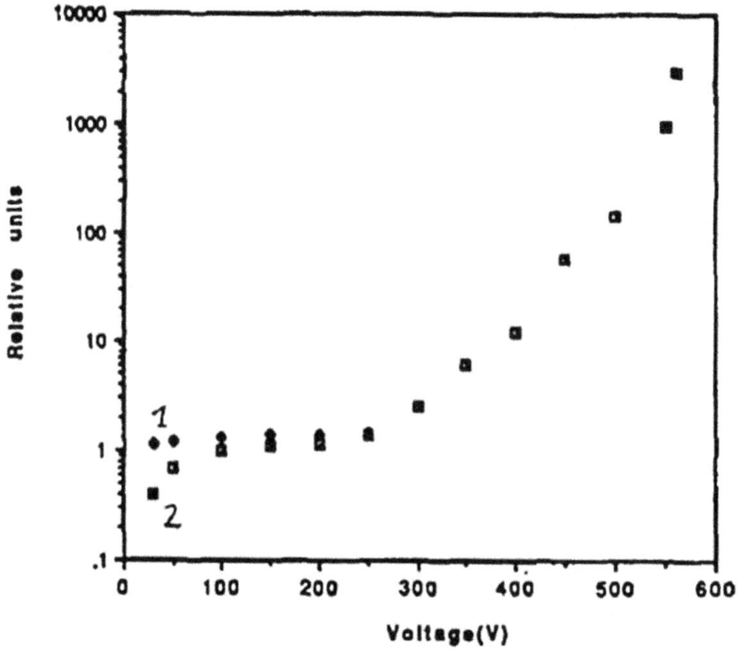

The highest gas gains that could be achieved, up to 3-4 x10^3, were obtained in DME (dimethyl ether) and isobutane. However, the quantum efficiency was typically 30-50% lower than the original value in vacuum. This is probably due to impurities in the gases. When using CH_4, the quantum efficiency was 80-90% of the vacuum value, but the maximum gain was only ~1000.

Figure 16 shows a typical gas gain as a function of operating voltage for a parallel plate detector with a Sb_3Cs photocathode.

The practical maximum gas gain of photosensitive detectors is often limited by photon and ion feedback rather than electrical breakdown. With the photocathodes described here 10%feedback was observed for gas gains around 100.Themaximum gain was flux dependent. At high fluxes it was lower. The maximum gain before discharge is quoted for rates $< 10^3$ Hz.

The ageing properties of the photocathodes in different gases are shown Figure 17 (Peskov, 1995).

No ageing was observed in noble gases (the test was done up to a collected charge of ~10 mC/cm^2). A rather strong ageing effect was observed in quenching gases. In mixtures of noble gases with quenchers the rate of the quantum efficiency degradation was proportional to the quencher concentration. One should note that the photocathodes treated by glow discharge or O_2, had far worse ageing properties than the one presented in Figure 17.

The next important contribution came from Breskin's group. This includes construction of a more sophisticated evaporation setup allowing better control of parameters as evaporation speed, temperature vs. time, the photocathode thickness, and also real time measurements of its quantum efficiency. The evaporation system was equipped with several magnetic manipulators for manufacturing sealed detectors.

Figure 17. Relative drop of the photocurrent (or quantum efficiency) versus collected charge for different conditions: 1-Ar, 100 Torr, 2-Ar + 10% DME, 3-Ar + 30% CH_4, 4-CH_4, 20 Torr 5-CH_4, 1 atmosphere; the Cs_3Sb photocathode is covered by 20nm CsI. All measurements were done at a gain of 10.

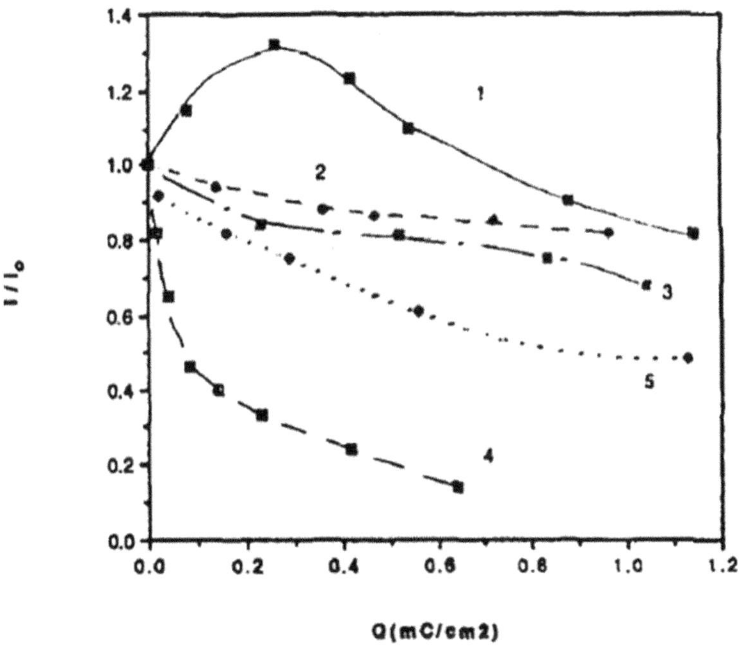

A similar system was built by MELZ research laboratory in Moscow (Shefer, 1998). It is shown schematically in Figure 18.

The setup was connected to a gas manifold and was equipped with a residual gas analyser.

It consists of a bakeable stainless steel vacuum chamber, in which either Cs_3Sb or K-Cs-Sb photocathodes can be prepared in a controlled way, and their quantum efficiency is measured in situ and in real time. The photocathode quality critically depends on many parameters, some of them are not clearly defined, resulting in large fluctuations in the photoemission.

One of the main features of this setup was high initial vacuum which can reach values of 10^{-9} Torr, by means of a turbo molecular pump. As a result, stable photocathodes with reasonably high quantum efficiency could be produced. Another important feature was that the photocathode substrate was located within a glass cylinder and mounted above a stainless-steel cylindrical anode, surrounded by a resistive oven made of tungsten wire, allowing local heating of the volume around the photocathode up to 270°C.

The photocathode evaporation site comprised three evaporation positions, one for antimony and two for alkali metals. It also contained a collimated halogen lamp for measurement of light transmission through the photocathode using a photodiode. This was used to control the antimony thickness. The same lamp was also used for monitoring the photocurrent during the photocathode activation.

The quantum efficiency was measured by illuminating the photocathode with light filtered by a monochromator and recording the photocurrent from the photocathode.

For the photocathode production alkali metal dispensers (SAES Getters Cs/NF/5.5/17FTlO + 10, K/NF/3,1/17FTlO + 10) and 99.999% pure elemental antimony (Alfa, antimony shots) were used. The latter previously melted in a Ta boat in the photocathode evaporation position.

Figure 18. The evaporation system used by Schefer et al; the photocathode substrate was located within a glass cylinder, installed inside the vacuum chamber and surrounded by a resistive oven. The photocathode evaporation site comprised three evaporation positions, shown schematically in the figure: one for antimony and two for alkali metals. The movement between the sites was performed by magnetic manipulators. The lamp inside the glass cylinder was used for several measurements: the light transmission through the photocathode, to control the antimony thickness and for monitoring the photocurrent during the photocathode activation.
Shefer, 1998.

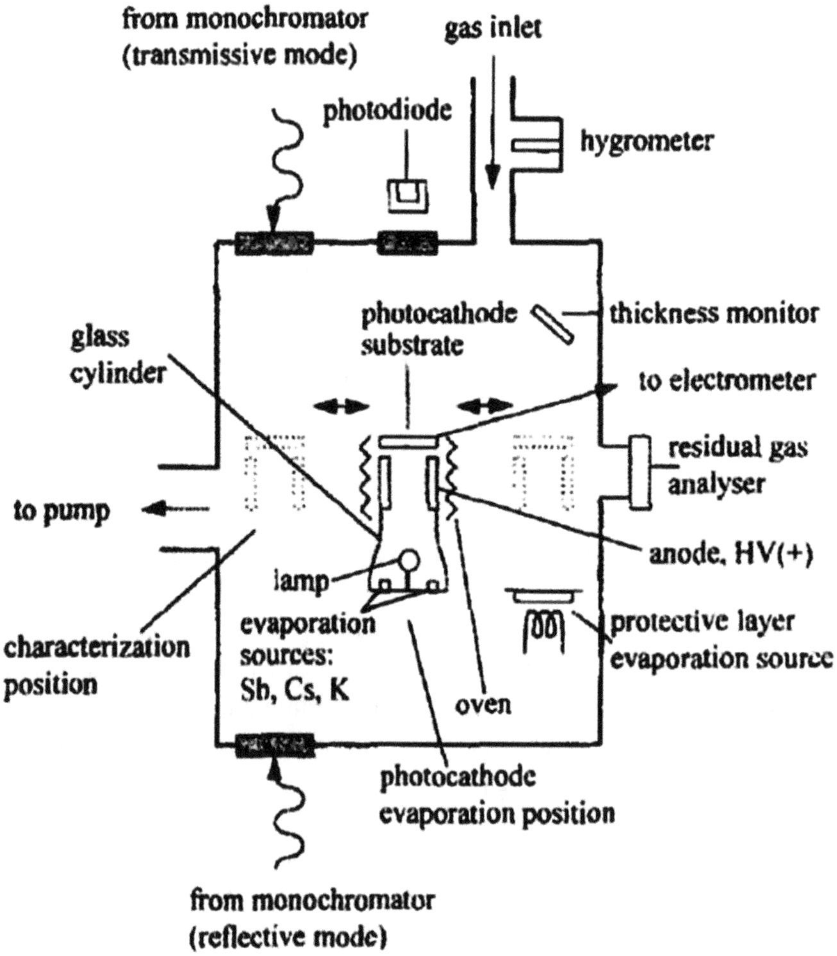

The substrates were optically polished suprasil quartz cleaned with NaOH and distilled water, followed by 10 minutes rinse in acetone and 10 min in ethyl alcohol in an ultrasonic bath. After that they were dried using clean nitrogen. Electrical contacts are provided by evaporation of a 100 nm thick aluminium film on the circumference of the substrate.

The evaporation chamber is then evacuated and heated to 210ºC for 36 h. The residual gas was monitored using the vacuum gas analyser. During this baking period, the activation oven was heated to 270ºC and small currents were applied to the alkali metal dispensers and to the antimony boat in order to outgas them from possible contamination. Finally, the chamber was cooled down to room temperature over a 5 hour period.

The first step in the production procedure of both photocathode, Cs_3Sb or K-Cs-Sb, was the evaporation of antimony, keeping the setup at room temperature. The antimony thickness was monitored by measuring the light attenuation in the substrate during the evaporation. Typically, the evaporation lasted for 15-30 s and was stopped after the light flux decreased to 70% of its initial value. The Cs_3Sb photocathodes were prepared by heating the antimony substrate to 160-170ºC before the Cs evaporation from the Cs generator. The photocurrent from the photocathode was continuously monitored throughout the activation process, and the Cs evaporation was terminated when it reached a maximal value. Then the photocathode was cooled down and the photocurrent was measured again. It was observed that stable photocathodes usually underwent an increase in photocurrent after being cooled to room temperature.

For more complicated alkali-antimonite photocathodes, the substrate was heated to 195-200°C before potassium was evaporated. The evaporation was terminated when a maximum photocurrent was reached, indicating that a K-Sb photocathode had been formed. To remove some of the potassium and thus to optimize the caesium reaction, the substrate was heated to 240°C for 10 sec and then cooled down to 195-200°C. At this point a decrease of the photocurrent to about 50% of the maximal value was usually observed. At the final stage of the photocathode production the substrate was cooled down to 175-180°C and caesium and antimony are alternately evaporated until the photocurrent reached a new peak value.

A typical quantum efficiency of Cs_3Sb and K-Cs-Sb photocathodes manufactured in this way are shown in Figure 19.

It is interesting to note that the quantum efficiency of these laboratory-made photocathodes in the spectral range of 250-450 nm was comparable to and in some cases even exceeded the quantum efficiencies seen in commercially produced photocathode for vacuum photomultipliers. However, they showed

Figure 19. Typical quantum efficiency vs. wavelength measured in vacuum for Cs_3Sb and K-Cs-Sb reflective photocathodes
Shefer, 1998.

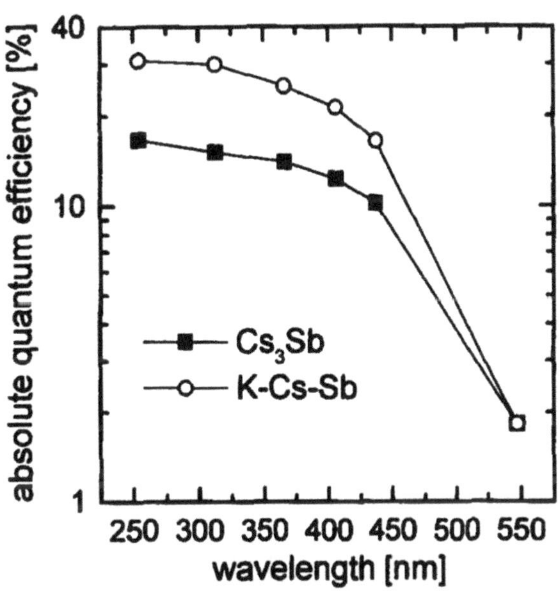

lower quantum efficiency at wavelengths around 550 nm. Both photocathodes were found to be stable in vacuum and in high-purity methane (99.9995) over periods of a few days.

The following procedure found to be useful for the photocathode which was stable in vacuum for four days. The chamber was filled with 1 atmosphere of CH_4 for 1 hour, and pumped before the quantum efficiency was measured again. The photocathode was then left in vacuum for another two days and after that again exposed to CH_4 for four days. After these exercises no strong changes in quantum efficiency was observed indicating that this photocathode can be used in a clean gas atmosphere.

4. SYSTEMATIC STUDIES OF PHOTOCATHODE PROTECTION

The Breskin group took up the task to evaluate other protective materials. They not only reproduced results obtained with CsI protective films (Borovik-Romanov, 1994; Peskov, 1994), but also studied some other alkali halides coatings. Figure 20 shows the quantum efficiency of reflective Cs_3Sb photocathodes vs. the protective film thickness, measured at a wavelength of 365 nm.

Figure 20. The quantum efficiency of reflective Cs_3Sb photocathodes vs. the thickness of different protective films: Csl, NaI, CsF, SiO, hexatriacontane (HTC) and calcium stearate (Cast); all measurements were performed at 365 nm (points). For comparison data for the Cast film on a reflective CsI photocathode, measured at 170 nm, are also shown.
From Buzulutskov, 1997.

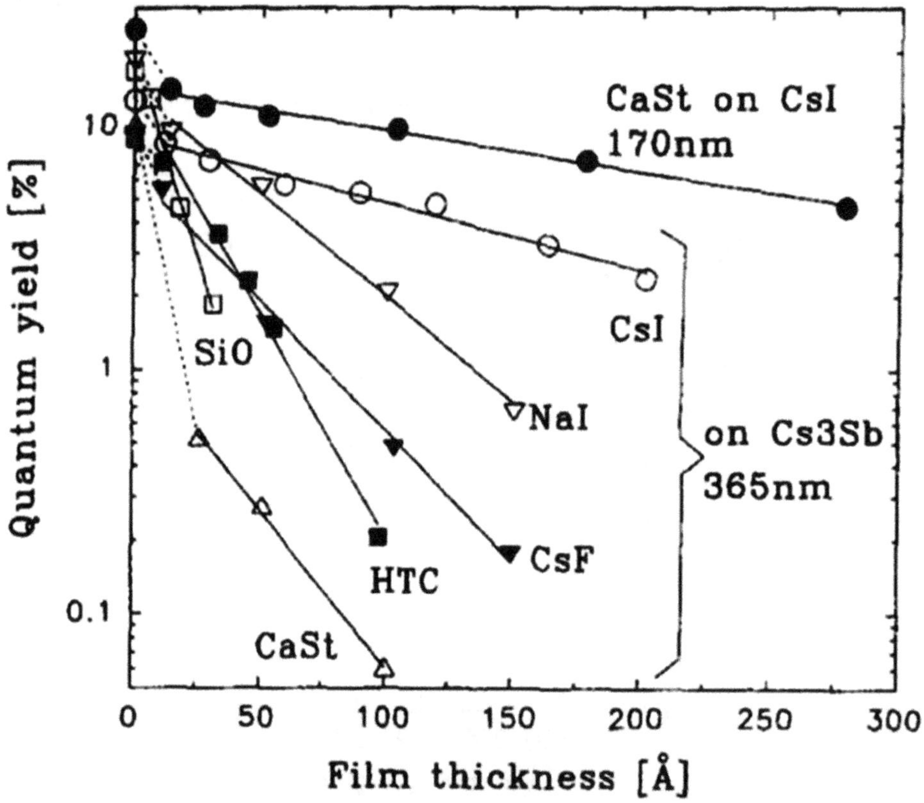

CsI and NaI films showed the highest protection capability and their thickness were found to be critical for the photocathode stability. They also showed good protection to limited exposure to oxygen, as shown in Figure 21.

The protection coatings were also successfully applied to bialkali photocathodes like K-Cs-Sb (Shefer, 1999), as shown in Figure 22.

The stability of coated photocathodes in gases was also investigated under avalanche multiplication. The study was done with a parallel plate avalanche chamber with 1 mm gap. Stable gains of 10^4 were achieved in 1 atmosphere of methane for both CsBr and CsI coated K-Cs-Sb photocathodes.

Figure 23 shows a gas multiplication curve measured for a semi-transparent K-Cs-Sb photocathode coated with30 nm thick CsBr films. The exponential behaviour of the gain vs. the applied potential indicates the absence of strong secondary effects such as photon and ion feedback.

The protection of all photocathodes studied with different coatings was reached at the expense of a reduction in quantum efficiency. The CsI-coated photocathodes offer a rather high spectral sensitivity over a wider energy range (cut-off at 2.8 eV) compared to TMAE or to CsI photocathodes.

Another interesting approach was to use removable films such as metalo-organic compounds, including EF (Charpak, 1992), TMAE (Anderson, 1994; Borovik-Romanov, 1994) or even solid Xe (Peskov,

Figure 21. Evolution of Cs_3Sb photocathodes quantum efficiency, coated with NaI films of different thicknesses (number in angstroms), after exposure to oxygen; exposure time at each data point was about 5 min, followed by measurement in vacuum. The measurements were performed at 365 and 436 nm.
From Buzulutskov, 1997.

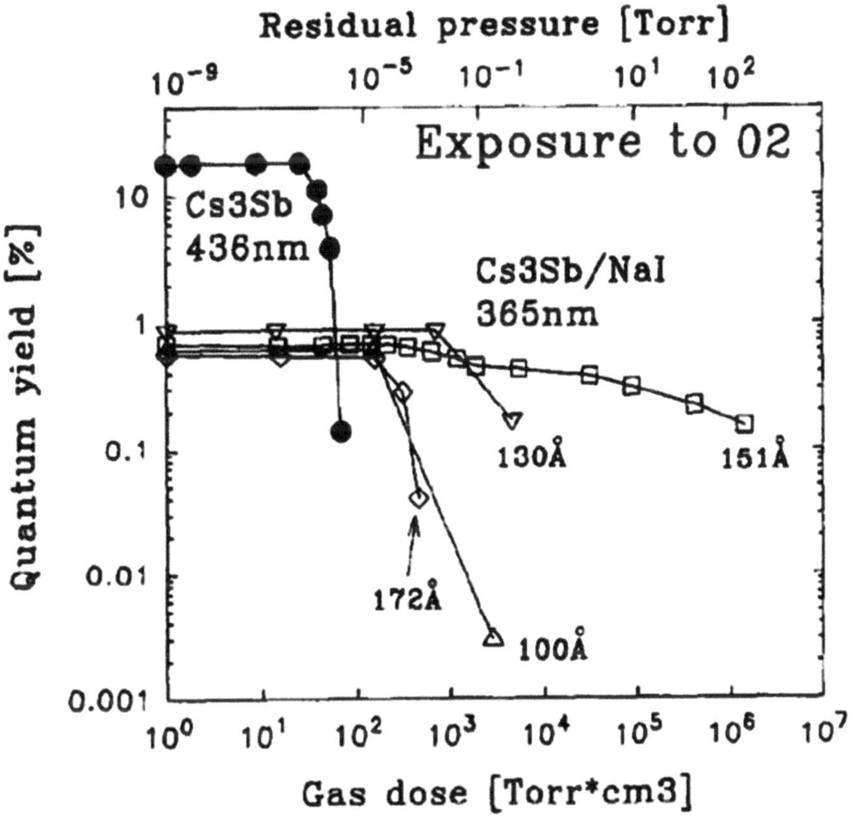

Figure 22. The quantum efficiency of bare and CsI-coated K-Cs-Sb photocathodes (20 and 25 nm thick) exposed to oxygen vs. its pressure; each data point represents 5 min of exposure to oxygen, followed by chamber evacuation and quantum efficiency measurement in vacuum.
From Shefer, 1999.

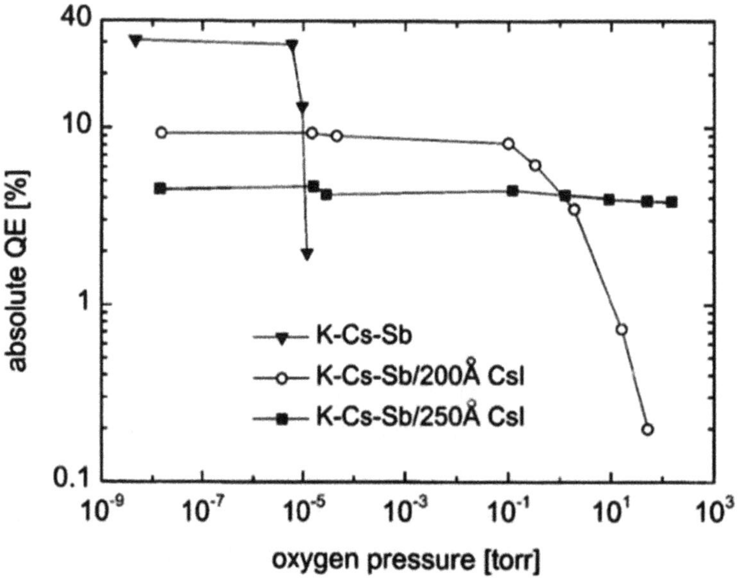

Figure 23. Gas gain as a function of voltage applied to a 1 mm gap parallel plate avalanche chamber coupled to a semi-transparent K-Cs-Sb photocathode coated with 30 nm thick CsBr and operated at 1 atmosphere of methane
From Shefer, 1999.

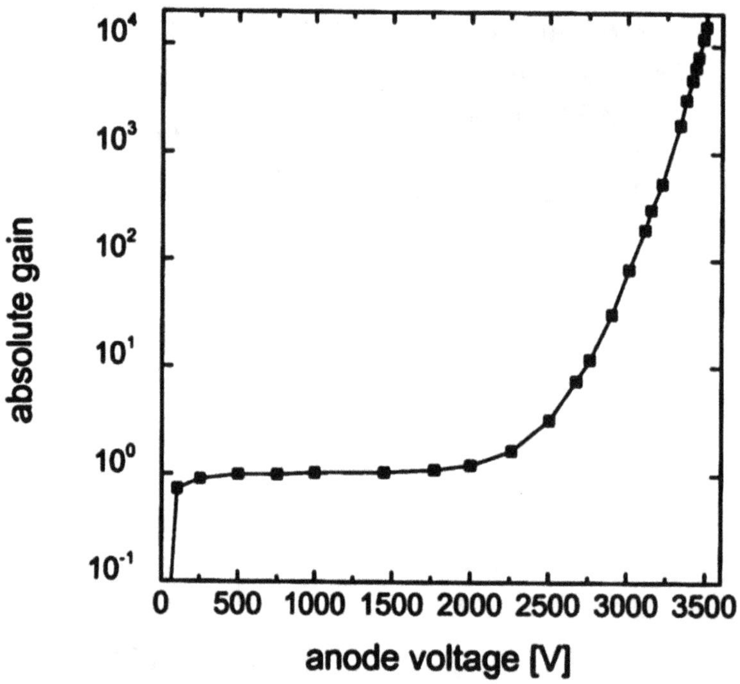

1995). The main attraction is that one can use such a coating when the photocathode is in contact with air or not clean gases. After the installation of the photocathode into the clean gas chamber, the coating can be removed allowing an increase in quantum efficiency. Breskin used paraffin hexatriacontane (HTC), n-$C_{36}H_{74}$, as a removable protective film (Breskin, 1998). The HTC had already been tested as a possible permanent protective coating for Cs_3Sb photocathodes (see Figure 20). Its photoelectron escape length is about 5-10nm which is much smaller than the escape length of a CsI film. However, the possibility to remove HTC coatings makes it possible to apply a thick, temporary coating.

These films were deposited on the Cs_3Sb photocathodes by thermal evaporation from a Ta boat. It begins to sublimate at temperatures higher than 120°C, which permits some baking of the vacuum chamber. The inversed procedure for removal of HTC films simply requires heating the photocathode. Results obtained with some cycles with this coating and its removal is shown in Figure 24.

Each cycle includes HTC deposition, followed by melting of the Cs_3Sb photocathode surface, cooling the protected photocathode to room temperature, exposure it to oxygen for some time and finally removal by heating. Melting of the evaporated HTC film (at about 80°C), makes it more uniform, thus improving its protective properties.

Figure 24. Evolution of a caesium antimony photocathode quantum efficiency coated with thick HTC films; the following treatments were applied: (1) initial photocathode deposition, (2) coating with 1.5 μm thick HTC, (3) HTC melting, (4) exposure to 25 Torr of oxygen for 5 min, (5) HTC removal by heating, (6) new coating with 0.8 μm thick HTC, (7) HTC melting, (8) exposure to 150 Torr of oxygen for 20 min, (9) HTC removal. The quantum efficiency were measured at 312 and 365 nm for a bare photocathode before HTC coating and after its removal.

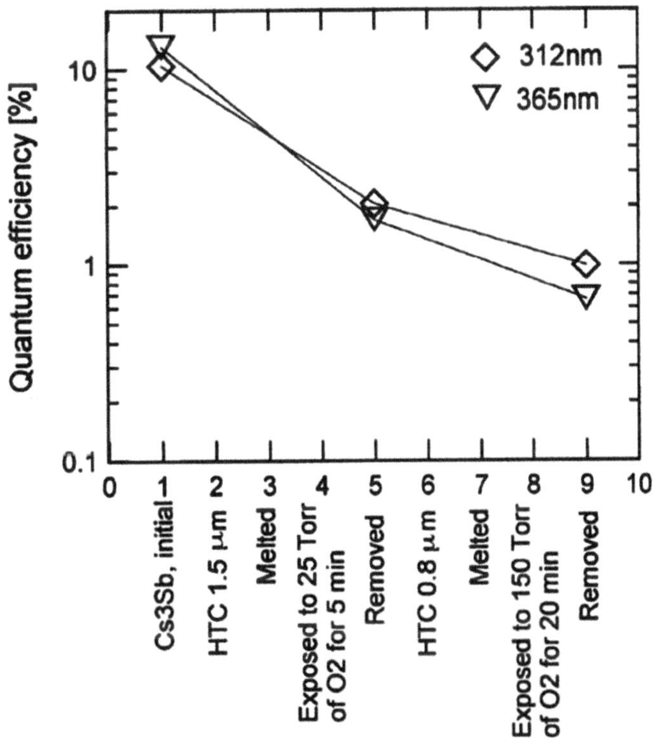

One can conclude that although this coating provides quite good temporal photocathode protection, deposition of HTC, and especially its melting procedure, as well as removal requires thermal treatments lasting over 20 min, which reduces the quantum efficiency of the Cs_3Sb photocathode. Again, protection comes at the expense of reduced quantum efficiency.

5. HOLE TYPE GASEOUS PHOTOMULTIPLIERS

The efforts of several research groups lead to mastering the production of efficient photocathodes which are capable of operating in a clean gas atmosphere and which withstand avalanche ion bombardment. However, the maximum achievable gain of parallel plate avalanche chamber combined with such photocathodes is below 1000 which is not sufficient for the single photoelectron detection with a high probability.

The maximum achievable gas gain is limited by photon and ion feedback. For highly efficient photocathodes both probabilities γ_{ph} and γ_+ are considerably increased. Parallel plate avalanche chamber and wire type detectors are open geometry detectors where all ions and a great part of the avalanche photons are able to reaching the photocathode and emit secondary photoelectrons.

To reach high gas gains one has to use detectors with a closed geometry, when only part of ions and a small portion of photons produced in the electron avalanche are able to reach the photocathode. An example of such amplification structure is the hole-type gaseous multipliers. The first attempt to combine photocathodes sensitive to visible light with glass capillary plates, operating in avalanche mode at atmospheric pressure, was performed by Peskov (Peskov, 1999). Glass capillary tubes are commercially available and are byproducts from microchannel plate production. Glass is a material which works well in vacuum, and matches all the requirements for photocathode manufacturing. Sakurai et al. had earlier demonstrated that one could get some gas avalanche multiplication inside capillary tubes using Ar-and Xe-based gas mixtures at atmospheric pressure (Sakurai, 1996). Their efforts were mostly concentrated on imaging of X-ray photons. The studies were oriented mainly towards avalanche-induced light emission, and not so much on detection of single photoelectrons.

In contrast, Peskov investigated the possibility to detect the charge from gaseous avalanches triggered by photoelectrons extracted from a photocathode sensitive to visible light (Peskov, 1999). The detectors used are presented in Figure 25.

The setup consisted of a gas chamber (not shown in the picture), which housed single or cascaded capillary plate electron multipliers combined with a semitransparent photocathode.

A quartz disc was used as a substrate for the semitransparent photocathodes. Prior to installation in the chamber, the quartz plate was covered with a thin film of Sb, GaAs or Ag. The thickness corresponded to 40-80% light absorption at 400 nm wavelength. The evaporation system used for the photocathode manufacturing was modified with respect to earlier setups used by the same groups (Peskov, 1994; Peskov, 1995). In particular the pumping system was improved allowing it to reach an initial vacuum of 10^{-7} Torr. As a result higher quantum efficiencies were obtained. Besides Cs_3Sb and GaAs-Cs photocathodes, some experiments were also done with AgO-Cs photocathodes. Unfortunately, the quantum efficiency of the latter photocathode was not reproducible due to difficulties to control the Ag oxidation process. In this setup the oxidization was done with a corona discharge initiated in oxygen at a pressure of 0.1 Torr before the activation of the Cs generator.

Figure 25. Single (a) and double (b) capillary plates combined with semi-transparent photocathodes sensitive to visible light and operating in avalanche modein Ar+ 5% CH_4 at atmospheric pressure; the diameter of the capillary plates is 20 mm, the thickness is 0.4 or 0.8 mm, the distance between the capillaries and between the capillary and the anode plate is about 3 mm.

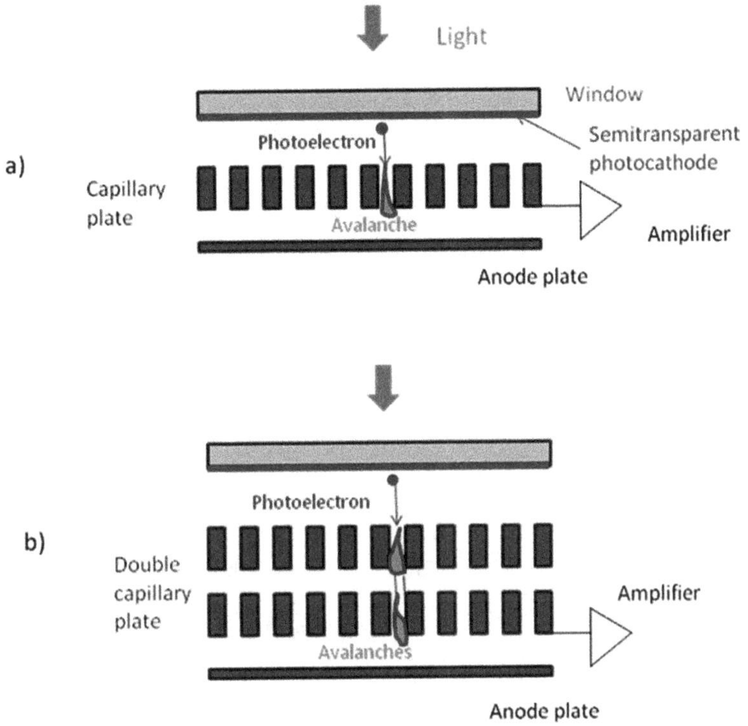

The quantum efficiency was first measured in vacuum. Then with the help of manipulators the photocathode was installed 5-7 mm above the capillary plates and the gas (Ar+ 5% CH_4 atmospheric pressure) was intrudes into the chamber. The photocathode was illuminated with a light source and the photocurrent was measured to calculate the gas gain, the collection efficiency and estimate possible contribution from photon feedback. In point-by-point QE measurements the photocathode was irradiated by monochromatic light, selecting lines from the Hg lamp with different filters. The results for a Cs_3Sb photocathode are presented in Figures 26 and 27.

By comparing the current values at low voltages when all capillary electrodes were connected together (plateau region) and at high voltage (multiplication region), one can conclude that with a single capillary plate one could reach gains of at least 10^3 without feedback. At higher gains the curve started to deviate from a pure exponential increase, indicating the appearance of feedback. The collection efficiency was estimated from the formula:

$$AI_1/I_2, \tag{1}$$

where I_1 is the current from the single or double step capillary plate when all electrodes were kept at the same potential, and I_2 is the current measured on the last (collection) electrode when the detector was operated in amplification mode with the overall gain A. It was found that the collection efficiency

Figure 26. Photocurrent vs. voltage for single and double-step capillary plates, combined with an Cs_3Sb photocathode: (1) a single capillary plate 0.4 mm thick, (2) two capillary plates 0.4 mm thick each in tandem (double-step capillary plates), (3) dark current for the double step capillary plate; the step in curve 2 is due to a reduction of the illumination light to avoid charging the cathodes of the capillary plates.

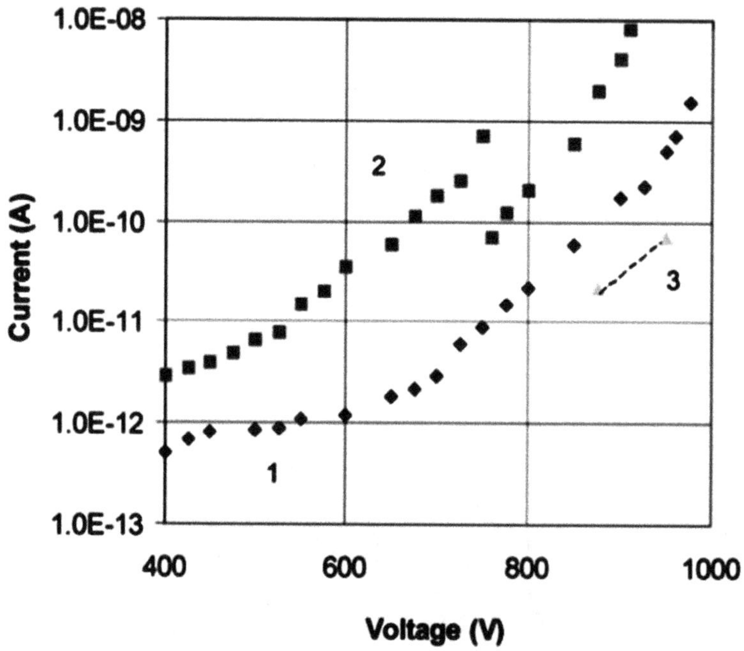

Figure 27. The net result of the quantum efficiency and the collection efficiency, measured for various photocathodes, combined with a double-step capillary plate; the gas mixture was Ar + 5% CH_4 at 1 atmosphere.
From Peskov, 1999.

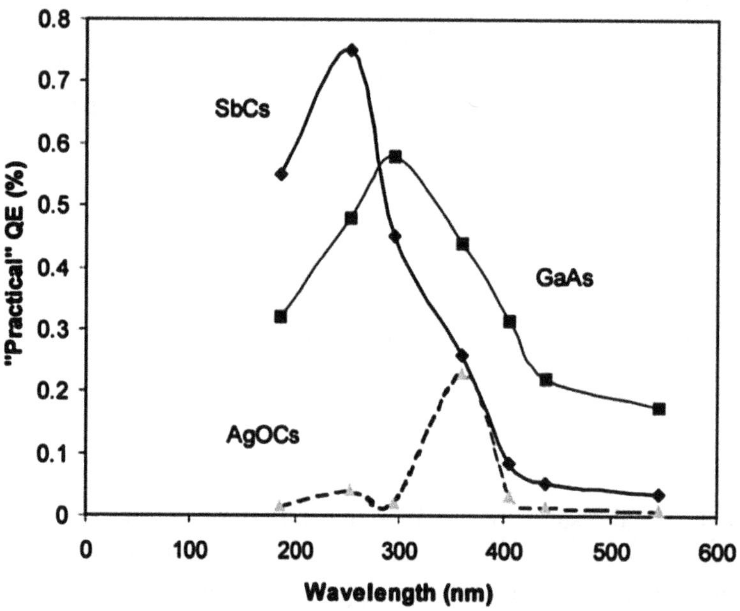

increased monotonically with multiplication and at high gas gains reaches 30-40%.With cascaded capillary plates total gas gains close to $5x10^4$ were achieved without signs of any feedback (which usually manifest itself as deviation from a pure exponential growth of the signal with voltage). The high gas gains were made possible due to the small solid angle from which the light from the avalanches may escape the capillary tube and reach the photocathode. Also only a fraction of the avalanche ions are reaching the photocathode. The other are collected on the cathodes of the capillary plate. At gas gains higher than $5x10^4$ the current curve starts deviating from the pure exponential growth, indicating the appearance of feedback.

At high gas gains the intensity of the filtered light (at 400 nm) from the Hg lamp was reduced by a factor of 10 at a current of10^{-9} A. This caused the "step" in current curve 2 in Figure 26.

The absolute value of the current I_2 was also used to calculate the net result of the quantum efficiency of the photocathodes and the collection efficiency of the capillary plates (the "practical" quantum efficiency). This was done by comparing value I_2/A with the current from calibrated photodiodes measured at the same wavelength. The results of these measurements are presented in Figure 27.

One can see that the "practical" quantum efficiency is actually not high, less than 1%.There are two reasons for that. The first one is back scattering of photoelectrons to the cathode, which appears in the gas media. The second reason is that not all electrons extracted from the photocathode enter inside the capillaries. Some of them are collected on the cathode of the capillary plate. As a result, when the photocathodes are combined with gaseous the capillary plates, their measured ("practical") quantum efficiency is lower than that originally obtained in vacuum.

In a comparative measurement a parallel plate avalanche chamber was used instead of the glass capillary plates. Its design is presented in Figure 28 (Peskov, 1999).The cathode of the parallel plate avalanche chamber was made of an 80% transparent stainless steel mesh stretched on a stainless steel ring with a diameter of 20 mm. The anode was a stainless steel disc. The gap between electrodes was 0.6 mm. The cathode and the anode were supported by ceramic spacers placed outside the active gap. To see the effect of light attenuation, in some tests a capillary plate with inclined capillaries was inserted in between the photocathode and the parallel plate avalanche chamber. This inclined capillary plate efficiently blocked all light produced in the electron avalanche in the parallel plate avalanche chamber from feeding back to the photocathode.

Figure 28. Schematic drawing of the parallel plate avalanche chamber combined with photocathode and capillary plate used in some comparative measurements

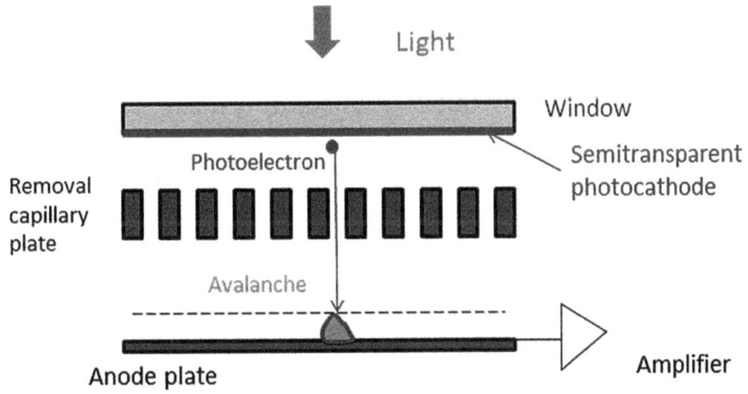

Figure 29. Photocurrent vs. voltage for a parallel plate avalanche chamber combined with a Cs_3Sb photocathode: (1) bare parallel plate avalanche chamber, (2) parallel plate avalanche chamber combined with capillary plate used as an avalanche light attenuator; the gas mixture was Ar + 20% CO_2 at atmospheric pressure.
From Peskov, 1999.

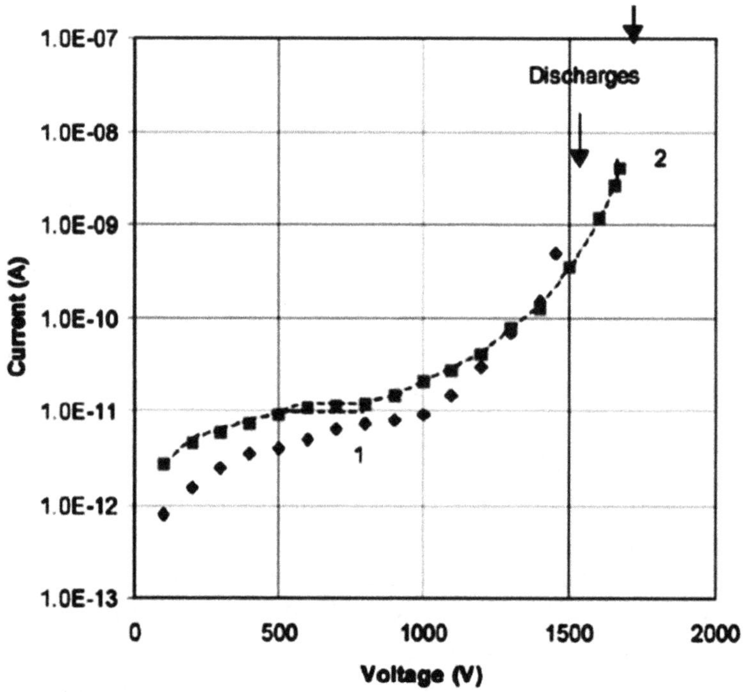

Photocurrent measurements performed with this device in Ar+20%CO_2 at atmospheric pressure are shown in Figure 29.

Ar + 20% CO_2 allowed gas gains 2-3 times higher Ar+5%CH_4. As in Figure 16, the photocurrent curve has a kind of "plateau" region at low voltages followed by the exponential growth due to avalanche multiplication. The plateau current occurs when all photoelectrons emitted by the photocathode are collected on the anode, but no gas amplification occurs in the parallel plate avalanche chamber. This corresponds to gas gain 1. The ratio of the current in the multiplication region to the plateau value gives the gas gain A.

Without the capillary plate photon and ion feedback limits the gas gain to only ~10^2. If a capillary plate with inclined capillaries is inserted between the photocathode and the parallel plate avalanche chamber the maximum gas gain is increased by an order of magnitude. At voltage across the capillary electrodes of more than 600 V more than 80% of the photoelectrons can drift through the capillary with losses. The remaining photoelectrons are collected on the upper electrode, a current that can be measured. At the same time, the capillary plate efficiently attenuates light and allowing higher gas gains (up to 10^3) to be achieved.

Earlier studies had showed that the ion feedback can be minimized in gas mixtures where $E_{in}>>E_{iq}$, where E_{in} and E_{iq} are the ionization potentials of the noble and quench gases, respectively (Peskov, 1975; Peskov, 1977). Thus, He ($E_{in}\approx24$ eV) with a quencher having a low ionization potential could be a good

Figure 30. Gain vs voltage for He + 5% isobutane (curve 1) and He + 5% Xe + EF mixture (curve 2)

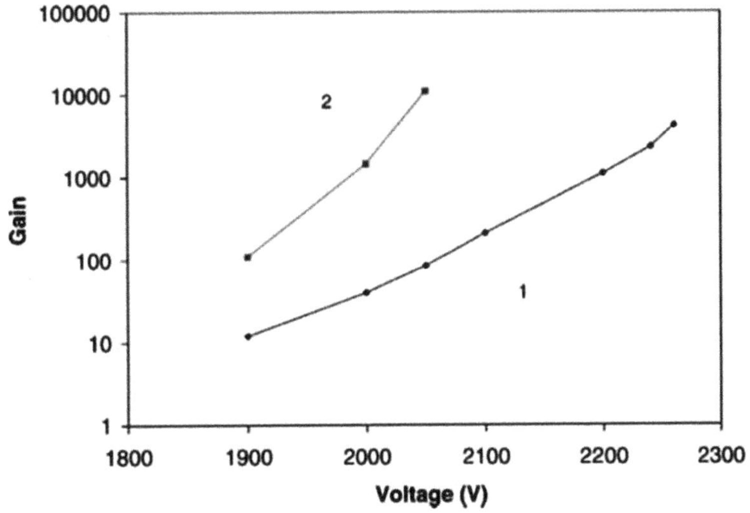

candidate. For this reason He+ 5% isobutane, He+ 5% isobutane + TMAE, He+ 5% Xe+EF and He+5% Xe + TMAE mixture gas mixtures were tested. The maximum gas gain, ~10^4, was achieved in He + 5% isobutane and He + 5% Xe + EF, as shown in Figure 30.

Figure 31 shows the quantum efficiency of a Cs_3Sb photocathode measured under different conditions. Curve 1 corresponds to measurements made in a vacuum and curves1 in He+ 5% isobutane. The same potential was applied to all electrodes to avoid electrostatic losses in the charge collection.

In the case of curve 3 a potential of ~600 V was applied to the capillary plate to make it fully transparent for drifting photoelectrons, but not sufficient to trigger electron multiplication.

Systematic studies of another photomultiplier prototype consisting for simplicity only of a single capillary plate operating in avalanche mode were performed by Guinji et al.(Giunji, 2002). The main focus of this work was not on achieving high gas gains, but rather on optimization of the photocathode manufacturing process in order to obtain quantum efficiencies close to commercial vacuum phototubes. The chamber where photocathodes were manufactured (evaporation chamber) and the chamber where they were combined with the detector (test chamber) were separated by a door. During the heating of the evaporation chamber and the photocathode manufacturing the door was kept closed. This prevented the penetration of Cs vapour to the test chamber and gave better results. After the manufacturing procedure was completed, the door was opened and the photocathode was moved to the test chamber and combined with a capillary plate. Two types of highly efficiency photocathodes were manufactured in this work: Cs_3Sb and bialkali.

In the case of the Cs_3Sb photocathode, a point-like Sb evaporator was used. It was a small sphere of Sb on a Ni wire. A few cm below the evaporator a spherical metallic mirror was placed for improving the evaporation quality and also used in the light measurements. After evaporation of Sb to a thickness corresponding to 60% transparency, the chamber was heated to 180°C and Cs was gradually evaporated. By applying 100 V between the cathode and the evaporator (or the mirror) one could monitor the dark current (a leakage current) due to Cs deposition on the electrodes. The first ~10 minutes, the dark current was quite unstable due to the migration of Cs inside the chamber and its sporadic deposition on

Figure 31. Quantum efficiency of a Cs_3Sb photocathode measured under various conditions: 1-measurements in vacuum, 2-He+5% isobutane; in cases 1 and 2 the same potential was applied to all electrodes 3-600 V applied across the capillary plate making it transparent to photoelectrons, 4-TMAE vapour added to the gas

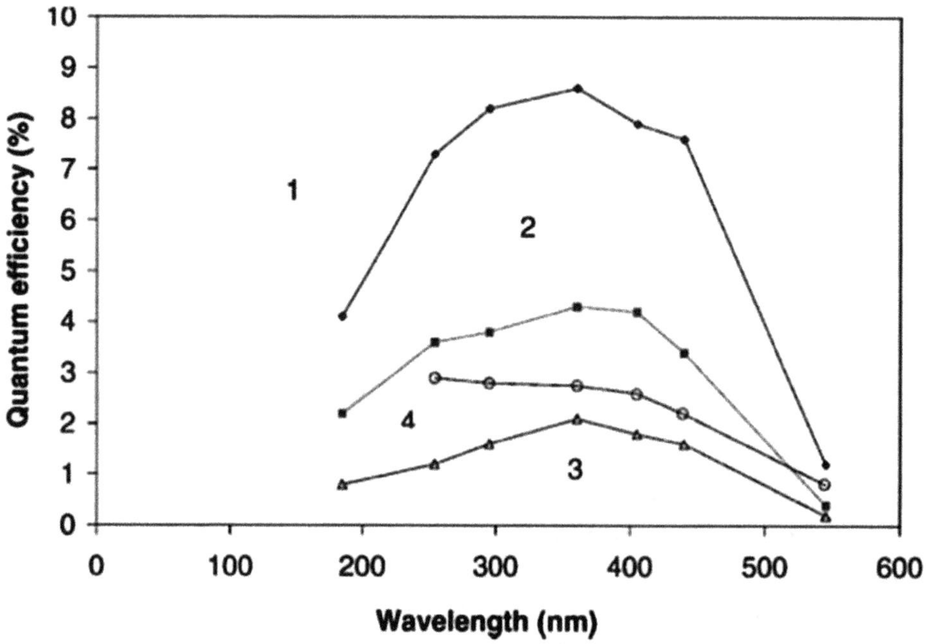

dielectric parts of the electrodes. After a strong decrease of dark current, the light source was switched on. At this moment the current sharply jumped in value due to the emission of photoelectrons from the cathode. When the current started decreasing the heating was stopped. Such a procedure guaranteed high efficiency and a good reproducibility in manufacturing of Cs_3Sb photocathodes (Giunji, 2002).

To manufacture bialkali photocathodes Giunji et al. first evaporated the Cs at room temperature. Then the vacuum chamber was heated to 180°C and the dark current was monitored. The heating was continued until the dark current dropped below <0.5 μA. The light source was then switched on and the photocurrent was monitored. When the photocurrent exceeded the leakage current with a few times, and reached some empirically defined value (~ 2 μA for their light source) the heating of the chamber was stopped. When the chamber was cooled down to room temperature, the Sb was evaporated to a thickness corresponding to 60% transmission. After that kalium (potassium) was evaporated (roughly 50% of the amount of Cs). Then the evaporation chamber was heated to 195°C and the dark current was monitored. After reaching a low value, the light source was switched on and the photocurrent was monitored. When it reached its maximum the heating was stopped and the photo cathode was ready.

Typical values of quantum efficiency of the Cs_3Sb and bialkali photocathodes manufactured and measured immediately after they were moved to the test chamber are shown in Figure 32. When a gas mixture of Ar + 5% CH_4 was introduced into the chamber, the quantum efficiency dropped by ~20% (curve 2).

In Figure 32 a "practical" quantum efficiency is also presented (curves 3). It is measured as a photocurrent on the anode at the maximum possible amplification voltage (1.32 kV) on the capillary electrodes, corresponding to a gas gain of 750. With a weaker light illumination giving 10 times lower current the

Figure 32. Quantum efficiency of Cs_3Sb (a) and bialkali (b) photocathodes, measured in vacuum (1) and in Ar+5% CH_4 (2) in charge collection mode; 3-"practical" quantum efficiency measured in amplification mode

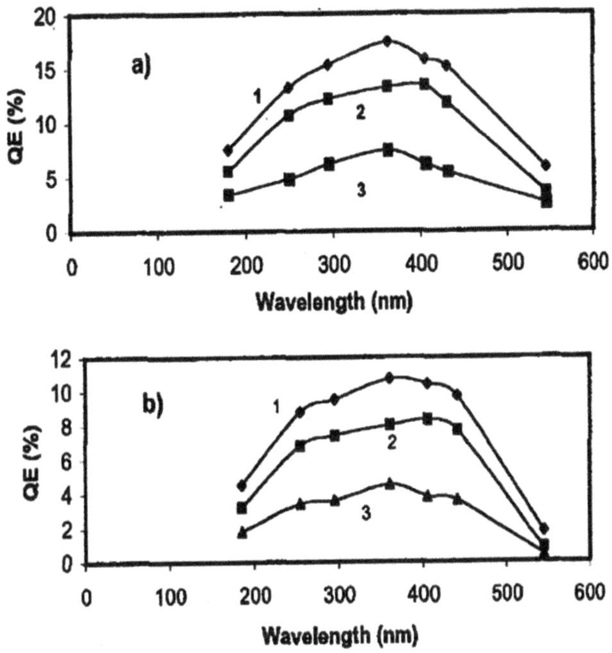

maximum achievable gain was 2.5 times higher. Since the maximum achievable gain is limited by ion and photon feedback, the difference in gain is attributed to a cathode excitation effect (described in the previous chapter) at high currents leading to an increase of the coefficients γ_{ph} and γ_+ which in turn reduces A_{max}. We should note that this was the first time such high gas gains were achieved with gaseous detectors with highly efficient photocathodes sensitive to visible light.

Figure 33 presents the results of the stability tests of the photocathodes. For such a large gas volume, in which the detectors were studied, the obtained stability shall be considered as good.

Looking back at Figures 7, 10, 15, 27, 31, 32, one appreciates how various research groups in a step-by step approach gained more and more experience in manufacturing highly efficient photocathodes sensitive to visible light. Using a sophisticated evaporation plant, quantum efficiencies close to the commercial vacuum devices were achieved.

An interesting concept was proposed by Rodionov et al. (Rodionov, 2002). Their photomultiplier consists of two planar parts: a sealed chamber with a semi-transparent bi-alkali or Cs_3Sb photocathode (the "scintillation chamber") and a position-sensitive gaseous detector with a CsI photocathode flushed with a gas at atmospheric pressure (the "readout detector"), as shown in Figure 34.

The main advantage of this approach is that the sensitive photocathode is kept in a sealed chamber, which has only 2–4 feedthroughs and does not contain any outgassing materials. This ensures a high degree of cleanliness. As a result, the photocathodes have high quantum efficiency, are stable in time and do not show any sign of aging.

Figure 33. Variation of the quantum efficiency with time in He+5%CH_4 gas mixture for a) Cs_3Sb photocathode, b) bialkali photocathode
From Giunji, 2002.

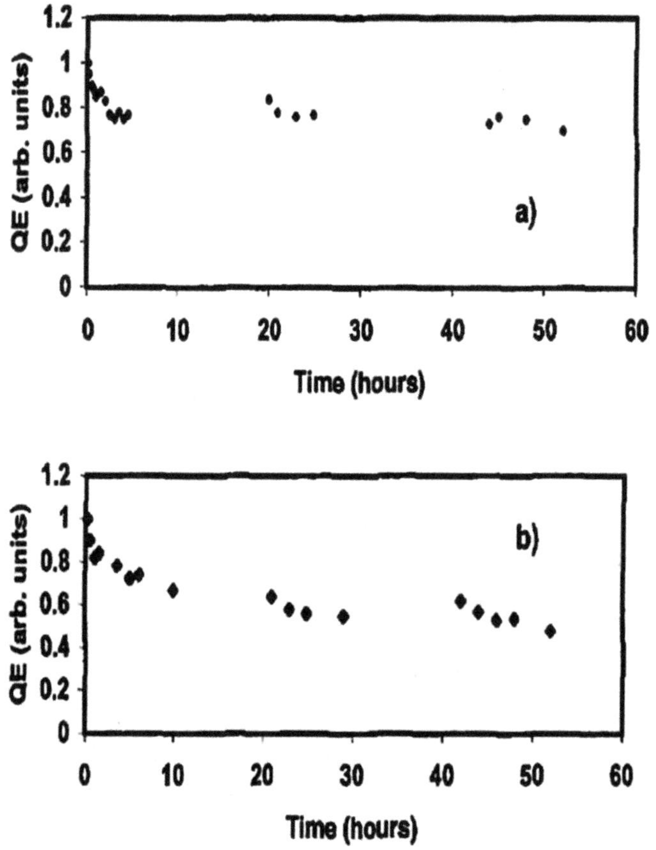

Figure 35 shows oscillograms of the signal from the vacuum PM (upper traces), detecting a weak pulse of the visible light hitting the Cs_3Sb photocathode and the corresponding charge signals from the MWPC, recording the Xe scintillation light.

A practical quantum efficiency of 3% was achieved at a wavelength of 400 nm (see Figure 36). The voltage on the MWPC was 1530 V (from Francke, 2002).

The position resolution of the detector was ~ 1 mm (Figure 37). The rate capability was better than ~10^4 Hz over the sensitive area and was limited mainly by the electronics used.

5.1 GEM-Based Detectors

The next decisive step in developing gaseous detector sensitive to visible light was done by Breskin's group. They manufactured sealed visible-light gas filled photomultipliers comprising multi-GEMs and K-Cs-Sb photocathodes (Balcerzyk, 2003). This group developed a sophisticated plant for photocathode manufacturing and their assembly with gaseous detectors.

The main modification compared to the relatively simple laboratory evaporators described earlier, was the use of three ultrahigh vacuum chambers integrated to one setup including, a load-lock where

Figure 34. Schematic drawing of the first type of a hybrid gaseous photomultiplier; the scintillation chamber (a 4.5 mm thick disc with a diameter of 30 mm) was filled with ultraclean Xe at atmospheric pressure. It consisted of a semi-transparent photocathode, a capillary plate and a parallel plate avalanche chamber with a 1mm thick gap and an exit window. Photoelectrons created by visible light from the Cs_3Sb photocathode drifted through the capillary plate (which acts in this design as a light attenuator) and produced a weak scintillation light during weak avalanche amplification in the parallel-plate mesh chamber. The position of the light burst was detected by a MWPC placed in a close contact to the exit window and having a semi-transparent CsI photocathode. The MWPC was not sealed. It was flushed with CH_4 or an $Ar+CH_4$ mixture at atmospheric pressure. The intensity of the visible light hitting the input window of the hybrid detector was calibrated with a PM-tube placed under the readout detector.

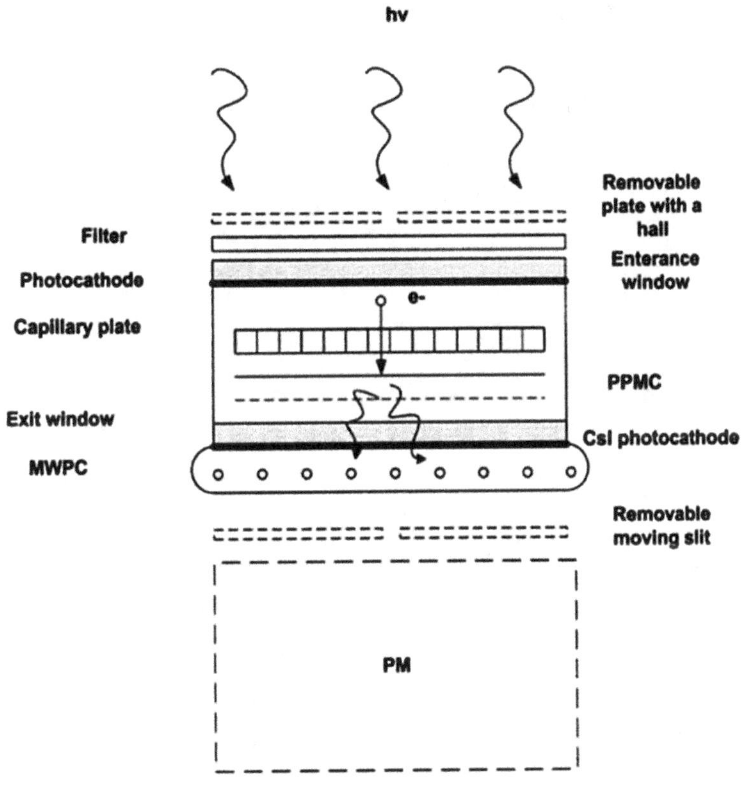

the glass window for the gas photomultiplier was initially baked, an activation chamber in which the semitransparent bialkali photocathode was deposited onto the window and its quantum efficiency evaluated, and an assembly chamber where the window was attached to the gas photomultiplier package containing the GEM multipliers. Each chamber was evacuated by its own turbo-molecular pump. An additional titanium sublimation pump could be used in the sealing chamber. Magnetic manipulators were used to move parts inside the chamber and from one chamber to another. The vacuum pressures in these chambers at room temperature were between $4x10^{-9}$ to and $3x10^{-10}$ Torr. All three chambers could be individually baked with internal quartz halogen lamps.

The glass window for the photomultiplier was first kept at 220°C for 24 h in the load-lock chamber and then was transferred to the activation chamber for photocathode deposition. The bialkali photocath-

Figure 35. Oscillograms of signals from the PMT (upper traces) and the MWPC (lower traces) detecting the secondary Xe scintillation light; the voltage between meshes of the parallel plate avalanche chamber was (a) 700 and (b) 800 V.

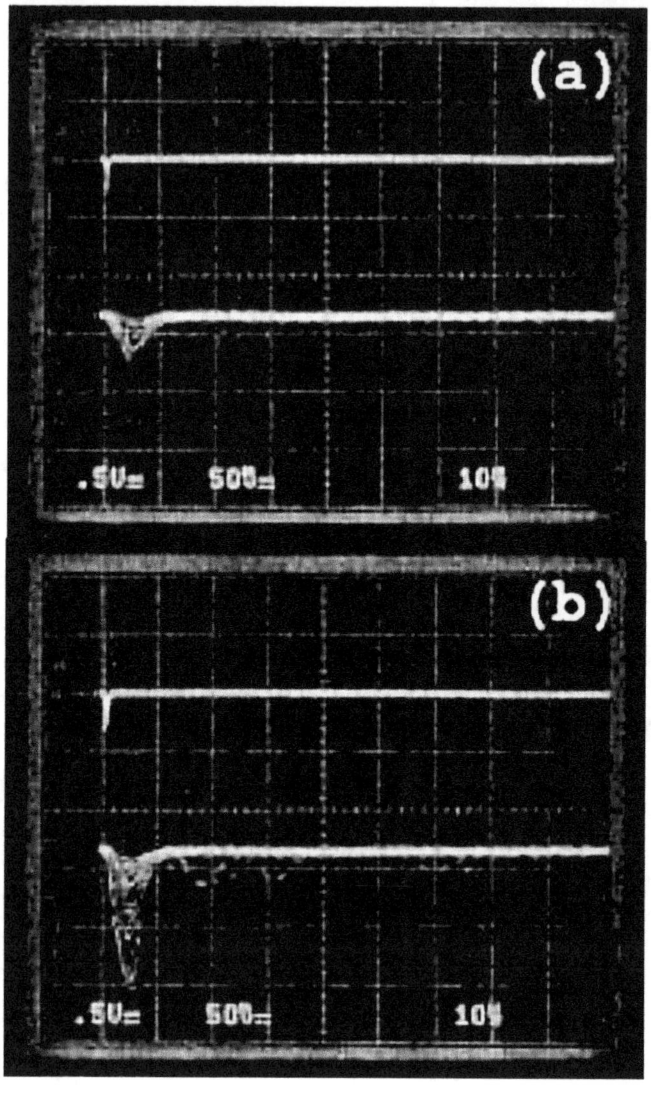

ode was prepared with the evaporation method described in the previous paragraph (see Shefer, 1999 for more details). After the evaporation, the photocathode transmission and its quantum efficiency were measured in situ using a calibrated light source and a monochromator. They succeeded manufacturing photocathodes with a vacuum quantum efficiency of 25% at its maximum. The photocathode was then moved to the sealing chamber in which the photomultiplier elements were installed in advance. The schematic drawing of the unsealed photomultiplier is shown in Figure 38.

The detector package had a cylindrical shape and was made of Kovar. It housed cascaded Kapton GEMs, 30 x30 mm^2 in size. The GEMs were mounted between ceramic spacers, defining the active area of GEMs to be 20 x20 mm^2. The distance between the photocathode and the top GEM was 5 mm and

Figure 36. The quantum efficiency of the Cs_3Sb photocathode measured 1) in vacuum, 2) in Xe at atmospheric pressure without a capillary plate, and 3) in Xe at atmospheric pressure with the capillary plate
From Francke, 2002.

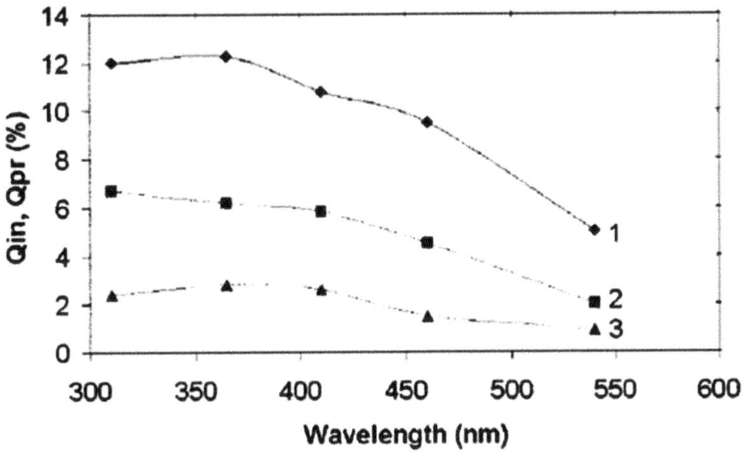

Figure 37. Charged signals from wires of the MWPC obtained with a 0.2 mm wide slit moving across the detector width
From Francke, 2002.

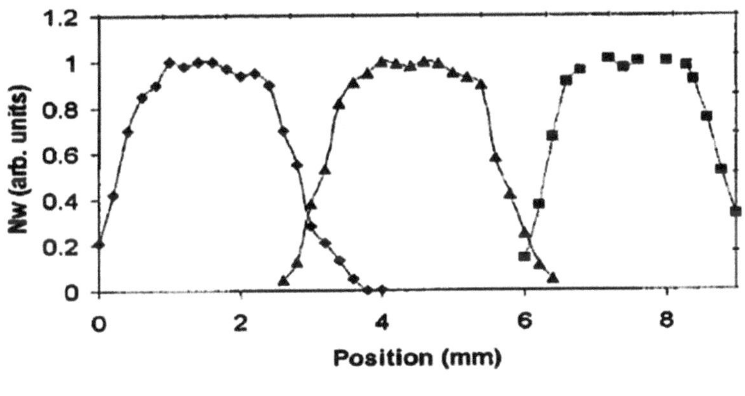

the distance between successive GEMs as well as between the last GEM and the anode was 1 mm. In the bottom of the package a SAES 172 getter was installed. When sealed the chamber with the assembled detector was baked for three days at 195°C. Then the getter inside the detector package was activated for 10 min by resistive heating to 450°C. After that an argon/methane gas mixture was let in to a pressure of ~700 Torr. The photocathode, previously held in the deposition chamber at the sealing temperature, was then transferred into the sealing chamber and brought into contact with the melted indium alloy on the top of the package. At this point, the package was rotated a few times by about to break the ineluctable oxide layer on the solder and the package was vibrated.

The chamber was sealed when the package was cooled to room temperature during 5–6 h. After that the sealed chamber was vented with dry nitrogen and the sealed photomultiplier was extracted to a glove box also flushed with nitrogen where it was additionally sealed with a silicone-based sealant

Figure 38. The internal parts of an unsealed gaseous photomultiplier

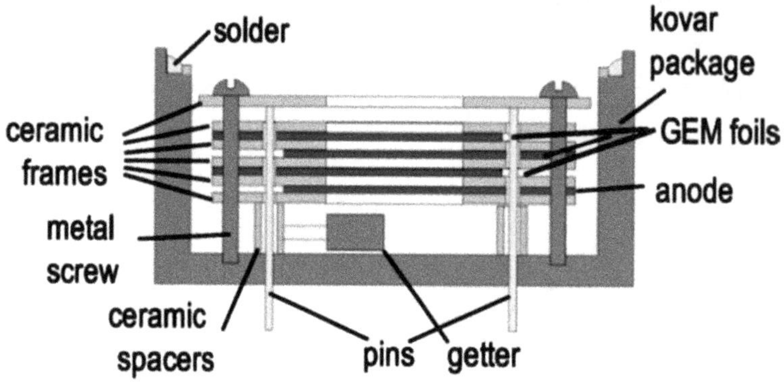

Figure 39. Photographs of a MHSP: top (left) and the bottom (right); the top one has the same electrode structure as an ordinary GEM whereas on the bottom rectangular dielectric openings between holes were created containing narrow (20 μm) metal strips in the middle.
From Lyashenko, 2007.

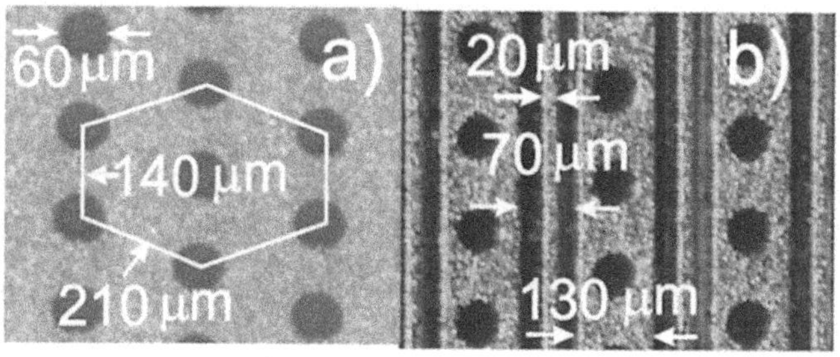

(Vacseal, Structure Probe, Inc.). The sealant was applied between the window and package, to ensure full protection against gas leak.

The highest quantum efficiency reached with this sealed device was 13% at 405 nm that falls to about 5% at 312 nm due to the transmission properties of the Kovar-glass window. The peak value was half of the quantum efficiency in vacuum and this was mainly due to back scattering in the Ar + 5%CH_4 gas mixture.

The gain curves measured in current mode with a single and double GEM operating in Ar + 5% CH_4 at 700 Torr. With double GEM a gas gain close to $2 \cdot 10^4$ was achieved. This is among the highest gas gains obtained, and with a quantum efficiency that is nearly an order of magnitude higher than in earlier studies, it was in important step forward.

A similar device with a UV-sensitive CsI photocathode have nearly as high gas gain with a visible photocathode as with a UV-sensitive CsI photocathode. It can be concluded that in both cases the ion and photon feedback were strongly suppressed resulting in gas gains sufficiently high for single photoelectron detection with a rather high efficiency, but not 100%.

With ordinary modern electronics, the gas gain needs to be increased 5-10 times more to have 100% efficiency for the single photoelectron detection. To achieve this Breskin's group focused on suppressing the ion and photon feedbacks even stronger. They combined a GEM and with two new hole type

Figure 40. Cross section of a MHSP: a) photoelectrons extracted from the cathode trigger two avalanches: 1) inside the holes and 2) near the anode strip on the MHSP back side; b) the avalanche electrons are quickly collected on electrodes. After their collection only positive ions remain in the region close to the initial avalanches. These ions are efficiently collected on the nearby electrodes so only a small portion of them will reach the photocathode.
From Veloso, 2000.

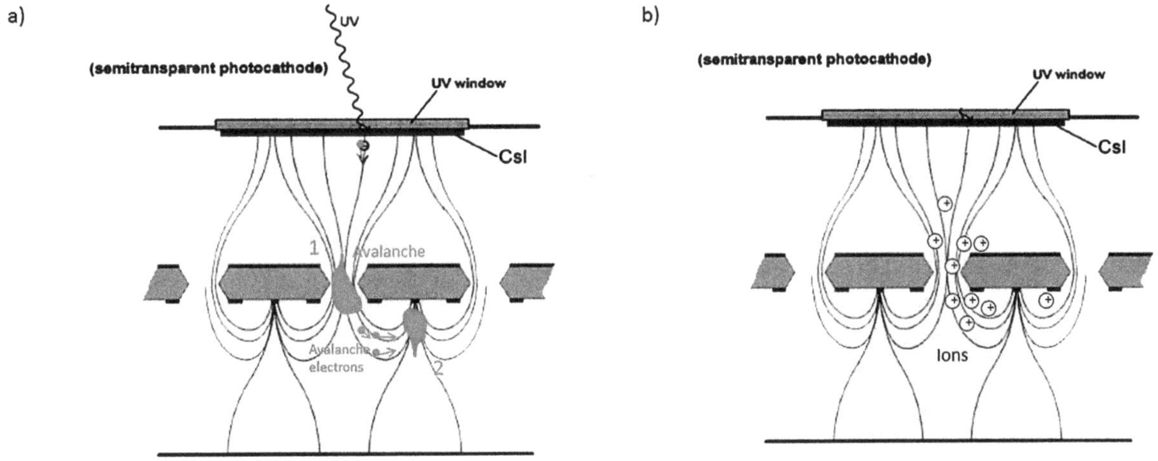

multipliers called microhole and strip chamber (or MHSP), proposed by Veloso et al., (Veloso, 2000). Basically it is a GEM made of Kapton having a different electrode pattern. The top and bottom of a MHSP is shown in Figure 39.

As can be seen, the top side has the same geometry as an ordinary GEM, a metalized surface with holes, whereas the bottom side has dielectric openings (70 μm wide) between the holes with metallic strips, 20 μm wide, manufactured in the middle.

The cross section of this device combined with a photocathode and the anode plate is shown in Figure 40. In MHSP a voltage is applied across the Kapton sheet (similar to GEM), ΔV_{GEM}. An additional positive voltage is applied to the metallic strips, $\Delta V_{AC\text{-}MHSP}$. The main idea behind this concept is that photoelectrons drifting from the photocathode experience two consequent multiplications: first in the holes (as in a usual GEM) and then part of avalanche electrons is extracted from the holes and directed by the electric field towards the strip, where they trigger a second multiplication.

Veloso et al. expected that this approach would allow increasing the overall gain, but it was also observed that the MHSP also significantly reduced the number of ions reaching the photocathode (Veloso, 2000). Later studies showed that, at properly tuned operation conditions in a cascaded electron multiplier comprising of different hole-multipliers with patterned surfaces, the fraction of ions back-flowing to the photocathode could be reduced down to 3×10^{-4} (Lyashenko, 2007).

A drawing of such a detector is shown in Figure 41. It consisted of a semitransparent photocathode (Cs_3Sb or bialkali) combined with a cascaded hole type multiplier and an anode plate. Three multiplication structures operating in tandem mode were used: F-R-MHSP (which is Flipped-Reversed-Micro-Hole & Strip Plate), GEM and MHSP. The first one was just an MHSP. It was placed inside the detector with the strip-coated surface facing the photocathode (flipped). Not a positive (as in MHSP), but a negative voltage was applied to its strips (reversed bias). The photoelectrons extracted from the photocathode are efficiently focused into the holes of F-R-MHSP (the negatively biases strips ensuring this), where they are multiplied. The avalanche electrons are then transferred into the GEM holes and finally to the MHSP for the last step of multiplications. The ions remaining after the multiplication avalanche (their paths are indicated by dotted arrows), relatively slowly (compared to electrons) drift back and most of them are captured by the system of properly biased electrodes. Only a minor fraction will reach the photocathode. This allowed much higher gas gains to be achieved than in any early detector.

Figure 42 shows the gain curves measured with this device. For comparison the results with double GEMs are also shown. As can be seen, with the F-R-MHSP/GEM/MHSP structure, due to the efficient

Figure 41. Drawing of an F-R-MHSP/GEM/MHSP gaseous photomultiplier combined with a semi-transparent photocathode; the main feature of this sealed detector is a special electron multiplier consisting of a cascaded hole-type structure, allowing to efficiently suppress the ion and the photon feedbacks. The trajectories of avalanche ion movement are schematically indicted by arrows. As can be seen, most of them are collected on the electrodes of hole-type multipliers and only a small fraction of them is able reaching the photocathode. As a result, the gas gain reached with this photomultiplier can be as high as 10^5 (see Figure 42).
From Lyashenko, 2009.

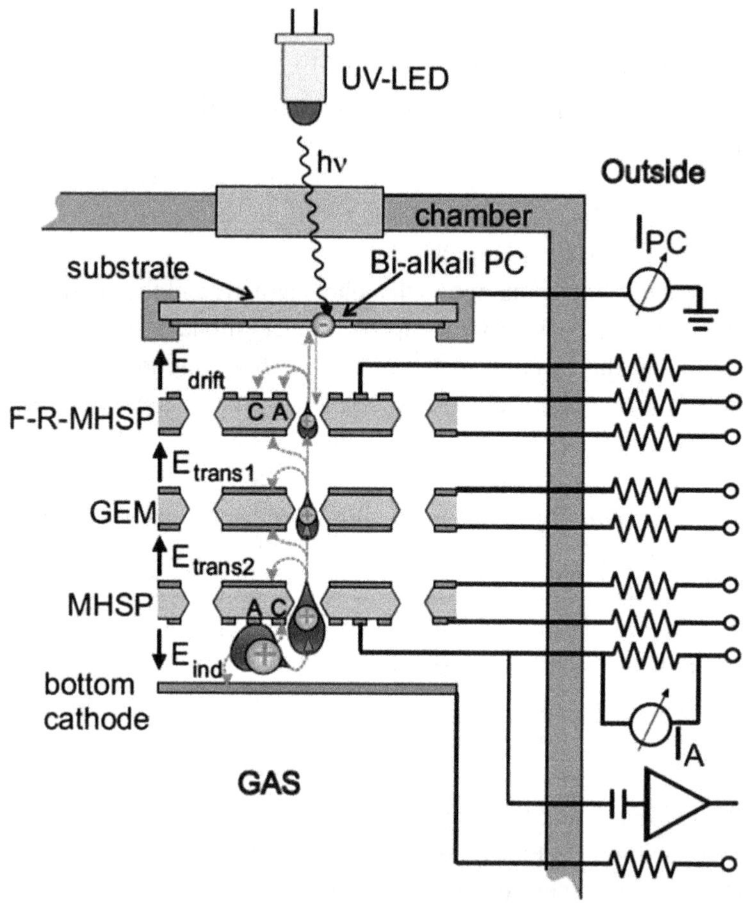

Figure 42. Gain curves measured with the F-R-MHSP/GEM/MHSP gaseous photomultiplier coupled to semi-transparent photocathodes: bialkali and CsI (upper curve); for comparison, data obtained with a double GEM are presented as well (lower curve). In both cases the gas mixture was Ar + 5% CH_4 at 700 Torr.
From Lyashenko, 2009.

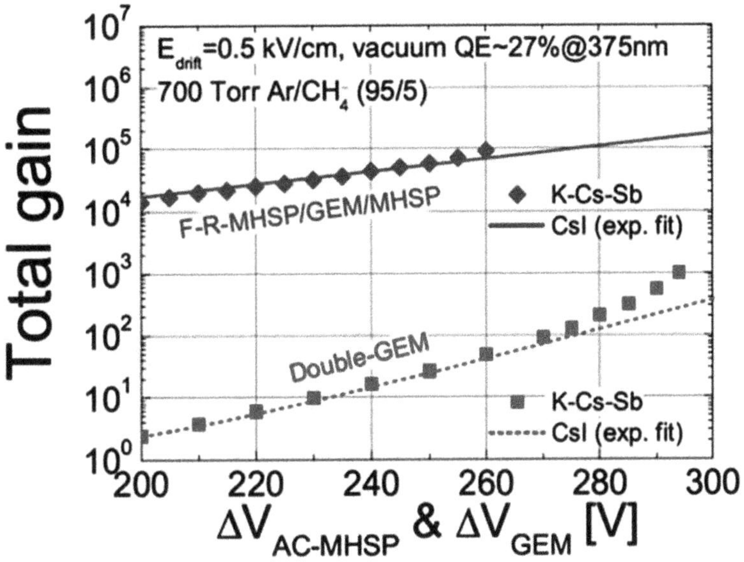

suppression of the ion and photon feedbacks, gas gains close to 10^5 were achieved, whereas with the double GEM the feedback becomes visible already at gains $>10^2$.

Another important result is the practical quantum efficiency measured at voltages optimized for efficient ion suppression. According to (Lyashenko, 2007) at some optimized voltages of across the F-R-MHSP, ΔV_{AC}, it operated with full collection efficiency (ε_{det}~1) of photoelectrons reaching its top surface (see Figure 43).

Their study indicates that the practical quantum efficiency is affected only by the backscattering of electron to the photocathode. Results of separate measurements of the quantum efficiency in the gas affected by the backscattering effect are shown in Figure 44.

The vertical axis on the right shows the calculated extraction efficiency of photoelectrons in the gas determined by the backscattering effect (from Lyashenko, 2009).

It was concluded that such a photomultiplier has quantum efficiency about 10% (Lyashenko, 2009).

5.2 Attempts to Develop Industrial Prototypes of Gaseous Photomultipliers

A Japanese group in collaboration with Hamamatsu Photonics made intense efforts to develop commercial prototypes of sealed gaseous photomultipliers sensitive to visible light. They tested a parallel-plate electron multiplier made of metallic meshes and combined with a bialkali photocathode (Tokanai, 2009). This device was similar to one described earlier in work (Peskov, 1995). However, the gap between the meshes was reduced allowing operation at lower voltages. It is shown in Figure 45.

Figure 43. Gas gain (left vertical axis), measured on the collection electrode placed under the F-R-MHSP- see Figure 41- (so-called visible gain) and the photoelectron detection efficiency ε_{det} (right vertical axis), which comprise the electron collection efficiency to the holes of F-R-MHSP (that affects the QE of the detector), as a function of the inter-strip voltage, ΔV_{AC}, of a F-R-MHSP at two conditions: a) E_{drift}= 0.2 kV/cm and b) E_{drift}= 0.5 kV/cm; the measurements were performed at different hole voltages.
From Lyashenko, 2007.

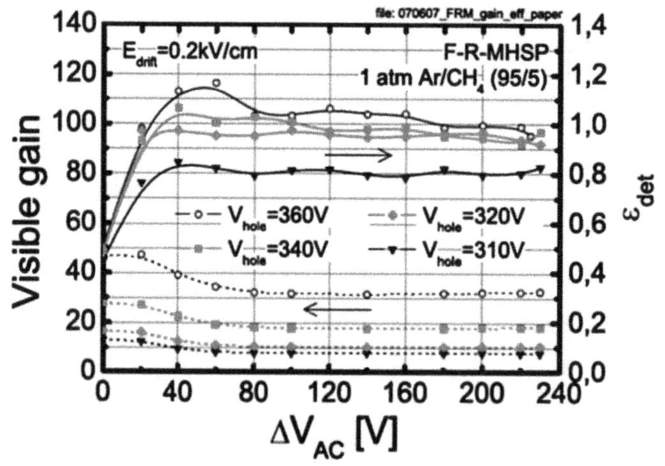

(a)

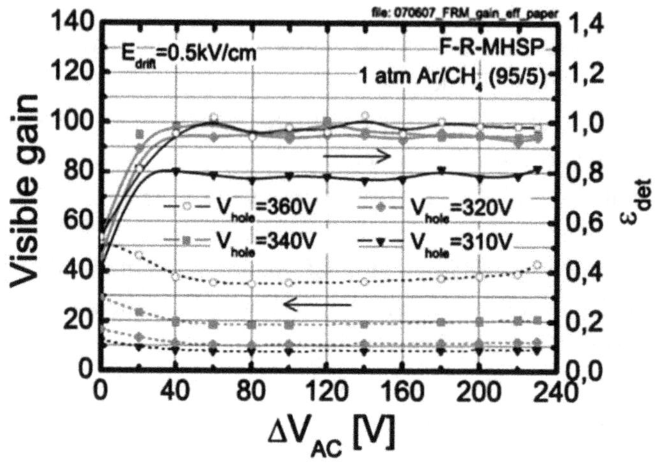

(b)

The quantum efficiency of this device is shown in Figure 46. They used Hamamatsu's evaporation facility and technology, giving high quantum efficiency, exceeding 10% at the peak of the sensitivity.

The maximum gas gain achieved with this detector was about 10^3 which is typical for the open-type geometry gaseous photodetectors (see Figure 47).

The same team tested later a single glass GEM as an amplification structure shown in Figure 48.

Figure 44. Results of quantum efficiency measurements for a bialkali photocathode in vacuum and in Ar + 5% CH_4 at 700 Torr (the lowest curve) at an electric field of 500V/cm

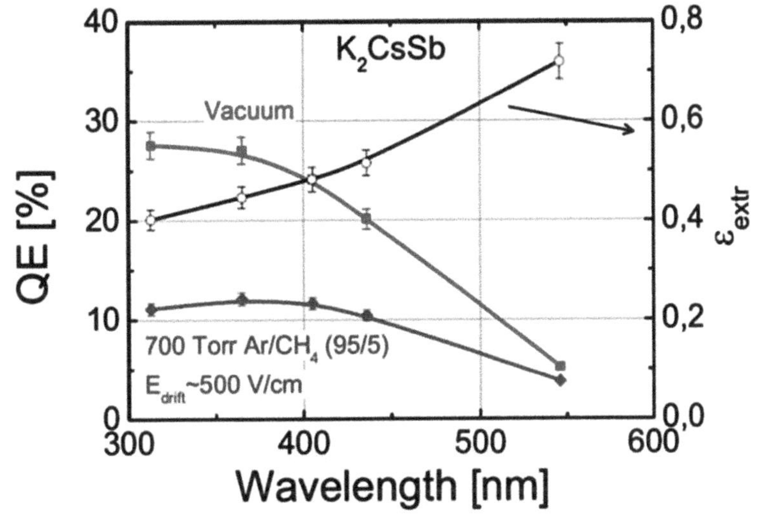

Figure 45. Photograph (a) and schematic view (b) of the sealed gaseous photomultiplier having a bialkali photocathode combined with a microgap parallel plate avalanche chamber
From Tokanai, 2009.

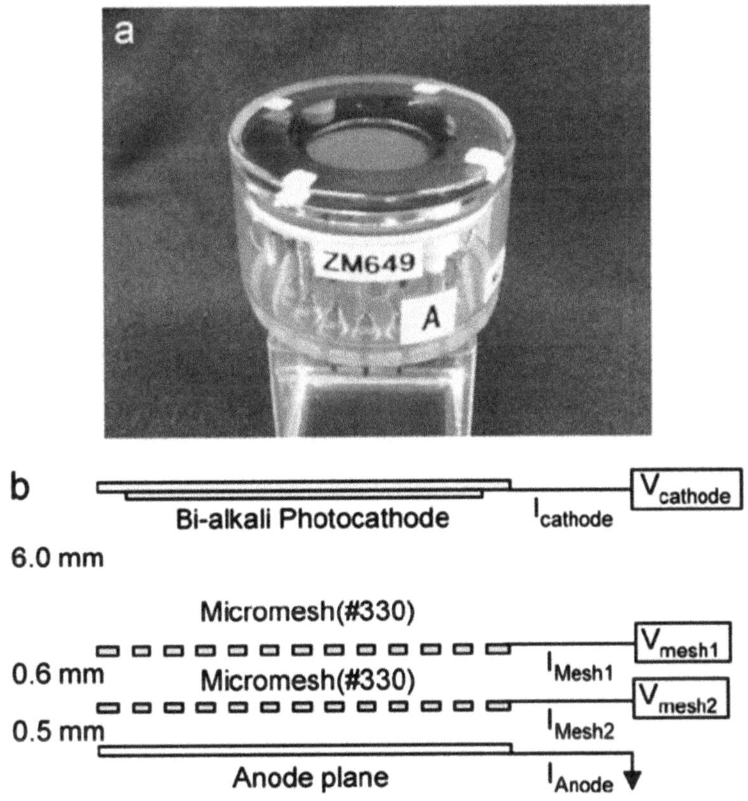

Figure 46. Quantum efficiencies of the first commercial sealed gaseous photomultipliers with a bialkali photocathode combine with a small-gap mesh parallel plate avalanche chamber; it was operated with Ar + 10% CH_4 and Ne + 10% CF_4 at atmospheric pressure.
From Tokanai, 2009.

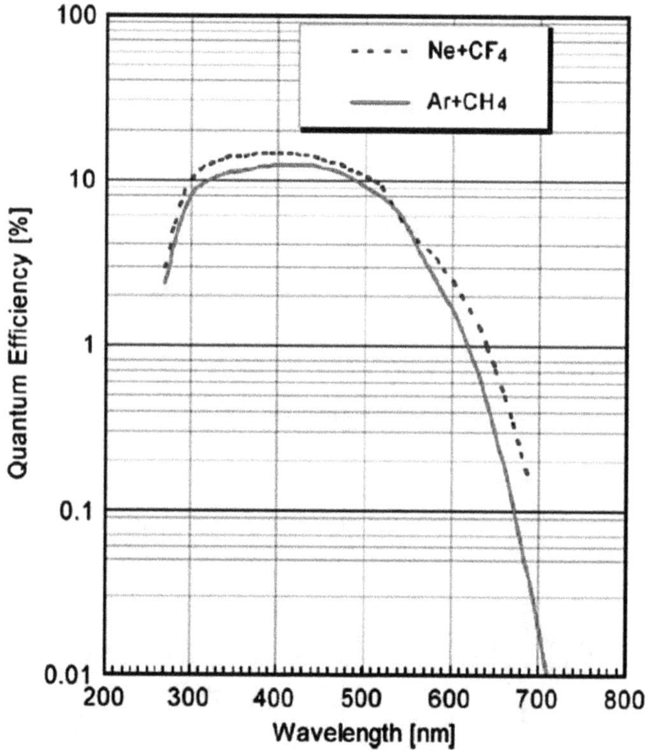

Figure 47. Gain vs the voltage applied across the meshes of the parallel plate avalanche chamber for the sealed gaseous photomultiplier with a bialkali photocathode operating in Ne + 10% CF_4 at atmospheric pressure
From Tokanai, 2009.

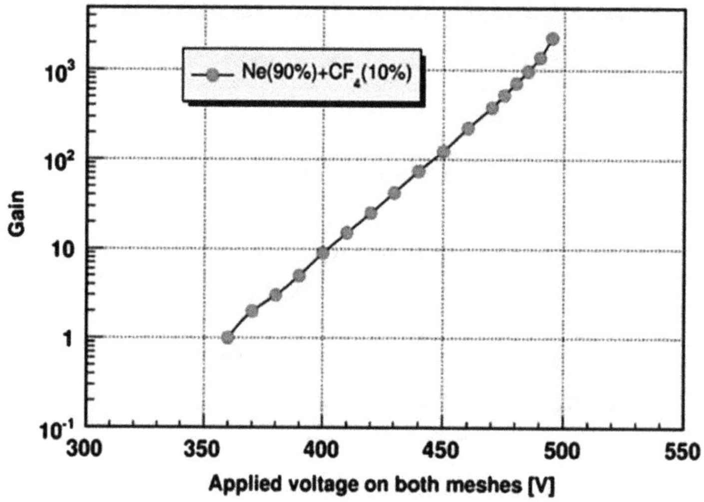

Figure 48. Structure of a bi-alkali photocathode MB-GP consisting from a glass GEM and a parallel plate avalanche chamber
From Sumiyoshi, 2011.

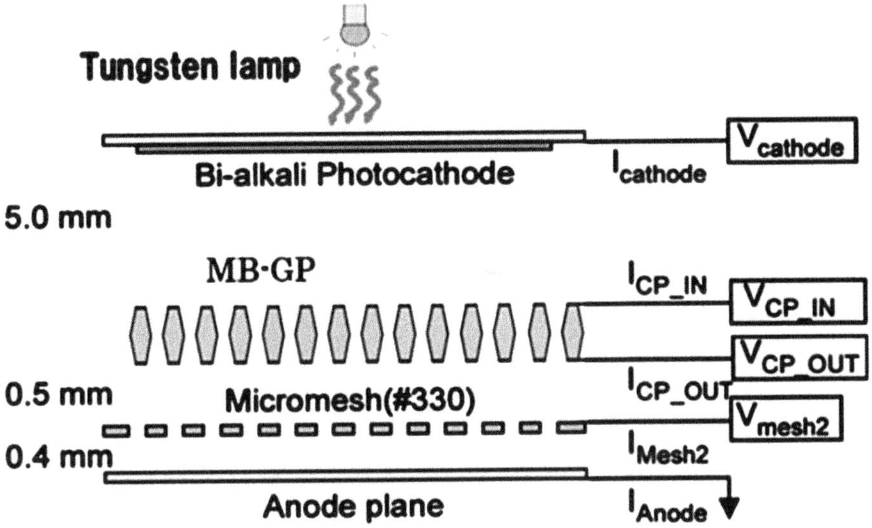

This GEM, called Micro-Blasting Glass Plate (or MB-GP) was manufacture from a 300 μm thick Pyrex glass sheet by a sand-blasting method. It used 230 μm hole pitch and 80 μm and 126 μm inner and outer hole diameters (as in standard GEM the holes had a double-conical shape).

Photographs of the MB-GP are shown in Figure 49.

In the gain measurements only the voltage across glass GEM electrodes was applied. The multiplication between the mesh and the anode were not exploited. Although a maxim gain of about 10^3 was achieved, which is similar to what has been observed earlier with a single capillary plate, one single glass GEM was not sufficient for good feedback suppression, as shown in Figure 50.

They also experimented with ordinary GEMs without great success. They observed that Kapton absorbs alkali vapour during the activation of the photocathode and the bi-alkali photocathode was not well formed if Kapton was present. As a result, the sensitivity of the Kapton GEM photomultiplier to visible light was only 3% of that of glass GEM.

Gas gains up to 10^4 with ion feedback was achieved with a photomultiplier design consisting of two small gap parallel plate avalanche chamber operating in cascade mode. The novelty was the use of meshes having different diameters of the rounded holes: 100 μm on the top mesh and 190 μmon the bottom, as shown in Figure 51.

Such a structure more efficiently suppresses the ion feedback than the normal meshes earlier. This can be illustrated by Figure 52 showing results of simulations of avalanche electrons and ions flow using the Maxwell 3D and GARFIELD simulation tools.

In the left part of Figure 52, the motion of the avalanche electrons in the double parallel plate avalanche chamber is presented. At this time frame ions can be considered as not moving since electrons are drifting much faster. In the right part of this figure the ion movement is shown when most of the avalanche electrons already have been collected on the mesh electrodes. As can be seen, due to the low

Figure 49. Photographs of the glass GEM manufactured by a micro-blasting technology (MB-GP): (a) MB-GP, 1inch in diameter MB-GP, (b) quality of holes produced by the micro-blasting method and (c) cross-section of the holes having a double conical shape
From Sumiyoshi, 2011.

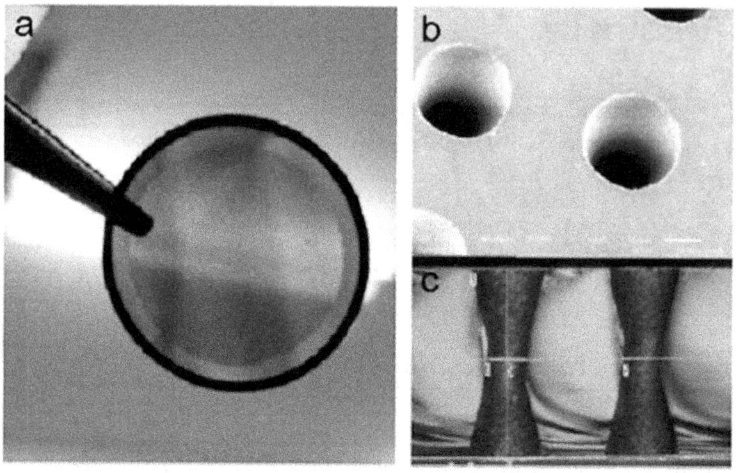

Figure 50. Gain vs. a voltage applied between the electrodes of the MB-GPs; clear deviation from an exponential growth is visible at gas gains >10, indicating the appearance of feedback effects.
From Sumiyoshi, 2011.

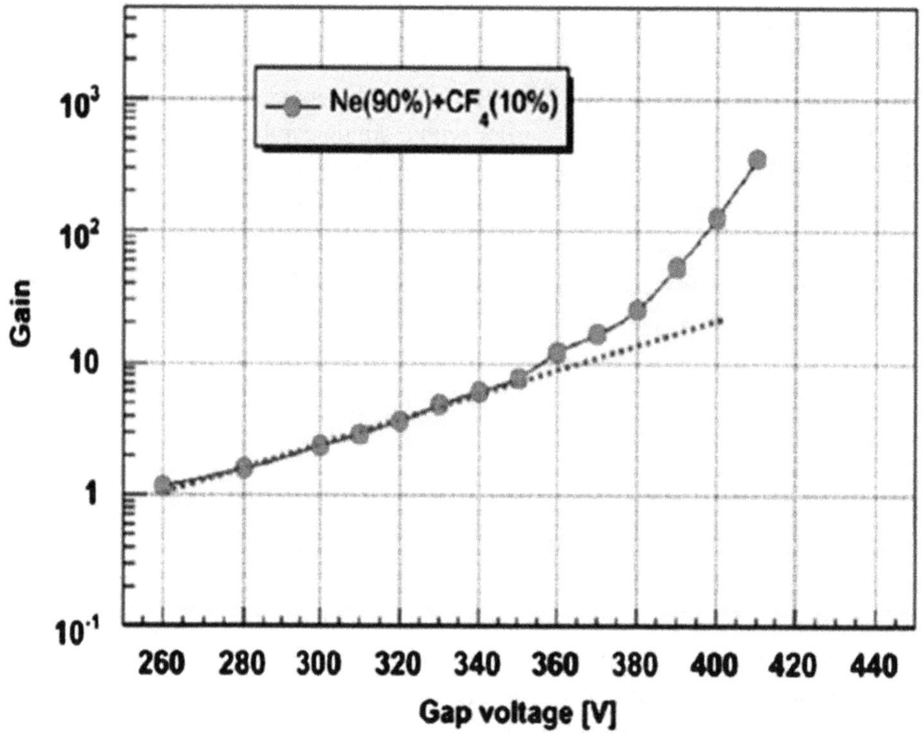

Figure 51. The schematic drawing of a sealed gaseous photomultiplier consisting from a semitransparent bialkali photocathode combined with small gap cascaded parallel plate avalanche chamber the cathode and the anode metallic meshes of which has round-hole opening with 100 μm and 190 μm diameters, respectively
From Tokanai, 2014.

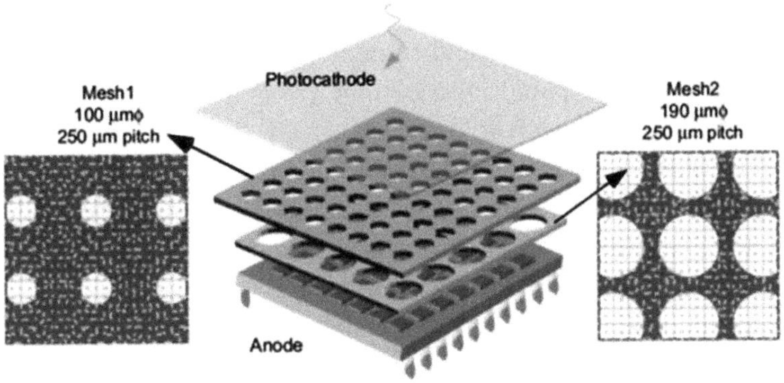

Figure 52. Results of simulations of an avalanche development triggered by a single electron; left-avalanche electrons tracks, right-ion backflow trajectories
From Tokanai, 2014.

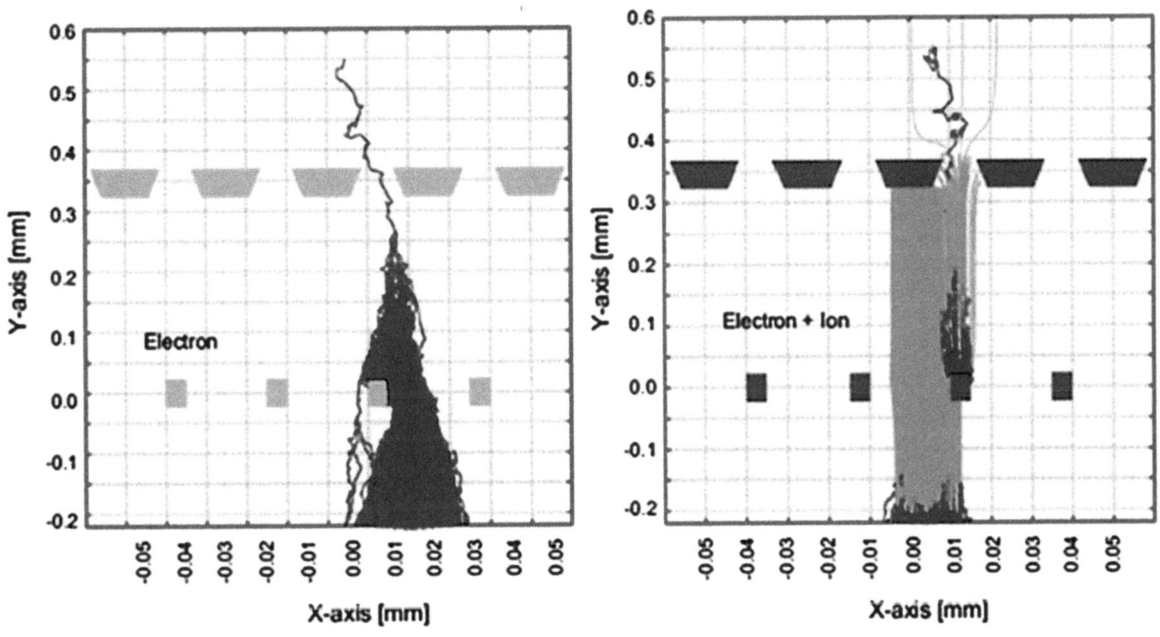

optical transparency of the first mesh (a small ration of hole diameters to their pitch), most of the ions are collected on the inner part of the first electrode. The ion back flow (IBF) can be expressed as:

IBF $=(I_{ipc}-I_0)-/I_{anode}$, (3),

where I_{ipc} is the current measured at the photocathode, I_0 is the photocurrent from the photocathode and I_{anode} is the avalanche current measured on the anode electrode. For a single parallel plate avalanche chamber:

$$IBF \sim E_{dr}/E_{PPAC},$$

where E_{dr} is the drift field and E_{PPAC} is the field strength inside the parallel plate avalanche chamber. Since the quantum efficiency of the photocathode in the gas is increasing with E_{dr}, there is a tradeoff between the quantum efficiency and the IBF for each gas gain.

Results of the IBF measurements are presented in Figure 53.

As can be seen, for these particular measurements the lowest IBF achieved was about $5x10^{-4}$. This indicates that gas gains higher than 10^4 could be reached with this detector.

The gas gain measurements are presented in Figure 54.

The maximum achievable gain was about 10^4 although no sign of the feedback is visible (Tokanai, 2014). In photodetectors breakdown occurs via a slow mechanism (a feedback loop). As no feedback could be detected at the breakdown gas gains, one can suspect that the breakdown in this detector appeared due to internal defects in the parallel plate avalanche chamber structure. Hence with a better designed detector one can expect higher gains to be achieved

The stability with time of various commercial prototypes of gas filled photomultiplier was also studied.

Two important observations were done:

Figure 53. IBF vs. the voltage applied across the gap between mesh 2 and the anode plate in Figure 51
From Tokanai, 2014.

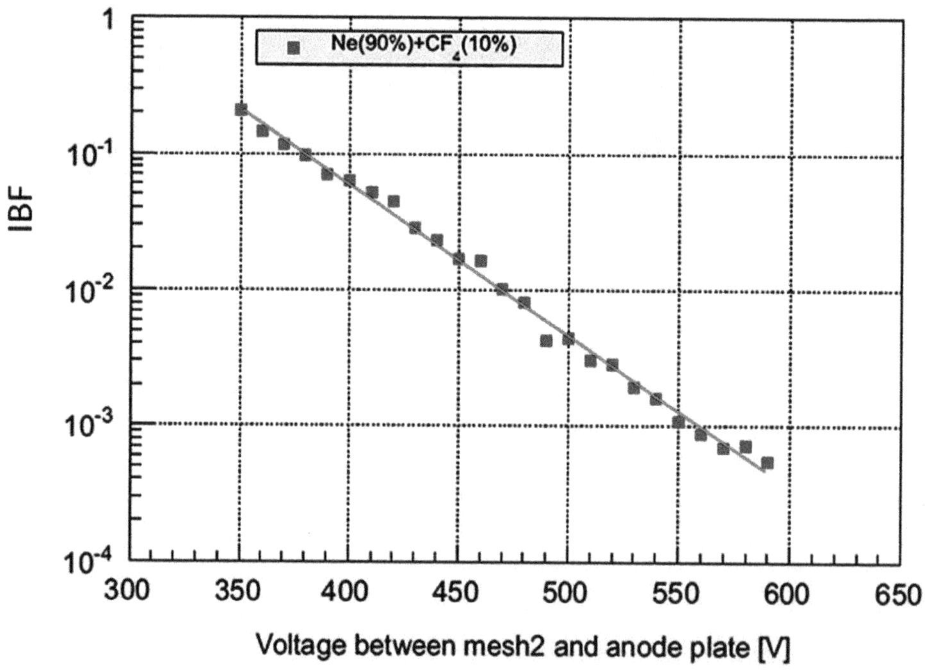

Figure 54. Gas gain curved measured the gaseous photomultiplier
Tokanai, 2014.

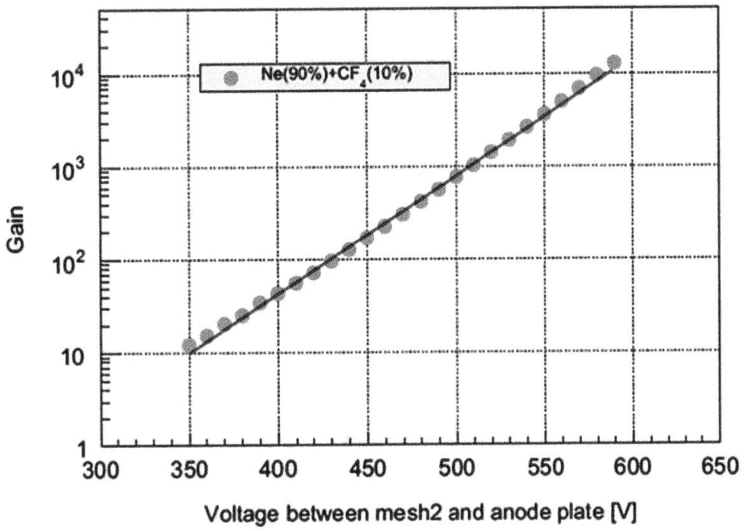

1. Bialkali and Cs_3Sb photocathodes which have not been exposed to the strong light fluxes are stable with time.
2. Strong light irradiation leads to some degradation of the quantum efficiency due to ion bombardment.

Figure 55 shows the results of a long-term monitoring a bialkali photocathode in a gas atmosphere. Practically no changes in the quantum efficiency happened after~1 year of contact with a clean gas.

In Figures 56, 57, and 58 the stability of the gas photomultiplier operating at various gas gains are illustrated.

Results obtained at a gas gain of 10 are shown in Figure 56.

In this test the gaseous photomultiplier was continuously irradiated with monochromatic light corresponding to an initial photocurrent current from the photocathode (I_0) of 20 pA. The figure shows the anode current (I_{anode}) at the applied voltage of 620 V across the glass GEM as a function of the operation time. It can be seen that the anode current is almost constant during the operation time of 16h.The accumulated charge on the photocathode was about 34nC/mm^2.

The quantum efficiency of the gaseous photomultiplier measured before and after 16 h of operation under light irradiation are shown in Figure 57.

The results obtained at a much higher gas gain, 670, are shown in Figure 58.

After 182 h of continuous operation, corresponding to an accumulated anode charge of about 5 mC, the anode signal had decreased with17%. The calculated density of the charge accumulated on the bialkali photocathode surface was 0.4 μC/mm^2.

The results obtained by the Japanese group proves that manufacturing of commercial gas filled photomultipliers capable operating at rather high gas gains and having excellent time stability is possible.

Figure 55. Stability with time of a gaseous photomultipliers with a bialkali photocathode exposed to Ar + 10% CH_4 at atmospheric pressure
From Tokanai, 2011.

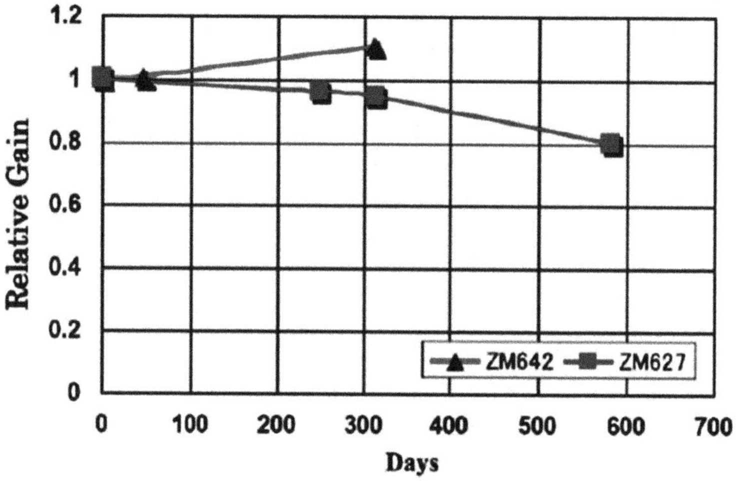

Figure 56. The anode current I_0 from the gaseous photomultiplier combined with bialkali photocathode as a function of operation time; the detector was filled with a gas mixture of Ne + 10% CF_4 at 0.9 atmospheres.
From Moriya, 2013.

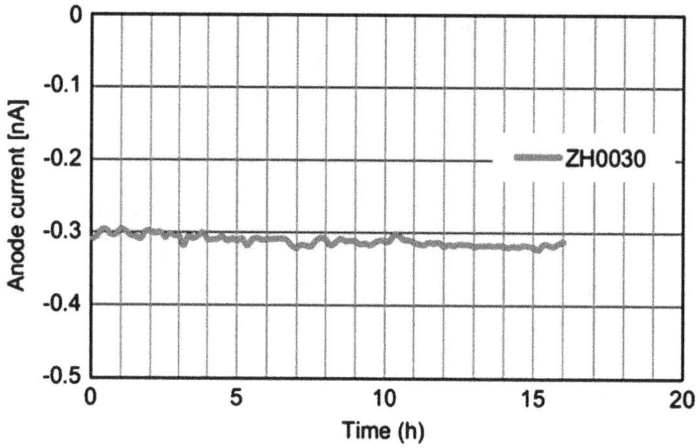

6. CONCLUSIVE REMARKS

Promising results have been obtained so far. Sealed gaseous photomultipliers sensitive to visible light have now high quantum efficiency and are stable in time. In contrast to vacuum devices, they have little sensitivity to magnetic fields, as shown in Figure 59, which is an important advantage allowing them to be used in many applications.

However, still some more work should be done before these detectors can be converted to commercial products. Since Kapton-based GEM and MHSP cannot assure long term stability, glass or metal-based

Figure 57. The quantum efficiency of the gaseous photomultiplier filled with Ar+10% CH_4 at 0.9 atmospheres combined with a bialkali photocathode and glass GEM amplification structure, measured before (0 h) and after16 h of continuous light illumination
From Moriya, 2013.

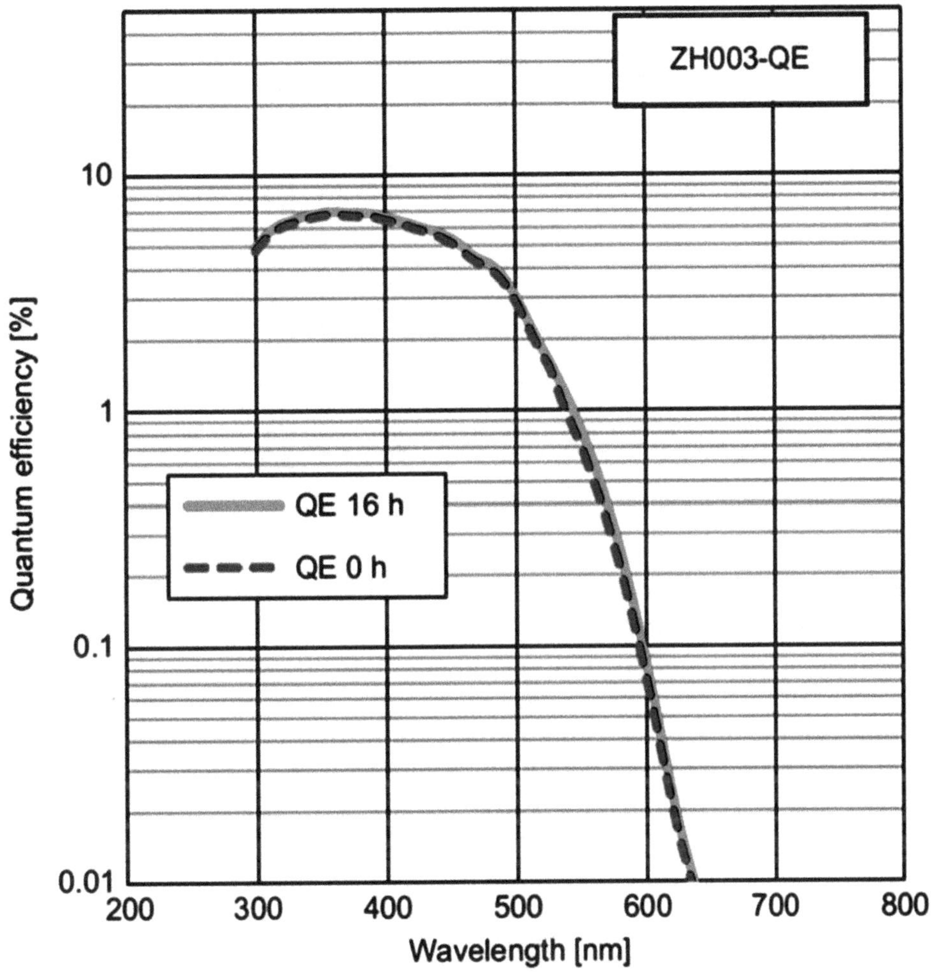

detectors should be developed allowing at the same time gas gains of about 10^5 and strongly suppressing the ion and photon feedback. Probably glass GEM and glass MHSP could be good candidates, but it remains to be proven.

In connection to this it will be relevant to quote a joke statement which Georges Charpak made at one of the RICH Workshops: "The next Noble prize in Physics in the field of gaseous detectors will be given to the people who developed gaseous photomultipliers sensitive to visible light". The future will show if this joke will be materialized or not.

Figure 58. Relative output current from the anode of the gaseous photomultiplier with bialkali photocathode and double parallel plate avalanche chamber detector as a function of operation time; the photomultiplier was filled with a gas mixture of Ne + 10% CF_4 at 0.9 atmospheres and operated at a gas gain of 670.
From Moriya, 2013.

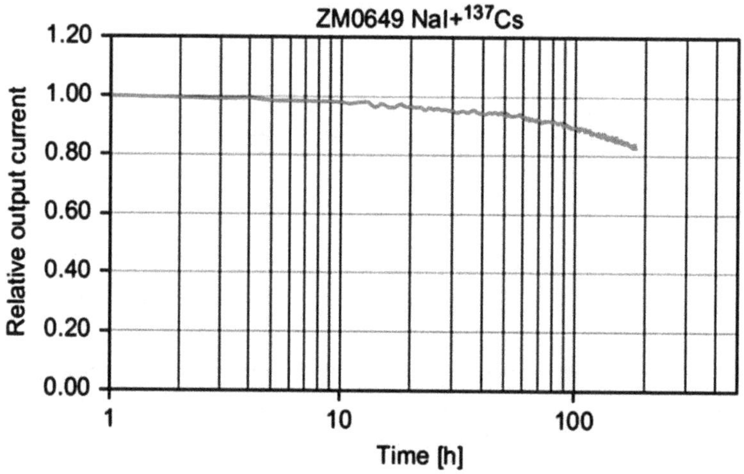

Figure 59. Comparative gain changes in magnetic fields for gaseous and vacuum (R5924 with fine meshes) photomultipliers oriented parallel (0°) and perpendicular (90°) to the magnetic field lines
From Tokanai, 2011.

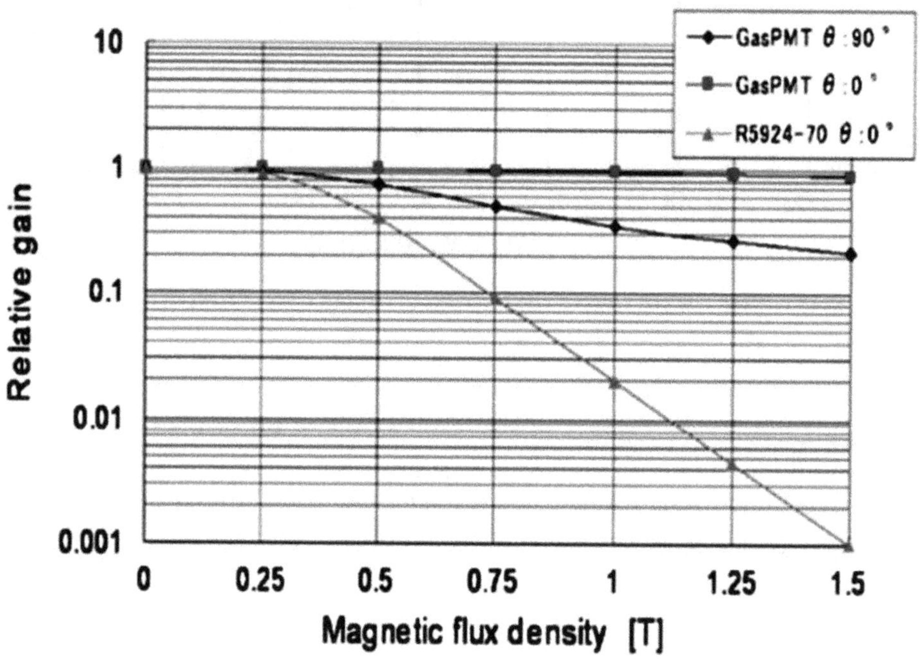

REFERENCES

Akkerman, A., Boutboul, T., Breskin, A., Chechik, R., & Giberhterman, A. (1994). Low-energy electron transport in alkali halides. *Journal of Applied Physics*, *76*(8), 4656–4662. doi:10.1063/1.357303

Anderson, D. F., Kwan, S., & Peskov, V. (1994). CsI and some new photocathodes. *Nuclear Instruments & Methods in Physics Research. Section A, Accelerators, Spectrometers, Detectors and Associated Equipment*, *343*(1), 109–116. doi:10.1016/0168-9002(94)90539-8

Balcerzyk, M., Mormann, D., Breskin, A., Sinh, B. K., Fritas, E. D. C., Chechik, R., & Rappaport, M. et al. (2003). Methods of preparation and performance of sealed gas photomultipliers for visible light. *IEEE Transactions on Nuclear Science*, *50*(4), 847–854. doi:10.1109/TNS.2003.815142

Biteman, V., Guinji, S., Peskov, V., Skurai, H., Silin, E., Sokolva, T., & Rodionov, I. (2001). Position sensitive gaseous photomultipliers. *Nuclear Instruments & Methods in Physics Research. Section A, Accelerators, Spectrometers, Detectors and Associated Equipment*, *471*(1-2), 205–208. doi:10.1016/S0168-9002(01)00967-6

Borovik-Romanov, A., & Peskov, V. (1994). Cs based photocathodes for gaseous detectors. *Nuclear Instruments & Methods in Physics Research. Section A, Accelerators, Spectrometers, Detectors and Associated Equipment*, *348*(2-3), 269–274. doi:10.1016/0168-9002(94)90745-5

Breskin, A., Buzulutskov, A., Shefer, E., Chechik, R., & Prager, M. (1998). Removable organic protective coating for alkali-antimonide photocathodes. *Nuclear Instruments & Methods in Physics Research. Section A, Accelerators, Spectrometers, Detectors and Associated Equipment*, *413*(2-3), 275–280. doi:10.1016/S0168-9002(98)00547-6

Buzulutskov, A., Breskin, A., Checik, R., Prager, M., & Shefer, E. (1997). Protection of caesium-antimony photocathodes. *Nuclear Instruments & Methods in Physics Research. Section A, Accelerators, Spectrometers, Detectors and Associated Equipment*, *387*(1-2), 176–179. doi:10.1016/S0168-9002(96)00984-9

Charpak, G., Dominik, W., Sauli, F. (1982).The gas photodiodes as a possible large-area photon detector. *Preprint* CERN-EP/82-169

Charpak, G., Peskov, V., & Sauli, F. (1992). Preliminary results of the study of gaseous detectors with solid photocathodes sensitive in the spectral region from 105 to 300 nm. *Nuclear Instruments & Methods in Physics Research. Section A, Accelerators, Spectrometers, Detectors and Associated Equipment*, *323*(1-2), 445–451. doi:10.1016/0168-9002(92)90330-7

Edmends, J. S., Miller, D. J., & Barlow, F. (1988). Investigation of prototype gas-filled photomultipliers. *Nuclear Instruments & Methods in Physics Research. Section A, Accelerators, Spectrometers, Detectors and Associated Equipment*, *273*(1), 145–156. doi:10.1016/0168-9002(88)90809-1

Enomoto, R., Sumiyoshi, T., & Fujita, Y. (1994). Test of various photocathodes. *Nuclear Instruments & Methods in Physics Research. Section A, Accelerators, Spectrometers, Detectors and Associated Equipment*, *343*(1), 117–120. doi:10.1016/0168-9002(94)90540-1

Francke, T., & Peskov, V. (2014). *Innovative applications and developments of micro-pattern gaseous detectors*. IGA Global. doi:10.4018/978-1-4666-6014-4.ch001

Francke, T., Peskov, V., Rodionov, I., & Sokolova, T. (2003). Novel position-sensitive gaseous detectors with solid Photocathodes. *IEEE Transactions on Nuclear Science*, *49*(3), 977–983. doi:10.1109/TNS.2002.1039600

Giunji, S., Peskov, V., Sakurai, H., Silin, E., Sokolova, T., & Rodionov, I. (2002). Micro capillary glass plates as a gaseous detector of photons and particles. *Nuclear Instruments & Methods in Physics Research. Section A, Accelerators, Spectrometers, Detectors and Associated Equipment*, *477*(1-3), 8–20. doi:10.1016/S0168-9002(01)01906-4

Gys, T. (2005). *Particle Detectors – Principles and Techniques*. Retrieved from: http://ph-dep-dt2.web.cern.ch/ph-dep-dt2/CAT2005_3b.pdf

Lyashenko, A. V., Breskin, A., Chechik, R., Velosos, J. F. C. A., Dos Santos, J. M. F., & Amaro, F. D. (2007). Further progress in ion back-flow reduction with patterned gaseous hole – multipliers. *JINST*, *2*(08P08004), 1–18. doi:10.1088/1748-0221/2/08/P08004

Lyashenko, A. V., Breskin, A., Chechik, R., Velosos, J. F. C. A., Dos Santos, J. M. F., & Amaro, F. D. (2009). Development of high-gain gaseous photomultipliers for the visible spectral range. *JINST*, *4*(07P07005), 1–22. doi:10.1088/1748-0221/4/07/P07005

Majewski, S., Kuzmiertz, W, Sauli, F. (1983). *Operation of gas photodiodes in strong magnetic fields.* Preprint CERN-EP/83-89.

Moriya, T., Tokanau, F., Okazaki, K., Sakurai, H., Gunji, S., Kawabata, H., & Kishimito, S. et al. (2013). A concise quantum efficiency measurements system for gaseous photomultipliers. *Nuclear Instruments & Methods in Physics Research. Section A, Accelerators, Spectrometers, Detectors and Associated Equipment*, *732*, 269–272. doi:10.1016/j.nima.2013.08.054

Peskov, V. (1976). On the Secondary Processes on Gas Counters-I. *Soviet Physics. Technical Physics*, *20*, 791–794.

Peskov, V. (1977). On the Secondary Processes on Gas Counters-II. *Soviet Physics. Technical Physics*, *22*, 335–338.

Peskov, V., (1988). Gaseous emitters and detectors of VUV radiation based on Cs cathodes. *Journal of Appl. Spectr.*, *48*(2), 316-317.

Peskov, V., Borovik-Romanov, A., Sokolova, T., & Silin, E. (1994). Gaseous detectors of ultraviolet and visible photons. *Nuclear Instruments & Methods in Physics Research. Section A, Accelerators, Spectrometers, Detectors and Associated Equipment*, *353*(1-3), 184–188. doi:10.1016/0168-9002(94)91634-9

Peskov, V., Charpak, G., Mine, P., Sauli, F., Scigocki, D., Seguinot, J., & Ypsilantis, Y. et al. (1988a). Liquid and solid organic photocathodes. *Nuclear Instruments & Methods in Physics Research. Section A, Accelerators, Spectrometers, Detectors and Associated Equipment*, *269*(1), 149–160. doi:10.1016/0168-9002(88)90872-8

Peskov, V., & Silin, E. (1995). High efficiency photocathodes for gaseous detectors. *Nuclear Instruments & Methods in Physics Research. Section A, Accelerators, Spectrometers, Detectors and Associated Equipment*, *367*(1-3), 347–352. doi:10.1016/0168-9002(95)00654-0

Peskov, V. E., Silin, E. T., Sokolova, T., & Rodionov, I. (1999). First attempts to combine capillary tubes with photocathodes. *Nuclear Instruments & Methods in Physics Research. Section A, Accelerators, Spectrometers, Detectors and Associated Equipment*, *433*(1-2), 492–501. doi:10.1016/S0168-9002(99)00346-0

Rodionov, I., Francke, T., Peskov, V., & Sokolova, T. (2002). Hybrid gaseous photomultipliers. *Nuclear Instruments & Methods in Physics Research. Section A, Accelerators, Spectrometers, Detectors and Associated Equipment*, *478*(1-2), 384–390. doi:10.1016/S0168-9002(01)01778-8

Sakurai, H., Tamura, T., Gunji, S., & Noma, M. (1996). A new type of proportional counter using a capillary plate. *Nuclear Instruments & Methods in Physics Research. Section A, Accelerators, Spectrometers, Detectors and Associated Equipment*, *374*(3), 341–344. doi:10.1016/0168-9002(96)00226-4

Séguinot, J., Charpak, G., Giomataris, Y., Peskov, V., Tischhauser, J., & Ypsilantis, T. (1990). Reflective UV photocathodes with gas-phase electron extraction: Solid, liquid, and absorbed thin films. *Nuclear Instruments & Methods in Physics Research. Section A, Accelerators, Spectrometers, Detectors and Associated Equipment*, *297*(1-2), 133–147. doi:10.1016/0168-9002(90)91359-J

Shefer, E., Breskin, A., Buzulutskov, A., Chechik, R., Klin, M., & Prager, M. (1998). Laboratory production of efficient alkali-antimonide photocathodes. *Nuclear Instruments & Methods in Physics Research. Section A, Accelerators, Spectrometers, Detectors and Associated Equipment*, *411*(2-3), 383–388. doi:10.1016/S0168-9002(98)00350-7

Shefer, E., Breskin, A., Chechik, R., Buzulutskov, A., Singh, B. K., & Prager, M. (1999). Coated photocathodes for visible photon imaging with gaseous photomultipliers. *Nuclear Instruments & Methods in Physics Research. Section A, Accelerators, Spectrometers, Detectors and Associated Equipment*, *433*(1-2), 502–506. doi:10.1016/S0168-9002(99)00358-7

Sommer, A. H. (1968). Photoemissive materials: preparation, properties, and uses. New York, NY: Wiley.

Sumiyoshi, T., Tokanai, F., Sugiyama, H., Okada, T., Ohishi, N., Ohmura, T., & Gunji, S. et al. (2011). Development of gaseous PMT with micropattern gas detector. *Nuclear Instruments & Methods in Physics Research. Section A, Accelerators, Spectrometers, Detectors and Associated Equipment*, *639*(1), 121–125. doi:10.1016/j.nima.2010.10.032

Tokanai, F., Moriya, T., Takeyama, M., Sakurai, H., Gunji, S., Sumiyoshi, T., & Kishimoto, S. et al. (2014). Newly developed gaseous photomultiplier. *Nuclear Instruments & Methods in Physics Research. Section A, Accelerators, Spectrometers, Detectors and Associated Equipment*, *766*, 176–179. doi:10.1016/j.nima.2014.05.014

Tokanai, F., Sumiyoshi, T., Sugiyama, H., Okada, T., Ohishi, N., Ohmura, T., & Gunji, S. et al. (2009). Development of a gaseous PMT with micro-pattern gas detectors. *Nuclear Instruments & Methods in Physics Research. Section A, Accelerators, Spectrometers, Detectors and Associated Equipment*, *610*(1), 164–168. doi:10.1016/j.nima.2009.05.111

Tokanai, F., Sumiyoshi, T., Sugiyama, H., Okada, T., Ohishi, N., Sakurai, H., & Kishimoto, S. et al. (2011). Sealed gaseous photomultiplier with CsI photocathode. *Nuclear Instruments & Methods in Physics Research. Section A, Accelerators, Spectrometers, Detectors and Associated Equipment*, *628*(1), 190–193. doi:10.1016/j.nima.2010.06.314

Veloso, J. F. C. A., dos Santos, J. M. F., & Conde, A. N. (2000). A proposed new microstructure for gas radiation detectors: The microhole and strip plate. *The Review of Scientific Instruments*, *A71*(6), 2371–2376. doi:10.1063/1.1150623

Winn, D. R. (1989). Large photocathode detector for large calorimetry. *IEEE Transactions on Nuclear Science*, *NS-36*(1), 128–131. doi:10.1109/23.34418

KEY TERMS AND DEFINITIONS

Garfield: The program for detailed simulation of drift chambers in 2-D. It can calculate and plot the electric field, the potential and drift lines of electrons and ions for some specific geometries of the gas chamber.

Gas Electron Multiplier (GEM): A planer gaseous detector in which avalanches are developed in a strong electric field formed inside holes manufactured in a thin metalized from both sides dielectric sheet.

Geiger Mode: Operation condition of gaseous or solid sate detectors in which primary ionization trigger a breakdown: type discharge quenched by a resistor circuit. Gaseous detectors filled with special gas mixture can operate in a self-quenched mode, when the discharge is interrupted by internal mechanism without the help of the external circuit.

Hole-Type Gaseous Photomultiplier: A gaseous detector in which avalanche multiplication of primary photoelectrons happening in high electric field created by flat electrodes in holes located between them.

Imaging Intensifier: An electronic vacuum device allowing to increase the intensity of a weak light flux.

Photocathode Protection: Coating of a photocathode with a thin protecting film preventing the photocathode from the direct contact with the ambient gas, but at the same time allowing a fraction of photoelectrons to penetrate through this layer and appear in the sensitive volume of the photodetector.

PM-Vacuum Photomultiplier Tube: A vacuum electronic detector of light consisting from a photocathode and a dynode system. Photoelectrons extracted from the photocathode experience multiplication in the dynode system allowing obtaining large electronic pulse on the output sufficient for detection singe electrons.

Practical Quantum Efficiency: Efficiency of photon detection in the given photodetector after loses in its electrode system some fraction of primary photoelectrons created from the photocathode.

TV Camera: A cathode ray vacuum tube designed to reproduce television images.

TV Tube: Television equipment optically focusing an image on photosensitive mosaic scanned by electron beam.

Chapter 8
Alternative Position Sensitive Photomultipliers

ABSTRACT

In this chapter, alternative position sensitive vacuum and solid state detectors are described. We start with a description of the vacuum photomultiplier tubes and how they have evolved into solid state detectors. These devices are promising, but still fall short compared to gaseous detectors.

1. INTRODUCTION

In the previous chapters various position sensitive gaseous photomultipliers, capable of efficiently detecting single photons, have been described in great detail. What is most significant of these developments? To answer this question it will be useful to look deeper into alternative position sensitive photodetectors.

The first photodetectors sensitive to single photons was the vacuum photomultiplier tube (PMT) developed in the 1930ies. It was a tremendous success from the beginning. PMTs are still the most widespread detectors sensitive to single photons. They are used practically in every kind of experimental studies including high-energy physics, astronomy, space research, archeology, medicine, geology, biology, art, metallurgy, chemistry, agriculture, etc. Among these applications, physics experiments, particularly high energy physics and astroparticle physics experiments, are the most common users of PMTs. The developments of PMTs are mainly driven by the needs in physics experiments.

L.A. Kubetsky was the first who proposed the new method to amplify weak photocurrents by dozens of dynodes which amplified incoming electrons a thousand times without using traditional electronic tubes. His USSR Author's certificate #24040, has the priority date 4 August 1930 (Lubsandorzhiev, 2006).The device consisted of a source of primary photoelectrons (a photocathode) and a consecutive system of secondary electron emitters (dynodes) with a certain electron multiplication coefficient k_{sec} each. The principle of this new electron multiplier was based on the secondary electron emission effect discovered in 1902 by L. Austin and H. Starke (Bruining, 1954). The physics behind this phenomenon is as follows: if an electron with a high kinetic energy hit a surface, it creates energetic electrons inside the material and some of them, especially those located close to the surface, may escape the surface

DOI: 10.4018/978-1-5225-0242-5.ch008

Figure 1. The secondary electron emission effect; a high energy primary electron hitting a surface will create several hot electrons inside the material. Some of them with high enough kinetic energy, especially those located close to the surface can escape into the vacuum. As a result one primary electron can create k_{sec} electrons.

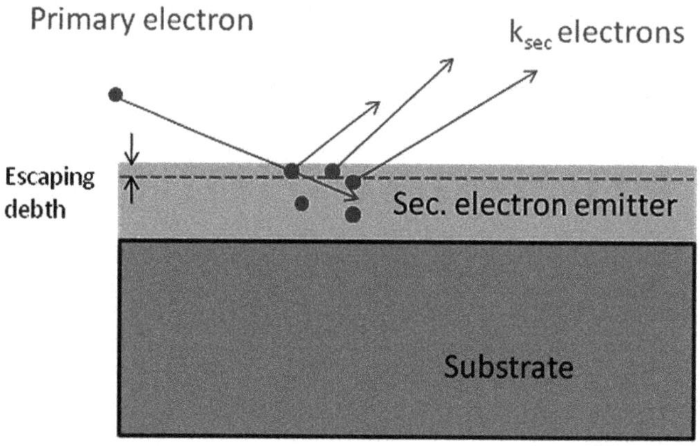

if their kinetic energy is exceeding the affinity or the work function of the material. This process is illustrated by Figure 1.

As in the case of photocathodes, the most efficient secondary electron emitters are semiconductors with small affinity, such as alkali antimonite. In metals the hot electrons rapidly lose their excess energy in collisions with other electrons and thus cannot efficiently escape into the vacuum (typically $k_{sec}<1$ in metals). In dielectrics and semiconductors with high affinity only a small fraction of the electrons will have high enough energy to cross the surface barrier. In dielectric materials the primary electrons will not be neutralized inside the material, but will cause the surface to charge up. For the best secondary emitting materials, such as Sb_3Cs and bialkali, the coefficient k_{sec} reaches values of 10 and 20, respectively, for primary electrons with kinetic energies close to 1kV (Hakamata,2007).

In the device proposed by Kubetsky the voltage applied to each subsequent dynode was higher compared to the preceding dynode. Photoelectrons created from the photocathode were accelerated to the first emitter knocking out secondary electrons from its surface. These electrons are in turn accelerated to the next emitter where they knock out new secondary electrons and so forth. The device can in principle have arbitrary number of emitters. The electron flux from the last emitter is collected on a collector or anode. The total gain of the device will amount to:

$$N_{sec} = k_{sec}^{\ z}, \tag{1}$$

where z is the number of emitters.

In 1933-34 L.A. Kubetsky developed a number of photomultiplier tubes with Ag-O-Cs photocathodes and circular secondary electron emitters made also from Ag-O-Cs. The photomultiplier tubes consisted of the photocathode and a multistage electron multiplier system including constant magnets for electron focusing. The total gain of the tubes reached 10^3-10^4.

According to W. Summer: “Kubetsky’s results have been picked up in USA (1934) and later in Germany and England” (Summer, 1957). As it sometimes happens, new ideas are generated almost at

Figure 2. Artistic view a PMT; light enters through the window covered with a semitransparent photocathode creating primary electrons, which are directed by the focusing electrode towards a system of dynodes, where they experience a multiplication by the secondary electron emission effect. The voltage on dynodes is supplied via a resistive voltage divider made from a resistive chain.
From Gys, 2005.

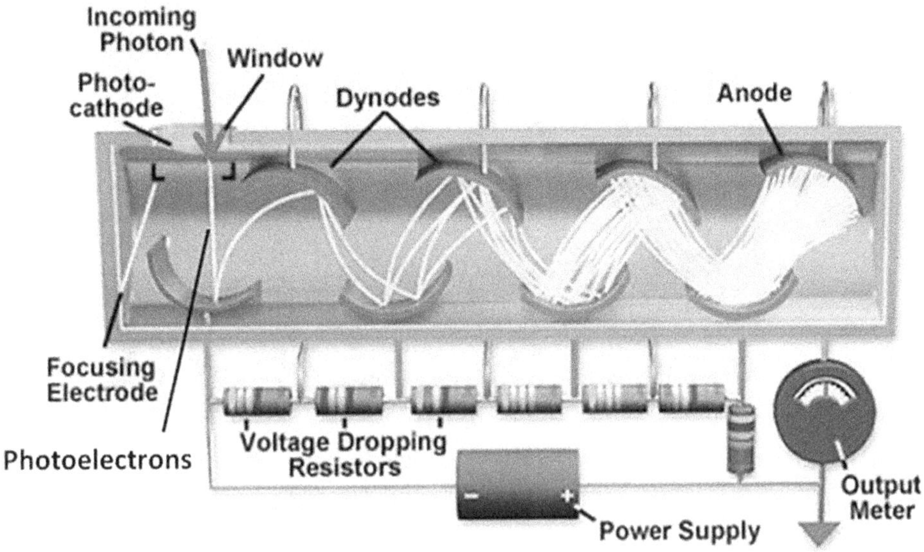

the same time and independently by several authors. It happens also that a bright idea of one person is immediately picked by other and rapidly implemented to a practical device. Whatever it was in case of the PMT, it is clear that the remarkable work of Zvorykin's group at RCA realized the first commercial devices of the PMTs. Since that early time the PMT designs have continuously been modified to improve their performance such as the maximum achievable gain by secondary electron emission, increasing its sensitivity, lowering the noise level, reduce undesirable secondary effects from positive ions etc.

A schematic drawing of a modern vacuum PMT is shown in Figure 2.

Incoming photons create photoelectrons in a reflective or semitransparent photocathode. The majority of them are directed by the focusing electrode to a system of dynodes covered by semiconductor secondary electron emitters. In each such dynode the electrons experience a sequence of secondary electron emission. Finally a large current pulse is created on the anode electrode. The total multiplication factor by secondary electron emission can reach values exceeding 10^7.

There are many designs of PMTs in the market. Each is optimized for a particular application. Some PMTs are designed for high quantum efficiency, reaching values up to 30% for certain wavelengths. Other have an optimized focusing electrode so that photoelectrons from the whole photocathode surface will arrive almost simultaneously on the first dynode surface in order to reach the best possible time resolution (~1 ns). The typical transit time from the photocathode to the first dynode is 20-80 ns. Some PMTs are designed specifically for applications requiring large sensitive area, with a photocathode surface as large as 1.6 m^2.

Figure 3. The pulse height distribution measured from the PMT under a weak light illumination, known as the single photoelectron response
From Gys, 2005.

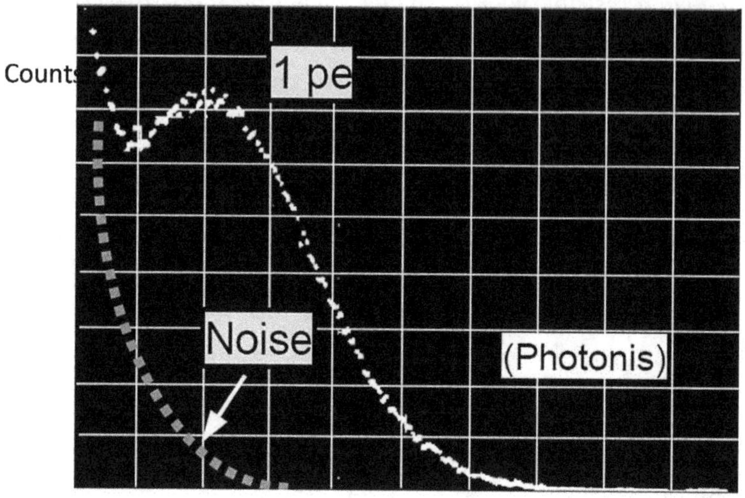

One of the main attractions of the PMTs is the possibilityto detect single photons, often referred as "photon counting". In this mode of operation each detected photon is separately time resolved at the anode of the PMT. This is illustrated in Figure 3 showing the pulse height distribution from the anode of the PMT when the intensity of the incident light is so low that there is no overlap between detected photons.

The multiplication process is statistical in nature. Events which all start as a single photoelectron at the photocathode produce a range of output pulse heights. A photon counting photomultiplier has a narrow signal distribution resulting in the well-defined single electron peak shown in Figure 3.The narrower the spread in the pulse heights are, the better suited is the photomultiplier for photon counting.

Another important characteristic of a good photon counting PMT is that increasing or decreasing the light intensity changes the area under the curve, but not the position of the peak. At high enough gains most of the events in the single photoelectron distribution is above the electronic threshold so that a counting plateau is reached.

Despite the great success of this device it has also several disadvantages, for example:

1. It is not a position sensitive device. One can only determine the position of the primary photons with an accuracy determined by the photocathode area.
2. It is extremely sensitive to magnetic fields, including the Earth magnetic field (only 30-60 μT). So a magnetic field shielding is usually required.
3. Some PMTs have after-pulses. These are spurious pulses that appear in the wake of the true pulses. They have two main causes: luminous origin, i.e. light emitted by the dynodes due to the electron bombardment, and ionization of residual gas traces.

The PMTs briefly described above (let's name them "traditional PMTs") were introduced for historical reasons in order to follow the progress in this field. With the focusing system used in the traditional PMTs it is quite difficult (and probably even impossible) to obtain position information better than the size of the photocathode.

In the last decades photonic companies worldwide have invested a lot of efforts in developing various position sensitive vacuum photomultipliers, which greatly enlarge the application of PMTs. The most important designs of these photodetectors will be described in the following paragraphs.

2. MULTI-ANODE PMTs

The multiple-anode PMT is equivalent to incorporating multiple PMTs in a single housing. One such detector is shown schematically in Figure 4.

In this photosensitive detector a specially shape dynode (covered with a photo-emissive material), makes it possible to localize the region where the secondary electron multiplication happens. The secondary electrons are collected on the anode plate segmented into isolated electrodes called a multi-anode. By measuring the signal from these electrodes, the position of the secondary electron avalanche can be determined with an accuracy equal or better than the size of each electrode.

The anode electrodes may have different geometry, but typically it is a 2D matrix. For example, the multi-anode PMT Hamamatsu H7546 has 8 x 8 square anode pads, each 2x2 mm^2. This particular detector has quantum efficiency of 20% at 400 nm and can operate at gain of $3x10^5$.

One limitation of such a device is "crosstalk" between the channels. The crosstalk indicates how accurately the light incident on a certain position on the photocathode surface is detected while still retaining

Figure 4. Schematic drawing of a multi-anode PMT with metal-channel dynodes; the primary photoelectron created at the photocathode experience secondary electron multiplication in the sequence of aligned channels and are collected on one or several anode electrodes.
From: https://www.hamamatsu.com/resources/pdf/etd/PMT_handbook_v3aE.pdf

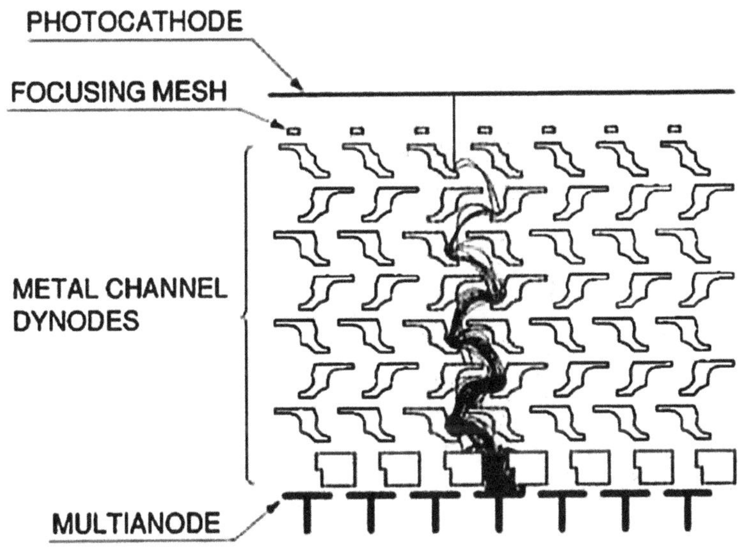

the position information. In the multi-anode PMTs crosstalk is mainly caused by spatial broadening of the secondary electron avalanche during its development. Light spread within the face plate may also cause the crosstalk.

In practice the crosstalk is determined as follows. The photocathode is irradiated with collimated light (typically 1mm in diameter) aligned in the middle of the one of the anode electrodes. To the output signal form this electrode is assigned a value of 100%.The output of each neighbouring anode is then expressed as a relative value with respect to this 100%. Usually the cross talk is less than a few percent in the neighbouring channels.

The position resolution of the multi-anode PMT with channel-type dynodes can be as good as~5 mm. Figure 5 shows the output signal distributions for each anode measured by scanning a pencil light beam (1mm in diameter) across the photocathode surface. As can be seen the anode pads structure is quite accurately resolved.

There are several dynode electrode systems to improve PMT's for a specific application. Figure 6 shows the electron trajectories in a PMT with grid-type dynodes.

Typically, this detector contains 15 to 19 stages of fine meshes mounted close to each other. This proximity configuration yields a relatively strong electric field and results in a high immunity to magnetic field. Such a device can stand magnetic fields up to 1 T. An interesting feature of this design is the wire signal readout system. Multiplied secondary electrons are repelled back from the last dynode and read out by perpendicularly arranged wires.

Figure 5. Results of a pencil light beam (diameter 1mm) scanned across the central readout pads of a multi-anode PMT

From: https://www.hamamatsu.com/resources/pdf/etd/PMT_handbook_v3aE.pdf

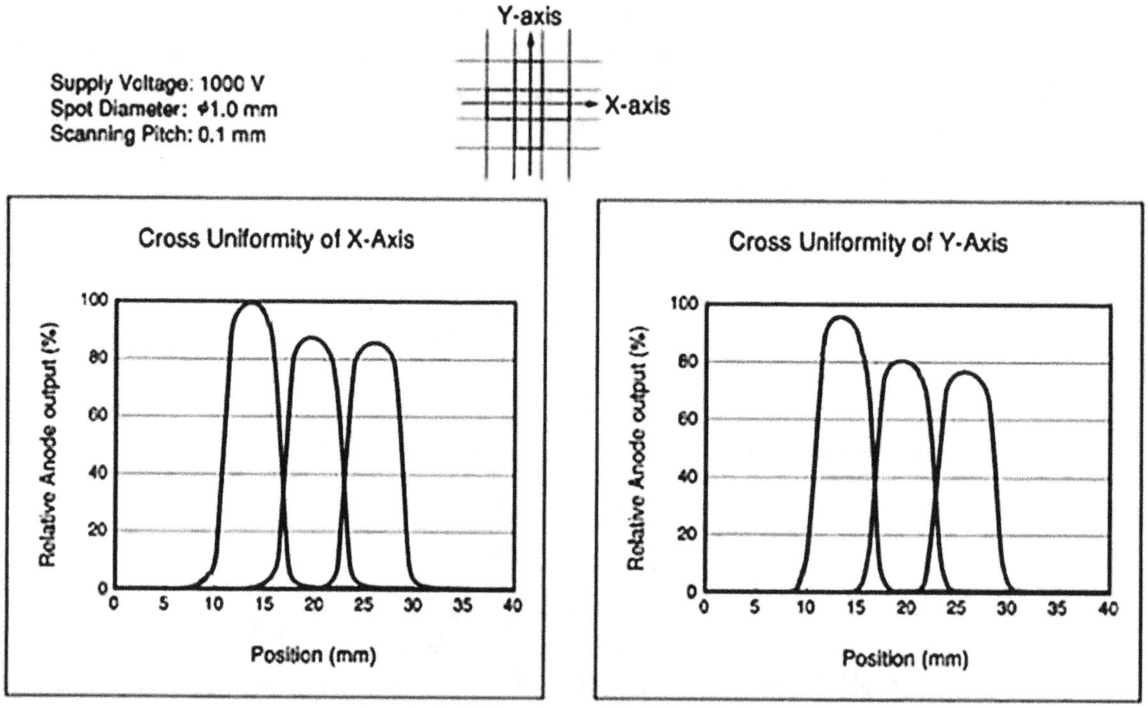

Figure 6. Schematic of a grid-type dynode photomultiplier with cross wire anodes
From: https://www.hamamatsu.com/resources/pdf/etd/PMT_handbook_v3aE.pdf

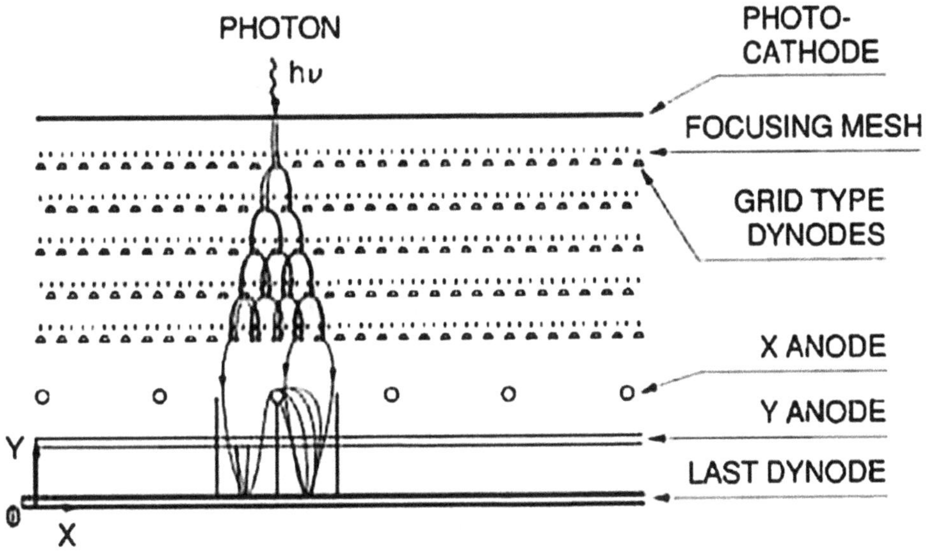

Figure 7 shows the signal distributions obtained by the x-y scan of the photocathode surface with a light spot, 50 μm in diameter.

As can be seen, the distributions are quite broad. By applying a centre of gravity method the space resolution can be dramatically improved. With a pulsed light source producing 4000 photons per pulse a spatial resolution 0.3 mm FWHM can be obtained (Hakamata, 2007). Due to the Poison statistics, the width of the output signal is inversely proportional to the square root of the amount of the incident light. Hence, for weak intensities the position resolution is worse than for high intensities.

In the position measurements presented above each anode pad or crossed wire were connected to an individual electronic chain consisting of preamplifier, amplifier and shaper. In real applications all anode signals are analogue-to-digital (AD) converted and are stored in an acquisition system for later data analysis and image processing.

The advantage offered by this method is a good image quality obtained by correcting the centroid position distortions due to gain non-uniformity and light distribution truncation.

The disadvantages, however, are count rate limitation and high cost due to the large number of electronic channels.

An alternative method is the use a resistive chain. In this method the anodes are connected through a resistive network and the position signals are read out from the two ends for both X and Y direction, as shown in Figure 8.

The obvious advantage is a simple and cheap electronic read out (only four channels). This method can handle large and small signals, and a high count rate. The disadvantage is that the resistive method introduces an extra noise (depending on the number of resistors in the network) which worsen the spatial resolution as well as the position linearity response.

In some applications these two methods are combined together: the resistive chain method is applied after amplification all signals from the anodes independently.

Figure 7. Signal distribution measured with wire type readout electrodes when the photocathode was scanned with a collimated light beam a) scan in the x-direction, b) scan in the y-direction
From: https://www.hamamatsu.com/resources/pdf/etd/PMT_handbook_v3aE.pdf

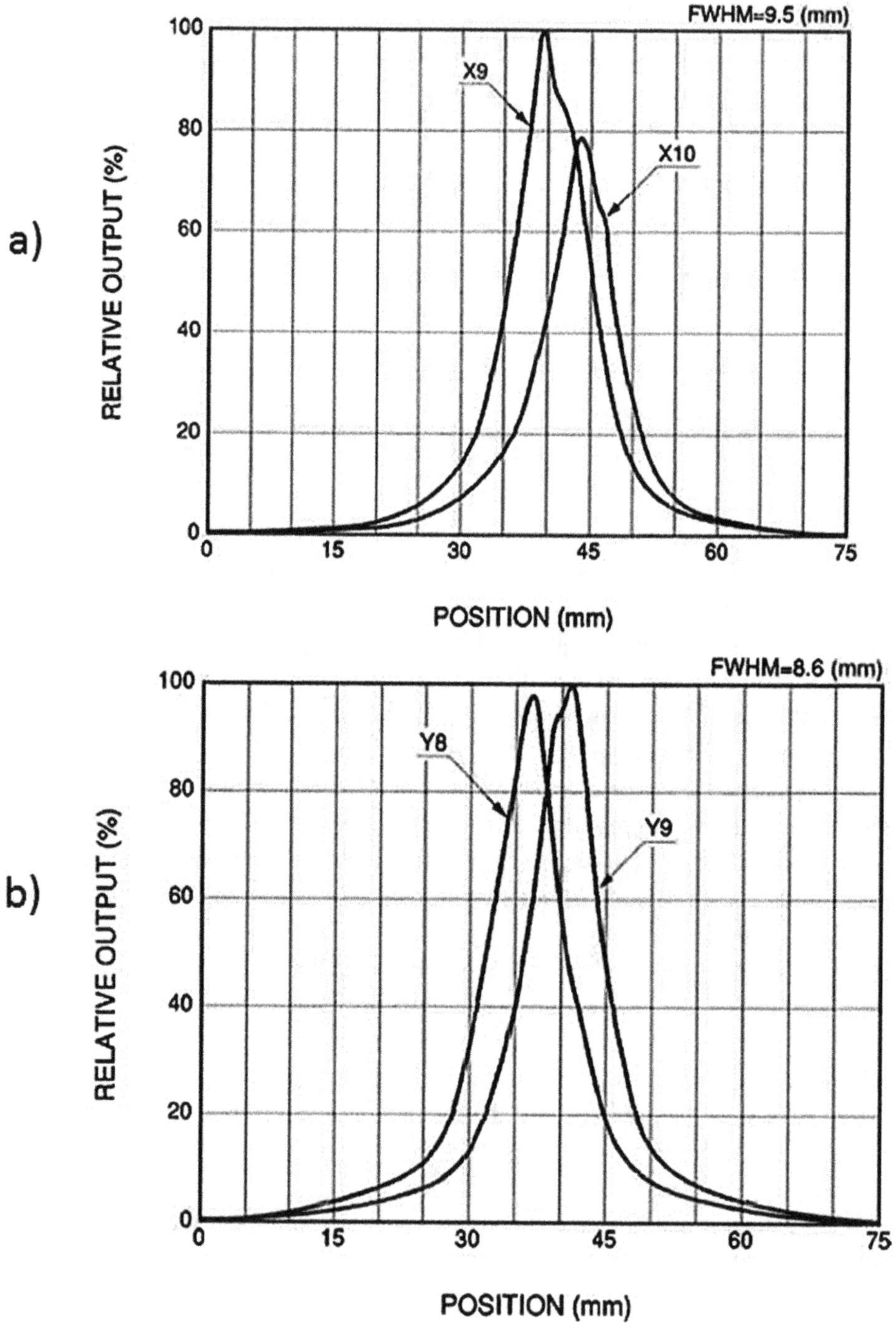

Figure 8. The principle of the resistive chain approach in which the coordinate of the secondary electron avalanche is determined by the charge division method
From: https://www.hamamatsu.com/resources/pdf/etd/PMT_handbook_v3aE.pdf

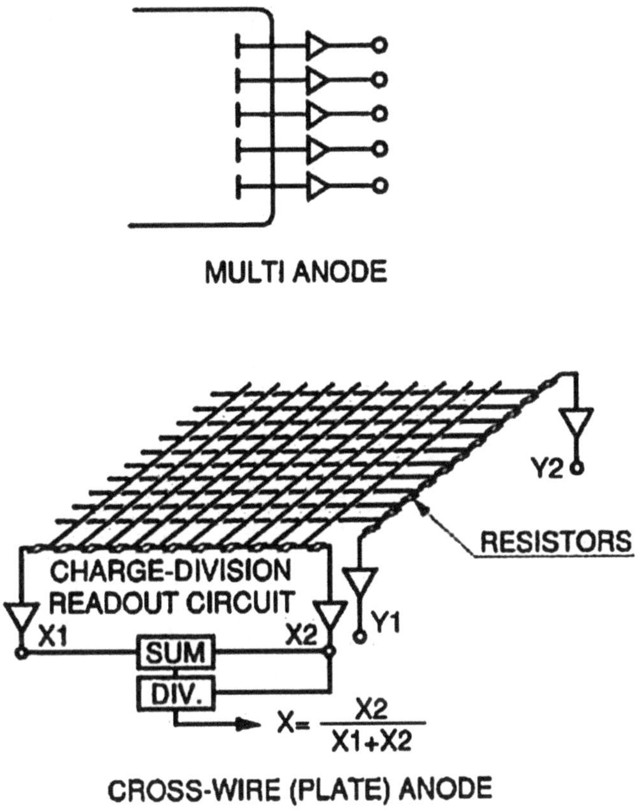

As any photosensitive detector, multi-anode PMTs exhibit some aging effect. It manifests itself by two main effects: the aging of the photocathode, which can lose quantum efficiency over time, and the aging of the multiplication chain, which causes a gain reduction. In most of applications the charge extracted from the photocathode is rather low and therefore the aging of the last dynodes (as in the case of the traditional PMTs) is much more significant than for the first ones.

The typical gain degradation of the multi-anode PMT due to the dynodes aging is shown in Figure 9.

With a current from the illuminated pixel was ~1 μA (which corresponds to single photoelectron emission at a rate of 2 MHz) a gain loss of about 60% can be seen in the first week of continuous operation (200 hours). Then the gain loss curve flattens, reaching 50% after two months of continuous operation (1600 hours). At this moment the total charge accumulated on each pixel reaches 0.5coulombs equivalent to 15 coulombs over the entire photocathode surface. Such aging is actually less than in the case of the traditional PMT. This is not astonishing since the dynode structures in these devices are completely different.

Concluding the paragraph, let us stress that the development of multi-anode PMTs was a great step forward. They not only offer some position resolution, but also preserve important features on traditional PMTs, such as clear single electron peak comparable to classical PMTs, fast electron transit time, and a small transit time spread (~ 0.5 ns FWHM).

Figure 9. Results of aging studies performed with a multi-anode PMT Hamamatsu R7600; the current from the illuminated pixel was about 1 μA.
From Calvi, 2011

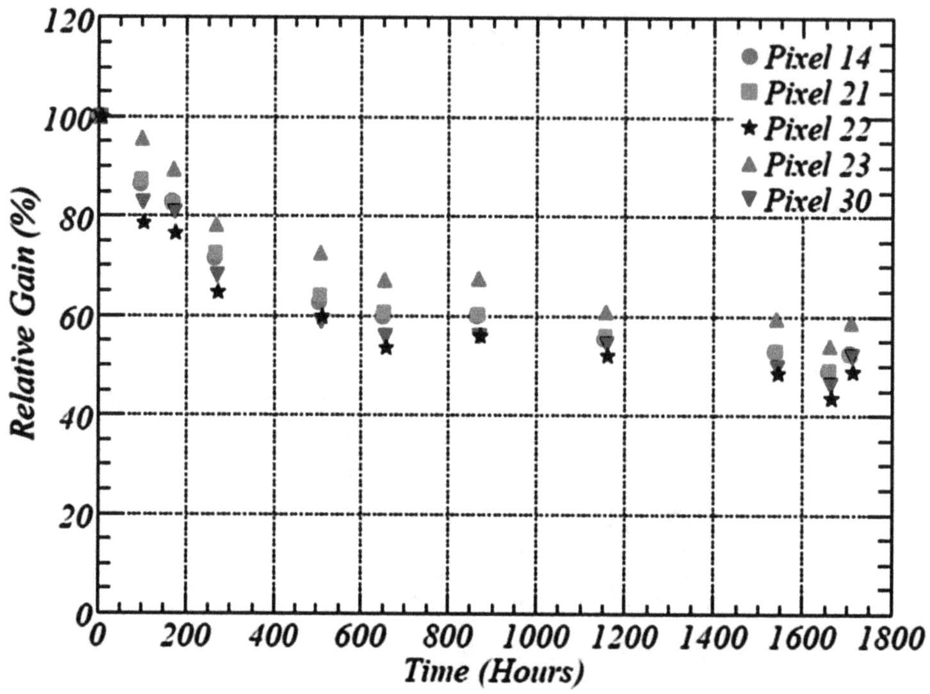

3. MCP-BASED POSITION SENSITIVE PMTs

Another popular position sensitive PMT, often called MCP-PMT, is based on glass microchannel plates (MCP). Glass capillary plates, having a capillary diameter of about 1 mm, were introduced in Chapter 2 as a gaseous detector. Their operation was based on gas avalanches (Townsend avalanches) developing in the strong electric field formed inside the capillaries.

The traditional capillary plate is operating in vacuum and exploiting another multiplication mechanism, secondary electron emission on the walls of the capillaries. Vacuum MCP is an array of micro capillaries (typical channel diameters are in the range of 10-100 μm) assembled together and oriented parallel to one another. The active surface of the MCP is typically between 3 and 100 cm^2. The channel axes are usually biased at a small angle (~8°) to the MCP surface.

This channel matrix is usually fabricated from a lead glass, treated in such a way as to optimize the secondary emission characteristics of each channel and to render the channel walls semiconducting so as to allow charge replenishment from an external voltage source. Each channel acts as continuous dynode structure which has its own dynode resistor chain. The common electrical contact to each channel is provided by deposition of a metallic coating, (usually Nichrome or Inconel) on the front and rear surfaces of the MCP, which then serve as anode and cathode electrodes. The total resistance between the metalized electrodes can be up to 10^9 Ω.

A schematic structure of a single MCP and the cross section of a channel are shown in Figure 10.

Figure 10. Schematic drawing of the MCP (a) and its principle of operation (b)
From: https://www.hamamatsu.com/resources/pdf/etd/PMT_handbook_v3aE.pdf

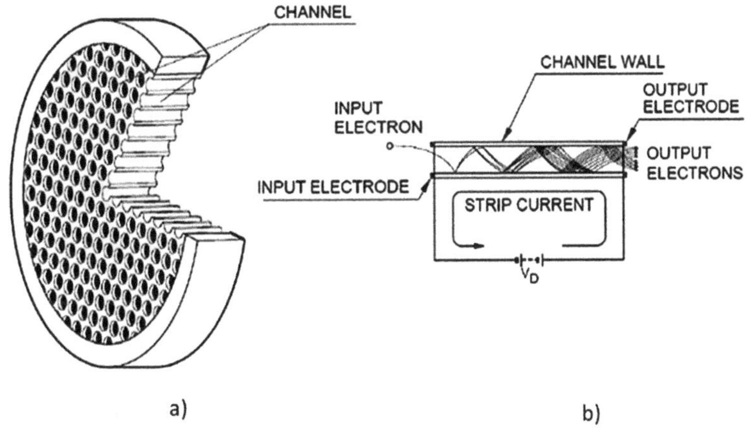

Figure 11. Schematics of double MCP photon detectors
From Dinu, 2005.

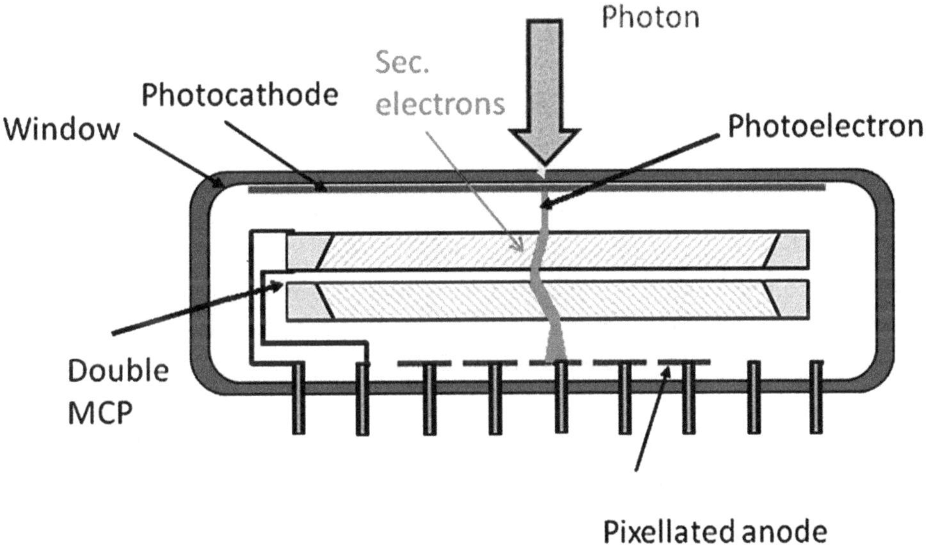

When a primary electron with a sufficient kinetic energy impinges in vacuum on the inner wall of the micro-capillary, secondary electrons are produced. Due to the strong electric field applied across the channel the secondary electrons are accelerated and produce an avalanche of secondary electrons.

Due to the small size of the amplification micro-capillaries, MCP's offer unprecedented high position resolution (limited only by the channel dimensions and spacing), fast time response with ultra-high time resolution (< 100 ps) and stable operation even in high magnetic fields. MCPs, used singly or in cascade (see Figure 11), allow electron multiplication factors of 10^4-10^7.

The concept of a continuous dynode electron multiplier was suggested by Farnsworth (Farnsworth, 1930). However, it was not implemented as a practical device until the1960swhen the understanding of

Figure 12. Schematic drawing showing the creation of positive ions in a double MCP and their interaction with the photocathode leading to photocathode aging as well as the creation of secondary electrons *From Gys, 2014.*

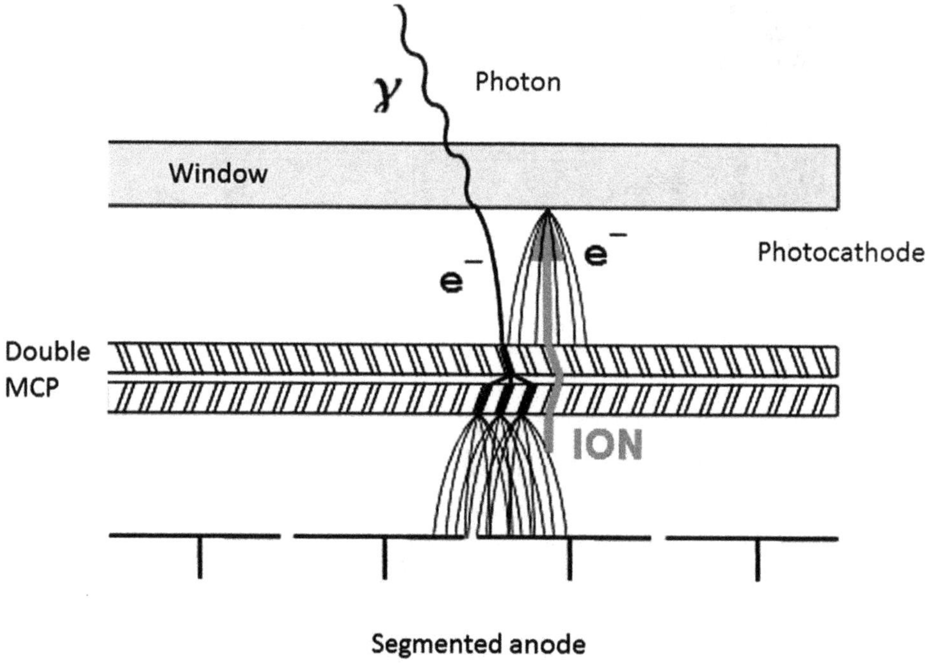

the processes of secondary electron emission and in the development of high surface resistance glasses had been made (Oschepkov 1960; Heroux, 1960; Goodrich 1962; Adams 1966).

Mainly used as an amplification element for image intensification devices, MCPs also have direct sensitivity to charged particles and energetic photons which have extended their applications to such diverse fields as X-ray and VUV astronomy, e-beam fusion studies, and nuclear science, where to date most applications have capitalized on the superior MCP time resolution characteristics.

The first prototypes of MCP-based photomultipliers sensitive to visible light were developed and tested at the middle of 1970 (see for example Bateman, 1976). At the end of the last century MCP-based photomultipliers combined with Ag-O-Cs and other photocathodes become largely available from various photonic companies.

As in other vacuum photodetectors, e.g. traditional PMT and multi-anode PMT, positive ions created due to the ionization of the residual gas and moving towards the photocathode may strongly effect the MCP performance, e.g. cause a photocathode degradation or create secondary electrons. The effect is illustrated in Figure 12.

The maximum gain that can be achieved without significant ion feedback in a straight-channel MCP is of the order 10^3 to 10^4. These gain levels are not high enough to detect single photoelectrons. For this reason special measures are implemented to suppress the ion back flow to be able to increase the gain.

A traditional approach is to use inclined or curved micro capillaries as shown in Figure 13.

Using a triple MCP the rate of the photocathode aging can be decreased by about order of magnitude (Barnyakov, 2009). Another "standard" approach is to use a thin film (5–10nm)of aluminum oxide in front of the first MCP stage. This film, sometimes called "ion barrier" prevents ions, generated during

Figure 13. Geometry of micro capillaries used in high gain (up to 10^6) MCPs for the suppression of ion back flow: a) Chevron MCP stack, b) so-called "Z-plate" MCP stack, c) curved-channel MCP; in this drawing N^+ stands for positive ions.
From Timothy, 2010.

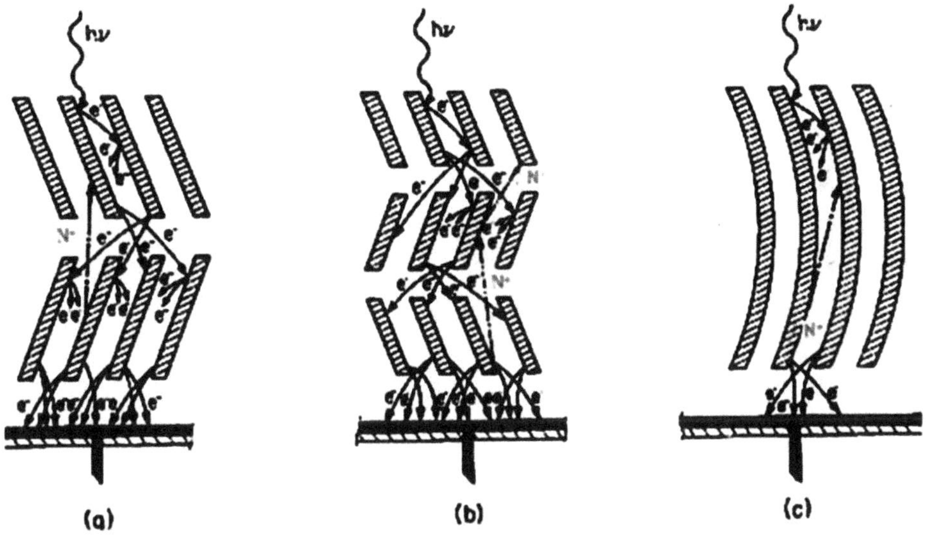

the secondary electron emission process, from penetrating to the cathode. However, the disadvantage of this technique is a ~40% reduction of the collection efficiency. To avoid this loss Hamamatsu recently developed PMTs in which the protection layer was placed between the two MCP stages.

Another possibility to reduce the ion back flow is by improving vacuum level inside the MCP-PMT and also to clean the MCP surfaces by electron scrubbing (Lehmann, 2013).

Recently the attention of researchers turned towards the atomic layer deposition (ALD) technique (Conneely, 2013). This is a chemical vapor process allowing to deposit atomic mono-layers on a substrate. It is used to improve the secondary emission yield and thus to achieve significantly higher gain at the same operating voltage across an MCP (see Figure 14).

The atomic mono-layers deposited by ALD coating prevents contaminants in the MCP glass to evaporate out. ALD coating dramatically improves the aging characteristics of the MCP-PMT. As can be seen from Figures 15 and 16, both the photocathode and the secondary electron emission gain remain quite stable up to a total accumulated charge of 5 C/cm^2.

This value is acceptable for most applications. Just like traditional and multi-anode PMT, the MCP-PMT also have a clear single electron peak, which can be even more pronounced due to the gain saturation a in photon counting mode.

The main attraction of the MCP-PMTs are their extremely high position and time resolutions. The MCP-PMT's intrinsic position resolution is theoretically determined by the capillary size and their pitch. In practice the actual position resolution is determined by the geometry of the anode segmentation and the readout scheme. For example, some commercial MCP-PMTs have anodes segmented into square pads with a size of a few mm (Hakamata, 2007). In this case the spatial resolution is determined by the pad size.

Figure 14. Detector gain for two MCP-PMTs: ALD coated and uncoated; both MCPs consisted of two identical chevron stacked MCPs. The gain is plotted as a function of the total MCP bias across both plates.
From Conneely, 2013.

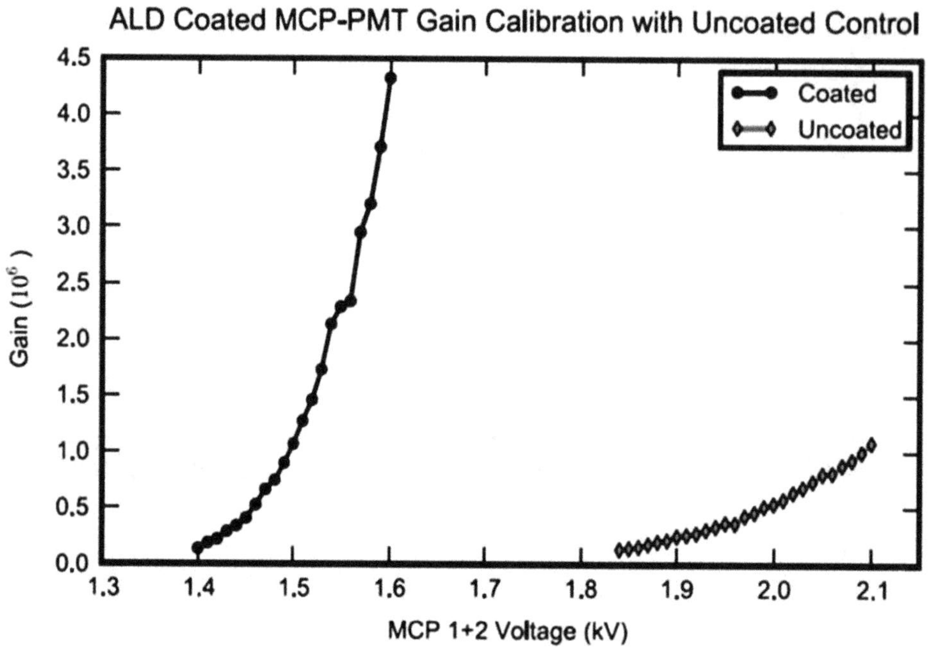

Figure 15. Quantum efficiency of the ALD-coated MCP-PMT measured at different total accumulated charges
From Conneely, 2013.

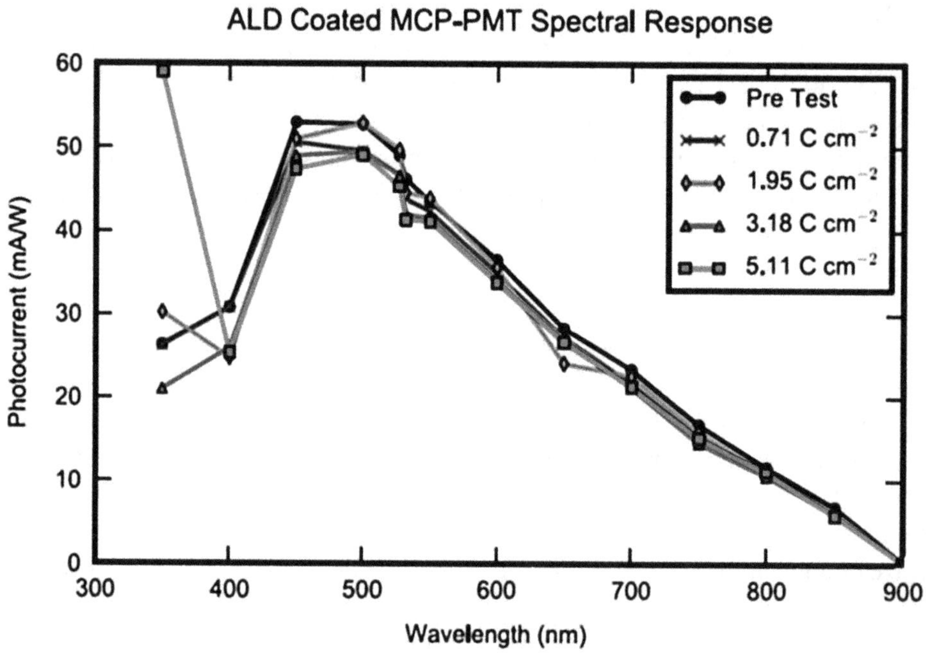

Figure 16. Gain variation of o the ALD coated MCP-PMT as a function total extracted charges, at three different initial gains/voltages
From Conneely, 2013.

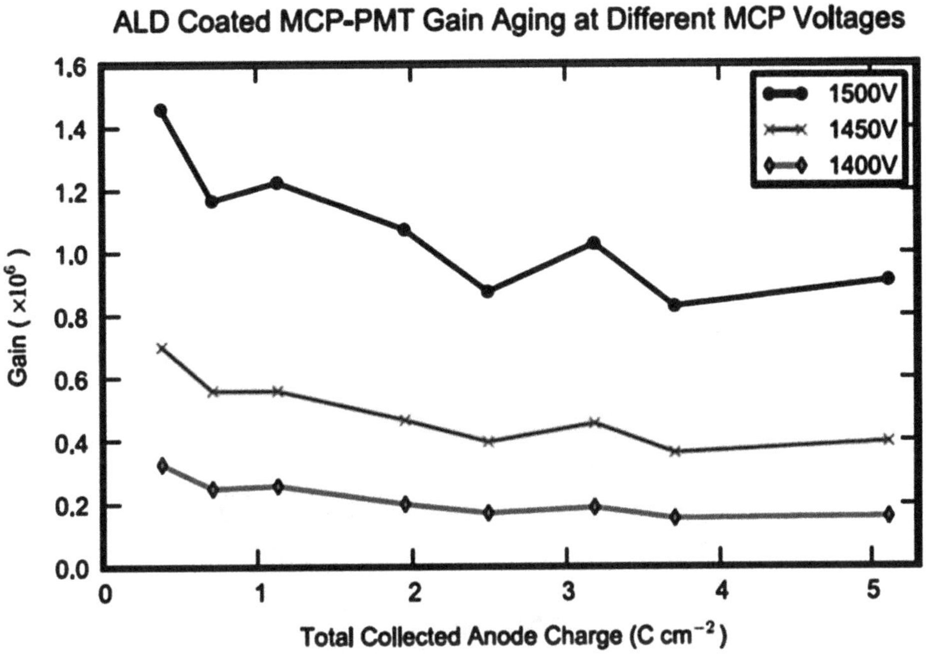

A review of various methods to achieve an extreme position resolution can be found in (Tomothy, 2010). Below we will highlight several state-of -art developments.

Figure 17 shows a schematic view of a double MCP–PMT, readout by an array of perpendicular strips located on the anode plate.

In this detector primary photoelectrons multiplied within the capillaries, resulting in an electron cloud which is accelerated and collected on two orthogonal sets of metal electrodes (isolated from each other) forming the anode. The primary photoelectron position is determined from the charge centroid across several strips. To ensure an accurate centroid centre reconstruction the strip pitches to be optimized so that the charge impinges on several neighbouring strips of the anode pattern.

For the MCP-PMT shown Figure 17 the pattern period was ~0.7 mm for both the upper and lower conductor strips (Siegmund, 2011). The distance between the anode readout plate and the double MCP stack was ~2.5 mm to reflect the charge cloud spread.

This readout system offers an excellent spatial resolution. Figure 18 shows the image of a 12 mm pore MCP pair. The pores are clearly resolved, dead pores are visible, and the MCP multi-fibre distortions of the pore alignment are also seen.

Figure 19 shows the spatial resolution (derived from a sharp edge) for several double MCP gains. As can be seen, a special resolution of about 17 µm FWHM at a gain of $1.4x10^6$ is possible.

In the work of Bellazzini the electron cloud from a single photoelectron in a single MCP was collected and encoded by a pixelated custom analogue ASIC consisting of many charge sensitive pixels, each 50 µm in size arranged on a hexagonal pattern as shown in Figure 20 (Bellazzini, 2008).

Figure 17. A schematic diagram of a sealed double MCP-PMT with a cross-strip readout anode plate; the charge distributions on the strips created by a single event are also shown.
From Siegmund, 2011.

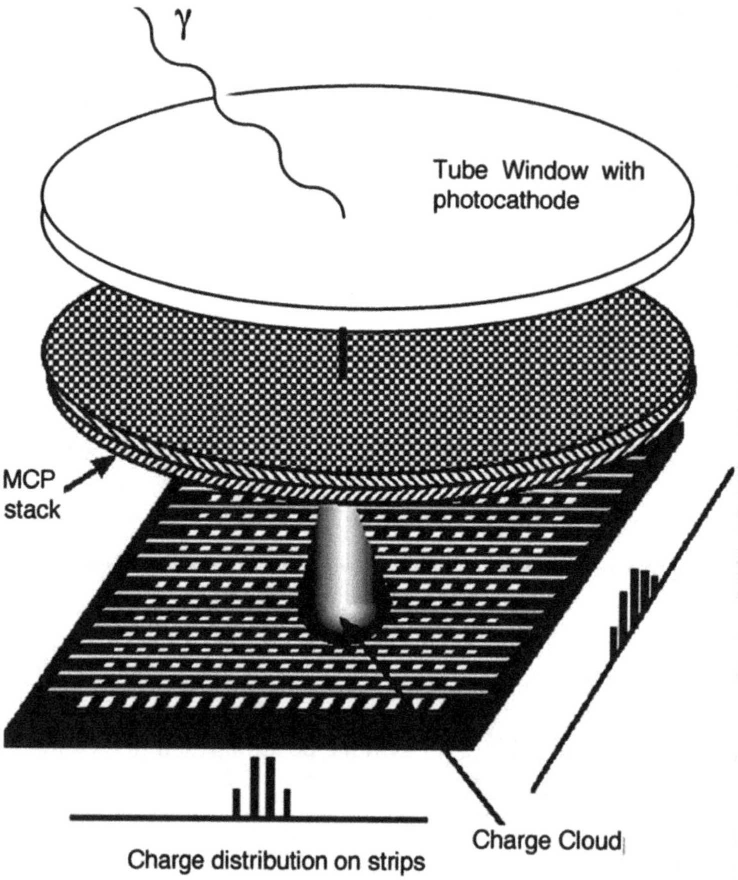

Figure 18. Image of a double MCP pair with 12 μm pores; the MCP gain was 5x10⁶; the size of the readout plate was 8x8 mm².
From Siegmund, 2003.

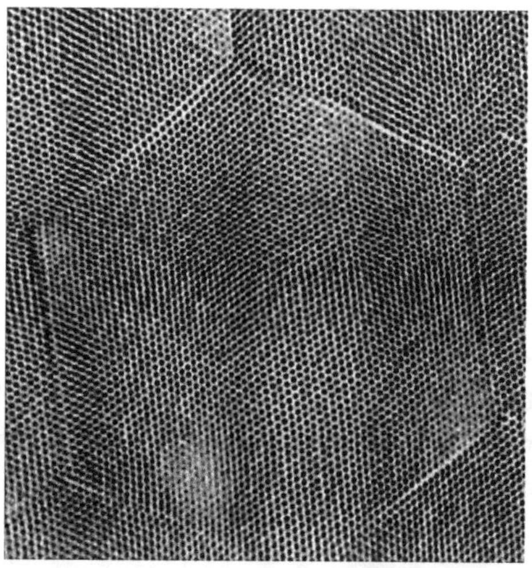

Figure 19. Space resolution (FWHM) as a function of the MCP gain
From Siegmund, 2011.

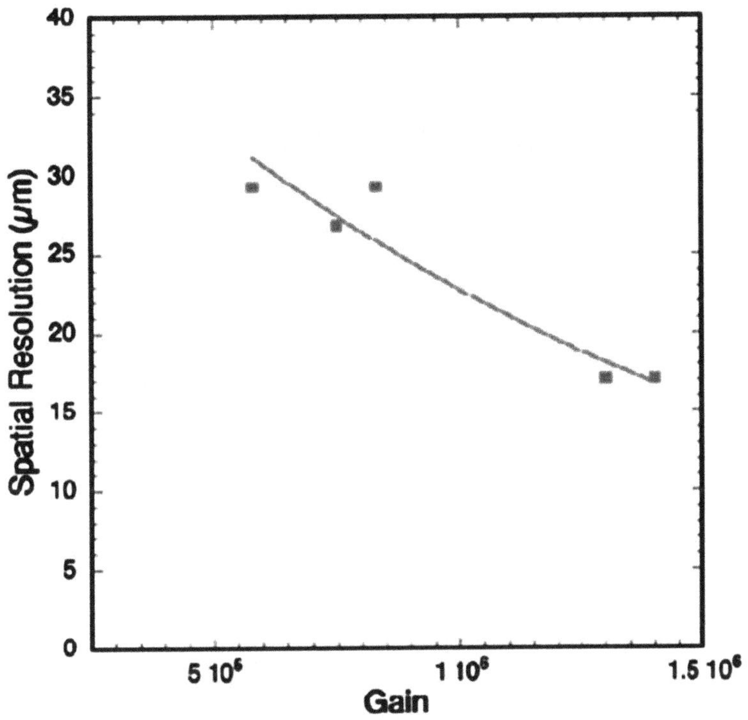

The ASIC samples the charge cloud and allows to measure with high precision the charge centroid on an event-by-event basis. A detailed description of this CMOS-based readout system can be found in (Bellazzini, 2006).

Figure 21 shows the reconstructed image of a standard Air Force Resolution Mask (USAF-1951) with a zoom around groups 6 and 7. The last element of group 6 has a width of only 4.4 μm and a line density of 114 line pairs/mm.

In Figure 22 a cut through of the first horizontal bar of group 6 shows a spatial resolution of 3 μm FWHM.

The MCP-PMTs also offer excellent time response characteristics, which are actually the best among currently available vacuum phototubes. In traditional PMTs the signal pulse is broadened due to several factors, e.g. due to the emission angle and initial velocity distribution of the primary photoelectrons, the velocity distribution of the secondary electrons, space non uniformity effect of the focusing lenses etc. In contrast, in MCP-PMTs, where the strong electric fields are applied to the photocathode region, the emission angle distribution and the initial velocity distribution of photoelectrons are almost negligible. Also the transit time of the primary photoelectrons is small, allowing a dramatic reduction in the transit time spread.

The time resolution of the MCP-PMT depends on the gain and the light pulse intensity. Usually, lasers, producing short light pulses (14 ps) are used for the measurements of the time resolution, the rate capability and the dependence of the gain inside magnetic fields. The light pulses are guided through a glass

Figure 20. Artistic view of the high-resolution detection system for single/multiple photon imaging consisting of a photocathode, a single MCP with 4 μm pores and CMOS-based pixelated readout
From Bellazzini, 2008.

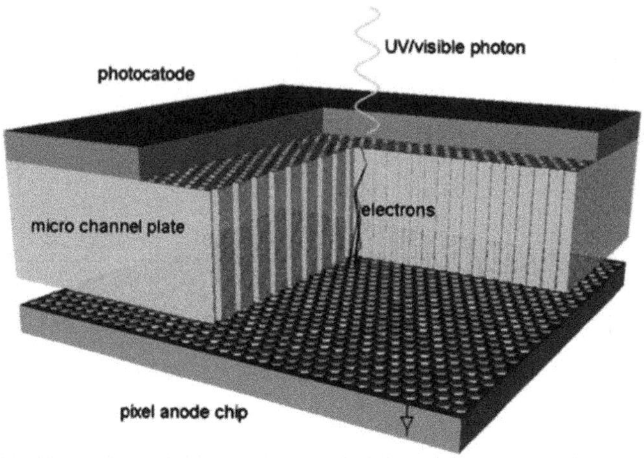

Figure 21. An image of ~1mm² region of the USAF mask around groups 4, 5, 6 and 7 obtained with a 4 μm pore single MCP; the structure of the micro-pores is clearly visible and all the bars of group 6 elements are distinguishable in the insertion with enlarged view.
From Bellazzini, 2008.

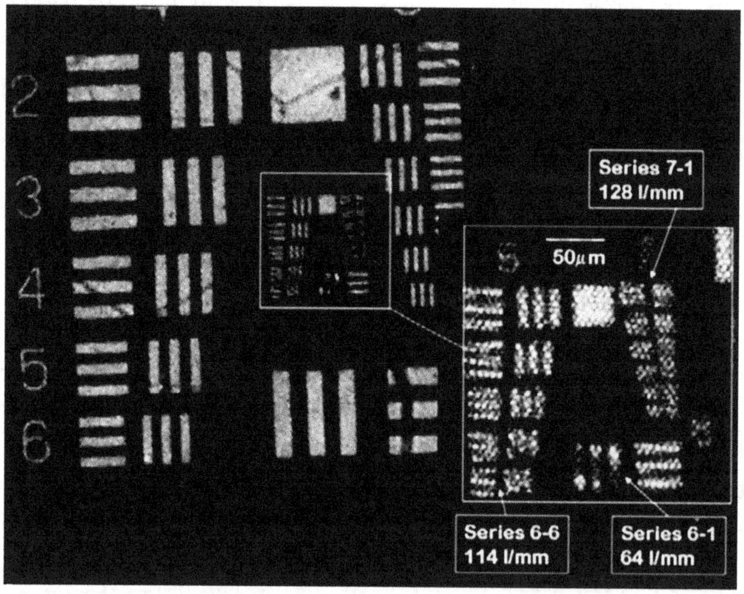

Figure 22. Result of as can through the first element of group 6; a spatial resolution of 3 μm FWHM was measured.
From Bellazzini, 2008.

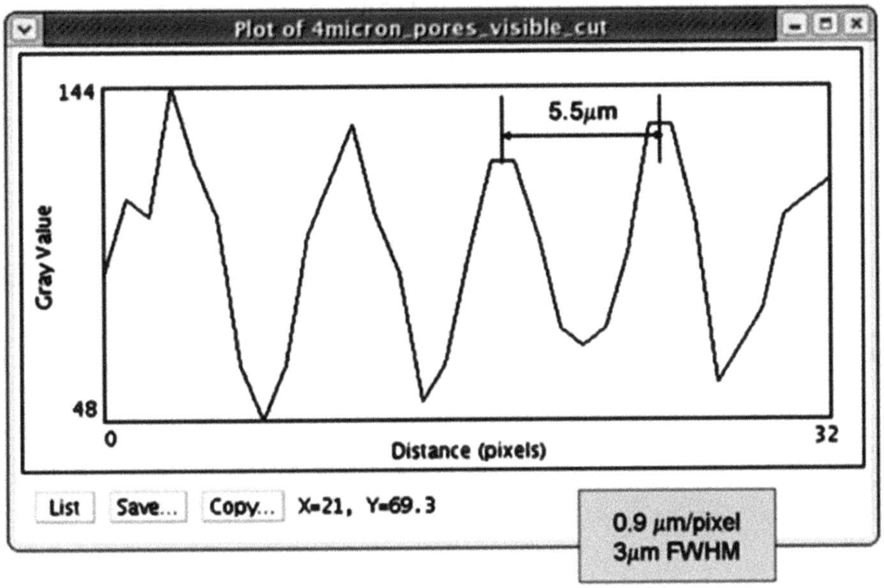

fiber to the detector and neutral density filters attenuate their intensity to a single photon level. The time resolution is obtained from the jitter between the start time (laser trigger) and stop time (MCP-signal).

The typical time resolution obtained with MCP-PMT detecting single photons with and without magnetic field is illustrated in Figure 23.

When a MCP was used to detect a pulse of the Cherenkov light a time resolution close to 5 ps was obtained (Inami, 2006).

MCP-PMTs also have superior characteristics in their (lack of) sensitivity to magnetic fields. The influence of the magnetic field on the electron trajectory is minimized when:

1. The distance between the photocathode, dynode systems and the anode is shortened to reduce the electrons deviation.
2. The electrodes should ensure a parallel electric field between the photocathode and the anode, so that all electrons travel parallel to the tube axis.
3. The value of the electric field should be as high as possible

The MCP-PMTs designs perfectly well meet all these requirements.

In Figure 24 the gain as a function of the tilt angle and the rotation angle is shown for two specific Hamamatsu and Photonics MCP- PMTs. In all cases with increasing the tilt or rotation angle a gain drop at magnetic fields stronger than of 0.5 T was observed.

In magnetic fields perpendicular to the MCP axes (the extreme case) the gain drops sharply at magnetic fields > 0.08T (Hakamata, 2007). At other angles the MCP stands significantly stronger magnetic fields.

Figure 23. The time resolution of some MCP-PMTs (XP85012 and Hamamatsu R10754); for the XP85012 also the dependence on the magnetic field is shown.
From Uhlig, 2012.

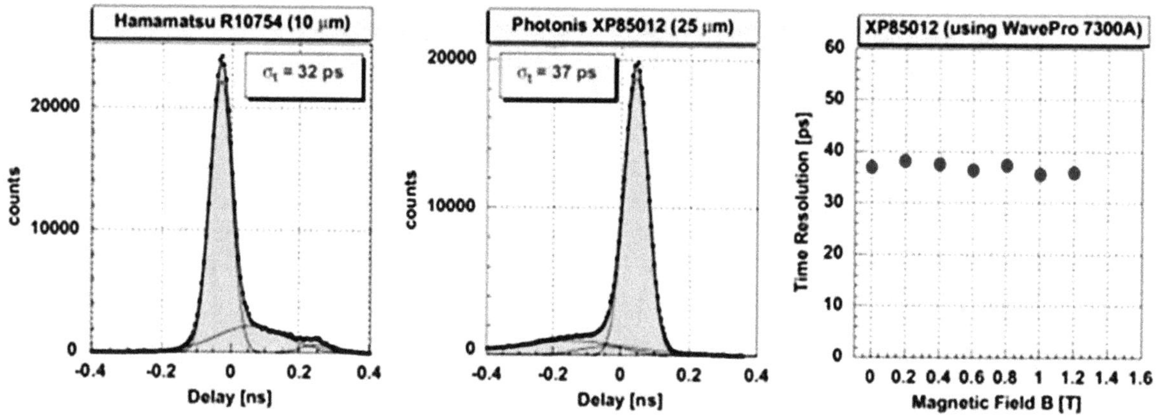

Figure 24. Magnetic field behavior of two selected MCP-PMTs: Photon is XP85012 and Hamamatsu R10754 at various tilt and rotation angles
From Uhlig, 2012.

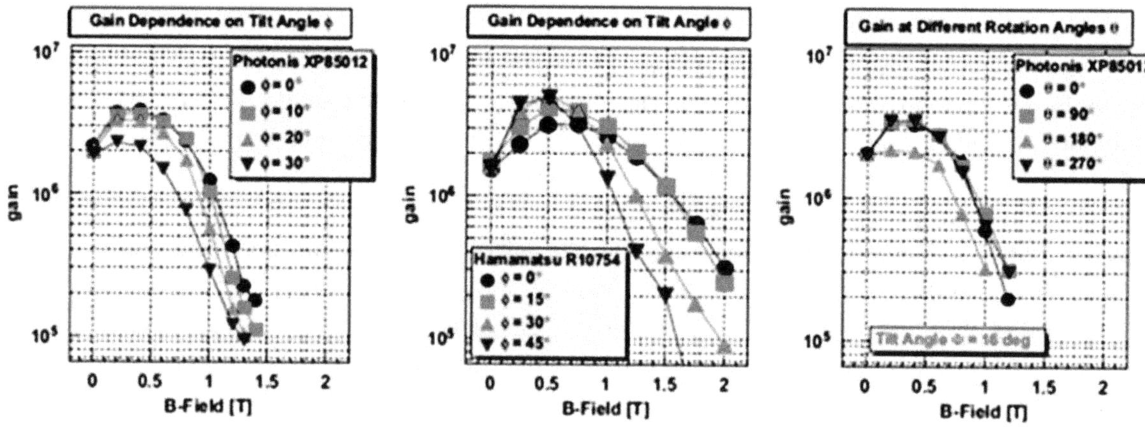

At the end of this paragraph it will be interesting to compare the main characteristic of multi-anode PMTs with MCP-PMTs (see Table 1). As can be seen from this table, MCP-MPTs are superior in many aspects, except the r dark current rate.

4. HYBRID VACUUM PHOTODETECTORS

Commercial multi-anode PMTs and MCP-PMTs have limited active areas, typically ≤ 25cm^2. There are efforts to develop an MCP-PMT with a size of about 20x20 cm^2 (Siegmund, 2012).

Another interesting approach to develop large area vacuum photodetectors is a so-called hybrid photodetector (HPD), which combines two or more detector principles in one design.

Table 1. Comparison of MCP-PMT with a multi-anode PMT

Characteristics	Multi-Anode PMT	MCP-PMT
Gain at zero magnetic field	10^6-10^7	10^6-10^7
Gain at 2 T	~0	>5x10^5
Position resolution(mm)	~5	~0.01
Time resolution (ps)	150	<50
Dark current rate (kHz/cm^2)	0.01	2
Stability with rate (MHz/cm^2)	Up to 10	5

From Lechmann, 2009.

Figure 25. The principle of the HPD; a photoelectron created from the photocathode accelerates in a strong electric field and gain sufficient kinetic energy to produce many electron-hole pairs in the Si sensor or some other type of detectors.
From Joram, 2003.

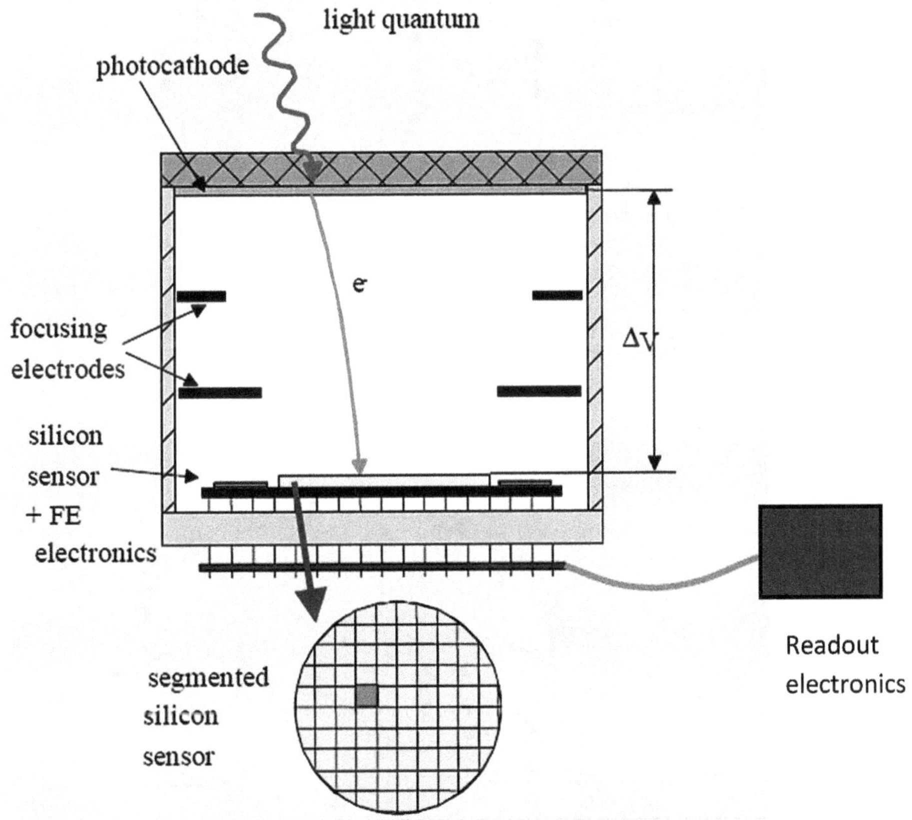

Like PMTs, the hybrid photodetector is a vacuum tube with a photocathode that reacts to light, an "electron multiplier" that multiplies electrons, and an output terminal that outputs an electrical signal. But whereas PMTs use multiple dynodes as electron multipliers, in most of the HPD a silicon sensor is used instead, as shown in Figure 25.

A photoelectron is accelerated in a strong electric field and produces many electron-hole pairs in a silicon sensor, which in the depleted silicon sensor results in a detectable current. In some designs of HPDs, the photoelectrons extracted from the large area photocathode are focused onto a small detector. Instead of a silicon sensor, other detectors can be used as well, for example a MCP.

Probably the first position sensitive HPD was realized in 1971 (Beaver, 1971). In this device the photoelectrons was accelerated to typically 30 keV. They hit a silicon substrate segmented into a linear array of 38 individual diodes (0.1 mm pitch). This HBD adopted the proximity focusing principle and comprised a solenoid magnetic field which allowed for scanning a 2D optical image.

Figure 26. Photoelectron trajectories in cross focusing (a) and fountain focusing (b) designs of HPD
From Joram, 1999.

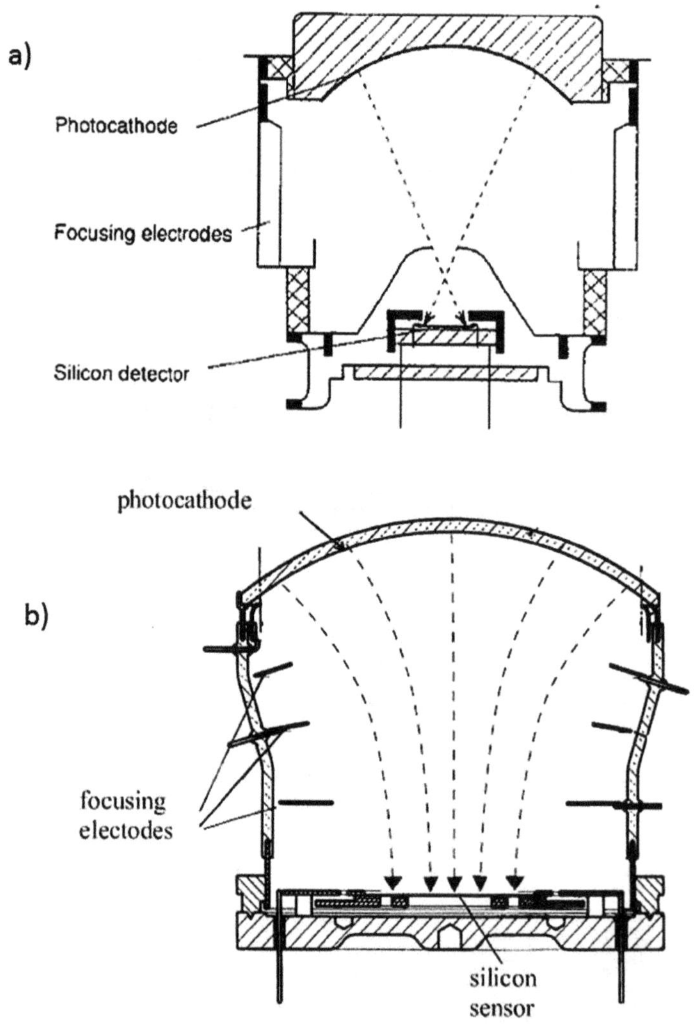

Later a similar concept was implemented in so called DIGICON tubes. In his detector, a linear diode array consisting of 512 elements (40 x 500 μm², 50 μm pitch), was used as a readout/"multiplication" element.

The development of HPD gained new momentum in the middle of the 1990ies in the CERN RD7 and RD19 programs. As a result of these efforts, an imaging silicon pixel array tube (ISPA) was built, in which for the first time the readout electronics was incorporated inside the vacuum tube. This allowed the use of a finely segmented pixel array sensor (64 x 16 = 1024 pixels of 75 x500 μm²) bump bonded to a binary readout chip. The proximity focusing design leads to compact and magnetic field tolerant detectors (because of the small gap between photocathode and silicon sensor). Since there is no demagnification, the photosensitive area of the detector is limited to the size of the silicon detector.

Besides proximity focused HPDs, other geometries were successfully implemented as well, so called "cross focusing" and "fountain focusing"(see Figure 26).

A cross focusing design offers a high resolution imaging because the electrostatic lenses largely compensate for the initial spread of the photoelectron velocity and emission angle at the photocathode. This geometry allows for strong demagnification, by the cost of a relatively large distance between cathode and the sensor, which results in pronounced magnetic field sensitivity.

Figure 27. An alternative HPD design; the sealed scintillation chamber was operated in vacuum. The photoelectrons extracted from the photocathode and accelerated in a strong electric field hit the scintillator emitting VUV light. The VUV light was detected with the gaseous readout detector.
Rodionov, 2002.

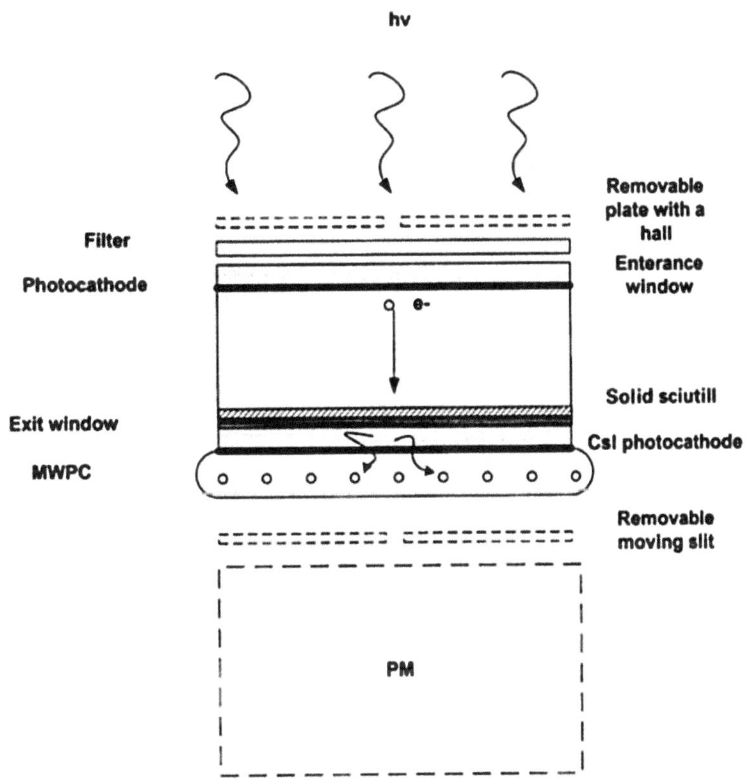

The fountain focusing design is an alternative to the cross focused HPD if a reduced spatial resolution is acceptable. The electron optics does not correct for the emission angle distribution. Its advantage is in a simple and compact tube design combined with a linear demagnification over the full acceptance. The sensitivity to magnetic fields is comparable to the cross focusing design.

Nowadays various designs of small size pixelated HPD are commercially available, for example from DEP (Delft Electronic Products).There were also intense attempts of various research groups to develop alternative designs of HPD. A few examples will be given below.

In the work of Rodionova hybrid gaseous photomultiplier was tested consisting of a vacuum chamber, comprisingba photocathode sensitive to visible light and a VUV scintillator, and a readout chamber based on a photosensitive MWPC (Rodionov, 2002). This apparatus is shown in Figure 27.

Electrons extracted from the semitransparent photocathode (two types of photocathodes were used: bi-alkali or Cs_3Sb) were accelerated in the proximity focused vacuum chamber and hit a thin layer (~200 μm) of a metalized (~0.05 μm Al) scintillator $KMgF_3$ (see Charpak, 1989) on top of a CaF_2 window, where VUV photons were emitted. The scintillation light from the other

(non-metalized) side of the scintillator was detected with the gaseous MWPC with a semitransparent CsI photocathode operating with CH_4 or an Ar+CH_4 mixture at 1 atmosphere. The position resolution of this prototype was about 2 mm FWHM.

Another interesting efforts pursued at CERN was the development of a large-area HPD by Braem et al. They built a device with a spherical borosilicate UV-glass entrance window, 127 mm in diameter, as

Figure 28. An artistic view of a large-area fountain focusing HPD developed at CERN
From Joram, 2003.

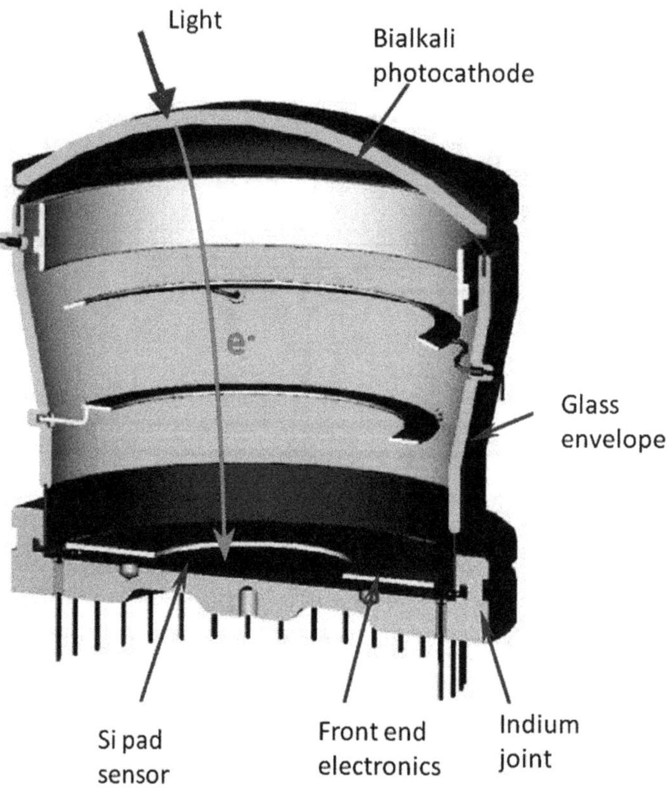

shown in Figure 28 (Braem, 2000).The fountain-focusing electron optics magnify the image onto a 50 mm diameter silicon sensor, containing 2048 pads with a size of 1x1 mm[2].

Besides large sensitive areas and rather good position resolutions, HPDs based on Si or other solid-state sensors, have also excellent energy resolution. The number of created electron-hole pairs are:

$$n_{eh} = E_k/W_{eh}, \tag{1}$$

where E_k is a kinetic energy of a photoelectron hitting the sensor and W_{eh} is the mean energy required for creating an electron-hole pair (for Si it is about 3.6 eV(Scholze, 1998)).

The variance of the signal charge is given by:

$$\sigma_{ch} = \sqrt{(Fn_{eh})}, \tag{2}$$

where F is the so called Fano factor (Fano, 1947).

In case of Si F~0.1, so:

$$\sigma_{ch} = 0.3\sqrt{n_{eh}}, \tag{3}$$

At a typical kinetic energy of the photoelectrons, Ek~25 keV, n_{eh}~7000 which ensures high statistics and therefore excellent accuracy in determining the number of primary electrons. Indeed, Si based

Figure 29. Photoelectron spectra obtained with a large area HPD with a pad Si sensor
From Joram, 2003.

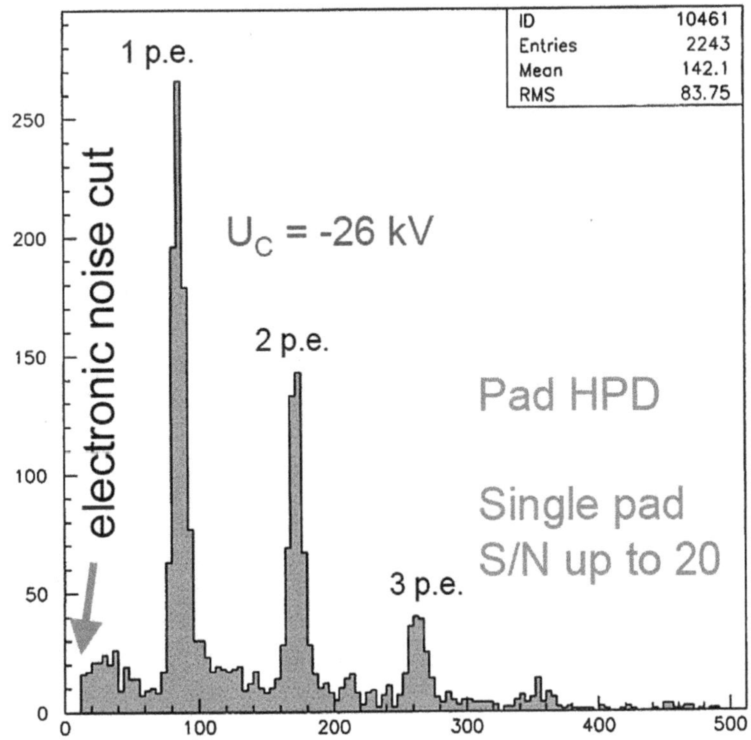

HPDs, including pad HPD, easily separate the signal produced by a single and a few photoelectrons, as shown in Figure 29.

A 10-inch in diameter HPD with a spherical entrance window covered with a highly efficient Rb_2Te semi-transparent photocathode has been developed (Braem, 2003). The detector also adopts the fountain focusing electron optics which allows for a linear demagnification of the image over practically the full tube diameter.

The exciting developments of large area vacuum photodetectors are continued by several groups. For example, another potentially interesting approach is to use a large area spherical photocathode with an MCPs in its centre (Wang, 2012) instead of the Si sensor as was used at CERN (Braem, 2003).

5. THE MAIN APPLICATIONS OF POSITION SENSITIVE VACUUM PMTs

Multi-anode PMTs and MCP-PMTs have many applications requiring parallel multichannel readout and time resolved photon counting, for example in spectroscopy, biomedical fluorescent detection, environmental monitoring, laser scanning detection including a so called LIDAR (light detection and ranging) a remote sensing method using the pulsed laser light to measure ranges and generate a three-dimensional information about the shape of the surface and is characteristics. Probably, the largest among them are in medicine and biology, e.g. position emission tomography (PET), small animal imaging, compact gamma camera, scinti-mammography, 2-D radiation monitoring etc.

In the last years there have been intense efforts aiming to develop MCP-PMT based time of flight PET allowing a significant image quality improvement.

Applications of multianode PMTs, MCP-PMTs and hybrid PMT in high-energy physics is also steadily growing. This will be illustrated by a few examples below.

The HERA-B (a particle physics detector of the HERA accelerator at the German Electron Synchrotron) Ring Imaging Cherenkov (RICH) collaboration was the pioneer in using a multi-anode PMTs (Arino, 2004). Their spherical mirror–based detector used 2240 multi-anode PMTs with bi-alkali photocathodes to detect Cherenkov rings. The quantum efficiency of these detectors at the peak of their sensitivity was about 20%, the single photoelectron detection efficiency reached 98%, the cross-talk did not exceed 0.2%, the background rate was below a few Hz per channel.

Up to 500 multi-anode PMTs were also implemented in the COMPASS (Common Muon and Proton Apparatus for Structure and Spectroscopy) RICH central region (Abbon, 2008), replacing the CsI-MWPC, which suffered at high fluxes from the cathode excitation effect. Due to the extended spectral sensitivity, the number of detected photons per Cherenkov ring was 60 and is to be compared with the 14 photons detected by the first generation of photon detectors made with CsI-MWPC. This extends the particle identification capability towards lower momenta.

The LHCb scintillating pad and pre-shower detectors use scintillating tiles readout by multi-anode 64 channel PMTs. In total of about 12.000 detector cells. The position of the scintillating pad and pre-shower detectors in the LHC-b calorimeter is shown in Figure 30.

They are made of plastic scintillator layer, divided in about 6000 cells of different size in order to obtain better granularity near the beam. The ionization generates a light pulse that is collected by a wavelength shifting fiber that is twisted inside the scintillator cell. The light is transmitted through a clear fiber to the readout system.

Figure 30. Schematics of the LHC-b calorimeter, consisting from a scintillating-pad detector (SPD), lead sheet, 2.5 radiation length, pre-shower detector (PS), an electromagnetic calorimeter (ECAL) and a hadronic calorimeter (HCAL); e, h and γ stand for electron, hadron and gamma photon, respectively. *From Philippov, 2014.*

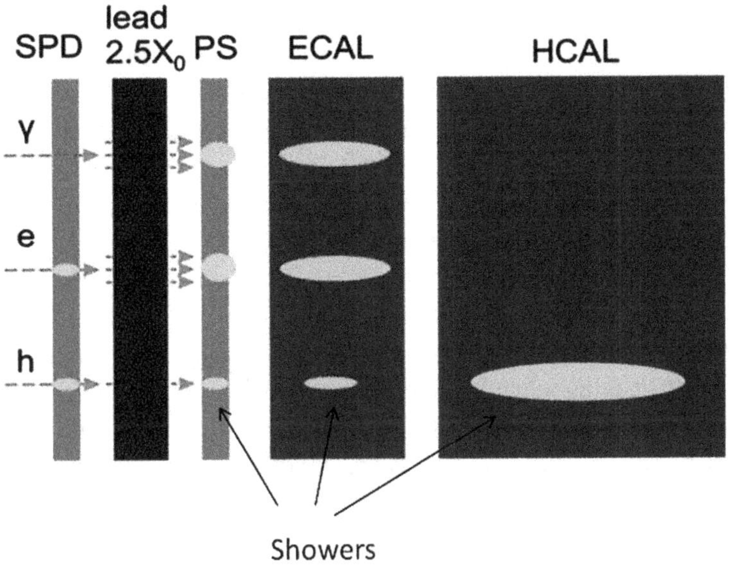

Multi-anode PMTs (680 detectors) were also successfully used in AMS-02 (Alpha Magnetic Spectrometer) space mission for the detection of Cherenkov rings. The Cherenkov light is produced in a radiator made of silica aerogel and sodium fluoride and is driven to the photocathodes by an array of 10,880 solid light-guides with different shapes to maximize the photon collection efficiency. The final matrix effective spatial granularity is of 8.5x8.5 mm^2.

Due to their superior time characteristics, MCP-PMTs are particularly suited for photon time-of-propagation (TOP) applications investigated for the future upgrade of particle identification devices in super B factories. In a TOP Cherenkov counter, the two-dimensional information of a Cherenkov ring generated in a quartz bar is represented by the TOP and horizontal angle of the Cherenkov photons after their propagation by total internal reflections into the bar (Figure 31).

MCP-PMTs are considered for several future experiments, e.g. PANDA (anti-Proton ANnihiliation at DArmstadt), Belle-II (an upgrade of the B factory experiment conducting by the Belle collaboration, Japan), CBM-RICH (RICH detector at the compressed baryonic matter experiment) etc. They are also

Figure 31. Artistic view of a TOP counter; the emitted Cherenkov photons are propagated inside quartz radiator and detected by the photon detector at the bar end. TOP counter measures the propagation time. *From Inami, 2014.*

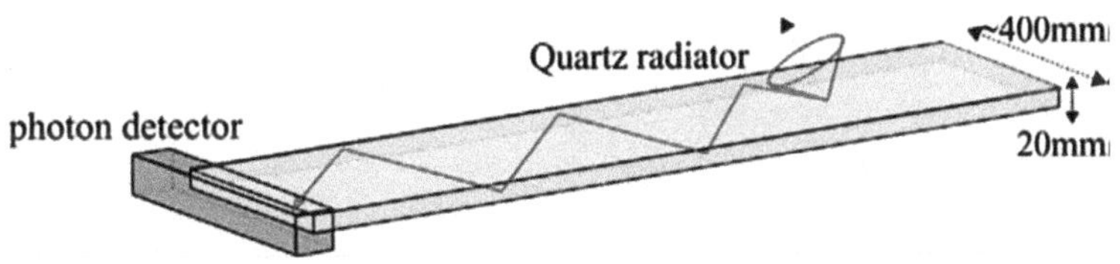

proposed for a time of internally reflected Cherenkov light system (TORCH) for the LHC-b upgrade to provide charged particle identification in the low momentum regime (up to10 GeV/c). The TORCH detector is supposed to supplement the current particle identification system after removal of the aerogel in RICH-1, which currently performs this task.

In some high-energy physics experiments the HPDs have advantages. Indeed, the internal demagnification enables close packing allowing to achieve an active area fraction of the system of 65%. Improved HPDs have high quantum efficiency in the UV region, about 30% at 270 nm. Finally, the low noise of the silicon pixel readout is leading to high signal/noise ratio. All these features make HPDs attractive in RICH detectors. About 500 HPDs (a total area of 3.3m^2) were used in HPD RICH during the Run I of the Large hadron collider (LHC) from 2010 to 2012. The tubes de-magnify the photocathode image by a factor of ~5 onto a small silicon detector array with ~500 x 500 μm^2 pixels, offering the real granularity 2.5 x2.5 mm^2.

Even in rather high magnetic fields, HPDs of proximity focusing type can be used. This option has been exploited in the hadronic calorimeter of the CMS (Compact muon solenoid) experiment at the LHC. In this case, the HPD having a small gap of 3.3 mm between the photocathode and the silicon pad detector, operate in a magnetic field of B~3.8 T, oriented in a such a way that the E field are parallel to the B field. Each HPD has 18 pixels, 20 mm^2 each, and detect the light from the scintillating fibers.

The examples described above show that vacuum position sensitive devices represent a completive technology in the photo-detection field. New coming ideas and developments, e.g. hybrid avalanche photodetector - HAPD - (Korpar, 2014), will further enlarge their applications.

6. SOLID-STATE DETECTORS

Solid state detectors, including pixelated ones, are used in various application since a long time. In this paragraph our focus will be on position sensitive solid state sensors capable of detecting single photons.

The development started with avalanche photodiodes. The first work in this direction was done in the 1960s at the RCA company by McIntyre (McIntyre, 1961) and by Haitz in the Shockley research laboratory (Haitz, 1964). Avalanche photodiodes operating in linear and in Geiger mode were in the sixties and early seventies of the last century an active field of experimental and theoretical research. A model of the operation was developed and verified with various test structures.

The schematic and operation principle of the avalanche photodiode (APD) are demonstrated with Figure 32.

This device is essentially p-n photodiodes operated under reverse bias near the breakdown voltage. The configuration consists of an n$^+$ layer followed, in sequence, by a thin p-layer, a depleted region and a heavily doped p-layer. The corresponding electric field distributions are also shown. The field strength has it is maximum at the n$^+$p junction.

Under the influence of the electric field the electron can gain sufficient energy for ionization, i.e. formation of an additional electron-hole pair. Such avalanche development is similar to the avalanche mode in gaseous detectors. The gain of this process (a so called linear or proportional mode) is:

$$G_e = \exp(\alpha_S d_s) \quad (4)$$

Figure 32. Incident photons create electron-hole pairs in the depletion layer of this structure and these moves towards the respective p-n junctions. If the external bias increases this localized electric field to above about 10^5 V / cm then the carriers in the semi-conductor collide with atoms in the crystal lattice, and the resultant ionization creates more electron-hole pairs. The avalanche development is similar to the happened in gaseous detectors.
From Fisher, 2012

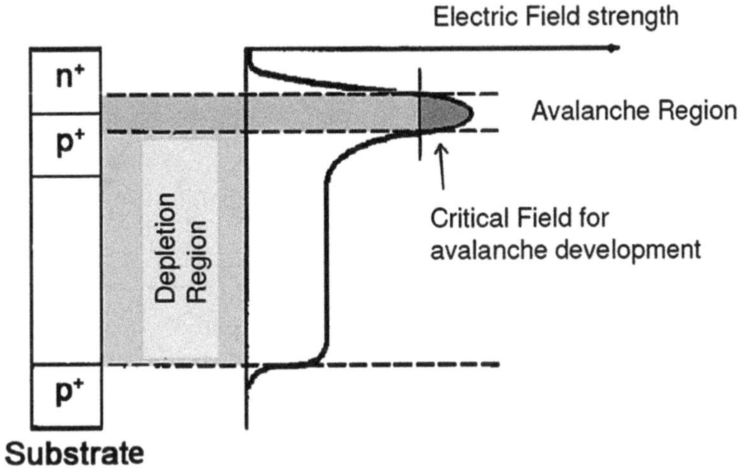

where α_S is the electron ionization coefficient (which is a function of electric field and also the temperature) and the d_s is the distance within which the multiplication happens.

The secondary hole can also ionize and form additional electron-hole pairs. However, since the hole mobility is less than the electron mobility, higher fields are required than for electron ionization.

When the partial gain due to holes:

$$G_h \geq 2, \tag{5}$$

the combined multiplication of electrons and holes leads to a sustained avalanche, i.e. a breakdown.

At some specific condition (e.g. low noise and specially designed electrode structures) it is possible to operate an APD in stable Geiger mode, in which, opposed to the analogue operation mentioned above, one can detect individual incident photons. This Geiger mode is also similar to the Geiger mode of the gaseous detectors. In this case the APD has to be operated at a few volts above its breakdown voltage and under well-controlled operating conditions such as the applied voltage and temperature.

A simple way to quench a Geiger discharge in an APD is similar to what used in gaseous Geiger counters, i.e. the use of a high ohmic resistor in series to the diode. After quenching, the resistor prevents an instantaneous recharge of the diode capacitance and allows resetting to the initial bias above breakdown.

Geiger APDs have the advantage of large output pulses, which depending on the overvoltage and diode capacitance, can reach values between 10^5 and 10^7electrons, which is sufficient for single photon detection. However, there are also several disadvantages. Besides free electrons being generated by the photoeffect, electron–hole pairs are constantly generated thermally. Traps are another source of charge carriers. In combination with the recombination time of the diode these effects limit the active area of

Figure 33. Simplified scheme of the array of pixelated Geiger APD
From Haba, 2007.

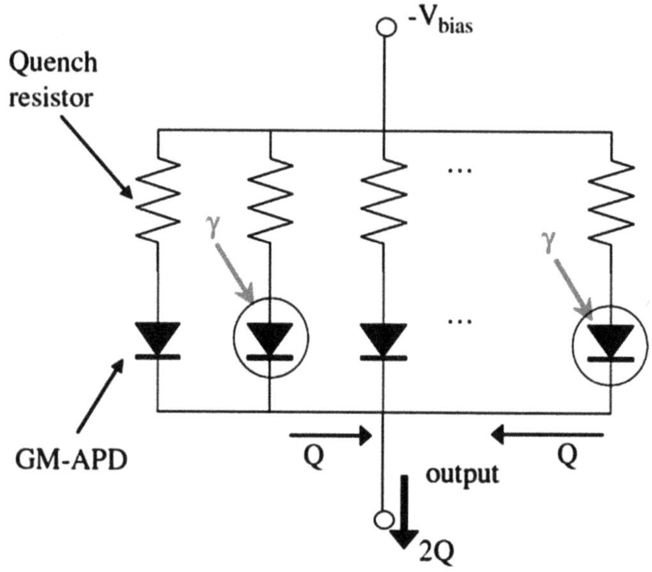

the Geiger APD. Currently commercially available APD from Perkin Elmer Optoelectronics, have a diameter of the active area of about 0.5 mm and requires cooling.

A breakthrough in these developments was the introduction of a micro-pixel Geiger photodiodes (Golovin, 1998; Sadygov, 1998).In this detector concept, a densely packed array of photodiodes (up to 10,000 per mm^2) is fabricated on the same substrate. Each single Geiger APD has its own micro quenching resistor (see Figure 33). The output signal of the device is the analog sum signal of all "fired" microcells.

This device, based on Si, has many names which depend on vendor, e.g. Si PM, multiphoton pixel counter etc. Below we will use an abbreviation SiPM - silicon photomultiplier.

Currently commercially available SiPMs have pixel sizes of about 50x 50 μm^2, as shown in Figure 34.

It has rather small sensitive surface (typically between 1and 9 mm^2) and high dark count rates (up to MHz for 3 x3 mm^2. To enlarge the sensitive surface and also to obtain spatial resolution usually a matrix of SiPMs is used, as shown in Figure 35.

To characterize the spectral sensitivity of SiPM often a so-called photon detection efficiency (PDE) is used instead of the quantum efficiency (QE).The (PDE) is defined as:

$$PDE = QE \times F_g \times P_a, \tag{6}$$

Here F_g is geometric factor and P_a is the avalanche probability. The PDE may have quite high values, as shown in Figure 36.

With a large number of pixels it is difficult to achieve high PDE because each pixel needs space for a quenching resistor, decreasing the active area, F_g, of the sensor as shown in Figure 37.

One of the features of a SiPM is its pulse-height response on the incident light. Each Geiger APD generate output pulses with equal amplitude and therefore the sum of the output signal is proportional

Figure 34. Magnified photo of the pixels of the Hamamatsu SiPMS10362-11-050C, consisting from small avalanche diodes; the individual pixel size is 50 x 50 μm².
From Xu, 2014.

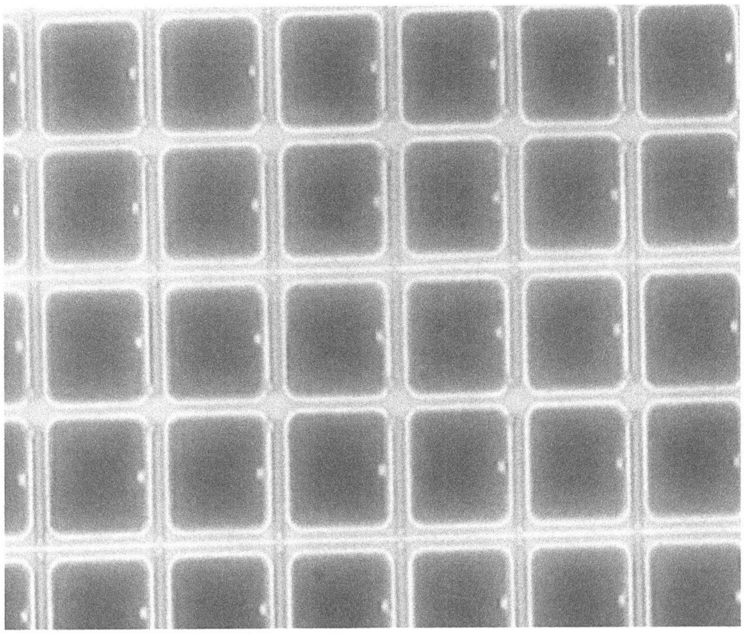

Figure 35. Photo of 4x4 arrays of 3x3 mm² SiPMs from Hamamatsu; the individual cells are marked by dashed lines.
From Quiang, 2013.

Figure 36. PDE measured with different types of SiPMs, mentioned in the insertion
From Musienko, 2007.

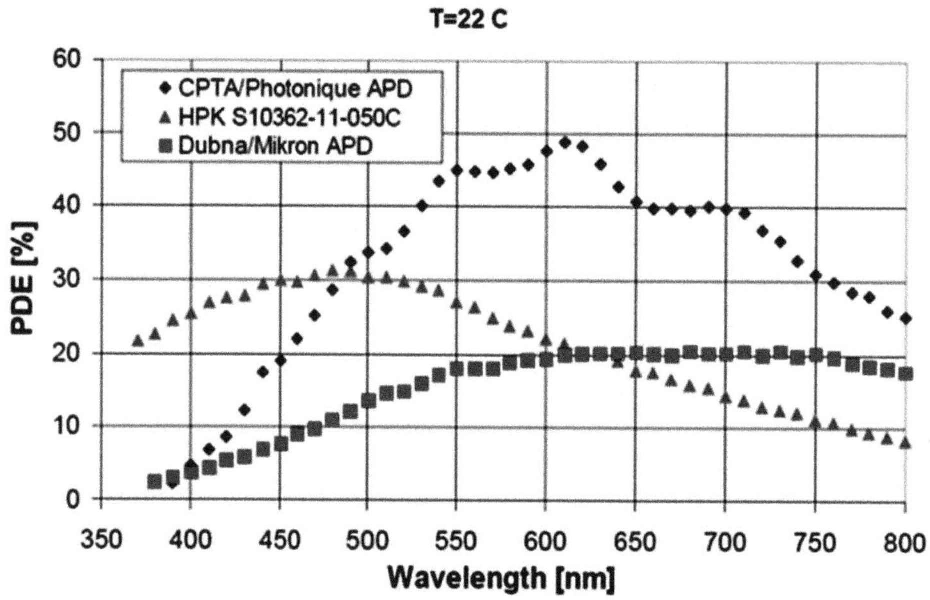

Figure 37. PDE vs. wavelength measured with SiPMs having different number of pixels; often more light is lost in the optical collection system than may be gained from the higher quantum efficiency of the detector itself. To increase the geometric acceptance, light concentrators have been employed in some detectors.
Korpar, 2014a.

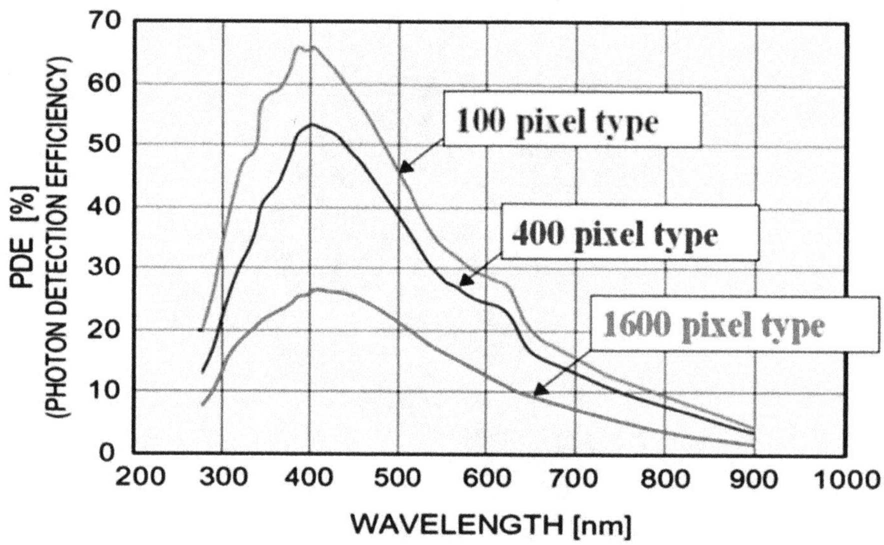

Figure 38. Pulse height spectra measured with a 1 x1 mm² HamamatsuSiPMS10362-11-025C for three different light intensities (the red and blue curves correspond to the lowest and strongest intensities, respectively) showing peaks corresponding to different numbers of photoelectrons generated
From Roncali, 2011.

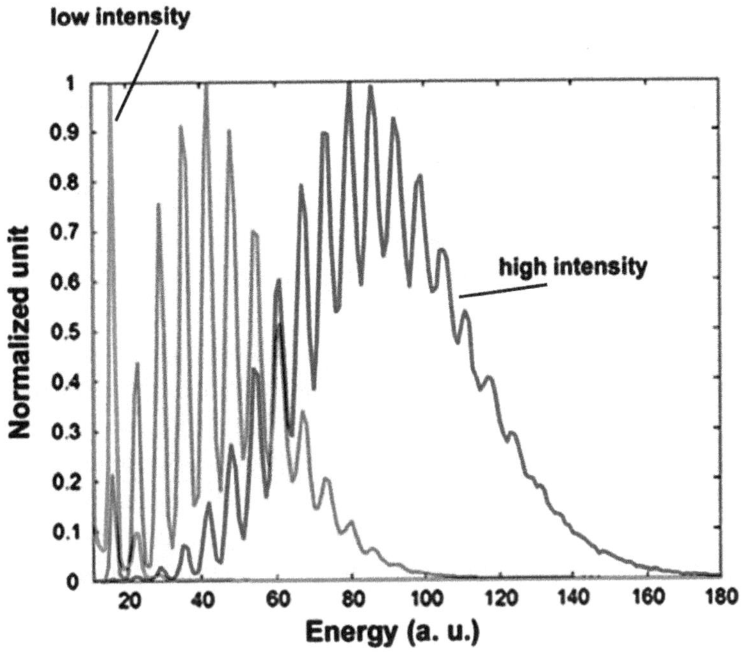

to the number of the incident photons. As a result, a APD is a good photon counting detector, as demonstrated in Figure 38.

In conclusion, the main advantages of SiPMs are:

- High quantum efficiency,
- Low operating voltages,
- Insensitivity to magnetic fields, AND
- Robustness and compactness.

Since the late 1990-ies the development of SiPMs has diversified and many prototypes and commercial devices exist. For example, there are promising efforts to develop avalanche Charged Coupled Devices (CCD) (Holl, 2006).

It should be noted that highly stable, highly sensitive APD systems are often more expensive than a comparable PMT-based system. For the majority of instrumentation based applications, the larger detection area, higher gain and superior signal to noise ratio of the PMTs make them still the detector of choice for many years to come.

7. APPLICATION OF SOLID-STATE DETECTORS

The only solid-state detector capable of detecting single photons with high efficiency is SiPMs. Their main field of application is detection of light from scintillators (crystals and plastics) and fibers.

From our point of view the most important among them is the use in nuclear medicine, where compact devices operating at low voltages are attractive. Many prototypes have been built and tested. It is beyond the scope of this paragraph to review all these promising developments, and we have chosen to highlight only a few examples.

In report (Dinu, 2014) was described a prototype of high resolution handheld radiation detector for radio-guided surgery. It consist of a gamma collimator, a $LaBr_3$ (Ce) scintillator and a 4x 4 array of SiPMs (Figure 39), allowing to track a radioactive isotope injected to the patient body for marking a region of interest. A modified device (without collimator and with an ultrathin light-tight window) was also developed for detecting traces of beta emitters.

Another promising application of SiPMs is small animal positron emission tomography (PET). PET in general is a nuclear medicine imaging technique which produces a three-dimensional image of functional processes in the body. To conduct the scan, a short-lived radioactive tracer isotope is injected into the body. The tracer is then chemically incorporated into a biologically active molecule. After some time the active molecule concentrate in tissues of interest. Then the patient is placed in the imaging scanner. As the radioisotope undergoes positron emission decay the emitted positron travels in the tissue for a short distance (typically less than 1 mm) until it interact with an electron and annihilate. The produced pair of gamma photons moves in opposite directions. These are detected when they reach a scintillating crystal in the scanning device, creating a burst of light which is detected by a photo-sensor, typically a PMT.

There were already intense efforts to implement SiPMs in compact PETs, which are mainly used for scanning small animals on which various medical test are performed (DelGuerra, 2011; Cabello, 2013; Lu, 2014). As an example we will present here some results obtained in work (Lu, 2014).

Their PET imaging system consists of a stationary detector gantry, a motor-controlled animal bed module, electronics modules, and power supply modules (see Figure 40).

The detector gantry is 655 mm high by 600 mm wide, and consists of 54 detector modules arranged in a ring with an inner diameter of 114 mm. Each PET detector has a LYSO scintillator (LYSO is crystals of $Lu_{1.9}Y_{0.1}SiO_5$(Ce) 0.5 mol%) directly coupled to a 4x4 SiPM array through the optical silicone oil (see Figure 41).

Figure 39. Schematics of the SiPM -based medical imaging (SIPMED) gamma camera
From Dinu, 2014.

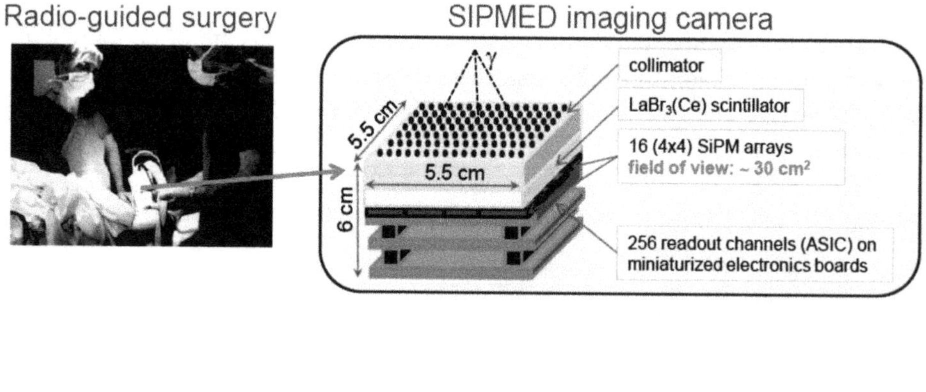

Figure 40. Photograph of the PET imaging system developed in work
LU, 2014.

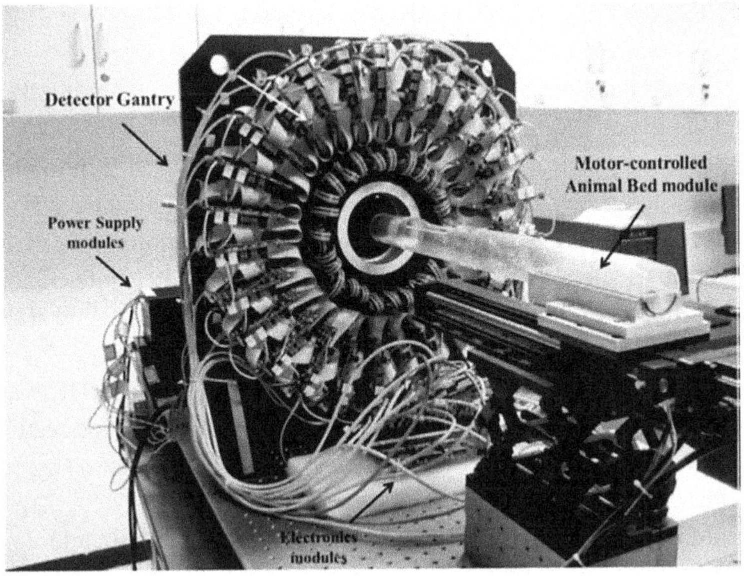

Figure 41. Geometry of the PET detector blocks: (a) each block has a 4x6 LYSO scintillator block coupled to a4x4 SiPM array, (b) left: schematic drawing of the detector block: middle: its axial view; right: a trans-axial view
From Lu, 2014.

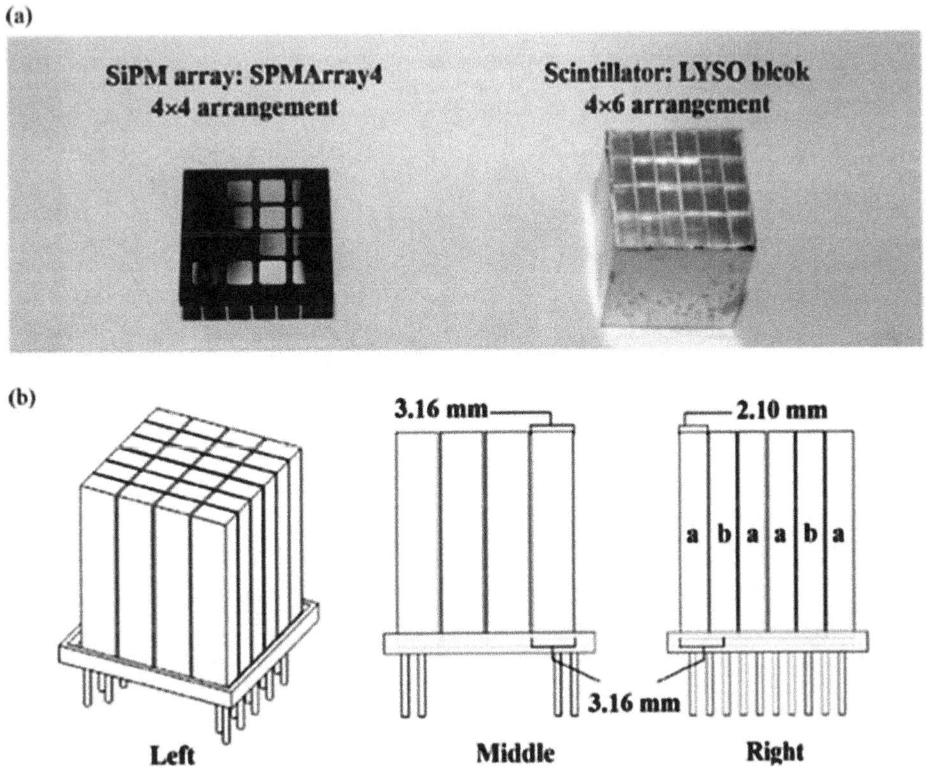

The SPM array is a16-element SiPM that has a3x3 mm^2 pixel chip area. Each pixel has a3x3 mm^2 active area that includes 4774 microcells. The SiPM detectors are operated at a voltage of 30 V to achieve a gain of about $2.4x10^6$. Each of the scintillator blocks has a4x6 matrix of crystals 2x 3x15 mm^3.

Examples of images obtained with such a PET are shown in Figure 42.

The radioactivity accumulation in the abdomen of the mouse was evident in all PET images, which also clearly indicated the tumor's location.

There are also active attempts from various groups to promote SiPM technology into high-energy physics and astrophysics experiments. Prototypes of various detectors have been built and tested.

The CALICE (calorimeter for linear collider experiment) collaboration investigated the possibility to use SiPM, which are insensitive to large magnetic fields, for the ILC (international linear collider) calorimeter. They have already installed and tested thousands of SiPMs attached at the end of the wave-length shifting fibers embedded in the scintillation tiles.

Besides readout of fibres and scintillators SiPMs are also considered for RICH applications.

In work (Korpar, 2010) SiPMs were used to detect Cherenkov photons produced in an aerogel radiator. To improve the light collection SiPMs were combined in a light collection system. In Figure 43 is shown an image of the Cherenkov ring recorded with SiPMs

Other promising studies are oriented on adopting SiPMs in the MAGIC telescope and other astrophysical instruments (Techima, 2008; Mazin, 2013).

Figure 42. PET images (top) of a mouse (bottom right) with an implanted tumor in its liver; the trans-axial, coronal and sagittal images of the mouse are illustrated. The 3mm diameter tumor tissue, which can be seen in the liver, is confirmed in the PET images.
From Lu, 2014.

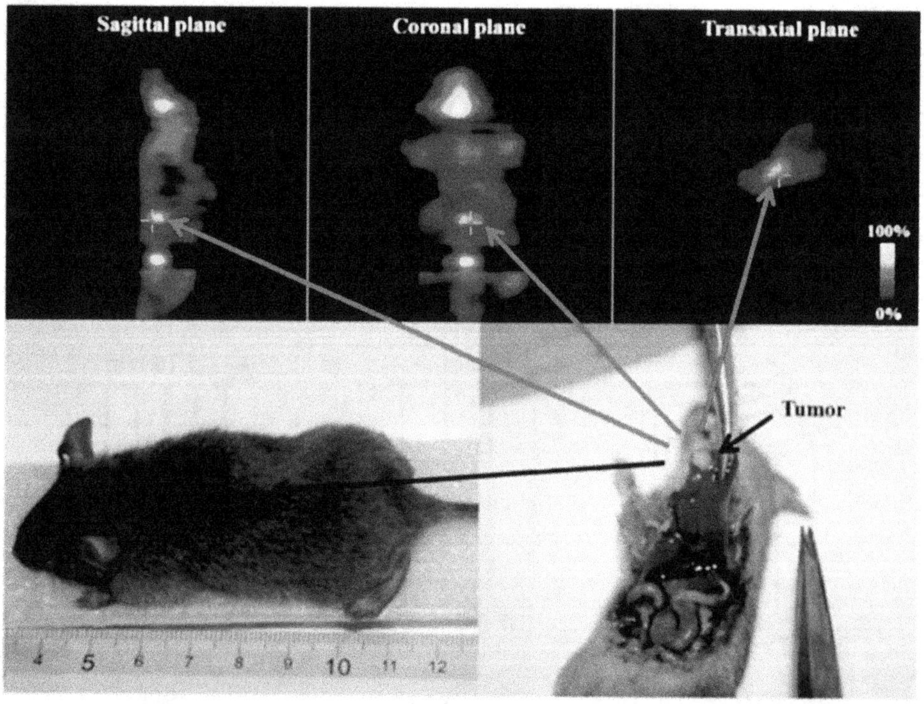

Figure 43. The accumulated SiPM hits represented in Cherenkov angle space for measurements without with the light collection system; the azimuthal distribution on the Cherenkov ring is uneven due to limitations in acceptance.
From Korpar, 2010.

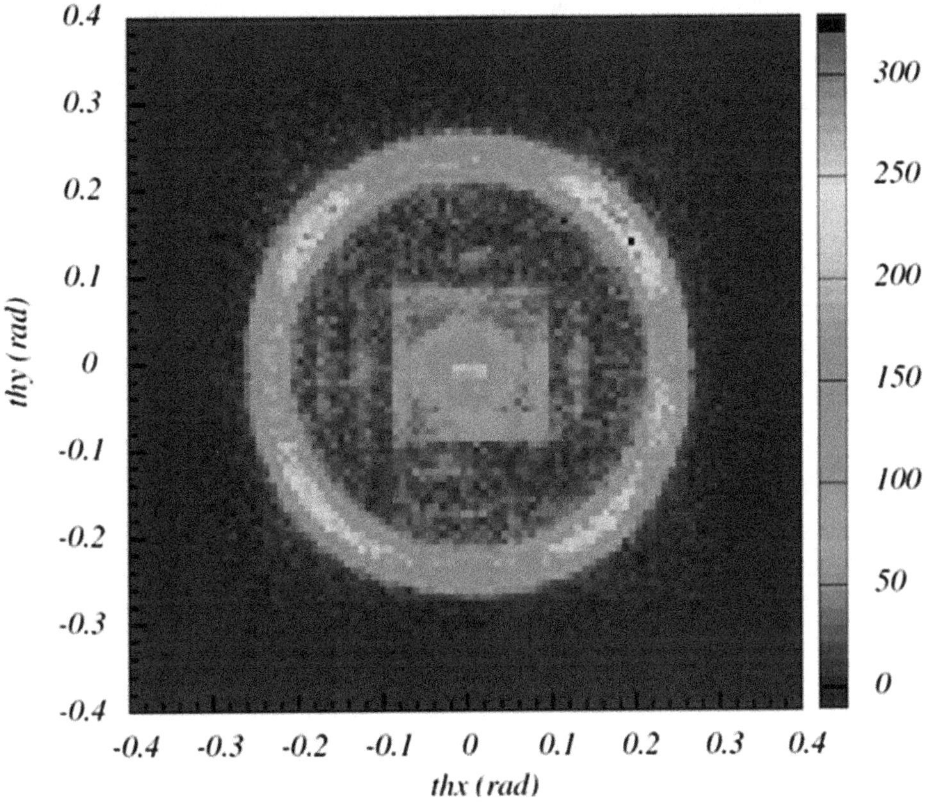

To conclude this paragraph we have to admit, that in spite of great potentials no real detector based on SiPM has been built so far. There is no doubt that in the nearest future SiPMs will be a widely used detector in many application and experiments.

8. CONCLUSIVE REMARKS

In this Chapter the state-of-the-art developments in vacuum and solid-state detectors for the detection of single photons have been described. Let us conclude by a comparison between the different detectors described, including gaseous detectors, to evaluate clearly their advantages and disadvantages. For completeness the traditional PMTs, from which the development started, are included in Table 2 as well.

As can be seen from the data presented in the table, gaseous detector have the similar characteristics in respect of quantum efficiency and position resolution as most other detectors. Their main advantage is that they can be built with high sensitive area.

Table 2. The main characteristics of the modern position sensitive photomultipliers capable of detecting single photons with high efficiency

Type of Single Photon detector	Largest Sensitive Area in a Single Unit Used in Practice, cm²	Position Resolution, μm	Time Resolutions, ns	Gain	Range of Sensitivity, nm	Quantum Efficiency at the Peak of Sensitivity, %	Sensitivity to Magnetic Field
PMT	1600	10^4-$5x10^5$	0.5-1	$>10^6$	120-900	35	high
Position sensitive PMT	~100	~$5x10^3$	~0.5	10^5-10^6	170-700	~20	reduced
MCP	~100	3-17	~0.035	10^4-10^7	170-750	~20	small
HPD	~100 Prototypes: 300	Potentially up to 100		~10^4	170-700	~25	sensitive
SiPM	~0.1 (up to 10 in matrix)	~300	0.1	$>10^5$		20-30	Insensitive up to 15T
Gaseous PMs	2800	30	~5	10^4-10^5	120-500	20-30	Very small

Therefore, gaseous detectors are preferable in applications requiring high surface coverage at low cost. The relevant example of such application could be the large RICH detectors built for high-energy physics experiments, e.g. the ALICE RICH detector based on CsI-MWPC has an active are of about 11 m².

REFERENCES

Abbon, P., Alekseev, M., Angerer, H., Apollonio, M., Birsa, R., Bordalo, P., & Wollny, H. et al. (2008). The COMPASS RICH-1 detector upgrade. *The European Physical Journal. Special Topics*, *162*(1), 251–257. doi:10.1140/epjst/e2008-00800-2

Adams, J., & Manley, B. W. (1966). The mechanism of channel electron multiplication. *IEEE Transactions on Nuclear Science*, *NS-13*(3), 88–99. doi:10.1109/TNS.1966.4324084

Ambrosio, C. D. H., & Leutz, H. (2003). Hybrid photon detectors. *Nuclear Instruments & Methods in Physics Research. Section A, Accelerators, Spectrometers, Detectors and Associated Equipment*, *501*(2-3), 463–498. doi:10.1016/S0168-9002(03)00431-5

Arino, I., Bastos, J., Broemmelsiek, D., Carvalho, J., Chmeissani, M., Conde, P., & Živko, T. et al. (2004). The HERA-B ring imaging Cherenkov counter. *Nuclear Instruments & Methods in Physics Research. Section A, Accelerators, Spectrometers, Detectors and Associated Equipment*, *516*(2-3), 445–461. doi:10.1016/j.nima.2003.08.173

Barnyakov, A. Yu., Barnyakov, M. Yu., Barutkin, V. V., Bobrovnikov, V. S., Buzykaev, A. R., Inami, K., & Yurikusa, Y. et al. (2009). Photomultiplier tubes with three MCPs. *Nuclear Instruments & Methods in Physics Research. Section A, Accelerators, Spectrometers, Detectors and Associated Equipment*, *598*(1), 160–162. doi:10.1016/j.nima.2008.08.084

Bateman, J. E., Apsimon, R. J., & Barlow, F. E. (1976). A new photomultiplier tube utilizing channel plate electron multiplier as the gain producing element. *Nuclear Instruments and Methods*, *137*(1), 61–70. doi:10.1016/0029-554X(76)90249-4

Beaver, E. A., Mc, C. E., & Ilwain, C. E. (1971). A Digital Multichannel Photometer. *The Review of Scientific Instruments*, *42*(9), 1321–1324. doi:10.1063/1.1685376

Bellazzini, R., Spandre, G., Minuti, M., Baldini, L., Brez, A., Cavalca, F., & de Oliveira, R. et al. (2006). Direct reading of charged multipliers with a self- triggered CMOS analogue chip with 105 k pixels and 50 μm pitch. *Nuclear Instruments & Methods in Physics Research. Section A, Accelerators, Spectrometers, Detectors and Associated Equipment*, *566*(2), 552–562. doi:10.1016/j.nima.2006.07.036

Bellazzini, R., Spandre, G., Minuti, M., Brez, A., Baldini, L., Latronico, L., & Siegmund, O. et al. (2008). Single photon imaging at ultra-high resolution. *Nuclear Instruments & Methods in Physics Research. Section A, Accelerators, Spectrometers, Detectors and Associated Equipment*, *591*(1), 125–128. doi:10.1016/j.nima.2008.03.040

Braem, A., Chesi, E., Dulinski, W., Filthaut, F., Go, A., Joram, C., & Ypsilantis, T. et al. (2000). Highly segmented large-area hybrid photodiodes with bialkali photocathodes and enclosed VLSI readout electronics. *Nuclear Instruments & Methods in Physics Research. Section A, Accelerators, Spectrometers, Detectors and Associated Equipment*, *442*(1-3), 128–135. doi:10.1016/S0168-9002(99)01210-3

Braem, A., Chesi, E., Joram, C., Séguinot, J., Weilhammer, P., Giunta, M., & Sartori, G. et al. (2003). Development of a 10-inch HPD with integratedreadout electronics. *Nuclear Instruments & Methods in Physics Research. Section A, Accelerators, Spectrometers, Detectors and Associated Equipment*, *504*(1-3), 19–23. doi:10.1016/S0168-9002(03)00744-7

Britting, A., Eyrich, W., & Lechmann, A. (2009). *Studies of MCP properties*. Retrieved from: http://www.google.fr/url?sa=t&rct=j&q=&esrc=s&source=web&cd=3&ved=0CDQQFjAC&url=http%3A%2F%2Fwww-hep.uta.edu%2F~ihowley%2FAFP%2FReadings%2Flehman-giessen.pdf&ei=ZT1_VM7HCNHrarOQgLAF&usg=AFQjCNEkTXwOt8a3gu-eW7ZR1woIU_A1mA&bvm=bv.80642063,d.d2s

Bruining, H. (1954). *Physics and Applications of Secondary Electron Emission*. London: Pergamon Press.

Cabello, J., Barrillon, P., Barrio, J., Bisogni, M. G., Del Guerra, A., Lacasta, C., & Llosá, G. et al. (2013). High resolution detectors based on continuous crystals and SiPMs for small animal PET. *Nuclear Instruments & Methods in Physics Research. Section A, Accelerators, Spectrometers, Detectors and Associated Equipment*, *718*, 148–150. doi:10.1016/j.nima.2012.08.094

Calvi, M., Giachero, A., Gotti, C. (2011). Aging and time resolution measurements for the Hamamatsu R 7600 multi-anode photomultiplier tube. *IEEE Nucl. Sci. Sympos. Conference Record*.

Calvi, M., Giachero, A., Gotti, C., Maino, M., Matteuzzi, C., & Pessina, G. (2013). Characterization of a Hamamatsu R7600 multi-anode photomultiplier tube with single photon signals. *JINST*, *8*(02P02012), 1–11. doi:10.1088/1748-0221/8/02/P02012

Charpak, G., Peskov, V., & Scigocki, D. (1989). New scintillators for photosensitive gaseous detectors. Proceedings of Sympos. on particle identification at high-luminosity hadron colliders. Batavia, IL: CERN-LAA-EC-89-003.

Conneely, T. M., Milnes, J. S., & Howorth, J. (2013). Extended lifetime MCP-PMTs: Characterization and life time measurements of ALD coated microchannel plates, in a sealed photomultiplier tube. *Nuclear Instruments & Methods in Physics Research. Section A, Accelerators, Spectrometers, Detectors and Associated Equipment*, *732*, 388–391. doi:10.1016/j.nima.2013.07.023

Del Guerra, A., Belcari, N., & Bisogni, M. G. (2011). Silicon Photomultipliers (SiPM) as novel photodetectors for PET. *Nuclear Instruments & Methods in Physics Research. Section A, Accelerators, Spectrometers, Detectors and Associated Equipment*, *648*, S232–S35. doi:10.1016/j.nima.2010.11.128

Dinu, N. (2005). *Silicon photomultiplier (SiPM) new technique for photon counting*. Presentation at College de France. Retrieved from https://www.google.ch/search?q=Silicon+photomultiplier+(SiPM)+new+technique+for+photon+counting&oq=Silicon+photomultiplier+(SiPM)+new+technique+for+photon+counting&aqs=chrome.69i57.4540j0j7&sourceid=chrome&es_sm=93&ie=UTF-8#q=Dinu%2C+Silicon+photomultiplier+(SiPM)+new+technique+for+photon+counting

Dinu, N. (2014). *Silicon Photomultiplier - characteristics and applications*. Presented at the Seminar in Geneva University. Retrieved from http://dpnc.unige.ch/seminaire/talks/dinu.pdf

Fano, U. (1947). Ionization yield of radiations. II. The fluctuations in the number of ions. *Physical Review*, *72*(1), 26–29. doi:10.1103/PhysRev.72.26

Farnsworth, P. T. (1930). *Electron multiplier*. US Patent number 1 969 399.

Filippov, S. (2014). *First years of running for the LHCb calorimeter system*. Retrieved from https://cds.cern.ch/record/1667194/files/LHCb-TALK-2014-035.pdf

Fisher, P. (2012). *Amplifying Photo Detectors: Avalanche Photodiodes Silicon Photomultiplier*. Retrieved from: http://sus.ziti.uni-heidelberg.de/Lehre/Detectors1213/Part7_PhotoDetectors.pdf

Golovin, V. (1999). *Avalanche Photodetector*. Russian Agency for Patents and Trademarks, Patent No. RU 2142175.

Goodrich, G. W., & Wiley, W. C. (1962). Continuous channel electron multiplier. *The Review of Scientific Instruments*, *33*(7), 761–762. doi:10.1063/1.1717958

Gys, T. (2005). *Particle Detectors –Principles and Techniques(3/5) Lecture 3b – Photodetection*. Retrieved from http://ph-dep-dt2.web.cern.ch/ph-dep-dt2/CAT2005_3b.pdf

Gys, T. (2014). *Micro-channel plate photomultiplier tubes for fast photon detection*. Presentation at CERN detector seminar. Retrieved from https://indico.cern.ch/event/288433/

Haba, J. (2008). Status and perspectives of Pixelated Photon Detector (PPD). *Nuclear Instruments & Methods in Physics Research. Section A, Accelerators, Spectrometers, Detectors and Associated Equipment*, *595*(1), 154–160. doi:10.1016/j.nima.2008.07.061

Haitz, R. H. (1964). Model for the Electrical Behavior of a Microplasma. *Journal of Applied Physics*, *35*(5), 1370–1376. doi:10.1063/1.1713636

Hakamata, T., Kume, H., & Okano, K. (2007). *Photomultiplier tubes. Basics and applications. Edition 3a*. Retrieved from https://www.hamamatsu.com/resources/pdf/etd/PMT_handbook_v3aE.pdf

Heroux, L., & Hinteregger, H. E. (1960). Resistance strip magnetic electron multiplier for the extreme ultraviolet. *The Review of Scientific Instruments*, *31*(3), 280–286. doi:10.1063/1.1716957

Holl, P., Andritschke, R., & Eckhardt, R. (2006). A New High-Speed, Single Photon Imaging CCD for the Optical. *2006 IEEE Nuclear Science Symposium Conference Record*. doi:10.1109/NSSMIC.2006.354202

Inami, K. (2014). TOP counter for particle identification at the Belle-II experiment. *Nuclear Instruments & Methods in Physics Research. Section A, Accelerators, Spectrometers, Detectors and Associated Equipment*, *766*, 5–8. doi:10.1016/j.nima.2014.07.006

Inami, K., Kishimoto, N., Enari, Y., Nagamine, M., & Ohshima, T. (2006). A 5ps TOF-counter with an MCP–PMT. *Nuclear Instruments & Methods in Physics Research. Section A, Accelerators, Spectrometers, Detectors and Associated Equipment*, *560*(2), 303–308. doi:10.1016/j.nima.2006.01.027

Joram, C. (1999). Large area hybrid photodiodes. *Nuclear Physics B - Proceedings Supplements*, *78*(1-3), 407–415. doi:10.1016/S0920-5632(99)00578-2

Joram, C. (2003). *Hybrid photon detectors (HPD) and multi anode photomultiplier tubes (MAPMT)*. Retrieved from: http://www.google.fr/url?sa=t&rct=j&q=&esrc=s&source=web&cd=1&cad=rja&uact=8&ved=0CCYQFjAA&url=http%3A%2F%2Findico.cern.ch%2Fevent%2Fa035690%2Fsession%2Fs10%2Fcontribution%2Fs10t1%2Fmaterial%2F0%2F1.pdf&ei=ojh_VMnEL9LUasS6gugK&usg=AFQjCNG9rhdupwqOLC7A1Hphv23INhzMOw

Korpar, S., Adachi, I., Hamada, N., Higuchi, M., Iijima, T., Iwata, S., & Yusa, Y. et al. (2014). A 144-channel HAPD for the Aerogel RICH at Belle II. *Nuclear Instruments & Methods in Physics Research. Section A, Accelerators, Spectrometers, Detectors and Associated Equipment*, *766*, 145–147. doi:10.1016/j.nima.2014.05.060

Korpar, S., Chagani, H., Dolenec, R., Križan, P., Pestotnik, R., & Stanovnik, A. (2010). A module of silicon photo-multipliers for detection of Cherenkov radiation. *Nuclear Instruments & Methods in Physics Research. Section A, Accelerators, Spectrometers, Detectors and Associated Equipment*, *613*(2), 195–199. doi:10.1016/j.nima.2009.11.043

Korpar, S., Tahirovic, E., Krizan, P., & Pestotnik, R. (2014a). Test of the Hamamatsu MPPC module S11834 as a RICH photon detector. *Nuclear Instruments & Methods in Physics Research. Section A, Accelerators, Spectrometers, Detectors and Associated Equipment*, *766*, 107–109. doi:10.1016/j.nima.2014.05.074

Lechmann, A., Britting, A., & Eyrich, W. (2013). Significantly improved life time of micro-channel plate PMTs. *Nuclear Instruments & Methods in Physics Research. Section A, Accelerators, Spectrometers, Detectors and Associated Equipment*, *718*, 535–540. doi:10.1016/j.nima.2012.11.109

Lu, Y., Yang, K., Zhou, K., Zhang, Q., Pang, B., & Ren, Q. (2014). Development of a SiPM-based PET imaging system for small animals. *Nuclear Instruments & Methods in Physics Research. Section A, Accelerators, Spectrometers, Detectors and Associated Equipment*, *743*, 30–38. doi:10.1016/j.nima.2014.01.010

Lubsandorzhiev, B. K. (2006). *On the history of photomultiplier tube invention*. ArXiv:physics/0601159

Mazin, D., Bangale, P., & Sitarek, J. (2014). *Towards SiPM camera for current and future generations of Cherenkov telescopes.* ArXiv: 1410.5070

McIntyre, R. J. (1961). Theory of Microplasma Instability in Silicon. *Journal of Applied Physics, 32*(6), 983–995. doi:10.1063/1.1736199

Musienko, Y. (2007). *Tests and performance of multipixel Geiger mode APD's and APD's for the CMSECAL.* Presented at the International workshop on new photon-detectors PD07 Kobe University, Kobe, Japan.

Oschepkov, P. K., Skvortsov, B. N., & Osanov, B. A. (1960). Application of a continuous secondary electron multiplication for amplifying small currents (Translation). *Pribory Tekh Eksper., 4*, 89–91.

Philippov, P. (2014). *First year of running for the LHC-b calorimeter system.* Retrieved from https://www.google.ru/url?sa=t&rct=j&q=&esrc=s&source=web&cd=1&cad=rja&uact=8&ved=0CBwQFjAA&url=https%3A%2F%2Findico.cern.ch%2Fevent%2F301616%2Fcontribution%2F7%2Fmaterial%2Fslides%2F0.pptx&ei=dh-YVdPKN4m0ygPntJyIDg&usg=AFQjCNHALpPVepyHDolqTAJjJj2b3ZEqTg

Qiang, Y., Zorn, C., & Fernando Barbosa, F. (2013). Radiation hardness tests of SiPMs for the JLab Hall D Barrel calorimeter. *Nuclear Instruments & Methods in Physics Research. Section A, Accelerators, Spectrometers, Detectors and Associated Equipment, 698*, 234–241. doi:10.1016/j.nima.2012.10.015

Rodionov, I., Francke, T., Peskov, V., & Sokolova, T. (2002). Hybrid gaseous photomultipliers. *Nuclear Instruments & Methods in Physics Research. Section A, Accelerators, Spectrometers, Detectors and Associated Equipment, 478*(1-2), 384–390. doi:10.1016/S0168-9002(01)01778-8

Roncali, E., & Cherry, S. R. (2011). Application of Silicon Photomultipliers to Positron Emission Tomography. *Annals of Biomedical Engineering, 39*(4), 1358–1377. doi:10.1007/s10439-011-0266-9 PMID:21321792

Sadygov, Z. (1998). *Avalanche Detector.* Russian Agency for Patents and Trademarks, Patent No. RU 2102820.

Scholze, F., Rabus, H., & Ulm, G. (1998). Mean energy required to produce an electron-hole pair in siliconfor photons of energies between 50 and 1500 eV. *Journal of Applied Physics, 84*(5), 2926–2939. doi:10.1063/1.368398

Siegmund, O. H. W., John, V., Vallerga, J. V., & Tremsin, A. S. (2011). Microchannel Plate Imaging Photon Counters for Ultraviolet through NIR Detection with High Time Resolution. *Proceedings of the Society for Photo-Instrumentation Engineers, 8033*, 80330V-1, 80330V–12. doi:10.1117/12.884271 PMID:23833700

Siegmund, O. H. W., McPhate, J. B., Tremsin, A. S., Jelinsky, S. R., Hemphill, R., Frisch, H. J., & Mane, A et al.. (2012). Atomic layer deposited borosilicate glass microchannel plates for large area event counting detectors. *Nuclear Instruments & Methods in Physics Research. Section A, Accelerators, Spectrometers, Detectors and Associated Equipment, 695*, 168–171. doi:10.1016/j.nima.2011.11.022

Siegmund, O. H. W., Tremsin, A. S., Vallerga, J. V., Abiad, R., & Hull, J. (2003). High resolution cross strip anodes for photon counting detectors. *Nuclear Instruments & Methods in Physics Research. Section A, Accelerators, Spectrometers, Detectors and Associated Equipment*, *504*(1-3), 177–181. doi:10.1016/S0168-9002(03)00816-7

Summer, W. (1957). *Photosensitors: a treatise on photo-electric devices and their application to industry*. London: Chapman and Hall Ltd.

Teshima, M., Dolgoshein, B., & Mirzoyan, R. (2008). SiPM development for astroparticle physics applications. *Proceedings of the 30th International Cosmic Ray Conference*. Universidad Nacional Autónoma de México.

Timothy, J. G. (2010). In M. C. E. Huber, A. Pauluhn, J. L. Culhane, J. G. Timothy, K. Wilhelm, & A. Zehnder (Eds.), Microchannel plates for photon detection and imaging in space. I n: Observing Photons in Space. ISSI Scientific Reports Series, ESA/ISSI (pp. 365–390). Academic Press.

Uhlig, F., Britting, A., Eyrich, W., Lehmann, A., Schwarz, C., & Schwiening, J. (2012). Performance studies of microchannel plate PMTs. *Nuclear Instruments & Methods in Physics Research. Section A, Accelerators, Spectrometers, Detectors and Associated Equipment*, *695*, 68–70. doi:10.1016/j.nima.2011.12.062

Xu, C., Klanner, R., Garutti, E., & Hellweg, W.-L. (2014). Influence of X-ray irradiation on the properties of the Hamamatsu silicon photomultiplier S10362-11-050C. *Nuclear Instruments & Methods in Physics Research. Section A, Accelerators, Spectrometers, Detectors and Associated Equipment*, *762*, 149–161. doi:10.1016/j.nima.2014.05.112

Yamamoto, K., Yamammura, K., Sato, K. (2007). Development of multi-pixel photon counter (MPPC). *IEEE Nucl. Sci. Conf. Record, 24*(292), 1511-1515.

KEY TERMS AND DEFINITIONS

Aerogel: A material consisting from grains of amorphous SiO_2 (with sizes from 1to 10 nm) packed together in a three-dimensional structure filled by a trapped air. Aerogel's refractive index, depending on its structure, covers the range between condensed stated (n=1.3) and gases (n=1.001) and for this reason is a very convenient Cherenkov radiator.

Calorimeter: A dedicated apparatus for measuring deposited energy in a dense material by photon, charged particles and showers. Most of calorimeters are position sensitive to measure energy depositions depending on their location.

Dynodes: A system of electrodes in a vacuum detector serves as an electron multiplier for primary photoelectrons through secondary electron emission.

Electromagnetic Calorimeter: A calorimeter designed to measure the energy of particles and photons which interact with the calorimeter matter via electromagnetic interactions.

Hadronic Calorimeter: One which measures the deposit energy of particles that interact with the calorimeter matter via the strong nuclear force.

Hybrid Detector: A detector which combines in one design two or more operational principles.

PMT: A photomultiplier tube-a general name for a vacuum apparatus for detection ultra-weak flux of light (UV, visible or infrared) through the use of photoemission from a photocathode and a system of dynodes in which the primary photoelectron is multiplied via a secondary electron emission.

Super B-Factory: A high luminosity electron-positron collider to be designed for producing vary large quantities of B mesons; this supposed to reveal new physics though studies of rare or suppressed decays.

Chapter 9
Cherenkov Light

ABSTRACT

Consider an electron approaching a sample of glass with a velocity close to the speed of light, c. As the electron moves through the glass light will instantly be emitted along its track, if its velocity is high enough. Even more, the electron will leave the glass sample before the light since the velocity of the particle inside the sample is larger than the speed of the light. At first sight, this seems to be in contradiction to Einstein's theory of special relativity, which states that nothing can travel faster than the speed of light, but one often forgets the important condition: in vacuum. This is the story of light emitted at particle travelling faster than the speed of light.

1. INTRODUCTION

The light emitted in a medium as a charged particle traverses it, is normally known as Cherenkov light after the Russian scientist who first studied it. The most important feature of this pale blue light is that it is emitted in a cone at a specific angle to the track, depending of the velocity of the incoming particle. This means that by measuring this angle one can determine the velocity of relativistic particles with good accuracy. Combining this information with information about the particle's momentum, the mass of the particle is known and, since the mass of each particle is unique, the particle is identified.

The Cherenkov effect has been used to identify particles ever since its discovery, and is becoming an even more important part of particle physics experiments of today, since new technology has provided better and better equipment to detect the faint light. These detectors are known as Cherenkov detectors.

2. CHERENKOV LIGHT

During the early years of the 1930's a Russian physicist, Sergei Ivanovich Vavilov, suggested to one of his students, Pavel Alexejevitj Cherenkov, to study the luminescence that occurred when gamma rays interact with different substances dissolved in water (Ekspong, 1958; Lennander, 1958).

DOI: 10.4018/978-1-5225-0242-5.ch009

Already in 1910 Marie Curie had found that bottles of concentrated radium- solution glow with an uncanny pale blue light. The first deliberate attempt to study the phenomenon was made by Mallet in 1929 (Mallet, 1929). He found that the light emitted from a wide variety of transparent bodies placed close to a radioactive source always had the same bluish white quality, and that the spectrum is continuous, not possessing the line or band structure characteristic of fluorescence (Jelley, 1958). He was the first to appreciate the generality of the effect and to notice that in a number of other aspects also, it is very different from fluorescence and other known forms of luminescence. Unfortunately Mallet did not pursue the work, nor did he attempt to offer an explanation for the origin of the light. The subject then lay dormant until Cherenkov began to study the phenomenon.

The delay in the study of the phenomenon was due to a number of causes. For instance, at the time of these early observations, much work was going on in the systematic study of the fluorescence and phosphorescence of materials irradiated by ultra-violet light, X-rays and the newly discovered radiation from the radioactive elements. The diverse and relatively complicated phenomena associated with these forms of luminescence only helped to postpone the discovery of Cherenkov radiation; the latter in any case was so weak that it was frequently masked by the presence of these other effects. The absence of really sensitive light detectors also contributed to the delay in the discovery of Cherenkov radiation; the early work had been carried out either by visual observation or by photographic recording with long exposure. Tamm has pointed out that the effect could have been predicted theoretically much earlier if a non-critical, oversimplified view of the implication of the theory of special relativity had not restrained thinking (Tamm, 1959). One often says that particles cannot travel faster than the speed of light and forgets the important addition: in vacuum. The maximum velocity for particles is the same(c), but the phase velocity of light in matter is lower (c/n), where n is the refractive index of the material.

Cherenkov started by examining the pure solvents of the substances. In 1934, his redistilled samples of water showed small luminescence when they were irradiated with gamma rays. Since water is not expected to fluoresce, this was a most remarkable result. Tamm has said that other physicists who were told about Cherenkov's results made jokes about the hallucinations of a young man who had spent too long in the dark, or if they believed in the results, explained them as normal fluorescence (Tamm, 1958). Vavilov, on the other hand, made the hypothesis that the effect does not originate from the gamma rays but is caused by Compton electrons (Vavilov, 1934). This hypothesis was later confirmed experimentally by Cherenkov, who also showed that the effect could be caused by fast electrons from a β source.

One of the greatest scientists of that time, Niels Bohr, did believe in the effect and visited Cherenkov in Leningrad. "Wunderbar, Wunderschön" he repeated several times during the demonstrations of the most essential properties of the radiation, namely its space asymmetry (Cherenkov 1986). Nevertheless, the atmosphere of distrust of the new effect from the scientific community remained. The most open, and sharp, manifestation of this distrust was the refusal by the journal Nature to publish a short paper submitted by Cherenkov, summarizing the essence of the phenomena and its main properties. "Receiving a negative answer from them and, fortunately, also the article, I consulted Vavilov and right then transferred it to another envelope and sent it to the editors of the American journal *Physical Review*. They were more attentive to it and, rather soon, published it in the next issue of the journal" (Cherenkov, 1986; Cherenkov, 1937). The letter from the editor of *Nature* is shown in Figure 1.

New experiments by Cherenkov showed that the effect was not due to fluorescence, e.g. measurements of intensity after heating and by adding substances that quench the fluorescence. None of the measurements showed any change in the intensity. Furthermore, he showed that the light is partially polarized. All together this meant that the effect was new and that it was not related to fluorescence.

Figure 1. The letter from the editor of Nature refusing to accept Cherenkov's original work on Cherenkov light; the editor rejected the paper, which later awarded Cherenkov the Noble Prize in Physics. Instead Cherenkov published his results in Physical Review which immediately accepted it.
From Cherenkov, 1986.

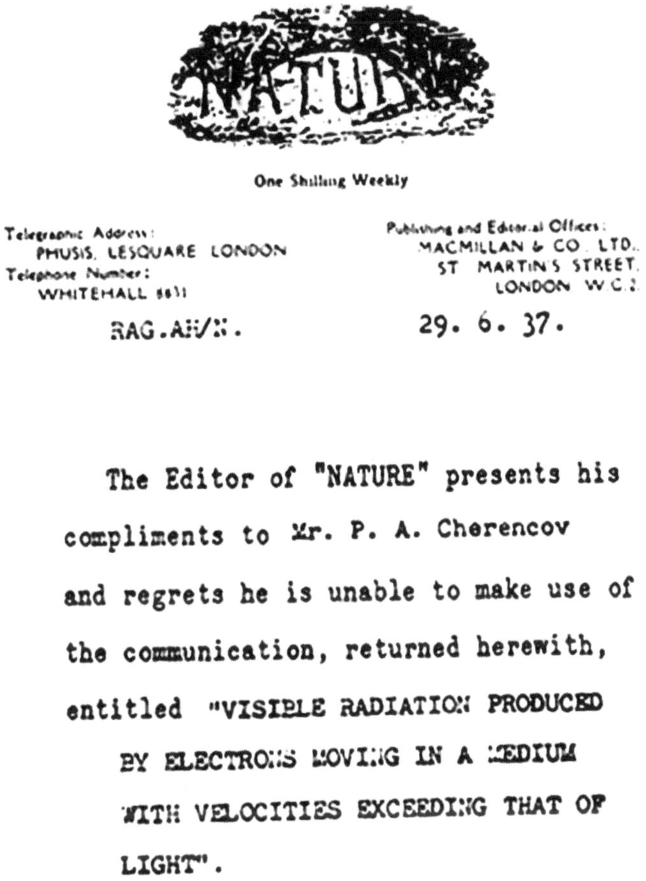

NATURE

One Shilling Weekly

Telegraphic Address:
PHUSIS, LESQUARE LONDON
Telephone Number:
WHITEHALL 8831

Publishing and Editorial Offices:
MACMILLAN & CO. LTD.,
ST. MARTIN'S STREET,
LONDON, W.C.2

RAG.AH/M.

29. 6. 37.

The Editor of "NATURE" presents his compliments to Mr. P. A. Cherencov and regrets he is unable to make use of the communication, returned herewith, entitled "VISIBLE RADIATION PRODUCED BY ELECTRONS MOVING IN A MEDIUM WITH VELOCITIES EXCEEDING THAT OF LIGHT".

The apparatus used by Cherenkov is shown in Figure 2 (Cherenkov, 1934).

The liquid was placed in a platinum crucible placed in a wooden block. A platinum vessel was chosen to eliminate background light from the vessel itself. The collimated gamma source was placed either below or beside the container. The luminescence could be studied through the optical system. A graded wedge (K) was used for relative intensity measurements and a filter (F) for crude spectral analysis. A Nichol prism was used to study the polarization when the gamma source was placed beside the container. He found a small effect of about 20% partial polarization where the electric vector of the field is parallel to the propagation direction of the gamma ray.

The fact that the effect is caused by Compton electrons, and not by the gamma rays, was proved later in a series of experiments where the source finally was exchanged for a β source. During these experiments he also made the very important discovery that the light is not emitted isotropic ally, but at a specific angle. This lead Tamm to the idea that the light does not originate from the collisions between the electrons and the molecules, but from the free path between the collisions (Frank and Tamm, 1937).

Figure 2. The first experimental set-up used by Cherenkov in 1934; Cherenkov placed a radioactive source under or beside a container of liquid solution. Through an optical system he observed the Cherenkov light produced. In this way he could make spectral, spatial and polarizing studies of the mysterious light. From Ekspong, 1958.

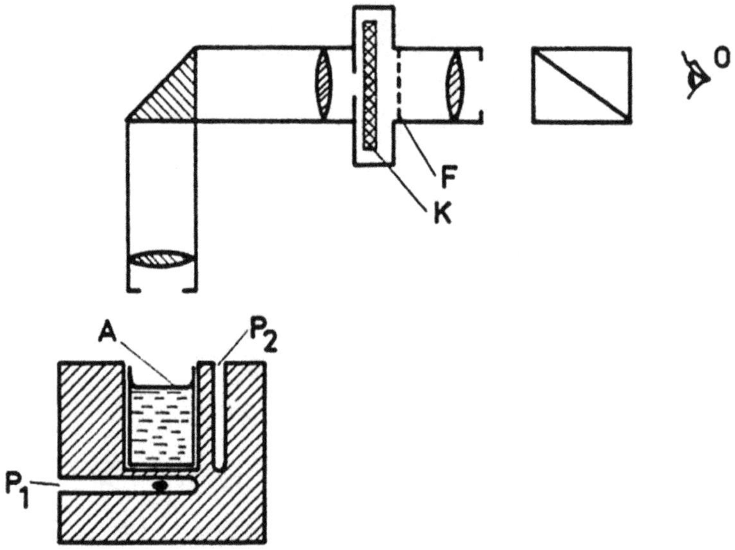

Cherenkov and Vavilov could not explain the new effect theoretically themselves. The theoretical interpretation of the effect is due to the two Russian theoreticians Ilja Michajlovitj Frank and Igor Jevgenevitj Tamm who presented a paper together in 1937 (Frank and Tamm, 1937). The explanation they gave is based on the classical theory of electrodynamics expressed in Maxwell's equations. The theory predicted new properties of the effect which Cherenkov verified experimentally, such as the absolute intensity, the spectral distribution and the angle of emission.

A refined theoretical explanation was presented in 1940 by Ginsburg (Ginsburg, 1940; Lennander, 1958), who produced a quantum theory of the phenomenon which was henceforth known as Cherenkov radiation.

In 1958 Cherenkov, Frank and Tamm were awarded the Nobel Prize in physics for their discovery.

3. GENERAL THEORY

3.1 Introduction

It is a well-known fact that electrically charged particles which travel with a constant velocity in vacuum cannot emit electromagnetic radiation. A particle cannot travel faster than the speed of light in vacuum, c, and therefore it cannot "run away" from the electromagnetic field. This was such a well-established fact that it delayed the theoretical prediction of the Cherenkov radiation until Cherenkov had started his experiments and Frank and Tamm could explain it through classical electrodynamics.

If the particle is not traveling in vacuum, but in a medium, the particle can indeed travel with a velocity larger than the speed of light in the medium, $v=c/n$, where n is the refractive index of the medium.

This is in some ways an analogous phenomenon to the V shaped shock wave from supersonic airplanes. When the airplane travels faster than the speed of sound in air, it loses energy because of the emission of shock waves. Cherenkov radiation can therefore in a trivial way be considered as a bow wave from a charged particle, traveling at a speed faster than the speed of light in the medium and emitting light while losing energy. (The analogy stops here since the physical laws behind the emission and the propagation of the two waves are completely different).

Suppose, for instance, an electron to be moving relatively slowly through a piece of glass or other transparent medium. Figure 3a) shows a section of the glass in the vicinity of the track of this electron, the circles representing the individual atoms composing the glass. Normally these will be roughly spherical in shape, and undisturbed.

However, in the region close to the passing electron, which at a particular instant in time is at the point P, the electric field of the particle distorts the atoms so that the negative charges of the electrons are displaced to one side of the heavier positive charges of the nuclei of these atoms. The medium thus becomes polarized about the point P. When now the electrons moves on to another point, say P', the elongated atoms around P return to their normal shape. While the atoms are distorted they behave like elementary dipoles, with the negative poles pointing away from the track if the passing particle is a negative electron, or vice versa for a positron or proton. Thus, as the particle passes through the medium, each elemental region of the glass along the track will in turn receive a very brief electromagnetic pulse. Owing to the complete symmetry of the polarization field surrounding the electron, there will be no resultant field at large distances and therefore no radiation. There is symmetry in both azimuth and along the axis, in this case.

Figure 3. Polarization effect when a charged particle traverses a dielectric media at: a) low speed, b) high speed; normally the charged particles polarize the surrounding neutral atoms in a symmetric way. It, however, the particle is moving close to the speed of light in the medium a residual polarization exist which can interact coherently and emit Cherenkov light when the atoms recombine.
From Séguinot, 1989.

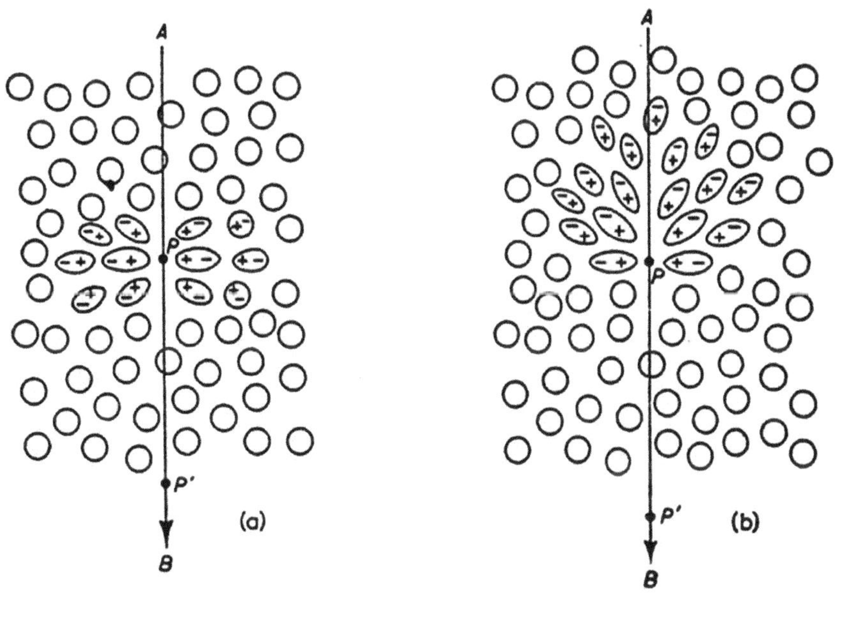

If however the electron is moving fast, that is at a speed comparable to that of light in the medium, the picture is quite different, as shown in Figure 3b). In this case the polarization field is no longer completely symmetrical. In a plane perpendicular to the trajectory of the particle, symmetry is preserved, but along the axis there is a resultant dipole field which will be apparent even at large distances from the track of the electron. Such a field will be momentarily set up by the electron at each element along the track in turn, each element then radiating a brief electromagnetic pulse.

In the general case, the radiated wavelets from all parts of the track interfere destructively so that, at a distant point, the resultant field intensity is still zero. However, if the velocity of the particle is higher than the phase velocity of the light in the medium, it is possible for the wavelets from all portions of the track to be in phase with one another so that, at a distant point of observation, there is now a resultant field.

It will be understood from the Huygen construction shown in Figure 4 that this radiation is only observed at a particular angleΘ with respect to the track of the particle, namely that angle at which the wavelets from arbitrary points on the track are coherent and combine to form a plane wave front. This coherence takes place when the particle traverses *AB* in the same time that the light travels from A to C. If the velocity of the particle is βc where c is the velocity of light in vacuum, in a time t the particle will travel a distance $AB = \beta ct$, and the light a distance $AC = tc/n$, where n is the refractive index of the medium. From this we obtain:

$$\cos\theta = \frac{1}{\beta n} \tag{1}$$

which is known as the "Cherenkov relation".

Figure 4. Huygen's construction to illustrate the coherence of wavelets forming a plane wave front when the velocity of the particle is larger than c/n; when the velocity is lower than c/n the wave front interfere destructively and no residual field exists far away from the particle. When the velocity is larger than c/n the wave front interferes constructively and a residual wave front is formed at a characteristic angle which depends on the velocity of the particle.
From Séguinot, 1989.

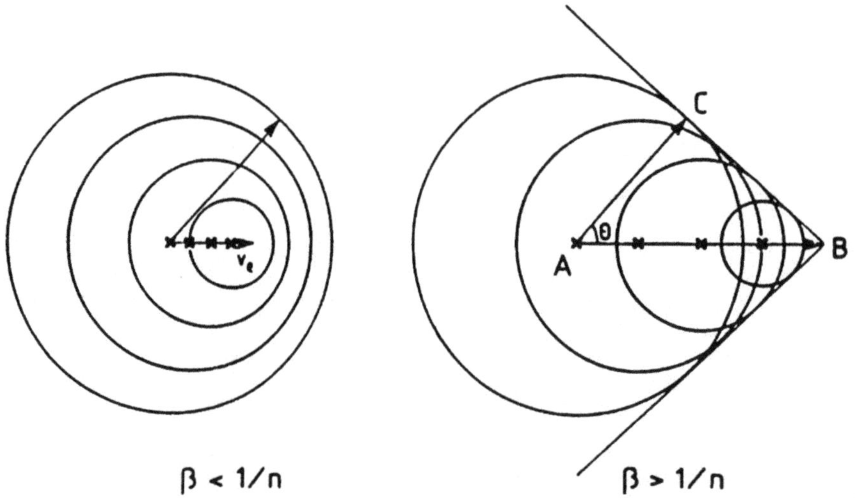

It is seen that:

1. For a medium of given refractive index n, there is a threshold velocity $\beta_{min} = 1/n$, below which no radiation takes place. At this crucial velocity the direction of radiation coincides with that of the particle.
2. For an ultra-relativistic particle, for which β-> 1, there is a maximum angle of emission, given by $\Theta_{max} = \cos^{-1}(1/n)$.
3. The radiation occurs mainly in the visible and near-visible regions of the spectrum, for which $n>1$. Emission in the X-ray region is impossible since n is less than unity and (1) cannot be satisfied.

Figure 4 has been drawn only in one plane. There is of course complete symmetry about the axis of the particle. The light originating from each element around the track is propagated along the surface of a cone whose apex is at this element, whose axis coincides with the track, and whose semi-vertical angle is the Cherenkov angle Θ. The distribution in Θ of the light intensity approximates to a δ-function, and the polarization is such that the electric vector, E is everywhere perpendicular to the surface of the cone, and the magnetic vector H tangential to this surface.

It will be realized that there are two further conditions to be fulfilled to achieve coherence, in addition to those stated previously. First, the length L of the track of the particle in the medium must be large compared with the wavelength, λ, of the radiation in question, otherwise diffraction effects will become dominant and the light distributed over an angle $\delta\Theta \sim (\lambda/L)\sin\Theta$, instead of appearing at only one angle Θ as in (1). Secondly, the velocity of the particle must be constant during its passage through the medium, or, to be more specific, the differences in the times for the particle to traverse successive distances λ, must be small compared with the period λ/c of the emitted wave.

It is appropriate here to emphasize that this radiation phenomenon should not be confused with either recombination or excitation radiation associated with ionization caused by the particle. Neither should it be confused with Bremsstrahlung, which is an acceleration radiation produced if the particle enters the sphere of influence of the electromagnetic field of an atomic nucleus (or electron). The two types of radiation have completely different spectral distributions (Bremsstrahlung dominates at very short wavelengths) and different angular distributions. Furthermore, Bremsstrahlung depends so strongly on the mass of the particle that it is only significant for electrons, while the Cherenkov radiation is independent of the mass of the particle.

3.2 The Classical Theory of Frank and Tamm

In 1937 Frank and Tamm presented a paper together (Frank and Tamm, 1937) where for the first time a theoretical explanation of the Cherenkov effect was given.

In their original treatment of the problem of radiation of an electron moving uniformly in a dielectric medium, Frank and Tamm made the following simplifying assumptions which have been summarized by Jelley (Jelley 1958):

1. The medium is considered as a continuum, so that the microscopic structure is ignored; the dielectric constant ε_r is then the only parameter used to describe the behavior of the medium.
2. Bremsstrahlung is neglected.

3. The medium is assumed to be a perfect isotropic dielectric, so that the conductivity is zero, the magnetic permeability μ_r=1, and there is no absorption of radiation.
4. The electron is assumed to move at a constant velocity; i.e. any slowing down due to ionization and multiple scattering is ignored.
5. The medium is unbounded and the track length infinite.

The theory for the Cherenkov effect is based on Maxwell's equations for the electromagnetic field, with the above assumptions. The medium is described by the permittivity $\varepsilon_r\varepsilon_0$ or the refractive index $n = \sqrt{\varepsilon_r}$.

All field variables (*E*, *H*, *j*, etc.) are composed of several frequency components $E_w d\omega$, $H\omega d\omega$, $j\omega d\omega$ etc. The sum of these components, the Fourier integral, forms each field (in a specific point, at a specific time):

$$E = \int_{-\infty}^{\infty} E_{\omega} e^{i\omega t} d\omega, H = \int_{-\infty}^{\infty} H_{\omega} e^{i\omega t} d\omega, etc. \tag{2}$$

The ω-dependent components are inversely obtained from the field variables as:

$$E_{\omega} = \frac{1}{2\pi} \int_{-\infty}^{\infty} E e^{-i\omega t} dt, H_{\omega} = \frac{1}{2\pi} \int_{-\infty}^{\infty} H e^{-i\omega t} dt, etc. \tag{3}$$

It is necessary to introduce the following connections between the field variables with the above assumptions:

$$D_{\omega} = \varepsilon_r \varepsilon_0 E_{\omega} \tag{4}$$

$$H_{\omega} = \frac{B_{\omega}}{\mu_0} \tag{5}$$

where *E* and *H* are the electric and magnetic field vector, *D* the electric displacement vector and *B* the magnetic flux density vector. The Maxwell equations for each frequency component in the medium are then:

$$\begin{aligned} divH_{\omega} &= 0 \\ curlH_{\omega} &= j_{\omega} + i\omega D_{\omega} \\ divD_{\omega} &= \rho_{\omega} \\ curlE_{\omega} &= -i\omega \mu_0 H\omega \end{aligned} \tag{6}$$

where *j* is the current density, and ρ is the density of free charges. Re-writing these equations as a function of the vector and scalar potentials *A* and ϕ leads to the following set:

$$H_{\omega} = \frac{1}{\mu_0} curl A_{\omega}$$
$$\nabla^2 A_{\omega} + \omega^2 \mu_0 \varepsilon_r \varepsilon_0 A_{\omega} = -\mu_0 j_{\omega}$$
$$\nabla^2 \varphi_{\omega} + \omega^2 \mu_0 \varepsilon_r \varepsilon_0 \varphi = -\frac{\rho_{\omega}}{\varepsilon_r \varepsilon_0}$$
$$E_{\omega} = -i\omega A_{\omega} - grad \varphi_{\omega} \tag{7}$$

where use have been made of the Lorentz condition between the vector and the scalar potentials:

$$div A_{\omega} + i\omega \mu_0 \varepsilon_r \varepsilon_0 \varphi_{\omega} = 0 \tag{8}$$

The current density j and the charge density ρ depend on the charge q, velocity v, and the direction of the particle. If an electron of charge e is moving through a medium along the z-axis with a constant velocity v, the corresponding charge and current density is equal to:

$$\rho = q\delta(x)\delta(y)\delta(z - vt) \tag{9}$$

$$j_x = 0$$
$$j_y = 0$$
$$j_z = ev\delta(x)\delta(y)\delta(z - vt) \tag{10}$$

where δ denotes Dirac's delta function. Expanding j_z into its Fourier components using (3) one gets:

$$j_z(\omega) = \frac{e}{2\pi} e^{-\frac{i\omega z}{v}} \delta(x)\delta(y) \tag{11}$$

(here the relation $\delta(z\text{-}vt) = 1/IvI\ \delta(z/v\text{-}t)$ has been used).

Introducing cylindrical coordinates ρ,ϕ and z, because of the cylindrical symmetry of the problem, this relation transforms into:

$$j_z(\omega) = \frac{e}{4\pi^2 \rho} e^{-\frac{i\omega z}{v}} \delta(\rho) \tag{12}$$

Introducing this expression in (7, 2nd equation), and putting $A_{\rho} = A_{\phi} = 0$, $A_z(\omega) = u(\rho)\ e^{-iwz/v}$ results in:

$$\frac{\partial^2 u}{\partial \rho^2} + \frac{1}{\rho}\frac{\delta u}{\delta \rho} + s^2 u = -\frac{e\mu_0}{4\pi^2 \rho}\delta(\rho) \tag{13}$$

where

$$s^2 = \frac{\omega^2}{v^2}\left(\beta^2 n^2 - 1\right) = -\sigma^2; \beta = \frac{v}{c}$$
$$\left(\varepsilon_r \varepsilon_0 \mu_0 = \frac{n^2}{c^2} since\ n = \sqrt{\varepsilon_r} and\ c = \frac{1}{\sqrt{\varepsilon_0 \mu_0}}\right) \tag{14}$$

Thus u is a cylindrical function satisfying the Bessel equation:

$$\frac{\partial^2 u}{\partial \rho^2} + \frac{1}{\rho}\frac{\delta u}{\delta \rho} + s^2 u = 0 \tag{15}$$

everywhere with the exception of the pole ρ = 0. To find the condition to be satisfied by u at ρ = 0, first replace the right hand side of (13) by f, such that:

$$f = \left\{ -\frac{2e\mu_0}{4\pi^2 \rho_0^2}, \quad if\ \rho < \rho_0 \right.$$
$$f = 0, \qquad if\ \rho > \rho_0 \tag{16}$$

then integrate (13) over the surface of a disc of radius ρ_0:

$$\int_0^{2\pi}\int_0^{\rho_0} \frac{\partial^2 u}{\partial \rho^2} \rho d\rho d\varphi + \int_0^{2\pi}\int_0^{\rho_0} \frac{1}{\rho}\frac{\delta u}{\delta \rho} \rho d\rho d\varphi + \int_0^{2\pi}\int_0^{\rho_0} s^2 u \rho d\rho d\varphi = \int_0^{2\pi}\int_0^{\rho_0} f \rho d\rho d\varphi \tag{17}$$

The ϕ dependence is only a constant for all terms and vanishes. The remaining terms look as follows:

$$\left[\frac{\delta u}{\delta \rho}\rho\right]_0^{\rho_0} - \int_0^{\rho_0} \frac{\delta u}{\delta \rho} d\rho + \int_0^{\rho_0} \frac{\delta u}{\delta \rho} d\rho + s^2 \left\{ \left[U\rho\right]_0^{\rho_0} - \int_0^{\rho_0} U d\rho \right\} = -\frac{2e\mu_0}{4\pi^2 \rho_0^2} \int_0^{\rho_0} \rho d\rho \tag{18}$$

Approaching the limit ρ_0 -> 0 all terms but the first vanish:

$$\lim_{\rho \to 0} \rho \frac{\delta u}{\delta \rho} = -\frac{e\mu_0}{4\pi^2} \tag{19}$$

Let us now distinguish between two different cases:

- **At Low Velocities:** First consider the case of an electron moving at low speed such that $\beta n < 1$, $s^2 < 0$ and $\sigma^2 = -s^2 > 0$, σ thus being a real quantity. In this case the solution of the homogeneous Equation (15), satisfying the pole ρ = 0 (17) and vanishing at infinity, is:

$$u(\rho) = \frac{ie\mu_0}{8\pi} H_0^{(1)}(i\sigma\rho) \tag{20}$$

where $H_0^{(1)} = J_0 + iY_0$ is the Hankel function of the first kind.

If $\sigma\rho >> 1$, which is the case for macroscopic distances, one can use the asymptotic values of $H_0^{(1)}$:

$$H_0^{(1)}(x) = J_0(x) + iY_0(x) \sim \sqrt{\frac{2}{\pi x}} e^{i\left(x - \frac{\pi}{4}\right)} \tag{21}$$

Using the asymptotic value the vector potential $A_z(\omega)$ is obtained according to 18):

$$A_z(\omega) = \frac{e\mu_0}{4\pi} \frac{1}{\sqrt{2\pi\sigma\rho}} e^{-\sigma\rho} e^{-i\omega z/v}; \left(\sigma\rho >> 1\right) \tag{22}$$

Thus in the case of small velocities, the field of the electron decreases exponentially with ρ, so that at a point remote from the track of the electron, no radiation is observed.

- **At High Velocities:** If however the velocity of the electron is so large that within a certain frequency range $\beta n = (v/c)\, n(\omega)$ is greater than 1, the parameter s in (14) is real and the general solutions of Equations (15) and (19) represents, at infinity, a cylindrical wave. Requiring u in this case to represent a outgoing and not ingoing wave, the solution of (15) satisfying (19) becomes:

$$u(\rho) = -\frac{ie\mu_0}{8\pi} H_0^{(2)}(s\rho); if\ \omega > 0 \tag{23}$$

and a complex conjugate expression if $\omega < 0$, s being assumed to be positive. Here $H_0^{(2)} = J_0 - iY_0$ is the Hankel function of the second kind. At macroscopic distances from the track, where $s\rho >> 1$, one can use the asymptotic value of $H_0^{(2)}$:

$$H_0^{(2)}(x) = J_0(x) - iY_0(x) \sim \sqrt{\frac{2}{\pi x}} e^{-i(x - \pi/4)} \tag{24}$$

Then the vector potential $Az(w)$ becomes, according to (21):

$$A_z(\omega) = -\frac{e\mu_0}{4\pi} \frac{1}{\sqrt{2\pi s\rho}} \exp\left(-\frac{i\omega z}{v} - i\left(s\rho - \frac{3\pi}{4}\right)\right); \left(\omega > 0, s\rho >> 1\right) \tag{25}$$

With the help of (14) the exponent can be transformed as follows:

$$A_z(\omega) = -\frac{e\mu_0}{4\pi}\frac{1}{\sqrt{2\pi s\rho}}\exp\left(-i\omega\frac{z\cos\theta+\rho\sin\theta}{c/n}+i\frac{3\pi}{4}\right) \tag{26}$$

where the angle θ is defined by the Cherenkov relation $\cos\theta = 1/\beta n$. Thus, if $\beta n > 1$ a wave is propagated to infinity along the direction θ.

Calculating the field intensity in the wave zone with the aid of (7) and (8), and re-transforming into the real fields, through (2), there are only three non-vanishing components of the field vectors:

$$\begin{aligned} H_\varphi &= -\frac{\alpha}{\mu_0\sqrt{\rho}}\int_{\beta n>1}\frac{1}{\sqrt{s}}\sqrt{\frac{1}{4\rho^2}+s^2}\cos\left(\chi+\arctan(2\rho s)\right)d\omega \\ &\approx -\frac{\alpha}{\mu_0\sqrt{\rho}}\int_{\beta n>1}\sqrt{s}\cos\left(\chi+\frac{\pi}{2}\right)d\omega;(s \gg 1/\rho) \end{aligned} \tag{27}$$

$$\begin{aligned} E_\rho &= -\frac{\alpha}{\sqrt{\rho}}\int_{\beta n>1}\frac{1}{\sqrt{s}}\frac{c}{\beta n^2}\sqrt{\frac{1}{4\rho^2}+s^2}\cos\left(\chi+\arctan(2\rho s)\right)d\omega \\ &\approx -\frac{\alpha}{\sqrt{\rho}}\int_{\beta n>1}\frac{1}{\sqrt{s}}\frac{\sqrt{\beta^2 n^2-1}}{\beta^2 n^2}\cos\left(\chi+\frac{\pi}{2}\right)\omega d\omega;\left(s \gg \frac{1}{\rho}\right) \end{aligned} \tag{28}$$

$$E_z = -\frac{\alpha}{\sqrt{\rho}}\int_{\beta n>1}\frac{1}{\sqrt{s}}\left(1-\frac{1}{\beta^2 n^2}\right)\cos\left(\chi-\frac{\pi}{2}\right)\omega d\omega \tag{29}$$

where

$$\alpha = \frac{e\mu_0}{8\pi}\sqrt{\frac{2}{\pi}} \;\; and \;\; \chi = \omega\left[t-\frac{z\cos\theta+\rho\sin\theta}{c/n}\right]+\frac{3\pi}{4} \tag{30}$$

The integration have to be carried out only over values of ω for which $\beta n(\omega) > 1$.

The electric vector E of the wave lies in the meridian plane (z,ρ), as shown in Figure 5. The magnetic field vector H is in the ϕ direction, i.e. tangential to the circular cut perpendicular to the direction of motion of the particle, meaning that the light propagates in the direction ExH at an angle θ to the track of the particle.

Figure 5. The polarization vectors of the Cherenkov light; the magnetic field vector H is in the ϕ direction, i.e. tangential to the circular cut perpendicular to the direction of motion of the particle, meaning that the light propagates in the direction ExH at an angle θ to the track of the particle.
From Séguinot, 1989.

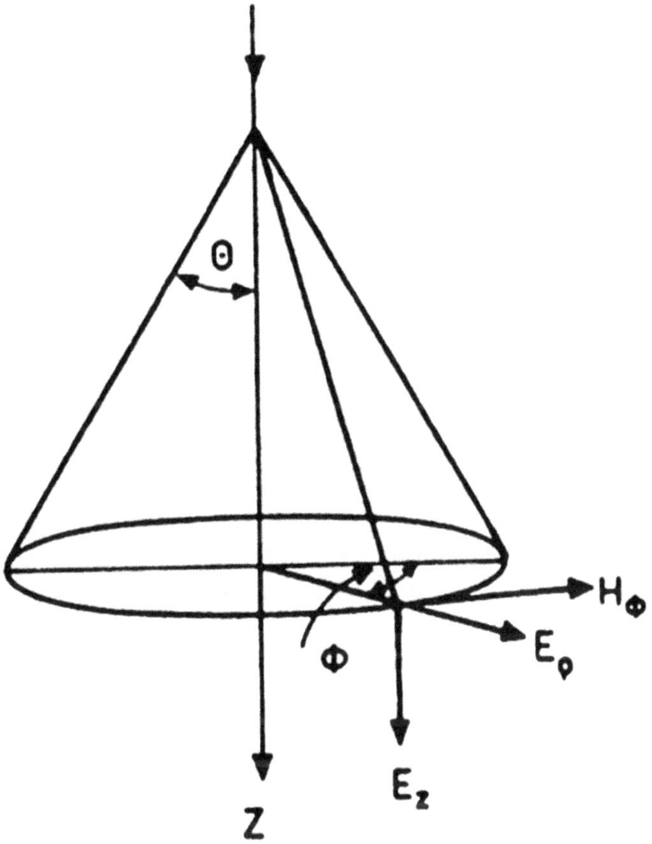

The total energy *W* radiated by the electron through the surface of a cylinder of length l whose axis coincides with the line of motion of the electron track, is equal to:

$$W = 2\pi\rho l \int_{-\infty}^{\infty} \left[E \times H\right]_{\rho} dt \tag{31}$$

With the aid of the formula:

$$\int_{-\infty}^{\infty} \cos(\omega t + \alpha)\cos(\omega' t + \beta) dt = \pi\delta(\omega - \omega') \tag{32}$$

one obtains the fundamental equation for the output of radiation:

$$\frac{dW}{dl} = \frac{Z^2 e^2}{16\pi^2 \varepsilon_0 c^2} \int_{\beta n>1} \left(1 - \frac{1}{\beta^2 n^2}\right) \omega d\omega \tag{33}$$

Here the charge Z is added to show the dependence of the charge of the particle.

A real medium is always dispersive, so that radiation is restricted to those frequency bands for which $n(\omega) > 1/\beta$. Media which are transparent at visible wavelength normally have $n > 1$ only in the near ultra-violet and longer wavelengths, and there may even be absorption bands in the infra-red. In the X-ray region $n(\omega)$ is always less than 1 and radiation is therefore forbidden.

The number of photons emitted is deduced from (33), using the relation between photon energy (E) and wavelength (λ), $E = hc/\lambda$, to be:

$$\frac{dN}{dl} = 2\pi Z^2 \alpha \int_{\beta n>1} \left(1 - \frac{1}{\beta^2 n^2(\lambda)}\right) \frac{1}{\lambda^2} d\lambda \tag{34}$$

where λ is the wavelength of the light ($\lambda = c/v$), and α is the fine structure constant, $2\pi e^2/4\pi\varepsilon_0 hc = 1/137.04$.

Consider, for instance, a high energy electron moving at $\beta = 1$ through 10 mm of water. In the visible region of the spectrum between 400 nm and 700 nm, water has a relatively constant refractive index of $n = 1.33$. In this region 200 photons are emitted, corresponding to a loss of 500 eV kinetic energy. This is negligible in comparison to the 2 MeV the particle loses through ionization of the water during the passage.

From (33) and (34) one can draw some important conclusions:

1. The energy loss is in the order of keV/cm.
2. The radiation of light depends on the medium only through the index of refraction n.
3. The emission is proportional to the square of the charge of the particle, but independent of the mass of the particle.
4. The number of photons emitted at a wavelength λ is proportional to $d\lambda/\lambda^2$. Thus, the Cherenkov light is mainly emitted in the ultra-violet region of the spectrum.
5. Cherenkov radiation appears as a continuous spectrum.
6. The number of emitted photons (per centimeter) in the visible region (λ: 400 -> 700 nm) of the spectrum is approximately $N = 500\, Z^2 \sin^2\Theta$, where Z is the charge of the incoming particle.

Note also that the light is emitted in extremely short pulses and the radiator does not have any after-glow, which allows very high counting rate. For a 10 mm thick radiator the pulse is only 30 ps long, i.e. the time it takes the particle to go through the radiator.

It is interesting to note that Sommerfeld in 1904 obtained an expression very similar to (1.2) for the radiation losses from an electron moving in vacuum with a constant velocity $v>c$. This of course was prior to the establishment of the theory of relativity which states that the condition v>c can never be realized.

Even more interesting to note is that the British scientist Oliver Heaviside in November 1888 published a paper where he deduced the Cherenkov effect theoretically from Maxwell's equations (Heaviside, 1888). This was long before the discoveries of cosmic rays and radioactivity, and explains why neither he, nor anyone else, could verify this effect experimentally. This paper was then forgotten until Kaisser in 1974 in Nature published the extract shown in Figure 6 (Kaisser, 1974).

When Maxwell died in 1879, no one had fully understood his new theory of electromagnetic phenomena nor was it presented in such a way that is was easily understandable. Heaviside was the first to present the theory in an understandable way when he eliminated all unnecessary variables, which complicated Maxwell's system of equations. Hertz later arrived at the same form of these equations. It is in this form that Maxwell's theory has been generally understood. During this work Heaviside learned a deeper insight into the theory than anyone else in those days, and among other things deduced the Cherenkov effect. This was even before Hertz made his epoch-making experiments demonstrating the existence of electromagnetic waves.

Figure 6. The original work of Heaviside in 1888 to investigate the effects of a charge moving through a dielectric medium with a speed of motion exceeding that of light; Heaviside actually predicted Cherenkov light emission from particles moving through a medium at velocities faster than the speed of light in the medium long before Einstein postulated that nothing can move faster than the speed of light in vacuum and the discovery of radioactivity. It was a purely theoretical interpretation of the Maxwell's original equations which Heaviside rewrote in a more understandable manner.
From Kaisser, 1974.

Returning to the case of a charge q at a point moving through a dielectric, if the speed of motion exceeds that of light, the disturbances are wholly left behind the charge, and are confined within a cone, AqB. The charge is at the apex, moving from left to right along Cq. The semi-angle, θ, of the cone, or the angle AqC, is given by

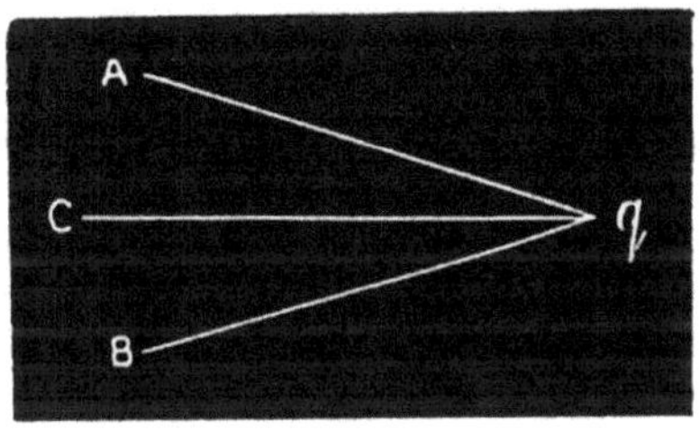

$$\sin \theta = v/u,$$

where v is the speed of light, and u that of the charge. The magnetic lines are circles round the axis, or line of motion. The displacement is away from q, of course, and of total amount q, but not uniformly distributed within the cone. The electric current is towards q in the inner part of the cone, and away from q in the outer.

It will be seen that the electric stress tends to pull the charge back. Therefore, applied force on q in direction Cq is required to keep up the motion. Its activity is accounted for by the continuous addition at a uniform rate which is being made to the electric and magnetic energies at q. For the motion at the wave-front, at any point on Aq or Bq, is perpendicularly outward, not towards q. Whilst the cone is thus expanding all over, the forward motion of q continually renews the apex, and keeps the shape unchanged.

Steady motion alone is assumed.

To avoid misconception I should remark that this is not in any way an account of what would happen if a charge were impelled to move through the ether at a speed several times that of light, about which I know nothing; but an account of what would happen if Maxwell's theory of the dielectric kept true under the circumstances, and if I have not misinterpreted it. [See footnote on p. 516, later.]

Nov. 18, 1888.

3.3 Corrections Introduced by Quantum Mechanics

In the classical theory of Cherenkov radiation discussed above, effects due to the reaction of the emitted radiation on the motion of the particle have been ignored. When these effects are taken into account, the expressions for the radiation condition and the energy production have forms which differ slightly from those obtained in the original theory. The corrections are naturally expected to be small, since the energy of the radiated quanta is small compared to the kinetic energy of the particle.

It is possible to arrive at the radiation condition by arguments about the conservation of momentum and energy, in a manner similar to that used in discussing the Compton effect.

Let a charged particle of rest mass m_0 be traveling through the medium at a constant velocity u. Assume now that at some stage along its track it emits a photon of energy hν at an angle Θ with respect to the original direction of the particle (see Figure 7). The latter will experience an instantaneous loss of energy and it will be assumed that it now travels with a velocity v at some angle $\phi$with respect to its original direction.

Conservation of momentum and energy leads to the expression:

$$\cos\theta = \frac{1}{\beta n} + \left(\frac{\Lambda}{\lambda}\right)\left(\frac{n^2-1}{2n^2}\right) \tag{35}$$

where Λ is the de Broglie wavelength of the particle:

$$\Lambda = \frac{h\sqrt{1-\beta^2}}{m_0 u} = \frac{\sqrt{1-\beta^2}}{\beta}\lambda_0 \tag{36}$$

λ_0 is the Compton wavelength (0.0024 nm for electrons).

Figure 7. Reaction effects on a charged particle emitting a photon; the emission of Cherenkov photons will cause the particle to deflect. The deflection is however small due to the small energy loss in Cherenkov light emission.
From Francke, 1991.

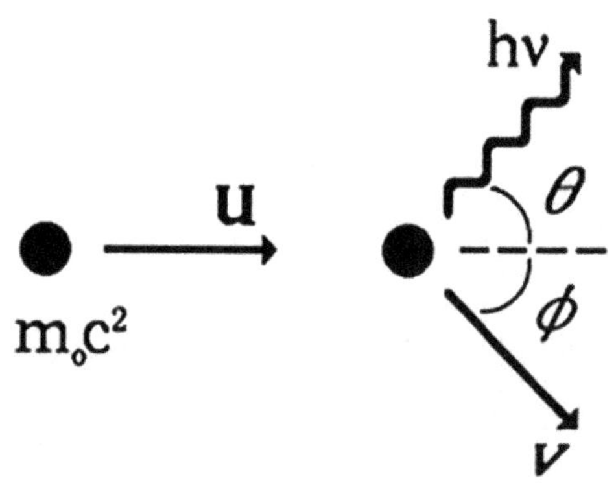

The second term in (1.4) differs from the classical radiation condition, $\cos\Theta = 1/\beta n$, and is extremely small in any practical case. For example, for electrons at the threshold in water (β =0.75), Λ is 0.0021 nm and the second term, $(\Lambda/\lambda)\,((n^2\text{-}1)/2n^2)$, is of the order of 10^{-6}.

Note that λ in (1.4) is the wavelength *in the medium*. The validity of these equations rests on the assumption that the momentum of a photon in the medium is hνn/c and not hν/c. (Normally it is expressed as the wavelength of the photon changes when it enters a medium with different index of refraction). While the medium has been assumed to modify the momentum of the photon, no such modification has been made for the particle. The justification for this is that the photon wavelength (in the optical region) is so large compared with interatomic distances that the radiation is influenced by the macroscopic properties of the medium, whereas the particle, owing to its small de Broglie wavelength, finds itself traveling most of the time through empty space, its dimensions being small compared with these interatomic distances. It is for this reason that the ordinary relativistic equations for momenta in free space are used to deduce (35).

Heitler discussed in 1944 the question about coherence in the quantum picture of Cherenkov radiation (Jelley, 1958). How are the phases of the individual quanta defined, so that the coherence of the Cherenkov light is preserved? His answer to this question is that in a light wave the uncertainty of the phase ϕ is related to a corresponding uncertainty of the number of quanta N in the wave, by the expression $\Delta N\,\Delta\phi \sim 1$. Thus for a number of quanta emitted from the track of the particle, the phases may be defined as closely as desired, in which case either the number of quanta in a given wavelet is entirely undetermined, or, the total intensity may be known but it is then not possible to say which quanta are associated with which part of the wave front.

In 1957, Sokolov and Loskutov showed theoretically that if the particle has spin, the polarization, the threshold condition and radiation yield are slightly modified (Jelley, 1958). A small fraction of circularly polarized light is added to the normally linearly polarized light. This contribution does not disappear at the threshold, whereas the linearly polarized portion of the radiation does. Although Harding and Henderson reported observations of nonpolarized light at the threshold in 1947, this cannot be attributed to spin effects since the effect is far weaker than what is detectable (Jelley, 1958).

4. CONCLUSION AND REMARKS

The physical effect, discovered by Cherenkov, leads to important practical devices allowing to detect and identify high energy charged particles. There are two main types of Cherenkov detectors: a threshold Cherenkov counter and a ring imaging (RICH) detector. The threshold counter records particles having velocity V_p greater that the speed of light in the give media, $V_p > c/n$.

By changing the media particles with various velocities can be selected.

The RICH detector is based on fact that angle of the emitting Cherenkov photons depends on the particle velocity:

$$\cos\theta = c/V_p n$$

By measuring the angle θ one can determine V_p and from the knowing some complimentary parameters, such as momentum, identify the particle. Both types of detectors are widely used in high-energy physics and astrophysics experiments. In more details these devices and the concrete applications will be described the following chapters.

REFERENCES

Cherenkov, P. A. (1934). The visible glow of pure liquids under action of γ-rays. *Soviet Physics, Doklady*, *2*, 451.

Cherenkov, P. A. (1937). Visible radiation produced by electrons moving in a medium with velocities exceeding that of light. *Physical Review*, *51*(4), 378–379. doi:10.1103/PhysRev.52.378

Cherenkov, P. A. (1986). At the threshold of discovery. *Nuclear Instruments and Methods*, *S248*(1), 1–4. doi:10.1016/0168-9002(86)90487-0

Ekspong, G. (1958). *Cherenkovstrålning,* Kosmos Fysiska Uppsatser, P.A. *Norstedt & Söner*, *36*, 6.

Francke, T. (1991). *A ring imaging Cherenkov counter using a solid radiator.* (PhD Thesis). Royal Institute of Technology, Stockholm, Sweden.

Frank, I. M., & Tamm, I. J. (1937). *Coherent Visible Radiation of Fast Electrons Passing through Matter*, Dokl. *Academie des Sciences de l'URSS*, *14*, 109.

Jelley, J. V. (1958). *Cherenkov radiation and its applications*. Pergamon Press.

Kaisser, T. R. (1974). Heaviside radiation. *Nature*, *247*(5440), 400–401. doi:10.1038/247400a0

Lennander, S. (1958). Nobelpriset i Fusik 1958, Kosmos Fysiska Uppsatser, P.A. *Norstedt & Söner*, *36*, 3.

Séguinot, J., *Les compteurs Cherenkov: applications et limites pour lídentification des particules. Dévelopments at perspectives.*, CERN preprint, CERN EP/89-92 (1989).

Tamm, I. (1959). *General Characteristics of Radiation Emitted by Systems Moving with Superlight Velocities with some Applications to Plasma Physics. In Les Prix Nobel en 1958* (p. 122). P.A. Norstedt & Söner.

Vavilov, S. I. (1934). On the possible cause of the blue γ glow in liquids. *Soviet Physics, Doklady*, *2*, 457.

ADDITIONAL READING

Landau, L. D., Liftshitz, E. M., & Pitaevskii, L. P. (1984). *Electrodynamics of Continuous Media"*. New York: Pergamon Press.

Smith, S. J., & Purcell, E. M. (1953). Visible Light from Localized Surface Charges Moving across a Grating". *Physical Review*, *92*(4), 1069. doi:10.1103/PhysRev.92.1069

KEY TERMS AND DEFINITIONS

A, Scalar Potential: A mathematical representation of a field on Euclidean space - time.

Bessel Equation: A differential equation often used for wave propagation in cylindrical or spherical coordinate systems.

Bremsstrahlung: Electromagnetic radiation produced by the deceleration of a charged particle when deflected by another charged particle, typically an electron by an atomic nucleus. The moving particle loses kinetic energy, which is converted into a photon, thus satisfying the law of conservation of energy.

Compton Wavelength of the Electron: The Compton wavelength of an electron is equivalent to the wavelength of a photon whose energy is the same as the rest-mass energy of the particle.

D, Electric Displacement Vector: A field that describes the flux density in a dielectric material caused by displacements of charges by an electric field.

de Broglie Wavelength: A quantum mechanical effect. The de Broglie wavelength of a particle is equivalent to the wavelength of a photon whose energy is the same as the rest-mass energy of the particle.

Dielectric Medium: A dielectric material (dielectric for short) is an electrical insulator that can be polarized by an applied electric field. When a dielectric is placed in an electric field, electric charges do not flow through the material as they do in a conductor, but only slightly shift from their average equilibrium positions causing dielectric polarization.

E, Electric Field Vector: A mathematical representation of an electric field generated by electric charge and time-varying magnetic fields.

Electromagnetic Field: An electromagnetic field (also EMF or EM field) is a physical field produced by electrically charged objects. It affects the behavior of charged objects in the vicinity of the field. The electromagnetic field extends indefinitely throughout space and describes the electromagnetic interaction. It is one of the four fundamental forces of nature.

Fourier Integral: A mathematical transformation from space coordinates to frequency coordinates.

H, Magnetic Field Vector: A mathematical representation of the magnetic field generated by electric charges in movement, or elementary particles.

Hankel Function: A special set of Bessel equations.

Maxwell Equations: Maxwell's equations describe how electric and magnetic fields are generated and altered by each other and by charges and currents.

Permittivity: ε, a measure of the resistance that is encountered when forming an electric field in a dielectric medium. $D = \varepsilon E$.

Polarization: The orientation of oscillations in the plane perpendicular to a transverse wave's direction of travel.

Relativistic Velocity: The velocity at which the kinetic energy of a particle cannot be considered to increase following Newton's equations of movement, but must be corrected by relativistic effects.

Speed of Light: The velocity of light in vacuum is 299,792,458 m/s. A velocity no matter or elementary particle can exceed.

β Source: A radioactive source emitting electrons and positrons.

δ, Dirac's Delta Function: A function which is zero everywhere except at the origin, where it is infinite, and which is also constrained to satisfy the identity that the integral over the delta function is unity.

γ Source: A radioactive source emitting gamma rays.

Chapter 10
The History of Cherenkov Detectors

ABSTRACT

As early as 1937 Pavel Alexejevitj Cherenkov wrote in one of his papers (Cherenkov, 1937), "...it is possible to pose the problem: for a given electron velocity find a liquid with such a refractive index, that the effect begins to show up. After some additional work this variant of experiments can be used for determination of electron velocities." Since then many different types of Cherenkov detectors have been successfully developed and used, without which particle physics would probably not have come as far as it has today.

1. CHERENKOV DETECTORS

All Cherenkov detectors consist of a radiator and some photosensitive device to detect the faint light. In the early experiments performed by Cherenkov the radiator was water, or some other liquid, and the photosensitive device either his eyes or a photographic emulsion.

The development of the photomultiplier tubes in the 1940's revolutionized the concept of particle physics, and made it possible to detect small light pulses electronically with much higher efficiency and in shorter times. It became possible to detect the Cherenkov light from a single particle.

Cherenkov detectors are normally designed to benefit from one or several of the following relations:

1. No light is emitted when the velocity$\beta < 1/n$.
2. Light is only emitted when the velocity$\beta > 1/n$.
3. Light is emitted in the forward direction.
4. $\cos \Theta_c = 1/\beta n$.
5. The number of emitted photons per meter radiator in the wavelength interval $(\lambda_{min}, \lambda_{max})$:

$$N \approx 2\pi\alpha Z^2 \left(1 - \frac{1}{\beta^2 n^2}\right)\left(\frac{1}{\lambda_{min}} - \frac{1}{\lambda_{max}}\right) \tag{1}$$

DOI: 10.4018/978-1-5225-0242-5.ch010

Normally Cherenkov counters are designed for some of these relations depending on the purpose of the detector. Threshold Cherenkov counters use a radiator with a refractive index chosen in such a way that one type of particle produce Cherenkov light while another does not. In this way they can be separated. Direction sensitive detectors use the property that light is emitted in the forward direction. Other detectors use the Cherenkov relation $\cos \Theta_c = 1/\beta n$ to detect light only in a specific angular interval. Finally, Ring Imaging Cherenkov (RICH) counters also use the Cherenkov relation to calculate the velocity of the particle from the Cherenkov angle Θ_c.

2. THE PIONEERS

The first really useful Cherenkov counter was developed and constructed by Jelley in 1951 (Jelley, 1951). It was direction sensitive and used the directional property of Cherenkov light. The construction is shown in Figure 1.

Figure 1. Jelley's direction sensitive Cherenkov detector from 1951; the aim of the detector was primarily to detect cosmic rays with a velocity high enough to produce Cherenkov light in water. It was also used to reject particles moving upwards instead of downwards. Only the light from particles traveling downwards will reach the PM tube as Cherenkov light is only emitted in the forward direction.
From Ekspong, 1958.

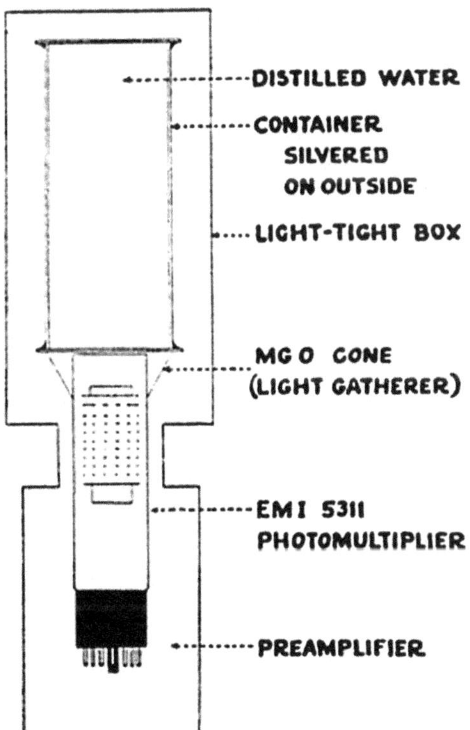

Figure 2. Cherenkov detector of the Getting type; the radiator is a cylindrical cone with top angle ϕ equal to the Cherenkov angle Θ_c. Only Cherenkov light emitted at an angle equal to the top angle of the cone will leave the base of the cone perpendicular to the surface which is required for the light to be focused by the lens through the collimator.
From Ekspong, 1958.

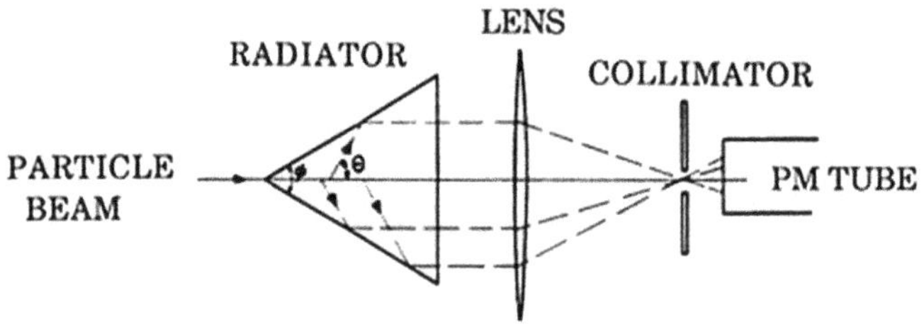

The radiator, distilled water, was contained in a cylinder, silvered on the inside and enclosed in a light-tight box. A photomultiplier tube was connected to one side with a light-collecting cone. Reflection at the other end was prevented by a black piece of paper. The counter was used to detect muons in the cosmic rays. With the counter placed vertically and the photomultiplier below the water radiator, it will register cosmic particles mainly coming from the zenith. Placed upside-down the light will not reach the PM tube.

The first Cherenkov detector used to detect single particles was designed by Getting at MIT in 1947 (Ekspong, 1958). His type of detector was used at the Birmingham proton-synchrotron in 1956 to detect950 MeV protons. The layout is shown in Figure 2.

The radiator was a cone of acrylic plastic (n = 1.5), 14 cm long. The cone was constructed to have the top angle equal to the Cherenkov angle. The light will then leave the cone perpendicular to the base. A lens focuses the parallel rays onto a small diaphragm which cuts away all light but that emitted at the desired angle. The light was then detected by a photomultiplier.

The most famous of the early Cherenkov detectors is probably the one designed and used by Owen Chamberlain, Emilio Segré, Thomas Ypsilantis and Clyde Wiegand when they discovered the antiproton at Berkeley, California, in 1955 (Chamberlain, 1955).

By 1954 the new Bevatron was ready and many physicists used it to hunt for the elusive antiproton that had been predicted by Dirac in 1930. In the beam used, the ratio between antiprotons and pions was 1:50 000, and the intensity of the beam was so low that only one antiproton per 1-5 minutes went through the detectors. The spectrometer included time of flight measurements over 12 m to separate the antiprotons from the pions, but because of the huge background it was considered necessary to have Cherenkov counters to veto accidental coincidences of two mesons. The schematic design of the Cherenkov counter is shown in Figure 3.

In the first experiment the Cherenkov radiator was a 6.3 cm long sample of fused quartz. A cylindrical mirror was arranged to reflect only light emitted in the right direction. This light was then focused within a small region where it could be detected by photomultiplier. The light was divided into three parts by three flat mirrors, to avoid having the beam of particles traverse the photomultiplier. This was the same design as had been proposed earlier by Samuel Lindenbaum (Chamberlain, 1960).

To further decrease the possibility of misidentification, a more conventional Cherenkov counter was also used. It was a container of liquid, a fluorocarbon with an index of refraction of 1.28, viewed directly

Figure 3. The Cherenkov detector used in the discovery of the antiproton in 1955; only particles in a limited velocity range produce light that reaches the PM tubes. The mirrors allow the PM-tubes to be placed outside of the particle beam.
From Chamberlain, 1960.

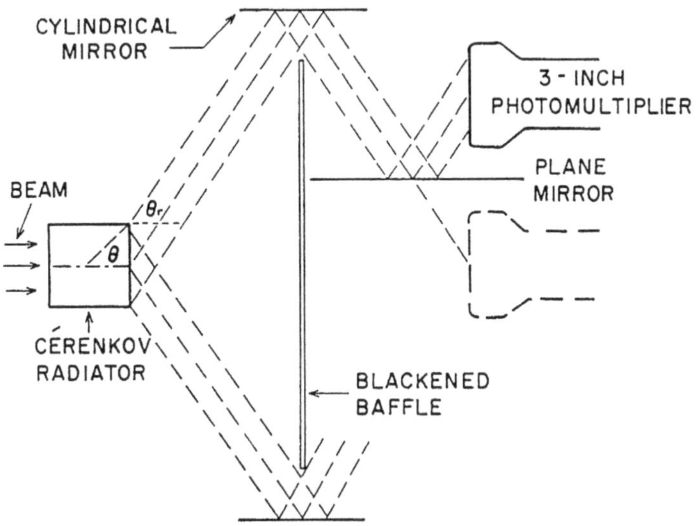

by a photomultiplier. Particles moving as slowly as the antiprotons would produce no Cherenkov light in this material. However, faster particles, such as pions and electrons, would produce Cherenkov light and would give rise to a pulse at the output of this counter.

The most copious production of antiprotons was expected to beat a velocity of 87% of the speed of light, while other background particles would be traveling at approximately 99% of the speed of light. To make the experiment somewhat easier, however, slightly slower antiprotons were chosen, with a velocity of 78% of the speed of light. The apparatus did indeed work and in 1955they could report the observation of 20 antiprotons, and within 3% they could verify that it has the same mass as the proton but opposite charge (Chamberlain, 1955).

In 1959 Owen Chamberlain and Emilio Segré were awarded the Nobel Prize in physics for the discovery of the antiproton. Thomas Ypsilantis, aged 27 years at the time of the discovery of the antiproton, did not share the prize with Chamberlain and Segré, but devoted his whole scientific life to the development of Cherenkov detectors.

3. THE BIRTH OF RICH COUNTERS

In 1976 Jacques Séguinot together with Thomas Ypsilantis proposed a design of a Cherenkov counter which would detect the ring of Cherenkov light itself, and not only look for emission or no emission of Cherenkov light (Séguinot and Ypsilantis, 1977). They were inspired by the work of Litt and Meunier (Litt and Meunier, 1973) on a "Differential Isochronous Self Collimating" type of detector (DISC) with which it was claimed that it was possible to have a resolution of $\delta\beta/\beta = 10^{-7}$. In a DISC counter the Cherenkov photons emitted at different points along the particle's straight line trajectory are focused by a reflective mirror (radius R, focal length f = R/2) to give a circular ring image of radius r at the mirror focal plane. In the small angle approximation the focal surface is a plane and the radius of the ring image is $r = f \tan\Theta_c$.

The angle determination in a DISC counter is obtained by using an annular diaphragm of radius r at the mirror focal plane. Photomultipliers located behind the diaphragm detect the transmitted photons. For optimum resolution additional optical elements are added to correct for chromatic dispersion due to the variation of refractive index with the energy of the emitted photons. Such counters are suitable for velocity (mass) selection in collimated (momentum analyzed) primary particle beams but completely unsuitable for velocity measurement of secondary particles emerging from an interaction. The phase space occupied by these particles is large whereas the phase space acceptance of the DISC counter is small.

To improve the performance Séguinot and Ypsilantis proposed a different particle detector, shown in Figure 4, consisting of a spherical mirror of radius R whose center is the source of secondary particles (target) and a spherical detector surface at radius R/2 with the Cherenkov radiating medium between R/2 and R.

Figure 4. Schematic layout of the first Ring Imaging Cherenkov detector for large phase space acceptance; the ring images produced by two different particles emerging from the target are shown. The Cherenkov light produced between the detector and the mirror are reflected back onto the photo-sensitive detector by the mirror. The spherical shape of the mirror also focuses the cone of Cherenkov light into a ring in the focal plane at the detector. During the years many RICH detectors based on this basic design has been successfully used in particle physics experiments.
Rom Séguinot and Ypsilantis, 1977.

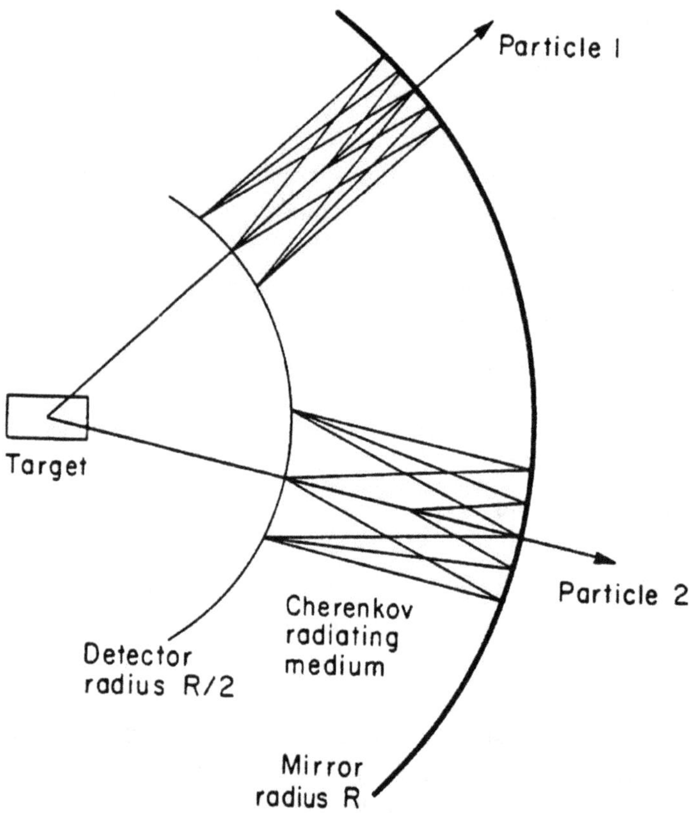

An ideal photo-detector in a RICH counter should have the following properties:

1. Large spherical surface sufficient to accept all ring images generated by all secondary particles which intercept the mirror.
2. Uniform sensitivity over a wide velocity range (i.e. no diaphragms).
3. Capability of localizing the spatial position of photons in the ring image.
4. High quantum efficiency.
5. Good time resolution.

To detect the ring image several different types of photo-detectors were proposed.

Multiwire proportional chambers were discarded since only one coordinate is obtained directly by the photoelectron, while the other coordinate has to be obtained through charge division or by induced pulses in strips in the cathode plane. Charge division was considered not to be possible since the signal from a single photoelectron is too weak, and the induced pulse method gives rise to ambiguities if the number of photoelectrons is too large.

A more elegant method was proposed consisting of drifting the photoelectrons of the ring image to an amplifying wire which induces pulses on a closely coupled electromagnetic delay line.

A third proposal was to use an array of needles at the focal plane of the mirror, as shown in Figure 5. Each needle is centered in a honeycomb structure of squares (or hexagons) and closed on the top by a transparent wire mesh. The honeycomb and mesh are maintained at negative voltage with the needle grounded through a resistance. Each needle is thus shielded from the others and acts as an elementary Geiger or proportional counter. Using needles, instead of wires, in the Geiger mode operation improves the timing, since in a discharge on a needle the avalanche propagates up to the needle tip and down only a short distance because of the rapid drop in electric field at the needle surface as the radius of curvature increases, and not all along as is the case for a wire in Geiger mode.

Figure 5. Proposal of a photo-sensitive detector for Cherenkov ring imaging used either in Geiger- or proportional mode; the major challenge in all Cherenkov detectors is to build a detector which is sensitive enough for the very faint Cherenkov light. Besides being sensitive to single UV-photons it shall also be position sensitive and fast.
From Séguinot and Ypsilantis, 1977.

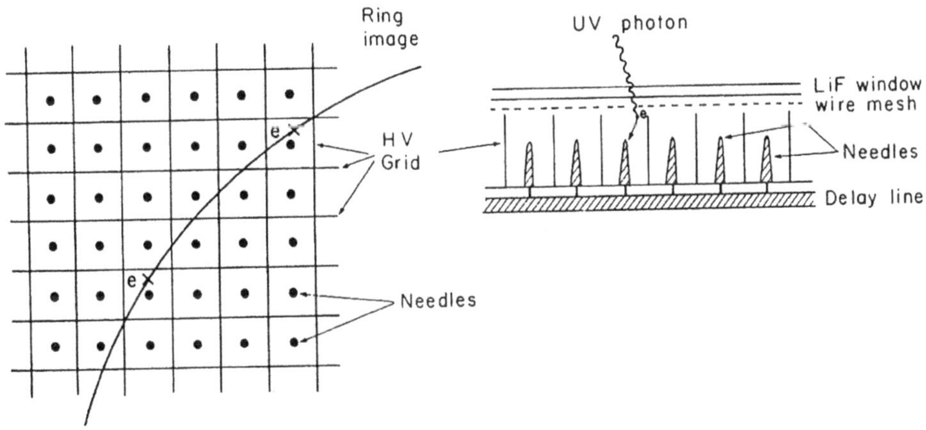

Several of these detectors can be cascaded so as to cover, with the required resolution, the full range of β values expected for a given experiment.

There are several severe problems connected with such a detector. First one has to find a photo-cathode which converts the photons into photoelectrons with high quantum efficiency. Secondly, the detector and the electronics has to be very sensitive to be able to detect a single photoelectron, and there has to be a way to read out the large number of channels that arises because of the fine granularity needed for determining the co-ordinates of each photoelectron. A radiator medium with the correct index of refraction for a specific purpose, with low dispersion, is the crucial point since the chromatic errors are the main limiting factors on the resolution of the velocity β.

4. MODERN RICH COUNTERS

Since the first proposal in 1976, a number of RICH counters have been successfully developed and built and they are now important parts of several high energy spectrometers.

The two large experiments SLD at SLAC and DELPHI at CERN have both included a RICH counter in their spectrometer. Both uses double radiators, i.e. a 45 cm gas radiator with z =1.0017 (in the visible region) (C_5F_{12}) and a 1 cm liquid radiator with n =1.26 (in the visible region) (C_6F_{14}), and a similar construction. They are designed to identify e, μ, π, K and p from 0.5 to 40 GeV/c momentum.

Figure 6 shows a schematic diagram of the mechanical layout of the SLD RICH (in the United States a RICH counter is commonly referred to as a Cherenkov Ring Imaging Detector, CRID). The

Figure 6. Schematic diagram of the mechanical layout of the SLD RICH counter; two charged tracks going through the Barrel RICH and the Forward RICH are indicated. The Cherenkov detectors cover nearly all angles a secondary particle can be emitted.
From Leith, 1989.

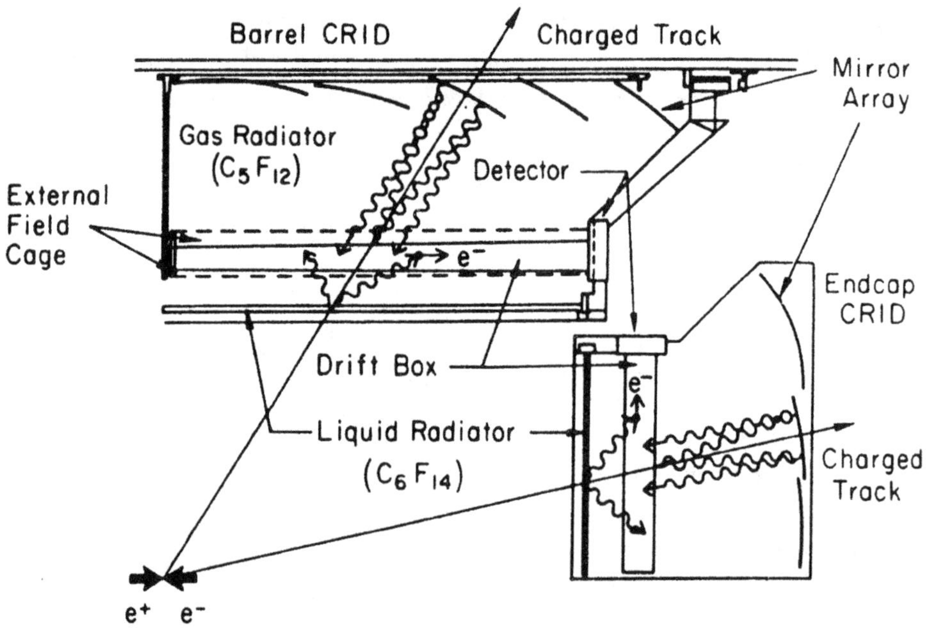

Figure 7. Conversion and drift volume of the RICH counter used in SLD; the ring of photoelectrons is drifted up to a meter to the end where they are detected in a MWPC. Cherenkov rings from one particle traversing the two radiators are indicated.
From Leith, 1989.

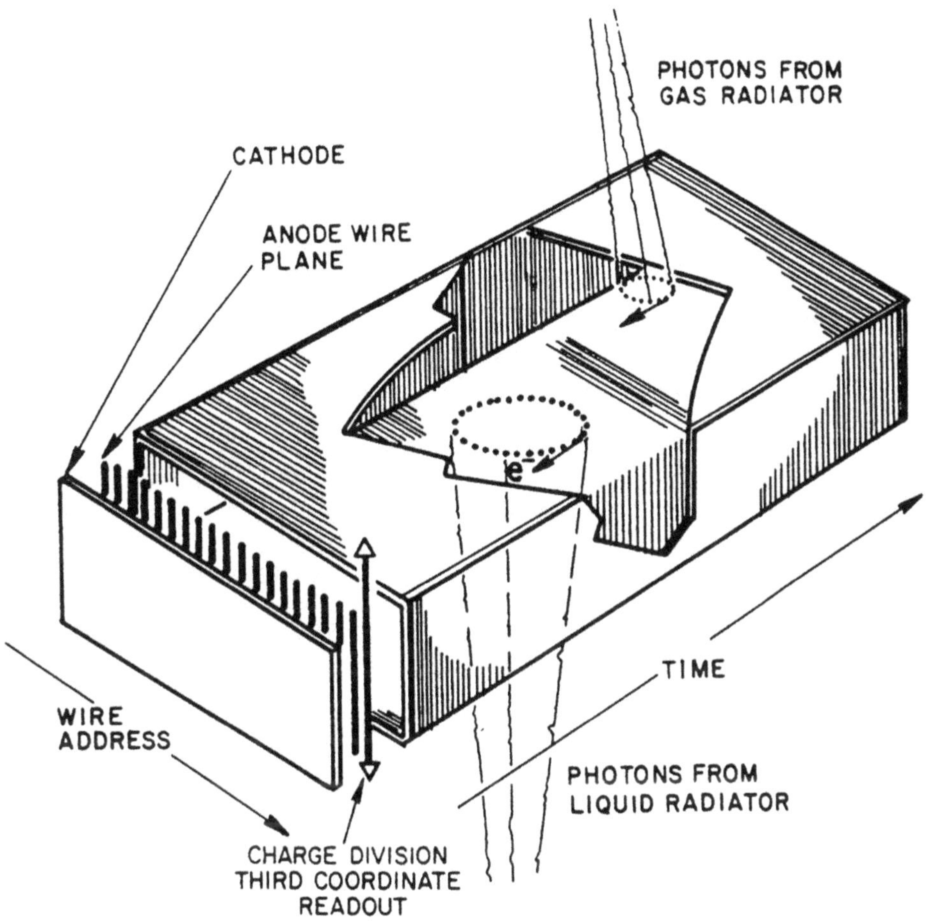

spectrometer is cylindrical with a Barrel RICH counter along the axis and a Forward RICH counter at each end. The secondary particles produced in the e^+e^- annihilation traverses the RICH detector where Cherenkov light is produced in the two radiators. The light from the gaseous radiator is reflected back by a spherical mirror (parabolic in DELPHI), into the conversion region in the drift box, while the light from the liquid radiator enters the conversion region directly. The photons are converted into photoelectrons which are drifted as far as one meter to the end where they are detected by a multiwire proportional chamber (MWPC) filled with a gas mixture C_2H_6+TMAE. The anode wires give the first co-ordinate of the conversion. The second component, i.e. along the axis of the cylinder, is given by the time it takes the photoelectron to drift to the wire. In SLD the third co-ordinate is obtained through charge division with a resistive anode wire, while in DELPHI it is measured through the induced pulse in the cathode strips perpendicular to the wires.

Figure 7 shows the drift volume and the read out system of SLD. DELPHI is similar, although it uses cathode strips to obtain the vertical co-ordinate instead of charge division. The photosensitive gas mixture also was slightly different: CH_4+C_2H_6+TMAE. Due to the different refractive index between the gas and the liquid the two rings, originating from the two radiators, have very different radius. The total thickness of the barrel is 75 cm.

5. CONCLUSION AND REMARKS

Cherenkov effect fascinated many scientists who tried to exploit it in practice for particle identification. Slowly, starting from first small prototypes, bigger and more sophisticated devices were built and used in some measurements. A real breakthrough happened after introduction a concept of a RICH detector. However, a full scale implementation of these new detectors would be impossible without a complimentary work: search for vapors having low ionization potential and compatible with gaseous detectors such as MWPC. As a result of these efforts two useful for the RICH application vapors were identified: TEA and TMAE. After solving technological problems associated with the chemical aggressively of TMAE the latter was successfully used in such a giant detector as DEPHI RICH as well as in many other detectors.

REFERENCES

Chamberlain, O. (1960). The early antiproton work. In *Les Prix Nobel an 1959*. P.A. Norstedt & Söner.

Chamberlain, O., Segré, E., Wiegand, C., & Ypsilantis, T. (1955). Observation of antiprotons. *Physical Review*, *100*(3), 947–950. doi:10.1103/PhysRev.100.947

Cherenkov, P. A. (1937). Izv. Akad. Nauk SSSR. *Ser. Fiz. OMEN*, *1*, 455.

Ekspong, G. (1958). Cherenkovstrålning, Kosmos Fysiska Uppsatser, P.A. *Norstedt & Söner*, *36*, 6.

Leith, D. (1989). A review of 4π Cherenkov Ring Imaging detectors. In *Proceedings ofSymposium on particle identification at high luminosity hadron colliders*. Fermi National Accelerator Laboratory.

Séguinot, J., & Ypsilantis, T. (1977). Photo-Ionization and Cherenkov ring imaging. *Nuclear Instruments and Methods*, *142*(3), 377–391. doi:10.1016/0029-554X(77)90671-1

ADDITIONAL READING

Apsimon, R. J., Flower, P. S., Freeston, K. A., Hallewell, G. D., Morris, J. A. G., Morris, J. V., & Thompson, L. F. et al. (1986). The Recent Operational Performance Of The Cern Omega Ring Imaging Cherenkov Detector. *IEEE Transactions on Nuclear Science*, *33*(1), 122–131. doi:10.1109/TNS.1986.4337063

Arnold, R., Baillon, P., Besch, H. J., Bosteels, M., Christophel, E., Dracos, M., & Ypsilantis, T. et al. (1988). A Ring Imaging Cherenkov Detector: The Delphi Barrel Rich Prototype. Part A: Experimental Studies Of The Detection Efficiency And The Spatial Resolution. *Nuclear Instruments and Methods*, *A270*(2-3), 255–288. doi:10.1016/0168-9002(88)90695-X

Ekelöf, T., Séguinot, J., Tocqueville, J., & Ypsilantis, T. (1981). The Cherenkov Ring Imaging Detector: Recent Progress And Future Development. *Physica Scripta*, *23*(4B), 718–726. doi:10.1088/0031-8949/23/4B/023

Ratcliff, B. (2003). "Imaging rings in Ring Imaging Cherenkov counters". Nucl. Instr. and Meth. in Phys. *Res.*, *A502*, 211.

KEY TERMS AND DEFINITIONS

Cherenkov Light: A new type of light emitted when charged particles traverse a transparent medium with a velocity higher than the speed of light in the medium. It is named after the Russian scientist Pavel Cherenkov who discovered it.

Gas-RICH: A RICH detector using a gaseous radiator to produce the Cherenkov light.

MWPC: Multi Wire Proportional Counter, a type of detector often used in particle physics experiments.

Photomultiplier (PM) Tube: A photosensitive device operating in vacuum.

RICH: Ring Imaging CHerenkov detector. A type of detector which measures the velocity of elementary particles by measuring the emission angle of the Cherenkov light produced when the particle traverses a medium with a velocity higher than the speed of light in the medium.

β: A measure of velocity expressed as the fraction of the speed of light (300.000 km/s). β is normally expressed in per cent of the speed of light.

Chapter 11
A NaF RICH Counter

ABSTRACT

RICH detectors are an important type of photosensitive detectors often used in particle and astroparticle physics for particle identification. They are quite complicated to construct and operate. The following chapters will describe in more detail examples of such detectors and their performance.

1. THE HISTORY OF THE NaF RICH PROJECT: THE CPLEAR SPECTROMETER

The initial impulse for the NaF RICH detector came originally from the CPLEAR experiment at CERN. The CPLEAR collaboration intended to measure the CP violating parameters in the neutral kaon system with initially-pure K^0 and K^0beams from the annihilation reaction p p-> p^{+-} K^{-+} K0 (K^0) at rest. To do so, a good separation of pions and kaons is necessary. The strangeness of the neutral kaon is identified by the sign of the charged kaon. This is determined by measuring the deflection in a 0.5 T magnetic field, while the separation of pions and kaons is performed by a Cherenkov counter. The Low Energy Antiproton Ring (LEAR) at CERN can deliver $2{\cdot}10^6$ antiprotons per second, and the branching ratio for the interesting reaction is 0.4%. Therefore, the particle identification device has to respond fast to reduce the rates at the first trigger level. It also has to be compact since an old magnet is used and the space allocated for the particle identification is small. Figure 1 shows a cross-section of the detector with a typical event.

The detector consists, moving from the center outwards, of the target, two proportional chambers, six drift chambers, a streamer tube system, scintillators (S1) used for time of flight, the Cherenkov counter, a second layer of scintillators (S2), an electromagnetic calorimeter and the coil of the magnet. The beam axis is perpendicular to the plane of the figure.

The requirements of the particle identifier are the following:

1. Highly efficient K-π separation (1:10^4) up to 1 GeV/c
2. **Fast:** A trigger signal has to be obtained in 250 ns.
3. **Compact:** Only 8 cm are totally available.
4. Low density material, not to convert γ-rays from π^0 decays.
5. Insensitive to magnetic field (0.5 Tesla).

DOI: 10.4018/978-1-5225-0242-5.ch011

Figure 1. A cross-section of the CPLEAR detector with atypical event; an antiproton annihilates with a proton in the hydrogen target in the center producing a charged pion, a charged kaon and a neutral kaon. The neutral kaon decays into a $\pi^+\pi^-$ pair on its way out of the detector. The overall diameter is 2 m and its length is 5 m. The spectrometer consists (from the center and outwards) of multiwire proportional chambers, streamer chambers, scintillators, liquid Cherenkov detectors and a lead electromagnetic calorimeter.
From Jonsson, 1988.

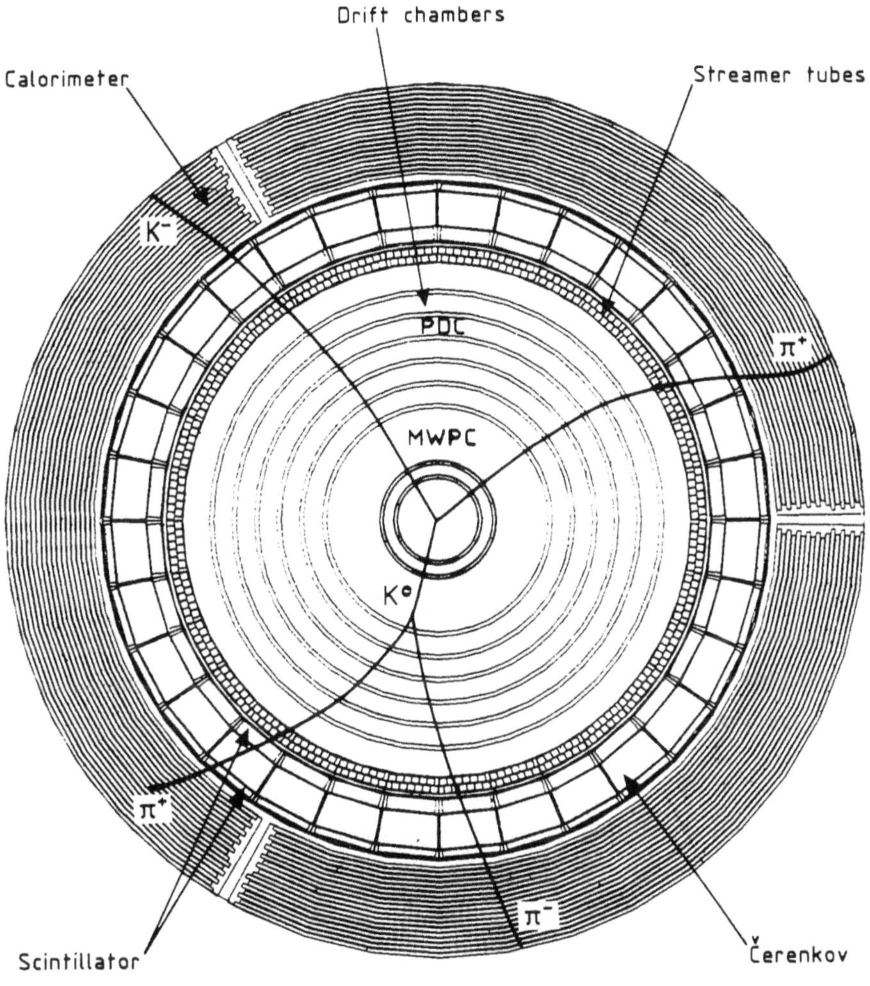

Two independent projects started to try to develop a detector which fulfils the requirements: a conventional threshold Cherenkov counter and a RICH counter. The idea was to have a sandwich consisting of a 4 cm conventional threshold Cherenkov and a 4 cm RICH counter.

An 8 cm conventional threshold Cherenkov counter turned out to be the cheapest and simplest solution which gives satisfactory results and was later used in the experiment.

Figure 2. The conventional threshold Cherenkov counter; Cherenkov light is produced in the liquid fluorocarbon, C_6F_{14}. The Cherenkov light at short wavelengths is absorbed and reemitted at a longer wavelength by a wavelength shifter dissolved in the fluorocarbon to better match the sensitivity of the photomultipliers used, and reduce the absorption in the plastic tubes. One track from a particle produced in the gaseous H_2 target is shown. The PM tubes have to be placed outside the magnet, and so the light guides have to be more than a meter long.
From Jonsson, 1988.

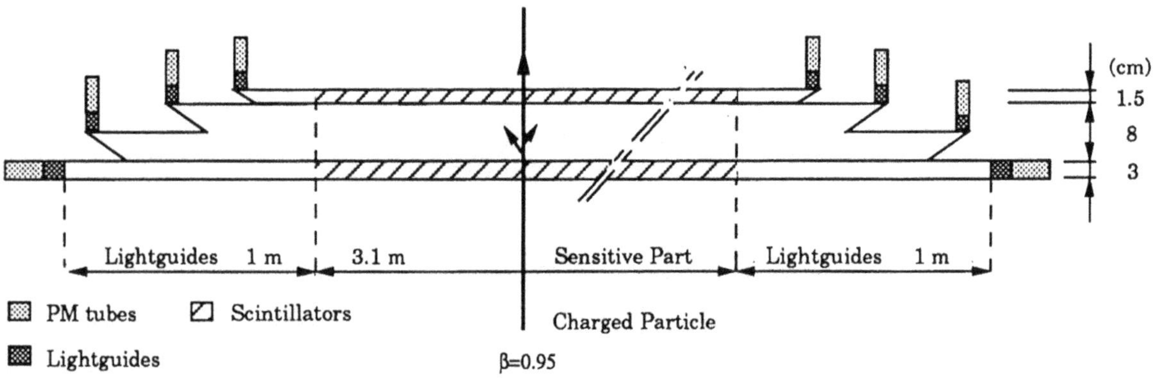

The development of the RICH counter continued anyway as a pure detector development project, and resulted in a fully tested prototype which performs better than ever expected in the beginning.

2. THE CONVENTIONAL CHERENKOV COUNTER

The conventional Cherenkov counter is a threshold Cherenkov counter and consists of 32 extruded tubes of UV transparent polymethyl methacrylate (PMMA) with a wall thickness of 2 mm (Rickenbach 1989). The ends of each tube are closed directly by light guides of the same material, and equipped with two PM tubes at each end. The Cherenkovs are filled with a liquid radiator C_6F_{14} (FC72, Fluorinert) with a refractive index of 1.26 in the visible region of the spectrum and a very low dispersion down to a wavelength of 200 nm. The whole tube is wrapped in aluminum foil and made light tight. The Cherenkov light produced in the liquid travels to the PM tubes at the ends through total reflection in the detector walls. One detector module is shown in Figure 2.

Since most of the Cherenkov light is produced in the far UV region where the transmission of the UV plastic is low, the efficiency of the Cherenkov counter can be largely improved by wavelength shifters. A suitable wavelength shifter was found in the form of 2,5 – diphenyloxazole (PPO) which absorbs light between wavelengths of 200 and 350 nm. The maximum solubility of PPO in FC72 is 22 mg/l at room temperature (when using an ultra-sonic bath to improve the solubility). At the used concentration of 20 mg/l the PPO absorbs more than 99V% of the light in one centimeter in the wavelength range of 280 to 310 nm. The absorbed light is re-emitted between the wavelengths of 300 and 450 nm (peak at 340 nm) with a quantum efficiency of about 70%. The emission peak is at a rather short wavelength compared with the transmission curve of UV transparent PMMA which makes thin counter walls necessary. Note that the total reflection occurs on the outside of the tube, so that the light has to go through the counter walls several times. The FC72 is practically 100% transparent for wavelengths down to at least 200 nm.

Owing to the reaction kinematics of the desired events the ideal Cherenkov threshold is $\beta_{thr} = 0.84$ corresponding to the maximum kaon velocity. This requires a refractive index of $n = 1/\beta_{thr} = 1.19$. However, FC72 is the only liquid available in suitable quantities with a sufficiently low refractive index ($n = 1.26$). Owing to energy loss in FC72 and the detectors in front of the Cherenkov detector, the effective threshold velocity of the incident particle is not given by $\beta_{thr} = 1/n = 0.79$ but by $\beta_{thr} = 0.81$ which is not far from the required threshold of $\beta_{thr} = 0.84$. This does not affect the separation power of the detector, but reduces the identification efficiency of high energy kaons, since some kaons will be identified as pions as they produce Cherenkov light.

Beam tests with a prototype of such a Cherenkov counter were carried out at PSI in Villigen, Switzerland. Measurements of the z and angular dependence of the photoelectron yield show that the response of the Cherenkov counter is improved significantly by painting the middle 80 cm of the detector on all four outer sides with white reflective paint (NE561). The Cherenkov light cannot be totally internally reflected in the central region, since the particles produced at the target enter the Cherenkov detector at nearly normal incidence. The white paint acts as a diffuser and scatters this part of the Cherenkov light isotropically, resulting in a factor of two increase in the photoelectron yield.

This simple construction fulfilled the requirements of the CPLEAR experiment even though the efficiency in the center of the detector is not as good as desired. The light guides have to come out of the magnet through small holes in the end caps in order to keep the PM tubes away from the magnetic field. This introduces light losses since the light guides have to be more than a meter long, but also mechanical problems when opening the magnet since all PM tubes have to be removed.

The counter is very fast and a primary trigger is available in100 ns. Good PM tubes that are available today make it possible to clearly put a threshold at only one photoelectron with nearly no random triggers from noise.

3. THE NaF RICH COUNTER

In parallel with the development of the conventional Cherenkov counter a RICH detector was also developed that would fulfil the properties. In the CPLEAR Cherenkov group two ideas came up which would render such a detector possible:

1. The use of a solid Cherenkov radiator.
2. To use a crystal both as radiator and window separating the drift and the conversion spaces.

Several prototypes were built and tested. The project started by using a low pressure multiwire proportional chamber (MWPC) to detect the light. The low pressure operation was chosen to minimize the background from ionization by the primary particle inside the MWPC itself. Tests were also made with different photocathodes, both gases and solids, to find the most efficient for this purpose. Different setups and windows were tested as well as different electronics for the read out.

Later operation at atmospheric pressure was chosen to be able to incorporate the detector into a real experiment, and pad readout was chosen to avoid problems with ambiguities in the reconstruction of the ring pattern. This included tests of a completely new MWPC where one of the cathode planes was replaced by a pad plane, together with VLSI electronics to be able to read it out.

The low pressure NaF RICH counter has two major drawbacks. First, in a real spectrometer low pressure operation is most undesirable, since it introduces unwanted material for the vacuum chambers. Second, strip readout gives raise to ambiguities in the reconstruction of the hit position of the photoelectrons. To overcome these problems, a RICH counter using a NaF radiator, operating at atmospheric pressure and using pad readout has been constructed and tested. This is a description of the design and performance of this detector, and an evaluation of the experimental results.

3.1 Detector Overview

To make a RICH counter useful for modern spectrometers, it has to be compact and fast. With a solid radiator there is no need for a UV-transparent vessel which makes the detector more compact and simplifies the construction. The detector is made fast by using a multiwire proportional chamber with short drift distance to detect the photoelectrons produced by the photosensitive compound. To be practical, it has to be operated at atmospheric pressure since a vacuum chamber will introduce unwanted extra material where gamma rays and electrons can produce electromagnetic showers. Since the number of photoelectrons detected per event must be fairly large to reconstruct a ring with good accuracy, strip

Figure 3. Set-up of the NaF RICH detector using pad readout and operated at atmospheric pressure; Cherenkov light is emitted in a NaF crystal. The cone of light expands in a helium filled volume before it enters into the photosensitive detector through a thin quartz window. The photosensitive detector is a MWPC operating with TMAE and ethane at atmospheric pressure. The Cherenkov light produced by a β = 0.82 particle at normal incidence is indicated.
From Francke, 1991.

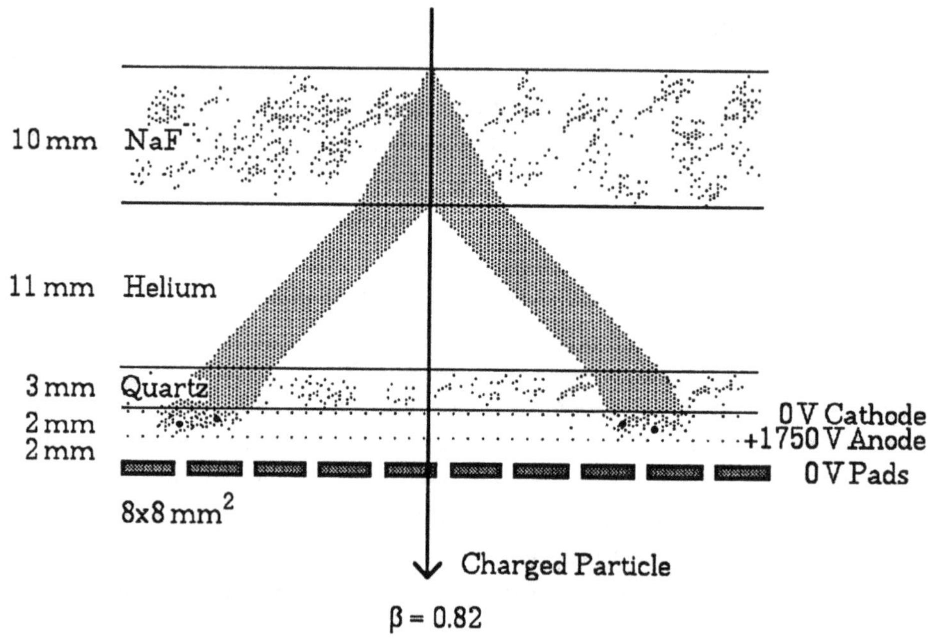

readout is not possible to use because ambiguities arise in the reconstruction of the ring. In this detector the problem is solved by reading out the induced signal in a cathode pad plane, where the two dimensional position is available directly from the pad position.

Two versions of the atmospheric NaF RICH, with and without an UV-mirror, have been tested. In the following, mainly the version without mirror will be described since the mirror version presents the major inconvenience of parasitic reflections. The principle of construction is shown in Figure 3.

The particles enter the detector from above, and the Cherenkov light produced in the NaF radiator refract out of the crystal and is expanded into a ring in the helium filled drift volume. This light then enters the photosensitive multiwire proportional chamber through a thin window. In the photosensitive detector, the Cherenkov photons are converted into photoelectrons by TMAE or TEA, which are amplified and detected by the multiwire proportional chamber. The position of each photoelectron is determined by the induced pulse in the cathode pad plane, where a two-dimensional readout is possible without ambiguities.

Several NaF radiators with different dimensions have been tested, but all of them are 10 mm thick. The smallest crystal used is 50 by 50 mm^2, and the largest is circular with a diameter of 120 mm.

The distance between the window and the NaF crystal can be adjusted to regulate the radius of the Cherenkov ring. Since the window can be as thin as 3 mm, compared to 10 mm if the NaF crystal itself acts as both radiator and window, the thickness of the detector is only increased by some 10 millimeters compared to using a mirror, even though the drift volume has to be twice as thick. When using TMAE as photosensitive compound, the window can be made of quartz, while TEA requires a CaF_2 or a LiF window to transmit light in the sensitive wavelength regions. The drift volume is filled with helium to minimize the absorption losses of the Cherenkov photons, without producing background photons in the drift volume gas. In helium at atmospheric pressure, only 0.04 Cherenkov photons are emitted per centimeter in the sensitive wavelength region (170 - 220 nm), compared to 300 in the NaF radiator when it is traversed by a $\beta = 1$ particle. For particles with a Lorentz factor $y < 120$ no Cherenkov photons are emitted at all in the helium. Furthermore, the overall transmission of the system is higher since there is no mirror and the window is thinner.

The detector can also be modified to use the NaF crystal both as radiator and window and a mirror to reflect the Cherenkov light back, i.e. the same set-up used for the low-pressure NaF RICH. In this way the performance of the two set-ups can be compared. A schematic view of this set-up is shown in Figure 4. The NaF crystal separates the two chambers of the detector, while the mirror is supported in the upper chamber. The distance between the NaF crystal and the mirror can be varied to adjust the radius of the Cherenkov ring. With this set-up, the particles enter the detector from below. The ring of Cherenkov light produced in the NaF radiator refracts out of the crystal and is expanded into a ring in the helium filled drift volume. The light is reflected back by the mirror and enters the photosensitive detector through the NaF crystal, this time acting as a window.

The use of a multiwire proportional chamber to detect the photoelectrons produced simplifies the construction of large area detectors. Multiwire proportional chambers have high single photoelectron detection efficiency, which is important since the number of photoelectrons produced is low. Furthermore, a MWPC is fast. The amplification process takes less than 1 ns and the drift velocity under these circumstances is of the order 1 mm per 20 ns. Using a 4 mm thick MWPC, where the photoelectric conversion takes place inside the amplification region and the anode wires are placed in the center, the signal develops in approximately 50 ns. Thinner detectors are difficult to construct for large area applications, although they would probably improve the performance of the detector. To reduce the ionization of the

Figure 4. Schematic view of the NaF RICH detector using a mirror to reflect the Cherenkov light back through the NaF crystal; an alternative design of the NaF RICH detector uses a mirror to reflect the Cherenkov light back into the detector. The charged particles enter the detector from below. Cherenkov light is emitted in the NaF crystal and is expanded twice in the helium volume before entering the photo-sensitive detector through the NaF crystal. The advantage is a more compact design and the elimination of the quartz window. The drawback is the partial reflection of Cherenkov photons in the NaF crystal which gives an additional ring of light. The Cherenkov light emitted when the detector is traversed by a β= 0.82 particle is indicated.

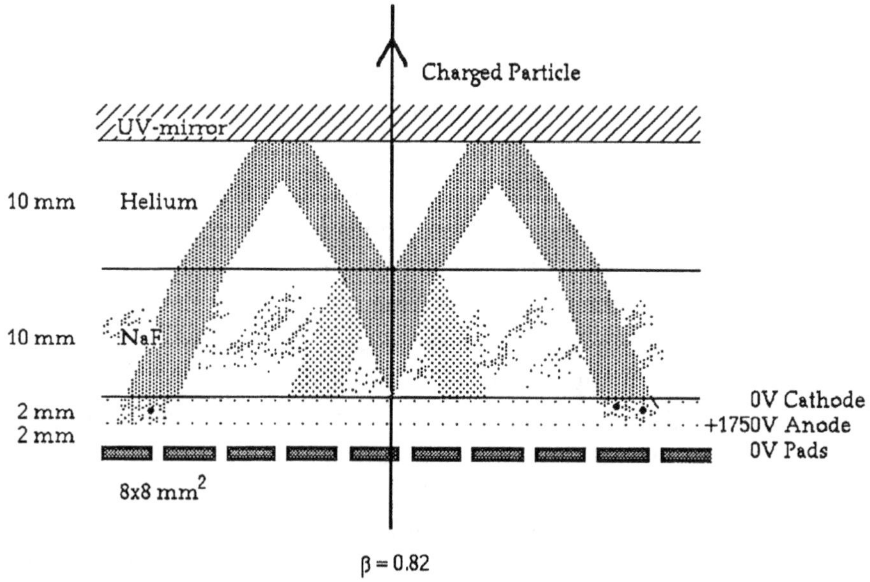

amplification gas by the primary particle is a second reason to keep the MWPC as thin as possible. In a 4 mm thick gas volume of methane and isobutane, approximately 50 electron-ion pairs are produced by minimum ionizing particles. They are all produced within a small area of the detector since they are produced along the track.

The anode wires are 4 mm apart and made of 20 μm diameter gold-plated tungsten. The upper cathode (cathode 1) is a stainless-steel mesh made of 50 μm diameter wires placed 500 μm apart. The lower cathode (cathode 2) is a printed circuit board with gold-plated square pads etched on the surface.

The position of the photoelectron is determined by reading out the induced pulse in the lower cathode (cathode 2). The cathode is segmented in gold-plated 8 by 8 mm^2 square pads which give anon-ambiguous, two-dimensional position by looking at the position of the pad fired. Smaller pads will give better position resolution, if the anode wires are placed closer to the cathode to reduce the spread of the induced pulse. But the signal on each pad will at the same time be smaller, and the number of electronic channels increases inversely with the square of the pad size. The number of pads is so large that they need to be read out by low-noise VLSI electronics placed close to the pad plane.

The anode wires are read out to give the projection of the ring, but also to control the detection efficiency and the gain variations of the chamber.

Operation at atmospheric pressure simplifies the construction of the detector since it only has to be made gas-tight, but not vacuum tight. The walls can be made of aluminum, and the chamber is made gas-tight using O-rings. The feedthroughs of cables can be made by normal connectors and the detector is light. On the other hand, the detector cannot be outgassed in a simple way, and hence impurities and contaminants stay in the chamber for a very long time, even if it is flushed with an inert gas. Therefore, the detector has to be flushed continuously with the gas mixture, and cannot be run as a closed system for a long time.

3.2 Experimental Results

The presentation of the experimental results will be separated into three different parts. First, a comparison between TEA and TMAE with the mirror set-up. Second, a comparison between the mirror set-up and the quartz window set-up using TMAE. Finally, a more detailed analysis of the performance of the NaF RICH counter using a quartz window and pad readout is presented. All results are compared with simulations, according to the simulation algorithm discussed above.

3.3 Analysis Algorithms

Some special algorithms are used in the analysis of the experimental data.

The scintillators defining the beam are 10 by 10 mm^2, but it is possible to obtain a better accuracy of the impact point of the primary particle by calculating the center of gravity of the pad amplitudes from the ionization signal. The analog information of each pad above threshold is available from the DRAMSs and the center of gravity without any weights on the pads seems to give the best resolution. The center of the fitted Cherenkov ring gives an independent reference position to the center of gravity. In this way, a resolution of ± 2 mm perpendicular to the anode wires, but even better parallel to the wires, is possible for the TMAE runs.

During the TEA runs, the gain is so high that the pad under the avalanche saturates. Due to some imperfections in the design of the mother board for the Amplex (a monolithic analog signal processor fabricated in 3 μm n-well CMOS), the baseline of the Amplex chip reading out the central 16 pads is shifted by the large pulse so all16 central pads gives a signal above threshold. This ruins all efforts to improve the impact position by center of gravity calculations, and the resolution of the impact point is ± 5 mm both perpendicular and parallel to the wires.

The result of the reconstructed impact points for a TMAE run is shown in Figure 5. The dashed lines correspond to anode wires while the squares illustrate the square pads. The beam is so wide that the incoming particles can be assumed to be uniformly distributed over the trigger scintillators. The two bands of hits coincide with the position of the anode wires, which shows that there is a small systematic error in the reconstructed impact point perpendicular to the anode wires. Parallel to the wires the distribution is more uniform.

The reconstruction of the position of a photoelectron is more complicated. During the TEA runs, where the number of photoelectrons per event is very small (about 4 per event), every photoelectron will form an isolated cluster of pads hit. The overlap between pads is also small. In this case, the position of the photoelectron can be calculated from the center of gravity of each cluster. Since a cluster very often is only one pad above threshold, the resolution in the reconstructed position is half the pad side, ± 4 mm.

When using TMAE, the number of photoelectrons detected per event is much larger and it is not possible to find isolated clusters of pads hit originating from a single photoelectron. Here, the center of each pad hit is used as a single photoelectron in the reconstruction of the Cherenkov ring. This is

Figure 5. Beam profile after reconstruction of the impact position using a center of gravity algorithm; the resolution is high. One clearly sees the shape of the 10 by 10 mm² scintillator defining the beam profile. In the y-direction (vertical) the resolution is slightly worse as the charges are attracted towards the anode wires spaced 4 mm apart before triggering a multiplication avalanche.

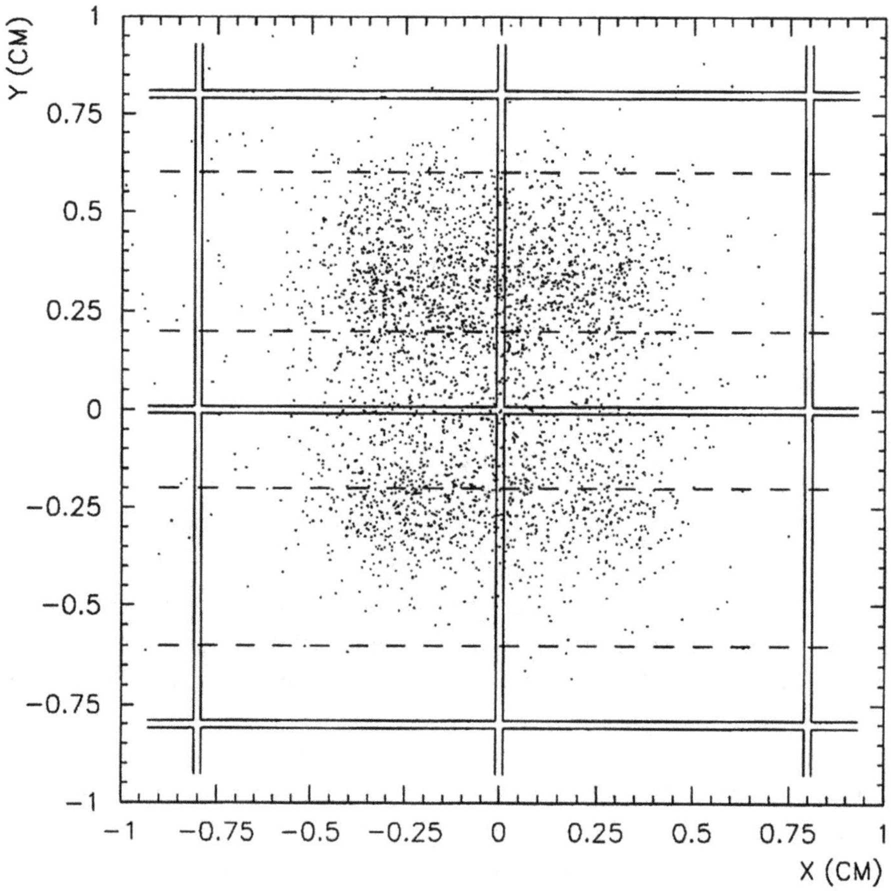

not correct if one wants to deduce the actual number of photoelectrons per event in which case a more rigorous comparison with the simulation is necessary.

The radius of the Cherenkov ring is calculated in two ways: either by calculating the mean distance between the reconstructed impact point of the primary particle and the reconstructed photoelectron, or by fitting a ring to all pads hit excluding the central hits from the ionization. These two methods give the same ring radius within one millimeter.

3.4 TEA/TMAE Comparison

The main difference between using TMAE compared to TEA is the spectral region that is used. When TMAE is used as photosensitive compound, the detector is sensitive from 220 nm (the threshold of TMAE quantum efficiency) down to 140 nm (the cut-off of NaF). TEA on the other hand is only sensitive to shorter wavelengths than 170 nm. The quantum efficiency of the TEA is also at least a factor of 2 lower

Figure 6. Single proton event (β =0.82) using a) TEA and b) TMAE; the higher quantum efficiency of TMAE is clearly visible as significantly more Cherenkov photons are detected. One can also see the effect of the lower amplification in the MWPC as the ionization signal in the center from the charged particle is significantly smaller in the TMAE operation compared to when using TEA.

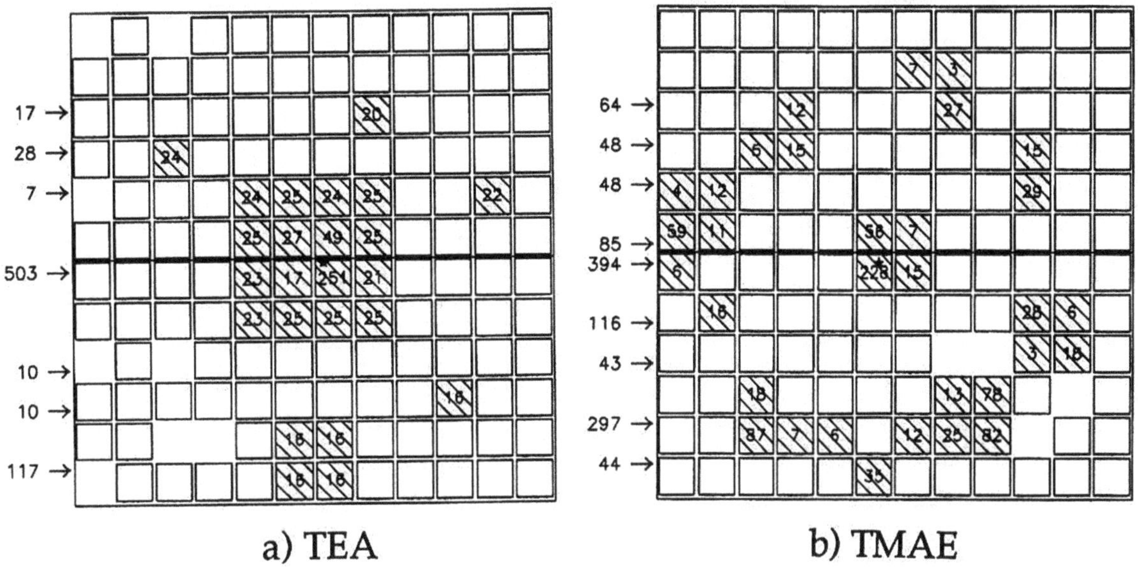

than of TMAE. Together, this implies that the expected number of photoelectrons when using TEA is at least a factor 5 less than with TMAE with the same set-up. In this comparison the mirror set-up is used with both photosensitive compounds to eliminate the influence of different windows. Hence, the NaF crystal acts as both radiator and window. For reasons discussed above, the gain during the TMAE runs is much lower than during the TEA runs. The TMAE is heated to 50°C to have short enough absorption length. The threshold used on the wires is 3 standard deviations of the noise distribution and on the pads 5 standard deviations if nothing else is stated.

A typical event using TEA and TMAE is shown in Figure 6 for 1.35 GeV/c protons, corresponding to $\beta = 0.82$. In the TEA event, the effect of the baseline shift is clearly visible as all 16 central pads are above threshold. In the TEA runs the number of photoelectrons detected per event can be estimated by just counting the number of clusters. Here, 5 photoelectrons form a ring. The numbers in the pads correspond to the signal amplitude on each pad, and the pads missing are dead channels. The arrows to the left represent the anode wires that are fired and their signal amplitudes. The star in the center illustrates the reconstructed center of gravity of a cluster.

For the TMAE event the ionization signal in the center is much smaller, only four pads hit in this event, and several more photoelectrons are detected and they form a nice ring.

The distribution of the number of pads hit per event for the two set-ups is shown in Figure 7. Figure 7a) is with TMAE and 7b) with TEA. The solid line shows the number of pads hit outside the central region (central 16 pads), and the dotted line the number of pads fired by the ionization. The contribution from parasitic reflections in the NaF crystal is negligible at these low velocities. The particles are 1.35 GeV/c protons ($\beta = 0.82$). In the TEA run the effect of the baseline shift is clearly visible as very often all 16central pads are above threshold. This is not the case in the TMAE runs, where the minimum ionizing

Figure 7. Number of pads hit per event for a) TMAE and b) TEA; the solid line illustrates the hits outside the central 16pads, and the dotted line the number of pads fired by the minimum ionizing particle. The higher quantum efficiency of TMAE is clearly visible by the large number of pads fired by Cherenkov light. The lower gain of the MWPC when using TME is shown by the smaller number of pads fired by the ionization.

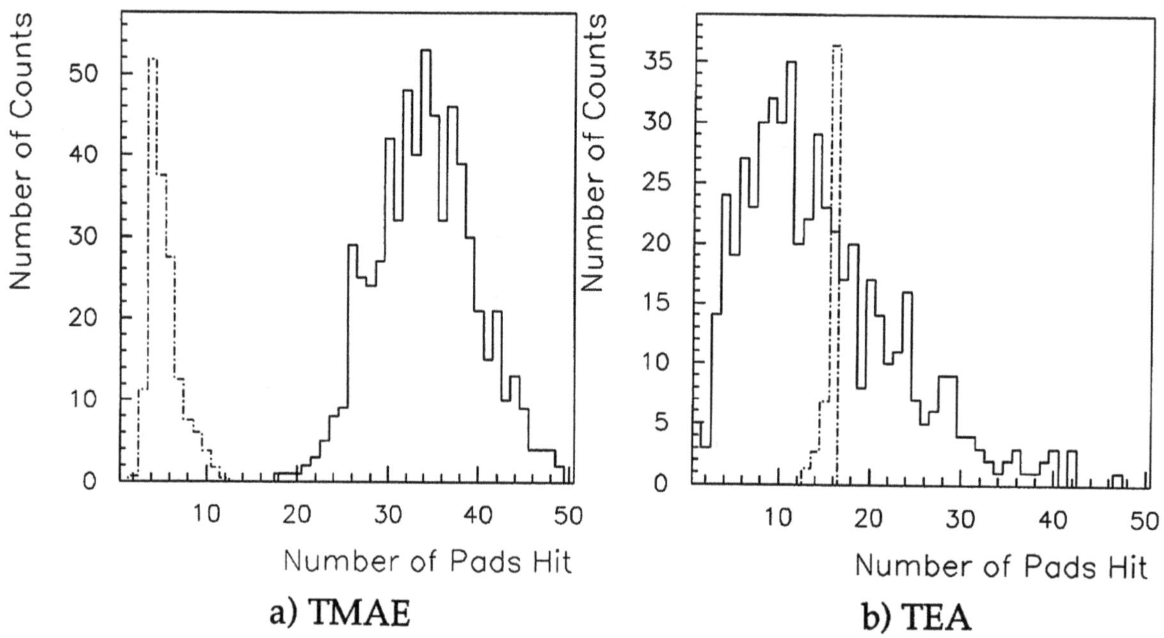

particle gives a detectable signal in 4 pads as an average. Note that the number of pads fired per event is significantly smaller in the TEA run than in the TMAE run even though the gain is much higher. This indicates that the number of photoelectrons produced with TEA must be a lot smaller than with TMAE.

The mean number of photoelectrons detected per event for$\beta = 0.82$ protons using TEA is shown in Figure 8. The number is estimated by counting the number of clusters per event. With this gain 4.9 photoelectrons are detected per event. The dotted line is a Poisson distribution with $<n>=4.9$ photoelectrons, normalized to the same number of events as in the measured distribution. The measured distribution is narrower than a Poisson distribution. This is expected since in events with a large number of clusters, the probability that two clusters will overlap and be counted as one cluster is larger than in events with a small number of clusters. In this way, a slightly too small number of clusters per event is estimated. A cluster is defined as a group of pads with a signal above threshold that are lying side by side, but not corner to corner.

The distribution of the ring radius calculated as the mean distance between the reconstructed impact position and the reconstructed position of each photoelectron is shown in Figure 9 for TMAE and TEA. Note that all events with at least three pads fired outside the central hits are included here. The small tail on the small ring radius side is due to pads influenced by the ionization signal outside the central 16 pads. With TMAE the standard deviation is 1.7 mm and with TEA 7.8 mm at a mean radius of approximately 36 mm. The particles here are also 1.35 GeV/c protons ($\beta = 0.82$).

Figure 8. The number of photoelectrons detected per event with TEA for β = 0.82 protons; the dotted line is a Poisson distribution with the same mean and number of events as the measured solid line. On average 4.9 photoelectrons are detected per event.

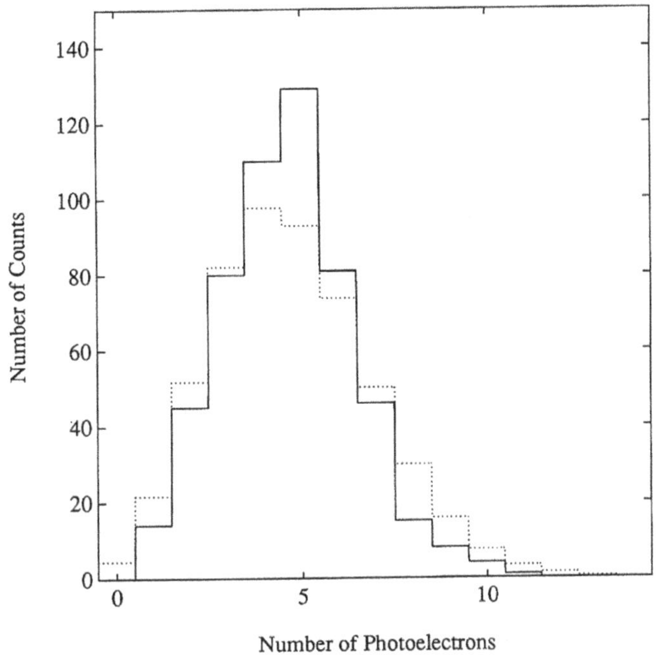

Figure 9. Distribution of the ring radius, calculated as the mean distance from the center of each hit, for β = 0.82 protons. a) is with TMAE and b) is with TEA. Due to the lower number of photoelectrons detected when operating the detector with TEA, the resolution of the reconstructed Cherenkov angle is significantly lower, 3.8 mm compared to 1.7 mm.

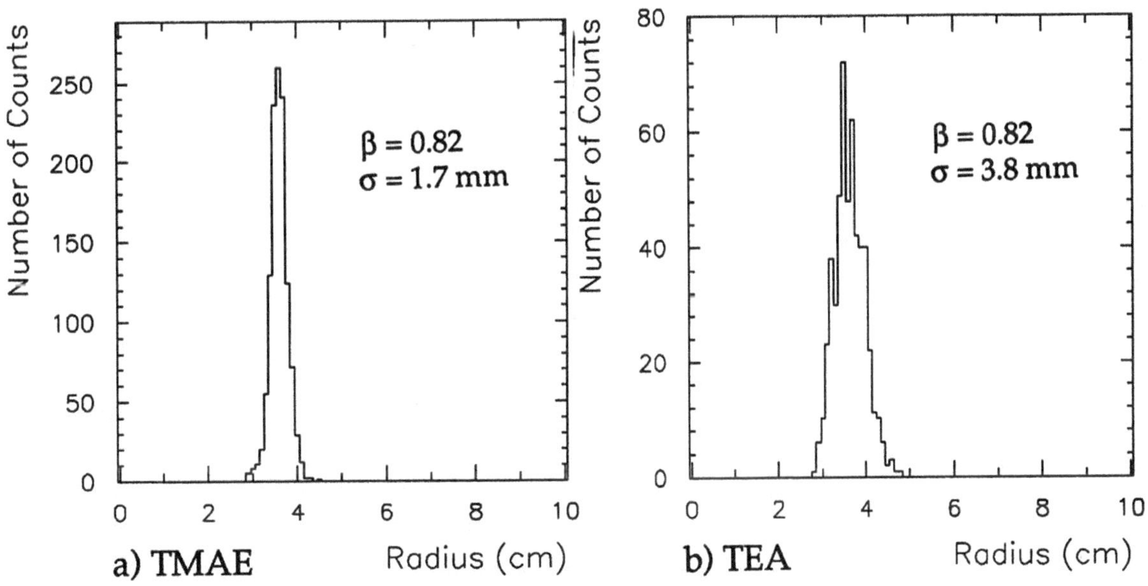

After this comparison TEA was abandoned since it gives too few photoelectrons per event to be able to build an efficient detector. Also no problems were observed with photon feedback when using TMAE even though the wires are unshielded. From here on, all measurements are done with TMAE.

3.5 Mirror: No Mirror Comparison

The main difference between using the mirror set-up compared to no mirror is the spectral region where the detector is sensitive. With the mirror set-up it is the transmission of the NaF that sets the limit at short wavelengths (140 nm) and the quantum efficiency of TMAE at long wavelengths (220 nm). Without the mirror the quartz window increases the short wavelength cut-off to 170 nm. The transmission of the system without the mirror is higher because there are no reflection losses in the mirror and the window is thinner. Concerning the number of photoelectrons per event, these two effects will approximately compensate each other, and the expected number of photoelectrons detected per event is expected to be approximately the same for large beta ($\beta > 0.84$). But, since the wavelength region is narrower without the mirror and shifted to longer wavelengths the chromatic errors due to the dispersion of NaF are expected to be smaller without a mirror than with. Furthermore, the parasitic reflections in the NaF crystal do not affect the reconstruction as much without the mirror as with.

The number of pads hit per event with and without a mirror is shown in Figure 10 for $\beta = 0.84$ and $\beta = 1$ particles. The solid line corresponds to measurements with the mirror and the dotted line to measurements without the mirror. Only the pads fired outside the ionization region are included. The number of pads hit per event is approximately the same with both set-ups. The gain of the wire chamber is approximately equal in the two set-ups, which implies that the detected number of photoelectrons is approximately the same.

Figure 10. Number of pads hit per event for a) $\beta = 0.84$ and b) $\beta = 1$ particles; the solid line corresponds to measurements with the mirror and the dotted line to measurements without the mirror. The increased number of pads hit for β=1 particles is due to the increased number of emitted Cherenkov photons.

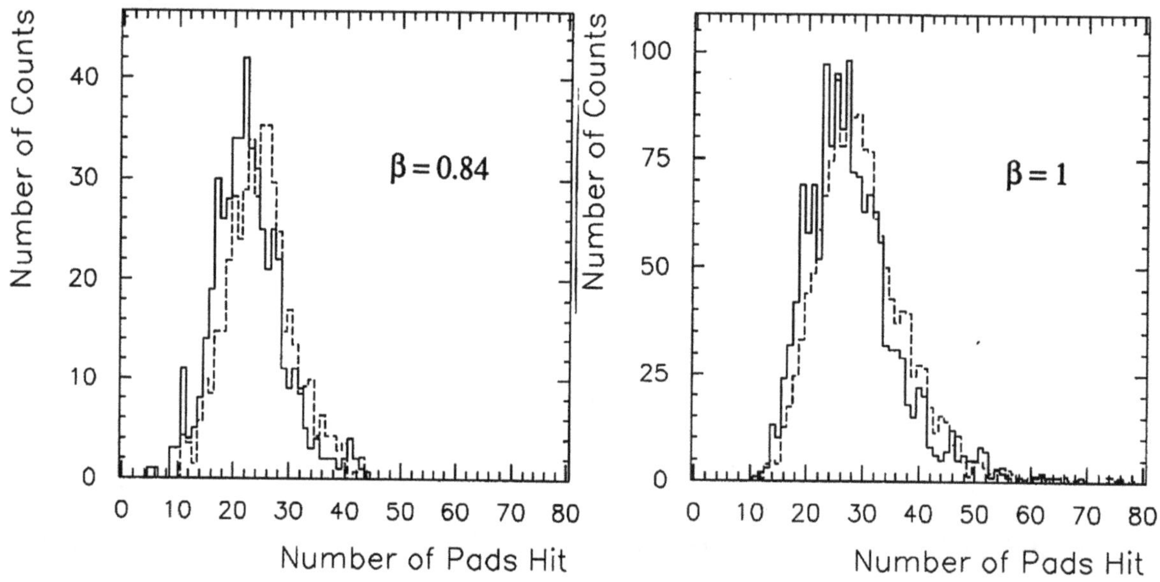

Figure 11. Ring radius distribution for a) $\beta = 0.84$ and b) $\beta = 1$ particles; the solid line corresponds to experimental results with the mirror set-up, and the dotted line to experimental results with the quartz set-up. The quartz setup gives a slightly better resolution of the reconstructed Cherenkov angle due to a better optical performance of the system resulting in a larger number of detected Cherenkov photons.

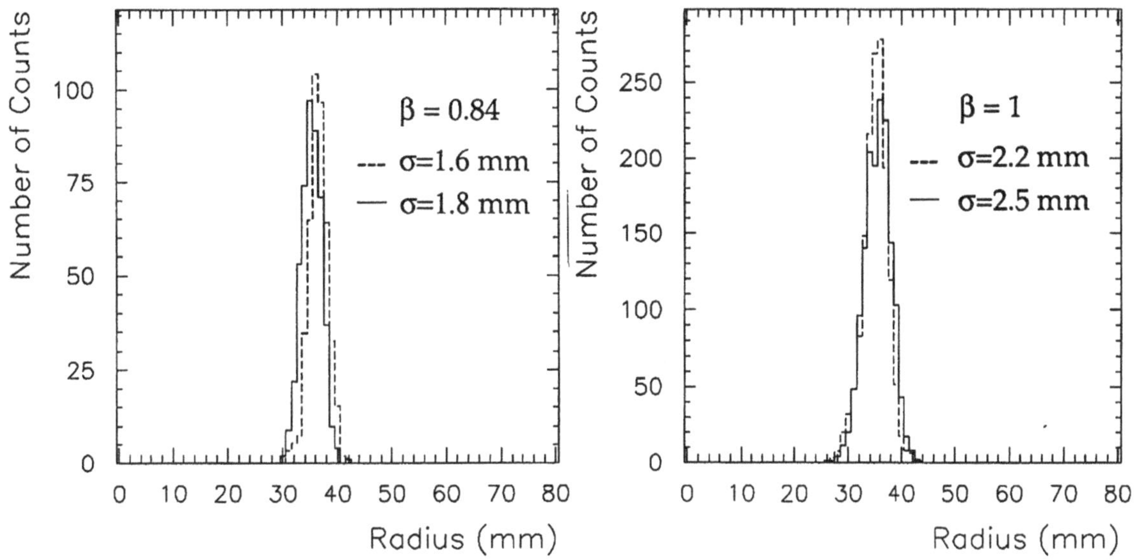

The difference in resolution is illustrated in Figure 11. It shows the distribution of the ring radius for the two set-ups for $\beta = 0.84$ and $\beta = 1$ particles. The ring radius of each event is calculated by the mean distance method described earlier. Also here, the solid line corresponds to measurements with the mirror and the dotted line to measurements without the mirror. The resolution is considerably higher without the mirror than with. At beta close to unity, this is mainly due to problems with parasitic reflections that produce a second ring inside the desired Cherenkov ring using the mirror set-up, which confuses the reconstruction of the ring-radius. At lower beta, the parasitic reflections coincide with the ionization signal, and it is mainly the chromatic errors of the NaF crystal that cause the reduced resolution with the mirror set-up.

4. THE ATMOSPHERIC PRESSURE NaF RICH DETECTOR USING A QUARTZ WINDOW AND PAD READOUT

This is a detailed description of the performance of the final version of the NaF RICH detector set-up shown in Figure 3. The10 mm thick NaF crystal is used as Cherenkov radiator. The Cherenkov photons refracts out of the radiator and are expanded into a ring in a helium filled drift volume before they enter the photosensitive detector through a 3 mm thick quartz window. The photosensitive multiwire proportional chamber uses TMAE as photosensitive compound and methane + 5% isobutane as amplification gas. The chamber is operated at atmospheric pressure and can be heated between 20° and 60°C to increase the vapor pressure of TMAE so that the absorption length is compatible with the thickness of the multiwire proportional chamber. It is necessary to convert as many photons as possible into photoelectrons to be able to reconstruct the Cherenkov ring with good resolution.

A non-ambiguous reconstruction of the position of each detected photoelectron is achieved by using pad readout. The pads are 8 by 8 mm^2 and gold plated.

4.1 Perpendicular Incidence

Typical single events at various velocities of the primary particle are shown in Figure 12. The NaF-quartz distance has been varied to give approximately the same ring radius for all β. The increase of the number of photoelectrons with increased β (up to β=0.94) is clearly seen. Above β = 0.9 a it decreases slightly again as β approaches 1. The central spot is due to the ionization signal and the star shows the center of gravity of the signal amplitudes of the pads fired by the ionization. This position is used as an impact point of the particle with the multiwire proportional chamber.

Figure 12. Single events for various values of β; the NaF-quartz distance has been adjusted to give approximately the same ring radius for all events. As the velocity increases one can clearly see how the number of Cherenkov photons detected increases. When the velocity, β, approaches unity some Cherenkov photons are trapped inside the NaF crystal due to total reflection, decreasing the number of detected Cherenkov photons.

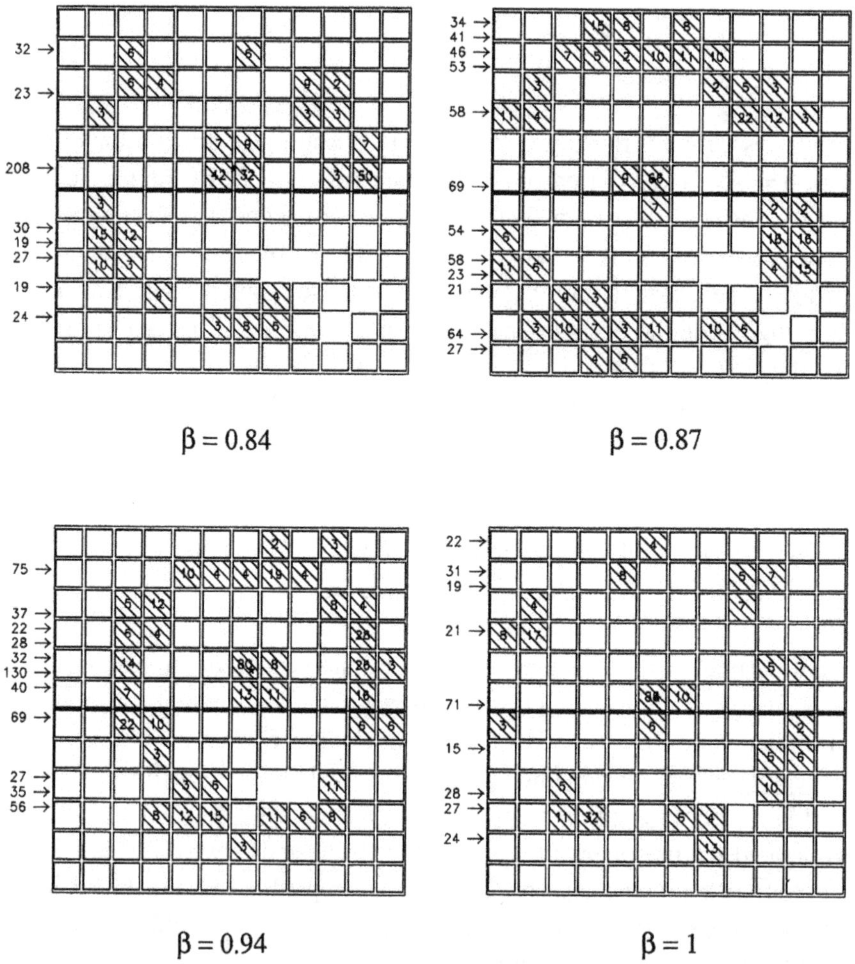

The numbers on the pads correspond to the amplitude of the signal on each pad above threshold. The pads missing are dead channels. The arrows to the left indicate the anode wires with signals above threshold with their signal amplitudes. Nice rings are detected for single events over a large velocity range (β:0.80 ->1) which indicates the good separation power of the detector.

Accumulated events with the same set-up as in Figure 4 are shown in Figure 13. In each figure, the analog signals of each pad for about 1000 events are accumulated. In the center, the large signals originate from the ionization of the amplification gas by the minimum ionizing primary particle. Around this large signal, a ring of Cherenkov light is observed. An estimation of the number of detected photoelectrons as a function of β is obtained by comparing the amplitude of the pads forming the Cherenkov ring with the ionization, since the ionization is approximately independent of β at these high velocities.

Figure 13. Accumulated events at various values of β; the NaF-quartz distance has been adjusted to give approximately the same ring radius for all runs. A maximum in the detected number of Cherenkov photons is observed around β = 0.94. The ionization signal is approximately the same for all velocities.

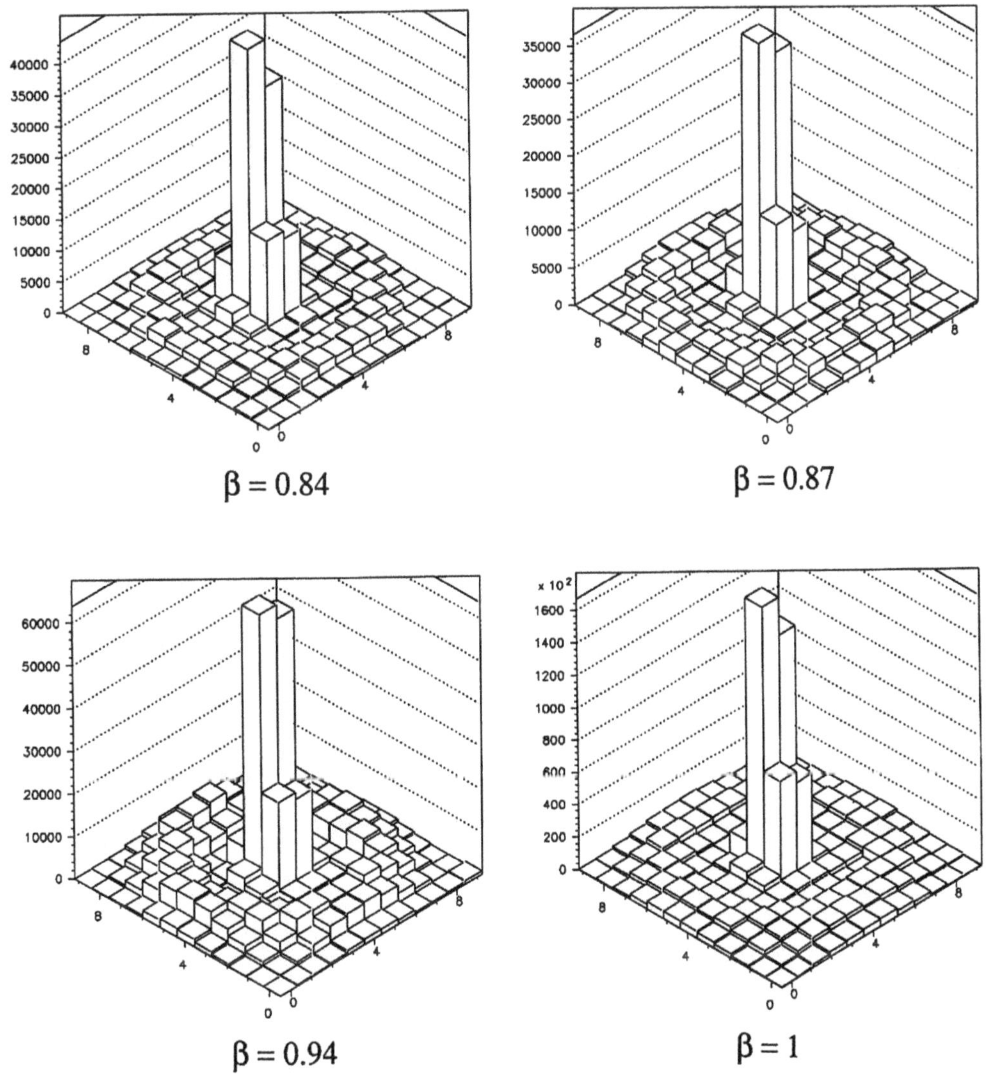

The mean number of pads hit per event as a function of β is shown in Figures 14 and 15. The two figures correspond to measurements with different gains. The solid line is the expected behavior from the simulations and the data points indicate the experimental results. The error bars indicate the width of the distribution (r.m.s.) of the measured number of pads hit. Also here, the NaF-quartz distance has been adjusted so that the Cherenkov ring radius at each β varies between 35 and 39 mm. The dotted line corresponds to the simulated number of photoelectrons detected as a function of β.

As can be seen in the figures, there is a factor of about 2.3 between the number of detected photoelectrons per event and the number of pads hit. This ratio turns out to be approximately independent of beta for a fixed ring radius.

The number of photoelectrons detected per event reaches a maximum at $\beta = 0.92$ and then decreases again. The increase for $\beta<0.95$ follows the increase in the number of emitted Cherenkov photons. Close to beta equal L the reflection losses in the NaF crystal and in the quartz together with the reduced transmission of the grid reduce the number of detected photons.

The radius of the Cherenkov ring can be calculated in different ways. One way is to calculate the mean distance between each pad hit and the reconstructed impact point of the particle. This results, to a high accuracy, in the same radius and resolution as fitting a ring to each event, which is mathematically much more complicated and hence takes a longer time on-line. Figure 16 shows the simulated ring radius as a function of β for a fixed NaF-quartz distance of 12 mm. The solid line corresponds to the mean distance

Figure 14. Simulations of the number of pads hit (solid line) and the number of detected photoelectrons per event (dotted line) as a function of β; the NaF-quartz distance has been adjusted so that the Cherenkov ring radius varies between 35 and 39 mm. The data points show the experimentally measured number of pads hit per event.

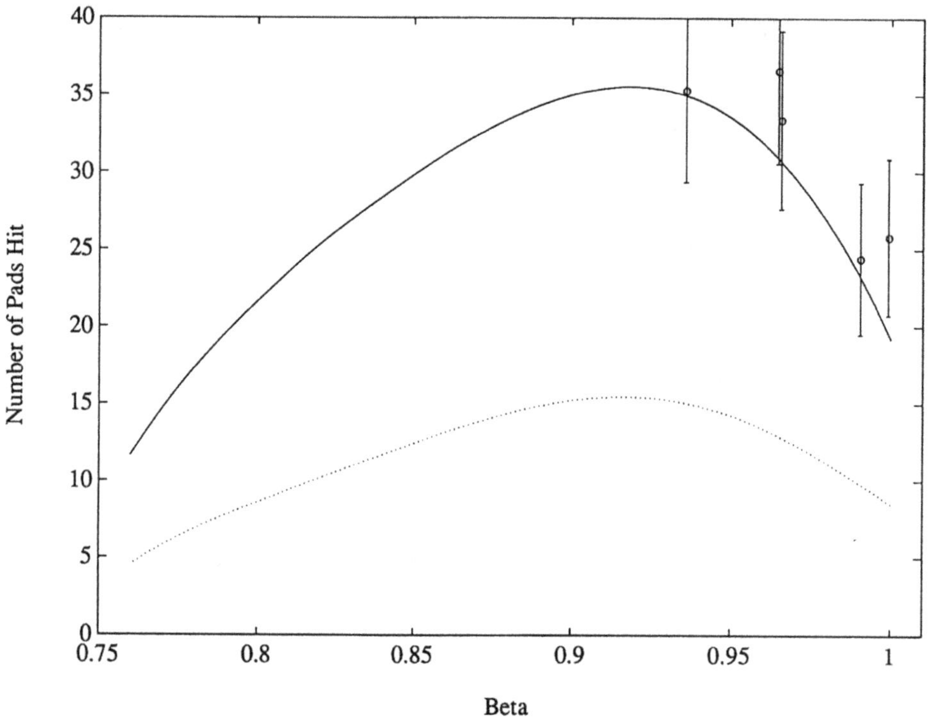

Figure 15. Simulations of the number of pads hit (solid line) and the number of detected photoelectrons per event (dotted line) as a function of β at a lower gain; the data points show the experimentally measured number of pads hit per event. The lower gain results in a) a lower probability to detect a Cherenkov photon, and b) in a lower number of pads fired per detected Cherenkov photon.

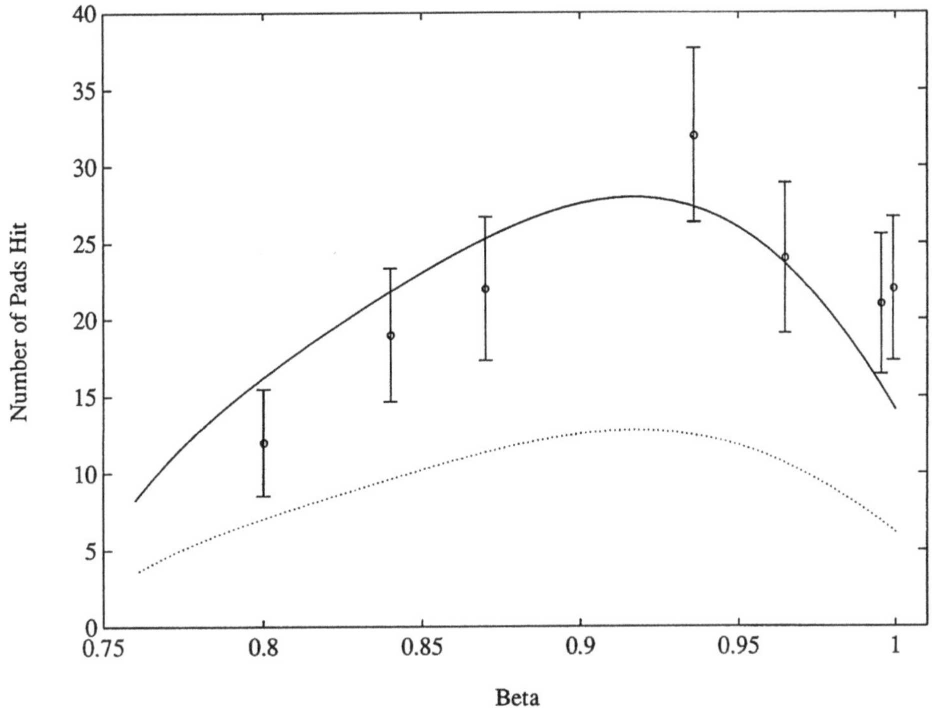

between each pad hit and the reconstructed impact point of the primary particle, while the dotted line shows the fitted ring radius as a function of β. The third line indicates the Cherenkov ring radius fitted to the true detection point of each photoelectron.

As can be seen in Figure 16, for ring radius larger than 15 mm the algorithms give the same result to a high accuracy. In a real detector, the ionization signal spreads out over a radius of at least 15 mm, and thus rings with a smaller radius than 15 mm cannot be separated from the ionization signal. Hence, the simple mean ring radius method is adequate to use for calculating the ring radius of a single event.

The uncertainty on the reconstructed ring radius depends strongly on the number of photoelectrons detected per event, and hence on the velocity of the primary particle. Figure 17 shows the distribution of the reconstructed ring radius for different β. Also here, the NaF-quartz distance has been adjusted so that the ring radius is approximately the same (R=36 mm) for all runs.

At low β, the number of photoelectrons produced is small and hence the distribution is broad. When beta increases, the resolution improves as the number of photoelectrons detected per event increases. Close to β equals 1, the resolution gets worse again as the number of photoelectrons decreases. Note that for a ring radius of about 37 mm and using 8 by 8 mm^2 pads the standard deviation is never larger than 2.5 mm in the velocity range between $\beta = 0.80$ and $\beta = 1.0$. It should be noted that this resolution includes all errors such as chromatic errors, uncertainty of the position due to finite pad size, the smearing of the ring due to the thickness of the radiator and the finite absorption length of TMAE, etc.

Figure 16. Simulations of the mean ring radius (solid line) and the fitted ring radius (dotted line) as a function of β; the NaF-quartz distance is fixed at 12 mm. The third line (dot,dash) corresponds to the correct ring radius obtained by fitting a ring to the detection point of each photoelectron. The deviation at low β is to due the difficulties to separate pads fired from a Cherenkov photon from those fired by the ionization.

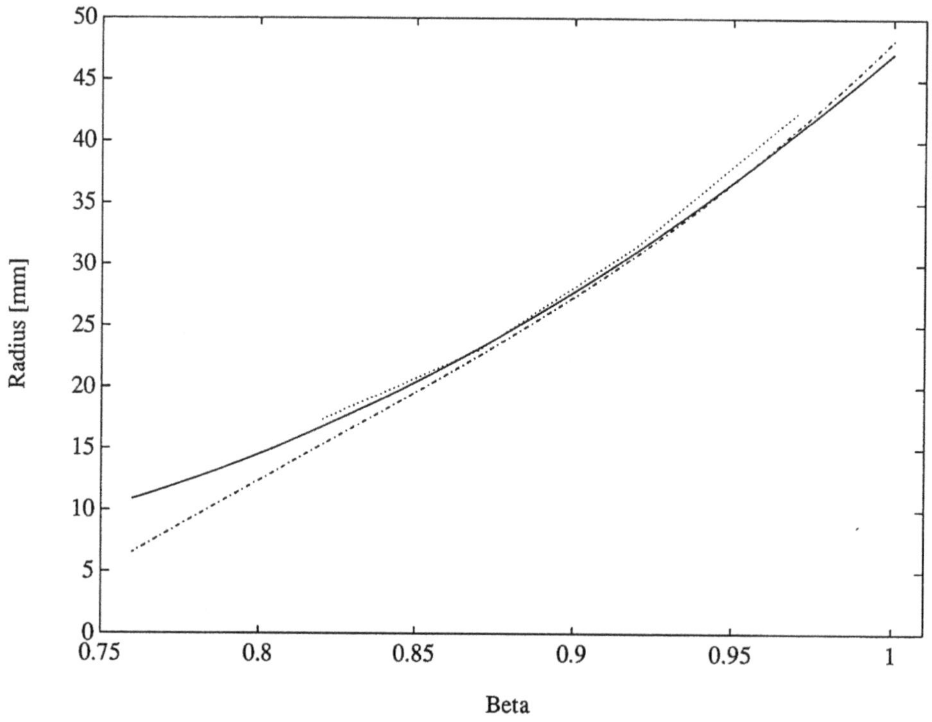

For a fixed ring radius of 36 mm the expected resolution (r.m.s.) as a function of β is shown in Figure 18. The different curves correspond to different TMAE temperatures between 20°and 60°C. This illustrates the influence of the TMAE absorption length on the resolution. The highest resolution is obtained forβ = 0.85-0.90. In this velocity interval the highest number of photoelectrons is detected and the chromatic errors are small since the Cherenkov angle is much smaller than the total reflection angle. Above a temperature of 40°-50°C, the temperature of the TMAE only influences the resolution to a small degree. Hence, an optimum working temperature is 50°C. Furthermore, what can also be observed from Figure 18 is that the resolution presents a plateau over a large velocity range at high temperatures. Higher temperature also decreases the variation of the resolution as a function of β.

The chamber has been tested between 30° and 60°C with good results, and the experimentally measured resolutions are indicated in the figure as data points.

One useful parameter in the analysis of the performance of the NaF RICH detector is the mean refractive index of NaF at a given set-up. The mean refractive index, n, is calculated from the measured ring radius R using the geometrical parameters and the temperature dependent mean absorption length of the TMAE through:

$$R = t_{NaF}/2 \tan \theta_C + d_{NaF\text{-}quartz} \tan \alpha + t_{quartz} \tan \gamma + L_{TMAE} \sin \alpha \qquad (1)$$

Figure 17. Reconstructed ring radius distribution for various values of β at a TMAE temperature of 40°C; the resolution depends primarily on the detected number of Cherenkov photons. As β approaches unity, more and more photons are trapped inside the NaF crystal by total reflection.

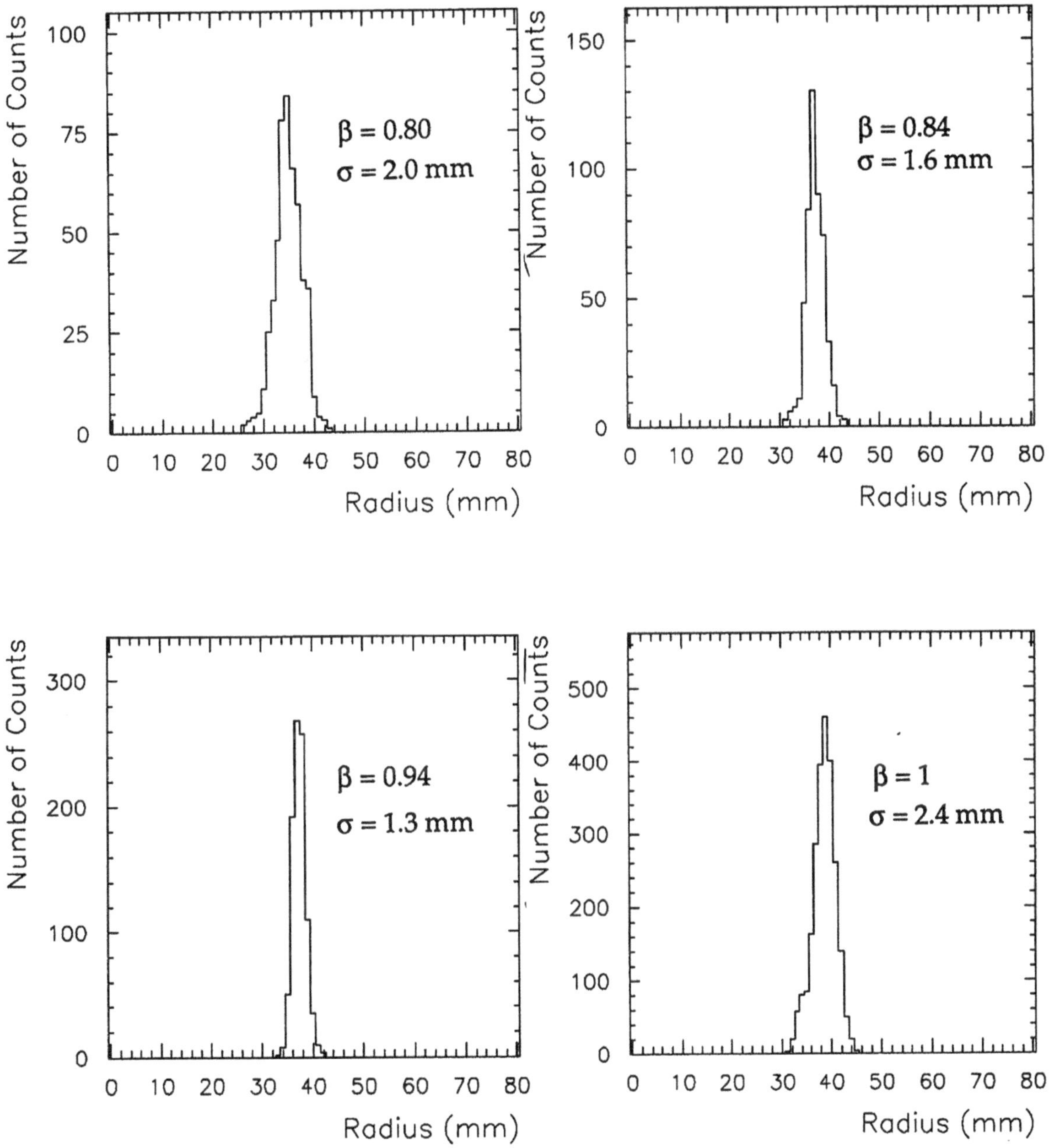

where t_{NaF} is the thickness of the NaF crystal, $d_{NaF\text{-}quartz}$, the NaF-quartz distance, t_{quartz} the thickness of the quartz and L_{TMAE} the temperature dependent TMAE absorption length. If n_{quartz} is the refractive index of quartz, then the angles are defined through:

Figure 18. Simulated resolution as a function of β at different TMAE temperatures; the NaF-quartz distance is adjusted to give an approximately constant ring radius of 36 mm for all β. The data points correspond to experimentally measured resolutions. The temperature of the TMAE has large impact on the resolution of the reconstructed Cherenkov angle. The higher the TMAE temperature, the higher is the quantum efficiency to detect a Cherenkov photon.

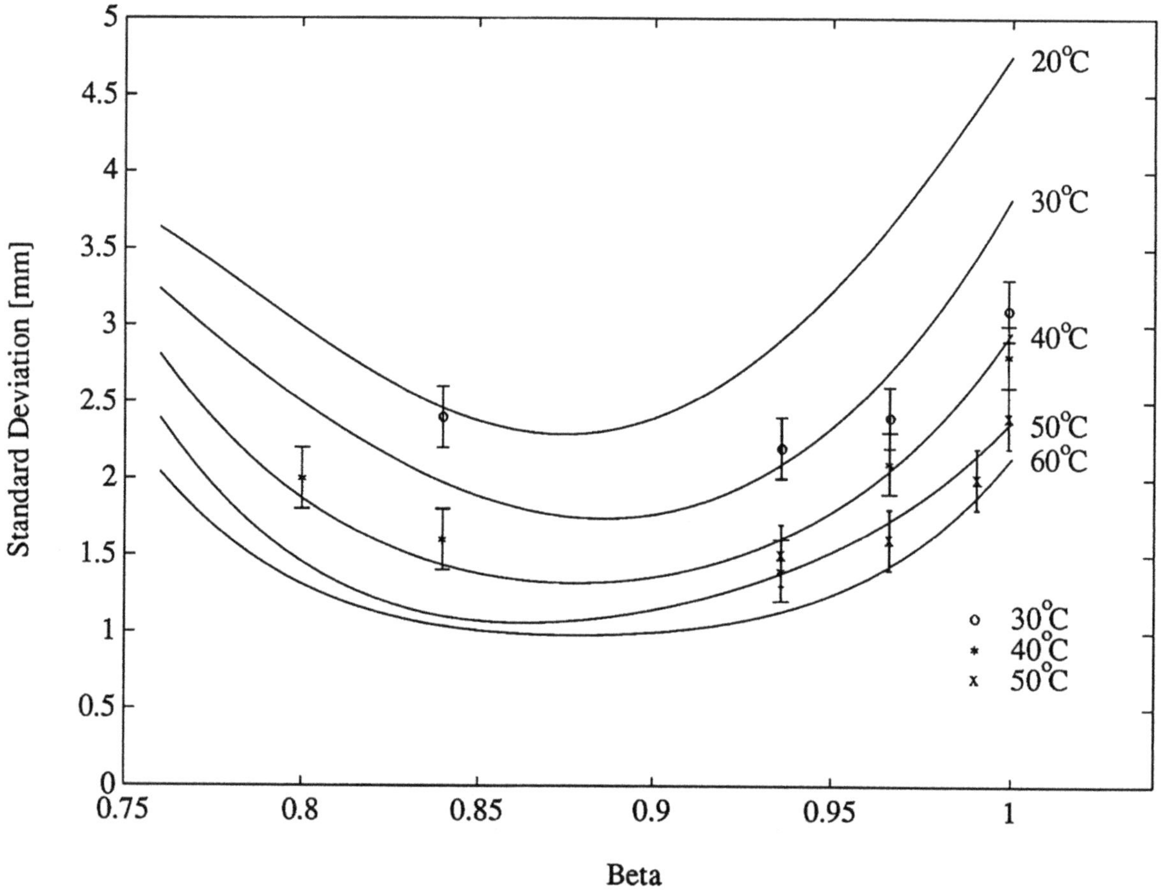

$$\cos \theta_C = 1/\beta n \tag{2}$$

$$\sin \alpha = n \sin \theta_C \tag{3}$$

$$n_{quartz} \sin \gamma = \sin \alpha \tag{4}$$

This is a way of comparing measurements at different beta with this small prototype, where the NaF-quartz distance has to be adjusted between every run to keep the ring inside the active area of the detector. If the mean refractive index as a function of beta is known, then the behavior of the detector over a large velocity range, and at a given set-up can be calculated. Figure 19 shows the simulated mean refractive index as a function of β at four different temperatures. The vertical error bars of the measured data points correspond to the influence of the uncertainty of +0.5 mm in the NaF-quartz distance and 2% uncertainty in beam momentum. The horizontal error bars correspond to a 2%uncertainty in the beam momentum. The drop in the mean refractive index at high β can be explained from the fact that photons of short wavelengths, where the refractive index of NaF is high, are lost by reflections in the NaF and quartz or absorbed by the grid since they refract out at a large angle. Therefore, photons with long wavelengths will have a larger probability of being detected than photons with short wavelengths. Hence, the mean refractive index will drop. This effect is only significant at high beta. At low beta, the momentum of the particle has been corrected for the energy lost in the material between the last bending magnet, defining the momentum, and the detector. The material is listed in Table 1 with an approximate energy loss at each stage for 1.5 GeV/c momentum protons, calculated using the Bethe-Bloch equation (Particle Data Group 1988). The total energy lost is about 34 MeV for 1.5 GeV/c protons, and about 26 MeV for minimum ionizing particles. The small discrepancy between the measured mean refractive index and that simulated at low β can be explained by a 2% too high momentum from the magnet settings of the last bending magnet defining the momentum of the beam. This corresponds approximately to the uncertainty in absolute momentum in the beam (Simon, 1990). In the data points at β = 0.84 the effect of 2% uncertainty in the beam momentum is large. At higher momentum this effect is significantly smaller, since a 2% change in momentum does not significantly change the particles at high momenta.

Figure 19. Reconstructed mean refractive index as a function of β at various temperatures; the solid lines correspond to simulations and the data points to experimental results. It is necessary to adjust the refractive index of the NaF with β in the reconstruction of the Cherenkov angle to compensate for the increasing losses of Cherenkov photons inside the NaF crystals by total reflection.

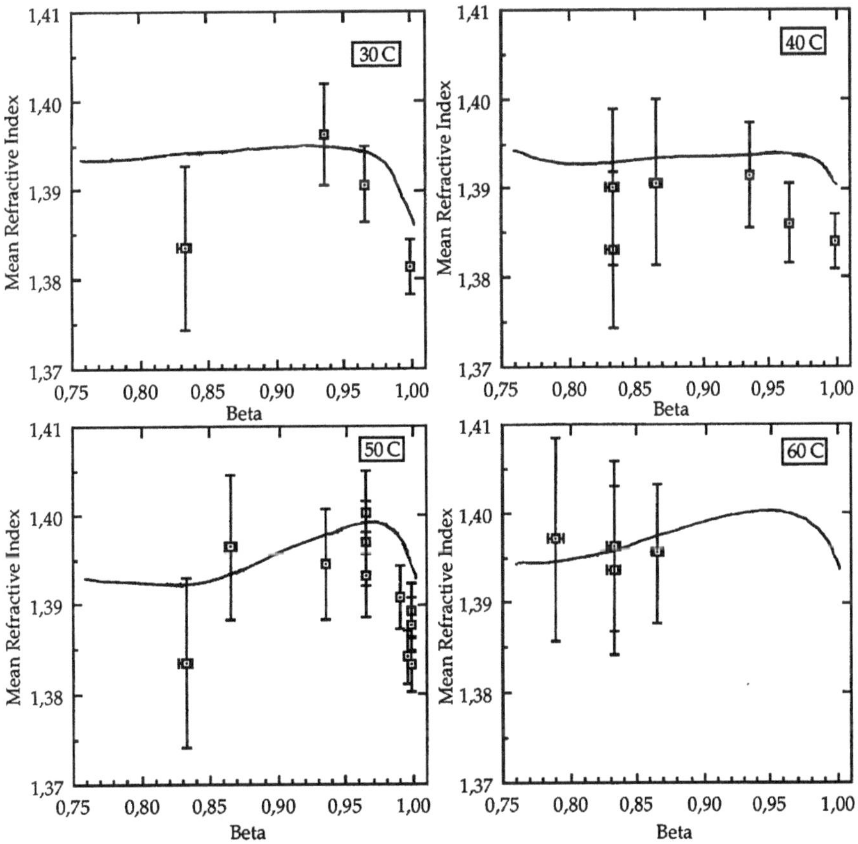

Table 1. Absorbing material before the detector

Object	Material	Thickness [mm]	Approx. Energy Loss* [MeV]
12 Scintillators	Plastic	70	18
Gas Cherenkov	N_2, 2 bar	2000	1
Air		10000	3
Microstrip	Silicon	10	5
JetSet RICH	Quartz + Al		5
Wire Chambers		1	Entrance Knob
Al	2	1	
Sum			**34 MeV**

* For 1.5 GeV/c protons.

4.2 Anode Readout

The signals from the anode wires can possibly be used to generate a fast trigger. To examine this, all anode wires are read out and analyzed. Accumulated $\beta = 1$ events from the anode signals are shown in Figure 20.

The projections are in accordance with the anode projections measured with the low pressure chamber, but the ionization signal is larger. The two horns of the ring are clearly seen, and the central peak is due

Figure 20. Anode projection of accumulated $\beta = 1$ events; the anode wires alone can be used to give an estimate of the radius of the Cherenkov ring.

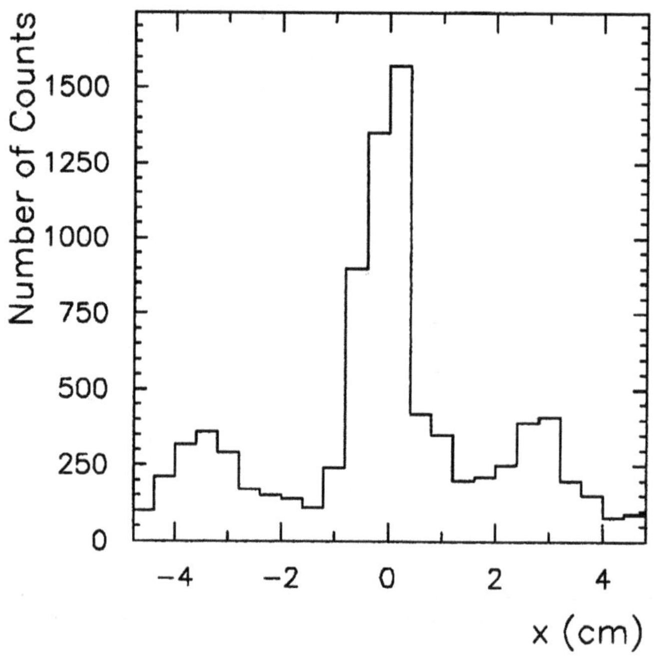

Figure 21. Correlation between the maximum anode radius and the reconstructed ring radius; the error bars for one measurement are shown and corresponds to the width (r.m.s.) of the two distributions. The errors in all measurements are approximately the same. The pad plane gives a more accurate measurement of the ring radius, while the anode wires can be used to get a fast trigger.

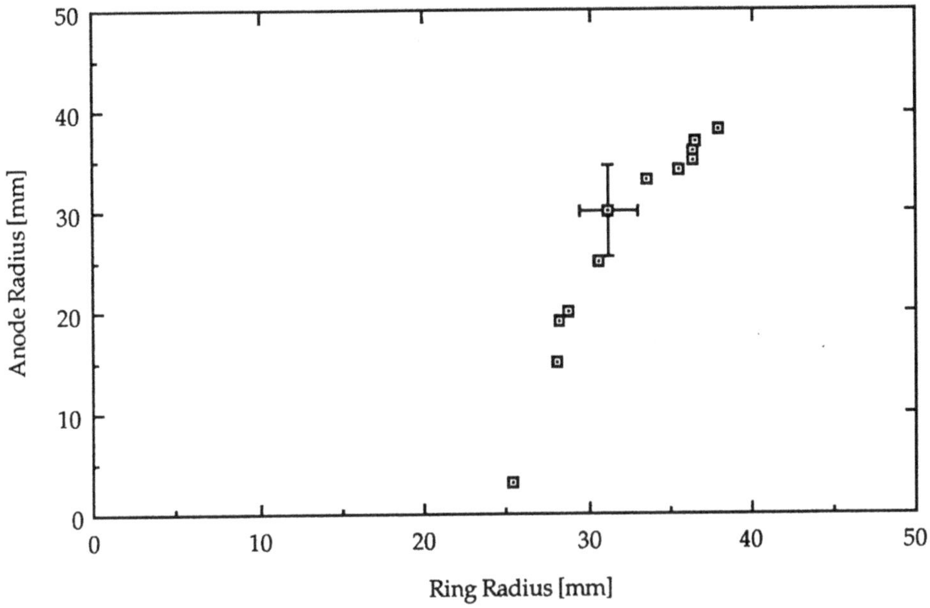

to the ionization. The distance between two adjacent wires is 4 mm. The small asymmetry in the figure is due to a small misalignment of about 0.5°.

From these horns, an accurate estimation of the ring radius is possible. For single events, the best estimation of the ring radius is possible by looking at the distance between the outermost wires fired by the event. The large signal in the center from the ionization can be used as a reference to find an event. The correlation between this maximum anode projection radius and the reconstructed ring radius is illustrated in Figure 21. The errors in all data points are approximately the same, and hence error bars are only displayed for one measurement.

The vertical error bars correspond to the width (r.m.s.) of the distribution of the maximum anode radius. The horizontal error bars correspond to the standard deviation of the reconstructed ring radius. Here the TMAE temperature is 60°C. The cut-off at a ring radius of 25 mm is artificial and is due to that only hits outside the central 16 pads are used in the reconstruction of the ring radius. Typically only one or two anode wires are fired by the ionization.

This property of the anode signals can be used to generate a fast trigger. It only gives a very rough estimation of the ring radius, but the information can be available in less than 100 ns if fast amplifiers are used on the anode wires.

4.3 Non-Perpendicular Incidence

Until now, all events have been due to particles at normal incidence to the detector. To be useful in a real experiment, the detector has to be able to discriminate between particles also at angular incidence.

Figure 22. Single β = 0.86 proton event at 9° incidence angle; only half of the ring is detected as the lower half is trapped inside the NaF crystal by total reflections.

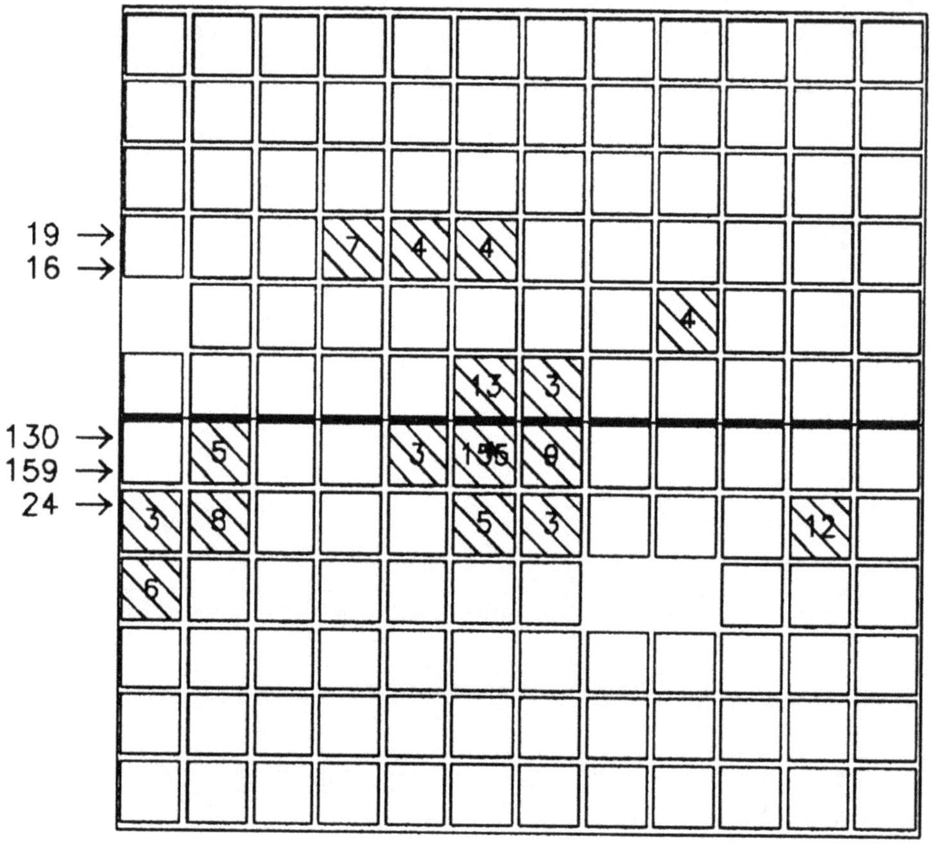

A typical single event is shown in Figure 22, corresponding to a $\beta = 0.86$ proton at 9° incidence angle.

The shape of the hit pattern is not an ellipse, but rather a drop at small angles and a wide u at larger incidence angles. The density of the photoelectrons is much higher in the center of the hit pattern, and decreases rapidly away from the center.

The simulated photoelectron hit pattern for β = 0.85 and β = 1 events at different incidence angles are shown in Figure 23. The central hits are due to Cherenkov light produced in the quartz window that can escape out of the quartz for non-normal angular incidence. At normal incidence, all these photons are trapped by total reflection inside the quartz if the β is larger than 0.84. The ionization signal is not included in the simulations for clarity.

The drop-like shape opens up to a wide U shape already at very small incidence angles for β = 1 events. This is due to the fact that the Cherenkov photons are emitted at an angle very close to the total reflection angle of the NaF, which is why only a small increase in the incidence angle of the photons towards the NaF surface is enough to cause total reflection and hence no light will come out on the far side of the Cherenkov cone. This is not a severe limitation of the detector, since, on the other hand, more Cherenkov light will refract out on the near side of the Cherenkov cone as the incidence angle of the photons towards the NaF surface is much smaller than the total reflection angle. This implies that on the near side of the Cherenkov cone, a larger number of photons will refract out compared to normal

Figure 23. Simulated photoelectron hit pattern at different angular incidence for β = 0.85 and β = 1 events; the effect of total reflection inside the NaF crystal is clearly visible even for small Cherenkov angles (low values of β).

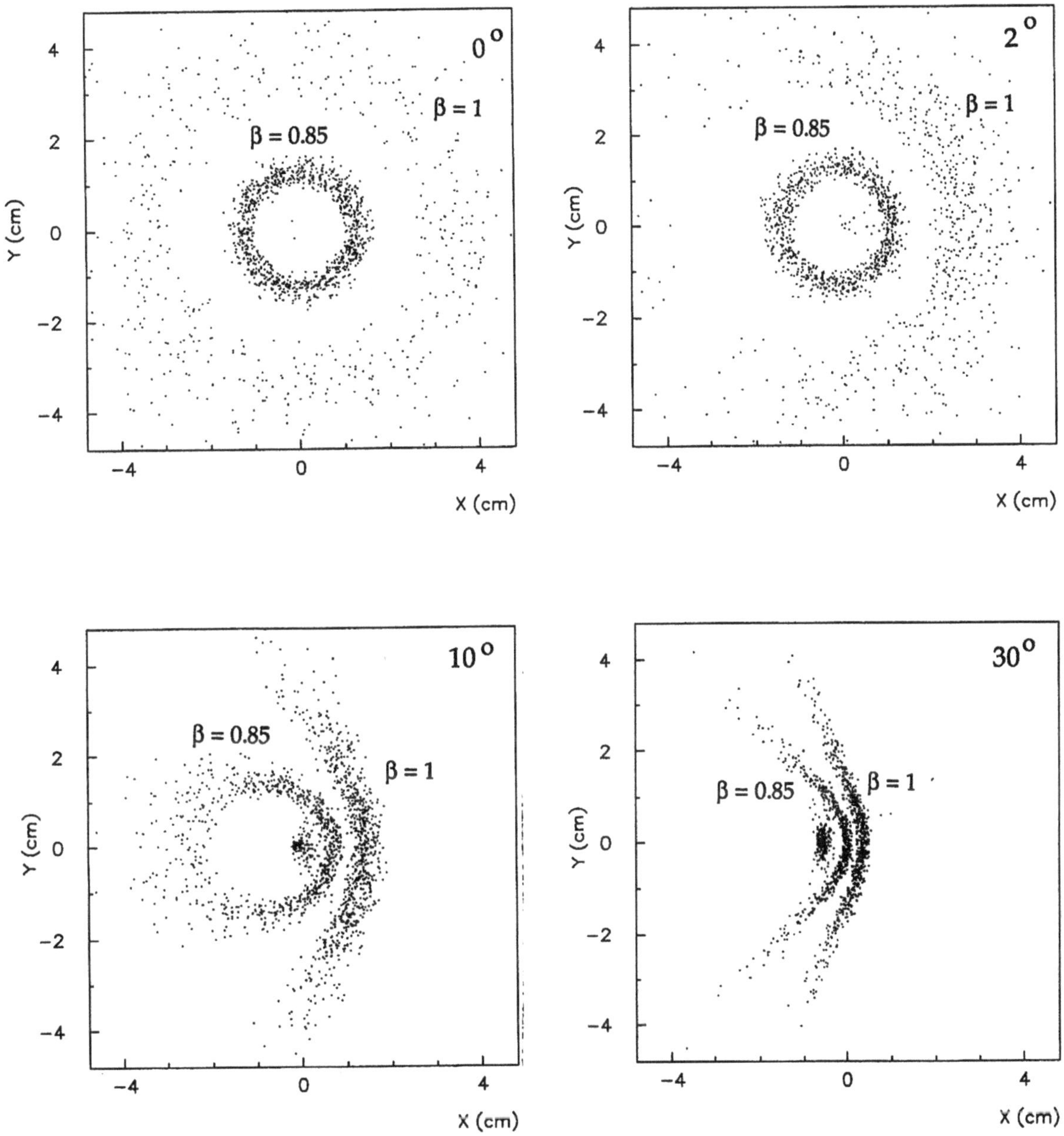

incidence angle, but also that the chromatic errors are reduced since the photons does not refract out close to the total reflection angle any longer. To fully benefit from these reduced chromatic errors, the pads have to be made very small since the change in the distance between the ionization center and the hit pattern will only vary slowly with the beta of the incidence particle. For large pads, the resolution decreases significantly at angular incidence compared to normal incidence.

Figure 24. Measured mean Cherenkov angles at different incidence angles for β = 0.97 protons; the loss of Cherenkov photons in the NaF crystal by total reflection results in a poorer resolution of the reconstructed Cherenkov angle.

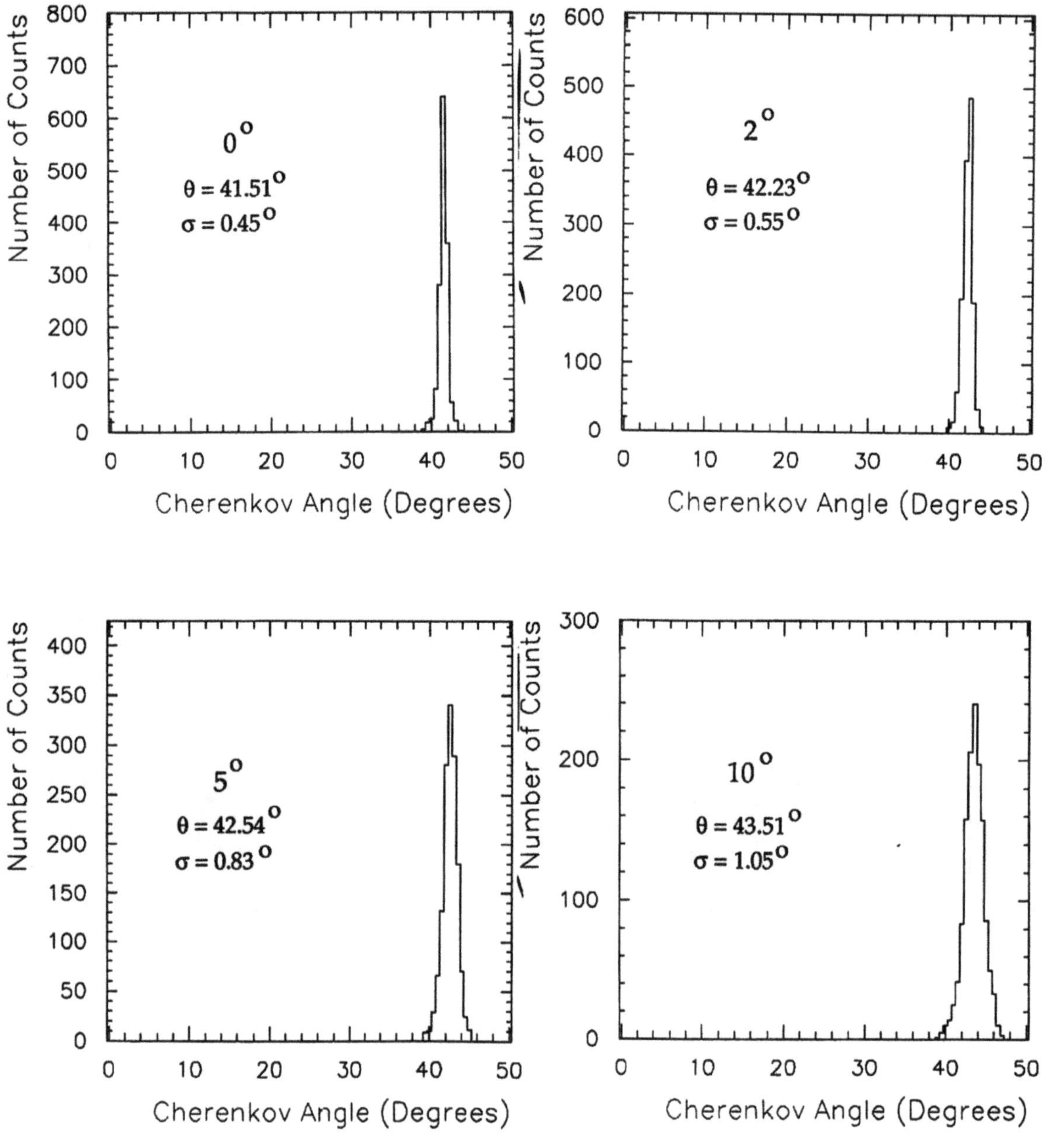

A fast and simple way of separating particles from one another is by measuring the distance between the bottom of the U to the impact point of the particle. The major drawback of this technique is that this distance changes very slowly with the velocity of the primary particle, and hence the separation power is small at high beta. A more severe discrimination is possible by calculating the mean Cherenkov angle of each event. The Cherenkov angle of the photon generating each hit is calculated from the hit position of the photoelectron using the same technique as when the mean refractive index of NaF was calculated for normal incidence, as described above. A mean Cherenkov angle of an event is then calculated as the mean of all Cherenkov angles from each hit. It is found that when calculating the mean, unit weights give the best resolution compared to using the signal amplitude or the logarithm of the signal amplitude as weights. Figure 24 shows the reconstructed Cherenkov angles for 5 GeV/c ($\beta = 0.97$) protons at different angles of incidence.

The resolution decreases rapidly as the angle of incidence increases, but it is still possible to have good separation power up to10° of incidence, and fair separation power up to 30° of incidence.

All this analysis demands that the incidence angle and the momentum of the particle are known with good accuracy. This is normally the case in a real spectrometer, since the particle identifier is often situated outside a tracking system in a magnetic field, which, with high accuracy, gives both the incidence angle and the momentum of the primary particle.

5. SEPARATION POWER

The separation power of the NaF RICH detector using a quartz window and pad readout will mainly be presented in terms of how well it can separate protons from pions, but the results can easily be transformed to be used for separation between other particles. This discussion is in two parts. First, the separation power for normal incidence particles (approximately the same results are valid for small incidence angles) which is the case in a fixed target experiment far away from the interaction region. Secondly, a more general discussion of the separation power at angular incidence.

5.1 Normal Incidence Angles

At normal incidence angles, the resolution of the detector can either be expressed in terms of ring radius or Cherenkov angle. The solid lines of Figure 25 show the expected ring radius as a function of momentum for electrons, muons, pions, kaons and protons for a NaF-quartz distance of 12 mm. The mean index of refraction dependence on beta is included in the curves. The measured resolutions (r.m.s.) at 50°C and different velocities (ring radii) are illustrated as error bars on the right side of the figure.

Using these ring radii and resolutions, the difference in ring radius between two particles as a function of momentum can be expressed in number of standard deviations of the reconstructed ring radius. Figure 26 shows the difference in ring radius between two particles expressed in standard deviations of the reconstructed ring radius. As standard deviation, the mean standard deviation of the two particles at each momentum is used, since the resolution varies with momentum. Also here, the NaF-quartz distance is 12 mm and the TMAE is kept at a temperature of 50°C.

At a ring radius of less than 18 mm, the hit pattern of the Cherenkov ring interferes with hit pattern of the ionization of the amplification gas by the primary particle, and hence it is not possible to determine ring radii smaller than 18 mm. In this region the detector acts as a threshold Cherenkov counter where only an upper limit of the particle velocity is determined by the detector. The threshold velocity is determined by the lateral spread of the ionization signal and the NaF-quartz distance, which determines the ring radius for a specific velocity. This corresponds to the slope at low momentum, where the ring radius of the slow (massive) particle is considered to be equal to the ionization radius as long as it can be resolved. At higher momentum, the ring radii of the two particles approach each other and

Figure 25. Ring radius as a function of momentum for different particles; the error bars on the data points illustrate the experimentally obtained resolutions. Hence, the NaF RICH detector can efficiently be used to identify elementary particles over a wide momentum range.

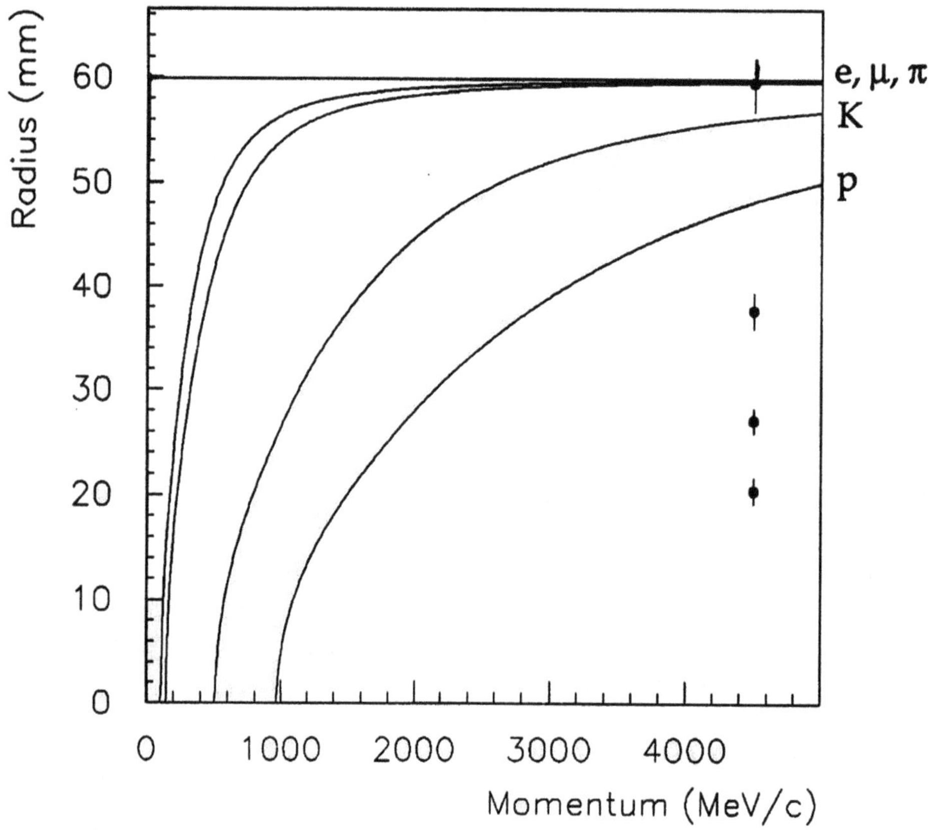

Figure 26. Difference in ring radius between two particles expressed in number of standard deviations of the reconstructed ring radius; a good particle separation is possible over a wide momentum range.

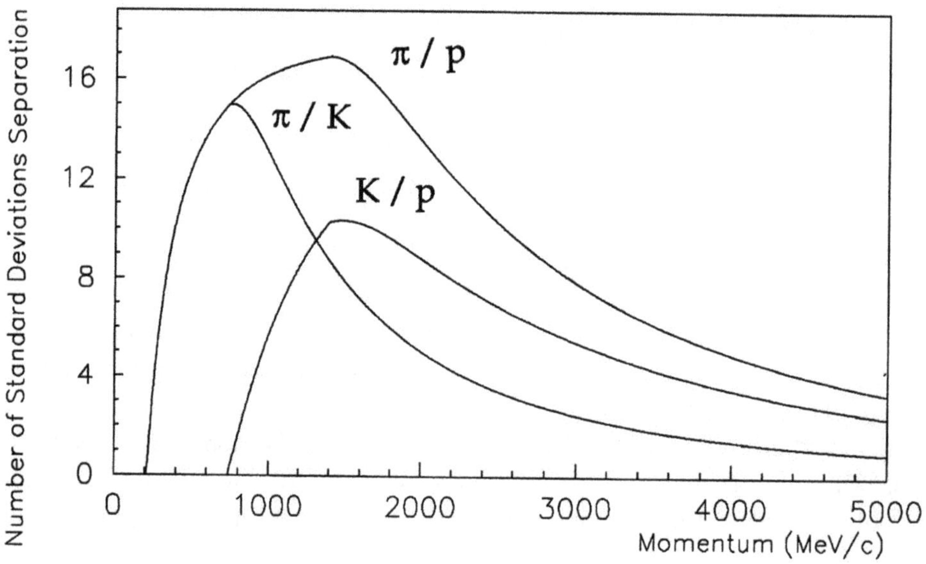

the separation power of the detector decreases. For normal incidence particles, a 3 standard deviation separation is possible between 0.2 and 5 GeV/c momentum between pions and protons, and between 0.2 and 2.5 GeV/c momentum between pions and kaons. In a real spectrometer, there is normally some kind of time of flight or dE/dx measurements that fulfil particle identification at low momenta where the RICH detector is not sensitive enough.

Figure 26 shows that the separation power of the NaF counter is superior to conventional methods like time of flight, dE/dx, etc.

5.2 Angular Incidence

At non-normal angles of incidence, the performance of the detector is determined by the accuracy to which the Cherenkov angle of the emitted photons can be determined. The resolution is limited by chromatic errors of the NaF and of the quartz, the pad size, the uncertainty in the impact point of the primary particle, the thickness of the radiator, the number of photoelectrons detected, the absorption length of TMAE and the uniformity of the gain in the electronics and the wire chamber, etc.

Simulations of the reconstructed Cherenkov angles as a function of the incidence angle of the particle are shown in Figure 27 for different beta. The smallest angle between a Cherenkov photon and NaF

Figure 27. Simulations of the reconstructed Cherenkov angles as a function of β for different angles of incidence; the NaF-quartz distance is fixed to 11 mm. The measured Cherenkov angle at $\beta = 1$ for 10° incidence is indicated with the measured resolution (r.m.s). At large incidence angles the index of refraction of NaF must be adjusted to compensate for the total reflection inside the NaF, especially for Cherenkov photons emitted at shorter wavelengths.

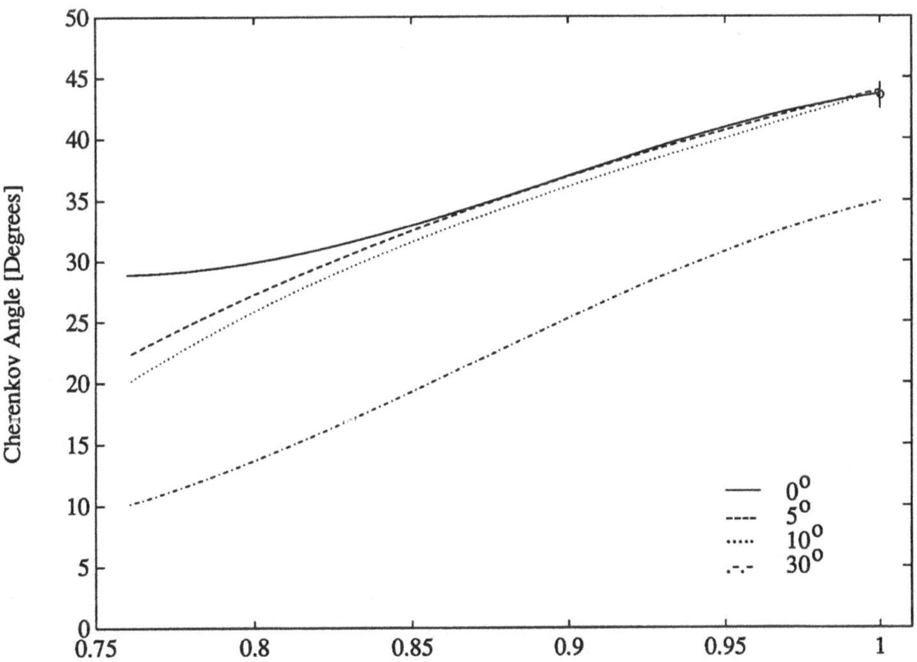

surface, inside the NaF, that can be reconstructed in the central region of the detector is about 30° if the NaF-quartz distance is 11 mm. This is because photons that refract out at smaller angles will convert inside the ionization region of the primary particle and hence cannot be distinguished from the ionization signal. This is also the main limiting factor on the resolution at large angles of incidence.

In the analysis, the central 16 pads around the impact point of the primary particle are excluded. This implies that events at low beta or at large incidence angles will not have the correct reconstructed Cherenkov angle, but only the outer tail of the hit pattern will be included in the reconstruction of the Cherenkov angle. Hence, the number of detected photoelectrons decreases rapidly, since most of them are produced in the central region of the hit pattern, which increases the width of the distribution of the reconstructed angle. Moreover, the reconstructed Cherenkov angle will be too large and hence the difference in angle between two particles of the same momentum, but different velocity, will be smaller than that expected if there were no ionization signal. If the NaF-quartz distance is increased this effect will decrease, but the Cherenkov ring radius at normal incidence and β equal 1, will be too large to be detected with the same set-up.

The shift to too small reconstructed Cherenkov angles at 30° incidence is a consequence to that the central 16 pads are excluded in the analysis. At 30° angle of incidence 90% of all hits are inside these 16 central pads, according to simulations. Therefore, only very few (1-5) pads are used in the reconstruction

Figure 28. Separation power between pions and protons of the NaF RICH detector expressed in number of standard deviations distance between the reconstructed Cherenkov angles as a function of momentum at various angles of incidence; the instrument is a powerful particle identifier, capable of separating pions from protons over a wide momentum range, even for quite large angles of incidence.

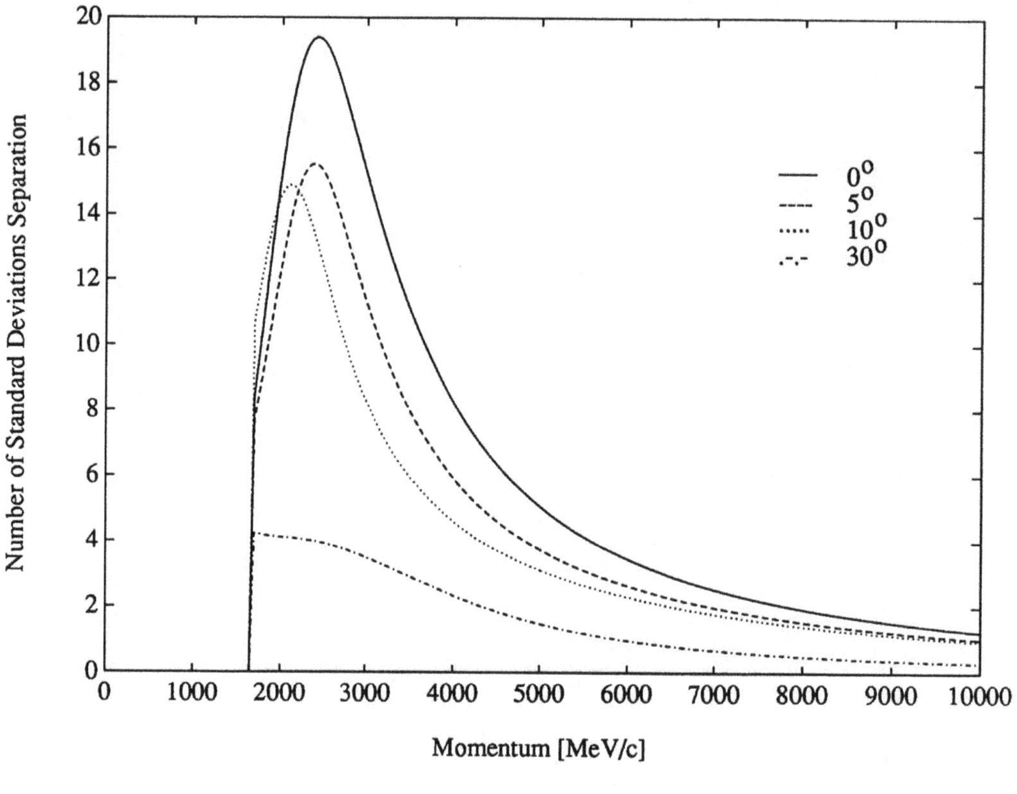

of the Cherenkov angle and the errors are hence very large. According to simulations, it is possible to slightly improve the resolution at large angles of incidence by reducing the excluded central area from 4 by 4 pads to 2 by 2 pads. In this way some pads fired by the ionization will be included in the reconstruction of the Cherenkov angle but the reconstructed resolution is still improved.

In the same way as at normal incidence, the separation power of the detector at non-normal angles of incidence can be illustrated by means of the number of standard deviations between the reconstructed Cherenkov angles of two different particles. Figure 28 shows this separation power as a function of momentum for different incidence angles for proton-pion separation. The standard deviation used to express the difference between the two reconstructed Cherenkov angles is the mean standard deviation of the two particles at each specific condition. The NaF-quartz distance used is 11 mm and the TMAE has a temperature of 50°C.

Comparing Figure 28 with Figure 26 shows that the method of calculating the mean Cherenkov angle gives a slightly superior separation power to the detector compared to calculating the mean ring radius. Very good separation (5 standard deviations) is possible between pions and protons between 1.8 and 5 GeV/c for normal incidence. The separation power is reduced at angular incidence, but up to 10° of incidence the same separation power is possible up to 7.5 GeV/c. For even larger angles of incidence the possibility of separating particles is strongly reduced, mainly because of the overlap of the hit pattern with the ionization signal.

6. CONCLUSION

The optimum performance of the NaF RICH detector is achieved if the NaF radiator is used together with a quartz window separating the drift volume from the photosensitive MWPC. The use of TMAE as photo converter is a necessity to have good separation power and high detector efficiency. To avoid having a large conversion region the TMAE has to be heated to 40°-50°C. Moreover, pad readout is used to give a non-ambiguous two-dimensional reconstruction of the hit pattern.

The use of TMAE results in at least five times more photoelectrons produced per event than the use of TEA. This implies that the detector can be operated at low gain which reduces the amplitude and spatial size of the ionization signal and the spatial size of the single photoelectron signal. What is even more important is that with the low gain operation there is no significant photon feedback and the anode wires need not to be shielded in any way.

The number of photoelectrons produced per event with the NaF-mirror set-up is approximately the same as with the NaF and quartz set-up. The resolution turns out to be slightly higher with the NaF and quartz set-up since the chromatic errors are reduced and the contribution from parasitic reflections does not disturb the reconstruction.

The ionization of the amplification gas by the primary particle puts a limit on the smallest ring radius that can be reconstructed. Typically 3 to 4 pads are fired by the ionization and the inactive area of the detector has a radius of approximately 20 mm. Each photoelectron fires about 2.5 pads as an average. Nice rings of up to 20 photoelectrons per event are reconstructed for single events which gives a good separation power and high efficiency to the detector.

For normal incidence the resolution of the reconstructed ring radius is less than 2.5 mm (r.m.s.) for a 37 mm ring radius. The highest resolution is obtained in the velocity range between $\beta = 0.85$ and $\beta =$

0.95, and the resolution is slightly reduced as beta approaches 1. This allows good particle separation between pions and protons up to 5 GeV/c and between pions and kaons up to 2.5 GeV/c momentum.

At angular incidence the discrimination power is slightly reduced. Good particle separation is possible between pions and protons up to 7.5 GeV/c momentum for incidence angles up to 10°. Up to 3 GeV/c a reduced pion-proton separation is possible for angles of incidence up to 30°.

The separation power of the NaF RICH counter is superior to conventional methods like time of flight, dE/dx, threshold Cherenkov counters, etc. It is fast and compact and represents only about 20% of a radiation length, and can therefore be used as an efficient particle identifier in the medium energy range ($p_{proton} < 5$ GeV/c).

REFERENCES

Francke, T. (1991). *A ring imaging Cherenkov counter using a solid radiator.* (PhD Thesis). Royal Institute of Technology, Stockholm, Sweden.

Jonsson, K. (1988). *Test of two PIDs with Cosmic Rays.* (PhD Thesis). KTH, Stockholm, Sweden.

Particle Data Group. (1988). Review of particle properties. *Physics Letters. [Part B], 204*, 64.

Rickenbach, R., Montanet, F., Backenstoss, G., Carlson, P., Chardalas, M., Franke, T., & Wildi, M. et al. (1989). Francke, T., ... Wildi, M., (The CPLEAR Collaboration), *The fast strangeness trigger of the CP-LEAR experiment. Nuclear Instruments & Methods in Physics Research. Section A, Accelerators, Spectrometers, Detectors and Associated Equipment, 279*(1-2), 305–309. doi:10.1016/0168-9002(89)91098-X

ADDITIONAL READING

Akkila, T., Francke, T., Lindblad, T., & Eide, Å. (1993). An Analog Neural Network Hardware Solution to a Cherenkov Ring Imaging Particle Identifier. *Nuclear Instruments & Methods in Physics Research. Section A, Accelerators, Spectrometers, Detectors and Associated Equipment, 327*(2-3), 566–572. doi:10.1016/0168-9002(93)90726-X

Backenstoss, G. (1989). The Fast Strangeness Trigger of the CP-LEAR Experiment. *Nuclear Instruments & Methods in Physics Research. Section A, Accelerators, Spectrometers, Detectors and Associated Equipment, 279*(1-2), 305–309. doi:10.1016/0168-9002(89)91098-X

Carlson, P. A RICH Counter for Antimatter and Isotope Identification in the Cosmic Radiation, *Proceedings of the 23rd International Cosmic Ray Conference*, Calgary, CanadaJuly 19 - 30, 1993, 2 (1993) 504.

Carlson, P., & Francke, T. (1994). *CAPRICE, Nouvelles de Saturne 1993-1994, Lab. nat.* Paris, France: Saturne.

Carlson, P., Francke, T., Löfgren, S., Weber, N., & Suffert, M. (1994). Results from Beam Tests of the CAPRICE RICH Detector. *Nuclear Instruments & Methods in Physics Research. Section A, Accelerators, Spectrometers, Detectors and Associated Equipment, 349*(2-3), 577–591. doi:10.1016/0168-9002(94)91229-7

Carlson, P., Francke, T., Weber, N., & Suffert, M. (1994). A RICH Counter for Identifying Antimatter and Light Isotopes in the Cosmic Radiation. *IEEE Transactions on Nuclear Science*, *41*(4), 1321–1327. doi:10.1109/23.322907

Francke, T., Lindblad, T., Eide, Å., Piuz, F., Williams, D., Martinengo, P., & Suffert, M. et al. (1991). Particle Identification by Cherenkov Ring Imaging using a Neural Network Approach. *Nuclear Instruments & Methods in Physics Research. Section A, Accelerators, Spectrometers, Detectors and Associated Equipment*, *307*(1), 47–51. doi:10.1016/0168-9002(91)90129-E

Francke, T., Montanet, F., Pavlopoulos, P., Rivoiron, C., Santiard, J. C., & Suffert, M. (1989). Compact and Fast RICH Detector using a Solid Radiator: Principle and First Experimental Results. *Nuclear Instruments & Methods in Physics Research. Section A, Accelerators, Spectrometers, Detectors and Associated Equipment*, *279*(1-2), 317–321. doi:10.1016/0168-9002(89)91100-5

Francke, T., Piuz, F., Williams, D., Martinengo, P., Ribeiro, R., & Suffert, M. (1991). A Fast and Compact Solid Radiator RICH Counter. *Nuclear Instruments and Methods*, *A310*(1-2), 146–149. doi:10.1016/0168-9002(91)91014-M

Francke, T., & Suffert, M. (1989). Yield of Photoelectrons Produced in TMAE and Solid TMPD by BaF2 Scintillation Light. *Nuclear Instruments & Methods in Physics Research. Section A, Accelerators, Spectrometers, Detectors and Associated Equipment*, *276*(1-2), 263–269. doi:10.1016/0168-9002(89)90643-8

Francke, T., Suffert, M., & Carlson, P. (1992, August). Identification of Antiprotons in Space Using a Compact RICH Counter. *IEEE Transactions on Nuclear Science*, *39*(4), 971–975. doi:10.1109/23.159743

KEY TERMS AND DEFINITIONS

Antiparticle: A group of elementary particles with similar characteristics as ordinary matter, except that charge and spin are opposite. When a ordinary elementary particle meets its antiparticle, they annihilate one another.

Antiproton: The antiparticle of the proton.

β: A measure of velocity expressed as the fraction of the speed of light (300.000 km/s). β is normally expressed in per cent of the speed of light.

C_6F_{14}: A fluorocarbon which is liquid at room temperature. Manufactured under the brand FC-72.

CAPRICE: The Cosmic AntiParticle Ring Imaging Cherenkov Experiments. A series of balloon borne experiments to measure the flux of antimatter and light isotopes in the cosmic radiation. In total CAPRICE made five successful flights between 1994 and 1999. The flights were performed from Lynn Lake, Manitoba, Canada, and Ft Sumner, New Mexico, USA.

Cherenkov Light: A new type of light emitted when charged particles traverse a transparent medium with a velocity higher than the speed of light in the medium. It is named after the Russian scientist Pavel Cherenkov who discovered it.

CPLEAR: A particle physics experiment at CERN aimed to measure the CP violating parameters in the neutral kaon system.

Kaon, Pion (π): Elementary particles.

LEAR: The Low Energy Antiproton Ring, a particle accelerator at CERN initially developed to store antiprotons.

MWPC: Multi Wire Proportional Counter, a type of detector often used in particle physics experiments.

NaF: The salt sodium fluoride. It is the solid material with the lowest index of refraction for visible and ultra violet light in room temperature.

Photomultiplier (PM) Tube: A photosensitive device operating in vacuum.

Positron: The antiparticle of the electron.

RICH: Ring Imaging CHerenkov detector. A type of detector which measures the velocity of elementary particles by measuring the emission angle of the Cherenkov light produced when the particle traverses a medium with a velocity higher than the speed of light in the medium.

TEA: Triethylamine. A transparent liquid which in vapor form is a photo-converter, i.e. it has a finite probability to absorb a light photon and emit a photoelectron.

TMAE: Tetrakis-(dimethylamino)-ethylene. A yellow liquid which in vapor form is an efficient photo-converter, i.e. has high probability to absorb a light photon and emit a photoelectron.

Chapter 12
Performance of the CAPRICE94 RICH Detector during the 1994 Balloon Flight

ABSTRACT

A RICH detector capable of detecting unit charged particles, e.g. antiprotons and positrons, was in 1994 used successfully for the first time in a balloon borne magnet spectrometer. The thin and compact CAPRICE94 RICH detector uses a NaF solid radiator, TMAE vapor as photo-converter and cathode pad readout in the photosensitive MWPC operated at low gain. 15 photoelectrons are detected per ring for $\beta = 1$, perpendicular incidence particles giving a resolution on the Cherenkov angle of 8 mrad, increasing to 14 mrad at 20^oC incidence angle. Besides particle identification on an event-by-event basis it efficiently rejects multiparticle events and albedo particles (Barbiellini, 1996).

1. INTRODUCTION

The Cosmic AntiParticle Ring Imaging Cherenkov Experiment (CAPRICE) has for the first time used a RICH detector in a balloon borne magnet spectrometer that is sensitive to unit charged particles. Previously, the Chicago University group (Buckley, 1992) has flown a gas-Rich detector sensitive to helium and higher charges.

The aim of the experiment is to measure the flux of antimatter, mainly antiprotons and positrons, and light isotopes in the cosmic radiation. Sophisticated particle identification detectors are needed to identify the rare antiprotons in a large background of electrons, muons and pions, and the positrons in an even larger background of protons.

The NMSU/WiZard CAPRICE94 spectrometer (Golden, 1991) is shown in Figure 1 and includes from top to bottom: a solid radiator RICH detector, scintillators for time-of-flight and dE/dx measurements, a 4 Tesla super conducting magnet with a tracking system of multiwire proportional chambers (MWPC) and drift chambers (Hof, 1994), and an electromagnetic Si-W calorimeter (seven radiation lengths deep) (Barbiellini, 1993).

DOI: 10.4018/978-1-5225-0242-5.ch012

Figure 1. The CAPRICE94 magnet spectrometer; it consists of a superconducting magnet giving an up to 4 T strong magnetic field in which the deflection of the cosmic particles is measured by a set of drift chambers. The particle identification is done by a solid radiator RICH detector, a W-Si electromagnetic calorimeter and a time-of-flight system.
From Francke, 1991.

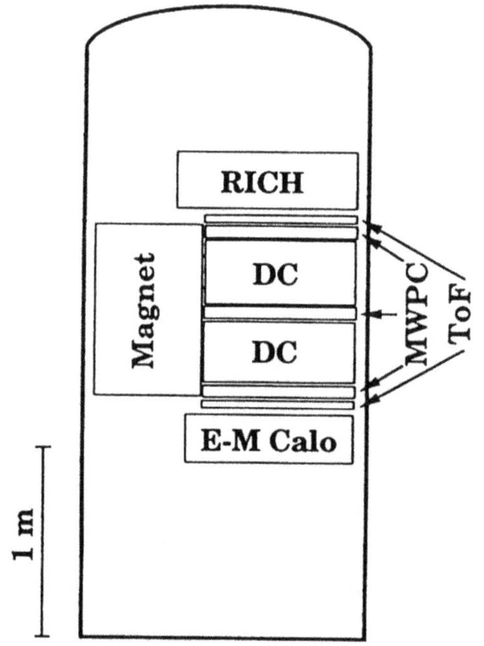

The flight took place from Lynn Lake, Manitoba, Canada, on August 8-9, 1994. Data was successfully collected during the 3.5 hour long ascent and for 23 hours at a float altitude above 36 km (less than 5 g/cm^2 residual atmosphere). Totally, data from more than 6 million cosmic rays were recorded. All detectors worked well during the flight.

2. THE CAPRICE94 RICH DETECTOR

The CAPRICE94 RICH detector is shown in Figure 2. Cherenkov light is emitted in a 10 mm thick solid radiator of NaF when a charged particle transverses the detector if $\beta > 1/n$ (β is the particle velocity relative to the speed of light, n is the refractive index of NaF). The cone of light refracts out of the crystal and is expanded in a nitrogen volume before entering into a photosensitive MWPC. The Cherenkov photons are converted into photoelectrons by the photosensitive compound tetrakis-(dimethylamino)-ethylene (TMAE) (Anderson, 1980), and are amplified in a MWPC. The induced pulse in the cathode pad plane is detected and gives an unambiguous image of the Cherenkov light. The MWPC is operated at low gain (a few times 10^4) with pure ethane as amplification gas. The size of the cathode pads are 8 by 8 mm^2.

The TMAE itself acts as an absorber of traces of oxygen and cleans the amplification gas from oxygen as it bubbles through the liquid TMAE before entering the detector. Any oxygen contamination would absorb the rare ultra-violet Cherenkov photons before they are detected. During the flight, a continuous

Figure 2. A schematic view of the CAPRICE94 RICH detector; a 1.7 GeV/c antiproton event is indicated. The Cherenkov light emitted in the NaF crystal is detected in a photosensitive MWPC with pad readout after expansion in nitrogen. A quartz window transparent to the ultra violet Cherenkov photons separates the nitrogen gas from the TMAE and ethane gas mixture in the MWPC. In the pad plane at the bottom is illustrated the two types of signals that occurs. At the edges are the signals from the Cherenkov photons. In the center are the signals from the ionization of the amplification gas by the charged particle.
From Francke, 1991.

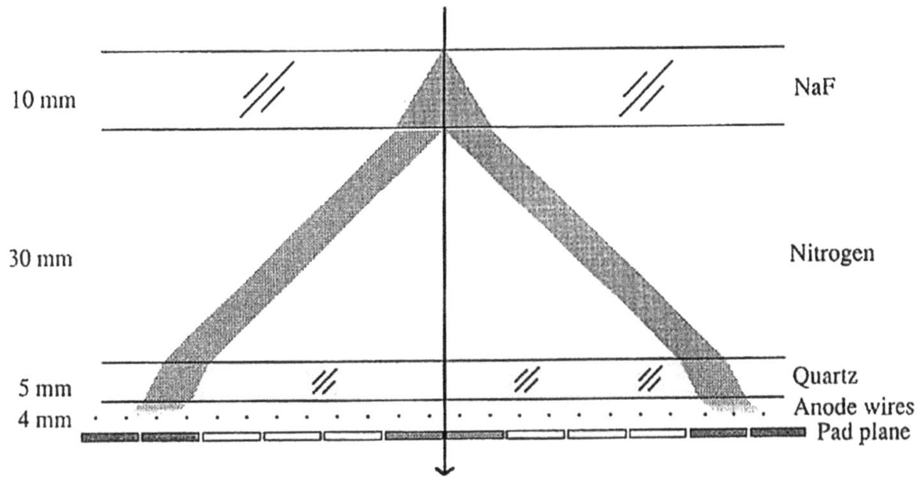

flow of ethane was used (3 l/h) and the TMAE was heated to 38°C to have a short enough absorption length. The detector was kept at 45°C to avoid condensation of TMAE.

The RICH detector and the results from particle beam tests are described in more detail in reference (Carlson, 1994).

3. EXPERIMENTAL RESULTS

3.1 Single Events

The image of the Cherenkov light is a ring at perpendicular incidence angle, which opens up to a wide U-shape at inclined incidence. Figure 3 shows four single events collected during the flight: a) a 1.5 GeV kinetic energy proton at 1° incidence angle, b) a 1.5 GeV proton at 10° incidence angle, c) a 1.1 GeV/nucleon ^{4}He at 2° incidence angle and d) a 1.5 GeV/nucleon ^{4}He at 20° incidence towards the detector, which is the maximum incidence angle during the experiment. The image of the Cherenkov light is well separated from the ionization of the chamber gas by the particle (the central spot) for particles with $\beta >$ 0.85 also for large angles of incidence.

3.2 Multiparticle Rejection

The induced pulse from the ionization of the chamber gas by the particle is used to get a redundant measurement of the impact point of the charged particle with the RICH, independent of the tracking system.

Figure 3. Single events from the flight: a) a 1.5 GeV proton at 1°, b) a 1.5 GeV proton at 10°, c) a 1.1 GeV/nucleon ^{4}He at 2°, and d) a 1.5 GeV/nucleon ^{4}He at 20° incidence angle; the ring, or ellipse at inclined angle of incidence, of Cherenkov light is clearly visible. The signals in the center are due to ionization of the amplification gas in the MWPC by the charged particle. Please note that the helium nuclei with their higher charge emit significantly more (4 times more) Cherenkov photons than the unit charged protons. The arrow indicates the direction of the particle as determined by the tracking drift chambers. The cross is the impact position as determined by the tracking drift chambers. The small ring is the impact position determined by the RICH itself.
From Barbiellini, 1996.

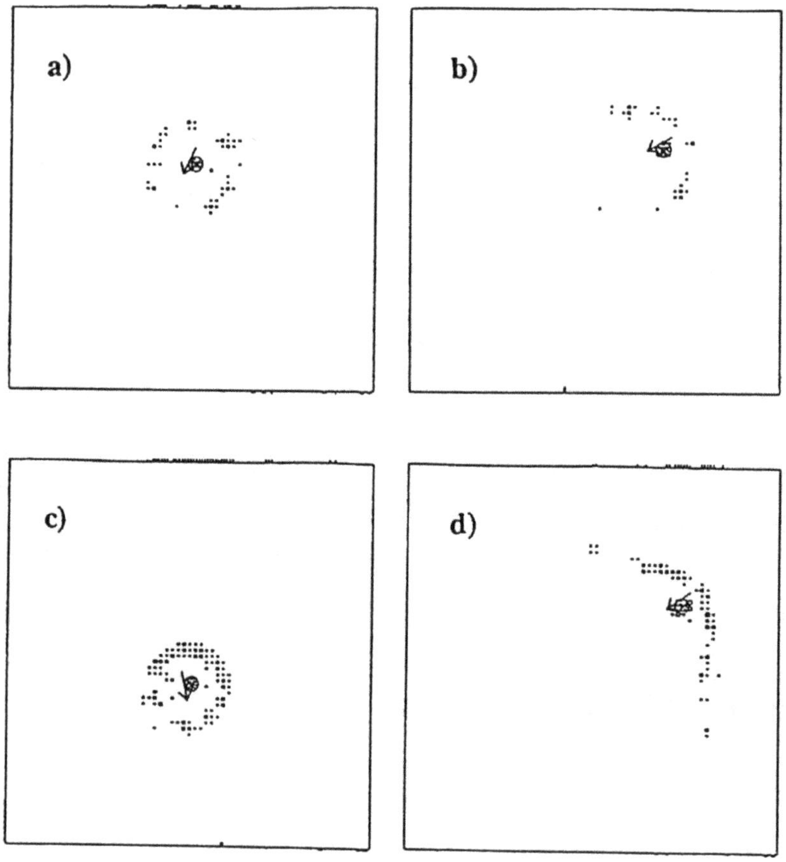

Figure 4 shows the distance between the predicted impact point from the tracking system (known with a precision better than 200 μm) and a center-of-gravity calculation of the central 25 pads around the pad with the largest signal, for all events with at least one reconstructable track.

The resolution of the center of gravity calculation in the RICH detector for clean single tracks is 1.4 mm independent of incidence angle. This quantity has shown to be a powerful tool for rejecting multiparticle events where two or more charged particles pass through the RICH close to each other, e.g. from interactions in the RICH itself. This will disturb the center-of-gravity calculation and give a value different from the tracking prediction.

Figure 4. Resolution of the impact position of the particle with the RICH using the data from the RICH alone; the resolution of the impact position as measured by the RICH itself is less than 2 mm. The tail in the distribution is due to several charged particles traversing the detector simultaneously, or from particles which have scattered between the RICH and the drift chambers. These particles are easily rejected by comparing the impact position determined by the RICH with the impact position determined by the tracking system.
From Barbiellini, 1996.

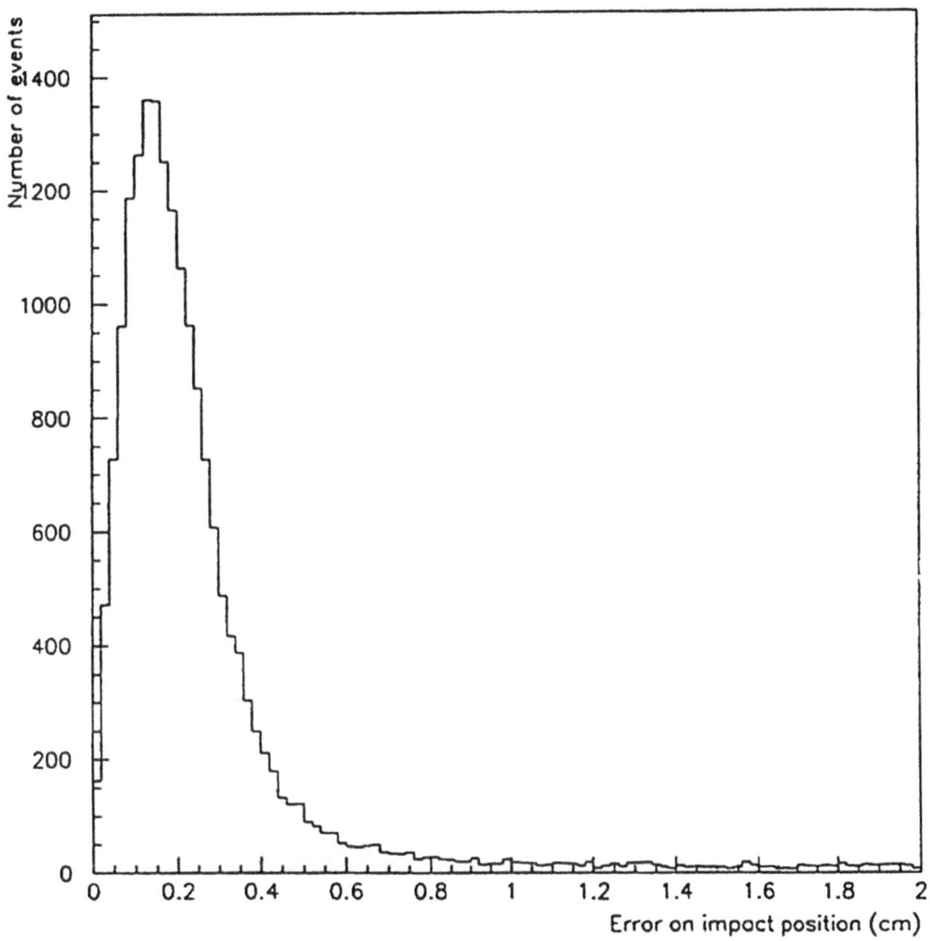

The amplitude in the pads is another useful tool for rejecting multiparticle events when the particles are more than 2 cm apart. A charged particle gives a pulse height close to saturation in the electronics while a signal originating from a Cherenkov photon is significantly smaller. This property is also useful for detector alignment with respect to the tracking system.

3.3 Particle Identification

The particle identification properties of the RICH detector are illustrated in Figure 5. The reconstructed Cherenkov angle versus rigidity (momentum/charge) for unit charged particles as selected by the energy

Figure 5. Reconstructed Cherenkov angle versus rigidity (momentum/charge) for a) Z = 1 and b) Z = 2 particles at float altitude; please note the good particle identification capabilities of the RICH detector over a wide momentum range. The RICH is capable to separate isotopes of both unit charge (^1H, ^2H, ^3H) and charge 2 (^3He and ^4He) from β = 1 particles such as electrons and positrons (e), pions (π) and muons (μ). The RICH detector is actually capable of separating isotopes of even higher charges (lithium, beryllium, boron, carbon, nitrogen, oxygen etc.). However, the flux of these elements in the cosmic radiation is too low to be measured during a 24 hour long balloon flight.
From Barbiellini, 1996.

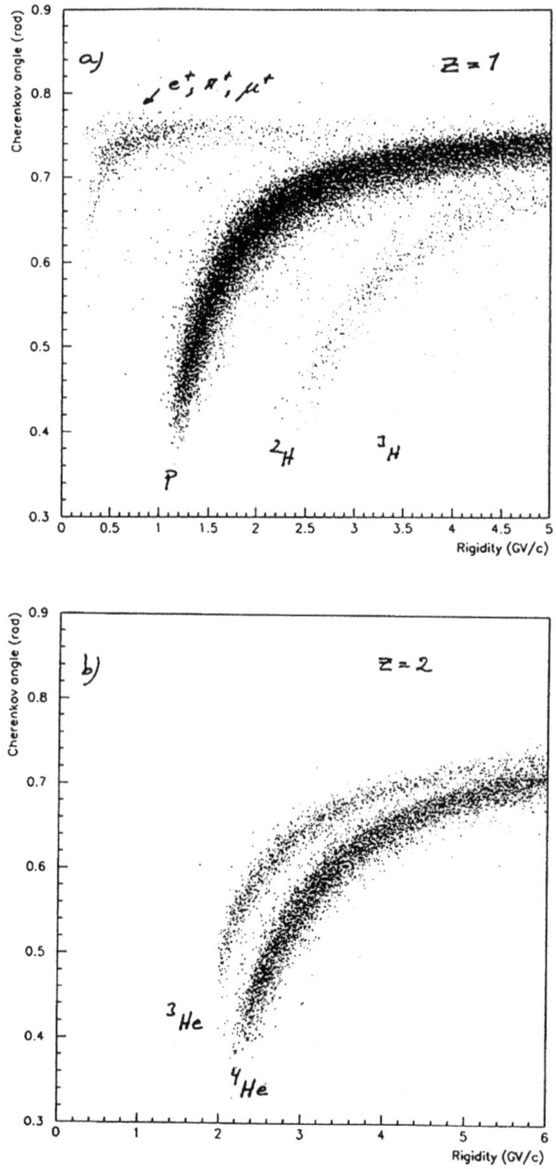

Figure 6. The resolution on the Cherenkov angle as a function of a) incidence angle of the particle for $\beta > 0.95$ particles, and b) β for particles of all incidence angles (0°-20°); the resolution deteriorates slightly at inclined incidence angles as the signals from the Cherenkov photons starts to overlap. The resolution improves rapidly with increasing velocity above the threshold velocity where Cherenkov light starts to be emitted ($\beta > 1/n$). This is due to the increasing number of emitted Cherenkov photons with increasing velocity.
From Barbiellini, 1996.

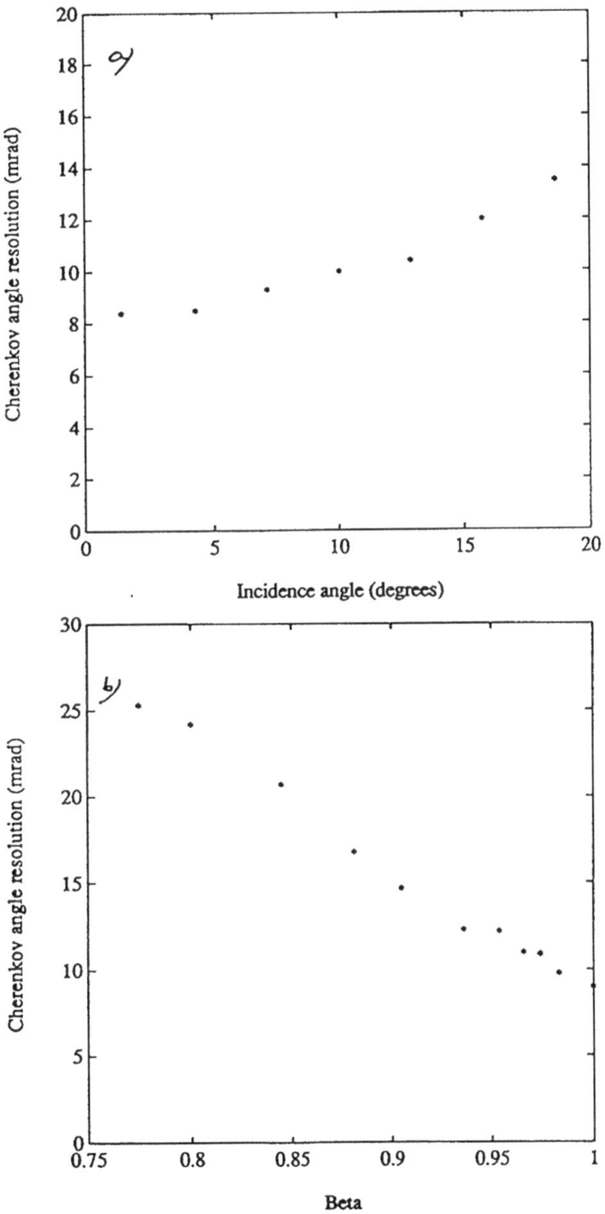

loss, dE/dx, in the two scintillators is shown in Figure 5 a) and ditto for charge two particles in Figure 5 b). As shown in the figure, the RICH is capable of isotope identification (^{1}H, ^{2}H, ^{3}H, ^{3}He and ^{4}He, etc.) as well as (anti)proton identification in a large background of $\beta = 1$ particles such as electrons and positrons (e), pions (π) and muons (μ).

The resolution of the Cherenkov angle during the flight is shown in Figure 6. The resolution (r.m.s) as functions of incidence angle for particles with $\beta > 0.95$ is presented in Figure 6a) and the resolution (r.m.s) as a function of β for particles of all incidence angles (0° -20°) in Figure 6b). The threshold for emission of Cherenkov light in NaF is $\beta = 0.75$, but the particles need to have $\beta > 0.85$ with the geometry used to clearly separate the image of the Cherenkov light from the ionization of the particle. The resolution of the Cherenkov angle during the flight was slightly worse than during the beam tests due to significant pressure changes (30%) inside the gondola during the flight. The anode voltage was adjusted continuously to compensate for the gain variations. The number of photoelectrons detected per event for $\beta = 1$ particles at perpendicular incidence is around 15 during the flight compared to 21 during the beam tests (Carlson, 1994).

3.4 Albedo Rejection

In space experiments it is important to separate between downward and upward (albedo) moving particles, since an albedo proton can mimic a downward moving antiproton when measuring the deflection in a magnetic field.

Traditionally this is done using time-of-flight between two scintillators, where the upper scintillator should give a signal before the lower. The directional sensitivity of a RICH detector has proven to be powerful for independent albedo discrimination as well. Out of 3245 albedo particles, with only one charged track through the spectrometer and with a velocity large enough to emit Cherenkov light, none has an image that could mimic a Cherenkov light image even with the loosest cuts. Note that the NaF crystals are polished on all sides, and not painted on the top to absorb Cherenkov light going upwards.

3.5 Cherenkov Angle Reconstruction Algorithm

For Cherenkov light imaging it is essential that the refractive index of the radiator is less than $\sqrt{2}$, not to trap the Cherenkov light inside the radiator by total reflection.

NaF is the only solid material with a refractive index less than $\sqrt{2}$ in the wavelength region where the CAPRICE94 RICH is sensitive (170-220 nm) (Carlson, 1994; Hempstead, 1991). However, the refractive index is close to $\sqrt{2}$ and part of the Cherenkov light will be trapped inside the NaF crystals by (total) reflections, especially photons of short wavelength and at inclined incidence angles of the particle. This implies that the mean wavelength of the Cherenkov photons that are detected in the photosensitive MWPC depends on the incidence angle and on the β of the incoming particle.

Furthermore, for small β and large incidence angles the Cherenkov light image partly overlap with the ionization of the chamber gas by the particle, and only the pads outside the ionization region can be used in the fit of the Cherenkov angle. This bias the Cherenkov angle in such a way that photoelectrons near the ionization will not be included in the fit of the whole image.

This has to be taken into account when reconstructing the Cherenkov angle of an event. One way to compensate for this effect is to use a "compensated" refractive index of NaF in the reconstruction algorithm that is a function of the particle incidence angle θ and the velocity β. Figure 7 shows the "com-

Figure 7. "Compensated" refractive index of NaF used in the reconstruction algorithm as a function of particle incidence angle θ and β; part of the Cherenkov light signals in the pad plane cannot be separated from the signals from the ionization by the charged particles, especially for small ring radius at low velocities and at large incidence angles. This cause some Cherenkov photons, especially those emitted at wavelengths where the refractive index of NaF is lower, to be excluded from the reconstruction of the Cherenkov angle. This is corrected for by using a predetermined "compensated" refractive index of NaF in the reconstruction algorithms.
From Francke, 1991.

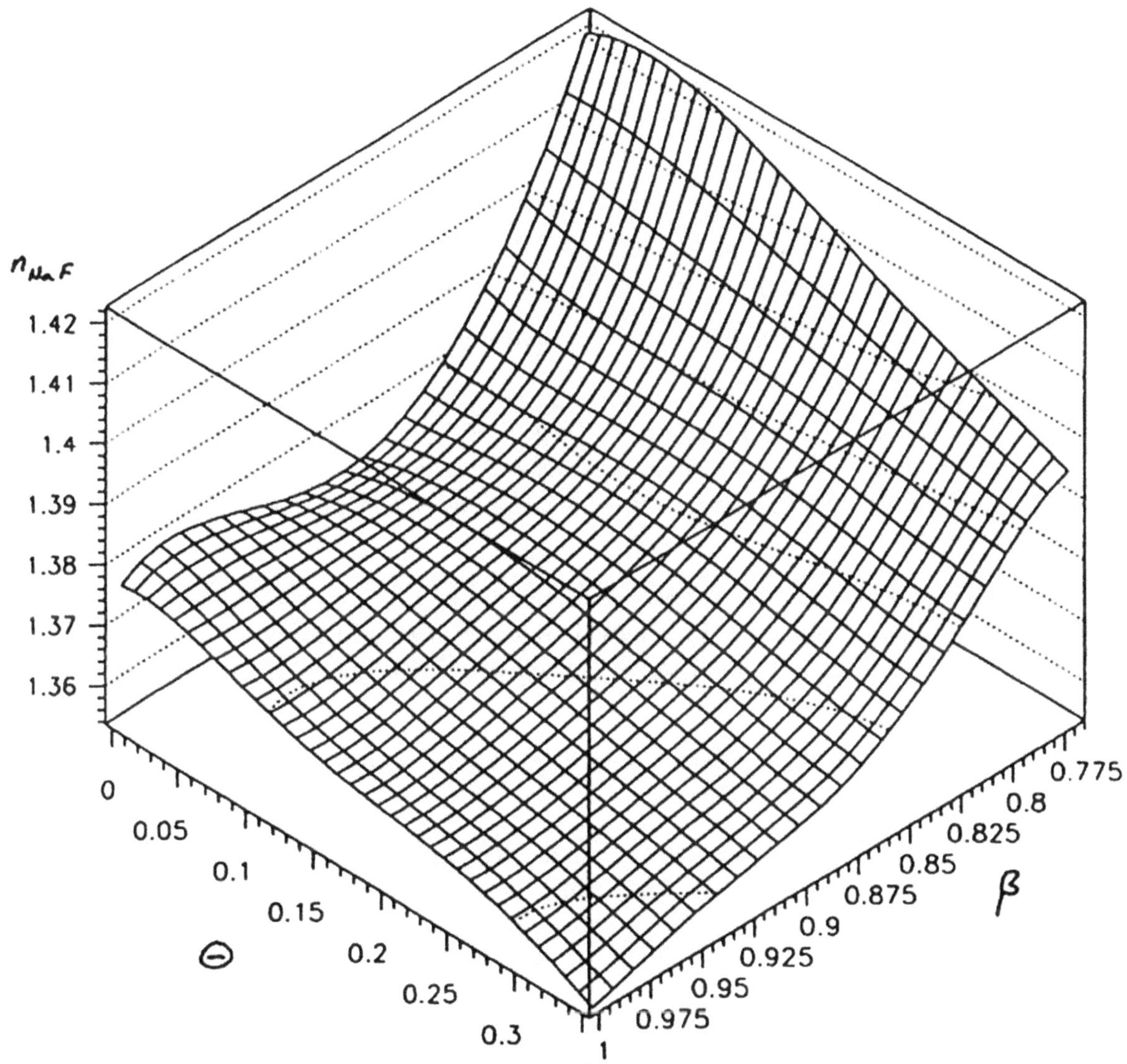

pensated" refractive index of NaF as a function of θ and β that best fits a clean proton sample. The effect is large for small β and inclined incidence angles. Using such a "compensated" refractive index instead of a constant value significantly increases the Cherenkov angle resolution. A different "compensated" refractive index is used for charge two particles.

4. CONCLUSION

A RICH detector sensitive to unit charged particles has been used for the first time in a balloon-borne magnet spectrometer. The detector performed well in the balloon borne environment despite large variations in pressure and temperature. The RICH technique allows for the first time reliable particle identification on an event-by-event basis necessary for identification of rare particles in a large background.

REFERENCES

Anderson, D. F. (1980). Extraction of electrons from a liquid photocathode into a low-pressure wire chamber. *Physics Letters. [Part B]*, *118*(1-3), 230–232. doi:10.1016/0370-2693(82)90634-7

Barbiellini, G., Basini, G., Bellotti, R., Bocciolini, M., Boezio, M., Massimo Brancaccio, F., & Zampa, N. et al. (1996). Performance of the CAPRICE RICH detector during the 1994 balloon flight. *Nuclear Instruments & Methods in Physics Research. Section A, Accelerators, Spectrometers, Detectors and Associated Equipment*, *371*(1-2), 169–173. doi:10.1016/0168-9002(95)01152-8

Bocciolini, M., Celletti, F., Grandi, M., Perego, A., Spillantini, P., & Basini, G. et al.. (1993). Silicon calorimeter for cosmic antimatter search. *Nuclear Physics*, *C32*, 77.

Buckley, J., Dwyer, J., Müller, D., Swordy, S., & Tang, K. K. (1992). A ring imaging Cherenkov telescope for observations of high energy cosmic rays. *Nuclear Instruments & Methods in Physics Research. Section A, Accelerators, Spectrometers, Detectors and Associated Equipment*, *323*(1-2), 380–385. doi:10.1016/0168-9002(92)90318-X

Carlson, P., Francke, T., Löfgren, S., Weber, N., & Suffert, M. (1994). Results from beam tests of the CAPRICE RICH detector. *Nuclear Instruments & Methods in Physics Research. Section A, Accelerators, Spectrometers, Detectors and Associated Equipment*, *349*(2-3), 577–591. doi:10.1016/0168-9002(94)91229-7

Francke, T. (1991). *A ring imaging Cherenkov counter using a solid radiator.* (PhD Thesis). Royal Institute of Technology, Stockholm, Sweden.

Golden, R. L., Grimani, C., Hull, R., Kimbell, B. L., Park, R., Stephens, S. A., & Picozza, P. et al. (1991). Performance of a balloon-borne magnet spectrometer for cosmic ray studies. *Nuclear Instruments & Methods in Physics Research. Section A, Accelerators, Spectrometers, Detectors and Associated Equipment*, *306*(1-2), 366–377. doi:10.1016/0168-9002(91)90343-O

Hempstead, M., Chauveau, J., Comby, G., Séguinot, J., & Ypsilantis, T. (1991). A measurement of the refractive index of sodium fluoride in the far ultraviolet. *Nuclear Instruments & Methods in Physics Research. Section A, Accelerators, Spectrometers, Detectors and Associated Equipment*, *306*(1-2), 207–211. doi:10.1016/0168-9002(91)90321-G

Hof, M., Bremerich, M., Menn, W., Pfeiffer, C., Reimer, O., & Simon, M. et al.. (1994). Performance of drift chambers in a magnetic rigidity spectrometer for measuring the cosmic radiation. *Nuclear Instruments & Methods in Physics Research. Section A, Accelerators, Spectrometers, Detectors and Associated Equipment*, *345*(3), 561–569. doi:10.1016/0168-9002(94)90516-9

ADDITIONAL READING

Barbiellini, G. (1993). A silicon imaging calorimeter prototype for antimatter search in space: Experimental results. *Nuclear Instruments & Methods in Physics Research. Section A, Accelerators, Spectrometers, Detectors and Associated Equipment*, *333*(2-3), 560–566. doi:10.1016/0168-9002(93)91208-5

Barbiellini, G. (1996). (The CAPRICE collaboration): *The cosmic ray positron to electron ratio in the energy range 0.85 to 14 GeV. Astron. Astrophys.*, *309*, L15.

Boezio, M., Carlson, P., Francke, T., Weber, N., Suffert, M., Hof, M., & Stochaj, S. J. et al. (1997). (The CAPRICE collaboration): *The cosmic ray antiproton flux between 0.62 and 3.19 GeV measured near solar minimum activity. The Astrophysical Journal*, *487*(1), 415–423. doi:10.1086/304593

Boezio, M., Carlson, P., Francke, T., Weber, N., Suffert, M., Hof, M., & Stochaj, S. J. et al. (2000). The Cosmic-Ray Electron and Positron Spectra Measured at 1 AU during Solar Minimum Activity. *The Astrophysical Journal*, *532*(1), 653–669. doi:10.1086/308545

Carlson, P., Francke, T., Weber, N., & Suffert, M. (1994). A RICH Counter for Identifying Antiprotons and Light Isotopes in the Cosmic Radiation. *Nuclear Instruments & Methods in Physics Research. Section A, Accelerators, Spectrometers, Detectors and Associated Equipment*, *343*(1), 198–202. doi:10.1016/0168-9002(94)90550-9

Carlson, P., Francke, T., Weber, N., & Suffert, M. (1994). A RICH counter for identifying antimatter and light isotopes in the cosmic radiation. *IEEE Transactions on Nuclear Science*, *41*(4), 1321–1327. doi:10.1109/23.322907

Francke, T., Lindblad, T., Eide, Å., Piuz, F., Williams, D., Martinengo, P., & Suffert, M. et al. (1991). Particle identification by Cherenkov ring imaging using a neural network approach. *Nuclear Instruments & Methods in Physics Research. Section A, Accelerators, Spectrometers, Detectors and Associated Equipment*, *307*(1), 47–51. doi:10.1016/0168-9002(91)90129-E

Francke, T., Montanet, F., Pavlopoulos, P., Rivoiron, C., Santiard, J. C., & Suffert, M. (1989). Compact fast rich detector using a solid radiator: Principle and first experimental results. *Nuclear Instruments & Methods in Physics Research. Section A, Accelerators, Spectrometers, Detectors and Associated Equipment*, *279*(1-2), 317–321. doi:10.1016/0168-9002(89)91100-5

Francke, T., Piuz, F., Williams, D., Martinengo, P., Ribeiro, R., & Suffert, M. (1991). A fast and compact solid radiator RICH counter. *Nuclear Instruments & Methods in Physics Research. Section A, Accelerators, Spectrometers, Detectors and Associated Equipment*, *310*(1-2), 146–149. doi:10.1016/0168-9002(91)91014-M

Francke, T., Suffert, M., & Carlson, P. (1992). Identification of antiprotons in space using a compact RICH counter. *IEEE Transactions on Nuclear Science*, *39*(4), 971–975. doi:10.1109/23.159743

Stochaj, S. (1997). The Cosmic-Ray Antiproton Flux between 0.62 and 3.19 GeV Measured Near Solar Minimum Activity. *The Astrophysical Journal*, *487*(1), 415–423.

KEY TERMS AND DEFINITIONS

Antiparticle: A group of elementary particles with similar characteristics as ordinary matter, except that charge and spin are opposite. When an ordinary elementary particle meets its antiparticle, they annihilate one another.

Antiproton: The antiparticle of the proton.

β: A measure of velocity expressed as the fraction of the speed of light (300.000 km/s). β is normally expressed in per cent of the speed of light.

CAPRICE: The Cosmic AntiParticle Ring Imaging Cherenkov Experiments. A series of balloon borne experiments aimed to measure the flux of antimatter and light isotopes in the cosmic radiation. In total CAPRICE made five successful flights between 1994 and 1998. The flights were performed from Lynn Lake, Manitoba, Canada, and Ft Sumner, New Mexico, USA.

Cherenkov Light: A new type of light emitted when charged particles traverse a transparent medium with a velocity higher than the speed of light in the medium. It is named after the Russian scientist Pavel Cherenkov who discovered it.

MWPC: MultiWire Proportional Counter, a type of detector often used in particle physics experiments.

NaF: The salt sodium fluoride. It is the solid material with the lowest index of refraction for visible and ultra violet light in room temperature.

Positron: The antiparticle of the electron.

RICH: Ring Imaging CHerenkov detector. A type of detector which measures the velocity of elementary particles by measuring the emission angle of the Cherenkov light produced when the particle traverses a medium with a velocity higher than the speed of light in the medium.

TMAE: Tetrakis-(dimethylamino)-ethylene. A yellow liquid which in vapor form is an efficient photo-converter, i.e. has high probability to absorb a light photon and emit a photoelectron.

Chapter 13
Performance of the CAPRICE98 Balloon Borne Gas-RICH Detector

ABSTRACT

A RICH detector using a gas radiator of C_4F_{10} and a photosensitive MWPC with pad readout has been developed, tested in particle beams at CERN and used in the CAPRICE98 balloon borne experiment. The MWPC was operated with a TMAE and ethane mixture at atmospheric pressure and used a cathode pad plane to give an unambiguous image of the Cherenkov light. The induced signals in the pad plane were read out using the AMPLEX chip and CRAMS. The good efficiency of the Cherenkov light collection, the efficient detection of the weak signal from single UV photons together with a low noise level in the electronics of the RICH detector, resulted in a large number of detected photoelectrons per event. For $\beta \sim 1$ charge one particles, an average of 12 photoelectrons per event were detected. The reconstructed Cherenkov angle of 50 mrad for a $\beta \sim 1$ particle had a resolution of 1.2 mrad (r.m.s). The RICH was flown with the CAPRICE98 magnetic spectrometer and was the first RICH detector ever used in a balloon borne experiment capable of identifying charge one particles at energies above 5 GeV. The RICH provided an identification of cosmic ray antiprotons up to the highest energies ever studied (about 50 GeV of total energy). The spectrometer was flown on 28–29 May 1998 from Ft Sumner, New Mexico, USA.

1. INTRODUCTION

A gas Ring Imaging Cherenkov detector (RICH) has been developed, tested in particle beams at CERN and flown in the CAPRICE98 balloon borne experiment (Ambriola, 1999). The detector was built to identify cosmic rays (Barbiellini, 1997; Bergström, 1999; Francke, 1999) in the rigidity range 2– 330 GV.

The CAPRICE98 spectrometer science objectives were to measure the antimatter component in the cosmic rays in the energy range 2–50 GeV and to study the cosmic ray composition in the atmosphere in the energy region between a few hundred MeV and 330 GeV. The main goal of the gas-RICH detector was to identify antiprotons in a large background of electrons, atmospheric and locally produced negative

DOI: 10.4018/978-1-5225-0242-5.ch013

pions and muons. Since the first observation by Golden (Golden, 1984) there has been several experiments measuring the energy spectrum of antiprotons in the cosmic rays. The energy spectrum has recently been determined with good accuracy below 4 GeV kinetic energy (Orito, 2000 and references therein). The observed spectrum indicates that the cosmic ray antiprotons are mostly of secondary origin, produced by the collision of cosmic ray nuclei with the interstellar gas. Above 4 GeV only two measurements exist (Golden, 1984; Basini, 1999), both with low statistics, and which give very different flux values.

The detection of antiprotons at high energies is an experimental challenge due to the difficulty in detecting and identifying the antiprotons against a vast background of other particles. Sophisticated particle identification detectors are therefore needed to identify the rare antiprotons. The information from the gas-RICH combined with the information from the other detectors in the CAPRICE98 magnetic spectrometer provided excellent particle identification. The RICH detector described in this paper is capable of mass resolving protons/antiprotons between 18 and 50 GeV (Bergström, 2000). Below 18 GeV it acts as an efficient threshold Cherenkov detector for proton/antiproton identification.

The spectrometer was one of the most complex instruments ever flown in a balloon borne experiment for cosmic ray studies. The apparatus included a tracking system with three drift chambers (Hof, 1994), a time-of-flight system, a gas-RICH detector, a silicon-tungsten imaging calorimeter (Bocciolini, 1996; Ricci, 1999) for particle identification and a 4Tesla super-conducting magnet (Golden, 1978). It was launched with a balloon from Ft Sumner, New Mexico, USA, on 28th May 1998 and floated at an atmospheric depth of about 5.5 g/cm^2 (corresponding to an altitude of about 36 km) for a period of 21 hours, at a mean vertical cutoff rigidity of about 4.5 GV. Information about more than 5 million cosmic ray particles was collected during the flight.

2. THE GAS-RICH DETECTOR

The RICH detector, shown in Figure 1, consisted of a photosensitive multiwire proportional chamber (MWPC) and a 1 m tall radiator box, filled with high purity C_4F_{10} gas at an average pressure of 911 mbar (Barbiellini, 1997; Bergström, 1999; Francke, 1999). As illustrated in Figure 1, charged cosmic ray particles ionized the gas in the MWPC before entering into the gas radiator, where Cherenkov light was emitted along the particle track. The Cherenkov light was reflected and focused by a spherical mirror back onto the photosensitive MWPC where the position of the Cherenkov light was accurately determined. The RICH was sensitive to light in the wavelength region 165–220 nm limited by the transmission of the quartz window (> 165 nm) and the quantum efficiency of the TMAE (tetrakis-dimethyl-amino-ethylene) (<220 nm) which was used in the amplification gas as photo-converter. The sensitive area of the detector was 480 x 480 mm^2 defined by the size of the quartz window.

2.1 The MWPC

The MWPC consisted of a frame with anode wires between an upper cathode pad plane and a lower cathode of conductive strips on a quartz window. The 128 gold-plated tungsten anode wires were 15 mm in diameter, 4 mm apart and 520 mm long. The cathode pad plane was a sandwich of two 0.5 mm thick Vetronite boards glued on the two sides of a 6 mm thick plate of Rohacell. Both Vetronite boards were plated with copper and gold. The lower board contained 4096 pads. Each pad (8 x 8 mm^2) was connected to a pin glued to the board and fed through the pad plane to the backside where it was used as a connec-

Figure 1. Schematic view of the CAPRICE98 RICH detector; the left upper expanded view illustrates how the gas in the MWPC is ionized by the cosmic ray particle. The position of the pads, wires and strips are indicated. The "-" and "+" in the figure are referred, respectively, to electrons and ions liberated in the ethane gas. In the lower part the Cherenkov light emitted in the C_4F_{10} gas close to the particle track is indicated by the shaded area. A spherical mirror reflects the light back up onto the MWPC. In the right upper expanded view, two Cherenkov photons (γ) enter the MWPC through the quartz window where they each liberate one photoelectron from the TMAE vapor.
Bergström, 2001.

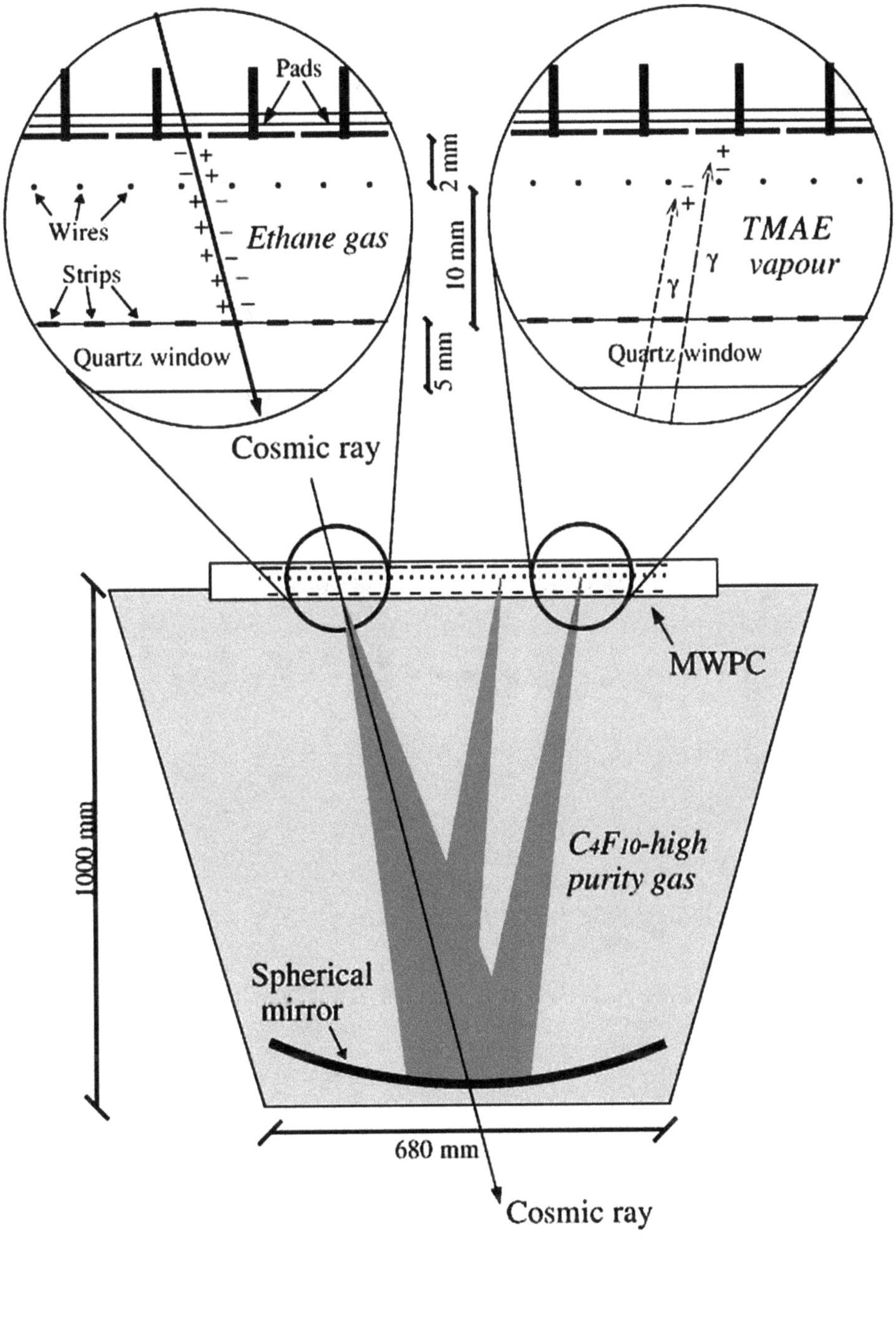

tor to the front-end electronics. The MWPC was operated with pure ethane saturated with tetrakis-(di-methyl-amino)-ethylene (TMAE), at moderate gain (4.4 x 10^5) in order to minimize photon feedback. A gas system was used to flush the MWPC with a low constant flow of TMAE saturated ethane gas.

To achieve the high detection efficiency of > 12 photoelectrons per event as an average for $\beta \sim 1$ particles (Bergström, 1999), the TMAE had to be heated to 35°C, giving a conversion probability of 80% in the 12 mm thick MWPC.

All parts of the gas volumes in the RICH detector that would be in thermal contact with the TMAE had to be heated at all times to prevent the gas from condensing on the inner surfaces of the detector. This required a regulated heating system, which was required to work at all times and during all possible operating conditions. The apparatus had to run with all systems active for several weeks continuously, during the integration and preflight testing at ground. The heating systems also had to take care of the temperature control during the flight when the temperature outside the payload could vary between -30°C and +40°C. The conditions for the instruments inside the gondola, which can be seen in Figure 2, were however not that extreme, since it was thermally insulated from the outside. The MWPC requires a stable temperature environment to be able to operate reliably. If the temperature of the MWPC was lower than the TMAE saturated ethane gas, then the TMAE would condensate. Discharges in the high voltage induced by the condensed TMAE could destroy the detector.

Figure 2. Temperature variations at float inside the gondola during the CAPRICE98 flight; the graph shows the temperature at different positions inside the gondola. The filled squares are the temperatures in the radiator of the RICH detector. The open crosses are the temperatures inside the payload. The stars are the temperatures of the TMAE liquid. The time is from the time of the launch.
Bergström, 2001.

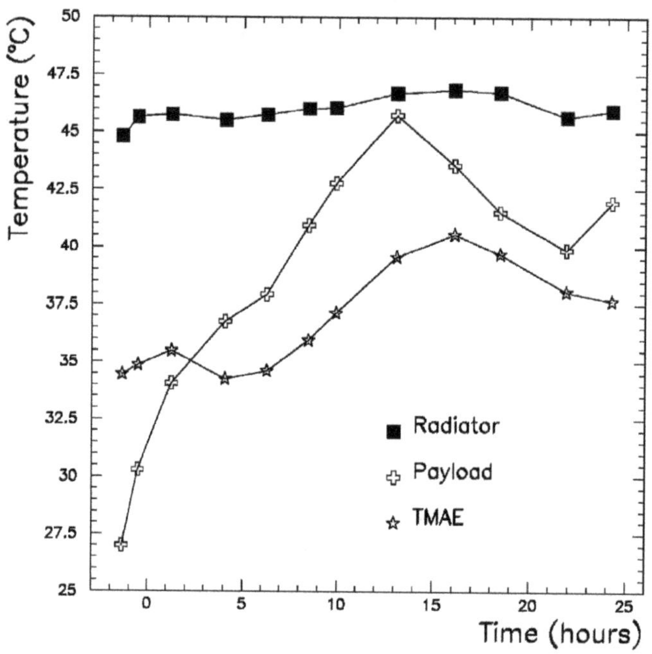

The 16 channel, low noise, VLSI (very large scale integration) chip AMPLEX (de Beauville, 1990) was used for the read out of the anode wires and the cathode pads. Each channel included a charge sensitive pre-amplifier, a shaping amplifier and a track and hold system where the charge was stored before being read out. The AMPLEX chips, 32 for the cathode pads (512 channels) and 8 for the anode wires (128 channels), were daisy-chained and readout in series (multiplexed) and fed to the VME based digital readout electronics, CRAMS (CAEN V550). A CRAMS (Card Random Access Memory) module has two independent analog to digital conversion blocks, which each can be programmed to read out between 32 and 2016 channels in steps of 32 channels. Therefore, four CRAMS (8 blocks) with 512 channels in each of the two blocks were used for the cathode pads and one CRAMS (one block) with 128 channels was used for the anode wires. In the CRAMS, each signal was digitized by a 10-bit flash ADC, pedestal subtracted and stored in a memory if the remaining signal exceeded a preset threshold value (usually five

Figure 3. Schematics of the read out electronics; the trigger logic in the center takes care of generating all the control signals (hold, start read and reset) and distributes them to the AMPLEX and VME read out. The multiplexed signals from the front-end electronics (AMPLEX) were sent to the CRAMS where the VME CPU module took care of the data handling and storage. The communication between the VME CPU and the trigger logic goes through a VME I/O module. After pedestal subtraction of the signal in the CRAMS, the VME CPU reads out the data. The data is then accessible for the central μVAX via a VME/Q-bus interface for storage on data disks and transmission to ground.
Bergström, 2001.

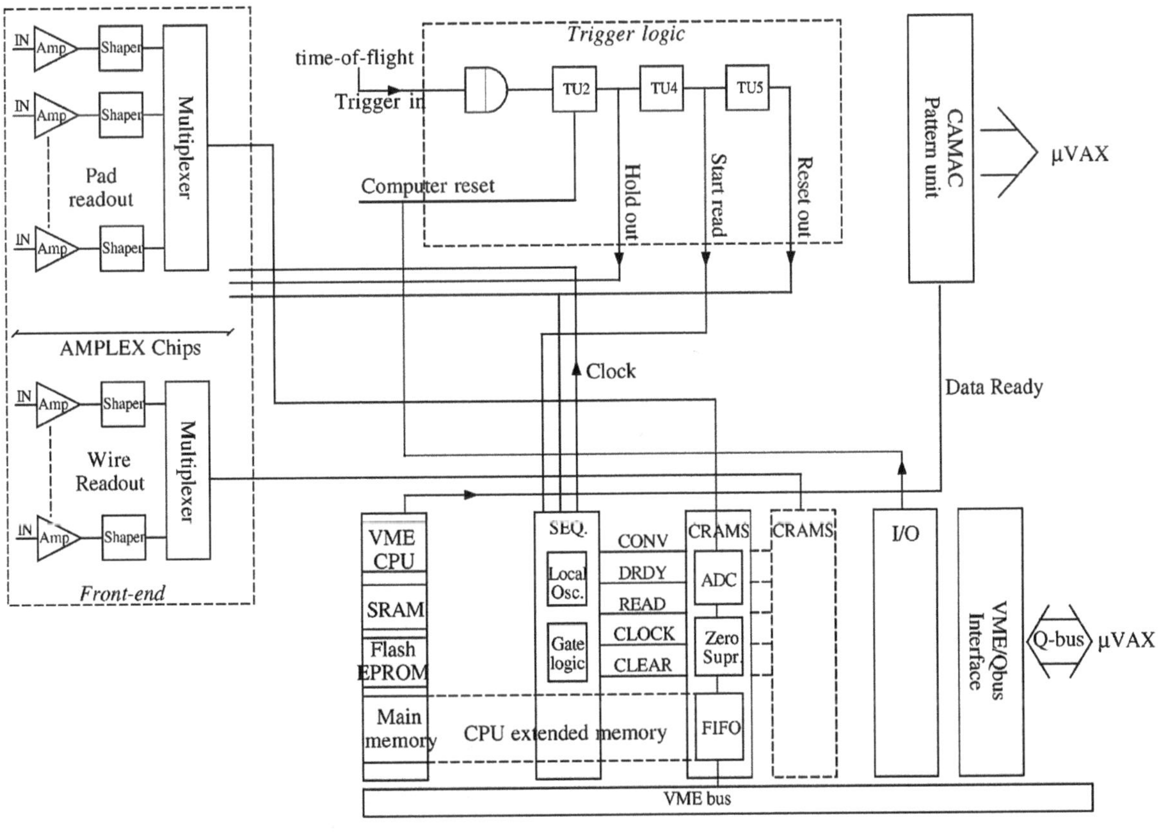

times the r.m.s noise). The readout time was 2 μs per channel giving a total readout time of approximately 1 ms. The trigger rate during the flight was 65 Hz. A typical signal from a single photoelectron was 70 fC (spread out over an average of 3.5 pads) and the equivalent noise of the electronics was 7 fC (r.m.s).

Figure 3 shows some details of the readout electronics and control system. The trigger logic generated all the control signals (hold, start read and reset) and distributed them to the AMPLEX and VME read out. The external trigger was given by the time-of-flight system. The multiplexed signals from the front-end electronics were sent to the CRAMS where a VME CPU (central processing unit) module (SYS68K/IBC-20 FORCE) managed the data handling and storage. After pedestal subtraction of the signal in the CRAMS the data was read by the VME CPU module and made accessible for storage on discs and transmission to ground by the central computer via a VME/Q-bus interface module. The memories in the CRAMS, where the signals were initially stored, could be considered as an extension of the VME CPU memory.

Figure 4. The output signal of the CRAMS as a function of the charge injected into the AMPLEX; note that above 400 fC the AMPLEX becomes saturated. The amplitude of the output signal and saturation limit for these two channels is typical for the response of all pad channels. The channel to channel variation is less than 10%.
Bergström, 2001.

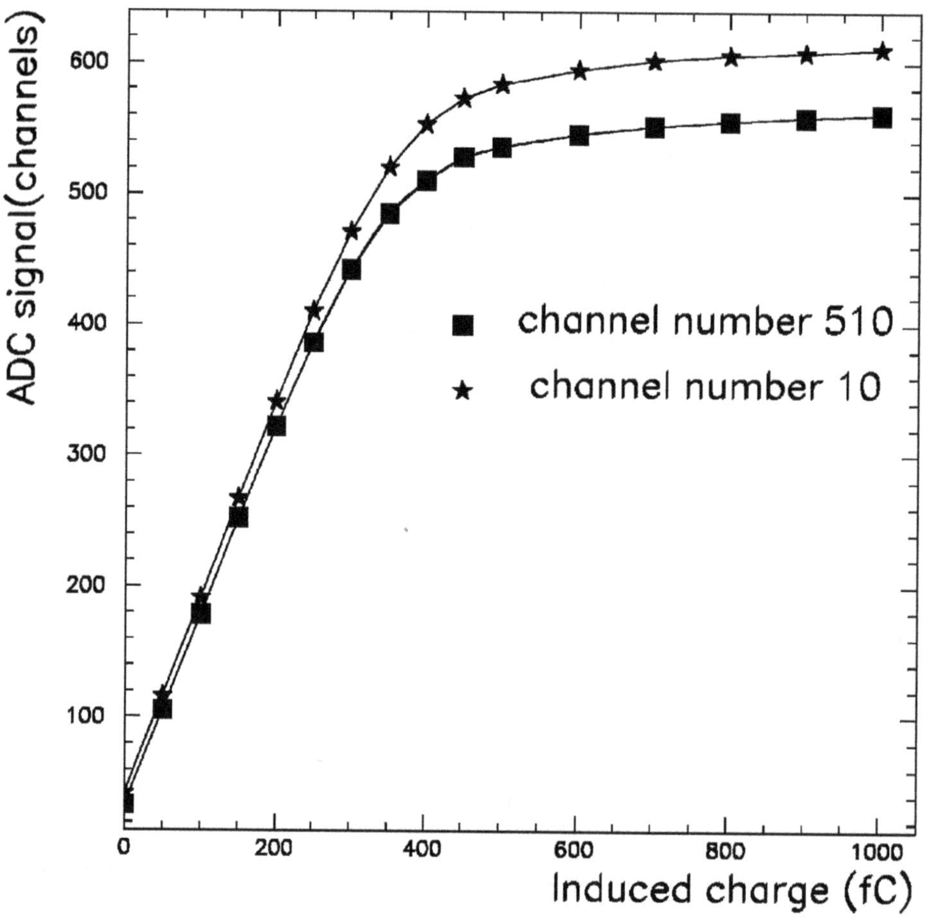

A pedestal and threshold calculation was performed at every start-up of the RICH detector. A computer-induced trigger was fed to the detector and the average noise level for each individual channel was determined. The noise measured was then later subtracted from the signal read out for each event before the data were saved on disks or transmitted via radio link to ground. Important effects that could require a recalculation were large temperature and pressure changes, electronic noise from other instruments, radio noise picked up by the wires in the MWPC, voltage fluctuations in the power feeding the detector, etc. The pedestal and threshold calculation could be performed at any moment when it was found to be necessary for a stable performance of the detector.

In Figure 4 the output of two CRAMS channels are shown as function of the induced charge in the AMPLEX. The channel to channel gain variations are less than 10%. Above 400 fC the AMPLEX becomes saturated, corresponding to the signal from six photoelectrons in a single channel.

The 5 mm thick quartz window (480 x 480 mm^2) was made of Suprasil 3, and had cathode strips evaporated onto it acting as the lower cathode. The strips were 150 µm wide, 3 mm apart giving a total transmission of 95%, which was independent of the particles angle of incidence since the strips were only 100 nm thick (10 nm nickel + 80 nm copper +10 nm nickel). The quartz window was glued into a frame of Invar which had the same thermal expansion coefficient as quartz and the frame could move freely with respect to the main frame holding the chamber, in order not to strain the quartz window by thermal expansions. The quartz window was transparent for wavelengths longer than 165 nm (WISAG 1988).

2.2 The Radiator

The radiator was built to have the largest possible acceptance and to withstand a vertical impact shock of about 10 g when the parachute opens. Because of weight constraints, aluminum was used as material for all the enclosure and for the support structure. The radiator box had to be absolutely airtight not to affect the purity of the C_4F_{10} gas. The internal volume of the radiator box was 800 liters with two inlets and two outlets. One inlet was only used during the first phase when the dry nitrogen was added to the system and was then disconnected and closed during the second (C_4F_{10} filling) and third (cleaning) phases. One inlet and one outlet were used for circulating the gas through the radiator. The gas flow was regulated so that the pressure in the radiator would at all times be a few mbar above the payload pressure. This was intended to prevent large pressure drops in the radiator, which could cause its thin aluminum box to implode, and also to prevent the payload gas (mostly nitrogen) to leak into the radiator. Due to the risk of failure of the radiator, there was an overpressure relief system connected to the second outlet with two oil filled bubblers with an intermediate balloon filled with cleaned gas. Thus, this dual system protected the gas system from contamination due to small changes of pressure differences between the radiator and the payload, while at the same time protecting the radiator box from exploding because of a sudden large input of gas into the radiator.

2.3 The Gas System

The overall purpose of this gas system was to separate and clean the C_4F_{10} gas used in the CAPRICE98 RICH radiator from all other gases and especially from water and oxygen, since oxygen and water vapor will absorb Cherenkov photons and thereby reduce the efficiency of the detector. The cleaning was performed in three steps. First, the radiator volume was flushed with several cubic meter of nitrogen. This had the purpose of drying out the whole system and gets rid of as much air and moisture as possible.

Figure 5. The complete gas cleaning system and the gas system for the ethane/TMAE mixture of the MWPC

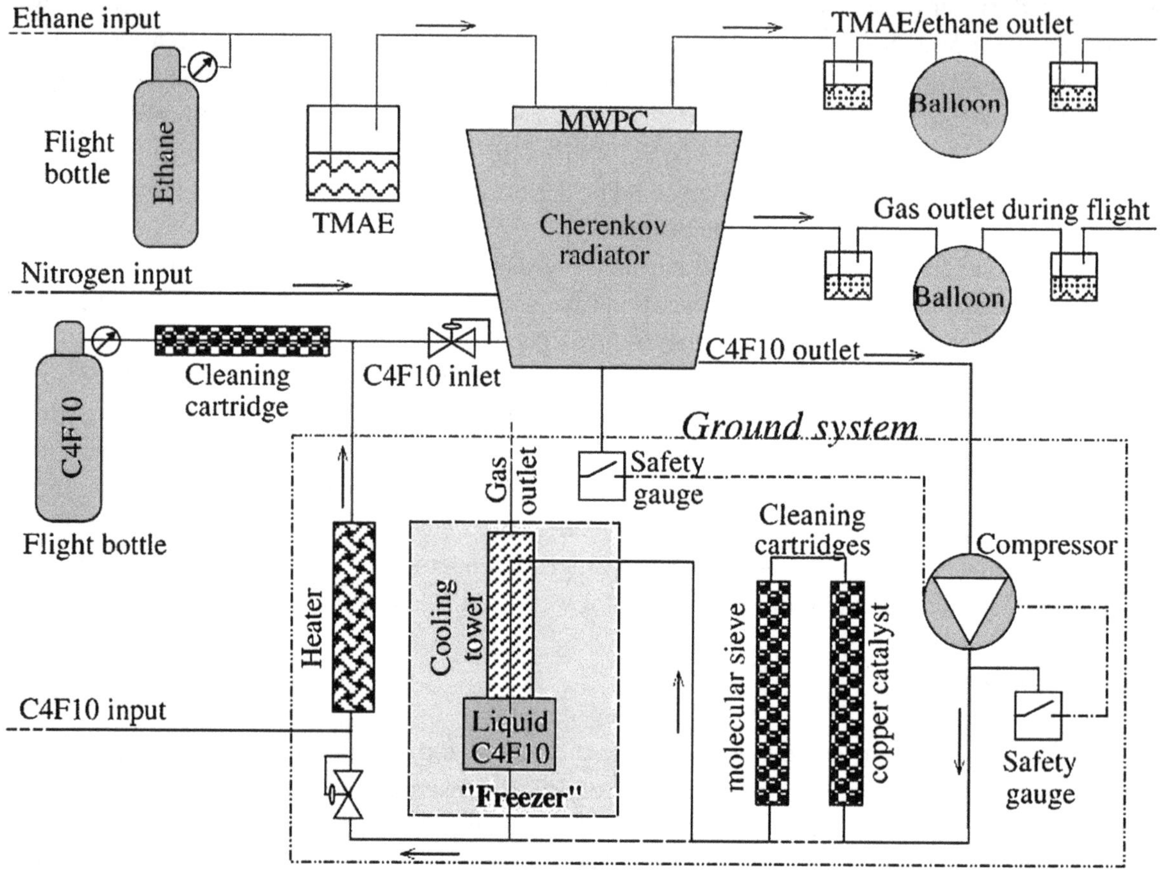

The following step was a process in which the C_4F_{10} (99.96% guaranteed by the manufacturer) was let into the system and at the same time the nitrogen was separated and expelled from the system using the combination of a freezer and a compressor. Finally, the C_4F_{10} gas was purified by filtering it through two cleaning cartridges. The C_4F_{10} gas system was designed to keep the impurities of oxygen below 0.12% (upper limit), which was sufficient for having a more than 50% transmission of Cherenkov light.

We show in Figure 5 the complete gas cleaning system and the gas system for the ethane/TMAE mixture of the MWPC. The system could be divided into two major parts, one internal part that was situated inside the CAPRICE98 payload and one external part only used during ground operations with most of the components mounted in a rack that had to be positioned close to the payload. All the gas tubes connecting these two parts of the system had to be fed through the payload shell with vacuum tight connectors without restraining the circulation of the gas.

The compressor, which was connected to the first outlet of the radiator, had the function of circulating the gas through the system. The gas was extracted from the radiator and the pressure was increased, so that the C_4F_{10} gas would easily liquefy in the freezer kept at -40°C. In case of a sudden decrease of the pressure in the radiator, a pressure sensor would automatically shut down the compressor. The C_4F_{10} that went back into the radiator was taken from this liquid, which guaranteed that no other gases, like nitrogen and oxygen which would still be in gas form at this pressure and temperature, could reenter

Figure 6. The operation principle of the ground and flight C_4F_{10} gas systems; the left side of this system was used during ground operations. The right side shows the flight system, which went into operation only during the flight.
Bergström, 2001.

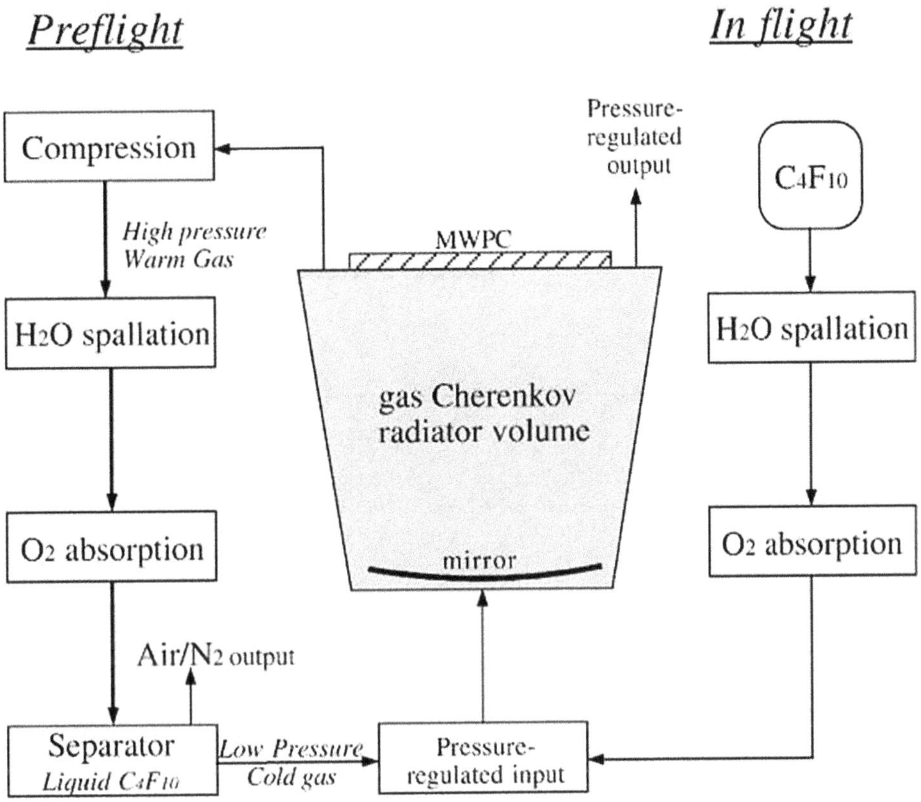

the radiator. These unwanted gases were instead expelled out of the system through an outlet valve. The remaining contamination of oxygen in the radiator, after usage of the cleaning system, was measured during the test at CERN to be as low as 230 ppm (Bergström, 1999). This was then a stable situation during all the experiment and was well below the limit that was needed to get the required efficiency for the Cherenkov light collection.

3. RESULTS FROM FLIGHT DATA

Figure 7 shows the payload pressure, which averaged at 911 mbar and Figure 2 the mean temperature, with an average of about 45.9°C in the radiator volume, during the CAPRICE98 flight. This resulted in a refractive index of about 1.0014, corresponding to a threshold Lorentz factor of $\gamma = 18.9$. The refractive index was estimated using ultra relativistic particles (electrons) and it was shown (Bergström, 1999) to follow precisely the change in the gondola pressure during the whole duration of the CAPRICE98 flight.

Figure 8 shows the number of pads used for the Cherenkov angle calculation, n_{eff}, for $\beta \sim 1$ particles during the flight. n_{eff} (Equation (2)) is directly proportional to the number of photoelectrons. The slow

Figure 7. Pressure variations inside the gondola during the CAPRICE98 flight at float altitude; the filled bullets are the pressures inside the payload and the stars are the pressure outside the payload. The times on both graphs are from the time of the launch.
Bergström, 2001.

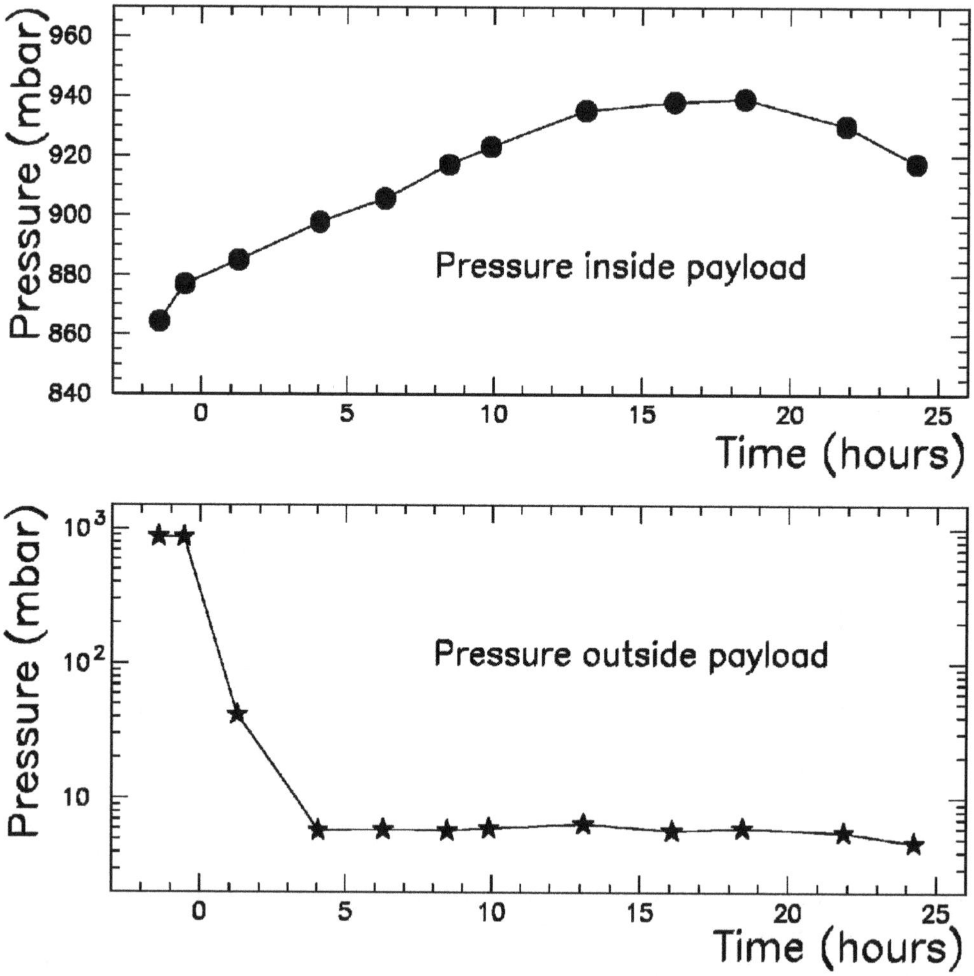

decrease with time that can be seen in the figure was due to increasing contamination in the radiator. This decrease was partly compensated by the increasing pressure during the flight.

Figure 9 shows a selection of events of different species, as detected by the RICH detector. The top left view shows an example of the Cherenkov ring in the pad plane from a proton, with a rigidity of 38.8 GV measured by the tracking system. Each square corresponds to one pad hit. The filled squares were selected as originating from Cherenkov light and open squares were either due to ionization or electronic noise. The big circle was obtained from the radius derived from the fitted Cherenkov angle y_c (Equation (1)). The "+" is the expected center of that circle, extrapolated from the reconstructed track given by the tracking system. A small circle and an "X" are located at the position where the particle crossed the multi-wire proportional chamber according to the RICH detector and the tracking system, respectively. The arrow shows the direction of the incident particle. The lower left figure shows an electron with a

Figure 8. The average number of pad hits that was used for the Cherenkov angle calculation, which is proportional to the number of photoelectrons; the time scale starts from the launch of the payload. The slow decrease in number of detected Cherenkov photons was due to an increasing contamination of the radiator gas during the flight. The decrease was partly compensated by an increasing pressure. Bergström, 2001.

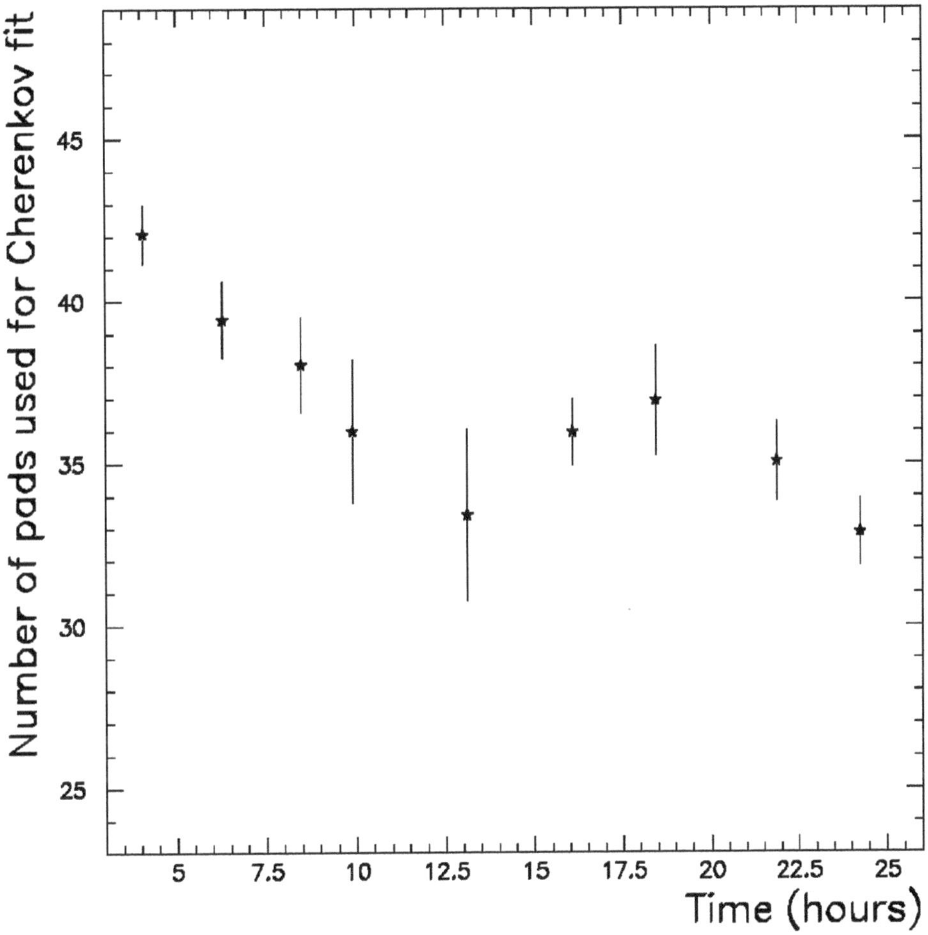

measured rigidity of 3.7 GV. The electron is fully relativistic and therefore the Cherenkov angle is at its maximum of about 51 mrad, corresponding to a ring diameter of about 11 cm. The upper right figure shows a helium nucleus with a measured rigidity of 95.6 GV. Important to notice is the much larger amount of Cherenkov light produced by this particle. The lower right figure shows an antiproton at a measured rigidity of 22.7 GV. The threshold rigidity to produce Cherenkov light for (anti)protons in the gas-RICH detector was about 18 GV. Already at rigidities of 25 GV, an average of six photoelectrons was detected.

A charged particle that crossed the multiwire proportional chamber induced a signal in several pads, of which at least one was a saturated signal. The particle impact position in the RICH can therefore be determined with a center of gravity calculation using the pads closes to the largest signal, which was the one most likely to be at the position of the track. A saturated pad has a signal that is typically 10 times

Figure 9. A selection of different hit patterns in the RICH pad plane from different particle species: the upper left figure shows the ionization and ring of Cherenkov light of a selected proton with a measured rigidity of 38.8 GV. The lower left figure is an electron at its maximum Cherenkov angle with a measured rigidity of 3.7 GV. The upper right figure shows a helium nucleus with a measured rigidity of 95.6 GV. The lower right figure shows an antiproton at a measured rigidity of 22.7 GV. The big circle in each figure was obtained from the radius derived from the fitted Cherenkov angle. The "+" is the center of the circle predicted by the tracking system. A small circle and an "X" are located at the position where the particle hits the multiwire proportional chamber according to the RICH detector and the tracking system, respectively. The size of the pad plane (indicated by the square border) is 51 x 51 cm². An arrow shows the direction of the incident particle as determined by the tracking system.
Bergström, 2001.

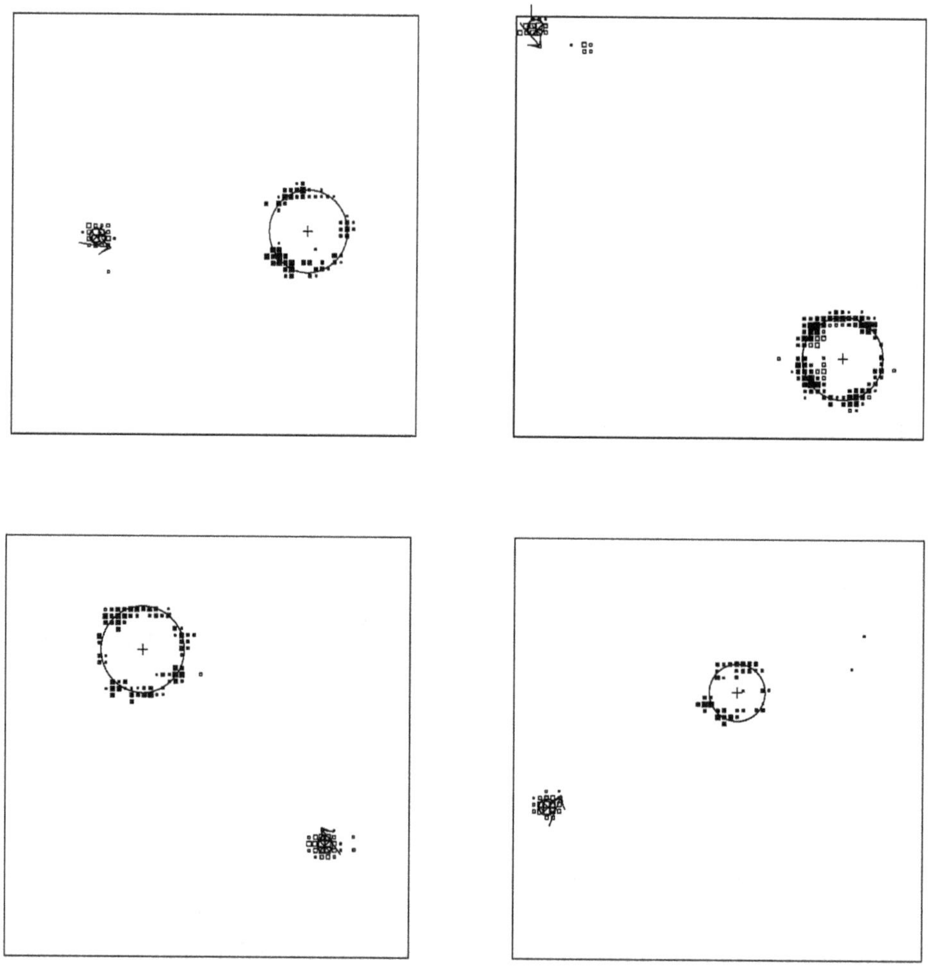

larger than a signal coming from a Cherenkov photon. Since the size of the MWPC was not covering the whole acceptance of the tracking system, only 33% of the events, which had the center of the Cherenkov ring in the MWPC, would have an ionization signal from the particle track going through the MWPC. Note also (in Figure 9), the clean appearance of the ring of Cherenkov light and the very low number of noise induced signals. This was important for particles with velocities below the threshold of Cherenkov light production, when the RICH was used as a threshold device. For identification of antiprotons between 2(3) GeV/c, which was the threshold for muons(pions), and 18 GeV/c, which was the threshold for antiprotons, this could be used since the high number of photoelectrons detected assured that muons and pions would produce a distinct Cherenkov light signal. An important feature, when using the RICH detector as a threshold device, was that it was not only using a "light/no-light" condition, but the tracking system gave information about where in the pad plane the Cherenkov light should have been detected, thereby greatly enhancing the detector immunity to noise. Less than one noisy pad out of 4096 channels per event was detected, allowing a stringent selection of antiprotons where no signals should appear in the predicted area, still maintaining a high identification efficiency. All 4096 channels were working during the entire flight.

Since the signal induced by one photoelectron was picked up by 3–5 pads, one approach would be to do a center of gravity calculation over these pads, weighing them by their signal value and from this deduce the Cherenkov angle for each individual Cherenkov photon. However, in most cases the probability that one pad could get part of its signals from at least two different photoelectrons was high because of the large number of photoelectrons. This made it difficult to estimate how much of the induced charge in a pad came from each photoelectron. Therefore, each pad was treated separately in the Cherenkov angle calculation. The signal amplitude, with its information of amount of induced charge, was not used for the same reason.

The Cherenkov angle was calculated for each pad that was not included in the 9 x 7 pad ionization area closest to the track, using an iterative geometrical method. This method used the "Fermat's principle" that light uses the path that gives the shortest possible time for travelling between two points. For this calculation we used: the position of the pad where the signal was induced, the position of the mirror and the direction of the particle (given by the tracking system) at the point where the Cherenkov light was emitted. Since the point of emission for the Cherenkov light was not determined, it was assumed that the light was emitted half-way between the MWPC and the mirror.

The Cherenkov angle y_c for the whole event was calculated using a so called "Gaussian potential method" (Agakichiev 1996). This is an iterative method that assigns a Gaussian weight:

$$\omega_i = e^{-(\theta_i - \theta_C)^2 / 2\sigma_i^2} \tag{1}$$

where θ_i and θ_c are the Cherenkov angle of each pad and for the whole event, to the Cherenkov angle calculated for each pad, depending on how much it differs from the fitted Cherenkov angle.

Figure 10 shows the reconstructed Cherenkov angle distribution for a large number of single pads. The data used for this distribution are from $\beta \sim 1$ electrons and muons selected from flight data. A Gaussian fit has been performed to the region 36– 64 mrad. This distribution gives a single pad resolution of 7.8 mrad. The tails of the distribution, which are not included in the fit, are due to the fact that the signal from each photoelectron spreads out over several pads. It has been shown in a previous work on the CAPRICE94 experiment that the Gaussian potential method gives the best result on the Cherenkov angle determination (Carlson, 1994). The multiwire proportional chamber and the calculation algorithms used for the CAPRICE98 experiment were the same as were used for the CAPRICE94 experiment. The Gaussian potential method had the benefit that pads with Cherenkov angles far from the mean value, with signals due to e.g. electronic noise, were suppressed.

Figure 10. The Cherenkov angle distribution for single pads for β~1 electrons and muons selected from flight data; the Gaussian fit gives a resolution (σ) of 7.8 mrad for a single pad.
Bergström, 2001.

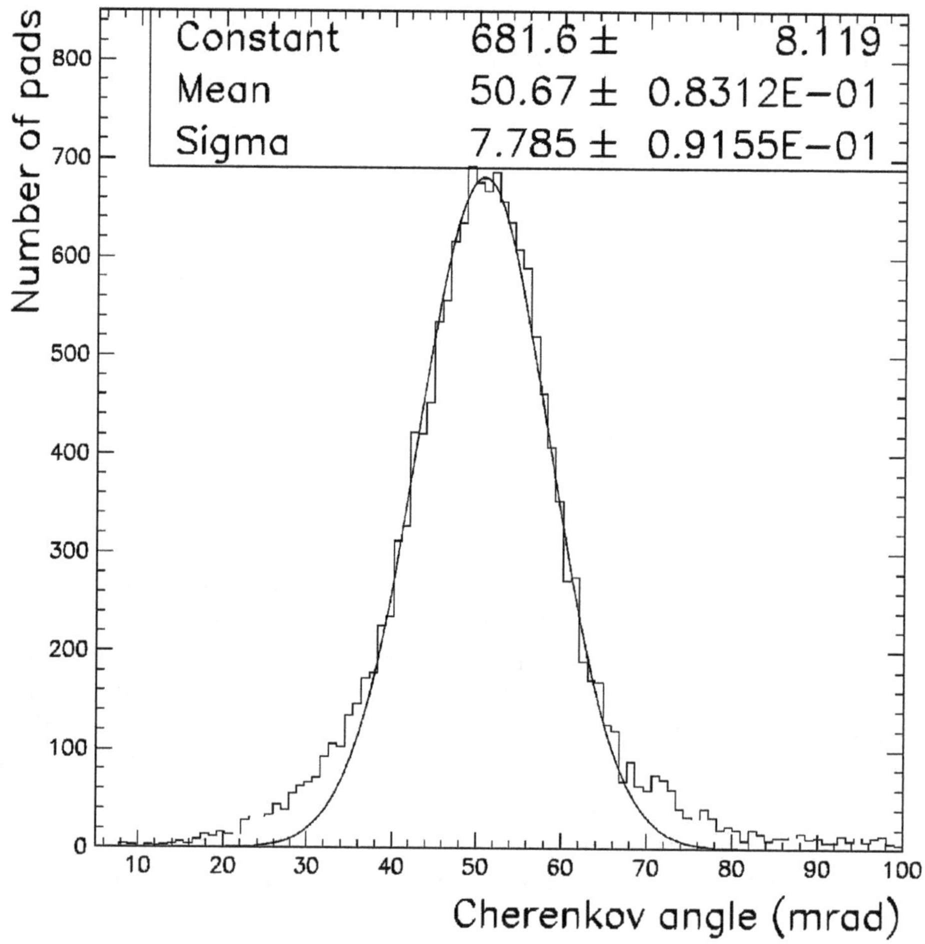

The measured Cherenkov angle for a large sample of data taken at ground a few days before the CAPRICE98 flight are shown in Figure 11. This sample of particles with a measured rigidity greater than 12 GV contains mostly muons. A Gaussian curve is fitted to the distribution and shows a resolution of about 1.2 mrad for β~ 1 particles with a Cherenkov angle of 50 mrad. The refractive index, which depends on the pressure, was measured to be 1.0012, which corresponds to a maximum Cherenkov angle of 50.0 mrad.

The measured resolution of 1.2 mrad of the Cherenkov angle means that the CAPRICE98 gas-RICH detector is capable of determining the velocity of a relativistic particle with a precision of less than 0.0057%. This corresponds to a particle moving with a velocity close to the speed of light, 299 792 km/s, will be measured with a precision of 1.7 km/s.

It is difficult to estimate the number of photoelectrons from the data, because of the over-lapping of the signal from different photoelectrons. The signal induced in a pad could be from more than one pho-

Figure 11. The Cherenkov angle distribution for selected muons with rigidity greater than 12 GeV/c from ground data; a Gaussian curve is fitted to the distribution. The resolution for $\beta = 1$ particles is less than 1.2 mrad. This means that the CAPRICE98 gas-RICH detector is capable of determining the velocity of the particles with a precision of less than 0,0057%. This corresponds to that a particle moving with a velocity close to the speed of light, 299 792 km/s will be measured with a precision of only 1.7 km/s. Bergström, 2001.

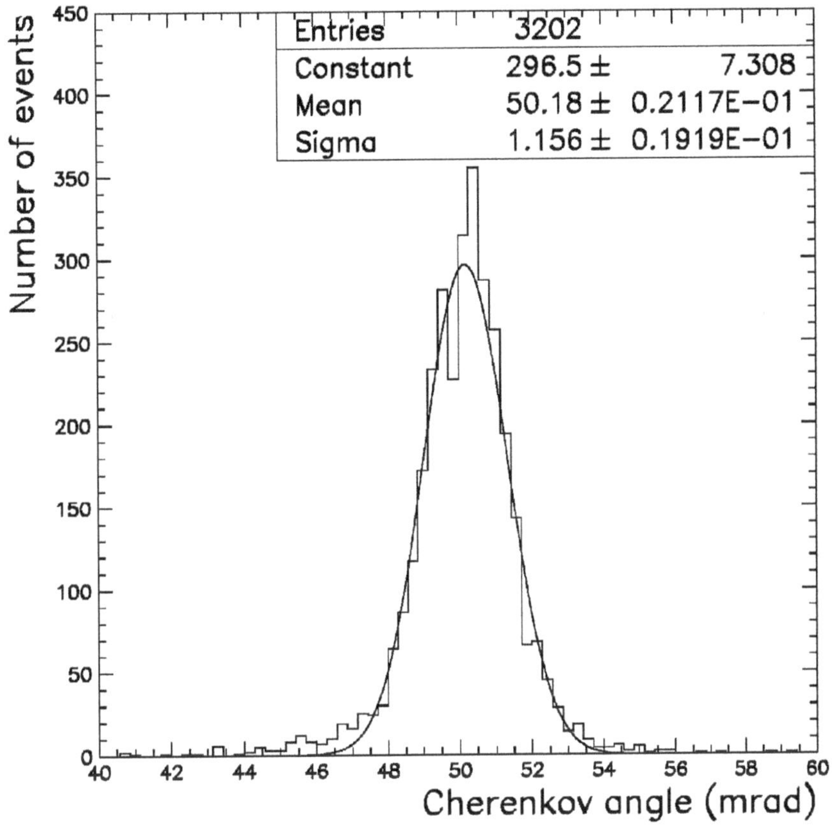

toelectron. A large number of photoelectrons, typically more than 20, would give so many overlaps that it would create a "full" ring. For most particles detected by the RICH detector this is not the case though.

We estimated the number of photoelectrons on an event-by-event basis. A number of electrons from a large sample of flight data were selected and the number of clusters, which should correspond to one photoelectron, was counted. The result was a mean value of 12 ± 2 photoelectrons per event. This result was cross checked by studying the dependence of the efficient number of pad hits used for the Cherenkov angle calculation:

$$n_{eff} = \frac{\left(\sum_{i}^{n_{pads}} \omega_i\right)^2}{\sum_{i}^{n_{pads}} \omega_i^2} \tag{2}$$

where ω_iEquation (1) is the weight of the Cherenkov angle of that pad. For the calculation a fixed value for σ_i was used, which was the average measured Cherenkov angle resolution for a single pad (7.8 mrad). n_{eff} is closely related to the number of pad hits due to Cherenkov light and therefore to the number of photoelectrons (N_{ph}), which is related to the Cherenkov angle (θ_c) through (Ypsilantis, 1994):

$$N_{ph} \sim N_0 L Z^2 \sin^2 \theta_c \tag{3}$$

where N_0 is the detector response parameter (given in cm^{-1}), L is the distance the particles travel in the Cherenkov light producing gas (in cm) and Z is the electric charge. A Gaussian fit was made to n_{eff} for a large number of $\beta \sim 1$ particles. It was then assumed that the full-width at half-maximum of this distribution divided by n_{eff} was proportional to $1/\sqrt{N_{ph}}$, where N_{ph} is the number of detected photoelectrons. This result in 15 detected photoelectrons per event. This also gives a response parameter, $N_0 = 57\ cm^{-1}$ for

Figure 12. As a function of wavelength are given: the quantum efficiency of TMAE, the transmission of the 5 mm thick quartz window and the reflectivity of the spherical mirror; a curve corresponding to the product of the TMAE quantum efficiency, the mirror reflectivity and the quartz window transmission is included. As can be seen from this product, the sensitive region of the RICH detector was 165–225 nm. In the wavelength interval 170 – 200 nm the average probability to detect a Cherenkov photon was 23%. *Bergström, 2001.*

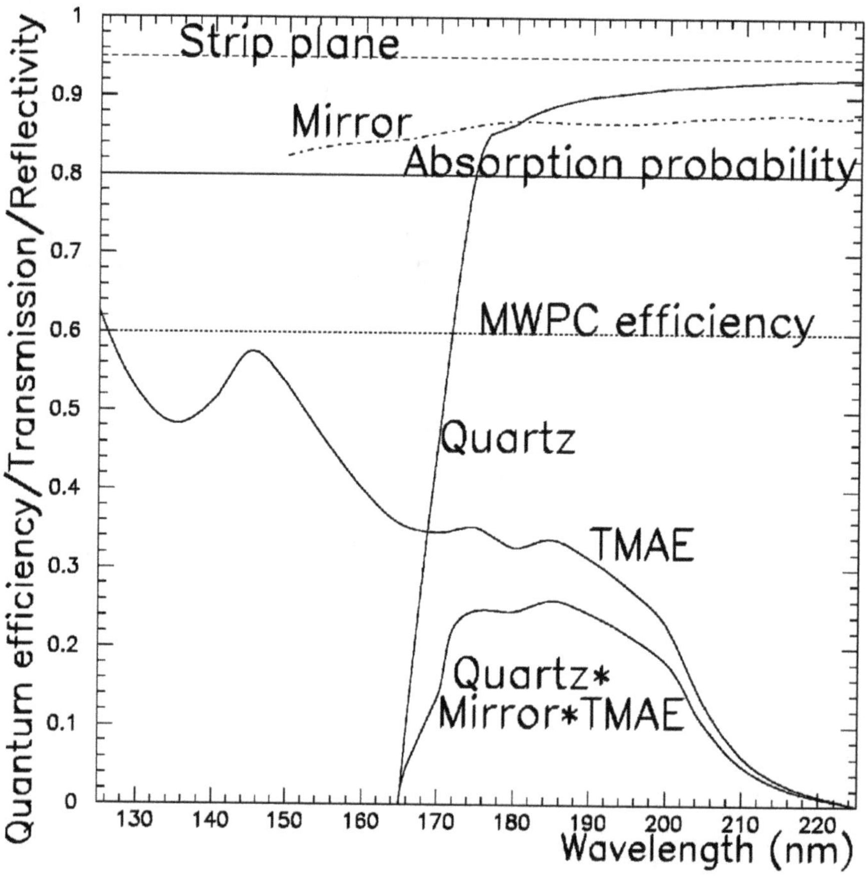

Figure 13. The measured Cherenkov angle resolution as a function of velocity (β) of the cosmic ray particle; the resulting resolution for β~ 1 particles was about 1.2 mrad corresponding to a precision in the velocity measurement of 0.0057%, or 1,7 km/s. The strong improvement in resolution for higher velocities is mainly due to the increasing number of photoelectrons produced. It is partly also due to the larger ring radius which gives each pad a better resolution in the reconstruction algorithm.
Bergström, 2001.

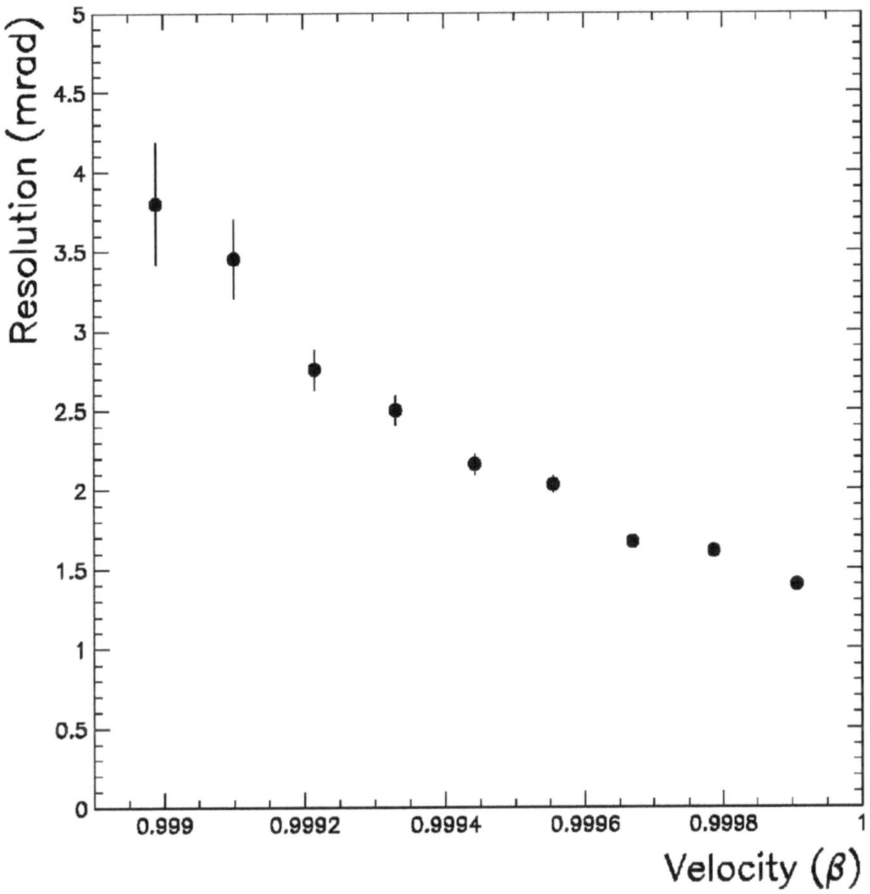

the average distance (L =98.5 cm) that the particles travel in the radiator of the RICH detector. For this study negative muons from data at ground were used, since they provide a clean sample of β~ 1 particles.

An estimation of the number of detected photoelectrons also was made by taking into account the measured reflectivity of the mirror (on average 87%), the quantum efficiency for TMAE, the quartz window transmission (85– 90%), the absorption probability (estimated to 80% in the 12 mm of ethane/TMAE gas at 35°C), the strip plane transmission (95%), the transmission of ultraviolet light in C_4F_{10} (100%) (Lenzen, 1994) and the efficiency to detect a photoelectron with the MWPC after the pedestal subtraction (60%). Integrating all of the above in the wavelength region 165–225 nm (see Figure 12) the result was that on the average 14 photoelectrons should be observed for β~ 1 particles. There was only one unknown factor left, the absorption of light in the remaining contamination of oxygen and water in the radiator. The transmission of light through the radiator depends strongly on the amount of residual oxygen and water contamination in the gas. By comparing the predicted and measured number of photoelectrons we have estimated the transmission to be greater than 85% in the average of 1.5 m that the Cherenkov photons had to travel in the radiator gas before reaching the MWPC.

The Cherenkov angle resolution was determined using a large sample of events. In Figure 13 it can be seen that the resolution as a function of velocity (β) varies from 4 mrad close to the threshold of β=0.9988 to about 1.2 mrad for fully relativistic particles. From the average number of photoelectrons (N_γ= 12), the measured Cherenkov angle resolution (σ_{tot} = 1.2 mrad for β ~1) and the relation $\sigma_{tot} = \sigma_{single}/\sqrt{N_\gamma}$ then gives a single photoelectron resolution of 4.5 mrad. This is in agreement with the angle resolution for each pad (7.8 mrad) and that each photoelectron on average gives a signal in 3–5 pads.

Besides particle identification on an event-by-event basis, the RICH efficiently rejected multi-particle events. The amplitude of the signal in the pads was used to decide if there were multiple particles passing

Figure 14. The reconstructed Cherenkov angle as a function of deflection; the dashed line indicates the expected Cherenkov angle for muons, pions and protons/antiprotons. These are selected charge one events passing the criteria track reconstruction and reconstruction of the Cherenkov angle. The negative section of the rigidity axis is for negatively charged particles. This experiment was the first to separate pions from muons and unambiguously identify high energy antiprotons.
Bergström, 2001.

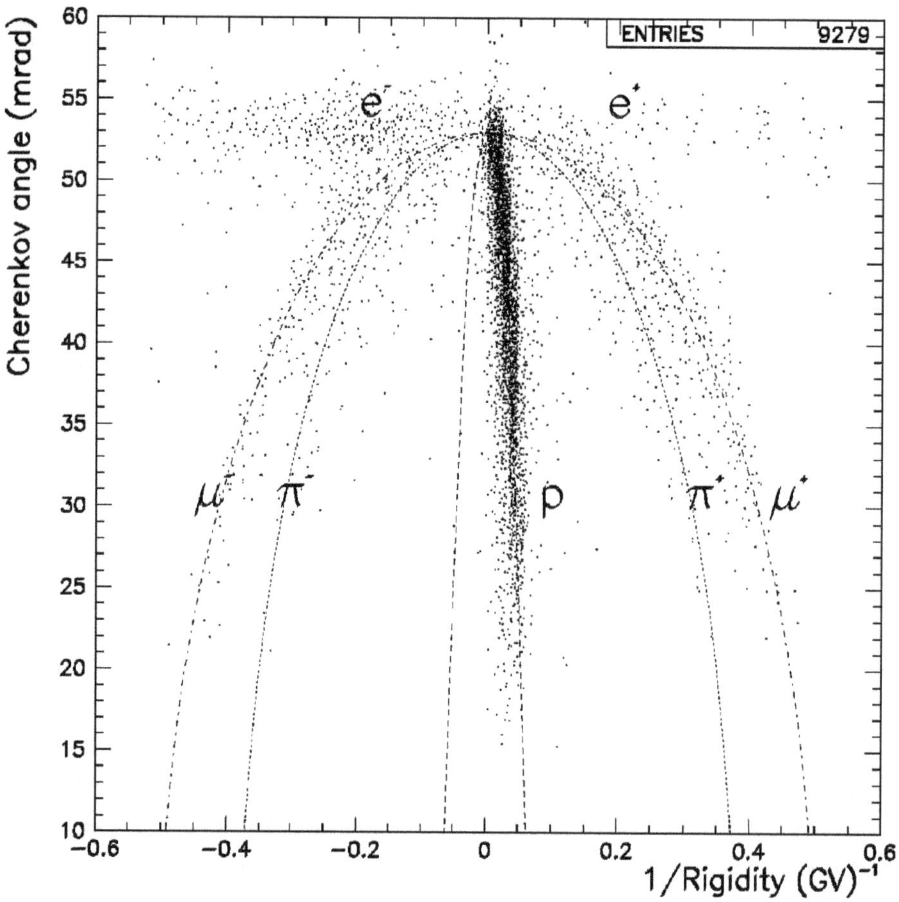

Figure 15. The Cherenkov angle as a function of rigidity for 5638 charge one particles measured by the CAPRICE98 experiment; to the right is a dense band of protons starting at approximately 18 GV and extending to higher energies and increasing Cherenkov angles. The main bulk of electrons and positrons are located at low energies (below 10 GeV) and at maximum Cherenkov angle. On the negative side, the location of five antiprotons between 20 and 50 GV are indicated by black squares.
Bergström, 2001.

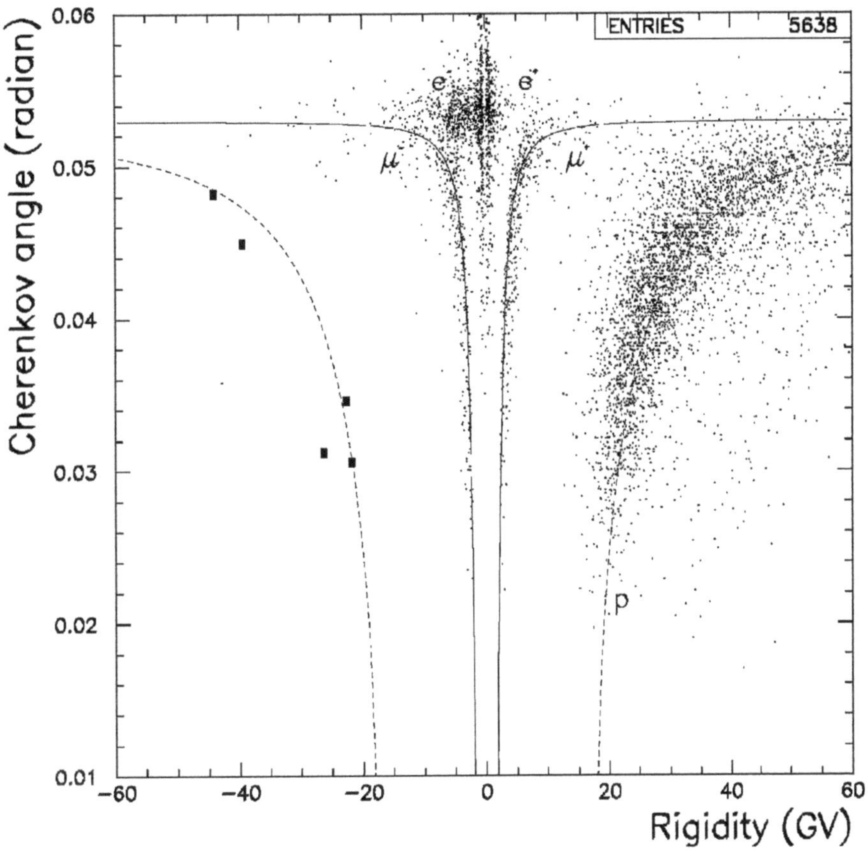

through the detector. The condition used was that there should not be two saturated pads more than 2 cm apart, because only the ionization by a charged particle could give a saturated signal. Albedo particles with a velocity high enough to produce Cherenkov light in the radiator were rejected by the RICH because the light from such a particle was detected in the wrong part of the MWPC and pad plane.

Figure 14 shows the reconstructed Cherenkov angle as a function of deflection $(\text{rigidity})^{-1}$up to 0.6 $(\text{GV})^{-1}$ for positive and negative charge one particles collected during the flight. These events were chosen for passing good track reconstruction conditions and the conditions on the reconstruction of the Cherenkov angle. The lines in the figure are the expected Cherenkov angles for protons, pions and muons. A dense band of about 6500 protons is extending from the Cherenkov light threshold at approximately 0.0056 $(\text{GV})^{-1}$ to the very highest rigidities (small deflections) and increasing Cherenkov angles. All of electrons and positrons are located at maximum Cherenkov angle as they are $\beta = 1$ particles at these rigidities (momentum). The RICH detector plays a crucial role in the CAPRICE98 experiment by the rejection of background muon and pion events. It is important to notice that muons and pions can be separated from one another in the rigidity region 2–6 GV. This is the first time a balloon borne experiment has been able to separate pions from muons. This also shows that the RICH detector allows to very precisely identifying muons in the atmosphere, which are important cross-check for the simulations carried out to calculate the atmospheric neutrino fluxes (Boezio, 1999).

Figure 15 shows the measured Cherenkov angle for 5638 particles in the rigidity region up to 60 GV measured by the CAPRICE98 experiment. These were selected charged one particles from flight data passing the criteria of the track and the Cherenkov angle reconstruction.

The momentum resolution of the RICH is comparable to the momentum resolution of the CAPRICE98 tracking system up to 50 GeV/c, and in the region 19–28 GeV/c even superior. The maximum detectable rigidity for the tracking system was 330 GV. The RICH could therefore be used to crosscheck, up to about 70 GeV/c, the momentum measurement done by the tracking system and to determine the efficiency of the tracking system. This gives the CAPRICE98 experiment a very precise understanding of the systematic uncertainties related to the tracking system.

4. CONCLUSION

A gas-RICH detector using C_4F_{10} gas as radiator has been developed and tested in particle beams at CERN. It has been an important part of a balloon borne magnet spectrometer experiment, CAPRICE98, which measured the antimatter component in the cosmic rays and the cosmic ray composition in the atmosphere. It allowed cosmic ray antiprotons with energies above 18 GeV to be mass resolved for the first time.

Because of the large number of detected photoelectrons and the low noise level in the MWPC, a good velocity (β) identification was possible by the induced signal on the pad plane originating from the Cherenkov light. By combining the RICH detector information on Cherenkov angle with the rigidity information in the magnetic spectrometer and the identification capability of the imaging calorimeter of the CAPRICE98 experiment, particles could be fully identified over large energy ranges.

In conclusion, the gas-RICH detector performed well during the CAPRICE98 flight and has shown to fulfil all the design requirements.

REFERENCES

Agakichiev, G., Drees, A., Glässel, P., Irmscher, D., Ososkov, G. A., Panebrattsev, Y., & Ullrich, T. (1996). Cherenkov ring fitting techniques for the CERES RICH detectors. *Nuclear Instruments & Methods in Physics Research. Section A, Accelerators, Spectrometers, Detectors and Associated Equipment, 371*(1-2), 243–247. doi:10.1016/0168-9002(95)01168-4

Ambriola, M., Barbiellini, G., Bartalucci, S., Basini, G., Bellotti, R., & Bergström, D. et al.. (1999). CAPRICE98: A balloon borne magnetic spectrometer to study cosmic ray antimatter and composition at different atmospheric depths. *Nuclear Physics B, 78*, 32.

Ambriola, M. L., Barbiellini, G., Bartalucci, S., Basini, G., Bellotti, R., & Bergstroem, D. (1998). *CAPRICE98: A balloon borne magnetic spectrometer to study cosmic ray at different atmospheric depths.* 26th International Cosmic Ray Conference, Salt Lake City, UT.

Barbiellini, G. (1997). A gas-RICH detector for cosmic ray studies. *Proceedings of the 25th International Cosmic Ray Conference*.

Basini, G. (1999). *Proceedings of the 26th International Cosmic Ray Conference*. Salt Lake City, UT: Academic Press.

Bergström, D. (2001). Performance of the CAPRICE98 balloon-borne gas-RICH detector. *Nuclear Instruments & Methods in Physics Research. Section A, Accelerators, Spectrometers, Detectors and Associated Equipment*, *463*, 161.

Bergström, D. (1999). *Performance of the CAPRICE98 gas-RICH detector*. (Lic. Thesis). Royal Inst. of Tech., Stockholm, Sweden.

Bergström, D. (1999). A gas RICH for cosmic ray studies. *Proceedings of the 26th International Cosmic Ray Conference*.

Bergström, D., Boezio, M., Carlson, P., Francke, T., Grinstein, S., Khalchukov, F., & Stochaj, S. J. et al. (2000). First Mass-resolved Measurement of High-Energy Cosmic-Ray Antiprotons. *The Astrophysical Journal*, *534*(2), 177. doi:10.1086/312665 PMID:10813676

Bocciolini, M., Celletti, F., Finetti, N., Grandi, M., Papini, P., Perego, A., & Zampa, N. et al. (1996). N. Zampa, *The WiZard/CAPRICE silicon-tungsten calorimeter. Nuclear Instruments & Methods in Physics Research. Section A, Accelerators, Spectrometers, Detectors and Associated Equipment*, *370*(2-3), 403–412. doi:10.1016/0168-9002(95)00841-1

Boezio, M., Carlson, P., Francke, T., Weber, N., Suffert, M., Hof, M., & Stochaj, S. J. et al. (1999). Stochaj, S. J., *New Measurement of the Flux of Atmospheric Muons. Physical Review Letters*, *82*(24), 4757–4760. doi:10.1103/PhysRevLett.82.4757

Carlson, P., Francke, T., Löfgren, S., Weber, N., & Suffert, M. (1994). Results from beam tests of the CAPRICE RICH detector. *Nuclear Instruments & Methods in Physics Research. Section A, Accelerators, Spectrometers, Detectors and Associated Equipment*, *349*(2-3), 577–591. doi:10.1016/0168-9002(94)91229-7

de Beauville, E., & Borer, K. (1990). Amplex, a low-noise, low-power analog CMOS signal processor for multi-element silicon particle detectors. *Nuclear Instruments & Methods in Physics Research. Section A, Accelerators, Spectrometers, Detectors and Associated Equipment*, *288*(1), 157–167. doi:10.1016/0168-9002(90)90481-K

Francke, T., Bergström, D., Boezio, M., Carlson, P., & Suffert, M. (1999). A gas RICH detector for space. *Nuclear Instruments & Methods in Physics Research. Section A, Accelerators, Spectrometers, Detectors and Associated Equipment*, *433*(1-2), 87–91. doi:10.1016/S0168-9002(99)00311-3

Golden, R. L., Badhwar, G. D., Jeffrey, L., Lacy, J., & Zipse, J. (1978). A magnetic spectrometer for cosmic ray studies. *Nuclear Instruments and Methods*, *148*(1), 179–185. doi:10.1016/0029-554X(78)90351-8

Golden, R. L., Horan, S., Mauger, B. G., Badhwar, G. D., Lacy, J. L., Stephens, S. A., & Zipse, J. E. et al. (1979). Evidence for the Existence of Cosmic-Ray Antiprotons. *Physical Review Letters*, *43*(16), 1196–1199. doi:10.1103/PhysRevLett.43.1196

Hof, M., Bremerich, M., Menn, W., Pfeiffer, C., Reimer, O., & Simon, M. et al.. (1994). Stochaj, S, *Performance of drift chambers in a magnetic rigidity spectrometer for measuring the cosmic radiation. Nuclear Instruments & Methods in Physics Research. Section A, Accelerators, Spectrometers, Detectors and Associated Equipment, 345*(3), 561–569. doi:10.1016/0168-9002(94)90516-9

Lenzen, G., Schyns, E., Thadome, J., & Werner, J. (1994). The use of fluorocarbon radiators in the DELPHI RICH detectors. *Nuclear Instruments & Methods in Physics Research. Section A, Accelerators, Spectrometers, Detectors and Associated Equipment, 343*(1), 268–272. doi:10.1016/0168-9002(94)90562-2

Orito, S., Maeno, T., Matsunaga, H., Abe, K., Anraku, K., Asaoka, Y., & Streitmatter, R. (2000). Precision Measurement of Cosmic-Ray Antiproton Spectrum. *Physical Review Letters, 84*(6), 1078–1081. doi:10.1103/PhysRevLett.84.1078 PMID:11017448

Ricci, M. (1999). *Proceedings of the 26th International Cosmic Ray Conference*. Salt Lake City, UT: Academic Press.

Ypsilantis, T., & Séguinot, J. (1994). Theory of ring imaging Cherenkov counters. *Nuclear Instruments & Methods in Physics Research. Section A, Accelerators, Spectrometers, Detectors and Associated Equipment, 343*(1), 30–51. doi:10.1016/0168-9002(94)90532-0

ADDITIONAL READING

Ambriola, M., Bartalucci, S., Bellotti, R., Bergström, D., Boezio, M., Bonvicini, V., & Zampa, N. et al. (2002). High-energy deuteron measurement with the CAPRICE98 experiment. *Nuclear Physics B - Proceedings Supplements, 113*(1), 88–94. doi:10.1016/S0920-5632(02)01826-1

Barbiellini, G., Bartalucci, S., Bellotti, R., Bergström, D., Bidoli, V., Boezio, M., & Zampa, N. et al. (2001). CAPRICE98: A balloon-borne magnetic spectrometer equipped with a gas RICH and a silicon calorimeter to study cosmic rays. *Nuclear Instruments & Methods in Physics Research. Section A, Accelerators, Spectrometers, Detectors and Associated Equipment, 461*(1-3), 269–271. doi:10.1016/S0168-9002(00)01222-5

Bergström, D. (2001). (The CAPRICE collaboration): *Performance of the CAPRICE98 balloon-borne gas-RICH detector. Nuclear Instruments & Methods in Physics Research. Section A, Accelerators, Spectrometers, Detectors and Associated Equipment, 463*, 161.

Boezio, M. (1999). (The CAPRICE collaboration): *The cosmic ray proton and helium spectra between 0.2 and 200 GeV. The Astrophysical Journal, 518*, 457. doi:10.1086/307251

Boezio, M. (2000). (The CAPRICE collaboration): *The cosmic ray electron and positron spectra measured at 1 AU during solar minimum activity. The Astrophysical Journal*, 531.

Boezio, M., Barbiellini, G., Bonvicini, V., Schiavon, P., Vacchi, A., Zampa, N., & Stochaj, S. J. et al. (2001). Measurements of cosmic-ray electrons and positrons by the Wizard/CAPRICE collaboration. *Advances in Space Research, 27*(4), 669–674. doi:10.1016/S0273-1177(01)00108-9

Boezio, M., Bonvicini, V., Schiavon, P., Vacchi, A., Zampa, N., Bergstrom, D., & Stochaj, S. J. et al. (2001). (The CAPRICE collaboration): *The Cosmic-Ray antiproton flux between 3 and 49 GeV. The Astrophysical Journal*, *561*(2), 787–799. doi:10.1086/323366

Boezio, M., Bonvicini, V., Schiavon, P., Vacchi, A., Zampa, N., Bergström, D., & Stochaj, S. J. et al. (2003). The Cosmic–Ray Proton and Helium Spectra measured with the CAPRICE98 balloon experiment. *Astroparticle Physics*, *19*(5), 583–604. doi:10.1016/S0927-6505(02)00267-0

Boezio, M., Bonvicini, V., Schiavon, P., Vacchi, A., Zampa, N., Bergström, D., & Stochaj, S. J. et al. (2003). Energy spectra of atmospheric muons measured with the CAPRICE98 balloon experiment. *Physical Review D: Particles and Fields*, *67*(7), 072003. doi:10.1103/PhysRevD.67.072003

Boezio, M., Carlson, P., Francke, T., Weber, N., Suffert, M., Hof, M., & Stochaj, S. J. et al. (1999). (The CAPRICE collaboration): *New measurement of the flux of atmospheric muons. Physical Review Letters*, *82*(24), 4757–4760. doi:10.1103/PhysRevLett.82.4757

Circella, M., Ambriola, M., Barbiellini, G., Bartalucci, S., Bellotti, R., Bergström, D., & Zampa, N. et al. (2001). Measurements of primary cosmic-ray hydrogen and helium by the WiZard collaboration. *Advances in Space Research*, *27*(4), 755–760. doi:10.1016/S0273-1177(01)00117-X

Circella, M., Ambriola, M. L., Barbiellini, G., Bartalucci, S., Bellotti, R., Bergström, D., & Zampa, N. et al. (2000). The WiZard collaboration cosmic ray muon measurements in the atmosphere. *Nuclear Physics B - Proceedings Supplements*, *85*(1), 355–360. doi:10.1016/S0920-5632(00)00529-6

Francke, T., Bergström, D., Boezio, M., Carlson, P., & Suffert, M. (1999). A Gas-RICH detector for space. *Nuclear Instruments & Methods in Physics Research. Section A, Accelerators, Spectrometers, Detectors and Associated Equipment*, *433*(1-2), 87–91. doi:10.1016/S0168-9002(99)00311-3

Kremer, J. (1999). (The CAPRICE collaboration): *Measurements of ground level muons at two magnetic locations. Physical Review Letters*, *83*, 4241.

Papini, P., Piccardi, S., Spillantini, P., Vannuccini, E., Ambriola, M., Bellotti, R., & Suffert, M. et al. (2004). *High-energy deuteron measurement with the CAPRICE98 experiment. The Astrophysical Journal*, *615*(1), 259–274. doi:10.1086/424027

KEY TERMS AND DEFINITIONS

Active Cupper: Pieces of pure cupper which have been heated up to 800°C and exposed to hydrogen. These cupper pieces are good absorbers of oxygen.

ADC: Analog to Digital Converter. An electronic circuit to convert analog digital signals to digital numbers.

AMPLEX: A sensitive electronic circuit developed by CERN for reading out sensitive particle physics detectors.

C_4F_{10}: Tetracarbondecafluoride. The densest liquid fluorocarbon, and hence the liquid fluorocarbon with the highest index of refraction for visible and ultra violet light at room temperature.

CAPRICE: The Cosmic AntiParticle Ring Imaging Cherenkov Experiments. A series of balloon borne experiments aimed to measure the flux of antimatter and light isotopes in the cosmic radiation. In total CAPRICE made five successful flights between 1994 and 1998. The flights were performed from Lynn Lake, Manitoba, Canada, and from Ft Sumner, New Mexico, USA.

CAPRICE94: The 1994 balloon borne flight by the CAPRICE experiment from Lynn Lake, Manitoba, Canada.

CAPRICE98: The 1997-1998 balloon flights by the CAPRICE experiment from Ft Sumner, New Mexico, USA.

Cherenkov Light: A new type of light emitted when charged particles traverse a transparent medium with a velocity higher than the speed of light in the medium. It is named after the Russian scientist Pavel Cherenkov who discovered it.

Gas-RICH: A RICH detector using a gaseous radiator to produce the Cherenkov light.

MWPC: Multi Wire Proportional Counter, a type of detector often used in particle physics experiments.

RICH: Ring Imaging CHerenkov detector. A type of detector which measures the velocity of elementary particles by measuring the emission angle of the Cherenkov light produced when the particle traverses a medium with a velocity higher than the speed of light in the medium.

TMAE: Tetrakis-(dimethylamino)-ethylene. A yellow liquid which in vapor form is an efficient photo-converter, i.e. has high probability to absorb a light photon and emit a photoelectron.

VLSI: Very large scale integrated electronic circuit.

β: A measure of velocity expressed as the fraction of the speed of light (300.000 km/s). β is normally expressed in per cent of the speed of light.

Chapter 14
CsI-RICH Detectors

ABSTRACT

The most successful application of CsI photocathodes are in Ring Imaging Cherenkov (RICH) detectors. Most RICH detectors used today in high energy physics experiments use a CsI photocathode coupled to a gaseous detector. They have shown to be efficient and stable over long periods of time. We will describe some of the most important RICH detectors used, and the technology behind them.

1. INTRODUCTION

CsI photocathodes have quantum efficiency in the UV region as high as TMAE vapour and thus represent a viable alternative in many applications. The main advantage of the CsI photo convertor is that it can be exposed to normal gases and even to dray air without loss in its quantum efficiency. This tremendously simplifies its handling such as transfer from the evaporation system to the gas chamber and also allows using cheap construction materials which strongly reduces the detector price.

Note also that in the case of solid photocathodes all photoelectrons are extracted from an extremely thin surface layer whereas in the case photosensitive vapour they are create within an orders of magnitude thicker gas volume. This causes a considerable jitter in arrival time of the primary electrons to the region of gas amplification. Therefore, CsI-based photosensitive detectors offer much better time resolution.

The first application of CsI photocathodes in a high-energy physics experiment was the CsI-MWPCs of the Threshold Imaging Cherenkov detector (TIC) at the CERN-NA44 heavy ion experiment (Braem, 1998). The schematic view of this device is shown in Figure 1.

The TICs main components are: the gas container which serves as a Cherenkov radiator where the UV photons are created, two V-shaped mirrors (inclined on 45°with respect to the chamber axis) and two CsI-MWPCs detecting the Cherenkov photons.

A particle of mass m_p, traversing a radiator with refractive index n, above the Cherenkov momentum threshold p_t, given by

$$p_t = m_p[\{1/(\sqrt{n}\{1-(1/n^2)\}], \quad (1)$$

DOI: 10.4018/978-1-5225-0242-5.ch014

Figure 1. Schematics of the threshold imaging Cherenkov detector in the NA44 experiment at CERN
From Braem, 1998.

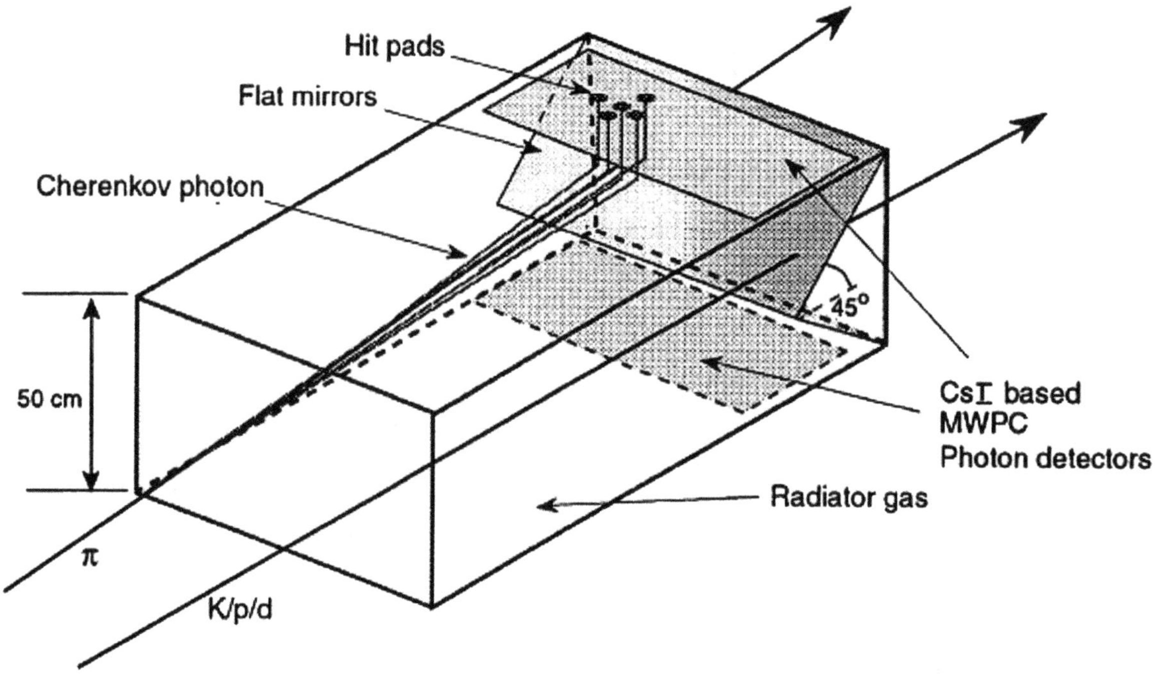

will result in a number of photons being created in the radiator.

In Figure 1, two particle tracks are shown for the illustration, one below and the other one (pion) above the Cherenkov threshold and emitting photons along its path. The Cherenkov light is reflected by the mirror and creates a well localized image of the Cherenkov photons emitted. The image is in the shape of a fiducial disc, which is the projection of the cone of Cherenkov light on the detector plane.

The V-shaped mirror geometry and the division of the photosensitive area in two parts minimizes the variation of the photon yield due to the different path lengths of particles and the Cherenkov photons through the gas radiator.

CsI-TIC were in use in the NA44 experiment during the 1995 and1996 physics runs with lead-lead collisions, allowing track-by-track identification of up to five simultaneous particles in the 3-8 GeV/c momentum range. No deterioration of the photocathode's quantum efficiency were observed during the experiment even though they were continuously ion bombarded corresponding to the total accumulated charge of about 100 μC/cm^2.Thesefirst results demonstrated that CsI photo cathodes are well suited for long term recording of Cherenkov photons.

2. CsI-MWPC RICH DETECTORS

Several high-energy physics experiments have been using CsI-MWPC operated with CH_4 at normal temperature and pressure, see Table 1.

Table 1. The main characteristics of Cherenkov detectors based in CsI MWPC

Experiment-Lab	3σ π/K Separation Momentum Range (GeV/c)	Max Interaction Rate (Hz)	Radiator (Length)	CsI Active Area (m^2)
ALICE-CERN/LHC	0.8-3	10^4	C_6F_{14} (15 mm)	11
STAR-BNL/RICH	0.8-3	10^4	C_6F_{14} (10 mm)	1.2
COMPASS-CERN/SPS	3-40	10^6	C_4F_{10} (3 m)	5.8
HALL A-TJNAF	0.8-3	10^6	C_6F_{14} (15 mm)	0.7
HADES-GSI	hadron blind	10^6	C_4F_{10} (0.4 m)	1.4

From Nappi, 2005.

Two of them, HADES (High Acceptance Di-Electron Spectrometer) and COMPASS (Common Muon and Proton Apparatus for Structure and Spectroscopy), use a gas (C_4F_{10}) radiator. The other three, STAR (Solenoidal Tracker at RHIC), ALICE (A Large Ion Collider Experiment) and JLAB-Hall A (Jefferson Lab Experiment), have a liquid (C_6F_{14}) radiator. All the projects have photon detectors equipped a similar front end electronics based on an ASIC developed by the ALICE Collaboration named GASSIPLEX, which was modified according to the specific needs of each experiment.

In 2006 the COMPASS RICH-1 was upgraded by replacing the GASSIPLEX with the APV25-S1 readout chip to withstand the increased interaction rates. Similarly, in 2008, the HADES RICH front end electronics was upgraded to the APV25-S1 chip to be able to reduce the peaking time to 0.2 μs and increase the trigger rate capability.

The CsI-MWPC used in these projects with the exception of HADES have a design similar to the one optimized by the ALICE-RICH group. For this reason we will focus on the ALICE RICH photosensitive detector, which also has the largest effective area, 11 m^2 (see for example Andres, 2002)

A photograph of the ALICE RICH detector taken during the installation in the tavern is shown in Figure 2.

It consists of seven individual modules located 4.7 m from the beam axis and mounted in a cradle where each of the modules is oriented to optimize the Cherenkov angle resolution. The size of each module (1.3 × 1.3 m^2) has been kept as large as possible, only limited by mechanical constraints. A special aluminium alloy was carefully selected as frame material to minimize outgassing into the chamber volume.

The schematics of the ALICE RICH photosensitive detector is shown in Figure 3. It is a MWPC having its bottom cathode covered with a thin CsI layer, acting as a photo conversion plane. The cathode is segmented into metalized pads, 8x8 mm^2, to give a two dimensional localization of the photoelectrons.

The Cherenkov radiator is a 15 mm thick layer of liquid C_6F_{14} with an index of refraction of n = 1.2834 at λ= 175 nm, corresponding to β_{min} =0.78. This corresponds to a threshold momentum for emission of Cherenkov light of p_{th}=1.26m_p, where m_p is the particle mass. The cone of Cherenkov light emitted in the liquid radiator expands in the proximity volume filled with CH_4 before reaching the CsI-MWPC photon detector thus allowing a better imaging of the Cherenkov pattern.

The signal recorded by the pad readout of the CsI-MWPC consists of the coordinates of the pads with a signal above a defined threshold, and the amount of charge induced on each of them enabling to improve the resolution of the impact position using a centroid method. The signal from each photoelectron is spread out over of 1-3 pads.

Figure 2. Photograph of the ALICE RICH detector made during its installation in the tavern. Individual modules are seen
From http://aliceinfo.cern.ch/Public/en/Chapter2/Chap2_HMPID-en.html.

Figure 3. The Alice RICH detector

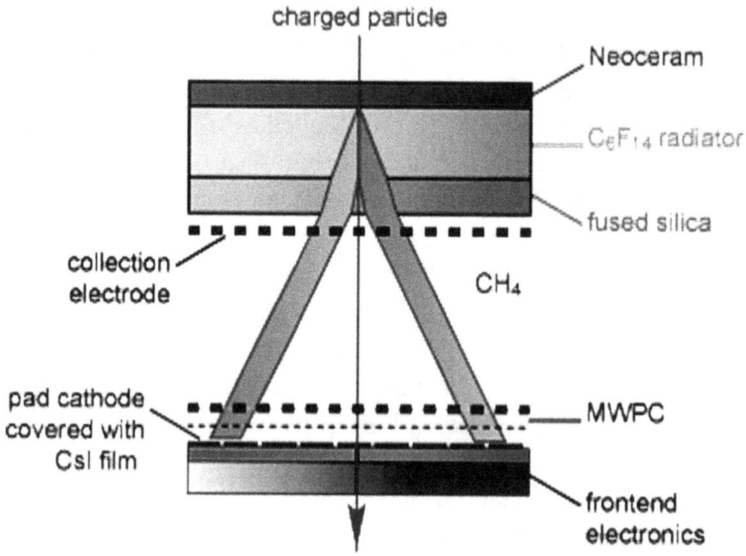

Figure 4. Cherenkov rings detected by the ALICE RICH module #3 for during the run with cosmic particles

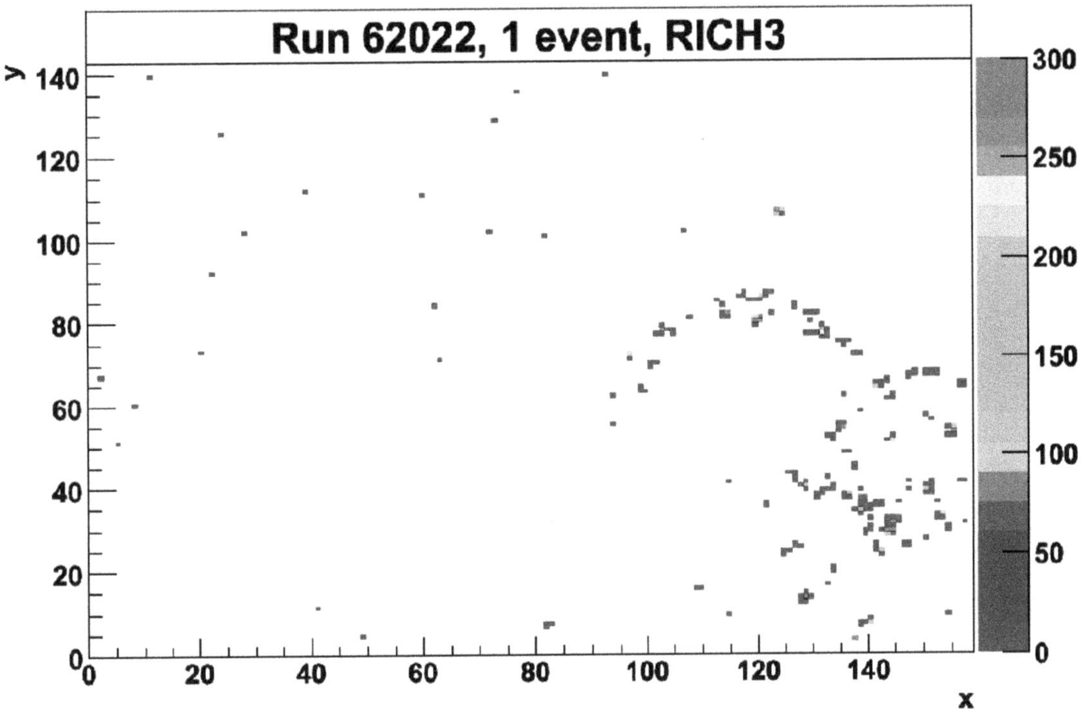

The use of a CsI-MWPC open geometry, i.e. without any measures to reduce ion and photon feedback from the avalanches, requires careful optimizations in order to maintain a stable detector operation in the high particle flux. Photon and ion feedback generate secondary photoelectrons emitted from the photocathode which gives spurious hits in the pad plane. These additional pads fired may be confused as Cherenkov photons or broadening the cluster of pads from a Cherenkov photon with false information. When the feedback is in the range 10-20% this effect can in some cases be considered as positive since it enlarges the cluster size which may help in identifying the Cherenkov rings. However, at higher levels of feedback the pad occupancy (i.e. the number of pads fired per Cherenkov photon) increases and the shape of the induced charge will be disturbed by such overlaps, degrading the centroid evaluation and the resulting localization accuracy.

In addition, at high gains sustained discharges due to the photon feedback may appear. In a single wire counter, when practically all photons produced in the avalanche reach the CsI photocathode, the discharge happens when $A\gamma_{ph} = 1$. In the MWPC geometry 30-50% of the photons produced in the electron avalanche reach the photocathode, and discharges will correspondingly appear higher gas gains. Therefore, there is a delicate a trade-off between the high gains necessary for efficient single photoelectron detection and the tolerable level of feedback.

Some results obtained with ALICE RICH are presented below. Figure 4 shows the display of one event in RICH chamber #3, where a clear Cherenkov ring pattern is present. In this particular case the data has been collected during the ALICE cosmic ray runs used for a first evaluation of the detector performance and alignment.

Figure 5. The Cherenkov angles vs particle momentum measured by the ALICE RICH in p-Pb collisions

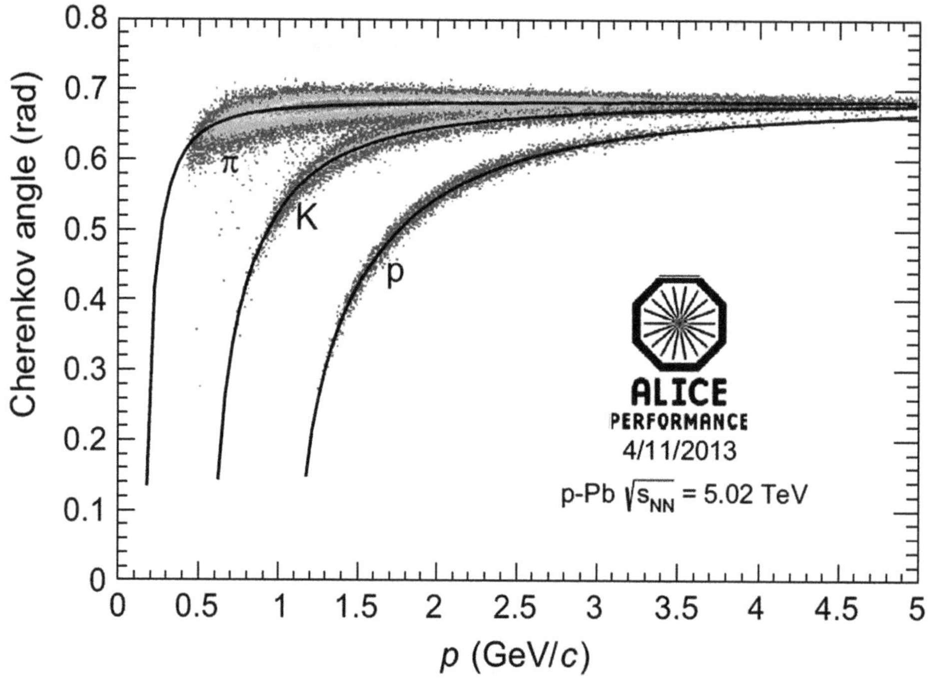

During real ALICE runs the particle identification is performed by determining the Cherenkov angle from the ring pattern. It is a difficult task (see, for example Linka, 1999, and especially in the high multiplicity environment of the most central Pb–Pb collisions. An algorithm has been implemented to determine the Cherenkov angle starting from the two-dimensional ring pattern in the photon detectors. It is based on the Hough Transform Method (HTM) to separate signal from background (Volpe, 2014). The ALICE RICH is useful in inclusive hadrons spectra measurements as well as in measurements where high purity particle identification is required.

In Figure 5 the reconstructed Cherenkov angle is shown as a function of the particle momentum for a large number of particles produced in p–Pb collisions. Three well separated bands are present, corresponding to pion, kaon and proton signals. The experimental data are in good agreement with the theoretical predictions (solid lines).

In the data presented in Figure 5 the background was negligible and the detector occupancy was only 0.1%, which is the case for p-Pb collisions. In central Pb–Pb collisions, the detector occupancy is ~30 times higher than in p-Pb. The particle density occasionally causes the pattern recognition algorithm based on the Hough Transform to fail and fake ring patterns are detected. Even so, the statistical particle identification used in inclusive hadrons production studies is still effective, and the background can be easily subtracted (Schukraft, 1998).

For track-by-track particle identification a different approach must be followed (De Cataldo, 2014; Volpe, 2014). From the knowledge of the expected Cherenkov angle and the expected theoretical standard deviation, it is possible to calculate the values of two particle identification estimators, the probability to be one of the charged hadron species and the difference between the value of the measured and the theoretically expected angle expressed in the number of standard deviations from the expected value.

Figure 6. Average number of reconstructed Cherenkov photons per track, and per CsI photocathode for β=1 particles, in pp and p-Pb collisions; the horizontal axis contains the different number of run periods per year to which the experimental points are corresponding
From De Cataldo, 2014.

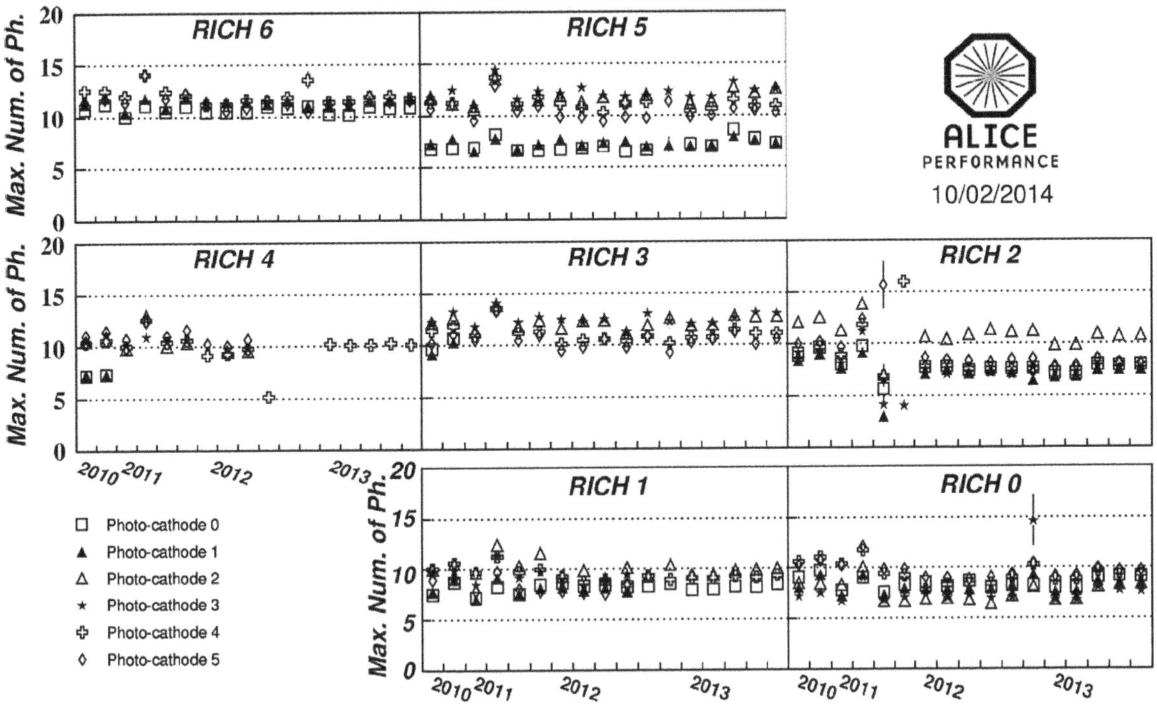

The stability of the CsI photocathodes was evaluated during the LHC (large hadron collider) runs. The CsI photocathode stability cannot be measured via a direct measurement of the CsI quantum efficiency extracted from the LHC data, because it is the convolution of several factors. An indirect way is to monitor the number of reconstructed Cherenkov photons per ring at the maximum emission angle (β=1 particles).

In Figure 6 the results from pp and p-Pb collisions for all the active photocathodes are presented. ~10 photons per Cherenkov ring are reconstructed. In RICH detector#5, one of the high-voltage sectors serving photo-cathode #1and #2 was off, and the mean number of the reconstructed photons is correspondingly smaller due to the not fully contained rings in the remaining active part of the CsI photocathode. The number of photons detected per ring remained constant over more than 3 years of operation.

In the COMPASS RICH1 the CsI photon detectors operate at high trigger rate than for ALICE, and in a high-intensity beam characterized by a huge dispersed halo of muons generated in upstream interactions.

In the original version of this detector Cherenkov photons were created in a 3 m long gas radiator of C_4F_{10}. The photons were focused by a large UV mirror wall (21 m^2), composed of 116 spherical mirrors, onto eight CsI-MWPC detectors, identical to the ALICE-RICH photon detectors (Figure 7a).

In 2006 the CsI-MWPCs in the central region (~25% of the detector surface) were replaced by a fast detection system based on multi-anode PMTs, as shown in Figure 7.b.The main reasons were to provide

Figure 7. The COMPASS RICH1: a) the first design, b) the present version
From Albrecht, 2002 and Tessarotto, 2014.

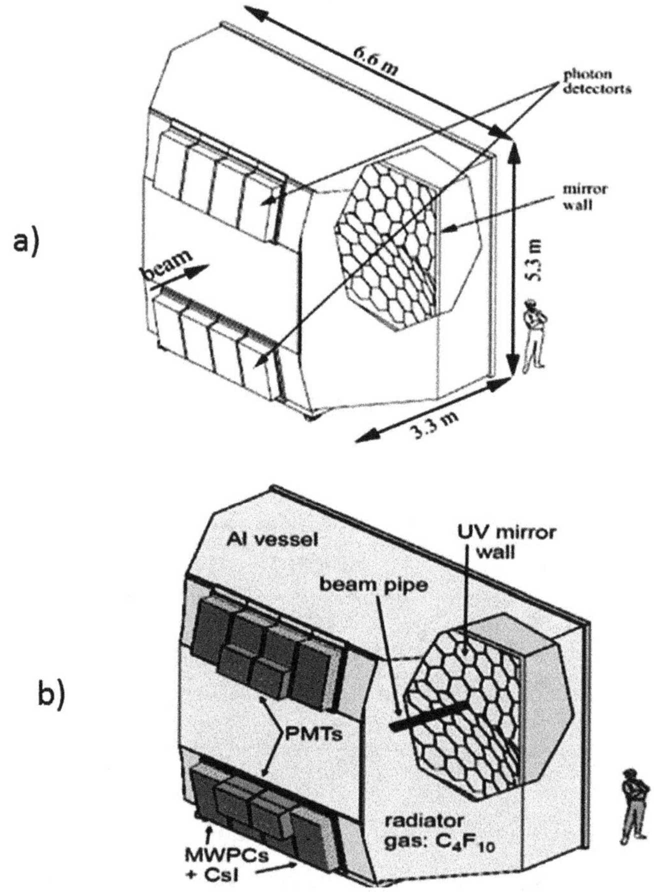

higher rate capability (one of the limitation came from the occasional instability mentioned in Chapter 3), and to get rid of the large uncorrelated background of mainly muons.

The upgraded detector showed excellent performance. A dedicated reconstruction and analysis software has been developed and used to perform data reduction, pattern recognition and hadron identification in COMPASS. For each particle a wide fiducial area of the photon detectors is defined where its Cherenkov photons could have arrived. This sometimes cover both part of the multi-anode photomultiplier tubes (MAPMT) and part of the CsI-MWPC, or part of the upper and part of the lower photon detector set. For each reconstructed hit inside the fiducial area an algorithm is used to compute the Cherenkov polar and azimuthal angles. These parameters were then used in the further analysis.

The reconstructed Cherenkov angle vs. the particle momentum is shown in Figure 8. The e, π, K and p band are clearly separated.

A schematic view of the HADES RICH is shown in Figure 9a and a photograph is presented in Figure 9b.

The idea of this detector is that leptons with 100 - 1500MeV/c momentum will produce Cherenkov light in the gaseous C_4F_{10}radiator, whereas all hadrons in the same momentum region have velocities far

Figure 8. The Cherenkov angle as a function particle momentum measured by the COMPASS RICH1
From Tessarotto, 2014.

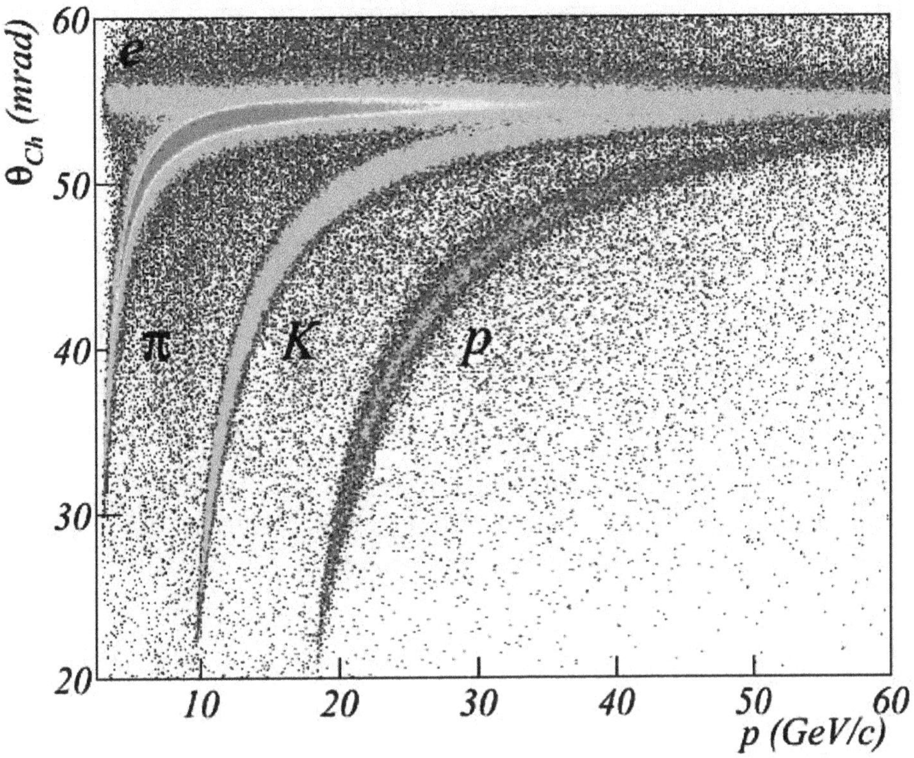

Figure 9. The HADES RICH: a) schematic layout, b) the photograph of the detector seen from the CaF_2 window side
From Zeitelhack, 1999.

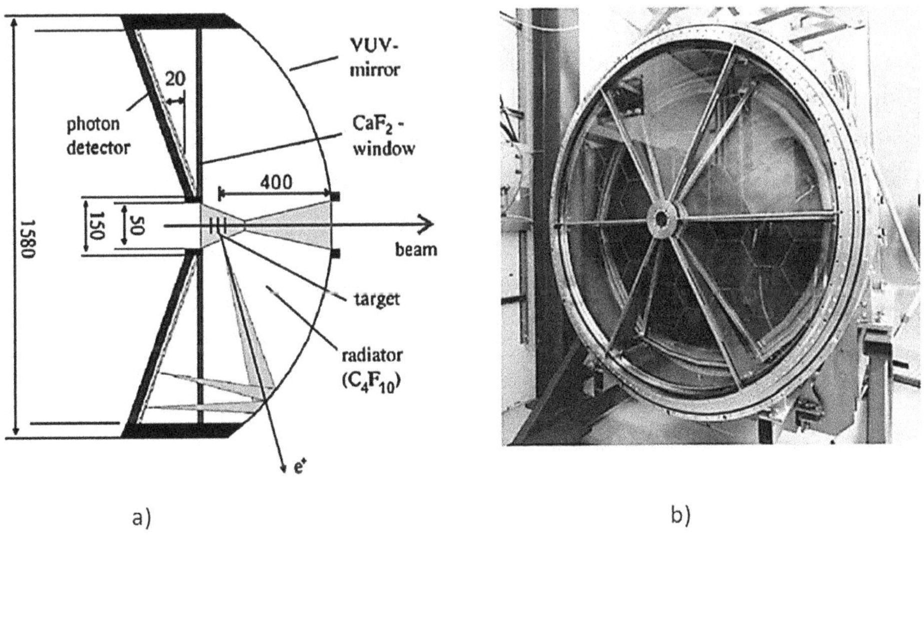

Figure 10. Schematic cross section of the HADES CsI-MWPC
From Zeitelhack, 1999.

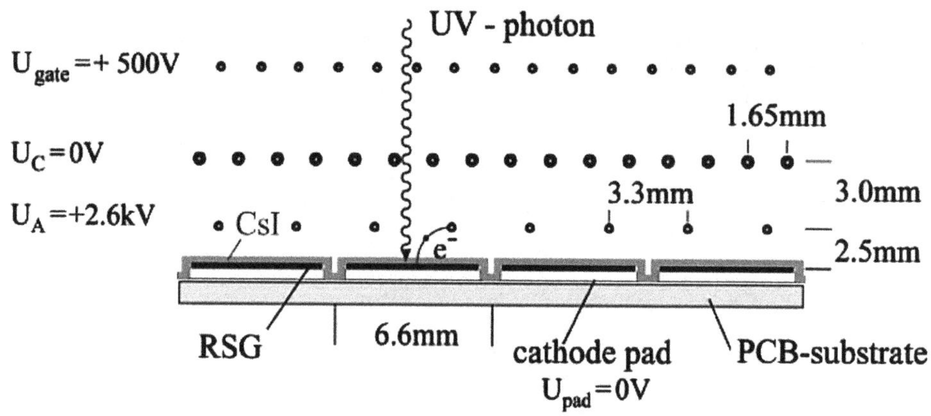

below the Cherenkov threshold. The emitted Cherenkov photons are reflected by a segmented spherical VUV-mirror and focused onto CsI-MWPCs, schematically shown in Figure 10.

The HADES photon detector design differs from the other RICH by the following features:

1. Substrate material.
2. Asymmetric gap in the chamber to improve the coupling between the cathode pads and the anode wires, and for reducing the mean number of pads on which the avalanche signals are induced.
3. Cathode plane segmentation into smaller size pads (6x6 mm^2) than "standard" CsI-MWPC used in RICH.

The detector has an unexpected high and stable value of the quantum efficiency of the CsI deposited on the printed circuit boards (Cu/Ni/Au-layers) with a pre-coated film of a graphite spray. The CsI quantum efficiency gradually increased within a few days and saturated after ~20 days. The conditioning effect is particularly efficient for long wavelengths. Initially, the quantum efficiency dropped ~10% and has since then been stable during the 14 years the detector has been in operation (Di Mauro, 2014).

Figure 11 shows the first rings detected with the HADES CsI-MWPC exposed to the reaction products of U-Pb collisions. For single lepton rings an average pad multiplicity~15 pads per ring was determined. The ring images were not obscured by charged particle background and thus the quality is sufficient for fast pattern recognition.

All the RICH detectors using CsI-MWPC during many years of operation demonstrates high quantum efficiency and excellent stability. This mature technology offers a unique possibility to build large-area photosensitive detectors at low cost, and contribute to obtaining important results in high-energy physics.

3. CsI-GEM RICH DETECTORS

The first application of CsI-GEM photosensitive detectors was in the PHENIX (pioneering high-energy nuclear interaction experiment) Hadron Blind Detector (HBD).The main purpose of this detector is to identify and reject γ conversions and π^0 decays. The rejection algorithm is based on the fact that the open-

Figure 11. Cherenkov rings produced by an electron-positron pair and detected in the HADES RICH detector
From Zeitelhack, 1999.

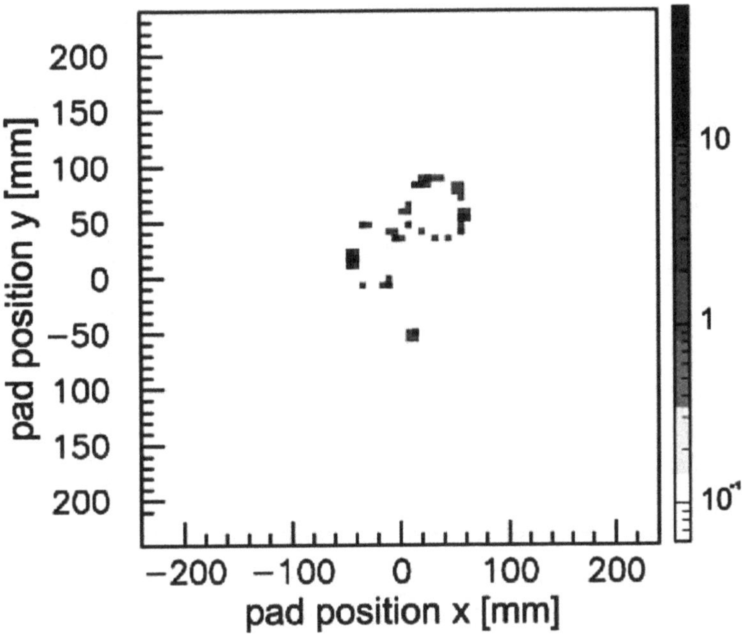

Figure 12. The HBD equipped with CsI-coated triple GEMs; since there is no expansion volume of the Cherenkov light, the image on the GEMs are blobs rather than rings
From Kamin, 2007.

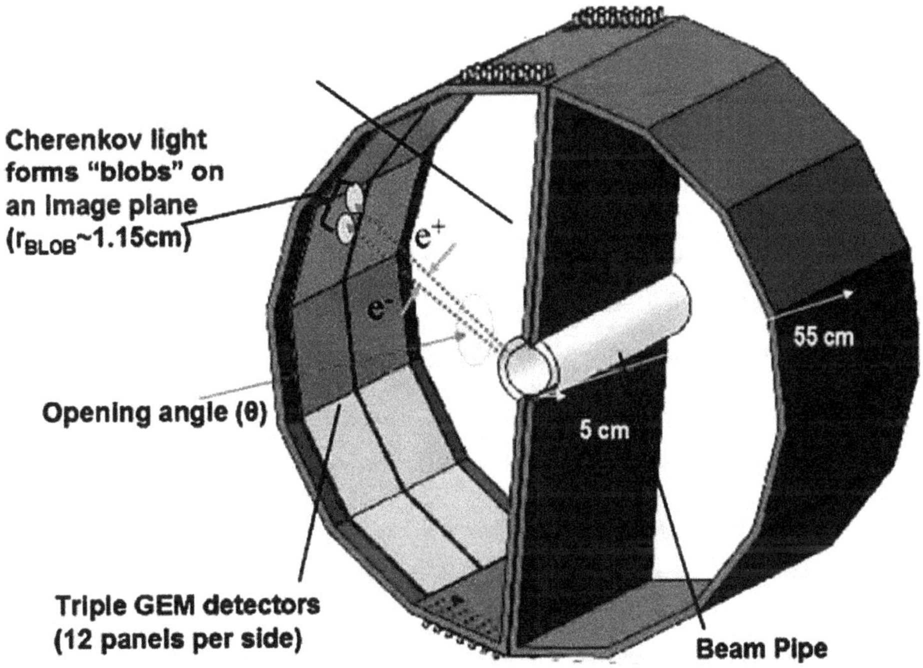

ing angle of the electron pairs from these sources is small compared to the pairs from light vector mesons and by applying an opening angle cut one can reject more than 90% of the conversions and π^0 decays.

The HBD is a windowless Cherenkov detector, using CF_4 as Cherenkov radiator and a reflective CsI photocathode and a triple GEM photosensitive detector. The schematic drawing of this detector is presented in Figure 12.

The detector consists of two half-cylinders. Twelve gas amplification modules are located on the inner side of each HBD vessel. Each module is 27×23 cm². The GEMs are subdivided into 28 HV modules to reduce the energy stored in them in case of an electric discharge. In this windowless configuration, the Cherenkov light from particles passing through the radiator is directly collected on a photosensitive cathode plane, forming a circular blob image rather than a ring as in a conventional RICH detector. The combination of a windowless detector with a CsI photocathode and a CF_4 Cherenkov radiator makes the detector sensitive to Cherenkov light over a wide energy range, 6 to 11.5 eV.

In the hadron-blind reverse bias mode shown in Figure 13 minimum ionizing particles (MIPs) generally produce a signal comparable to one Cherenkov-induced photoelectron and which is nearly always confined to a single pad. Contrary, the Cerenkov photons emitted by each traversing particle yield an average of 35 photoelectrons distributed across 2–3 pads. The ionization signal from charged hadrons (which do not produce any Cherenkov light) are easily distinguished from the Cherenkov light produced by the lighter particles both by signal height and the number of fired pads.

The readout pads have hexagon shape with area of 6.2 cm², which is slightly smaller than the size of the Cherenkov light blob (typically 10 cm²). An event display from p-p collisions in one of the detector arm is shown in Figure 14.

Figure 13. The HBD RICH detector; the top surface of the first GEM is coated with a CsI photoconvertor. Red and blue lines are electrons and photon tracks, respectively. Cherenkov light induced photoelectrons are accelerated through the GEM holes while the ionization electrons from MIPs are "swept" upwards towards the wire mesh
From Milov, 2006.

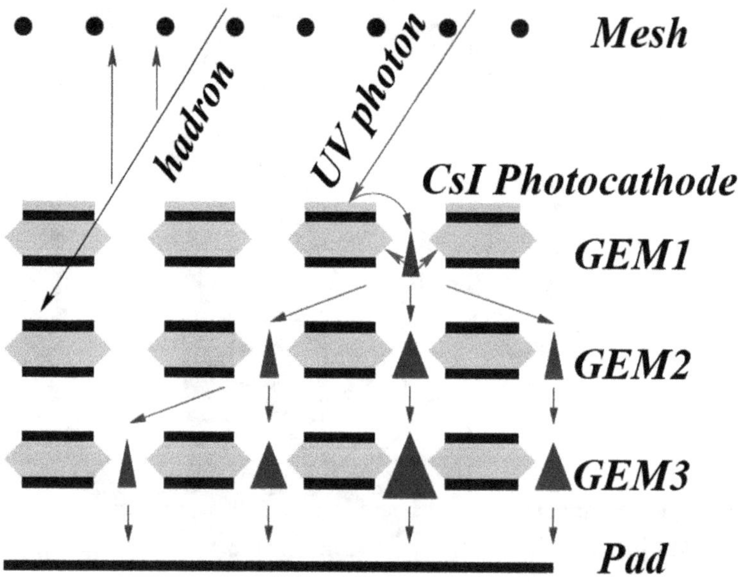

Figure 14. Display of one of the events of a p-p collision in the HBD detector
From Anderson, 2011.

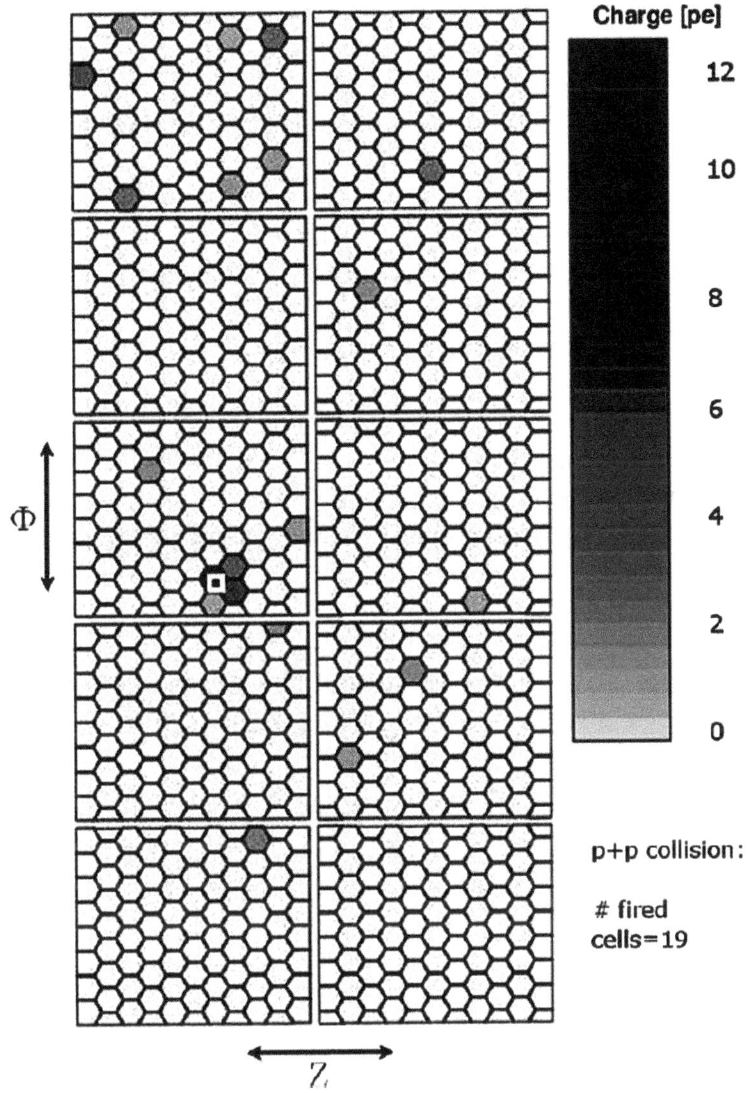

The event shows a few single pad hits from hadrons or scintillation photons and a clear cluster of four pads (indicated by the square in the figure) matched to the projection point of an electron track identified in the PHENIX central arm.

In the RHIC runs, which started in 2009, the HBD demonstrated good performance in terms of stability, single electron detection efficiency, position resolution, hadron rejection and single vs. double-hit recognition.

4. CsI-TGEM/RETGEM RICH DETECTORS

The success of the HBD detector started an interest also in CsI-coated TGEMs (thick gas electron multiplier) and RETGEMs (Resistive Electrode Gas Electron Multiplier). It is important to note that not all RICH detectors require the high position resolution offered by GEMs. Some RICH detectors like the ALICE RICH or COMPASS can perform well at rather modest spatial resolutions which make TGEM/RETGEMs attractive options.

With MWPC geometry the CsI photocathode is fully exposed to the feedback of avalanche light as well as to avalanche positive ion bombardment. To reduce the photon feedback one has to operate the CsI-MWPC in pure CH_4 and at a relatively low gas gains, ~ $3x10^4$. Even at this gain, if the avalanche ion bombardment is intense, as in high particle flux environments, the photocathode maybe excited which can cause breakdowns followed by a long recovery time.

The advantages of CsI-TGEM/RETGEMs detectors are:

1. They can operate at higher gains than CsI-GEMs or CsI-MWPCs.
2. They have the capability to supress photon and ion feedback.
3. They can operate and preserve high quantum efficiency in inflammable gases, e.g. mixtures of noble gases with quenchers.
4. Resistive coating makes them spark protected.

The first successful tests of CsI-RETGEMs and CsI-TGEMs (CsI coated RETGEM and TGEM respectively) were performed in works (Martinengo, 2011; Peskov, 2012; Alexeev, 2012). They used solid Cherenkov radiators like CaF_2 (Martinengo, 2011), fused silica (Alexeev, 2012). The schematic drawing of the setup used in studies of (Martinengo, 2011) is shown in Figure 15.

Figure 15. Layout of a CsI-TGEM/RETGEM RICH with a CaF_2 radiator

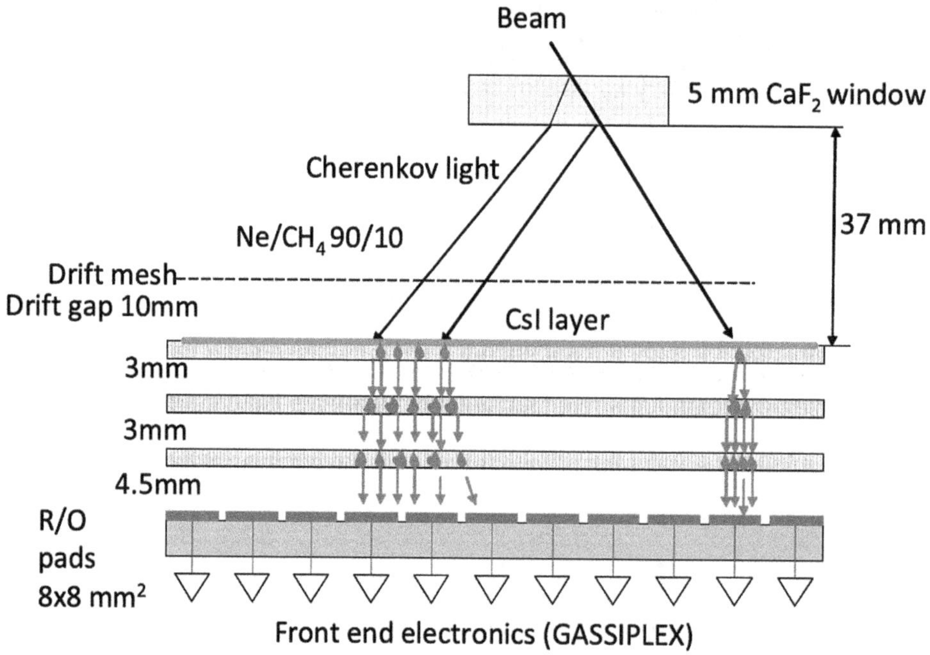

It consisted of a 5 mm thick CaF_2 window also acting as the Cherenkov radiator. The photosensitive detector was a triple CsI-TGEM or CsI-RETGEM, flushed with Ne+10%CH_4 or Ne+10%CF_4 at atmospheric pressure. Each TGEM/RETGEM had an active area of 10x10 cm^2 and was optimized for high gain operation. They had a thickness of 0.45 mm, a hole diameter of 0.4 mm with a pitch of 0.8 mm. The distance between each detector was 4.5 mm. The holes had 10 μm wide dielectric rims. Below the bottom TGEM/RETGEM a pad readout electrode was installed, which was similar to one used in ALICE CsI-MWPC. It had a pad size of 8x8 mm^2, connected to the standard ALICE RICH electronics (GASSIPLEX and DILOGIC chips).

The TGEM or RETGEM top electrode was coated with a CsI film. After the evaporation of a 0.3mm thick CsI layer, the response of the TGEM/RETGEM was measured using a VUV scanning system. Figure 16 shows the mean ratio between the TGEM/RETGEM photo-currents to the reference PM as a function of the time after the evaporation.

Figure 16. Ratio of the photocurrent values measures for different CsI-coated GEM, TGEM, and RETGEM (marked RTG1 and RTG2) immediately after the evaporation and 15 h latter.The quantum efficiency enhancement a few hours after the coating is similar to that obtained with photocathodes prepared for CsI-MWPCs, due to the heat treatment. The final photocurrent ratio ~3.5 is also close to one measured for the ALICE RICH photocathodes. This demonstrates that resistive Kapton or resistive paste serving as substrates in the CsI-RETGEMs, as in the HADES detector, do not cause any short-term degradation of the CsI quantum efficiency.

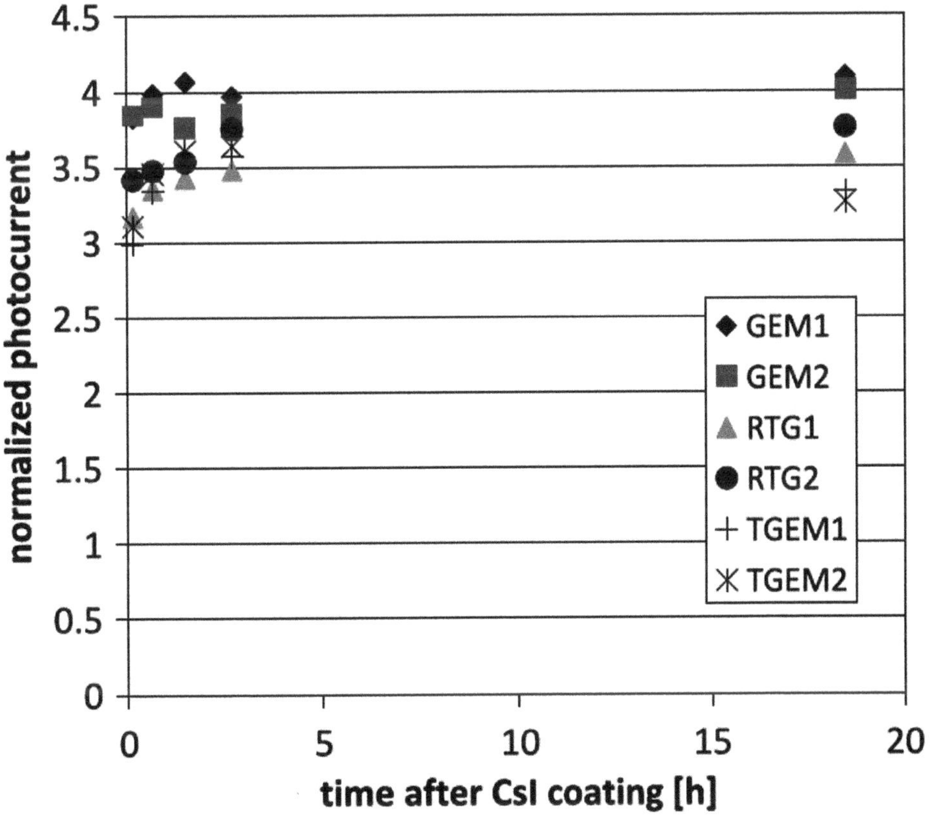

Because the quantum efficiency of a CsI-TGEM and a CsI-RETGEM are almost the same it is not astonishing that all results obtained with these two detectors during particle beam test under identical conditions are actually the same.

After the installation in a RICH detector, the quantum efficiency of the CsI-TGEM and the CsI-RETGEM has been evaluated using a Hg lamp combined with a narrow band filter with a peak transmission at 185 nm. The induced photocurrent was measured between the top electrode of the first TGEM (or RETGEM) and the drift mesh as a function of the voltage applied across this gap. The saturation value of this current was then compared with the saturation value of the current produced by the same light beam in a reference detector, a TMAE filled single-wire counter (Charpak, 2008). From these measurements the quantum efficiency of the CsI photocathode was evaluated to be 17 ± 2% at185 nm. This value is lower than for the ALICE CsI-MWPC (~21%). However, the ALICE measurements were performed in CH_4 whereas the TGEM used Ne + 10% CH_4 or Ne + 10% CF_4 gas mixtures in which the back diffusion of photoelectrons is stronger.

In the first runs at the particle beam CERN PS/T10, the CsI-TGEM/CsI-RETGEM was operated with a normal drift field, reaching stable gains up to ~10^5. In Figure 17a a typical image of overlapped events is shown obtained in a6 GeV/cπ^- beam entering perpendicular to the detector. The high index of refraction of CaF_2 traps all Cherenkov photons produced inside the radiator by total reflection. Hence, only the ionization signal from the minimum ionizing particle (the MIP- signal) is present in the figure and no Cherenkov light is visible.

At this run the CsI-TGEM operated with a reversed drift field (200V) at an overall gain of ~10^5.

Tilting the detector by an angle larger than 20° with respect to the particles direction, part of the Cherenkov cone escapes the CaF_2 radiator and falls in the acceptance of the top CsI-TGEM/RETGEM (Figure 17b). Therefore, only a fraction of the produced Cherenkov radiation can be detected. From this

Figure 17. Images of typical events obtained with triple CsI-TGEM/RETGEMs during PS/T10 beam tests: a) 2D-display of 1238 overlapped events when the RICH detector with the CaF_2 radiator was perpendicular to the particles beam; in the case the Cherenkov light was trapped inside the crystal and one can see only the image of particle beam (red spot in the center of the display), b) the RICH detector is oriented ~32° with respect to the beam. The spot visible at the top of the lower figure is the image of the MIPs and the horizontal band in the middle corresponds to the detected Cherenkov photons which escape the internal reflection in the CaF_2.

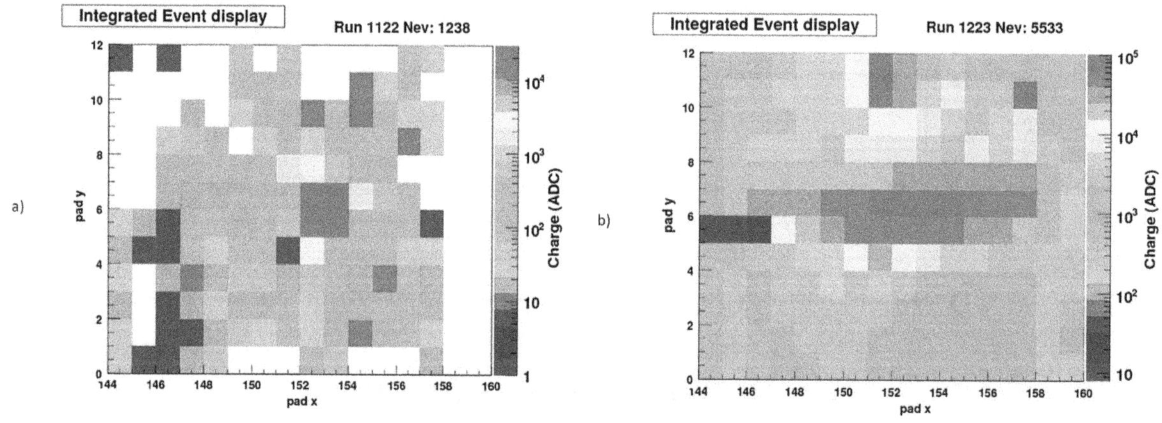

Figure 18. The RICH detector used by the COMPASS group at beam test; the triple CsI TGEM detectors as well as the reference multi-anode photomultiplier are shown. The hemispheric quartz radiator focuses the Cherenkov light into the ring sketched in the figure.
From Alexeev, 2012.

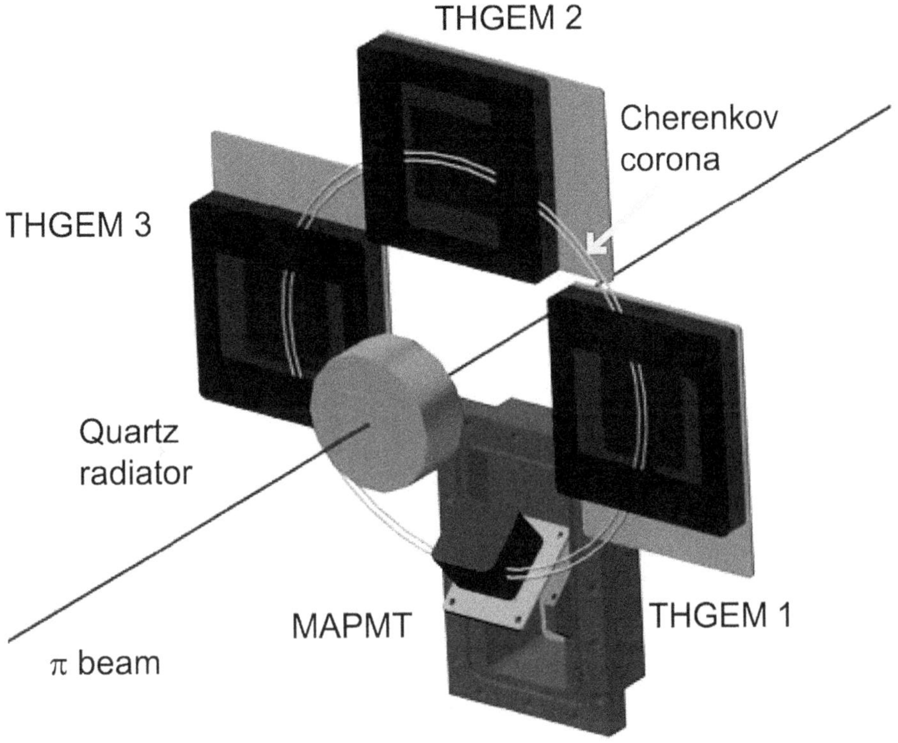

limited data it is quite difficult to directly evaluate the quantum efficiency of the detector operated in photoelectron counting mode. Monte Carlo simulations confirm that the intrinsic quantum efficiency of the CsI-TGEM/RETGEM in avalanche mode is comparable with that of the ALICE CsI photocathodes (Martinengo, 2011).

At a gain of the first TGEM of more than 200 all photoelectrons are fully collected in the TGEM/RETGEM holes producing avalanches there. The main factors affecting the quantum efficiency are then back diffusion and geometrical losses in the holes.

The first CsI-TGEM/RETGEM-based RICH detectors which produced Cherenkov rings are described in (Alexeev, 2012¸Peskov, 2012). In the detector shown in Figure 18 the beam particles traversed a hemispheric fused silica radiator focusing the Cherenkov light into a ring illuminating simultaneously 2-3 CsI-TGEM-detectors (see Figure 19) and central pixels of a multi-anode PMT. The TGEM geometry has a 3x3 cm^2 active surface, 0.4 mm thickness, 0.4 mm hole diameter and 0.8 mm pitch. The CsI-TGEMs operated in Ar/CH_4 50/50% mixtures at gains of ~10^5.

In the tests of a large area CsI-TGEM based RICH detector (Peskov, 2012) the same liquid radiator was used in ALICE RICH. This allows direct comparison between two detectors without complicated simulations. The design of this detector is shown in Figure 20. It consists of a 15 mm thick liquid C_6F_{14} radiator and six triple CsI-TGEMs (in some tests RETGEMs), each having an active area of 10x10 cm^2. The geometrical characteristics of each TGEM are: thickness 0.45 mm, hole diameter 0.4 mm, rim thick-

Figure 19. Superposition of event images; the dotted ring represents the expected Cherenkov ring.
From Alexeev, 2012.

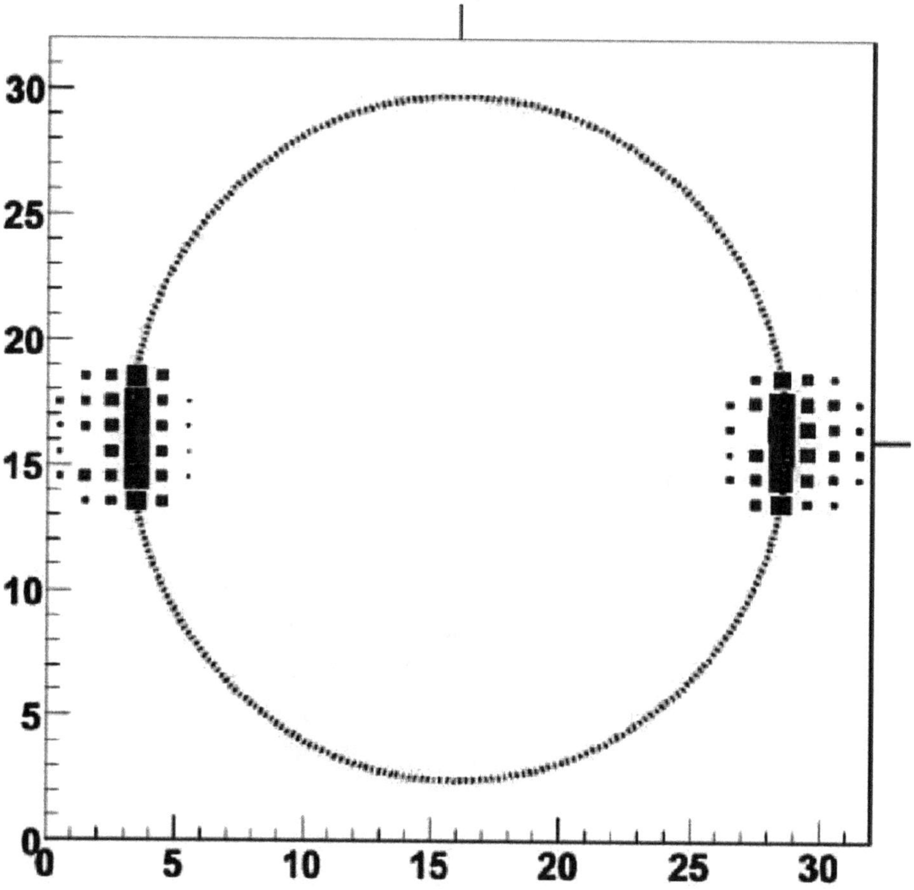

ness 10 μm and 0.8 mm pitch. A wire grid drift electrode was positioned 3 cm above the top electrode of the TGEMs. The gap between the TGEMs in each stack was 3 mm. One triple TGEM was placed behind the liquid radiator in order to detect the beam particles, whereas the other five were positioned along a circle in order to collect the Cherenkov light.

A pad read-out plane (pad size 8x8 mm^2) was placed 3 mm behind the triple TGEMs combined with a front-end electronics based on the standard ALICE-RICH GASSIPLEX and DILOGIC chips.

Figure 21 shows the superposition of Cherenkov ring images (or more precisely, the charge signal from the pads) obtained in Ne+10%CF_4 gas mixture. The ''spot'' in the center represents the MIPs (6 GeV/c pions) detected by the central TGEM. These results were obtained at overall gains of the triple TGEMs (measured on the bottom electrode of the triple TGEM) of 3x10^5.

From the presented data, taking into account the geometrical factors, it was determined that on average 9 Cherenkov photons per event can be recorded with this technology. In the ALICE MWPC operating in CH_4 it was 14-15 per particle.

From these data one can draw some conclusion about the quantum efficiency of the CsI layer deposited on the top of TGEMs/RETGEMs. For example, in Ne+10%CF_4 at a voltage across the top TGEM

Figure 20. Schematic drawing of a large-area RICH prototype, consisting of a C_6F_{14} liquid radiators, six triple CsI- TGEMs located in the area of Cherenkov rings and readout electronics: (a) cross section of the device, showing the direction of incident beam particles and (b) its top view, showing how the Cherenkov ring overlaps with an active area of the CsI-TGEMs.
Peskov, 2012.

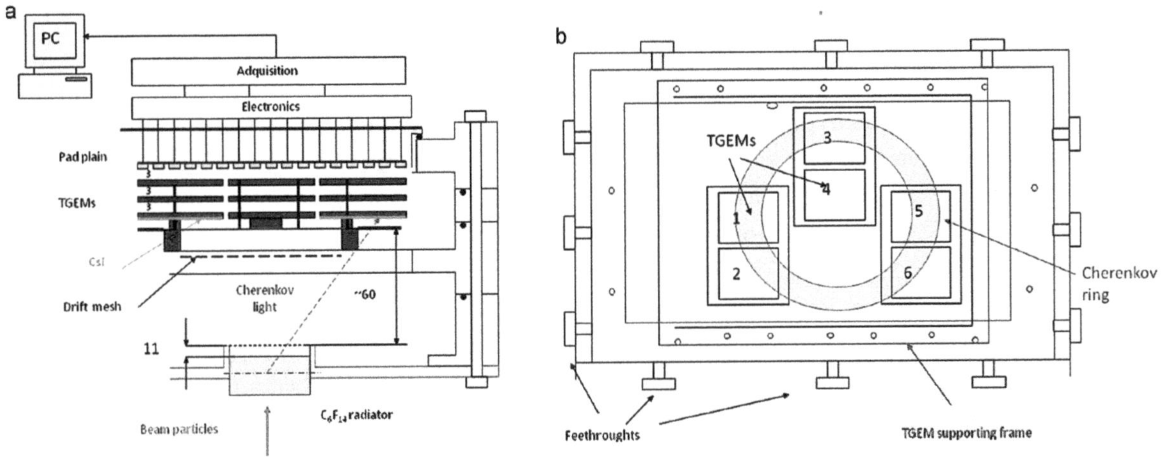

Figure 21. Cherenkov ring (superposition of events, raw data, and no noise subtraction) recorded with a large-area CsI-TGEM-based RICH detector, shown in Figure 20 and operated in Ne + 10%CF_4; the red spot in the center id the image of the particle beam.
From Peskov, 2012.

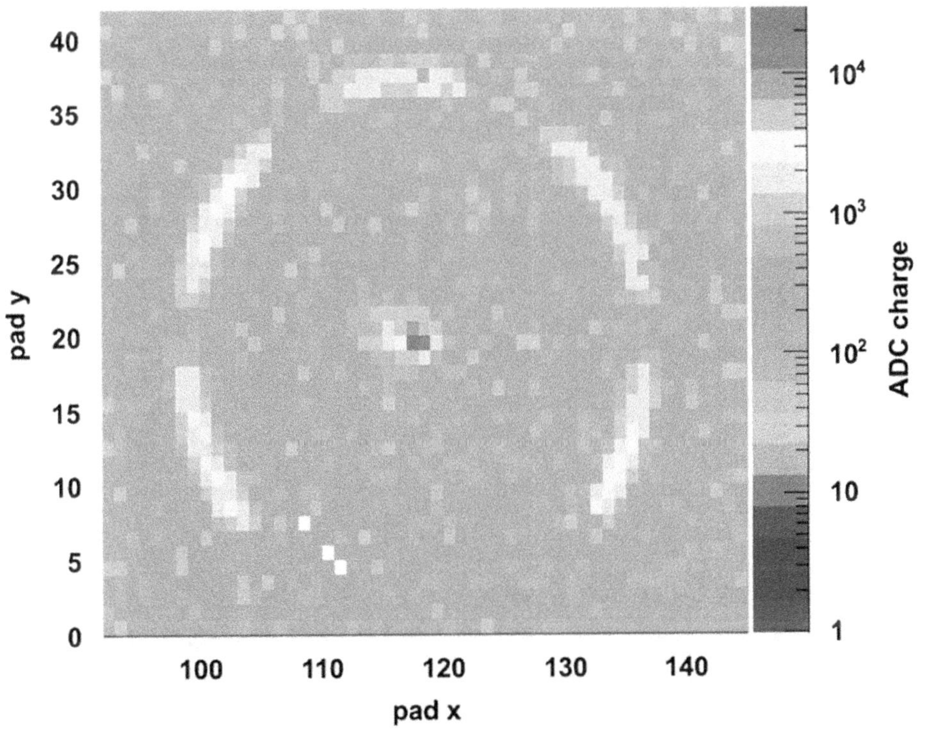

of550V, the extraction electric field strength between holes is about 1.5 kV/cm. At this electric field the photoelectron extraction efficiency from the CsI photocathode is 0.79, which is about $\eta_{rel}\approx0.9$ compared to the extraction efficiency in the pure CH_4 used in CsI-MWPC. The collection efficiency of the photo-electrons into the TGEM holes at 550 V across the TGEM is $\eta_{col}=0.95$. Keeping in mind that the active area of the TGEM is $S_{eff}\approx75\%$ of the total active area (15% are holes), the total correction factor will be:

$$K_{tot} = \eta_{rel} \text{ x } \eta_{col} \text{ x } S_{eff} = 0.9 \text{ x } 0.95 \text{ x } 0.75 = 0.64, \tag{2}$$

Hence, in the case of the hypothetical photodetector operating in pure CH_4 and having no holes the expected number of detected Cherenkov photons will be9/0.64=14 photons, which is close to the best CsI-MWPC data. By optimization of the gas mixture, the geometry and the gas gain of the first TGEM one can reach slightly higher values of η_{rel}, S_{eff} and $\eta_{col,}$ so that K_{tot} may approach 0.8.

Therefore, CsI-TGEM/RETGEM seems to be an interesting alternative to the MWPC technology especially since it allows the use of non-flammable gases and thus resolve a safety issue in some experiments such as the ALICE RICH or COMPASS detectors. In order to compete with CsI-MWPC in RICH applications, large area CsI-TGEMs should be developed and tested. An important step in this direction was done in work (Alexeev, 2014) in which the operation of 30x30 cm^2 active area CsI-TGEMs was investigated. With this detector (see Figure 22) a continuous Cherenkov ring (not separated patches as earlier) was possible to observe (Figure 23). There are attempts to develop TGEMs with an active area of 60x60 cm^2 and even larger.

Figure 22. Photograph of a TGEM with a large active area (30x30 cm^2), divided in six sectors; the eight large holes serve to position the support pillars.
From Alexeev, 2013.

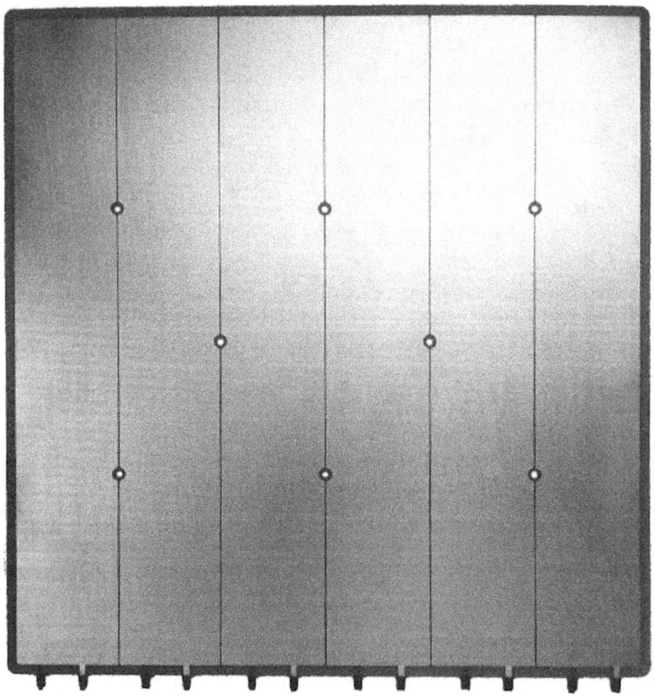

Figure 23. Accumulated Cherenkov rings and an image of charged particles in its center recorded with a CsI-TGEM with an active are of 30x30 cm^2 at a gas gain of $2x10^4$
From Alexeev, 2014.

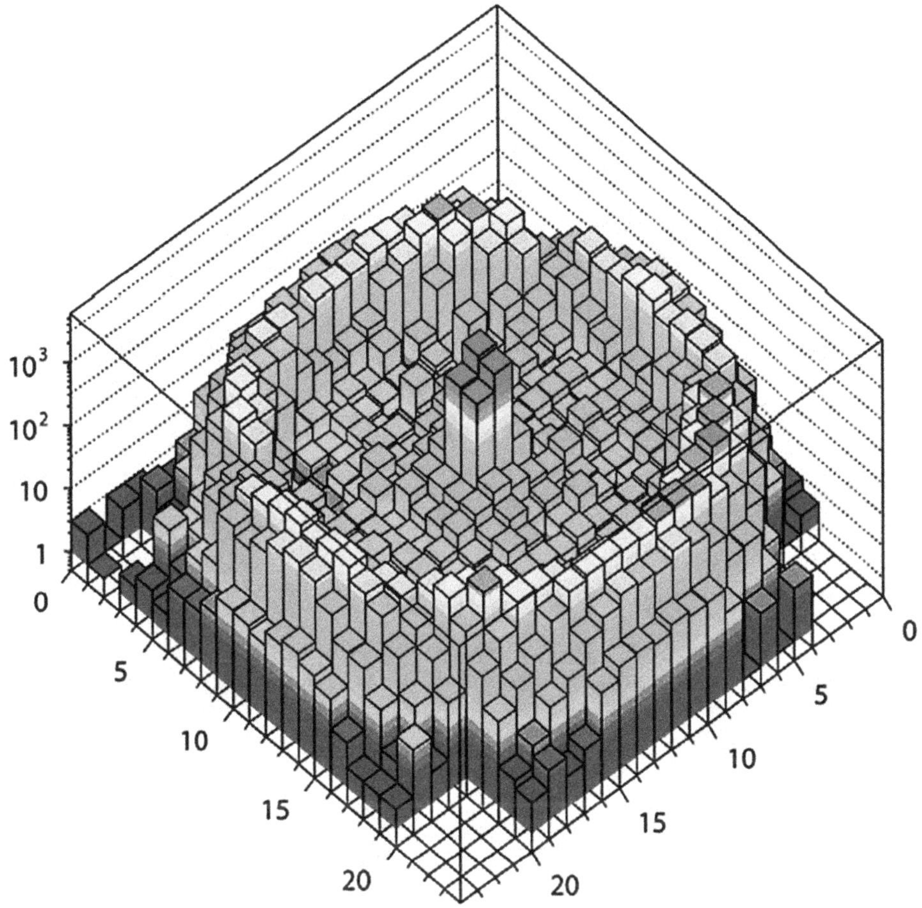

5. THE COMPASS RICH UPGRADE

A practical outcome of all these studies CsI-TGEMs are now included in the COMPASS upgrade program. The COMPASS RICH works in a high rate environment. The intense ion bombardment of the CsI photocathode may cause cathode excitation effects.

A single photon detector based on cascaded TGEMs does not suffer from photon feedback. Thanks to its "closed" geometry architecture only part the avalanche ions can propagate back to the photocathode surface by following the electric field lines. Measurements shows that with a triple TGEM structure the ion back flow fraction (IBF) is about 30% of the total ions generated, while the remaining70% is collected by the intermediate layers (Alexeev, 2013a). Lower IBF, close to 5%, was achieved with a triple TGEM having misaligned holes as shown schematically in Figure 24.

The idea is that positive ions, which have relatively low lateral diffusion during their drift, almost follow the electric field lines. In the case of misaligned holes most of the field lines starting from the holes of TGEMs #3 (or #2) will end up on the blind surface of the upper TGEM, and the ions which

Figure 24. Schematic drawing showing relative alignment of the holes of TGEM #1 and #3 in (red) and TGEM 2 (blue) in the staggered configuration-a triple THEM with misaligned holes
From Alexeev, 2013a.

follow these lines will be lost (Sauli, 2006). Contrary, electrons moving in opposite direction, are much stronger affected by lateral diffusion and have high probability to enter the holes of lower TGEM and generate avalanches there.

Besides using triple TGEMs, another option is currently under the evaluation by the COMPASS RICH group. It is a hybrid detector consisting from two staggered TGEMs and a MICROMEGAS (micromesh gas chamber) detector (see Figure 25).

MICROMEGAS provides an efficient ion collection by the cathode micro mesh for sufficiently large ratio of the values of electric field above and below this micromesh (Giomataris, 1996). Preliminary measuremens show that an IBF below a few % is possible to achive with this configuration. The latest protope has an active area of 30x30 cm^2 and there are ongoingefforts to build larger detectors, which shoud fit the main requiements of the upgraded COMPASS RICH.

6. CONCLUSIVE REMARKS

Photosensitive gaseous detectors based on CsI photocathodes, introduced at the works (Charpak, 1989; Séguinot, 1990; Dangendorf, 1990), are after many years of studies and developments implemented in several large-scale RICH detectors.

Figure 25. A hybrid detector, combining two technologies: TGEM and MICROMEGAS; the CsI is deposited on the first TGEM layer. The two TGEMs are aligned such that the center of a hole in one TGEM corresponds to a missing hole in another TGEM.

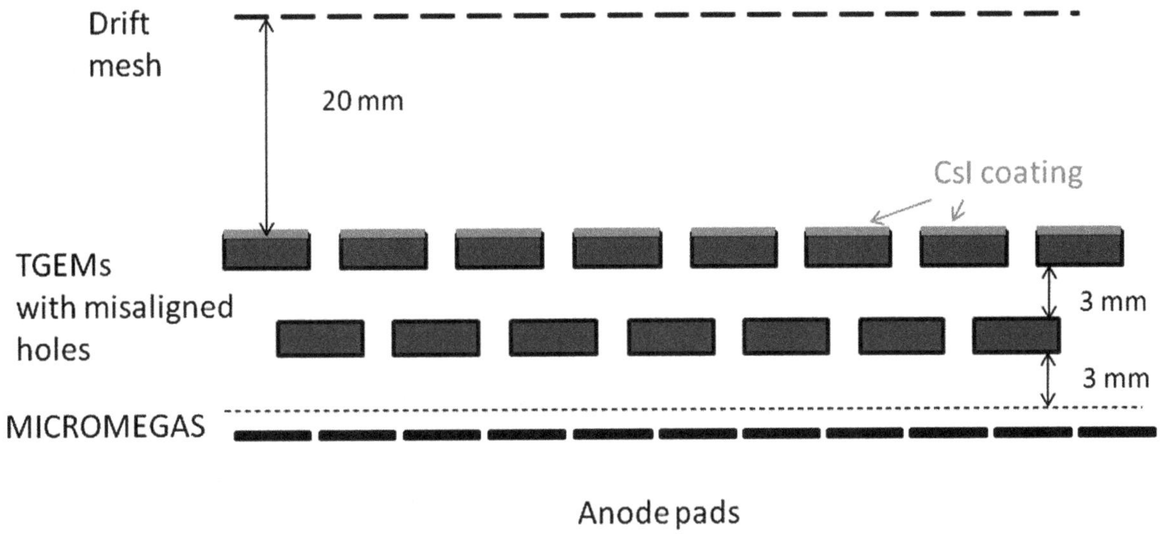

The confidence in this technology grew after the first successful operation of a CsI-MWPC in the STAR experiment, when one of the ALICE RICH modules was installed and tested there.

After the invention of micropattern gaseous detectors CsI- based photo-sensors gained new momentum and large area CsI-GEMs were implemented in the PHENIX hadron blind detector. CsI-TGEMs will be installed in 2017 at the COMPASS upgrade RICH.

CsI-based gaseous detectors, including the micropattern one must operate at high enough gas gains (close or above 10^5) to ensure high efficiency for the single photoelectron detection. At these gains heavily ionizing charged particles may cause sparking due to the Rather limit. One way to coop with this almost unavoidable sparking problem is to use detectors with resistive electrodes. A detailed description of resistive micropattern detectors, their main design and operation principle can be found in (Francke, 2014). Since these detectors are robust and withstand sparks, the COMPASS experiment is considering this approach as well (Alexeev, 2014a).

REFERENCES

Albrecht, E., Baum, G., Birsa, R., Bosteels, M., Brandamante, F., & Braem, A. et al. (2002). COMPASS RICH-1. *Nuclear Instruments & Methods in Physics Research. Section A, Accelerators, Spectrometers, Detectors and Associated Equipment*, *478*(1-2), 340–343. doi:10.1016/S0168-9002(01)01823-X

Alexeev, M., Birsa, R., Bodlak, M., Bradamante, F., Bressan, Buchele, M., … Veloso, J.F.C.A. (2014). MPGD-based counters of single photons developed for COMPASS RICH-1. *JINST, 9*, 1-9.

Alexeev, M., Birsa, R., Bradamante, F., Bressan, Buchele, M., Chiosso, M., …Veloso, J.F.C.A. (2013). Status and progress of novel photon detectors based on THGEM and hybrid MPGD architectures. *JINST,* 8, 1-18.

Alexeev, M., Birsa, R., Bradamante, F., Bressan, A., Büchele, M., & Chiosso, M. et al. (2014). Status and progress of the novel photon detectors based on THGEM and hybrid MPGD architectures. *Nuclear Instruments & Methods in Physics Research. Section A, Accelerators, Spectrometers, Detectors and Associated Equipment*, *766*, 133–137. doi:10.1016/j.nima.2014.07.030

Alexeev, M., Birsa, R., Bradamante, F., Bressan, A., Chiosso, M., Ciliberti, P., … Veloso J.F.C.A. (2013a). Ion backflow in thick GEM-based detectors of single photons. *JINST*, *8*, 1-17.

Alexeev, M., Birsa, R., Bradamante, F., Bressan, A., Chiosso, M., & Ciliberti, P. et al. (2012). Development of THGEM-based Photon Detectors for COMPASS RICH-1. *Physics Procedia*, *37*, 781–788. doi:10.1016/j.phpro.2012.02.422

Anderson, W., Azmoun, B., Cherlin, A., Chi, C. Y., Citron, Z., & Connors, M. et al. (2011). Design, construction, operation and performance of a Hadron Blind Detector for the PHENIX experiment. *Nuclear Instruments & Methods in Physics Research. Section A, Accelerators, Spectrometers, Detectors and Associated Equipment*, *646*(1), 35–58. doi:10.1016/j.nima.2011.04.015

Andres,Y., Barbosa, J., Braem, A., Carrone, E., Cozza, D., Davenport, M., ... Williams, T.D. (2002). R&D in ALICE: The CsI-based RICH high momentum particle identification detector. *EPJ Direct C*, *4*(S1), 1-7.

Braem, A., Fabjan, C. W., Franz, A., Kaneta, M., Paić, G., & Piuz, F. et al. (1998). A threshold imaging Cherenkov detector with CsI photocathodes. *Nuclear Instruments & Methods in Physics Research. Section A, Accelerators, Spectrometers, Detectors and Associated Equipment*, *409*(1-3), 426–431. doi:10.1016/S0168-9002(98)00118-1

Breskin, A., Alon, R., Cortesi, M., Chechik, R., Miyamoto, J., & Dangendorf, V. et al. (2009). A concise review on THGEM detectors. *Nuclear Instruments & Methods in Physics Research. Section A, Accelerators, Spectrometers, Detectors and Associated Equipment*, *598*(1), 107–111. doi:10.1016/j.nima.2008.08.062

Breskin, A., Peskov, V., Cortesi, M., Budnik, R., Chechik, R., & Duval, S. et al. (2011). CsI-THGEM gaseous photomultipliers for RICH and noble –liquid detectors. *Nuclear Instruments & Methods in Physics Research. Section A, Accelerators, Spectrometers, Detectors and Associated Equipment*, *639*(1), 117–120. doi:10.1016/j.nima.2010.10.034

Charpak, G., Benaben, P., Breuil, P., Di Mauro, A., Martinenego, P., & Peskov, V. (2008). Development of New Hole-Type Avalanche Detectors and the First Results of Their Applications. *IEEE Transactions on Nuclear Science*, *NS-55*(3), 1657–1663. doi:10.1109/TNS.2008.924084

Charpak, G., Peskov, V., Scigoscki, D., & Valbis, J. (1989). Mew scintillators for photosensitive gaseous detectors. *Proceedings of the Symposium on particle identification at high-luminosity hadron colliders*. Retrieved from: http://lss.fnal.gov/conf/C890405/

Dangendorf, V., Breskin, A., Chechik, R., & Schmidt-Bocking, H. (1990). A gas-filled UV –photon detector with CsI photocathode for the detection of Xe light. *Nuclear Instruments & Methods in Physics Research. Section A, Accelerators, Spectrometers, Detectors and Associated Equipment*, *289*(1-2), 322–324. doi:10.1016/0168-9002(90)90276-C

De Cataldo, G. (2014). The ALICE–HMPID performance during the LHC run period 2010–2013. *Nuclear Instruments & Methods in Physics Research. Section A, Accelerators, Spectrometers, Detectors and Associated Equipment*, *766*, 19–21. doi:10.1016/j.nima.2014.07.039

Di Mauro, A. (2014). Status and perspectives of gaseous photon detectors. *Nuclear Instruments & Methods in Physics Research. Section A, Accelerators, Spectrometers, Detectors and Associated Equipment*, *766*, 126–132. doi:10.1016/j.nima.2014.07.008

Francke, T., & Peskov, V. (2014). *Innovative applications and developments of micro-pattern gaseous detectors*. Hershey, PA: Engineering Science Reference.

Giomataris, Y., Rebourgeard, P., Robert, J., & Charpak, G. (1996). MICROMEGAs: A High granularity position sensitive gaseous detector for high particle flux environments. *Nuclear Instruments & Methods in Physics Research. Section A, Accelerators, Spectrometers, Detectors and Associated Equipment*, *376*(1), 29–35. doi:10.1016/0168-9002(96)00175-1

Kamin, J. A. (2007). A hadron blind detector for the PHENIX experiment at RHIC. *The European Physical Journal C*, *49*(1), 177–180. doi:10.1140/epjc/s10052-006-0100-y

Linka, A., Picek, J., Volf, P., & Ososskov, J. (1999). New solution to circle fitting problem in analysis of RICH detector data. *Czechoslovak Journal of Physics*, *49*(S2Suppl), 161–168. doi:10.1007/s10582-999-0099-z

Martinengo, P., Peskov, V., Breskin, A., Di Mauro, A., Mayani Paras, D., & Molnar, L. et al. (2011). R&D results on a CsI-coated triple thick GEM-based photodetector. *Nuclear Instruments & Methods in Physics Research. Section A, Accelerators, Spectrometers, Detectors and Associated Equipment*, *639*(1), 126–129. doi:10.1016/j.nima.2010.10.033

Milov, A. (2006). *Hadron Blind Detector for the PHENIX Experiment at RHIC*. ArXV: nucl-ex/0609014v1

Nappi, E., & Séguinot, J. (2005). Ring Imaging Cherenkov Detectors: The state of the art and perspectives. *Rivista del Nuovo Cimento*, *28*, 1–130.

Peskov, V., Bencze, G., Di Mauro, A., Mayani, D., Molnar, L., & Nappi, E. et al. (2012). First observation of Cherenkov rings with a large area CsI-TGEM-based RICH prototype. *Nuclear Instruments & Methods in Physics Research. Section A, Accelerators, Spectrometers, Detectors and Associated Equipment*, *695*, 154–158. doi:10.1016/j.nima.2011.11.045

Piuz, F., Andres, Y., Braem, A., Davenport, M., Mauro, A. D., & Goret, B. et al. (1999). The CsI-based ring imaging detector for the ALICE experiment: Technical description of a large prototype. *Nuclear Instruments & Methods in Physics Research. Section A, Accelerators, Spectrometers, Detectors and Associated Equipment*, *433*(1-2), 222–234. doi:10.1016/S0168-9002(99)00293-4

Santiard, J. C. (2004). The ALICE HMPID on-detector front-end and readout electronics. *Nuclear Instruments & Methods in Physics Research. Section A, Accelerators, Spectrometers, Detectors and Associated Equipment*, *518*(1-2), 498–500. doi:10.1016/j.nima.2003.11.068

Sauli, F., Ropelewski, L., & Everaerts, P. (2006). Ion feedback suppression in time projection chambers. *Nuclear Instruments & Methods in Physics Research. Section A, Accelerators, Spectrometers, Detectors and Associated Equipment*, *560*(2), 269–277. doi:10.1016/j.nima.2005.12.239

Schukraft, J., Piuz, F., Klempt, W., Leistam, L., De Groot, J., Beoli, S., & Tustonic, T. (1998). Technical Design Report of the High Momentum Particle Identification Detector. *CERN / LHCC 98–19 ALICE TDR 1*. Retrieved from: http://www.google.co.uk/url?sa=t&rct=j&q=&esrc=s&source=web&cd=1&ved=0CCEQFjAA&url=http%3A%2F%2Fcern.ch%2Falice-hmpid%2Fpublicns%2FTDR%2FTDRFirstpage.pdf&ei=IW2IVJHNO4fuaK2kgvAD&usg=AFQjCNEs8miubRdegKxtCzkqSVDfDIgEJQ&bvm=bv.81456516,d.d2s

Séguinot, J., Charpak, G., Giomataris, Y., Peskov, V., Tischhauser, J., & Ypsilantis, T. (1990). Reflective UV photocathodes with gas-phase extraction: Solid, liquid and adsorbed thin films. *Nuclear Instruments & Methods in Physics Research. Section A, Accelerators, Spectrometers, Detectors and Associated Equipment*, *297*(1-2), 133–147. doi:10.1016/0168-9002(90)91359-J

Tessarotto, F., Abbon, P., Alexeev, M., Birsa, R., Bordalo, P., Bradamante, F., …Takekawa, S. (2014). Long term experience and performance of COMPASS RICH-1. *JINST, 9*, 1-27.

Volpe, G., (2009). *Results from cosmic rays and first LHC beam with the ALICE HMPID detector.* ArXV:0907.4579v2

Volpe, G. (2014). Pattern recognition and PID procedure with the ALICE-HMPID. *Nuclear Instruments & Methods in Physics Research. Section A, Accelerators, Spectrometers, Detectors and Associated Equipment*, *766*, 259–262. doi:10.1016/j.nima.2014.05.031

Von Halleer, B., Roukoutakis, F., Chapeland, S., Altini, V., Carena, F., & Carena, W. et al. (2010). The ALICE data quality monitoring. *Journal of Physics: Conference Series*, *219*(022023), 1–9.

Zeitelhack, K., Elhardt, A., Friese, J., Gernhauser, R., Homolka, J., & Kastenmuller, A. et al. (1999). The HADES RICH detector. *Nuclear Instruments & Methods in Physics Research. Section A, Accelerators, Spectrometers, Detectors and Associated Equipment*, *433*(1-2), 201–206. doi:10.1016/S0168-9002(99)00371-X

KEY TERMS AND DEFINITIONS

CsI-MWPCs: A multiwire proportional chamber combined with a CsI photocathode.

CsI-RETGEMs: Resistive Electrode Gas Electron Multiplier with a top electrode coated with a CsI layer.

CsI-TGEMs: A thick gas electron multiplier combined with a CsI photocathode.

CsI-TIC: A threshold imaging Cherenkov detector in which the CsI-MWPCs was employed.

DILOGIC Chip: A sparse data scan readout processor developed for the ALICE RICH detector. It can handle up to 64 channels by group of 16, 32, 48 or 64 channels.

GASSIPLEX: An integrated front-end analog processor developed for the RICH and Dimuon spectrometer used in ALICE detector.

MICROMEGAS (Micromesh Gas Chamber): A parallel-plate avalanche chamber with a very small gap (50-150 µm) between the cathode mesh and the anode plate.

NA Experiment: Means the 44th experiment at CERN. It ran from 1989 through 1996.

TIC: Threshold Imaging Cherenkov detector.

Chapter 15
Other Applications of Photo-Sensitive Detectors

ABSTRACT

In this chapter the main applications of gaseous photomultipliers beyond RICH detectors will described. They include applications in spectroscopy, plasma diagnostic, astrophysics, flame detection, readout gaseous and solid scintillators, and cryogenic detectors. Their advantages will be described and compared with alternative techniques.

1. INTRODUCTION

Currently the main application of position-sensitive photomultipliers is in RICH detectors. In this Chapter review of the most promising applications beyond the RICH detectors, e.g.:

- Plasma diagnostics,
- Spectroscopy,
- Readout of UV scintillators,
- UV visualization in daylight condition,
- Forest fire detection,
- The use of photosensitive substances in X-ray imaging.

2. PLASMA DIAGNOSTICS

The first prototypes of MWPCs filled with photosensitive gases were developed at the same time by two independent groups, one located at CERN (Séguinot, 1977) and the other one in P.L. Kapitza laboratory in Moscow (Bogomolov, 1978).

The work of (Séguinot, 1977) is well known and led to development various RICH detectors. The application of photosensitive MWPCs for plasma diagnostic is much less known, since most of the

DOI: 10.4018/978-1-5225-0242-5.ch015

publications were in Russian, and this is why we dedicate this paragraph to this important subject (Bogomolov, 1978).

A stationary plasma is induced by ultrahigh frequency microwaves (λ_{mw}= 25 cm or 50 cm) in a cylindrical resonator filled with H_2, D_2 or noble gases at pressure of 1-30 atmospheres (see Figure 1). An interesting feature of this ultra-high frequency (UHF) plasma is sporadic appearance of ionization instability leading to the creation of dense short-lived (a few μs) filaments inside the plasma volume. The typical dimensions of the stationary plasma column depend on the microwave wavelength λ_{mv} and the generator power. The length can be between 12 to 30 cm and the diameter between 1.6 and 6 cm. The typical diameter of the dense filaments is 2-4 mm. The initial dense granularity is created in the center of the plasma. Due to the positive feedback between the resonator and the UHF generator, they immediately start to grow in both directions along the resonator axis until they reach the region of low electric field where they finally decay.

The stationary plasma column has an electron temperature about 6000°C and in H_2 and D_2 it emits a strong molecular spectra (H_2/D_2) as well as a so-called H^- continuum (Diatroptov, 1972), making it difficult to observe the instability in the UV or in the visible region of spectra (λ>220 nm). In contrast, the electron temperature of the short living dense filaments in the head regions was about few eV and they emitted bursts of bremsstrahlung radiation in the VUV region (λ<160 nm).

Figure 1. Illustration of a UHF plasma generator (called "Nigotron") coupled to a cylindrical resonator and diagnostic instruments used for the study of the ionization instability of the plasma

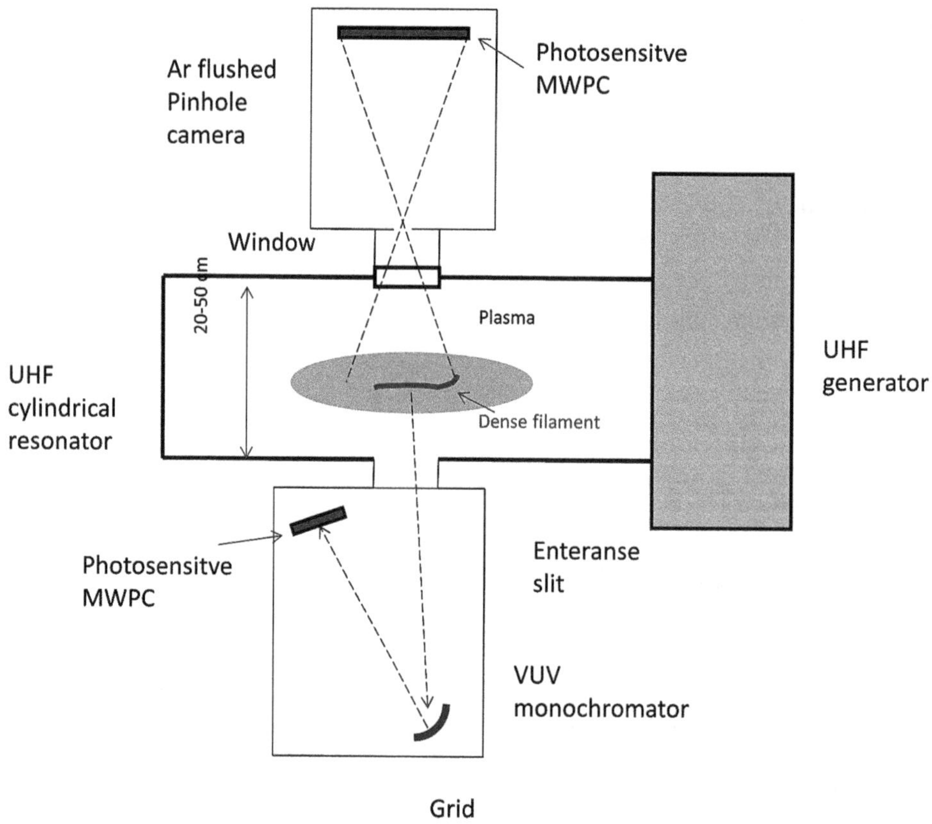

The most appropriate technique to visualize this instability and to detect the spectrum emitted from the filaments is the use a photosensitive MWPC, which practically is not sensitive to the UV and visible part of the bright H^- continuum. This allowed instability investigation practically without interference from the background (Peskov, 1980).

The apparatuses used for the plasma diagnostics in the VUV region are shown in Figure 1.

It includes several Ar-flushed pinhole cameras (for the simplicity only one of them is shown in the figure) inside which MWPCs, sensitive in various narrow band spectral intervals, were installed. These devices allowed the visualization of the filaments in the VUV region. There was also a vacuum monochromator allowing recording the spectrum of the filaments in coincidence with signals from the MWPCs installed in the pinhole cameras. As a detector the focal pane of the monochromator was either a photosensitive single-wire counter (Diatroptov, 1972) or a MWPC with a CuI photocathode (Peskov, 1980).

Each anode wire of MWPCs was connected to its own charge-sensitive preamplifier and the signals from all wires were readout by a multiplex and recorded on the oscilloscope screen. The stationary plasma profile and its spectrum were also measures in the visible and infrared range of spectra.

Some typical results obtained with these innovative diagnostic tools are shown in Figures 2-5. In Figure 2a the plasma profile (its cross-section) measured in the visible range of spectra is shown with the help of a one dimensional TV tube (dissector). In Figure 2b the signals (multiplex readout) from the wires of the MWPC sensitive in the spectral interval of 105-145 nm recorded at the moment of instability

Figure 2. Signals from the one-dimensional TV tube and the photosensitive MWPC measured the plasma profile in the visible and the VUV regions of spectra respectively at the moment of appearing the instability. In this particular case the visible plasma diameter in D_2 was about 15 mm, while the diameter of dense filament was between 1 and 2 mm.
From Bogomolov, 1978.

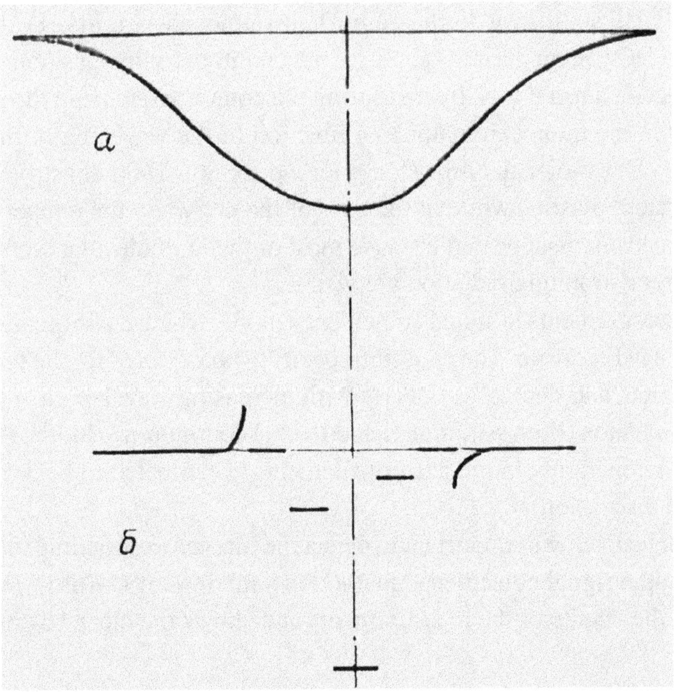

Figure 3. A digital 2D image of the dense filament appeared for a few μs inside the volume of the stationary UHF plasma: 1- the top view of the array of open ended-type MWPC flushed with toluene vapors, 2 –a superimposed projection of the visible image of the stationary plasma, 3-a line identifying anode wires, signal from which at least five times exceeded the given electronic threshold; this actually the digital image of the short leaving filament.
From Bogomolov, 1987.

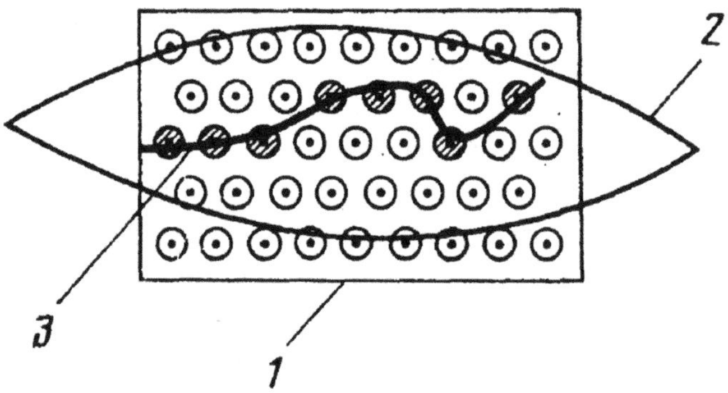

appearance, are presented. As can be seen, the diameter of the dense filament is almost 15 times smaller than the visible diameter of the stationary UHF discharge.

A two dimensional (2D) image of the short living filament detected with a special design of a photosensitive multiwire counter (for details see Karabadzhak, 1983) is shown in Figure 3.

Since the brightness of the filaments in the VUV region was much higher than the background, it was possible also to record their spectrum using only the signals from the gaseous photomultiplier located in the focal plane of the monochromator. They coincide in time with the signals from the MWPC installed in the pinhole camera. Some results of such measurements are presented in Figures 4 and 5. In Figure 4a the spectrum of the UHF stationary discharge in D_2 at 1atmosphere (curve1) and for comparison the spectrum of laboratory DC (direct current) arc also induced in the same gas (curve 2) are shown. Both spectrums were recorded in a usual way, by measuring the counting rate from the photosensitive single-wire counter, installed in the monochromator as a function of the wavelength. In the case of the UHF plasma one can see a power molecular emission with a peak at a 1600 angstroms (160 nm), which is typical for low-temperature plasma, while in the case of the arc when the temperature is higher (about 1 eV) the molecular spectrum disappeared because most of the D_2 molecular were dissociated and only "H^- continuum" and bremsstrahlung radiation are visible.

The results of the measurements in signal coincidence mode, which corresponds to the emission from the filaments, are shown in Figure 4b. The emission spectrum is corrected for the monochromator and the detector response function, and decreases linearly with increasing wavelength. This spectrum is a sum of the recombination radiation, bremsstrahlung, and the "H^-continuum". Judging from the slope of the spectrum, the electron temperature in the head of filaments is estimated to be 3-5 eV (see Bogomolov, 1987 for more detailed discussion).

To confirm this conclusion it was important to detect the lines corresponding to highly ionized atoms. This was done by the same signal coincidence method in a mixture of D_2 (or H_2) with a few percentage of He. Figure 5 shows the results of the measurements and shows the clear 164 nmHe II line (approxi-

Figure 4. Spectra of stationary discharges in deuterium measured by different methods: a) measurements of the counting rate: 1-microwave plasma spectrum; 2--dc-arc spectrum, b) filament spectrum measured by coincidences with a MWPC installed in pinhole camera. The ordinates represent percentages with respect to the counts measured by the first method. The gaps in the spectra at 121.6 nm is due to the light absorption in the dissociated D_2 surrounding the UHF discharge. The deuterium pressure in the microwave resonator was 1.3 atmospheres. The power dissipated in the discharge was ~10 kW.
From Bogomolov, 1987.

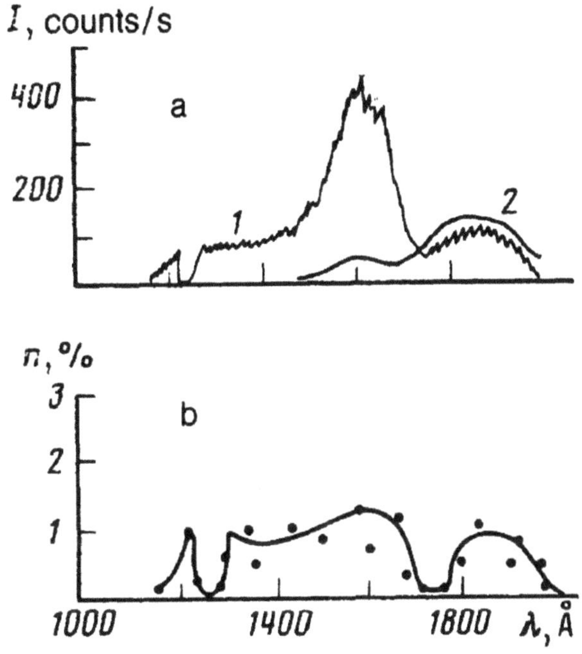

Figure 5. Results of measurements of He II line by coincidence with a MWPC; the gas pressure in the plasma installation was 1.5 atm. Open circles: D_2+ 0.5% He; filled circles: H_2+ 0.5% He, triangles: pure D_2. The power in the discharge is~9 kW. The spectral resolution is 0.03 nm.
From Bogomolov, 1987.

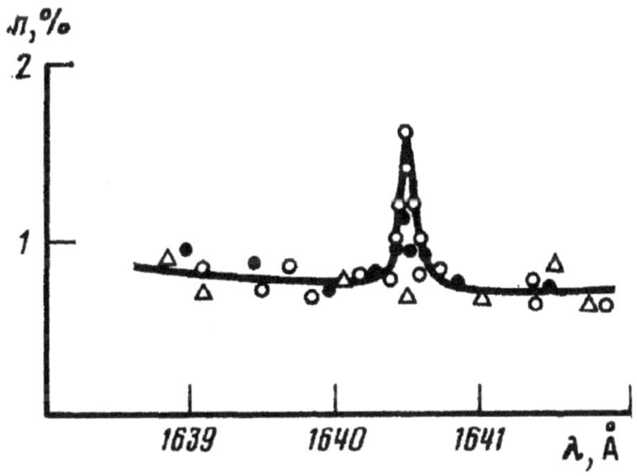

mate excitation energy 40 eV with a spectral resolution of 0.03 nm). It follows from the results that the filaments indeed have hot electrons with energy of at least 40 eV.

Later similar instability was observed by several other groups in various UHF discharges (see for example Skovoroda, 2001), indicating that it is a frequent phenomenon occurring in plasma generated in microwave electromagnetic fields.

3. SPECTROSCOPY

Photosensitive single-wire counters have been used for various spectroscopic measurements for quite a long time. Their important advantages compared to vacuum PMTs are extremely low noise courting rate (mainly cosmic rays, which can be easily rejected in proportional mode when the signal amplitude exceeds a certain value), and narrow sensitive band, which make them insensitive to the scattered wide-band lights. Both these features enables recording of VUV radiation with a high spectral resolution. The spectral resolution is determined not only by the spectrograph resolving power, but also the detectors to

Figure 6. Spectrum of stationary deuterium plasma (thick curve) recorded with a TMA filled gaseous photomultiplier; the dashed curve is the spectrum of a low-pressure discharge in hydrogen used to measure the transmission of the plasma in the resonator (see Figure 1) and the thin curve is the spectrum of the transmitted light. The ratio of these two curves determines the transmission vs. the wavelength. As can be seen that for wavelengths longer than 120 nm the gas has a transmission between 80 and 100%. All these measurements were performed with a spectral resolution of 0.03 nm. The power dissipated in the stationary discharge was 9.3 kW with a pressure of 1 atmosphere.
From Diatroptov, 1972.

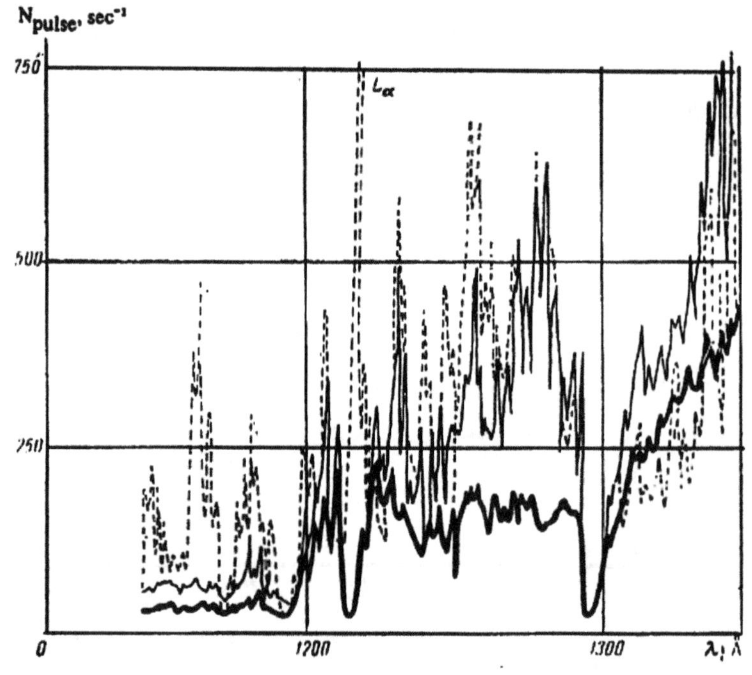

Figure 7. A H_2 spectrum detected with a one dimensional RPC; the H_2 lamp intensity was comparable to one used in Diatroptov, 1972. Note that some channels of the ASIC did not work, thus, one can see zero counts in these channels.

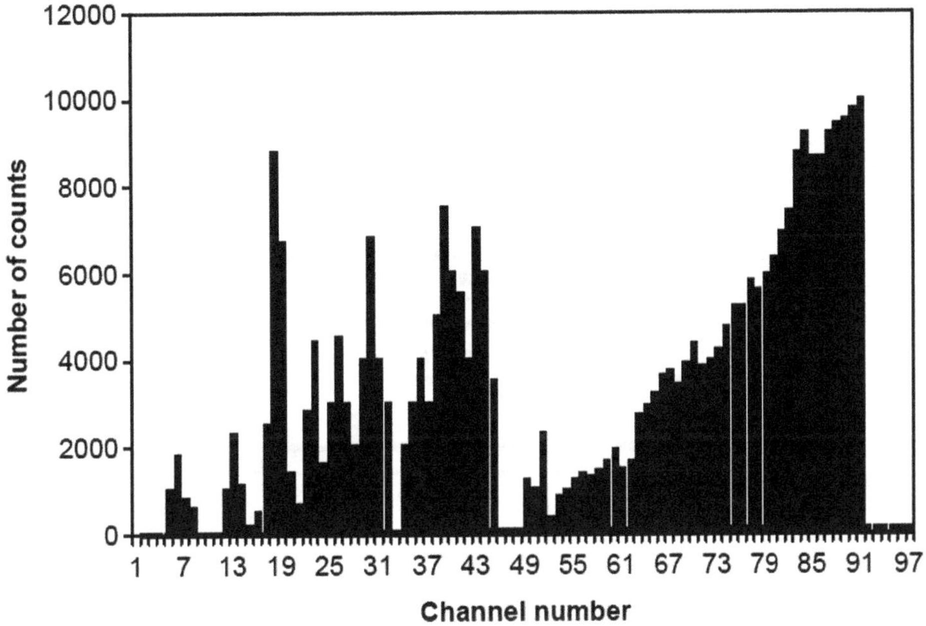

noise level. A high-resolution spectrograph delivers a narrow spectral interval and thus a small number of photons, which can be reliably recorded only with a detector having a high signal to noise ratio.

Figure 6 shows the VUV spectrum of the stationary UHF plasma in D_2 recorded with a TMA filled single wire counter with a high spectral resolution of 0.03 nm, thanks to practically zero noise level of the detector. To obtain such a spectrum takes ~15 -20 min, since the monochromator's grid is rotated by a step motor with a speed ensuring accumulation of enough counts from the gaseous photomultiplier at each grid position.

The same spectrum can be detected much faster if a position-sensitive detector is used in the focal plane of the monochromator. In this case the grid angle is fixed and array of light beams with various wavelength reflected from the grid are projected on of the active surface of the detector.

In the first attempt a MWPC with a CuI photocathode was used in the focal plane of the spectrograph to detect the spectrum of individual short-lived dense filament. The spectral resolution was low due to the large wire pitch (3 mm) (Peskov, 1980).

Later, a photosensitive RPC with a spatial resolution of 30 μm was used in the focal plane of a monochromator (Fonte, 2004).To minimize the number of electronic channels each group of 10 anode strips of the RPC were electrically connected together. Figure 7 shows the spectrum of a H_2 lamp detected with this RPC during a 0.1 s exposure (which is 9000 times faster than for the spectra presented in Figure 6).

Some spectrographs (called hyper spectrographs) allow detecting simultaneously the spectrum of the object and its one dimensional image. By moving the hyper spectrographs one can make 2-D scans of the surfaces. In one coordinate one obtains a one dimensional image of the object and in another coordinate one records its spectra. For a 2-D detector placed in the focal place of the hyper spectrograph one can

Figure 8. A thin-gap RPC structure developed for hyperspectroscopy
From Fonte, 2005.

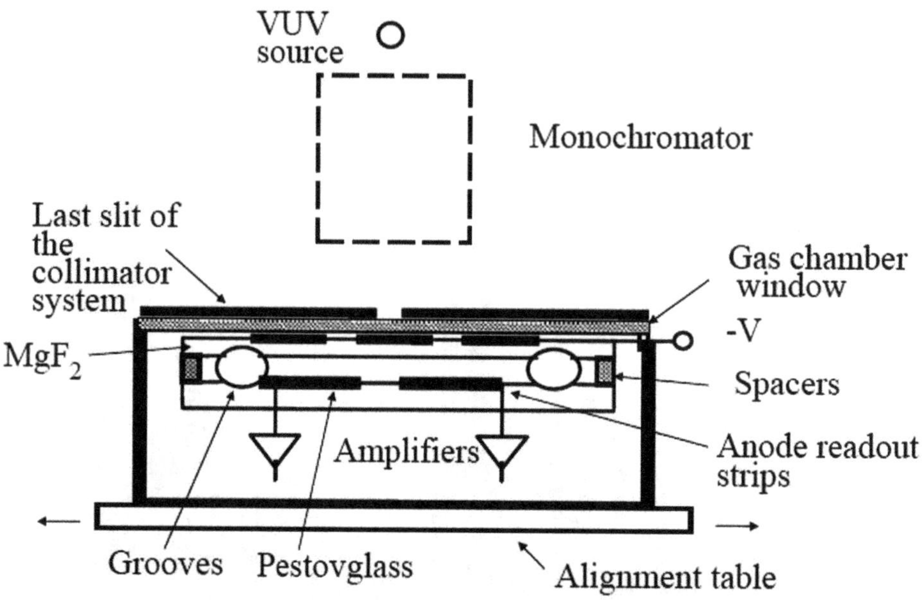

measure at the same time the image of the object with a rather high position resolution and its spectra at each point of the object's surface. This gives enormous resolving power since such a method allows one to make some conclusions about the chemical compositions of the surface.

It is attractive to extend the hyper spectroscopic method to the UV region of spectra, 185-280 nm. This region is unique for the hyper spectroscopic measurements because the sunlight with a wavelength of 185-280 nm is blocked by absorption in the upper layers of the atmosphere, but the atmosphere is transparent for this radiation at ground. This offers new technical possibilities in achieving materials and pattern recognitions. For example, it allows one to make unique hyper spectroscopic measurements using artificial UV light sources installed on a flying helicopter. The main problem with such an approach is that the extremely low intensity of the scattered and reflected UV light should be detected on the strong background of long wavelength sun light (with wavelengths longer than 280 nm). Existing MCPs are too noisy for these measurements.

A prototype of a gaseous photomultiplier developed for this purpose is shown schematically in Figure 8.

It is essentially a thin-gap RPC combined with a CsI photocathode operating in a sealed gas envelope (Rodionov, 2005).The cathode is made of a 2 mm thick MgF_2 plate with an area of 4 x 4 cm^2. The inner surface, facing the anode, is coated with a 20 nm thick photosensitive CsI layer. The outer surface is in contact with a 0.2 mm thick Al mesh, manufactured by a microelectronic technology, forming cells with an open area of 23 x 23 μm^2 at 30 μm pitch. The anode of the RPC is made of Pestov glass covered by Cr strips placed with 30 μm pitch, and the gas gap is 0.4 mm wide. The anode strips are connected individually to a counting ASIC readout chip.

The position resolution was measured by illuminating a single anode strip with a 30 μm diameter VUV beam. Figure 33 in Chapter 6 shows the response of this detector using different gases. The gas mixture He +0.8%CH_4+ EF (where EF stands for Ethylferrocene) offers the best compromise between position resolution and gas gain.

Figure 9. Counting rate vs. strip number measured with a collimated UV beam of 30 mm diameter for different gains and gas mixtures: He +0.8%CH_4+ EF and (triple mix.), CH2H2F4+ 10%SF_6 + 5% iso-butane; the position resolution is comparable to the strip pitch of 30 μm, except for the He +0.8% CH_4+ EF mixture at gains above 10^5 (b), where the resolution is spoiled by photon feedback.

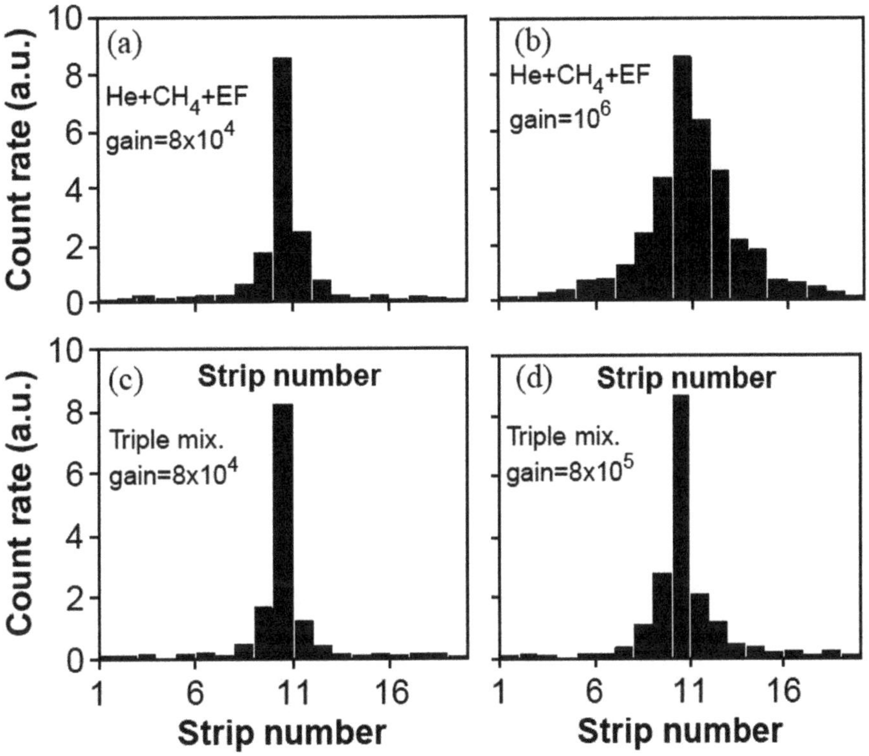

To obtain 2-D images in the focal plane of the hyper-spectrograph the following procedure was used. First the readout strips of the detector were oriented along the axis corresponding to the spectral measurements and then we rotated the detector 90° and measured the one dimensional image of the object at the selected wavelength.

Various tests were done in which hyper spectroscopic images were taken with this photosensitive RPC, and for comparison with MCPs as well. Figure 10 shows a photograph of two yellow stickers used as tests objects: a "dark" yellow (upper sticker) and "light" yellow (lower sticker).

The two stickers are hardy distinguishable by eye. However, by measuring the sticker's UV spectra one could easily distinguish between them as shown in Figure 11.

Figure 12 shows the image of the border between the stickers (a scan made in the direction perpendicular to the border) measured at 194 nm.

The UV hyper-spectroscopy can find a lot of applications from security to medicine. In medicine it may allow one to develop early diagnostics of some illnesses, as for example skin cancer.

Figure 10. A photograph of the two stickers having similar color: "dark yellow (upper part) and "light" yellow (lower part)
From Rodionov, 2005.

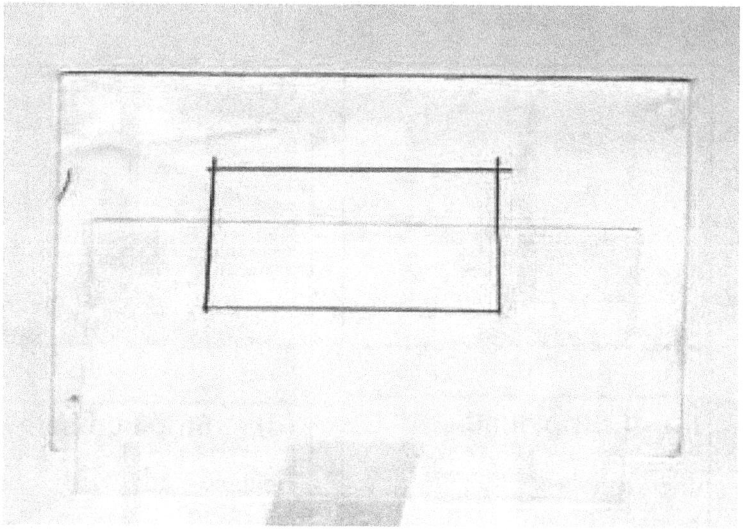

Figure 11. On line digital image of the spectra of the yellow stickers (a-"light" and b-"dark") measured with the photosensitive RPC in the spectral interval 185-200 nm
From Rodionov, 2005.

a)

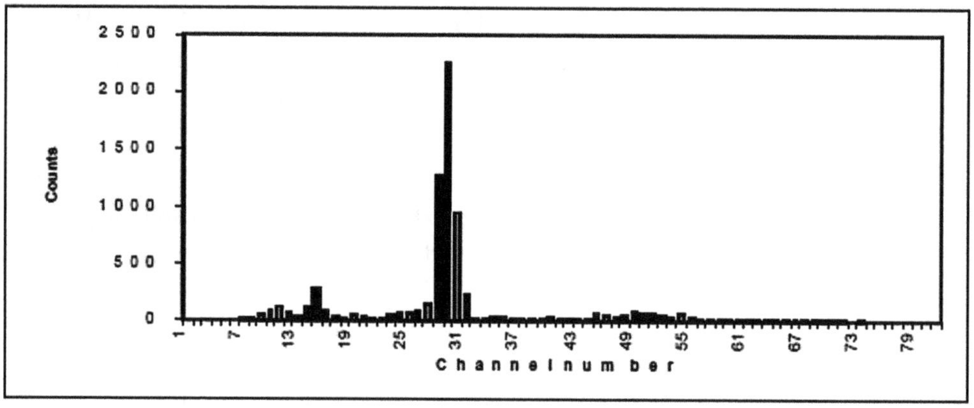

b)

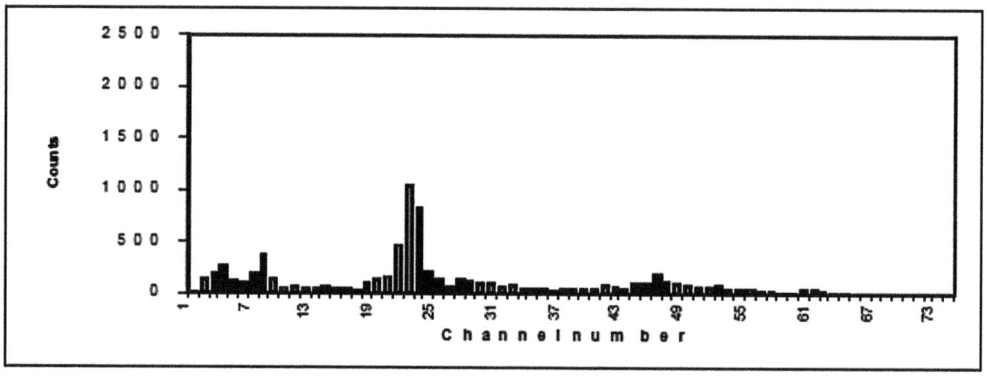

Figure 12. On line digital image of the border between two yellow stickers measured with an RPC at 194 nm; the border is clearly visible around channel 21.
From Rodionov, 2005.

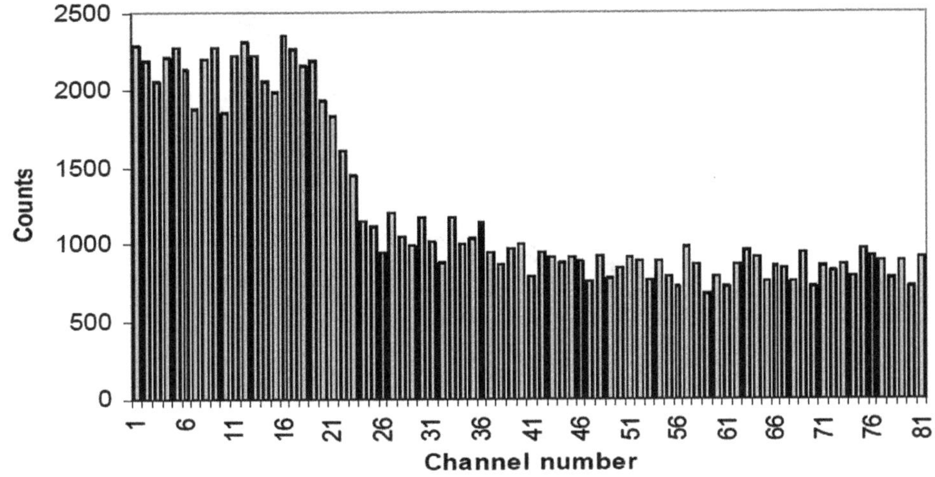

4. ASTROPHYSICS

The detection of the Cherenkov light emitted by secondary muons in atmospheric showers generated by high energy cosmic rays is a well-known detection technique of high energy cosmic rays, which has been used for many years (Holder, 2013). This method is practically unique in the energy range between 300 GeV and 30 TeV. Above 30 TeV the secondary particles produced in the atmosphere reaches the ground and permit the use of charged particle detector arrays located on the ground surface. Below 300 GeV the cosmic rays do not deposit enough energy in the atmosphere to be detected from the ground.

In the range 300 GeV to 30 TeV the secondary muons and electrons produced in air showers triggered by the high energy cosmic rays, produce Cherenkov light in the atmosphere which can be detected by ground based photosensitive detectors. Its main disadvantage is the necessity to work only during moonless nights, hence with a low duty cycle. To reduce the noise from night sky- or moon-light, it was proposed to work in the UV range since the amount of Cherenkov light emitted increases in the UV. It the wavelength range 200-280 nm it could even be possible to detect Cherenkov light from cosmic ray showers in full daylight. The ozone in the high atmosphere totally blocks out the sunlight at these wavelengths.

An early attempt was made to measure the Cherenkov light produced with a photosensitive TMAE (Tetrakis(dimethylamino)ethylene) filled MWPC (see Figure 12) located in the focal plane of aluminized parabolic mirrors (Bartoli, 1991).The photoelectrons were detected via charge induction in the back cathode plane which consists of 16 x 16 square conductive pads on a copper clad printed circuit board.

The advantage of the photosensitive MWPC compared to solar blind photomultipliers is the low cost, allowing covering a much larger active area with high granularity and good quantum efficiency. Another argument is that solar blind photomultipliers have residual quantum efficiency (10^{-3}) in the visible which may create problems in some cases, for instance one cannot point directly to the Moon.

Figure 13. Drawing of a TMEA filled MWPC with a pad cathode used for position measurements
From Anassontzis, 1992.

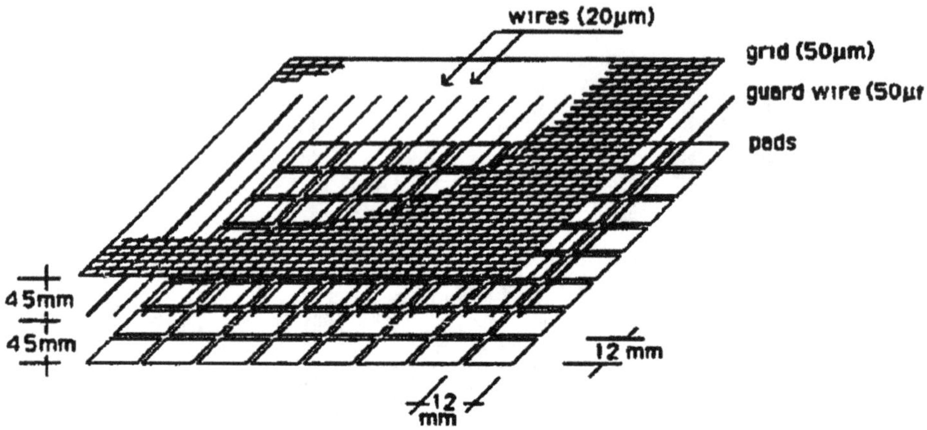

A typical air shower as seen on the on line computer screen is shown in Figure 14. Various tests show that TMAE-MWPC are sufficiently reliable for a full scale experiment, being able to work in open air in a rather severe environment of strong winds, large thermal variations and humidity changes.

5. DETECTION OF SPARKS AND FLAMES

5.1 Indoor Applications

5.1.1 Single Wire Counters

In many safety applications it is needed to detect sparks or flames under daylight conditions in houses, plants, storage buildings, refinery or chemical plants, oil production and drilling platforms, textile and paper manufacturing, underground and surface mining etc. Recent studies show that it can be efficiently achieved with UV sensitive detectors. Most flames in air, almost independently of their nature, emit UV-light (185–300 nm) whereas the lamps used to illuminate rooms and buildings practically do not emit in this spectral interval. The UV emission from the sun is fully blocked by glass windows. Thus one can detect the UV light from the flames with high signal to noise ratio. Commercial detectors exploiting this feature are already available from various companies, for example Spectrex Inc., (http://spectrex-inc.com/), ESP safety Inc. ((http://www.espsafetyinc.com/UV-flame-detector.php), Hamamatsu Photonics (http://www.hamamatsu.com/jp/en/3007.html), etc. The cheapest among them and one of the most efficient is the Hamamatsu sealed gaseous detector with metallic a photocathode R2868, shown schematically in Figure 15.

In a series of research work, (Carlson, 2002; Carlson, 2003; Rodionov, 2005; Bidault, 2006; Bidault, 2007; Periale, 2007; Di Mauro, 2007; Charpak, 2008) it was shown that gaseous detectors with CsI or CsTe photocathodes offer almost 1000 times higher sensitivity to flames compared to the best commercial sensors.

Figure 14. Typical air shower event as seen on the on-line data acquisition computer; the grid reproduces the 16 x 16 pads of the chamber cathode plane.
From Bartoli, 1991.

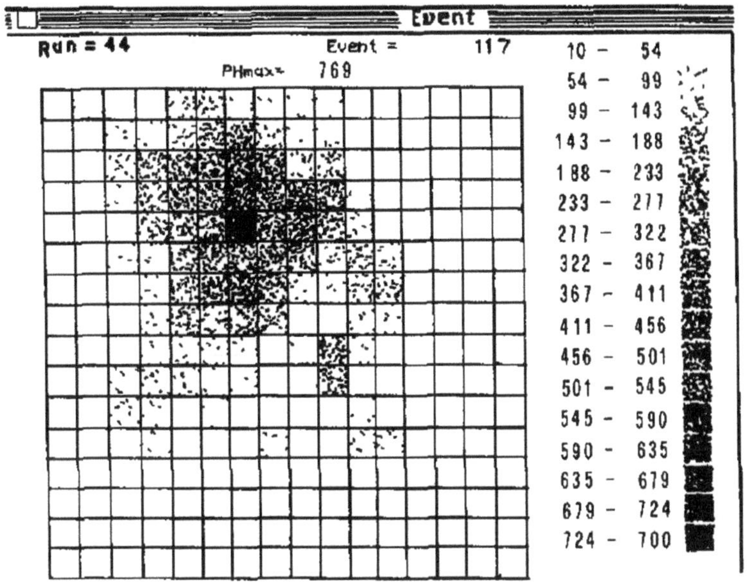

Figure 15. Principle of operation of the Hamamatsu UVtron R2868 filled with low pressure Ar; UV photons from the flame emits photoelectrons from the metallic cathode. The liberated photoelectrons trigger a glow discharge in the detector which is quenched by the external resistor. In this way the output signals have equal amplitudes, so this detector operates in digital mode. The maximum counting rate is limited by the glow discharge duration and is about ~1 ms.

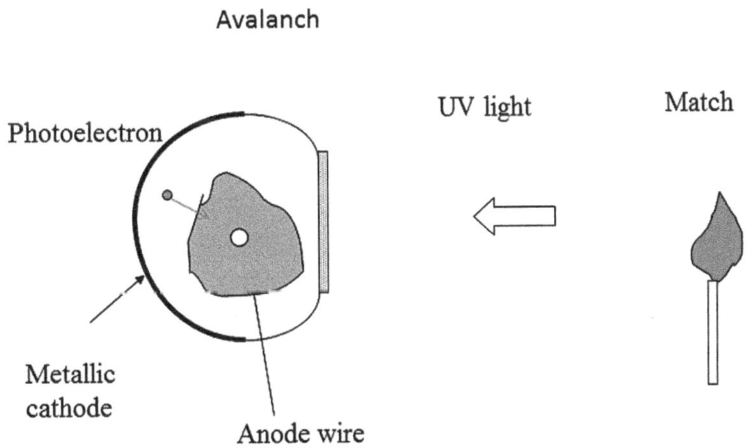

Figure 16. A sensitive gaseous photodetector for flams and sparks operating in proportional mode; the metallic photocathode is replaced by a CsI photocathode having quantum efficiency in the UV region orders of magnitudes higher than the metallic one. The window is made of MgF_2, the anode is a thin wire (50 μm in diameter), and the gas filling is Ar + (10-20)% CO_2 or Ar + (10-20)% CH_4. Photoelectrons extracted from the CsI photocathode trigger Townsend avalanches. The detector operates in proportional mode and the maximum counting rate at which the detection efficiency still remains high is about 10^5 Hz. *Francke, 2002; Carlson, 2002.*

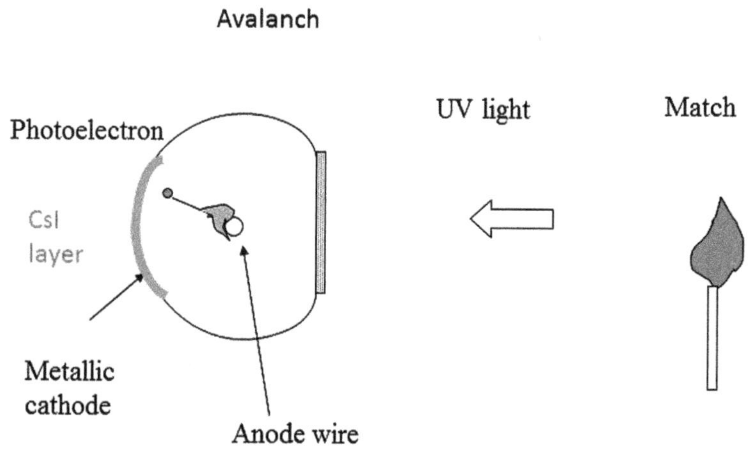

This is actually not so astonishing since the CsI quantum efficiency in the UV region is orders of magnitude higher than the quantum efficiency of the metallic photocathode used in the Hamamatsu UVtron R2868. Moreover the Hamamatsu UVtron operates in glow discharge mode (digital mode), whereas a CsI single-wire counter is operating in proportional mode allowing to discriminate signals produced by a flame and background signals due to e.g. cosmic rays. Another important advantage of the proportional mode is the possibility to detect transient events such as sparks, while in the UVtron a signal caused by a spark is undistinguishable from a signal produced by spurious noise photoelectron or by cosmic rays.

Such a detector is shown in Figure 16.

The results of the quantum efficiency measurements on the flame detector using different gases are shown in Figure 17. In the same figure are also presented the typical spectra of flames in air (red triangles) and the spectra of the sun measured at ground (light blue crosses), both in arbitrary units. The latest data were taken from the Hamamatsu data sheet for the flame sensor R2868 (http://www.hamamatsu.com/jp/en/3007.html). The curve for the spectra of the flames in air was verified in our measurement performed with help of the monochromator. As can be seen, the best overlap of the detector efficiency with the flame emission spectra is with a MgF_2 window.

Beside the laboratory prototype of the flame detector with a CsI photocathode, several sealed commercial products have been developed in collaboration with Oxford Instruments, Finland, shown in Figure 18.

The CsI photocathodes are manufactured at the CERN evaporation facility and then shipped to Oxford instruments for mounting inside the single-wire counters. They are installed into the detector after being in contact with air for one week or even more. After sealing the detector quantum efficiency and its operating voltage are periodically monitored. Some of them were studied under long time in continuous operation. No changes in their characteristics have been observed so far, demonstrating that the detectors remain stable for at least 10 years.

Figure 17. Results of the quantum efficiency measurements for some flame detectors: stars: a single wire counter with a CsI photocathode and with a quartz window, filled squares: the same type of the detector, but with a MgF_2 window, open rhombs: the detector filled with EF vapor; in the same figure are presented the typical spectra of flames in air (triangles) and the spectra of the sun at ground (crosses), both in arbitrary units.

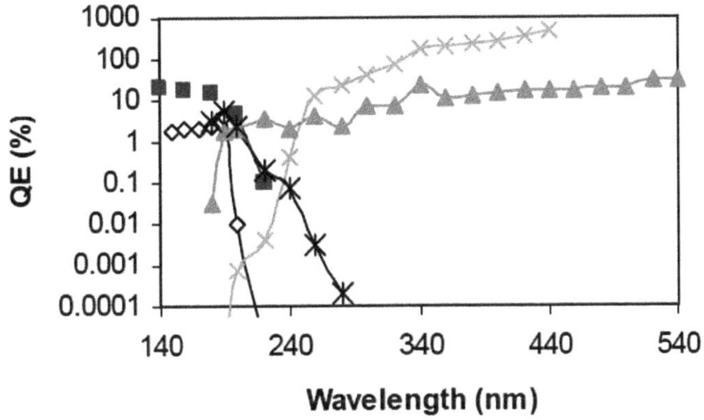

Figure 18. Photo of a sealed gaseous flame detector with CsI photocathode
Developed in collaboration with Oxford Instruments Inc.

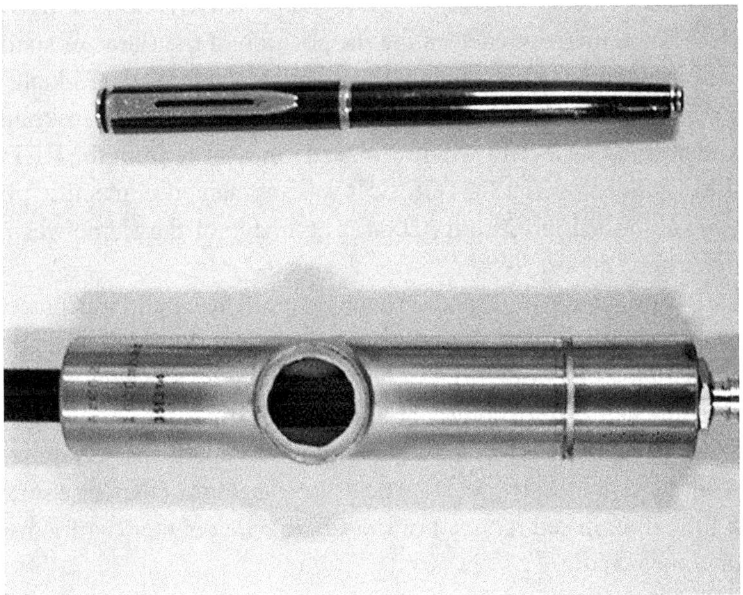

A comparison of the efficiency of the CsI-based detector for flames with the most sensitive commercial detector on the market, R2868, is presented in Table 1.

As can be seen from the data presented in Table 1, the laboratory prototype, having a MgF_2 window is more than 1000 time more sensitive than the best commercial flame detector existing on the market today, Hamamatsu R2868. The sealed commercial detector with a quartz window has sensitivity about 100 times higher than Hamamatsu R2868. Its lower efficiency is attributed to the fact the CsI photo-

Table 1. Counting rates produced by a candle placed at different distances from the flame detectors: Hamamatsu R2868, laboratory prototype with CsI photocathode and Oxford instrument industrial product

Distance (m)	Hamamatsu R2868	Laboratory Prototype (MgF_2 Window)	Industrial Product (Quartz Window)
1	705	Overloaded-continuous current	81.579
3	76	87.574	9.015
10	6	7.902	811
30		876	92

cathode used in this detector was exposed to ambient air for one week or more before being introduced into the detector.

5.1.2 Flame Detector Based on Micropattern Detectors Combined with CsI Photocathodes

The next generation of flame detectors with CsI photocathodes was based on micropattern electron multipliers: RPC (Rodionov, 2005), resistive electrode GEM or RETGEM (Bidault, 2006; Bidault, 2007; Di Mauro, 2007) and MICROMEGAS (Peyaud, 2015). This approach offers several advantages, for example the possibility to build planar detectors having relatively large active area with high sensitivity, and the capability to obtain UV flame images to determine the position of the flame or spark.

The first prototype of a flame imaging chamber was developed in work (Bidualt, 2006) and is sown in Figure 19. It was a gas chamber housing a RETGEM combined with a semitransparent CsI photocathode, 20nm thick, deposited on a CaF_2 window placed7 mm away from the RETGEM. The readout plate was placed 0.4 to 1 mm below the RETGEM. It was segmented in metallic pixels with 1.27 mm pitch similar to the one described in (Östling, 2004).The image of the flame was projected to on the chamber window be means of a lens.

Figure 20 shows a 1-D projection of a candle flame image. The candle was placed 1.5 m away from the detector and the counts were accumulated during0.1 s. When the candle was placed 30 m away it produced counting rates on the order of a kHz, to be compared with a background of less than10 Hz in a room fully illuminated by natural or fluorescent lights.

In work (Bidualt, 2006) was also performed a quantitative comparisons between the RETGEM and a single-wire counter with a CsI photocathode. Both detectors had almost the same sensitivity to the flame and could operate in fully illuminated rooms. For the single wire counter (with no segmented readout) in sunlight the signal to noise ratio:

$$k_s = S_F/N_{nF}, \quad (1)$$

(where S_F is the excess counting rate produced by the flame and N_{nF} is the counting rate without the flame) was small ($k_s < 1$). For the RETGEM the signal to noise ratio was much:

$$S_F/N_{nF} \sim k_s n_{pix}, \quad (2)$$

Figure 19. RETGEM-based imaging chamber with a semitransparent CsI photocathode develop in work Bidult, 2006.

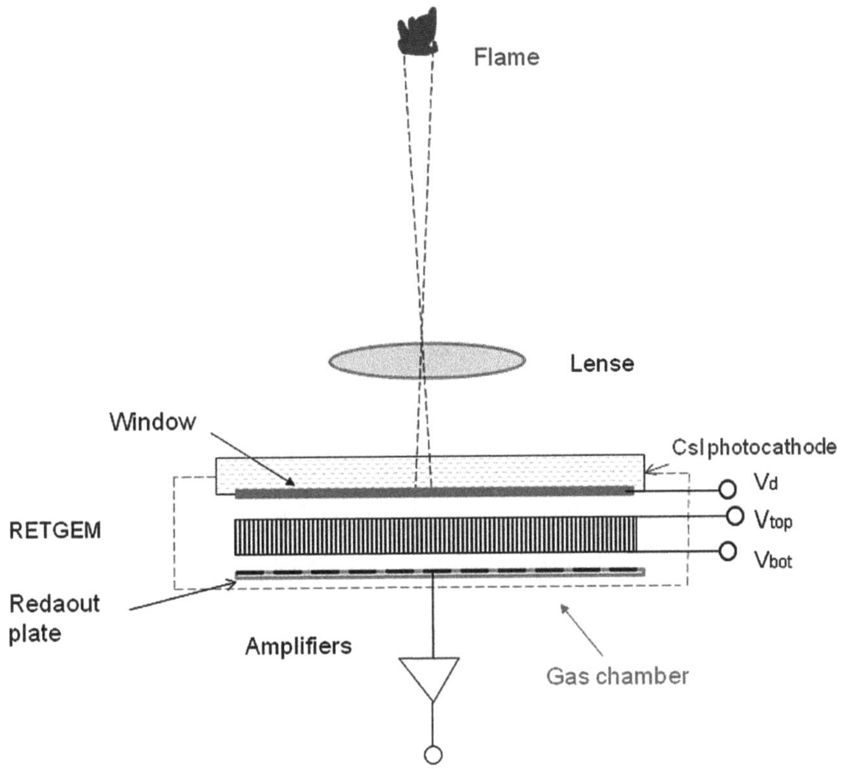

Figure 20. A digital image of a candle flame obtained with the RETGEM combined with a semitransparent CsI photocathode; the rows are separated by 1.27 mm. The acquisition time is 0.1 s. From Bidualt, 2006.

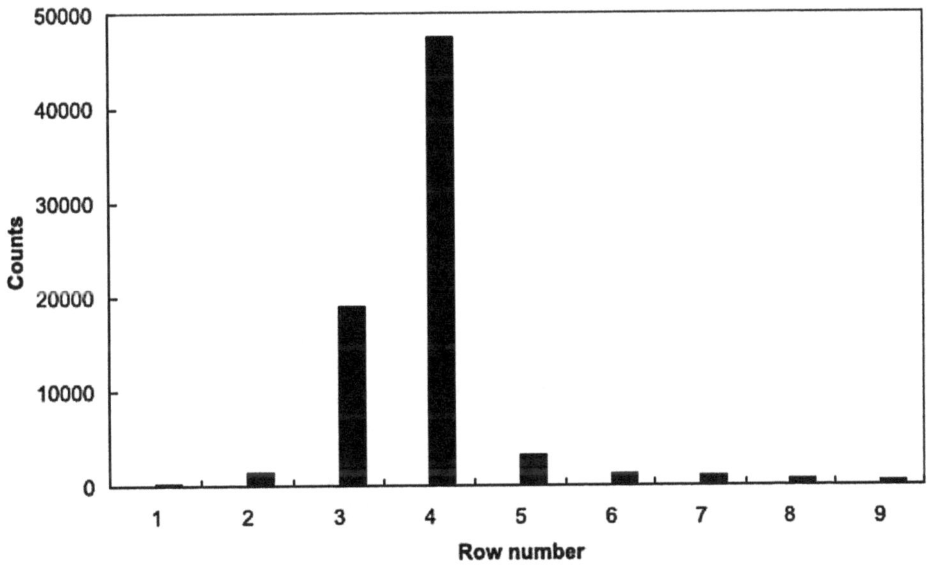

Figure 21. Layout of the flame imaging detector with reflective CsI photocathode

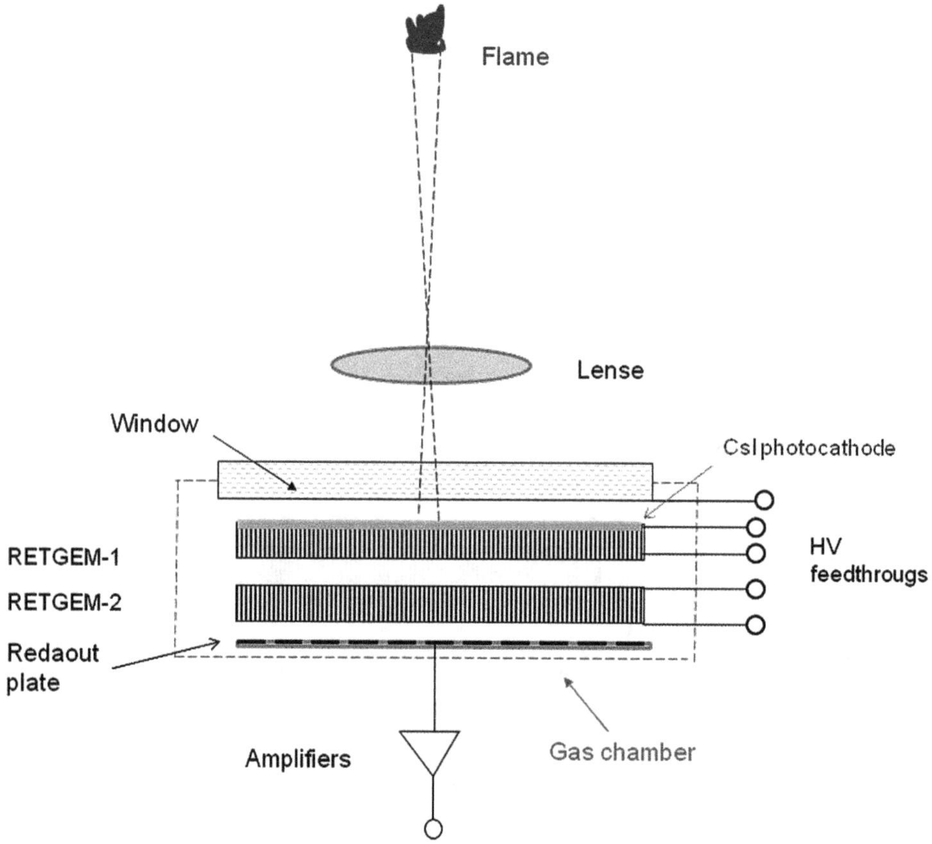

where n_{pix} is the number of pixels in the readout plate. This is because the image of small flames occurring far from the detector occupied one or few pixels, whereas the diffuse sunlight is distributed overall pixels. When looking at the image the few pixels illuminated by the flame clearly stand out over the uniform small background of sunlight.

Later in work (Di Mauro, 2007) was tested a double RETGEM with a reflective CsI photocathode deposited on the top surface of the first RETGEM, as shown in Figure 21.The detector operated at the reversed bias drift field doubling the sensitivity for UV light and at the same time strongly reduces the sensitivity to radioactive background. The images of a flame obtained with this detector can be found in (Di Mauro, 2007).

In 2012-2014, in the framework of the European project for fire, was developed a prototype of a flame detector in which MICROMEGAS combined with a reflective CsI photocathode (Peyaud, 2015). The schematic drawing of this device is shown in Figure 22. Its working principle is the following: UV photons extract electrons from the CsI photocathode deposited on top of the MICROMEGAS drift electrode. These electrons experience a two-step multiplication, first in the drift region and later in the avalanche gap of the MICROMEGAS. This multiplication concept was proposed in work (Peskov, 2001) in order to increase the diffusion of electrons triggering the avalanche in the final step of the multiplication. This approach allows increasing of the Raether limit and thus the maximum achievable gain of the MICROMEGAS.

Figure 22. A flame detector based on MICROMEGAS combined with a CsI photocathode; to increase the maximum achievable gain this device operates in double-step amplification mode.
From Peyaud, 2015.

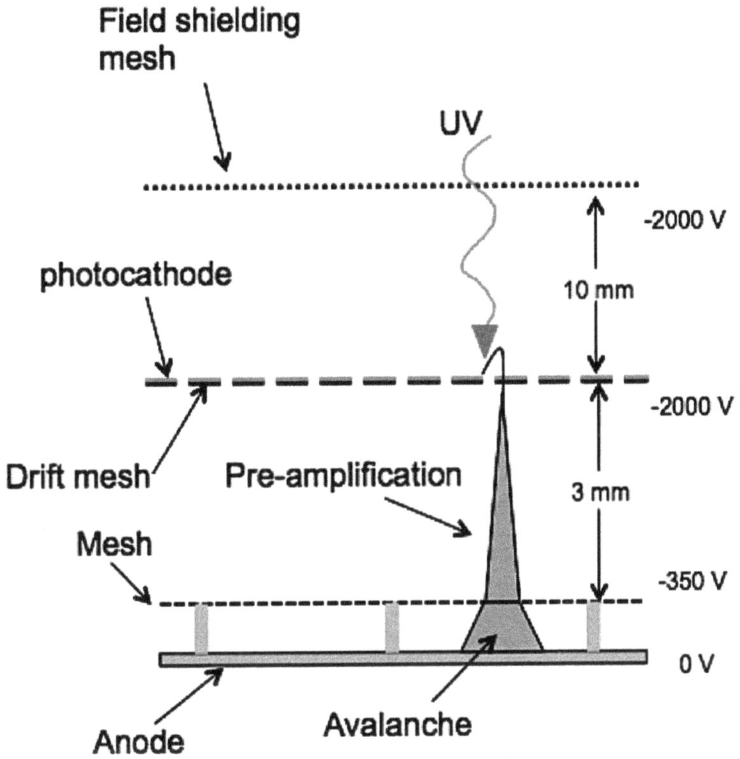

The sensitivity of this device to flames was comparable to the sensitivity of the CsI-RERGEM.

5.2 Outdoor Applications

5.2.1 Single-Wire Counters Filled with Photosensitive Vapors

UV light produced by flames can be detected in the presence of strong sunlight, since the UV light from the sun with wavelength $\lambda<280$ nm is fully adsorbed by the ozone in the upper layer of the atmosphere, while on ground level the atmosphere is quite transparent the radiation in the interval 185-280 nm. The tests performed with various designs of gaseous photomultipliers with CsI photocathodes demonstrated that they are able to detect flames in outdoor conditions, although the background rate N_{nF} produced by the sunlight is rather high. This was because the quantum efficiency of the CsI for $\lambda >280$nm is not really zero whereas the sun's emission in this spectra interval is extremely strong. As a result the convolution of the sunlight spectrum I_{sun} (λ) with the CsI quantum efficiency (Q_{CsI}) in the spectral interval of 280–500 nm is not zero either:

$$N_{nF} = \int_{280}^{500} I_{sun}(\lambda)\, Q_{CsI}(\lambda)\, d\lambda > 0, \tag{3}$$

Figure 23. Schematic drawing of a single-wire counter filled with EF vapors; in this design to the UV flame radiation enters the counter along the anode wire which ensures sufficient light absorption inside the detector.

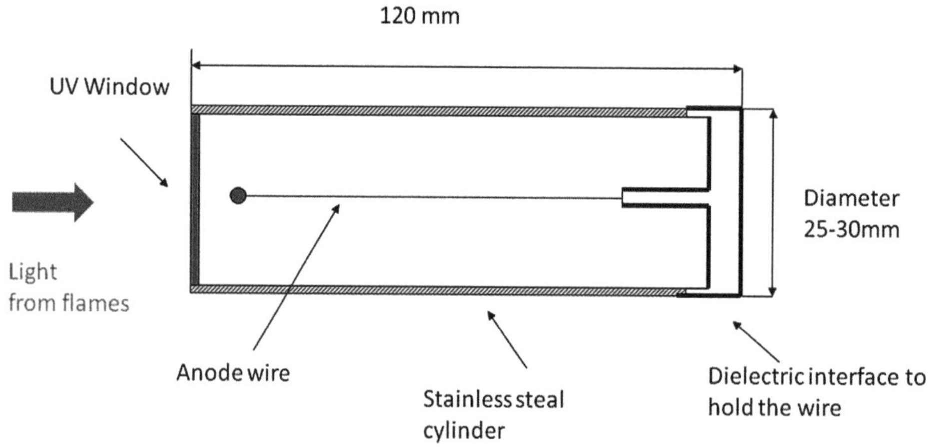

Much better results were obtained with gaseous detectors filled with photosensitive vapors like EF or TMAE. The quantum efficiency of the EF (Q_{EF}) and TMAE (Q_{TMAE}) are practically zero for $\lambda > 200$ and 220 nm, respectively so that:

$$N_{nF} = {}^{500}\int_{280} I_{sun}(\lambda)\, Q_{EF}(\lambda)\, d\lambda = 0, \tag{4}$$

and

$$N_{nF} = {}^{500}\int_{280} I_{sun}(\lambda)\, Q_{TMAE}(\lambda)\, d\lambda = 0, \tag{5}$$

Measurements show for example that in the case of EF vapors the background counting rate in direct sunlight N_{nF} is about 0.25 Hz. This counting rate is mainly caused by the cosmic radiation, and not by the sunlight. With such low background it is easy to detect small flames in the presence of direct sun light (Peskov, 2007). The single-wire counters filled with EF and TMAE and photograph of their commercial prototypes, manufactured at Oxford Instruments, are presented in Figures 23-26.

In Table 2 experimental results are presented comparing the sensitivity of flame detectors filled with photosensitive gases with the sensitivity of the commercial sensor Hamamatsu R2868. As can be seen from these data, the sensitivity of the EF and TMAE filled detectors are 100 and 1000 times, respectively, higher than the sensitivity of Hamamatsu R2868.

Usually, gaseous detectors filled with EF or TMAE vapors lose their sensitivity at temperatures below 10-15°C due to vapor condensation. To minimize this effect a special design of a single-wire counter was recently (2010-2012) developed in the frame work of the CERN Technology Transfer project. The modified detector has a cross-shaped cylindrical cathode with two windows aligned parallel to the anode wire (see Figure 27).

At low temperatures the photosensitive vapor condenses on inner surface of the detector including its windows which then act as liquid photocathodes. As was shown in Chapter 4, the quantum efficiency of these liquid photocathodes have reasonably high values even in long wavelengths, so the overlap with

Figure 24. Photograph of a commercial prototype of a sealed single-wire counter filled with Ar +20%CO_2+EF at 1atmosphere

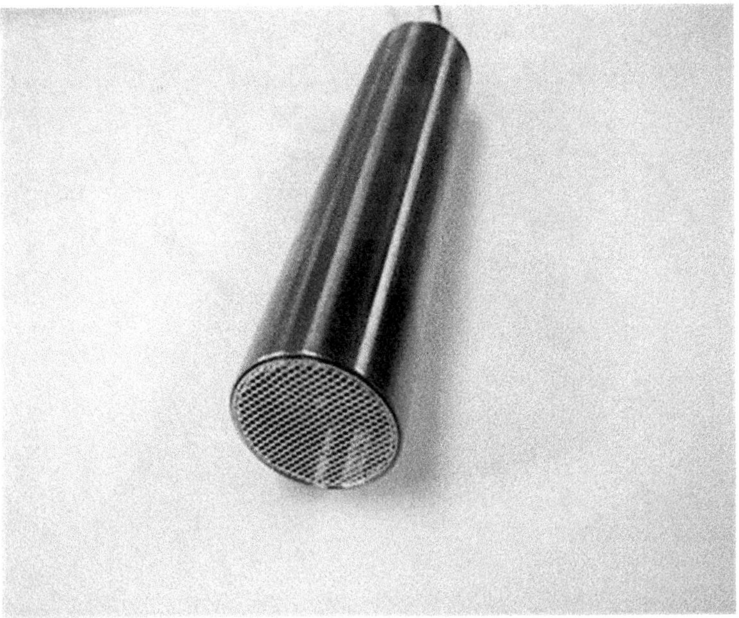

Figure 25. A TMAE filled single-wire counter; to minimize photoelectron production on the detector cathode (recall that TMAE-coated cathodes have an enhanced sensitivity to UV and visible light, see Chapter 4) the counter has a large diameter of the cathode (ensuring low electric field on its surface) and is supplied by a light-collimating tube.

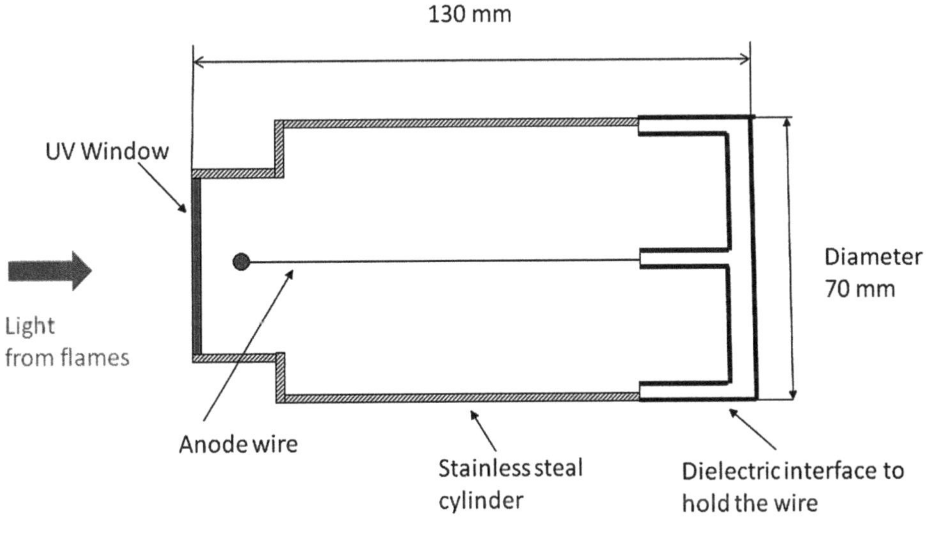

Figure 26. Photograph of a sealed single -wire counter filled with Ar +20%CO_2+TMAE at 1atmosphere

Table 2. Comparison of the sensitivity of sealed single-wire counters filled with EF and TMAE vapors and capable operating in direct sunlight, with the sensitivity of the most efficient commercial flame detector Hamamatsu R2868

Hamamatsu R2868		Our TMAE Detector at 23°C		Our EF Detector at 25°C	
Distance (m)	Mean Number of Counts per 10 sec: N_H	Mean Number of Counts per 10 sec: N_{tmae}	Ratio N_{tmae}/N_H	Mean Number of Counts per 10 sec: N_{ef}	Ratio N_{ef}/N_H
1,1	583	690.747	1.18 x 10^3	75.613	1.3 x 10^2
3	76	91.013	1.19 x 10^3	11.052	1.4 x 10^2
10	6	7.820	1.30 x 10^3	643	1.1 x 10^2
30	0,1	873	8 x 10^3	68	6 x 10^2
85		51		4	

the flame emission spectra remains relatively high and this partially compensate the loss of sensitivity due the pressure drop of photosensitive vapor.

Results of the efficiency measurements of such a detector filled with TMAE as a function of temperature are show in Figure 28. As can be seen, although the efficiency of this detector begins to sharply decrease at temperature below 10°C, it still remains ~ 100 times higher the sensitivity of Hamamatsu R2868.

Note also that the cross-shaped cathode helps to suppress the surface photoelectric effect both due to the geometry and also due to the low electric field in the cathode region exposed to the external light.

Figure 27. Schematics of a sealed single-wire counter filled with photosensitive vapors developed in the framework of CERN Technology Transfer project

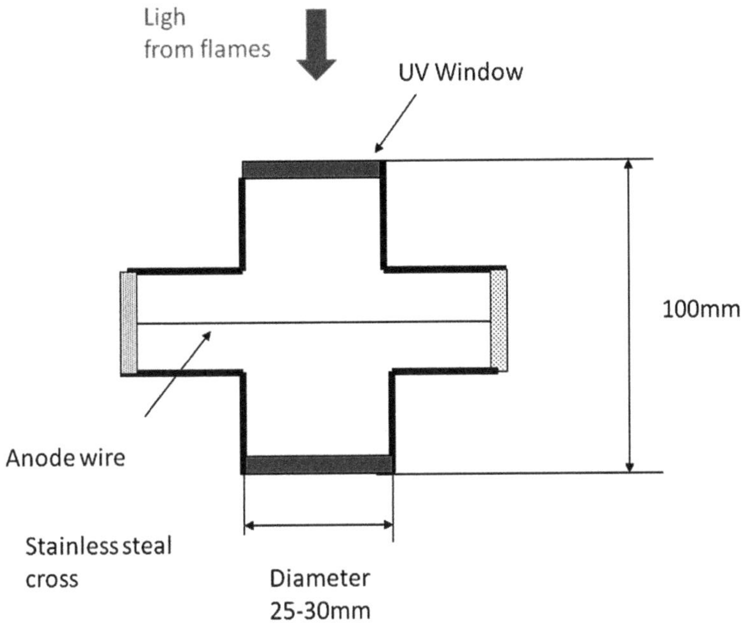

Figure 28. Counting rate vs. temperature for the single-wire counter design shown in Figure 27, detecting UV light from a candle located 10 m from the detector

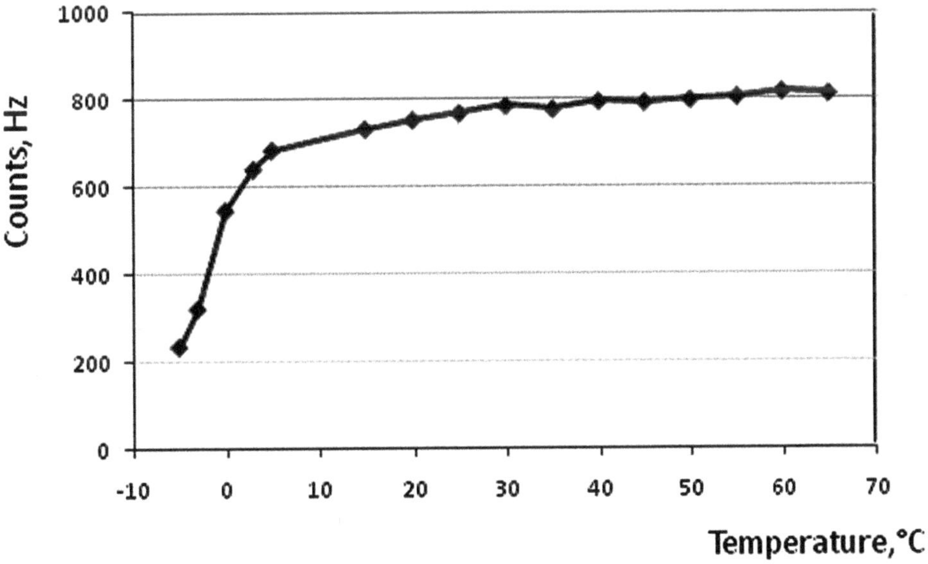

Figure 29. Possible arrangement of flame hazard monitoring with imaging detectors placed on watch towers
From Peskov, 2007.

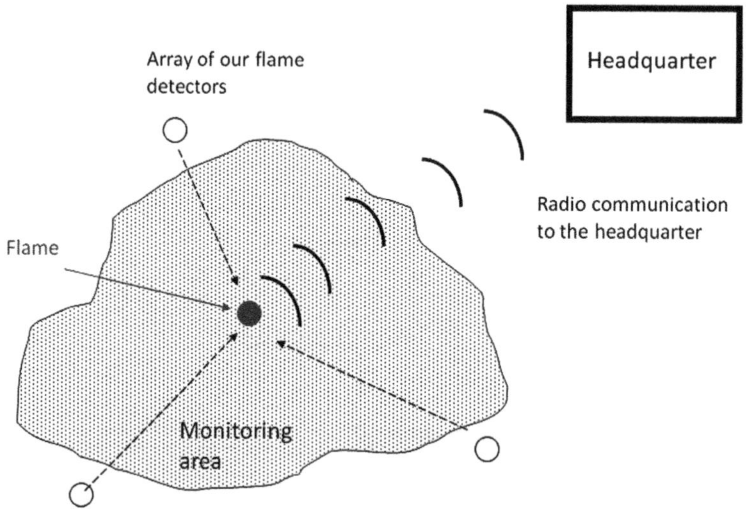

5.2.2 Imaging Flame Detectors Based on Photosensitive Gases

Further developments were focused on the construction of flame imaging devices filled with photosensitive vapors. The advantages of these detectors are that they can determine the position of the flame and also its intensity (see Figure 29).

This work was performed in a collaboration between the CERN ALICE RICH group, XCounter AB, Sweden, École des Mines in St Etienne, France and Pole University Leonard de Vinci, Paris, France and was led by Georges Charpak (Francke, 2005;Charpak, 2007; Bidault, 2007).

It was an exciting time:

Charpak's last area of research pertained to the development of flame detectors. In collaboration with colleagues from CERN and École des Mines Superior in St Etienne, we developed and tested several prototypes of position-sensitive gaseous detectors capable of visualizing flames by detecting their UV emission. We directed our initial efforts on MICROMEGAS combined with a CsI reflective photocathode. However, the MICROMEGAS alone, without a preamplification structure and when operated in single photon counting mode often sparked. What's more, this prototype required the use of a special filter to suppress the sunlight lowering its sensitivity to flames. After analyzing data derived from repeated tests, we opted for ethylferrocene (EF) as a photosensitive element because in the range of its sensitivity in air (185-200nm) the sunlight on the Earth's surface marginally exists. It is blocked by the ozone in the upper layer of the atmosphere. We determined that thick GEM with resistive electrode operated in mixture of Ne with ethylferrocene vapors was the optimal choice for our final version. This detector tracked down a flame of 1.5x1.5x1.5m^3 on a distance of ~1km. This sensitivity is 100- 1000 times higher than any commercial infrared or UV detector. Georges clearly maintained a preference for MICROMEGAS which he developed earlier with Peskov, Giomataris and others. In his practical way, he finally advocated a GEM approach and commented, "I have nothing against Sauli's GEM. If GEMs work better, *we will use GEM." The GEM–based flame detector operated so well that Georges decided to patent this device for flame detection. Georges was highly enthusiastic about this project, and believed in its future development. He called me (V.P.) two to three times per day at CERN, where most of the studies were undertaken or at St. Etienne. His ambitious plan involved the development of a network of cheap gaseous detectors for the early warning of forest fires. These detectors are designed to be installed in high risk regions where forest fires tend to erupt. They should operate in conjunction with compact battery powered radio transmitters to send warning signals to the fire brigade.*

Late one evening, George called and asked me to show up the next morning in St. Etienne: "We have to make a demonstration how our flame detector works for very important people". I don't know quite how he did it, but he beat me to St. Etienne. In a couple of hours, we managed to have our prototypes operating just in time for the arrival of "the important people" at the lab. These visitors included; a rector of the St. Etienne École des Mines, the president of the French Academy of Sciences, and the President of the Swedish Academy of Sciences. On cue, I ignited a cigarette lighter to demonstrate our detector's reaction to flame at various distances. Responding to the visitors' astonishment, Georges joked, "Vladimir and I will soon be as rich as Russian oligarchs, and I'll write a request to the French government to give him a title of Chevalier of Legion of Honor." The French president replied, "Write me a letter, I'll support your application…" The President of the Swedish Academy of Science was quite impressed yet reserved, and inquired, "Can the detector be used to measure the UV dose which people receive from the Sun? Swedish people who travel for vacation are often burned by the Sun." I replied. George added, "Yes, but then we have to use another version-with a CsI photocathode. We can do everything, but we need some funds.... soon we will get funds…your opinion and support is important. "These guest were not inclined to discuss financial backing, and they left the laboratory. Georges, remaining optimistic, was undeterred by the lack of discussion about actual funding. He remained in the laboratory and asked my colleagues and me to conduct various tests. When we demonstrated to him that our detector happens to be almost thousand times more sensitive than any commercial UV flame sensor and capable of detecting small flames even in direct sunlight, Georges blurted out some of the most colorful Russian expletives I have ever heard. I do not know where he learned these nasty words, but he continued to bellow them in rapid succession.

To his credit, Georges always believed that "tomorrow money will come;" despite the chronic lack of funds that constrained our research.

In 2009-2010, as a result of the absence of funding, we were forced to construct a cheap flame detector built with plastic boxes and filled with a mixture of Ne, EF vapors and air. This approach worked surprisingly well. With that in mind, Georges' dream of developing even simpler detectors operating in air saturated with EF vapors became a reality. We completed a successful prototype after hard work. It proved to be ten times more sensitive than the best commercial UV flame detectors, despite its simple and cheap design. Georges was elated; but he's health was declining at this time. He lost his mobility and could no longer travel to St Etienne or CERN. We stayed in contact through phone conversions and e-mail messages. During one of these phone conversations, Georges, who was by now strained and weak, said, "It was my dream throughout all my life to develop gaseous detectors operating in air. Now, the dream is materialized and I can die quietly". He said it in jest, but it truly touched my heart.

The flame imaging detector developed by the team lead by Charpak (Charpak, 2007, Bidault, 2007) is presented in Figure 30. It consists of a gas chamber filled with a mixture of Ar with EF or TMAE vapor at a total pressure of 1 atmosphere and combined with a lens (focal length F= 50 cm). Two RETGEMs operating in cascade mode, and a readout plate are installed inside the gas chamber. The RETGEMs were chosen because they are robust detectors withstanding sparks and capable to operate under poor gas quality conditions, thus leading to a competitive production cost. The RETGEMs have an active area of 4x4 or 10x10 cm^2 and a thickness of0.4 mm. The diameter of the holes is 0.3 mm, the pitch0.7 mm.

The schematic drawing of the readout plate is shown in Figure 30b. It was made of a printed circuit board with a fan-out of 10–30 Cu strips converging at one point, and with the center of the lens placed 50 cm away from the middle of the detector plane.

Each strip is individually read out. The image of any UV emitting object placed at a distance from the lens much larger than F is formed near the focal plane of the lens. If F>>d, where d is the size of the RETGEM used (typically d =10 cm), the UV photons from the flame are projected along one or two strips and cause the ionization of the photosensitive vapors along their passage.

The photoelectrons created in the drift region trigger Townsend avalanches in the holes of the RET-GEM, which in turn induce a signal on the readout strips. The typical overall gain at which the double RETGEMs operated was 10^5. Due to the fact that the photosensitive vapors employed have zero efficiency for λ>230nm it was possible to operate such a detector with a small background, even in presence of direct sunlight, an important improvement for practical applications.

By measuring the counting rate from strips, one can obtain a 1-D digital image of the flame. As an example, Figures 31a and b show the image of a flame of approximately 5x5x5 cm^3, placed 70 m away from the detector, obtained in open air on a sunny day. Note that for human eyes, such a small flame under the same light conditions is practically invisible.

As can be concluded from comparison of Figures 31a and b, the sensitivity of the detector filled with TMAE vapors was only a few times higher than with EF vapors. This was because the TMAE vapors were introduced at pressure much less than the saturated one to ensure more stability at temperatures below the 20°C.

Note that the EF vapor remain stable even in air and this allows manufacturing cheap UV sensitive flame detectors filled with air (Charpak, 2007).The drawback of this design is that the detector efficiency to flame radiation remains constant only in the ambient temperature interval 10–60 °C. At lower temperatures it is sharply dropping. This imposes certain conditions for exploiting these detectors in outside buildings.

5.2.2 Towards the Forest Fire Detection

Wild fires, for example bush fires or forest fires mainly occur in hot weather, so gaseous detectors filled with photosensitive vapors still remain attractive for these applications.

The measurements with the S-RETGEM shown in Figure 31 were performed at a distance of 70 m and without discernable UV flame absorption. But quite importantly, the S-RETGEM must demonstrate its ability to record flames at greater distances in order to present a realistic application for forest fire detection. Relevant tests were undertaken simultaneously with three detectors: RETGEM, TMAE-filled single-wire counter and Hamamatsu UVtron R2868 (Charpak, 2007).

Figure 32 shows the results obtained with a wooden fire (an approximate size 1.5 x 1.5 x 1.5 m^3) on a sunny day and recorded the first set of measurements at R_d= 50 m (a reference point) without an expec-

Figure 30. Schematic drawing of the RETGEM for visualization of flames: a) side view, b) top view of the readout plate; for simplicity, only 5 readout strips are shown. As can be seen, in the case of F>>d, the image of a small flame placed at a distance much larger than F will be projected along one or two strips. A 1-D digital image of the flame can be obtained by counting pulses from these strips.

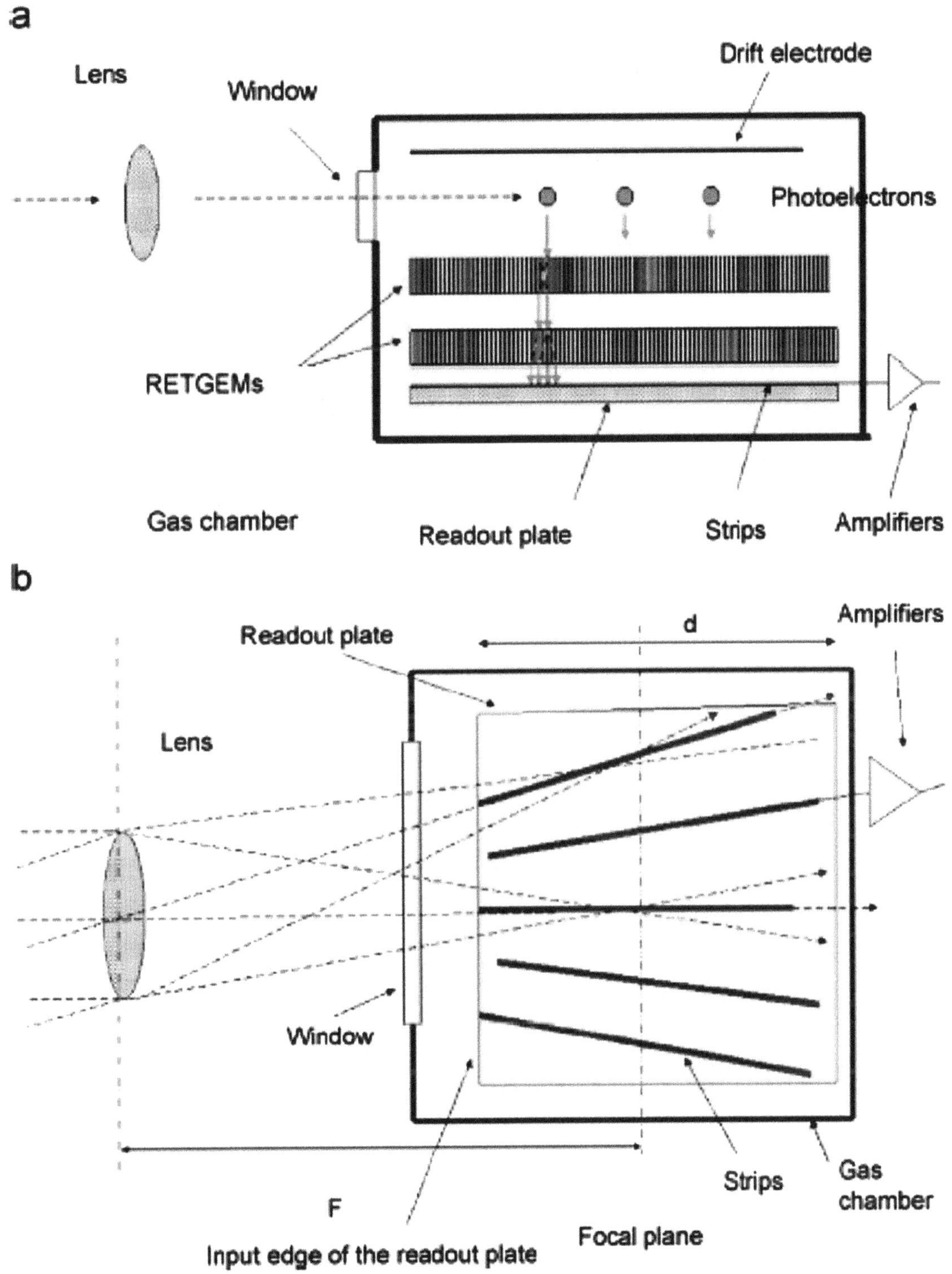

Figure 31. (a) A digital image (number of counts vs. the channel/strip number) of an alcohol flame of ~5x5x5 cm^3 located 70 m from the detector outside the building and exposed to direct sunlight; the detector was filled with a mixture of Ar + EF vapor at a total pressure of 1 atmosphere. The size of readout plate was 10x10 cm^2, and the pitch of the strips at the "input" edge of the readout plate (an edge close to the window) was 0.75 mm. (b) The digital image of the flame obtained under almost the same conditions, with a detector filled with Ar + TMAE vapors at a total pressure of 1 atmosphere

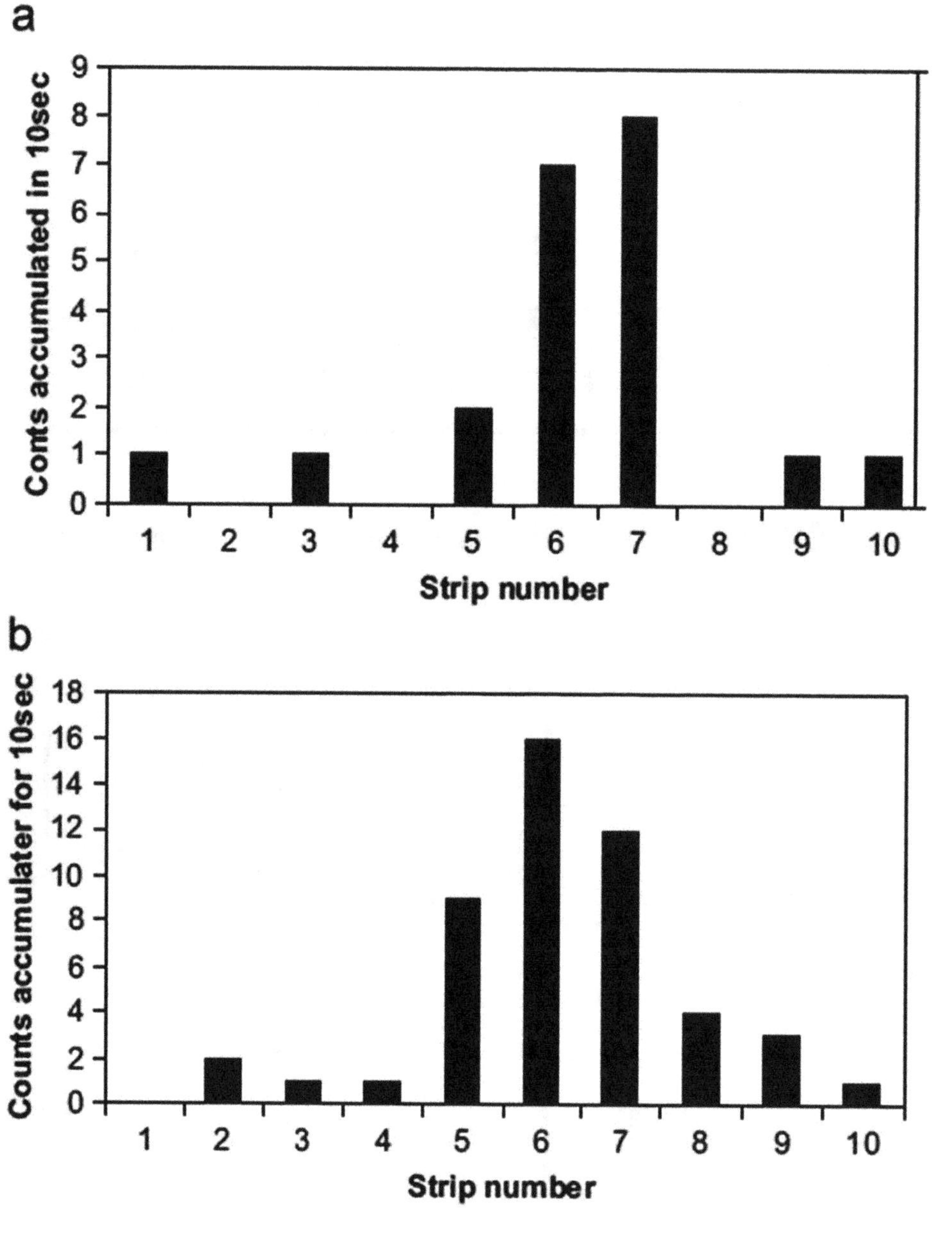

Figure 32. Counting rate vs. distance measured with three detectors: S-RETGEM (filled triangles), single wire counter filled with TMAE vapors (filled circles) and Hamamatsu R2868 (filled square) at 27°C from a flame 1,5 x 1,5 x 1,5 m³ in a sunny day. Open symbols indicate the calculated counting rate vs. distance with the assumption that it obeys $1=R_d^{-2}$ law.

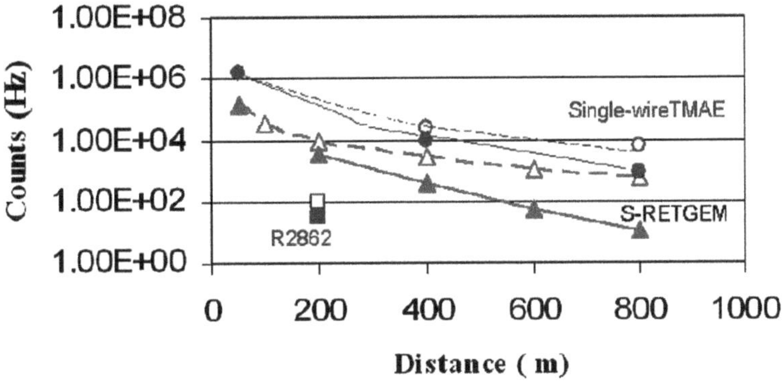

tation of strong light absorption. Then the distance was gradually increased up to 800 m (the maximum size of the area available for these measurements).

Open symbols in this figure indicate the calculated counting rate vs. distance with the assumption that it obeys the $1/R_d^2$ law. It is evident that both the S-RETGEM and the TMAE-filled counter efficiently detect the flame at a distance up to 800 m.

The counting rate deviates from the $1/R_d^2$ dependence. This allows for a rough estimate of the air absorption happening at great distances. Numerical calculations performed in work (Charpak, 2007) indicate that the detector response function is sharply peaked at λ_{EF}~195 nm for EF and λ_{TMAE} ~ 215 nm for TMAE. This is because the air sharply cuts the radiation in the interval $\lambda < 185$ nm for short distances and $\lambda < 220$ nm at longer distances, whereas the QE(λ) is zero at $\lambda > 200$ nm for EF and $\lambda > 220$ nm for TMAE and the flame emission steadily increases with the λ.

From the analysis of the measured at $N_{EF}(R_d)$ and $N_{TMAE}(R_d)$ the following values of the "characteristic" absorption coefficients were determined: 1/250m and 1/400 m for 215 nm an195 nm respectively. Applying these absorption coefficients, one can estimate that a 5x5x5 m³ fire can be detected with EF-filled S-RETGEM at a distance of 1.5 km if the counts are accumulated during 20 sec. By increasing the recording time (the interval between when the fire erupts and when the alarm sounds) to 100 sec the distance may be increased even further.

Note that at temperatures below 10°C gaseous detectors with CsI photocathodes can be used as well, since their sensitivity remained unchanged till cryogen temperatures. However, they should be combined with filters suppressing the sunlight radiation, which obviously increases the cost.

Both types of RETGEMs-based detectors (filled with photosensitive vapors or with CsI photocathodes combined with filters) can be installed on towers in risk areas including bushes and open areas and optimally implemented in combination with other types of flame detectors including CCD (Charged Coupled Devices), infrared (IR) sensors and others. The most advantageous application involves joint operation with IR detectors in small patrol airplanes since the transmission of the atmosphere for UV rapidly increases with the altitude enabling S-RETGEM detectors to identify small fires at a distance of a few kilometers.

6. PHOTOSENSITIVE GASEOUS DETECTORS FOR THE READOUT OF SCINTILLATORS

Another field of application for position-sensitive gaseous photomultipliers is readout of scintillators emitting light in the UV spectrum- The scintillators can be gaseous, solid and liquid. This technique allows to determine simultaneously the location where an event occurred in the scintillator (interaction with a high energy photon, charge particles, recoils etc.) and the energy deposited by the particle.

6.1 Gaseous Scintillators

The concept of the gas scintillation detector readout by the photosensitive MWPC was proposed in work (Policarpo, 1978), as shown in Figure 33.

Policarpo suggested a parallel plate avalanche chamber with electrodes made of metallic meshes, flushed with a pure noble gas, for example Xe. When an X-ray photon is absorbed in the drift/absorption region, it creates primary electrons which drift towards the parallel plate avalanche chamber and produces a burst of the scintillation light in the gap between its electrodes. The voltage between the electrodes is adjusted such that the electrons cannot reach sufficient kinetic energy to ionize the gas but only excite the molecules, emitting UV photons when they recombine, as shown in Figure 34. Policarpo proposed to detect this radiation with a MWPC filled with photosensitive vapors.

This approach offers excellent energy resolution (~8-10% FWHM for 5.9 keV X-rays) limited only by the statistics in in the number of primary electrons. At the same time the position where the scintillation light is emitted can be determined with a high accuracy, between 0.3 and 6 mm.

The proposed detector was almost immediately materialized in several parallel experimental works (Peskov, 1980; Charpak, 1980; Anderson, 1980) and later by many other authors. The main application of these detectors was astrophysics, where they were used in numerous flight experiments. Some prototypes were also built for medical applications.

Figure 33. A photo-ionization proportional scintillation detector; the dashed lines represent crossed wire meshes, having wires 50 μm in diameter and 450 μm apart. Grid 4 is laminated. The larger dots represent cathode wires (diameter of 100 μm) and the smaller dots anode wires (50 μm in diameter). The electric field in the scintillation region is high enough for a primary electron to gain enough energy between collisions to excite but not ionize the xenon atoms, producing VUV scintillation light as a result of the gas de-excitation processes. The scintillation light from the parallel plate avalanche chamber is detected by a photosensitive MWPC.
From Charpak, 1980.

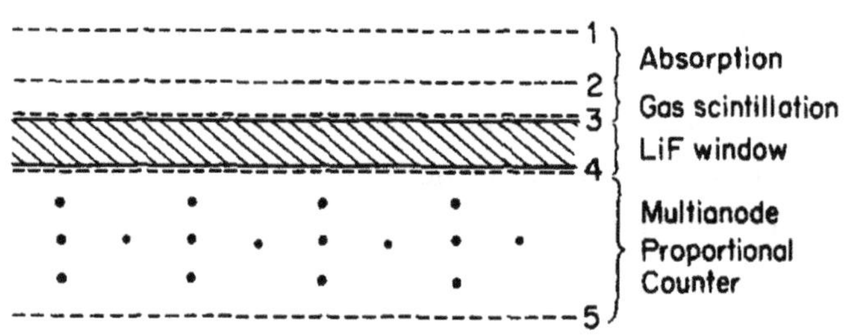

Figure 34. Spectral distribution of the scintillation light emission of pure noble gases, in moderate electric fields, together with the ionization potentials of some vapours
From Charpak, 1980.

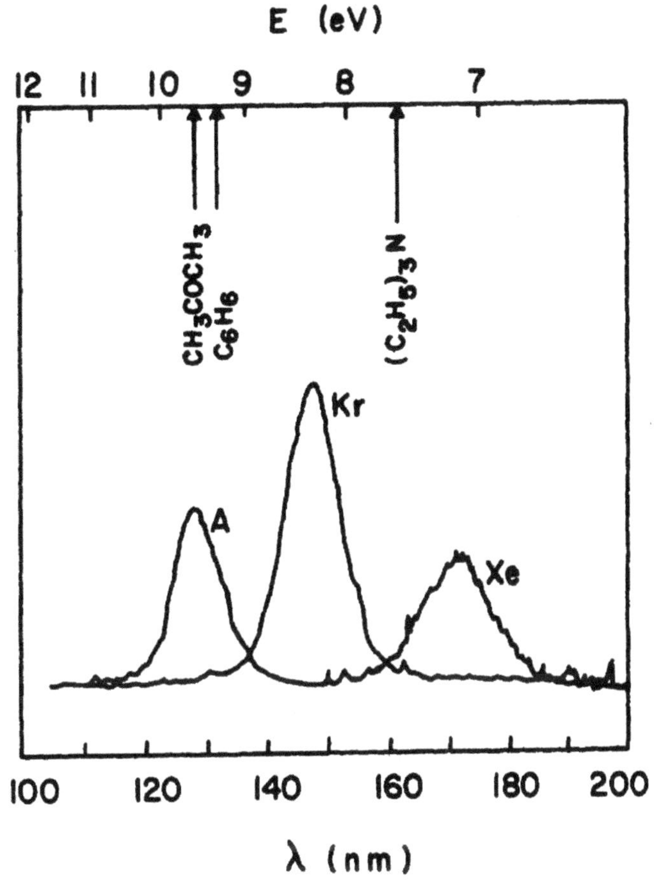

Starting from work (Dangendorf, 1990) CsI-MWPC began to be used for the readout of the gas scintillation chambers instead of MWPC filled with photosensitive vapors, see for example (Brauning, 1994; Rodionov, 2002).

There were also attempts to implement other multiplication structures combined with CsI photocathodes, e.g. CsI-TGEM (Periale, 2002; Periale, 2003), CsI- GEM (Lopes, 1999) or microstrip gas counters (Coelho, 2006).

The principle of operation CsI-GEM and CsI-TGEM was already described in previous Chapters. For this reason we will describe here in more details only the work (Coelho, 2006) in which the microstrip gas detector was used.

The schematic of this device is shown in Figure 35. An electron cloud is formed from an X-ray interaction in the absorption region. This electron cloud drifts under the influence of a weak electric field towards the scintillation region.

The VUV scintillation photons illuminate the CsI-film deposited on the surface of the microstrip gas counter. It is made of alternating anode and cathode metallic strips manufactured on a glass plate. When a voltage is applied between these strips, an electric field schematically shown in Figure 35 is

Figure 35. The operation principle of the gas scintillation chamber, readout by a CsI-coated microstrip counter; the VUV scintillation photons, incident on the CsI-film deposited on the MSP (micro strip plate) surface, induce the emission of photoelectrons from the active areas, the cathode strips. The photoelectrons undergo charge avalanche multiplication in the intense electric field around the anodes, being the resulting charge signal proportional to the energy of the absorbed X-ray.
From Coelho, 2006.

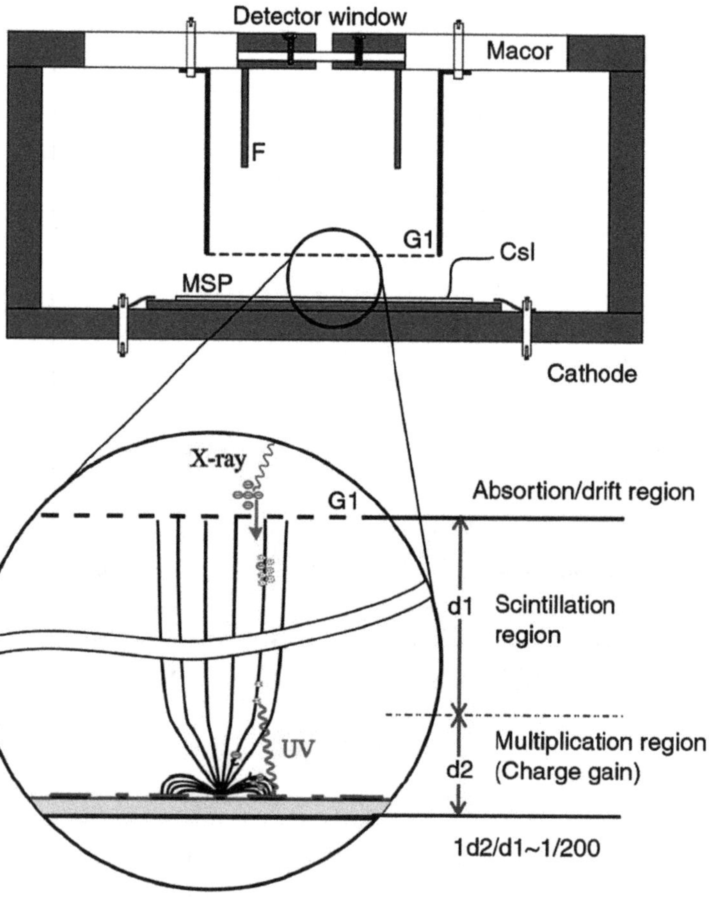

formed around each anode-cathode region. The photoelectrons experience avalanche multiplication in the intense electric field around the anodes. The resulting charge signal is proportional to the energy of the absorbed X-ray.

The CsI-coated microstrip plate serves simultaneously as the collection electrode for the primary electron cloud, as well as the photo-sensor for the VUV scintillation and as the amplification stage for the photoelectrons.

Nowadays most applications use solid state detectors as they offer both better energy and spatial resolutions.

Figure 36. Schematics of the test detector used in Anderson, 1993; it consisted from a BaF_2 scintillator coupled to a MWPC flushed with a gas mixture containing TMAE vapors.

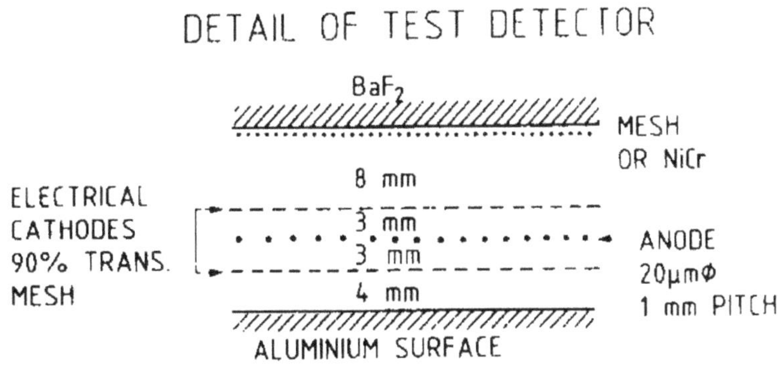

6.2 Solid Scintillators

Probably, the first attempt to record UV scintillation light from a solid scintillator was made in work (Anderson, 1993). In this work a BaF_2 crystal was coupled to the MWPC containing TMAE vapor, as shown in Figure 36. This detector was similar to one shown in Figure 33. The main difference was that the gaseous scintillator was replaced with the solid one.

The BaF_2 emission spectra have three components: one at 300 nm with a decay constant of 620 ns (a so-called slow component), and two peaks at 225 and 195 nm, respectively, with fast decay constants of about 0,6 ns (fast components). Sensistivity curves ofMWPCs, containingphotosensitive vapors such as TMAE or EF, quite well overlap with the emission spectra of the fast components, and they are not sensitive to the slow component. This opens the possibility of building fast detectors based on only the fast component of this scintillator.

A BaF_2 crystal has relatively high density, 4.9 g/cm^3, a radiation length of 2.1 cm and thus it is attractive for applications in calorimetry or in positron emission tomography (PET). A prototype of a BaF_2 electromagnetic calorimeter was built and extensively tested in work (Bouclier, 1998). It is made of fourteen cylindrical BaF_2discs, 12.6 cm diameter and 1, 2 .5 or 5 cm thick, shown in Figure 37.

The total BaF_2 depth was 40.5 cm, which represents 19.3 radiation lengths. Each crystal is preceded by a low pressure MWPC, which detects the BaF_2 scintillation produced by the development of the electromagnetic shower. Thanks to their low density, the MWPCs do not interfere with the shower development in the detector. Hence, the detector behaves as a homogeneous calorimeter, despite its segmentation.

The longitudinal segmentation was chosen to visualize the shower profile. Tests of this prototype performed in a 1-9 GeV/c beam gave the following results: an energy resolution better than $3.9\%/\sqrt{E_{in}}$ (where E_{in} is incoming energy in GeV), a position resolution of ~1 mm, and a time resolution better than 1 ns.

Several groups tried to implement a system consisting of BaF_2 scintillators coupled to TMAE-filled MWPCs in PET (position emission tomography) devices. Below only two examples will be given. In work (Bruyndonckx, 1996), was built and characterized a prototype of a small PET scanner optimized for experimental studies with small laboratory animals. Its main feature was the use of a large number of small crystals (2958 BaF_2 scintillators, 3 x 3 x20 mm^3 each) readout with one wire chamber, thus

Figure 37. Schematics of the BaF_2 calorimeter prototype consisting from fourteen cylindrical BaF_2 crystals with a TMAE filled MWPCs between them; a simulated 9 GeV electromagnetic shower in the calorimeter, displaying electron and positron tracks is also shown.
From Bouclier, 1998.

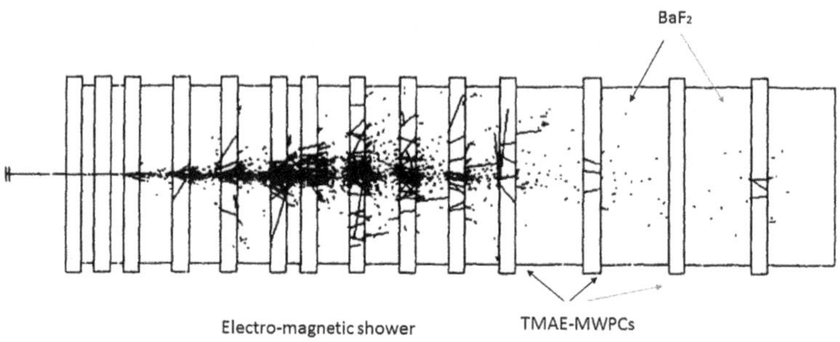

allowing for an improved resolution or an increased sensitive detector area. This approach offers an interesting and potentially low cost alternative to instruments using photomultiplier tubes.

In work (Ott, 2003) two large-area detectors (60x40 cm^2 each, mounted on a rotating gantry) which form part of a new type of whole body positron scanner, called PETRA, was investigated. In Figure 38 is shown a photograph of a clinic prototype of this PET camera.

Special efforts were done to suppress, by several methods, the scattered photons spoiling the PET performance. The scatteringcomes from photons scattering in the patient and the detector support structure.

Figure 38. Photograph of PETRA camera; in this design shielding to minimize counts in the detectors from activity outside the camera field-of-view was implemented.
From Ott, 2003.

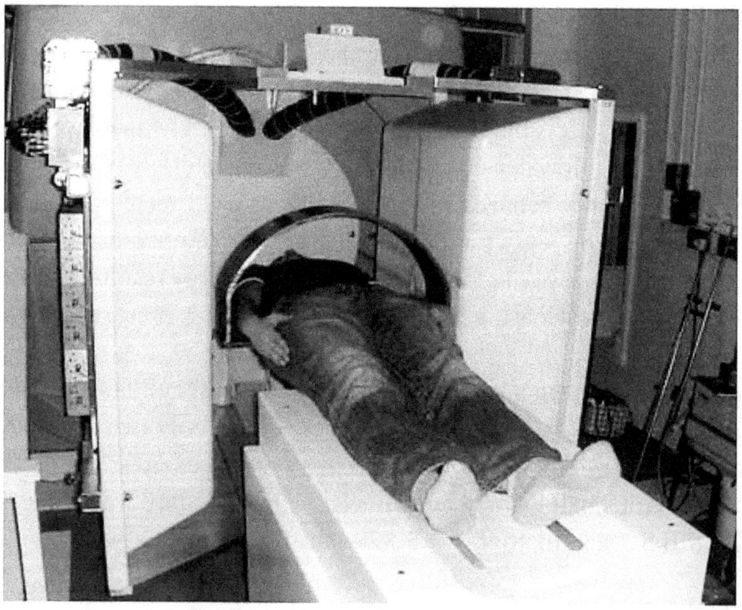

In spite of a lot of efforts from various groups, due to the complexity and strong completion from other PET designs, the works on BaF_2-TMAE PETs are practically abounded now.

However, the introduction of position-sensitive photomultipliers with CsI photocathodes, triggered new attempts d to develop practical devices based on solid UV crystal scintillators.

Studies in this direction were initiated by works (Charpak, 1991; Charpak, 1993) where parallel plate avalanche chambers with CsI photocathodes were used for the readout of BaF_2 crystals. This combination allowed to fully exploiting the fast scintillation properties of BaF_2 since the time resolution of the parallel plate avalanche chamber is much better than for a MWPC, 1 instead of the MWPC's ~ 500 ns. In the same work (Charpak, 1993) was also studied a $KMgF_3$scintillator, which emits at even shorter wavelengths than BaF_2 (between 140 and 220 nm with a decay time of about 2 ns).

Additional work on other fast crystal UV scintillators coupled to gaseous detectors with CsI photocathodes were investigated, for example $LiBaF_3$ (Peskov, 1991), LaF_3 doped with $Nd,^+$ (Kononenko, 1997), etc.

One should admit however that beside the laboratory prototypes that all these efforts did not lead yet to any massive applications. The present interest in application of gaseous detectors with solid photocathodes is shifted towards the readout of noble liquid scintillators.

6.3 Noble Liquid Scintillators

Noble liquids such as LAr, LKr and LXe (L stands for liquid) are unique detecting medias:

- Their stopping power is high enough for many applications,
- They are excellent scintillators, emitting in the VUV region of spectra,
- Primary electrons created inside the liquid could easily be drifted for long distance and collected on an electrode structure providing a charge signal,
- If necessary, primary electrons could even be extracted from the liquid to the gas phase (vapours above the liquid layer) and collected on electrodes placed there.

These properties make them attractive for several applications, for example noble liquid scintillating calorimeters (Chen, 1993), cryogenic TPCs-Time projection chamber- (Montanari, 2004), cryogenic PETs (Chepel, 1997). Expensive PMs are used for the detection of the scintillation light in these devices which considerably increases the overall cost of the noble liquid detector (Chepel, 2013). There have been several efforts to replace PMTs by avalanche solid-state detectors. However, the cost of these new devices is still high.

In several works (see for example Peskov, 2007; Di Mauro, 2009; Breskin, 2009; Badertscher, 2011) it was recently demonstrated that costly PMs and solid-state devices could be replaced by gaseous photosensitive detectors, i.e. detectors with a window (which could be immersed inside the noble liquids) or windowless, able to operate in pure noble gases or vapours above the noble liquids. The latter feature was based on the observation, made in work (Bondar, 2004) that GEM detector can operate in pure noble gases and at cryogenic temperature. At the same time the first sealed and windowless photosensitive detectors capable operating at cryogenic temperatures was developed (Periale, 2004).

6.3.1 Detectors with Windows

The sealed detectors developed were wither a single-wire counter shown in Figure 39 or with a CsI coated capillary plates (CP) and GEMs, shown in Figure 40.They were tested inside a special cryostat, which allows cooling to any temperature between room temperature and liquid nitrogen (LN_2) temperature with a selected time profile (see Periale, 2005 for more details).The CsI gaseous photomultipliers were directly immersed inside the noble liquids and were used to detect primary scintillation light produced by radioactive sources.

Other photocathodes were also studied at low temperatures, e.g. Sm, TMAE and TMAE + neopentane (NP) (Periale, 2004).

Figure 41 shows the signal amplitude from the single-wire counter produced by the scintillation light in noble gases (induced by a ^{241}Am source) vs. temperature for the CsI, TMAE, Sm (Samarium), and NP+TMAE (NP stands for neopentane) reflective photocathodes and for a GEM combined with a CsI photocathode.

At temperatures of T>120 K the detectors were filled with an Ar+CH_4gas mixture. In this gas the single-wire counter operated stable at gains of 10^4–$10^5$and the CsI GEM at a gain ~10^3. At T<120K the gas mixture was He+H_2. In this mixture, due to the strong back-diffusion effect, the "effective" quantum efficiency drops by a factor of 5–10 (compared to the Ar+CH_4 mixture), depending on the gas density. This in turn causes a drop in the signal amplitude to take place. This effect is clearly seen from Figure 41.Note that for a stable operation at T<100 K, the inner part of the detector and the gas should be very clean, otherwise impurities (especially water) can condensate on the photocathode's surface reducing its quantum efficiency and also giving rise to charge up effects. This particularly concerns the CsI photocathode. It should not be exposed to air and should be well outgassed to minimize charge up

Figure 39. Schematic drawing of a semi-cylindrical (radius ~15 mm) single wire counter with CsI photocathode developed for the detection of scintillation light from noble liquids

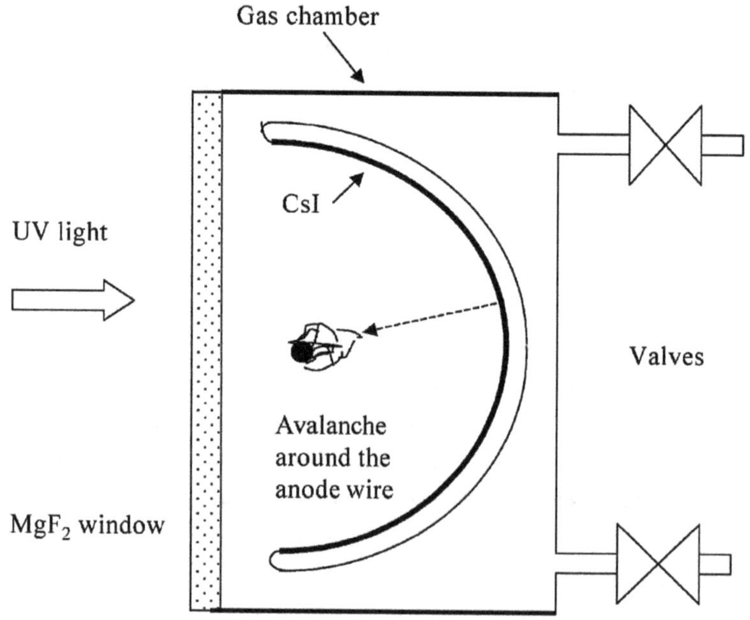

Figure 40. Schematic drawing of a hole-type detector installed inside a sealed gas chamber used for the detection of scintillation light from noble liquids

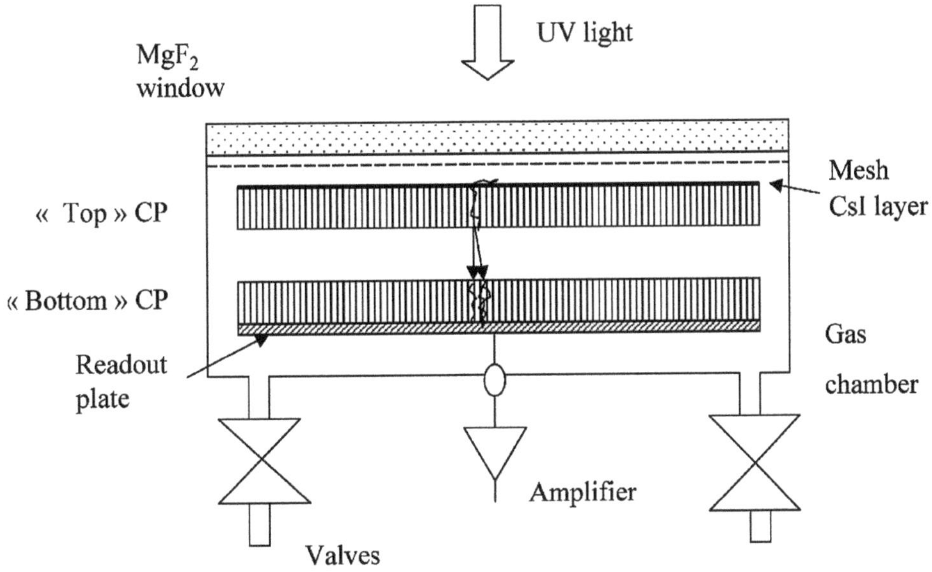

Figure 41. Measured signal amplitude (or amplitude recalculated from experimental data and normalized to the same gain) produced by the [241]Am scintillation light vs. temperature for various photocathodes and detector designs at a gas gain of 100; filled rhombus: a single wire counter (SWC) with CsI photocathode filled with Ar+CH_4 (for T>130 K) and He+H_2 (for T<130 K) and detecting Ar scintillation light, open squares: the same detector filled with Ar+$CH_4$and detecting Xe scintillation light, open triangles: GEMs detecting Ar scintillation light, open circle: SWC with a Sm photocathode detecting Ar scintillation light, crosses: SWC with a TMAE photocathode, dashes: SWC with a TMAE+NP photocathode
From Periale, 2004.

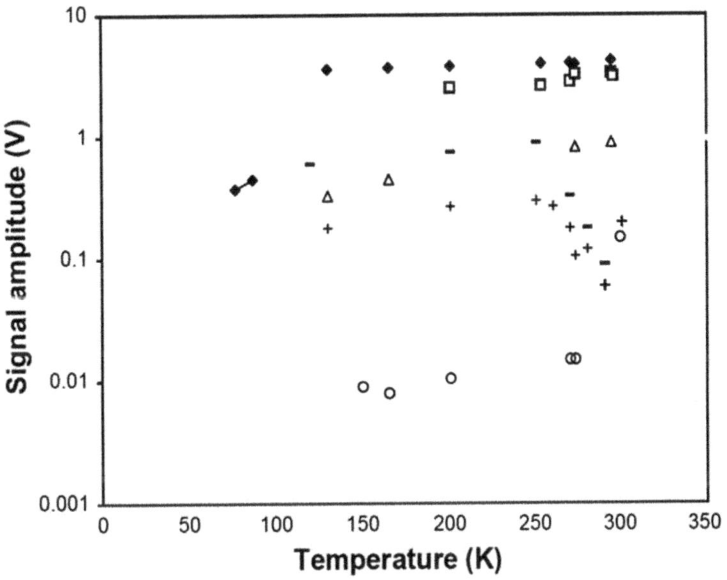

Table 3. Quantum efficiency (%) of various gaseous detectors, measured at room temperature

Detector Type	Photocathode				
	CsI at λ=120-130 nm (Ar Light)	CsI at λ=175 nm (Xe Light)	CsI at λ=165 nm (H_2 Light)	TMAE at λ=120-130 nm (Ar Light)	TMAE+NP at λ=120-130 nm (Ar Light)
Single wire (Ar + CH_4)	19 (1) 37.4 (2)	18.3 (1) 26 (2)	23.5 (3)	0.4 (1)	0.6 (1)
Single wire (He + H_2)	2.3 (1) 3.1 (2)	2(1)	3.2 (3)	0.15 (1)	0.25 (1)
GEM (Ar)	6 (1)		4 (3)		

From Periale, 2004.

Notes in parentheses indicate the method of measurement and calculations: (1) amplitude of signal produced by ^{241}Am in Ar or Xe, when the total number of emitted photons is well known, (2) with respect to the QE of EF, (3) with respect to CFM-3 counter.

effects. The data shown in Figure 41 shows the relative quantum efficiency of various photocathodes in different gases and at different temperatures. The absolute quantum efficiency was measured at room temperature, and is presented in Table 3.

As can be concluded from the data presented in Figure 41 and Table 3, the quantum efficiency of a CsI photocathode remains high down to at least 120 K. It has the potential to become an alternative to vacuum PMTs. The lower quantum efficiency observed with GEM can be related to the poor electron collection into the GEM holes since the first GEM is operated at a relatively low gas gain.

Tests with GEMs and capillary plates reveal that both can operate stable down to T~ 80K in He+10% H_2. For example, the maximum achievable gain of a bare capillary plate was 10^4at all temperatures from 300 to 80 K. The maximum achievable gain of CsI coated capillary plates were depending on temperature (it varied between 70 and 200). Probably, one can achieve a relatively high efficiency even for T < 120 K by operating the CsI coated capillary plates in a gas mixture with high concentration of H_2.

Further investigations revealed that TGEM (Perivale, 2001; Periale, 2002) and RETGEM (Bidault, 2006, Oliveira, 2007) are also suitable for cryogenic application (Periale, 2007a; Peskov, 2007 s). These detectors are easy to manufacture and they are capable of operating stable and at high gas gains at low temperatures.

In Figure 42 a setup used for systematic studies of these detectors at various temperatures is shown.

It consisted of a gas chamber with a CaF_2 window inside which a TGEM or a RETGEM was installed. It also included a monochromator combined with a Hg lamp, and a gas system allowing pumping the chamber and filling or flushing it with various gases, e.g. Ne, Ar or He or their mixtures with various quenchers (H_2, CO_2 or CH_4) atapressureof1atmosphere.The cathodes of the detectors were coated with CsI.

CsI-coated TGEM/RETGEM show high gains not only in mixtures with quenched gases, but also in pure noble gases (a feature which was later exploited in windowless photodetectors). As an example, in Figure 43 the gain curves measured in Ne and Ar at various temperatures are presented. The maximum achievable gains in these gases are approaching those obtained with a bare triple GEM (Bondar, 2004).

Figure 44 presents the results of the stability monitoring at different temperatures, when the detector was periodically exposed to low intensity radiation (~10 Hz/cm^2). It is evident that the RETGEM does not exhibit any strong charging up effect. Note that the RETGEMs used in these measurements had a modified design. They had strips instead of conventional unsegmented electrodes, so the avalanche

Figure 42. Schematic drawing the sealed gas chamber housing a CsI-RETGEM

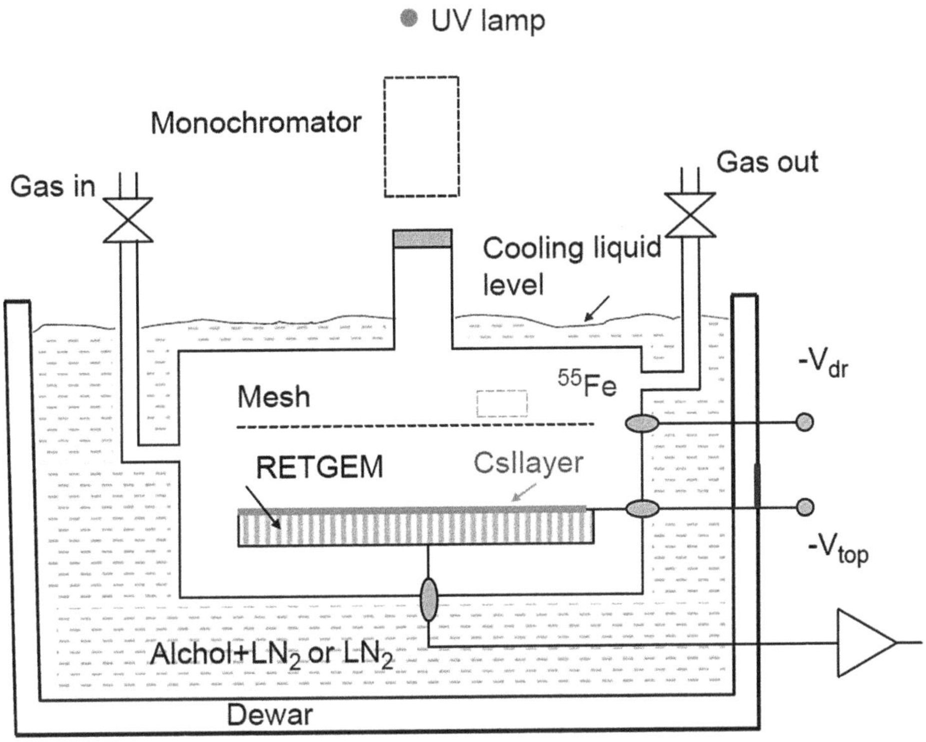

Figure 43. Gains versus voltage curves for single RETGEM measured in Ne at 300 and 78 K, and in Ar at 300 K, and 165 K

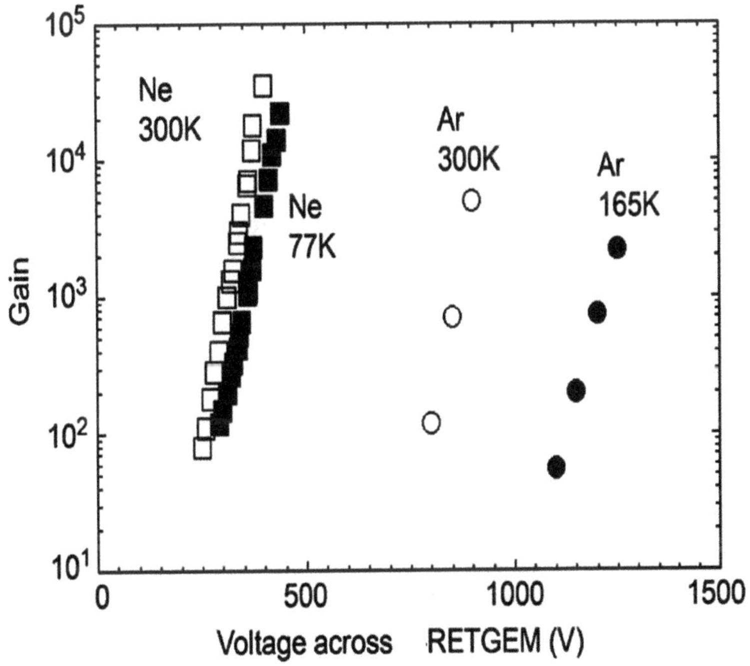

Figure 44. Gain versus time measured with RETGEM at various temperatures in Ne and Ar

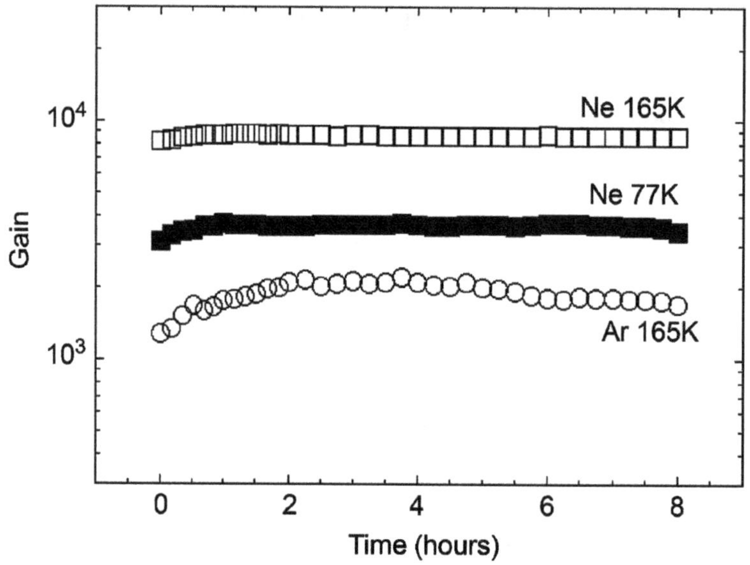

charge was collected near the hole where the avalanche happened which improved their performance (DiMauro, 2009).

The results of further systematic studies of a sealed CsI-TGEM photomultiplier operation at cryogenic temperatures can be found in (Breskin, 2011). The preliminary conclusion from these studies is that gaseous photodetector may challenge vacuum PMTs. They have a comparable sensitivity to scintillation light from noble liquids, they can be manufactured from materials having low radioactive background and they are practically insensitive to magnetic fields which can be an important advantage in some experiments.

The research in this direction still continues by several groups. For example, an interesting cascaded cryogenic detect was proposed and investigated in work (Duval, 2011).

6.3.2 Windowless Detectors

The first tests of windowless cryogenic photodetectors, e.g. GEM and capillary plates coated with CsI layers, were described in (Periale, 2004). Unfortunately, in this mode of operation the CsI cathode efficiency (per unit of area) is low compared to the detectors with windows filled with quenched gases due to the strong back diffusion of photoelectrons in pure noble gases. The sensitive area of windowless detectors could be rather large and thus the overall efficiency could be reasonably high.

The excellent characteristics of windowless TGEMs and RETGEMs made them especially interesting to study (see for example Periale 2006, Peskov,2007).The challenge was to reach stable operation in the gas phase above noble liquid, where quite high temperature gradient is possible. The setup used for these studies is shown in Figure 45 (Peskov, 2007).

It consisted of a cryostat with a test chamber placed inside it. The temperature in the cryostat was controlled by a computer and could be changed from 300 to 77 K. Depending on the measurements, either a single or a double TGEM/RETGEM (1 mm or 1.5 mm thick), with the top electrode coated with a CsI

Figure 45. The setup used for measurements with windowless TGEM/RETGEMs at cryogenic temperatures

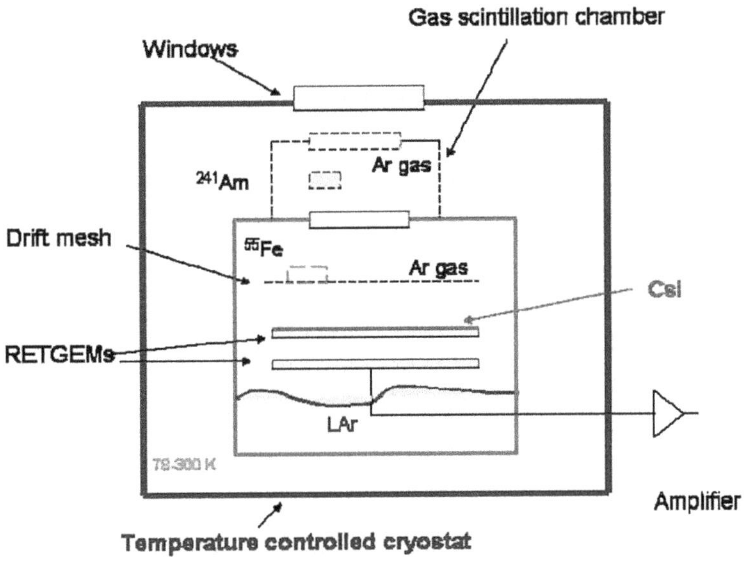

layer, was installed inside the chamber as well as a radioactive source ^{55}Fe for gas gain measurements. In some experiments, an additional scintillation chamber was attached to the test chamber and flushed by Ar at a pressure of 1 atmosphere. It contained an ^{241}Am alpha source. An important feature of this setup was the possibility to condense liquid argon (LAr) in a controlled way, which allowed tests to be carried out at different levels of noble liquid. Results of gain measurements in this condition are shown in Figure 46.

One can see that compared to the case where the RETGEM operated in Ar at 100 K, the operating voltage of the RETGEM placed 1–2 cm above the LAr level was higher. This could either be due to the higher gas density around the RETGEM or due to a thin layer of LAr formed on the surface of the RETGEM. Beside this difference the detector operated quite stable.

Figure 46. Gain vs. voltage for a single (filled symbols) and for a double (open symbols) RETGEM (1 mm thick) as measured in Ar at room temperature and pressure p = 1 atmosphere and at 100 K, p = 1 atmosphere (rhombus) and finally in the case when the RETGEM was 1–2 cm above LAr level at T ~ 89 K, p = 1.1 atmospheres (circles)
From Peskov, 2007.

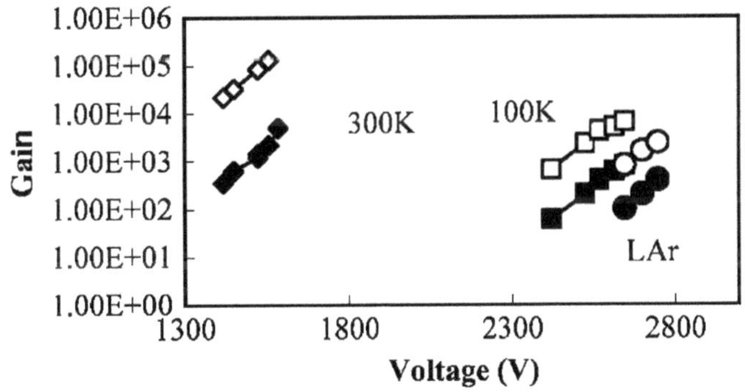

As in the case of sealed cryogenic detectors the quantum efficiency (Q_{pr}) of the windowless CsI-TGEM/RETGEMs at various temperatures was estimated from the amplitude of the signal S_{sc} (in electrons) from the detectors recording the scintillation light produced by alpha particles. It is based on the following relation:

$$S_{cs} = AN_{ph}\Omega Q_{pr}, \tag{6}$$

where A is a gas gain, N_{ph} is the number of UV photons emitted by the alpha source, and Ω is the solid angle at which the scintillation light reaches the CsI cathode. The value of S_{sc} (in electrons) was calculated from the measured amplitude of the signal and the known response of the amplifier on the given injected charge to its input circuit. The number of photons produced by the alpha particles was calculated from the following expression:

$$N_{ph} = E_{\alpha}/W_{ph}, \tag{7}$$

where E_{α} is the energy of alpha particles and W_{ph} is the energy required to produce a UV photon. The value of W_{ph} (W_{ph}~50 eV) was taken from the experimental work (Saito, 2002). One should note that when the value of W_{ph} is well known this method of the detector's quantum efficiency calibration is rather precise and was successfully used by several authors for the calibration of avalanche photodiodes (see e.g. Chandrasekharan, 2006 for more details).

Assuming that W_{ph} is independent on the temperature, the calculated Q_{pr} was then 28% and 17% at room temperature and 100 K, respectively. These results demonstrate that TGEM/RETGEMs could be an attractive alternative to PMTs or any other type of photodetectors for noble liquid TPCs.

This conclusion was later confirmed by the Novosibirsk group (Bondar, 2007). Figure 47 shows the setup used by this team. It was a CsI coated triple GEM, mounted in a cryogenic chamber. The detector was operated in Ar in a dual-phase mode close to the triple point, at a temperature of 84 K and a vapour pressure of 0.70 atmospheres. At this point, the liquid layer thickness was 10 mm (one more GEM foil was placed in the middle of the liquid layer and acted as a cathode). The authors successfully observed both the primary scintillation light and the ionization signals, produced by beta-particles in liquid. The amplitude of the scintillation signal is estimated to be about 30 photoelectrons per 600 keV of deposited energy.

This group also experimented with TGEM and confirmed that this is a very robust detector capable of operating stably at cryogenic temperatures up to 78 K. However, they failed to achieve a stable operation with RETGEM presumably due to the LAr condensation within the holes. Indeed, under the warming-up conditions, when LAr evaporated from the holes, the detector begun to operate with some gas gain. Note that this was an old version of the screen-printed RETGEM which indeed suffered from the charging up effect at temperatures below 100K (Di Mauro, 2007). This drawback was overcome with a new design described in (Di Mauro, 2007).

The studies described above identify potentials of gaseous photomultipliers (with and without windows) in cryogenic applications. The most important among them are: low cost, a large sensitive area, the possibility of choosing construction materials with low radioactivity levels and the practical insensitivity to the magnetic fields. Due to these properties gaseous detectors are under consideration as an option for XENON (Aprile, 2001) and ZEPLIN – (ZonEd Proportional scintillation in LIquid Noble gases) – see (http://www.shef.ac.uk/physics/research/pppa/research/ukdmc/ZEPLIN.html) WIMPs) weakly interacting massive particles) detectors and cryogenic PETs (Grignon, 2005).

Figure 47. Schematic view of the experimental setup to study the performance of the triple CsI-GEM in a dual phase LAr detector
From Bondar, 2007.

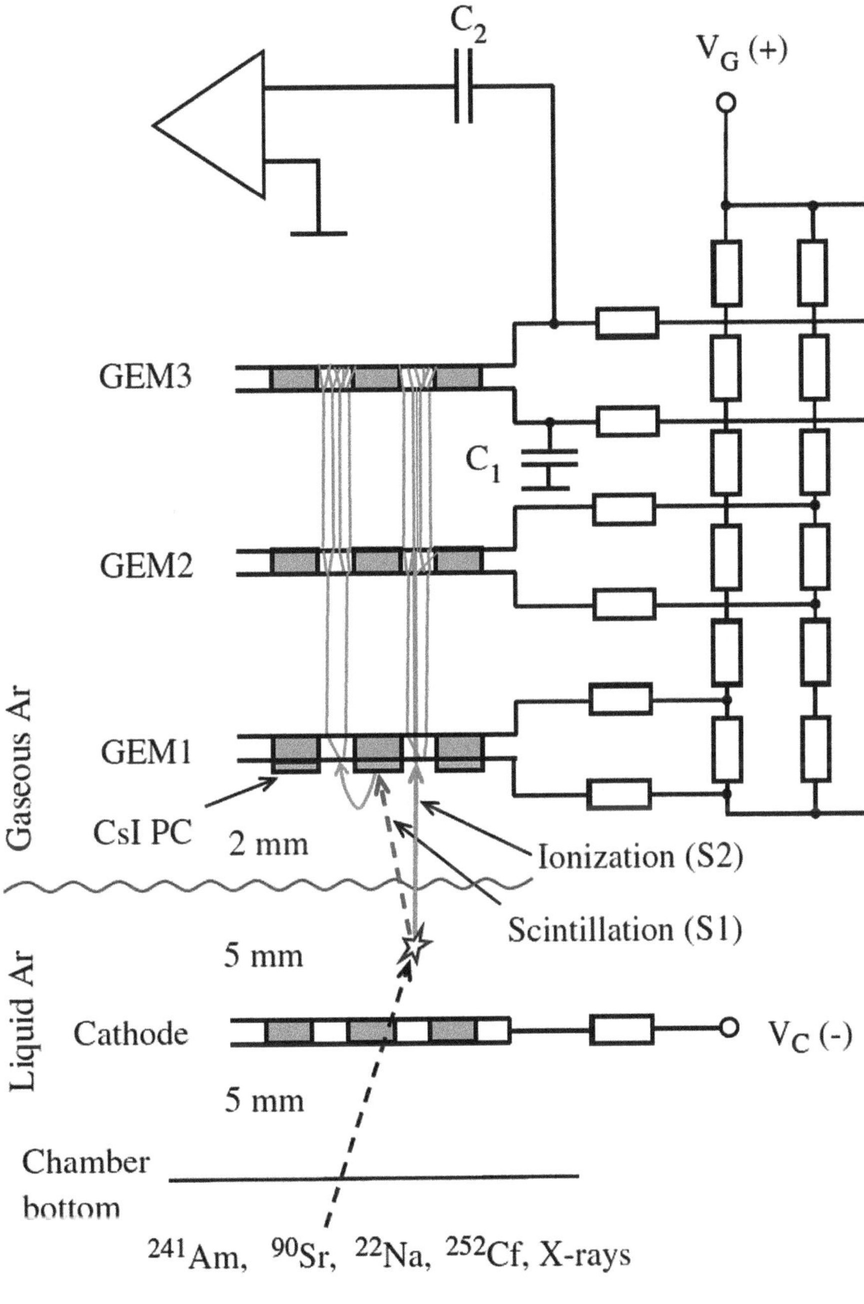

Figure 48. Schematic drawing illustrating a system for simultaneous detection of smoke and flames; the gaseous photomultiplier, shown in the middle, continuously records periodical signals from the UV sources. In case smoke appears these signals will be attenuated (see Figure 49) and a warning signal will be issued.

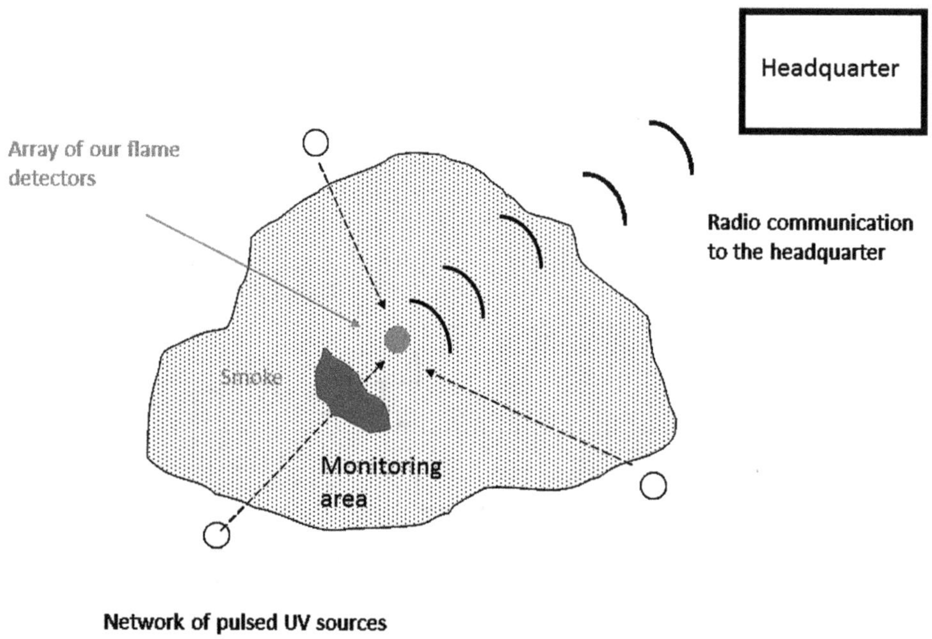

Figure 49. The signals from the periodical external UV/VUV source and the pulses from the flame which can be detected in the time intervals between the big pulses

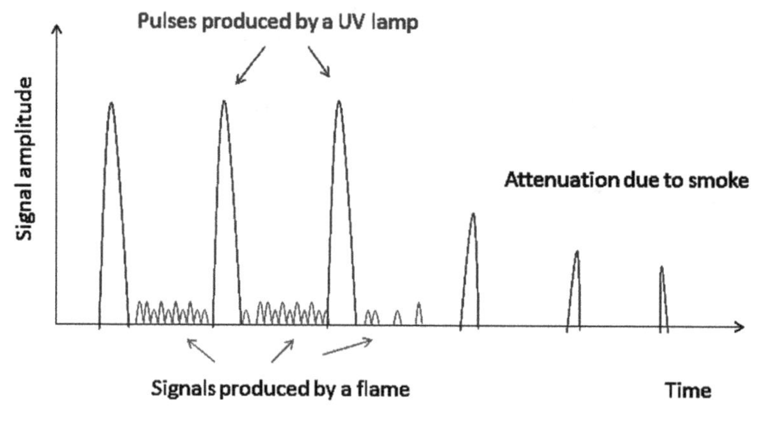

7. CONCLUSIVE REMARKS

In this chapter the main applications of gaseous photomultipliers beyond RICH detectors were described: spectroscopy, plasma diagnostic, astrophysics, flame detection, readout gaseous and solid scintillators, and cryogenic detectors.

More developments are in progress. For example, gaseous photomultipliers, in combination with pulsed UV/VUV source scan be used for simultaneous detection of smoke (Peskov, 2007) and dangerous gases (Carlson, 2003). The principle can be understood from Figures 48 and 49. The external pulsed sources produce quite large signal amplitudes in the photosensitive detector. If smoke appears on the way of any of these light beams, the signal will be attenuated, as shown in Figure 49, and this will trigger an alarm. In between these periodical pulses the detector is capable to detect a flame or a spark (if it appears).

In conclusion let us note that the photosensitive substances (e.g. TEA and TMAE vapors, CsI, etc.) were/are used also in other type of gaseous detectors, e.g. light imaging chambers (see for example Charpak, 1988) and in gaseous multipliers exploiting a secondary electron emission effects (Anderson, 1994; Ceron-Zeballos; 1995; Breskin, 1995; Breskin, 1996; Ceron-Zeballos, 1997; Fonte, 2000; Danielson, 2004).

REFERENCES

Agocs, G., Clark, B., Martinengo, P., Oliveira, R., Peskov, Pietropaolo, F., & Picchi, P. (2008). Developments and the preliminary tests of resistive GEMs manufactured by a screen printing technology. *JINST*, *3*, 1–10.

Anassontzis, E., Resvanis, L., Voulgaris, G., Ypsilantis, T., Tripp, R., Bral, A. R. P., & Milotti, E. et al. (1992). Background-free detector for cosmic ray showers in the atmosphere. *Nuclear Instruments & Methods in Physics Research. Section A, Accelerators, Spectrometers, Detectors and Associated Equipment*, *315*(1-3), 267–271. doi:10.1016/0168-9002(92)90714-F

Anderson, D. F. (1980). A xenon gas scintillation proportional counter coupled to a photoionization detector. *Nuclear Instruments and Methods*, *178*(1), 125–130. doi:10.1016/0029-554X(80)90866-6

Anderson, D. F., Bouclier, R., Charpak, G., Majewski, S., & Kneller, G. (1983). Coupling of a BaF_2 scintillator to a TMAE photocathode and a low pressure wire chamber.D. *Nuclear Instruments and Methods*, *217*(1-2), 217–223. doi:10.1016/0167-5087(83)90137-0

Anderson, D. F., Kwan, S., & Peskov, V. (1994). High counting rate resistive–plate chamber. *Nuclear Instruments & Methods in Physics Research. Section A, Accelerators, Spectrometers, Detectors and Associated Equipment*, *348*(2-3), 324–328. doi:10.1016/0168-9002(94)90755-2

Anderson, D. F., Kwan, S., & Salomon, M. (1994). A low-pressure, microstrip gas chamber operated with secondary-electron emission. *Nuclear Instruments & Methods in Physics Research. Section A, Accelerators, Spectrometers, Detectors and Associated Equipment*, *346*(1-2), 102–106. doi:10.1016/0168-9002(94)90692-0

Aprile, E. (2001). XENON: a Liquid Xe Experiment for Dark Matter. Proposal # 0201740 for National Science Foundation. Columbia University.

Bartoli, B., Peruzzo, L., Sartori, G., Bedeschi, F., Betrolucci, E., & Mariotti, M. (1991). Detection of ultraviolet Cherenkov light from high energy cosmic ray atmospheric showers: A field test. *Nuclear Instruments & Methods in Physics Research. Section A, Accelerators, Spectrometers, Detectors and Associated Equipment*, *302*(3), 515–521. doi:10.1016/0168-9002(91)90369-2

Bidault, J. M., Crotty, I., Di Mauro, A., Martinengo, P., Fonte, P., Galy, F., & Zanette, O. et al. (2007). The first applications of newly developed gaseous detectors with resistive electrodes for UV imaging in daylight conditions. *Nuclear Instruments & Methods in Physics Research. Section A, Accelerators, Spectrometers, Detectors and Associated Equipment*, *580*(2), 1036–1041. doi:10.1016/j.nima.2007.06.061

Bidault, J. M., Fonte, P., Francke, T., Galy, P., Peskov, V., & Rodionov, I. (2006). A novel UV photon detector with resistive electrodes. *Nuclear Physics B - Proceedings Supplements*, *158*, 199–203. doi:10.1016/j.nuclphysbps.2006.07.009

Bogomolov, G.D., Dubrovski, Y., & Peskov, V.D. (1978a). Multiwire Gas Counter for Coordinate Measurements in the VUV Region. *Instr. Exp. Tech*, *21*, 778-782.

Bogomolov, G. D., Dubrovski, Yu. V., Letunov, A. A., & Peskov, V. (1987). Experimental investigation of non-stationary streamer formations in a microwave plasma column at high pressures. *Soviet Physics, JETP*, *66*(2), 295–302.

Bogomolov, G. D., Dybrovski, U. V., & Peskov, V. D. (1978). Position-sensitive gaseous counters for spectral investigations in vacuum ultraviolet. *Prib. Tekhn. Eksp.*, *4*, 210–213.

Bondar, A., Buzulutskov, A., Grebenuk, A., Pavlyuchenko, D., Snopkov, R., & Tikhonov, Y. (2007). First results of the two-phase argon avalanche detector performance with CsI photocathode. *Nuclear Instruments & Methods in Physics Research. Section A, Accelerators, Spectrometers, Detectors and Associated Equipment*, *581*(1-2), 241–245. doi:10.1016/j.nima.2007.07.086

Bondar, A., Buzulutskov, A., Grebenuk, A., Pavlyuchenko, D., Tikhonov, Y., & Breskin, A. (2008). Thick GEM versus thin GEM in two-phase argon avalanche detectors. *JINST*, *3*(07P07001), 1–20. doi:10.1088/1748-0221/3/07/P07001

Bondar, A., Buzulutskov, A. L., Shekhtman, L., Snopkov, R., & Tikhonov, Y. (2004). Cryogenic avalanche detectors based on gas electron multipliers. *Nuclear Instruments & Methods in Physics Research. Section A, Accelerators, Spectrometers, Detectors and Associated Equipment*, *524*(1-3), 130–141. doi:10.1016/j.nima.2004.01.060

Bouclier, R., Charpak, G., Gao, W., Million, G., Mine, P., Paul, S., & Suffert, M. et al. (1988). Test of an electromagnetic calorimeter using BaF_2 scintillators and photosensitive wire chamber between 1 and 9 GeV. *Nuclear Instruments & Methods in Physics Research. Section A, Accelerators, Spectrometers, Detectors and Associated Equipment*, *267*(1), 69–86. doi:10.1016/0168-9002(88)90631-6

Brauning, H., Breskin, A., Chechik, R., Dangendorf, V., Demian, A., Ullmann, K., & Schmidt-Bocking, H. (1994). A large volume 3D imaging gas scintillation counter with CsI-based wire chamber readout. *Nuclear Instruments & Methods in Physics Research. Section A, Accelerators, Spectrometers, Detectors and Associated Equipment*, *348*(2-3), 223–227. doi:10.1016/0168-9002(94)90734-X

Breskin, A. (1995). Secondary emission gaseous detectors: A new class of radiation imaging devices. *Nuclear Physics B*, *44*(1-3), 351–363. doi:10.1016/S0920-5632(95)80054-9

Breskin, A., Chechik, R., & Frumkin, I. (1996). On atmospheric pressure operation of secondary electron emission gaseous X-ray imaging detectors. *Nuclear Instruments & Methods in Physics Research. Section A, Accelerators, Spectrometers, Detectors and Associated Equipment*, *380*(3), 562–567. doi:10.1016/0168-9002(96)00579-7

Breskin, A., Peskov, V., Cortesi, M., Budnik, R., Chechik, R., Duval, S., & Veloso, J. F. C. A. et al. (2011). CsI-THGEM gaseous photomultipliers for RICH and noble-liquid detectors. *Nuclear Instruments & Methods in Physics Research. Section A, Accelerators, Spectrometers, Detectors and Associated Equipment*, *639*(1), 117–120. doi:10.1016/j.nima.2010.10.034

Bruyndonckx, P., Xuan, L., Rajeswaran, S., Smolik, W., Tavernier, S., & Shuping, Z. (1996). Design and physical characteristics of a small animal PET using BaF_2 crystals and a photosensitive wire chamber. *Nuclear Instruments & Methods in Physics Research. Section A, Accelerators, Spectrometers, Detectors and Associated Equipment*, *382*(3), 589–600. doi:10.1016/S0168-9002(96)00804-2

Carlson, P., Francke, T., Lund-Jensen, B., & Peskov, V. (2003). Gaseous photomultipliers with solid photocathodes for the detection of sparks, flames and dangerous gases. *Nuclear Instruments & Methods in Physics Research. Section A, Accelerators, Spectrometers, Detectors and Associated Equipment*, *505*(1-2), 207–210. doi:10.1016/S0168-9002(03)01053-2

Carlson, P., Iacobeaus, C., Francke, T., Lund-Jensen, B., Periale., Peskov, V., & Rodionov, I. (2003). Beyond the RICH: innovative photosensitive gaseous detectors for new fields of applications. *Nucl. Instrum. and Methods in Phys. Res.*, *A502*, 189-194.

Ceron-Zeballos, E., Crotty, I., Fonte, P., Hatzifotiadou, D., Lamas Vlaverde, J., Peskov, V., … Zichichi, A. (1995). New developments of RPC: secondary electron emission and microstrip readout. In S.P. Ratti, & M. Merio (Eds.), *Proceedings of the International Workshop on RPCs and Related Detectors*. INFN.

Ceron-Zeballos, E., Crotty, I., Hatzifotiadou, D., Lamas, D., Valverde, J., Williams, M. C. S., & Peskov, V. et al. (1997). Resistive plate chambers with secondary electron emitters and microstrip readout. *Nuclear Instruments & Methods in Physics Research. Section A, Accelerators, Spectrometers, Detectors and Associated Equipment*, *392*(1-3), 150–154. doi:10.1016/S0168-9002(97)00256-8

Chandrasekharan, R., Messina, M., & Rubbia, A. (2006). Detection of VUV light at high quantum efficiency with large area avalanche photodiodes (LAAPDs). *Nuclear Instruments & Methods in Physics Research. Section A, Accelerators, Spectrometers, Detectors and Associated Equipment*, *567*(1), 45–47. doi:10.1016/j.nima.2006.05.058

Charpak, G., Banaben, P., Breuil, P., Di Mauro, A., Martinenego, P., & Peskov, V. (2008). Development of new hole-type avalanche detectors and the firts rsulsts of their applications. *IEEE Transactions on Nuclear Science*, *55*(3), 1657–1663. doi:10.1109/TNS.2008.924084

Charpak, G., Dominik, W., Fabre, J. P., Gaudaen, J., Peskov, V., Sauli, F., & Sauvage, D. et al. (1988). Some applications of the imaging proportional chamber. *IEEE Transactions on Nuclear Science*, *35*(1), 483–486. doi:10.1109/23.12770

Charpak, G., Gaudean, I., Giomataris, Y., Peskov, V., Scigocki, D., Sauli, F., & Stuart, D. (1993). Some studies of the applications of CsI photocathodes in gaseous detectors. *Nuclear Instruments & Methods in Physics Research. Section A, Accelerators, Spectrometers, Detectors and Associated Equipment*, *333*(2-3), 391–398. doi:10.1016/0168-9002(93)91181-L

Charpak, G., Peskov, V., Breuil, P., Benaben, P. Patent: Electron multiplication device and systems for detection ionizing radiation, FR20070001202, WO2008129159, 2007.

Charpak, G., Policarpo, A., & Sauli, F. (1980). The photo-ionization proportional scintillation chamber. *IEEE Transactions on Nuclear Science*, *NS-27*(1), 212–215. doi:10.1109/TNS.1980.4330830

Chen, M., Mullins, M., & Pelly, , D., Shotkin, S., Sumorok, K., Akyuz, D., ...Sugimoto, S. (1993). Homogeneous scintillating LKr/Xe calorimeters. *Nuclear Instruments & Methods in Physics Research. Section A, Accelerators, Spectrometers, Detectors and Associated Equipment*, *327*, 187–192.

Chepel, V., & Araujo, H. (2013). Liquid noble gas detectors for low energy particle physics. *JINST*, *8*(04), 1–114. doi:10.1088/1748-0221/8/04/R04001

Chepel, V., Lopes, M. I., Kuchenkov, A., Ferreira Marques, R., & Policarpo, A. J. P. L. (1997). Performance study of liquid xenon detector for PET. *Nuclear Instruments & Methods in Physics Research. Section A, Accelerators, Spectrometers, Detectors and Associated Equipment*, *392*(1-3), 427–432. doi:10.1016/S0168-9002(97)00196-4

Coelho, L. C. C., Veloso, J. F. C. A., Covita, D. S., Requicha Ferreira, L. F., & Dos Santos, J. M. F. (2006). High-pressure operation of a xenon-GPSC/MSGC hybrid detector for hard X-ray spectrometry. *Nuclear Instruments & Methods in Physics Research. Section A, Accelerators, Spectrometers, Detectors and Associated Equipment*, *569*(3), 796–802. doi:10.1016/j.nima.2006.09.061

Crotty, I., Fonte, P., Lemenovski, D., & Peskov, V. (1996). A new resistive plate chamber with secondary electron emitters and two dimensional microstrip readout. *IEEE Nucl Sci. Conf. Record*, *1*, 362–366.

Dangendorf, V., Breskin, A., Chechik, R., & Schmidt-Bocking, H. (1990). A gas filled UV-photon detector with CsI photocathode for the detection of Xe light. *Nuclear Instruments & Methods in Physics Research. Section A, Accelerators, Spectrometers, Detectors and Associated Equipment*, *289*(1-2), 322–324. doi:10.1016/0168-9002(90)90276-C

Danielsson, M., Fonte, P., Francke, T., Iacobaeus, C., Osling, J., & Peskov, V. (2004). Novel gaseous detectors for medical imaging. *Nuclear Instruments & Methods in Physics Research. Section A, Accelerators, Spectrometers, Detectors and Associated Equipment*, *518*(1-2), 406–410. doi:10.1016/j.nima.2003.11.038

Di Mauro, A., Lund-Jensen, B., Martinengo, P., Nappi, E., Oliveira, R., Peskov, V., & Santiard, J. C. et al. (2007). Development of innovative micro-pattern gaseous detectors with resistive electrodes and first results of their applications. *Nuclear Instruments & Methods in Physics Research. Section A, Accelerators, Spectrometers, Detectors and Associated Equipment*, *581*(1-2), 225–231. doi:10.1016/j.nima.2007.07.083

Diatropto, D. B., & Peskov, V. (1972). Vacuum ultraviolet spectrum of a free plasma filament in a microwave high-pressure discharge. *Soviet Physics JETF*, *34*, 554–558.

Duval, S., Breskin, A., Budnik, R., Chen, W. T., Carduner, H., Cortesi, M., & Thers, D. et al. (2011). On the operation of a micropattern UV-photomultipliers in liquid Xe. *JINST*, *6*(04), 1–12. doi:10.1088/1748-0221/6/04/P04007

Fonte, P., Francke, T., Pavlopoulos, P., Peskov, V., & Rodionov, I. (2005). Novel single photon detectors for UV imaging. *Nuclear Instruments & Methods in Physics Research. Section A, Accelerators, Spectrometers, Detectors and Associated Equipment*, *553*(1-2), 30–34. doi:10.1016/j.nima.2005.08.002

Fonte, P., & Peskov, V. (2000). Micro-gap parallel-plate chambers with porous secondary electron emitters. *Nuclear Instruments & Methods in Physics Research. Section A, Accelerators, Spectrometers, Detectors and Associated Equipment*, *454*(1), 260–266. doi:10.1016/S0168-9002(00)00838-X

Francke, T., & Peskov, V. (2002). *Patent: Flame and spark detector, automatic fire alarm and methods related thereto*. WO02097757, IPC: G08B17/12,H01J47/02.

Francke, T., Peskov, V. (2005). *Patent: Detector assembly, WO2006135321 (A1)*. EP1894179, IPC: G08B17/12H01J47/022006.

Grignon, C., Breskin, A., Carlier, T., Couturier, O., Cussonneau, J. P., & Ferrer, L. ... Vasseur, A. (2005). Simulation of a High Performance γ-Camera Concept for PET based on LXe and Gaseous Photomultipliers. *Dielectric liquids.2005 IEEE International Conference on Dielectric Liquids*, (pp. 357-360). IEEE.

Holder, J. (2013). Atmospheric Cherenkov telescopes for high-energy γ-ray astronomy. *Particle Data Group Review of Particle Properties*. Retrieved from: http://pdg.lbl.gov/2014/reviews/rpp2014-rev-particle-detectors-non-accel.pdf

Karabadjak, G. F., & Peskov, V. D. (1981). Multiwire X-Ray Counter with High Energy Resolution. *Prib. Tekhn. Eksp.*, *6*, 174–176.

Karabadjak, G. F., Peskov, V. D., & Podoliak, E. R. (1982). Gaseous counters for detection plasma radiation in the region 3-200 nm with spatial and time resolutions. *Journal of Applied Spectroscopy*, *3*, 509–516.

Karabadjak, G. F., Peskov, V. D., & Podolyalk, E. R. (1983). Coordinate gas counters with super high energy resolution for plasma diagnostics. *Nuclear Instruments & Methods in Physics Research. Section A, Accelerators, Spectrometers, Detectors and Associated Equipment*, *217*, 56–64.

Kononenko, W., Heinrich, J. G., Lockyer, N. S., Miller, J. M., Woody, C., & Kwan, S. (1997). Photon emission from 5 11 keV gamma rays incident on BaF_2 and LaF_3: Nd^{3+} crystals using a caesium iodide photocathode detector. *Nuclear Instruments & Methods in Physics Research. Section A, Accelerators, Spectrometers, Detectors and Associated Equipment*, *396*(1-2), 172–180. doi:10.1016/S0168-9002(97)00776-6

Lopes, J. A. M., dos Santos, J. M. F., Conde, C. A. N., & Morgado, R. E. (1999). A new integrated photosensor for gas proportional scintillation counters based on the gas electron multiplier (GEM). *Nuclear Instruments & Methods in Physics Research. Section A, Accelerators, Spectrometers, Detectors and Associated Equipment*, *426*(2-3), 469–476. doi:10.1016/S0168-9002(98)01419-3

Martinengo, P., Nappi, E., Oliveira, R., Peskov, V., Pietropaolo, F., & Picchi, P. (2009). A new generation of GEM detectors and their applications. *Nuclear Instruments & Methods in Physics Research. Section A, Accelerators, Spectrometers, Detectors and Associated Equipment*, *604*(1-2), 8–10. doi:10.1016/j.nima.2009.01.043

Montanari, C. (2004). The ICARUS project. A 3000 ton detector for neutrino and matter stability search. *Nuclear Instruments & Methods in Physics Research. Section A, Accelerators, Spectrometers, Detectors and Associated Equipment*, *518*(1-2), 216–219. doi:10.1016/j.nima.2003.10.064

Oliveira, R., Peskov, V., Pietropaolo, F., & Picchi, P. (2007). First test of thick GEM with electrodes made of resistive Kapton. *Nuclear Instruments & Methods in Physics Research. Section A, Accelerators, Spectrometers, Detectors and Associated Equipment*, *576*(2-3), 362–366. doi:10.1016/j.nima.2007.03.010

Östling, J., Brahme, A., Danielsson, M., Iacobaeus, C., & Peskov, V. (2004). A radiation-tolerant electronic readout system for portal imaging. *Nuclear Instruments & Methods in Physics Research. Section A, Accelerators, Spectrometers, Detectors and Associated Equipment*, *525*(1-2), 308–312. doi:10.1016/j.nima.2004.03.079

Ott, R. J., Flower, M. A., Divoli, A., & Binnie, D. (2003). Performance of the PETRRA positron camera: Reduction of scatter and random events. *Nuclear Instruments & Methods in Physics Research. Section A, Accelerators, Spectrometers, Detectors and Associated Equipment*, *513*(1-2), 61–64. doi:10.1016/j.nima.2003.08.002

Periale, L., Peskov, V., Braem, A., Di Mauro, A., Martinengo, P., Picchi, P., & Sipila, H. et al. (2007). Development of new sealed UV sensitive gaseous detectors and their applications. *Nuclear Instruments & Methods in Physics Research. Section A, Accelerators, Spectrometers, Detectors and Associated Equipment*, *572*(1), 189–192. doi:10.1016/j.nima.2006.10.272

Periale, L., Peskov, V., Carlson, P., Francke, T., Pavlopoulos, P., Picchi, P., & Pietropaolo, F. (2001). *Detection of the primary scintillation light from dense Ar, Kr and Xe with novel photosensitive gaseous detectors.* ArXive, Physics/0106048.

Periale, L., Peskov, V., Carlson, P., Francke, T., Pavlopoulos, P., Picchi, P., & Pietropaolo, F. (2002). Detection of the primary scintillation light from dense Ar, Kr and Xe with novel photosensitive gaseous detectors. *Nuclear Instruments & Methods in Physics Research. Section A, Accelerators, Spectrometers, Detectors and Associated Equipment*, *478*(1-2), 377–383. doi:10.1016/S0168-9002(01)01779-X

Periale, L., Peskov, V., Carlson, P., Iacobaeus, C., Francke, T., Pavlopoulos, N., & Picchi, P. et al. (2003). Evaluation of various planar gaseous detectors with CsI photocathodes for the detection of primary scintillation light from noble gases. *Nuclear Instruments & Methods in Physics Research. Section A, Accelerators, Spectrometers, Detectors and Associated Equipment*, *497*(1), 242–248. doi:10.1016/S0168-9002(02)01918-6

Periale, L., Peskov, V., Iacobaeus, C., Francke, T., Lund-Jensen, B., Pavlopoulos, N., & Pietropaolo, F. et al. (2004). The development of gaseous detectors with solid photocathodes for low-temperature applications. *Nuclear Instruments & Methods in Physics Research. Section A, Accelerators, Spectrometers, Detectors and Associated Equipment*, *535*(1-2), 517–522. doi:10.1016/S0168-9002(04)01719-X

Periale, L., Peskov, V., Iacobaeus, C., Francke, T., Lund-Jensen, B., Pavlopoulos, P., & Pietropaolo, F. et al. (2006). A study of the operation of especially designed photosensitive gaseous detectors at cryogenic temperatures. *Nuclear Instruments & Methods in Physics Research. Section A, Accelerators, Spectrometers, Detectors and Associated Equipment*, *535*, 517–522. doi:10.1016/S0168-9002(04)01719-X

Periale, L., Peskov, V., Iacobaeus, C., Francke, T., Lund-Jensen, B., Pavlopoulos, P., & Tokanai, F. et al. (2005). The Successful Operation of Hole-Type Gaseous Detectors at Cryogenic Temperatures. *IEEE Transactions on Nuclear Science*, *52*, 927–931.

Periale, L., Peskov, V., Iacobaeus, C., Lund-Jensen, B., Picchi, P., Pietropaolo, F., & Rodionov, I. (2007a). Photosensitive gaseous detectors for cryogenic temperature applications. *Nuclear Instruments & Methods in Physics Research. Section A, Accelerators, Spectrometers, Detectors and Associated Equipment*, *573*(1-2), 302–305. doi:10.1016/j.nima.2006.10.264

Peskov, V. (1980). *Complex/multiparameter plasma diagnostics in UV and X-ray regions*. (Doctor of science thesis). Kapitza Institute for Physical Problems, Moscow, Russia.

Peskov, V., Baibussinov, B., Centro, S., Di Mauro, A., Lund-Jensen, B., Martinengo, P., & Ventura, S. et al. (2007a). Development and first tests of GEM-like detectors with resistive electrodes. *IEEE Transactions on Nuclear Science*, *54*(5), 1784–1791. doi:10.1109/TNS.2007.905160

Peskov, V., Fonte, P., Danielsson, M., Iacobaeus, C., Ostling, J., & Wallmark, M. (2001). The Study and Optimization of New Micropattern Gaseous Detectors for High-Rate Applications. *IEEE Transactions on Nuclear Science*, *48*(4), 1070–1074. doi:10.1109/23.958725

Peskov, V., & Zichichi, A. (1991). A new scintillator, $LiBaF_3$, and its basic properties. In L. Cifarelli & T. Ypsilantis (Eds.), *New Technologies for Supercolliders* (pp. 49–55). New York: Plenum Press. doi:10.1007/978-1-4684-1360-1_4

Peskov, V., & Zichichi, A. (2007). *A New Supersensitive Flame Detector and it's Use for Early Forest Fire Detection* arXiv: 0709.2819

Peskov, V. D. (1980). Coordinate Gas Scintillation Counters of VUV and X-Ray Photons. *Instruments and Experimental Techniques*, *23*, 507–511.

Peyaud, A., Angelopoulos, A., Chemlis, C., Costopoulos, V., Chica, M., Giomataris, Y., & Quinlan, F. et al. (2015). The ForFire photodetector. *Nuclear Instruments & Methods in Physics Research. Section A, Accelerators, Spectrometers, Detectors and Associated Equipment*, *787*, 102–104. doi:10.1016/j.nima.2014.11.044

Policarpo, A. J. P. L. (1978). Coupling the gas scintillation proportional counter to photoinization detectors. *Nuclear Instruments and Methods*, *153*(2-3), 389–391. doi:10.1016/0029-554X(78)90977-1

Rodionov, I., Bidault, J-M., Crotty, I., Fonte, P., Galy, F., Peskov, V., & Zanette, O. (2005). Advanced Gaseous Photodetectors for Hyperspectroscopy and Other Applications. *IEEE Nucl., Sci., Conf. Record.*

Rodionov, I., Francke, T., Peskov, V., & Sokolova, T. (2002). Hybrid gaseous photomultipliers. *Nuclear Instruments & Methods in Physics Research. Section A, Accelerators, Spectrometers, Detectors and Associated Equipment*, *478*(1-2), 384–390. doi:10.1016/S0168-9002(01)01778-8

Saito, K., Tawara, H., Sanami, T., Shibamura, E., & Sasaki, S. (2002). Absolute number of scintillation photons emitted by alpha particles in rare gases. *IEEE Transactions on Nuclear Science*, *49*(4), 1674–1680. doi:10.1109/TNS.2002.801700

Séguinot, J., & Ypsilantis, T. (1977). Photo-ionization and Cherenkov ring imaging. *Nuclear Instruments and Methods*, *142*(3), 377–391. doi:10.1016/0029-554X(77)90671-1

Skovoroda, A. A., & Zvonkov, A. V. (2001). A stationary microwave discharge at high pressure of Argon. *Journal of Experimental and Theoretical Physics*, *92*(1), 78–85. doi:10.1134/1.1348463

KEY TERMS AND DEFINITIONS

Atmospheric Showers: A large scale (many km) cascade of ionizing particles and electromagnetic radiation triggered in atmosphere by a high energy cosmic ray.

Charged Coupled Devices (CCD): An imaging semiconductor circuit invented by W. Boyle and G. Smith, awarded in 2009 by the Noble prize in physics. The idea was to readout an array of pixels by rows. Stored electrons slide off the array onto a conveyor belt and van be translated to digital values: 1 or 0. The digital output makes it easy to transfer and treat the images.

Electromagnetic Shower: A cascade process triggered in a material by high energy electrons, positrons or photons. It occurs via two processes: pair production and bremsstrahlung.

GEM: Gas Electron Multiplier.

Plasma Diagnostics: Experimental methods allowing to measure plasma properties such as ion's concentration and their distribution in space and in time, electrons and ions temperature, etc.

Plasma: An ionized gas. It can be created either by heating of the gas or by applying to the gas volume a strong electric field, or a strong electromagnetic radiation (microwave, laser, etc).

Position Emission Tomography (PET): An imaging technique widely used in nuclear medicine and biology. It is bases on a positron emitting radio nuclides introduced into the investigating body. The surrounding detectors records gamma rays emitted from the area occupied by the radio tracers and after the computer analysis a three -dimensional images of the tracer concentration is obtained.

Resistive-Plate Chambers (RPC): A gaseous detector in which avalanche multiplication is happening between two parallel electrodes made of resistive material. The voltage to the detector is applied via metalized layers located on the outside surface of the resistive plates.

RETGEM: Resistive electrode GEM.

Samarium (Sm): A chemical element with an atomic number of 62.

Solid Scintillators: Solid materials (plastic, crystals) capable emitting visible or UV light after excited by ionizing radiation.

TEA (Triethylamine): Photosensitive vapors, used in the past in some RICH detectors.

TGEM: Thick Gas Electron Multiplier. This is a GEM, but made of a thicker plate.

Time Projection Chambers: A detector allowing to visualize particle tracks in three dimensions. It consists from a drift volume and a 2-D position sensitive collector. The latter determine the track projection whereas the third coordinate is determined from the drift time of primary electrons created by the passing particle in the drift region.

TMAE Tetrakis(dimethylamino)ethylene: Photosensitive vapors with the lowest ionization potential-about 5.28 Ev.

Triple CsI-GEM: A detector of UV photons, consisting from three cascaded GEMs in which the top electrode of the first GEM is coated with a CsI layer.

XENON Experiment: Experiment attempting to detect dark matter in the form of weakly interacting massive particles (WIMPs) via created by them recoils in a liquid xenon.

ZEPLIN: A dark matter experiment aiming to detect Galactic WIMPs using a liquid Xe target.

APPENDIX: WINDOWLESS GASEOUS DETECTORS FOR PLASMA DIAGNOSTICS

Besides the classical detectors of UV photons having entrance windows, in some applications, e.g. plasma diagnostics, windowless detectors were used as well which operated in special gas flush mode. They were designated for the detecting of plasma bremsstrahlung radiation with wavelength below the window transmission cut off, this lowest one is ~ 104 nm.

A design of one of such detectors is shown in Figure 50. In a simple version it was a single- wire counter separated from the plasma chamber (UHF resonator) via a special differential pumping/ flushing gas chamber having small holes, allowing to maintain pressure there lower than in the counter and the resonator. The role of this chamber is similar to one used in differential pumping. Recall that the latter is a special technique allowing to have different stages in a vacuum system between each a large difference in pressure is maintained. This could be between one part of the system which operates in ultrahigh vacuum and the rest of the volume kept at some non-negligible gas pressure. The pressure difference is maintained by means of a small aperture between these two parts.

The similar concept was implemented in the case of the gas-flushed system used for plasma diagnostic, operated, however at pressures between 2 and 30 atm (Bogomolov, 1978; Karabadjak, 1982). The idea was to separate the gas volume of the VUV detector from the plasma chamber using proper gas flows rather than a window. As can be seen from Figure 50, this was achieved by arranging two continuous gas flows: one from a single wire counter to the separation chamber and the other one from the plasma chamber to the separation chamber. The separation chamber in turn was kept at a pressure lower that in the VUV counter and the plasma chamber by a controllable releasing of the incoming gas to the atmosphere.

Figure 50. Schematic drawing of a windowless UV sensitive single-wire counter operating in gas mixture different from the one in which a plasma was induced

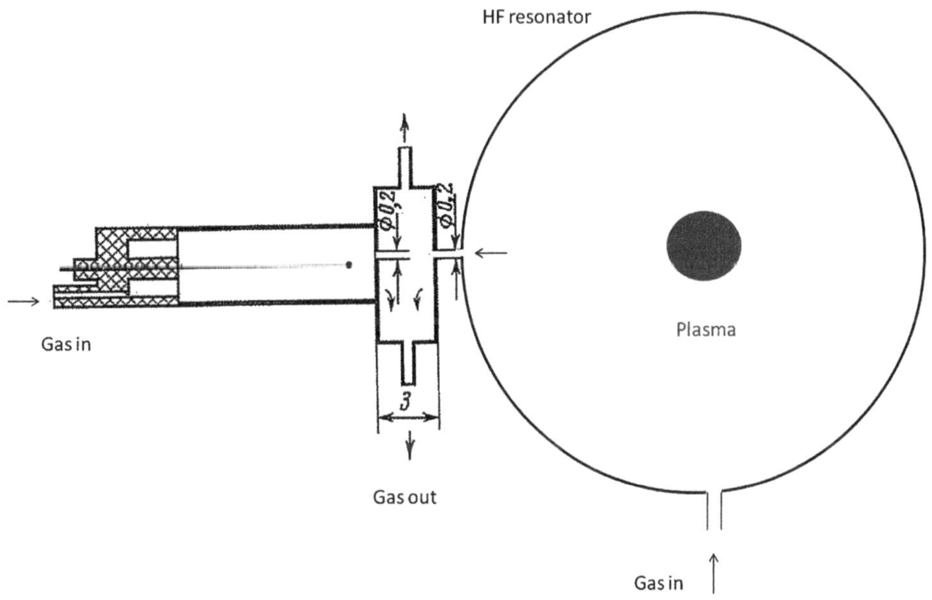

An automatic system continuously added the gas to the detector and the resonator keeping their pressure stable. Due to this the UV detector operates very stably. By adding various gases with ionization potentials below the ionization potential of the gas in the resonator one can perform measurements of absolute intensity of the VUV radiation in selected spectral intervals. For example, if the plasma chamber operates in He, than the gas in the resonator is transparent almost from 50 nm and by adding to the VUV detector consequently with mixture of He with Ne, Ar, Xe, Kr on can evaluate the absolute intensity of the bremsstrahlung plasma radiation in the wavelength intervals of: 50-57 nm, 50-79 nm, 50-88 nm and 50-102 nm respectively.

Since the intensity of plasma radiation in VUV region was quiet strong, the gas gain between 10 and 50 was sufficient for its reliable detection. Due to the small diameter of the aligned holes in the separation chamber the VUV entered the detectors almost parallel to its axis and as a result the contribution of the surface photoelectric effect to the signal was negligible.

However, the drawback of this method was that the gas mixture in the separating chamber, due to the photo absorption, may introduce some badly controlled attenuation of the plasma radiation. To avoid this, a complimentary method was used as well in which the differential chamber not only separated the gas flows, but also played a role of a gas filters. In this case the VUV counter operated in He with a small concentration of quenching gas (<1%) and Ne, Ar, Kr and Xe gas filters were introduced to this chamber by means of a special gas injector made of micro tubes having small holes (Karabadjk, 1981) In Figure 50, for simplicity, only one windowless detector is shown. In practice, an array of such detectors was used in plasma diagnostics VI.51.

Figure 51. Scematic drawing of a windowless matrix of miniature VUV sensitive single-wire counters used for detection of plasma radiation with a spatial resolution

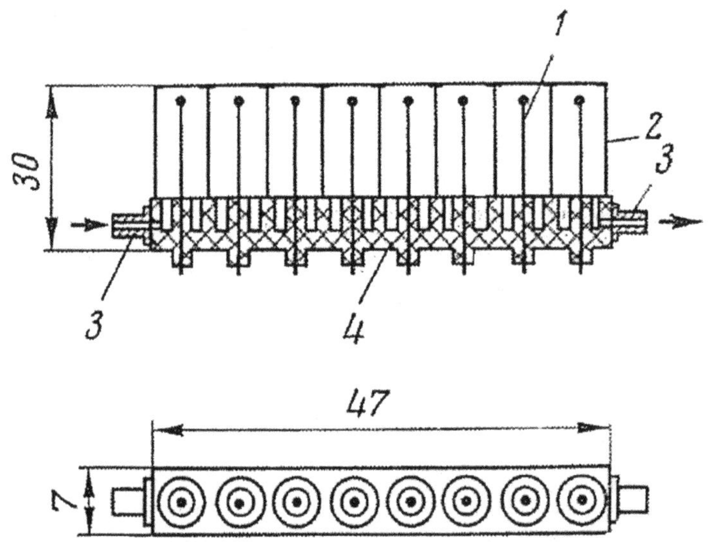

Each individual detector in this array, due to the aligned holes in the separation chambers, had a narrow solid angle of view and this allows to record VUV plasma radiation, depending on the detector matrix orientation, either from a well located regions along its axis or along it radius.

Some physical results obtained with windowless VUV detectors can be found in (Peskov, 1980).

Conclusion

In this book special position-sensitive photomultipliers based on avalanche gaseous detectors have been described. These detectors have several important features, of which some are really unique:

1. They can be combined not only with traditional semitransparent and reflective solid photocathodes, used e.g. in vacuum devices, but also with gaseous and even liquid/adsorbed layer photocathodes.
2. They have high QE from the VUV to the visible region of the spectra, comparable to vacuum photodetectors.
3. They operate in gas avalanche mode and the amplification factor can be up to 10^6 (or even up to 10^8 in Geiger mode).
4. The have excellent position resolutions, as small as 15 μm for MCPs.
5. Their QE and position resolution do not degrade in the high magnetic fields used in some high-energy physics experiments.
6. Some designs of gaseous photomultipliers have time resolutions as short as a few ns, close to that of vacuum PMTs.
7. They can have very large sensitive areas. For example, the active area of the ALICE RICH detectors is $11 m^2$.

Let's discuss these points in more detail.

1. WHAT IS THE ADVANTAGE OF USING GASEOUS PHOTOMULTIPLIERS FILLED WITH PHOTOSENSITIVE VAPORS OR COMBINED LIQUID PHOTOCATHODES?

Photosensitive vapors offer sensitivity in a well-defined spectral interval:

$$\Delta\lambda = \lambda_i - \lambda_w, \tag{1}$$

where λ_i is the threshold wavelength for the photoinization process and λ_w is the transmission cut off of the entrance window. Outside this interval the sensitivity of the detector is practically zero.

Such detectors have advantages in cases when one have to detect VUV radiation in the presence of a strong long-wavelength background. Example could be UV sensitive flame detectors which should operate in daylight conditions. In other applications it is important to perform various measurements in narrow VUV spectral intervals. For this the photodetectors are usually used in combination either with a monochromator or with narrow bandpass filters.

A gaseous photomultiplier filled with photosensitive vapors offer an alternative method. By a proper choice of λ_i and λ_w, one can sometimes perform this task directly without the use of filters or spectroscopic devices. The latter method was used, for example in plasma studies (Peskov, 1980) for imaging of short leaving filaments simultaneously in various spectral intervals. The simple approach was to use several MWP s filled with different photosensitive vapors.

Another important advantage of gaseous detectors filled with photosensitive vapors is the extremely low rate of noise pulses, only originating from the radioactive background and can be discriminated from the useful signals by the pulse amplitude

Liquid photocathodes offer several interesting features, e.g. the photo ionization threshold of non-polar liquids is lower than the corresponding gas. However, to handle a liquid photocathode in practices not so easy. This is overcome by using an absorbed layer of the photosensitive liquid deposited on the surface of a solid photocathode. An absorbed layer of some photosensitive gases, e.g. TMAE, form a double electric layer on the cathode and this lead to two important effects:

1. Enhancement of the QE of any solid photocathode.
2. Extension of the photocathode sensitivity threshold to longer wavelength.

These unique feature can only be realized using gaseous photomultipliers.

2. THE HIGH QE OF GASEOUS PHOTOMULTIPLIERS

In VUV (104-180 nm) the QE of vapors is comparable or even higher than in vacuum device or solid state detectors have (see, for example Day, 2004, Monteiro, 2005).

To achieve high QE in the UV and visible spectral interval (200-550 nm) gaseous photomultiplier are usually combined with the same photocathodes (e.g. CsI, CsTe, SbCs, bialkali) as vacuum devices. In principle one can expect comparable sensitivity at least when measured in vacuum. Indeed, this was proven by several research groups experimenting with photocathodes in their laboratories. However, there are two factors affecting the "practical" QE (measured at the given operational conditions). In the case of gaseous detectors the practical QE is less than in vacuum due to photoelectron back scattering and in the case some vacuum devices, for example vacuum PMTs, not all photoelectrons are collected and directed by the focusing system to the dynodes.

The values of practical QEs for gaseous photomultipliers were carefully measured and published data are widely available.

On the other hands, photonic companies usually do not publish the practical QE and instead present their photocathodes efficiencies measured at full photoelectron collections conditions (in current/diode mode when all dynodes and the focusing system are kept at the same potential). As follows from the on line Hamamatsu catalog hamamatsupmt handbook:

ver2.pdf: http://www.coseti.org/www.boseti.org/pdf/hamamatsu%20pmt%20handbook%20ver2.pdf.

The collection efficiency of some vacuum phototubes could be higher than 80% which can never be realized in a practical detector.

The main conclusion is that the practical QEs (in the spectral region 220-550 nm) of normal designs of gaseous photomultipliers, for example PPAC, operating in gases having low back diffusion, are close to vacuum PMTs. Unfortunately, the maximum achievable gain of such detectors is limited by feedback process and is about 50-100.

A later generation of gaseous photodetectors is based on cascaded electron multipliers which have the capability to strongly suppress the feedback processes and as a result the overall maximum achievable gains are increased on several orders of magnitude. In order to operate such detectors at relatively low voltages, gases with small a concentration of quencher are usually used resulting in noticeable losses of photoelectrons due to back scattering. For example, in the work (Lyashenko, 2009), the QE of SbK_2Cs photocathode measured in vacuum was 45% at peak. However, in Ar + 5% CH_4 at a pressure of 700 torr the QE dropped almost by a factor of two reaching 25% at 400 nm.

Therefore, although the cascaded detectors have potentials to have a QE in the region 200-550 nm very close to the best vacuum PMTs, due to the use of not optimized gas mixtures their QE is approximately factor two lower.

3. HIGH GAS GAINS

All detectors of single photons mentioned in this book are operating on the principle of electron multiplication. In vacuum photodetectors the electron multiplication occurs on the dynode system and is based on a secondary electron emission process. This phenomenon allows achieving a multiplication factor G sufficient for single photoelectron detection (typically vacuum devices operate at G = 10^5 - 10^6).

In most of vacuum photodetectors the pulse-height spectrum of the signals created by single primary electrons have a clear peak, which is essential in photon counting mode of operation since it ensures a counting plateau (i.e. close to 100% detection efficiency).

In gaseous and solid-state detectors electron multiplication is bases on avalanche ionization processes (Townsend avalanches) inside these media.

Modern solid-state photo detectors are operating in Geiger mode, where the output signals have almost equal amplitudes. In contrast, in most of applications gaseous photodetectors are used in proportional mode.

At moderate avalanche gains A (typically below 10^5) the pulse height distribution at the input of the front-end electronics has an exponential shape (called Furry distribution).

However, due to the low noise of modern electronics one can still detect single electrons with high efficiency. For example, in the case of ALICE RICH detectors at gain of A~ 4 x10^4 the detection efficiency is about 95-98% (Piuz, 2011).

By increasing the gas gain A one observes the development of a peak which is a favorable factor in achieving both high efficiency and a counting plateau. This transition is observed not only in wire-type detectors, but also in PPAC (Fonte, 1999) and is well understood theoretically (see for example Genz, 1973).

Numerous studies show that methane and ethane are the best gases to be used for single-photoelectron counting. They are significantly better than pure noble gases such as helium or argon or their mixtures with various quenchers.

Proportional mode of operation have several advantages. For example, it offers low operating voltages and the possibility to discriminate (by amplitude) useful signals caused by single photons from backgrounds. In some rare application gaseous photomultipliers, especially filled with photosensitive vapors are operating in Geiger mode.

4. HIGH POSITION RESOLUTION

As follows, the best position resolution achieved among various single photon detectors is so far with MCPs, about 15 μm (FWHM).

Where are gas detectors in this ranking?

One should note that gaseous detectors used for charged particle tracking in general have a spatial resolution approaching 17 μm (see for example Derre, 2002). However, the specifics of single photoelectron and track detection are slightly different. In particle tracking telescopes, several detectors are usually used and the particle track is fitted assuming that it is an infinitely thin straight line or a well know curved line (in the given magnetic field). In measurements with photons collimators are used so the determined position resolution often includes the collimator width.

As was reported in (Francke, 2004), with a1-D sensitive RPC photodetector a spatial resolution of about 30 μm was achieved in a digital form (including the collimator width). A similar position resolution was later measured in work (Fonte, 2005) with a 2- D photosensitive RPC.

Remarkable results were obtained in the work (Bellazzini, 2007). This team develop a detector consisting of a CsI-GEM charge multiplier combined a self-triggering CMOS analogue pixel chip with 105 k pixels with 50 mm pitch. This detector allowed to image the map of the GEM holes, arranged on a triangular pattern at 50 μm pitch (see Figure 1).

In work (Melai, 2011) a CsI–MICROMEGAS combined with a CMOS pixel array of the Timepix chip was investigated. Such a readout permitted high resolution imaging of single UV-photons with a resolution of approximately 7 lp/mm (71 μm).

5. INSENSITIVE TO MAGNETIC FIELDS

Vacuum PMTs are very sensitive even to weak magnetic fields, e.g. the Earth magnetic field. This serious drawback is strongly decreased in some advanced designs of position-sensitive PMTs (e.g., multianode PMTs and MCPs), in which the trajectories of secondary electrons are confined inside tiny multiplication channels.

Solid photon detectors are practically insensitive even to strong magnetic fields.

Gaseous detectors occupy an intermediate place. The effect of the magnetic fields depends detector design, orientation of magnetic field with respect to the electric field, and the readout scheme.

For perpendicular electric E and magnetic B fields the electrons drift with a reduced drift velocity at an angle (called Lorentz angle) to the electric field,

$$\tan \alpha_L = w_L \tau \quad (2)$$

where is the cyclotron frequency (w_L=eB/m), τ- is the collision frequency.

Figure 1. A cumulative map of several hundreds of thousands of single-photon events produces a kind of self-portrait of the GEM amplification structure with individual dots only 50 μm apart.
From Bellazzini, 2007.

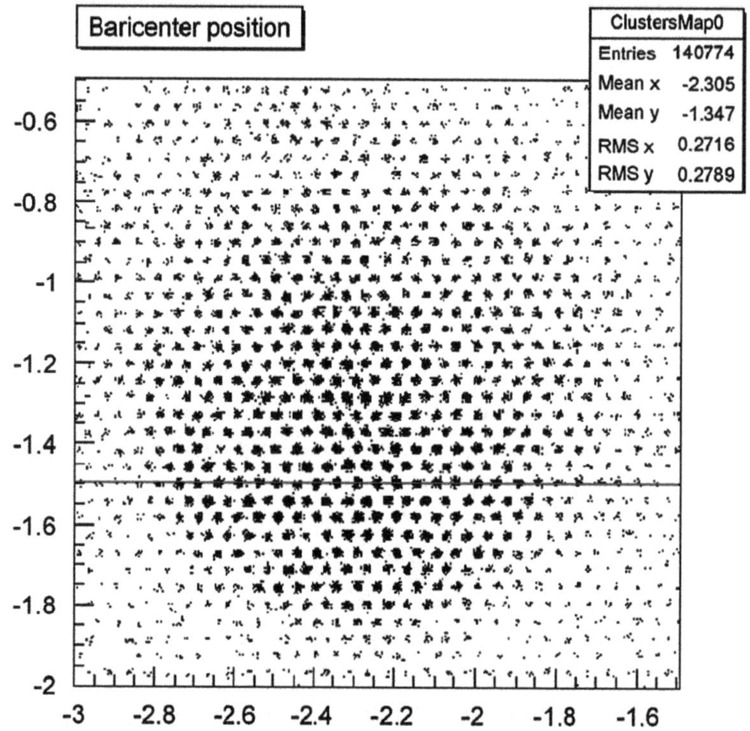

For parallel E and B fields the drift velocity and the longitudinal diffusion are unaffected, while the transverse diffusion, perpendicular to the fields, is reduced by a factor that depends on $\omega_L \tau$.

Some degradation of the position resolution was observed in early studies of MWPCs when the wires were parallel to the direction of the field (see for example (Charpak, 1970)). However, in many cases for the know Lorentz angle proper corrections could be applied allowing to preserve good spatial resolution.

In some gaseous photodetectors, for example RICH detectors, where the required spatial resolution is a few mm the effect of the magnetic field is negligible.

Figure 3 shows the performance of a CsI-MWPC (an ALICE RICH module) operating in magnetic fields up to 0.87 T. Figure 3a shows the variations of the number of pads hit, the mean number of resolved Cherenkov light clusters and the mean number of electrons as a function of the magnetic field measured at the same gas gain with pions of 350 GeV/*c*.

Figure 3b shows the variation of the mean number of resolved clusters of versus the gas gain for various magnetic field values. It can be seen that the curves remain parallel at different magnetic fields down to the minimum gain. Part of the drop with respect to no field value originates from an increase of the common noise induced by the magnet, not the field it creates.

As another example let's mention CH_4 + TEA filled MWPC used in CLEO III/CESR RICH detector, which operated successfully in a magnetic field of 1.5T.

Figure 2. A horizontal cut through the bi-dimensional map of the central positions allowing to estimate the intrinsic resolution of the CsI-GEM detector; the peak-to-peak distance corresponds to the GEM pitch. A gaussian fit of an individual peak distribution gives a width (rms) of ~4 μm.
From Bellazzini, 2007.

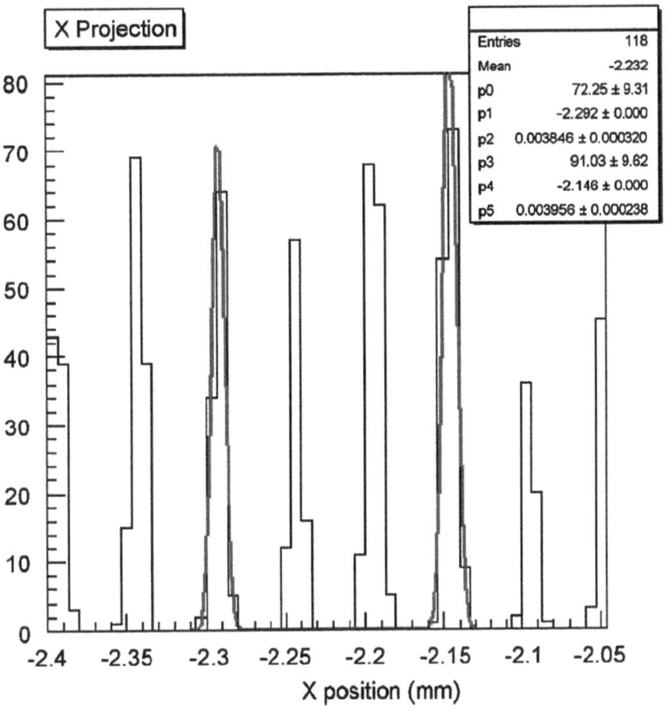

Figure 3. (a) The performance of a CsI -MWPC placed parallel to a uniform magnetic field; (b) mean number of Cherenkov electrons per ring as a function of the CsI-MWPC gain at different magnetic fields; experimental conditions: 350 GeV/c pions at normal incidence, radiator: 10 mm C_6F_{14}, one ADC channel corresponds to 0.17 fC.
From Piuz, 1999.

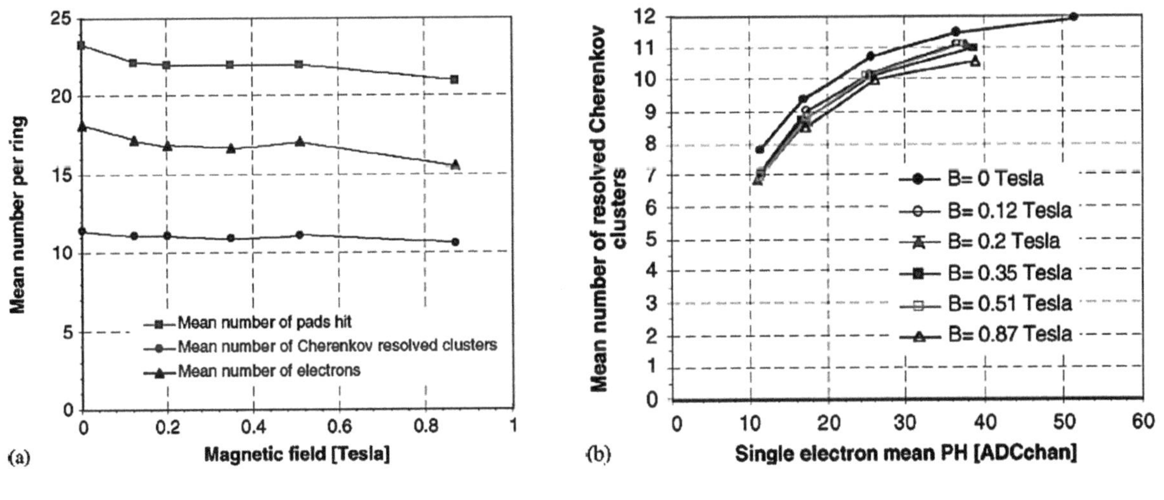

6. TIME RESOLUTION

The time resolution of a gaseous photomultiplier (as any other of its characteristics) depends on the type of the photo-convertor, the detector design and many other parameters, such as gas gain, electronic noise and threshold, output signal shape, number of photoelectrons per light pulse, etc.

The last two parameters are especially critical for gaseous detectors since their signals consist of two components: a fast one which is determined by the collection time of avalanche electrons and a slow one originating from the slow motion of positive ions.

Let us consider an ideal gaseous photomultiplier which has zero time jitter in the creation of primary photoelectrons, i.e. all electrons are created instantly as soon as the photon enters the detector. An example of such detector could be a PPAC. Suppose at time zero a primary photoelectron was created from its cathode. This electron will create an avalanche, electrons of which will arrive to the anode after the drift time of t_0. However, for a given pulse amplitude the arrival time recorded by the electronics will depend on their threshold. For a low threshold, the electronics will detect an arrival time which is identical to t_0. For a high threshold, the arrival time recorded by the electrons will be t_0 plus the time it takes for the signal to rise to the threshold level in the electronics, i.e. it depends on the signal amplitude and on the shaping time of the electronics. Hence, the distribution of signal amplitudes will affect the time resolution of the detector.

This is not a problem if all pulses have absolutely the same amplitude and the shape. In this ideal case the value of the threshold will only affect the recorded arrival time.

However, in reality pulse amplitudes are fluctuating and especially the signals from single electrons. In this case, for a fixed threshold the recorded arrival time will have a strong electronic time jitter spoiling the intrinsic time resolution of the detector. Note the ion drift time may be up to μs.

Figure 4. Schematic representation of the pulse from the PPAC recorded by a charge- sensitive pre-amplifier; it consists from a fast component (collection of the electron avalanches), a slow component (positive ions movement) and the pulse RC decay. The front of the pulse is delayed with respect of the creation of the primary photoelectron on a drift time t_0.

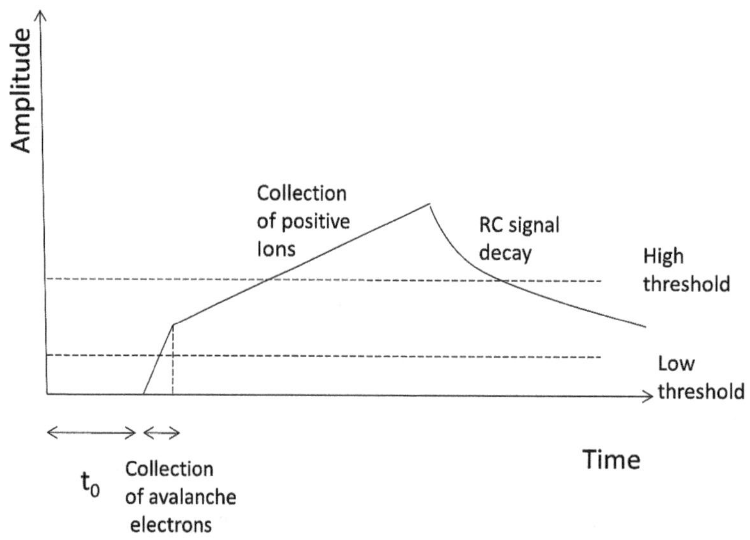

Obviously, this effect is much less critical in the case of very short pulses, typical for vacuum photodetectors, where only electron movement contribute to the pulse formation. If, for simplicity, one neglect the time jitter in the creation time of the primary photoelectrons, the maximum spread in the recorded arrival time will be around half of the pulse width (see Figure 5a). To further diminish the spread, a constant fraction discrimination is often used. Its principle is illustrated in Figure 5b. In this case the electronic threshold is automatically set by the electronics for each pulse. The electronics find the maximum of the pulse and set the threshold at the given ratio with respect to the maximum. Technically this can be implemented by various methods. In any case a signal that has a wide time variance will using a leading edge discriminator be narrowed significantly.

If many primary electrons are released simultaneously from the detector photocathode, the measured time resolution is considerably improved due to the several reasons, e.g. less fluctuation in the pulse amplitude, less contribution of the electronic noise.

The considerations in Figure 5 were done for the specific case when the jitter in the creation of primary electrons is zero. In practice this is almost true only for the gaseous photomultiplier based on a PPAC or a RPC combined with a solid photocathode. All other designs of gaseous detectors have a significant time jitter of the emission time of the primary eectron. Obviously, the highest jitter is found in detectors with photosensitive vapors. Especially if their pressure is low and the mean free path of photoabsorption is large.

Below a few examples of time resolution measured with various designs of gaseous detectors combined with CsI photocathodes are given.

In the work (Anderson, 1992) a timing resolution of 0.55 ns was achieved for single photoelectrons with a low pressure (20 Torr of CH_4) solid photocathode CsI-PPAC, see Figure 6. A constant fraction discriminator was used in these measurements.

(rms). If it is assumed that both chambers contribute equally to the resolution, this would imply a timing resolution for a single chamber of 0.55 ns.

In the work (Charpak, 2002) a time resolution of about 0.7 ns for single photoelectrons was measured with a solid photocathode CsI-MICROMEGAS operated at 1 atm.

The time resolution of a double CsI-TGEM measured with different number of primary photoelectrons is shown in Figure 7. The hole pitch of the TGEMs was 0.7 mm and this is why the jitter in arriving primary electrons created in various points on the CsI-coated top electrode of first TGEM was quite big, 8 ns for single photoelectrons

Better results were achieved with a triple CsI-GEM having a smaller hole pitch of 140 . The time resolution was improved to 2 ns for single photoelectrons (see Figure 8)

The time resolution depends on may parameters, including the detector gain and number of primary photoelectrons. Hence, it is not always straight forward to compare the time resolutions of different photodetectors, e.g. vacuum, gaseous and solid-state detectors.

Below an attempt is made to summarize some essential data obtained with several types of photodetectors.

Vacuum PMTs have a time resolution for single photoelectrons comparable to gaseous detectors: ~0.5 ns (for more details see, for example (Szczesniak, 2009). Other photodetectors usually have much better resolution, for example the time resolution for single photoelectrons of a multipixel MCP is about 30 ps (Va'vra, 2007).

The time resolution of a small gap RPC (0.1-0.3 mm) operating in heavy quenchers (such as $C_2H_2F_4$ + 10% SF_6 + 5% isobutene, see Fonte, 2002) and detecting minimum ionizing particles has a time reso-

Figure 5. a) Schematic explanation of the jitter in recorded arrival time in the case of the very narrow pulses; b) constant fraction discrimination having the dynamic threshold determined from the given fraction with respect to the pulse maximum

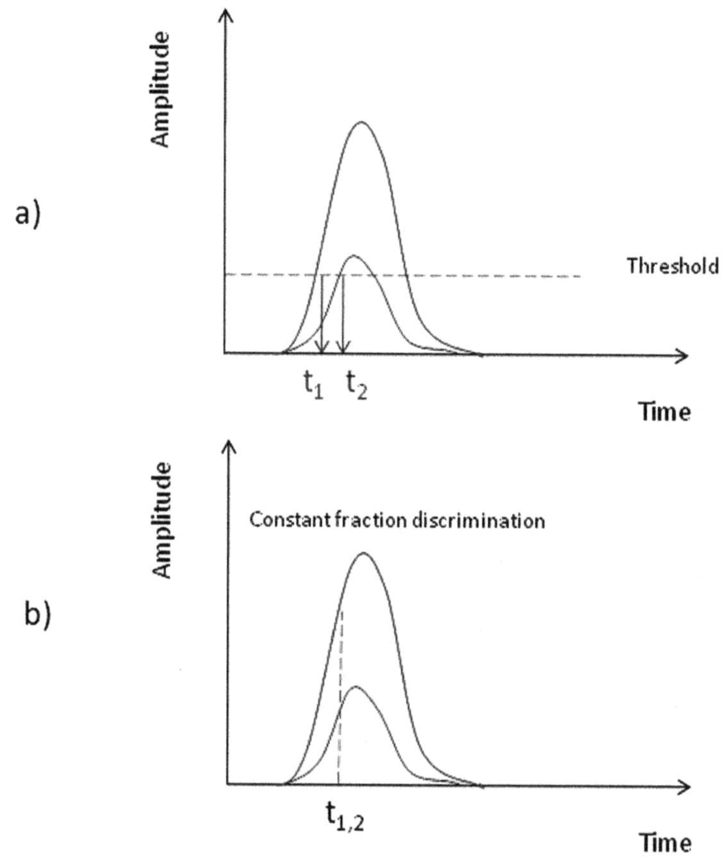

lution of 70-100 ps. These detectors are recording a cluster of a few primary electrons created near the cathode (Rieger, 2003, Lipmann, 2009).

7. LARGE AREA DETECTORS

The most important feature of gaseous photomultiplier is the possibility to build detectors with a very large sensitive area. Let us clarify this statement. Certainly, if necessary, one can construct a large area detector consisting from a mosaic of small modules. This approach is widely used in the case of vacuum and solid-state photodetectors.

Probably the most impressive example is Super-Kamiokande, the large water Cherenkov detector in Japan, used for the discovery of neutrino oscillation which was awarded 2015 Nobel Prize in physics. It is a stainless-steel tank, 39 m diameter and 42 m tall, filled with 50,000 tons of ultra-pure water with about 13,000 photo-multipliers are installed on the tank wall, see Figure 9. The detector is located at 1000 meter underground in the Kamioka-mine, Japan.

Figure 6. The timing spectrum obtained with single photoelectrons; measurements were performed with two identical CsI-PPAC combined with a Cherenkov radiator exposed to a muon beam. One chamber was used for the start and the other for the stop of the acquisition system. The total single-photoelectron resolution is 0.77 ns.
From Anderson, 1992.

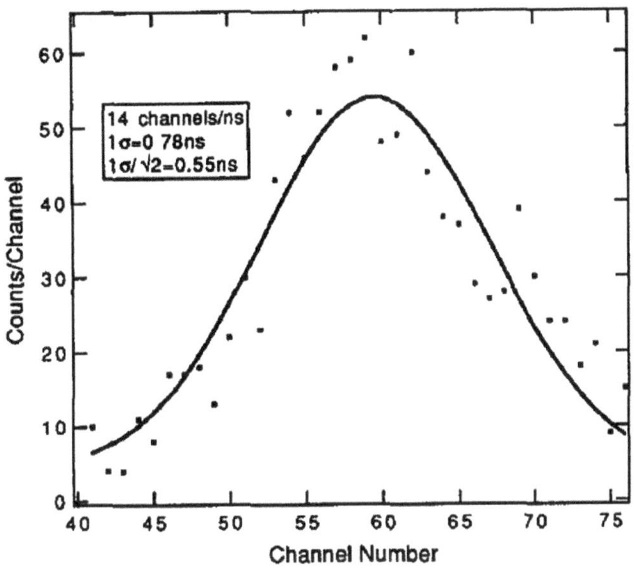

This approach cannot be considered as universal since it has several drawbacks, one of them is the "dead" space between the active areas of the detectors. This creates difficulties in pattern recognition when the Cherenkov ring contains a small number of photons in a large noise background.

Figure 7. Time resolution of a double CsI-TGEM vs. the number of primary photoelectrons
From Breskin, 2008.

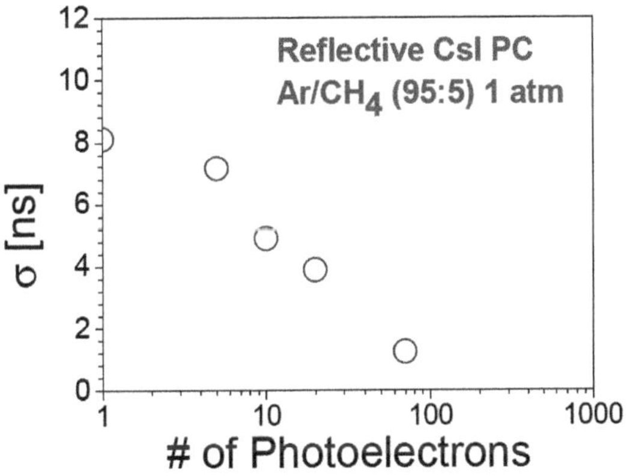

Figure 8. The time resolutions measured for 1, 10 and 150 photoelectrons released from the photocathode per light pulse; the data were obtained at a gain of $1.2x10^6$.
From Breskin, 2002.

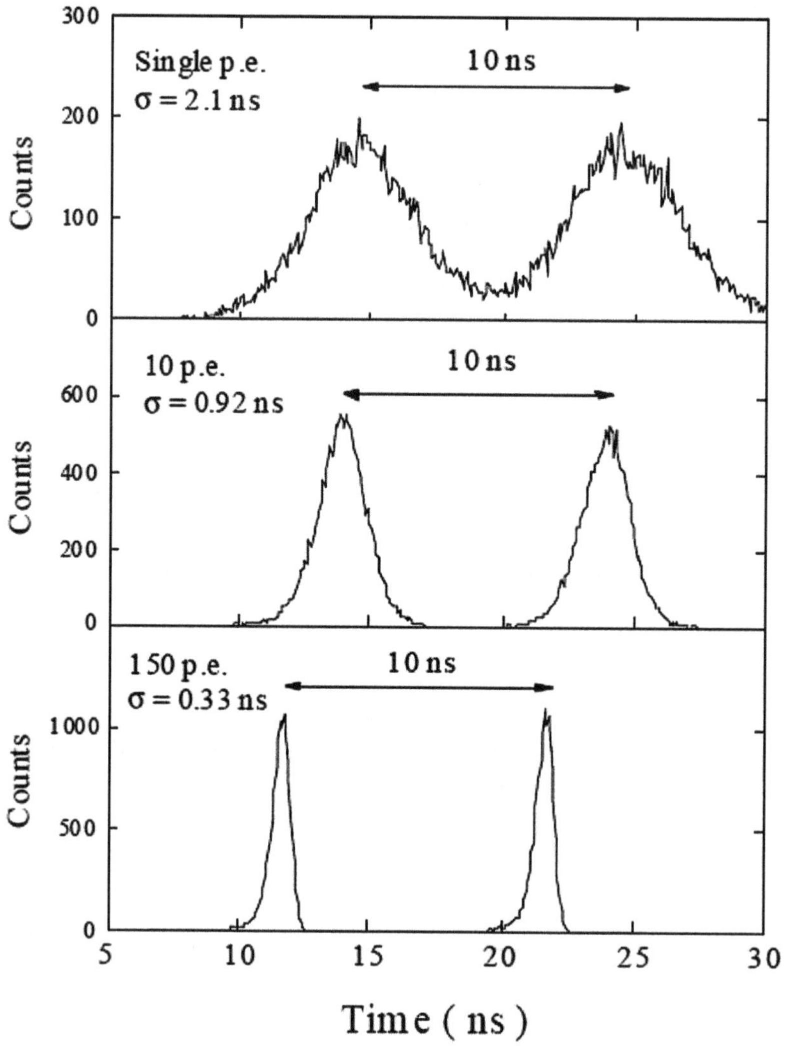

To minimize this problem some vacuum photodetectors, especially multianode MPTs, have squared geometry. Similarly solid-state photomultipliers are often assembled from a matrix of small individual detectors.

Gaseous photomultipliers offer the largest area of a single module detector (in the ALICE RICH each individual module has a size of 64x40 cm^2) and with a high position resolution (pad size of 8x8.4 mm^2). In a mosaic detector the dead zone between such "super"-modules is minimized. Moreover, in such an application as RICH each individual large-size modules are able to fully accommodate a Cherenkov ring, so the problem of the dead zone between such modules becomes not so important.

Another important advantages of gaseous photomultipliers, especially UV sensitive, is their low cost compared to alternative detectors.

Figure 9. Photograph of a Super-Kamiokande water Cherenkov detector the wall of which contains 13 000 large-size (50 cm in diameter) light collectors-photomultipliers; the Cherenkov light is projected as a ring on the wall of the detector and recorded by the PMTs.

To conclude this book, let us stress the main features of gaseous photomultipliers: insensitivity to magnetic fields, low cost, the possibility to choose construction materials with low radioactive background and best cost-effectiveness for large area coverage, good time and position resolution.

They are employed in the largest RICH detectors ever built, the DELPHI RICH and the SLD CRID, successfully operated between 1987 and 2000. Such systems were based on 3-D imaging of Cherenkov photons using a TPC approach: the photoelectrons, created by photoionization in a thick (~cm) layer of a gas mixture saturated with TMAE, drift towards a MWPC in a uniform electric field.

Later, the development of large-area CsI photocathodes made another breakthrough in Cherenkov photo detection allowing the transition from the photoionization in gas to photoelectron extraction from solid photocathodes in respectively faster MWPCs. Nowadays the majority of large RICH systems, presently in operation in high energy physics, are still based on this technology. The necessity to increase the counting rate capability for future applications (for example for the COMPASS RICH) has produced a further evolution of gaseous photomultiplier and all new developments are based on micro-pattern gaseous counters, combining fast detector operation with low cost per unit of surface. Promising results have also been obtained from R&D on micro-pattern photomultipliers using photocathodes sensitive to visible light.

Finally, it will be interesting to present a guess about the future of gaseous photomultipliers. Following the experience of the COMPAS experiment one can assume that in the near future most of the gaseous photomultipliers will be based on various micropattern multiplication structures. The main reason is that such structures efficiently suppress both photon and ion feedbacks, allowing to reach at the same time high gain, excellent rate and time characteristics and unprecedentedly high spatial resolution.

We think that the main applications of these detectors will be RICH and various UV sensitive devices capable to identify and visualize flames and sparks for fire safety (including homeland security) and various technical inspections.

REFERENCES

Anderson, D. F., Kwan, S., & Peskov, V. (1992). Properties of CsI and CsI-TMAE photocathodes. *Nuclear Instruments & Methods in Physics Research. Section A, Accelerators, Spectrometers, Detectors and Associated Equipment*, *323*, 626–634.

Bellazzini, R., Spandre, G., Brez, A., Minutii, M., Baldini, L., Lantronico, L., & Braem, A. et al. (2007). Imaging with the invisible light. *Nuclear Instruments & Methods in Physics Research. Section A, Accelerators, Spectrometers, Detectors and Associated Equipment*, *581*, 246–263.

Breskin, A., Buzulutskov, A., & Chechik, R. (2002). GEM photomultiplier operation in CF4. *Nuclear Instruments & Methods in Physics Research. Section A, Accelerators, Spectrometers, Detectors and Associated Equipment*, *483*, 670–675.

Breskin, A., Cortesi, M., Alon, R., Miuamoto, J., Checkik, R., Dangendorf, V., . . . Dos Santos, J. M. F. (2008). *Recent advances in thick-GEM (THGEM) detectors*. Paper presented at the RD-51 Workshop, NIKEF, Amsterdam.

Charpak, G., Derrie, J., Giomataris, Y., & Rebourgeard, P. (2002). Micromegas, a multipurpose gaseous detector. *Nuclear Instruments & Methods in Physics Research. Section A, Accelerators, Spectrometers, Detectors and Associated Equipment*, *478*, 26–36.

Charpak, G., Rahm, D., & Steiner, H. (1970). Some developments in the operation of multiwire proportional chambers. *Nuclear Instruments and Methods*, *80*, 13–34.

Day, D. (2004). *LAAPD performance measurements in liquid xenon*. Retrieved from http://www.nevis.columbia.edu/reu/2004/david.pdf

Derrie, J., & Giomataris, Y. (2002). Recent experimental results with MICROMEGAS. *Nuclear Instruments & Methods in Physics Research. Section A, Accelerators, Spectrometers, Detectors and Associated Equipment*, *477*, 23–28.

Fonte, P., Ferriera-Marques, R., Peskov, V., & Policarpo, A. (1999). Single-electron pulse-height spectra in thin-gap parallel-plate chambers. *Nuclear Instruments & Methods in Physics Research. Section A, Accelerators, Spectrometers, Detectors and Associated Equipment*, *433*, 513–517.

Fonte, P., & Peskov, V. (2002). High-resolution TOF with RPCs. *Nuclear Instruments & Methods in Physics Research. Section A, Accelerators, Spectrometers, Detectors and Associated Equipment*, *477*, 17–22.

Fonte, P., Peskov, V., Rodionov, I., & Fonte, P. (2004). High rate (up to 10^5 Hz/cm^2), high position resolution (30 μm) photosensitive RPCs. *Nuclear Instruments & Methods in Physics Research. Section A, Accelerators, Spectrometers, Detectors and Associated Equipment*, *533*, 163–168.

Genz, H. (1973). Single electron detection in proportional gas counters. *Nuclear Instruments and Methods*, *112*, 83–90.

Lippmann, C., Vincke, H., & Riegler, W. (2009). Simulation of RPC performance for 511 keV photon detection. *Nuclear Instruments & Methods in Physics Research. Section A, Accelerators, Spectrometers, Detectors and Associated Equipment*, *602*, 735–739.

Lyashenko, A. V., Breskin, A., Chechik, R., Veloso, J. F. C. A., Dos Santos, J. M. F., & Amaro, F. D. (2009). Development of high-gain gascous photomultipliers for the visible spectral range. *JINST*, *4*, 1–22.

Melai, J., Breskin, A., Cortesi, M., Bilevych, Y., Fransen, M., Van der Graaf, H., & Schmitz, J. et al. (2011). A UV sensitive integrated Micromegas with Timepix readout. *Nuclear Instruments & Methods in Physics Research. Section A, Accelerators, Spectrometers, Detectors and Associated Equipment*, *628*, 133–137.

Monteiro, C. M. B., Fernandes, L. M. P., Lopes, J. A. M., Veloso, J. F. C. A., & Dos Santos, J. M. F. (2005). Detection of VUV photons with large-area avalanche photodiodes. *Applied Physics. B, Lasers and Optics*, *81*, 531–535.

Peskov, V. (1980). *Complex/multiparameter plasma diagnostics in UV and X-ray regions*. (Doctor of science thesis). Kapitza Institute for Physical problems, Moscow, Russia.

Piuz, F., Braem, A., Davenport, M., Di Bari, D., Di Mauro, A., Elia, D., & Williams, T. D. et al. (1999). Final tests of the CsI-based ring imaging detector for the ALICE experiment. *Nuclear Instruments & Methods in Physics Research. Section A, Accelerators, Spectrometers, Detectors and Associated Equipment*, *433*, 178–189.

Riegler, W., & Lippmann, C. (2003). Detailed models for timing and efficiency in resistive plate chambers. *Nuclear Instruments & Methods in Physics Research. Section A, Accelerators, Spectrometers, Detectors and Associated Equipment*, *508*, 14–18.

Szczesniak, T., Moszynski, M., Swiderski, L., Nassalski, A., Synfeld-Kazuch, A., Dehaine, A. G., & Kapusta, M. (2009). A comparative study of fast photomultipliers for timing experiments and TOF PET. *IEEE Transactions on Nuclear Science*, *56*, 1017–1023.

Va'vra, J., Benitez, J., Leith, D. W. G. S., Mazaheri, G., Ratcliff, B., & Schwienig, J. (2007). A 30 ps timing resolution for single photons with multi-pixel Burle MCP-PMT. *Nuclear Instruments & Methods in Physics Research. Section A, Accelerators, Spectrometers, Detectors and Associated Equipment*, *572*, 459–462.

Appendix

AUTHORS' CONTRIBUTIONS TO THE DEVELOPMENT AND APPLICATIONS OF POSITION-SENSITIVE GASEOUS PHOTOMULTIPLIERS

This appendix describes the authors of this book, their scientific background and their personal contribution to the field of photodetectors.

We started collaborating with each other in early 1980, when both of us worked at CERN. At this time Tom Francke investigated various photosensitive elements for gaseous detectors (Francke, 1991) and later was deeply involved in construction and tests of TMAE filled MWPCs for RICH detectors. Vladimir Peskov already actively used photosensitive gaseous detector for spectral measurements and for plasma visualization in the UV region of spectra. He had several times visited Charpak's group at CERN and from 1986 become a CERN scientific associate.

From 1998 we worked in the same group in the Royal Institute of Technology, Stockholm and later in the photonic company XCounter in Sweden, led by Tom Francke. Together we performed many interesting developments and studies in the field of gaseous electron multipliers.

The authors' contributions can be summarized as follows:

1. **Development of Gaseous Detectors Filled with Photosensitive Vapours:** To our best knowledge the first use of photosensitive gaseous detectors for plasma diagnostic is dated by 1972 (Diatroptov,1972), when single wire counters filled with ethylferrocene and other photosensitive vapours were used for spectral investigation of plasma emission. Later Peskov's group developed photosensitive MWPCs filled with toluene vapours sensitive to wavelengths $\lambda<146$ nm. Their publication (Bogomolov, 1978) was sent to the publisher only two months later than the famous work of Séguinot (Séguinot, 1977) in which benzene vapours sensitive to wavelength $\lambda<135$nm were used. These two publications lead to a new era in development of photodetectors. Practically immediately photosensitive MWPCs began to be used in RICH detectors, for example in SLD at SLAC and DELPHI at CERN and for visualization of plasma instabilities allowing important discoveries to be made (Peskov, 1980a). MWPCs flushed with TMA vapours were used for the first time in a photodetector to detect secondary UV light from a gas scintillation counter (Peskov, 1980b, Bogomolov, 1987).
2. **Studies of Liquid Photocathodes:** The concept of a liquid TMAE photocathode was introduced by D. Anderson, and his idea triggered intense investigations performed by various groups. We (T. Francke and V. Peskov) also contributed to these studies (Peskov, 1988). Besides TMAE other liquid photocathodes were investigated, e.g. ethylferrocene (Charpak, 1989; Séguinot, 1990). It was discovered that absorbed layers of TMAE or EF deposited on metallic or organic photocathodes strongly enhance their quantum efficiency.

3. **CuI and Metalloorganic Photocathodes:** The first MWPC combined with a CuI photocathode is described in (Peskov 1980a). Such a detector operated on the principle of a surface photoelectric effect and for this reason was sensitive to longer wavelength than ones filled with photosensitive gases (Peskov 1980a; Charpak, 1991). Another important advantage of gaseous detectors combined with solid photocathodes is their excellent time resolution (better than 1 ns) since there are no space jitters in creating photoelectrons. They all come from the well-defined cathode (Peskov, 1992).

Other photocathodes were investigated as well: tetramethyl-p-phenylenediamine (TMPD) (Peskov, 1988; Francke Thesis), cyclopentadienyl-iron-hexamethyl-benzene (CPIHMB) (Peskov, 1989), diferrocenyl mercury (Charpak, 1991; Imrie, 1992), etc. A special focus was on photosensitive material stable in air which simplify their use in practice (Imrie, 1991).

4. **Development of Gaseous Photomultipliers Combined with CsI Photocathodes:** The greatest success came from the CsI photocathode. The first attempt to combine gaseous detector with a CsI photocathode was reported at Fermilab (Fermi National accelerator laboratory) in 1989 and triggered interest in the high energy physics community to this approach (Charpak, 1989). Inspired by these initial results other groups started to actively investigate CsI photocathodes, for example systematic studies were performed in work (Séguinot, 1990). Further tests revealed that one of the key procedures is heating the photocathode above 60°C, (Anderson, 1991). Another observation was that the highest CsI quantum efficiency was achieved when the evaporation process used freshly cleaved CsI crystals (undoped) rather than CsI powder which had been grinded long time ago (Anderson, 1994). In some brief laboratory studies, when one do not aim to reach the highest possible quantum efficiency, another very simple technique can be used often referred to as the "spray-on technique" when the CsI is dissolved in a solvent and spayed in air on a heated substrate (Anderson, 1992; Anderson, 1993). By this technique large area photocathode can be produced without access to a vacuum deposition system.

A further important step was developing the technology of transferring the CsI photocathode from the evaporation system into the gaseous detectors without exposure to air (Peskov, 1995). This ensured high quantum efficiency and the stability of the CsI photocathode after being combined with a gaseous detector.

In another series of studies the feedback problem in the CsI-PPAC was investigated and an optimum gas mixtures were found allowing achieving gas gains $\geq 10^6$ (Charpak, 1991a; Giomataris, 1992).

Based on the studies mentioned above, the authors of this book developed several designs of CsI-gaseous photomultipliers having not only high quantum efficiency, but also excellent position resolution (Iacobaeus, 2003; Francke, 2004; Fonte 2005; Bidault 2006).

A special place among them belongs to the hole-type gaseous photomultipliers which, due to their geometry, are capable to efficiently supress photon and ion feedbacks achieving higher gains even in poorly quenched gas mixtures (Iacobaeus, 2004; Iacobaeus, 2004a).

The technology of the CsI photocathodes, developed with the participation of the authors of this book, was carefully verified by the RD-26 collaboration established at CERN and was later accepted for the production of the CsI photocathodes for various large-scale experiments (in which we participated as well- see Hoedlmoser, 2006).

5. **Sealed Gaseous Detectors with CsTe Photocathodes:** The main advantage of the CsTe photocathode compared to other solid photocathodes mentioned above is that it has high quantum efficiency in a much wider spectral interval, 105- 350 nm. Unfortunately, this photocathode is easily damage by even a short contact with some gases: oxygen, water vapours etc. Hence, it cannot be exposed to air (Charpak, 1992a). This creates difficulties in its handling and exploitation, for example stable operation was achieved only with sealed detectors filled with ultraclean gases (Francke, 2002; Carlson, 2003; Fonte, 2005).

Such photomultiplier can be useful in many applications: RICH, flame detection, UV visualization in day light conditions etc.

6. **Development of Gaseous Photomultipliers Sensitive to Visible Light:** The gaseous detectors sensitive to visible photons represent the next generation of photomultipliers. The authors of this book made great contribution to these developments. For example in work (Peskov, 1988) an attempt was made to combine a Cs-based photocathode which was sensitive up to 450 nm with a single–wire counter filled with a mixture of a noble gas and hydrogen or methane (Peskov, 1988). Other photocathodes were also manufactured and combined with gaseous detectors, in particular GaAS(Cs) and AuCs, and their operation was investigated in various quenching gases (Borovik-Romanov, 1994; Peskov, 1994).

In the same work it was proposed and investigated coating of photocathodes sensitive to visible light with a thin film of CsI and some other materials. The systematic studies show that an additional thin layer on the top of the photocathode prevent or slow down the diffusion of the gas from the chamber whereas photoelectrons created from the photocathode have sufficient kinetic energy to penetrate through this layer and be recorded in gas detector Charpak, 1992a; Anderson, 1994).

After these first encouraging results, serious efforts were invested in developing highly efficient photocathodes. Maximum precautions were taken to ensure that the system is clean enough and made from appropriate materials having low outgassing properties (Peskov, 1995a). The highest quantum efficiencies (30 to 50%) and the most stable photocathode operations were obtained with glass chambers.

However, the maximum achievable gain of the first gaseous detectors combined with photocathodes sensitive to visible light is below 1000 which is not sufficient for the single photoelectron detection with a high probability. The maximum achievable gas gain was limited by photon and ion feedback. To reduce the feedbacks, in work (Peskov, 1999) the photocathodes were combined with glass capillary plates, operating in avalanche mode at atmospheric pressure. With such a device high quantum efficiencies and gas gains close to 10^5 was possible to achieve at the same time (Giunji, 2002). Nowadays, all prototypes of gaseous photomultipliers sensitive to visible light are based on cascaded hole-type structures.

Later another concept was propose, a hybrid detector (Francke, 2002; Rodionov, 2002). One of its version consisted of two planar parts: a sealed chamber with a semi-transparent bi-alkali or Cs_3Sb photocathode (the "scintillation chamber") and a position-sensitive gaseous detector with a CsI photocathode flushed with a gas at atmospheric pressure (the "readout detector").

7. **Aging and Stability Studies:** One of the key issues in applications of gaseous photomultiplier is not only their high quantum efficiencies and spatial resolutions, but also stability with time and aging. These characteristics were carefully studied by us for each photocathode and photodetector we had developed. Corresponding results are spread in the references presented above and also presented in some Chapters of this book. An example of these studies can be found in (Francke, 2002).
8. **Applications of Gaseous Photomultipliers:** As was already mentioned above, gaseous detectors filled with photosensitive vapours or combined with CuI photocathodes were used by us in spectroscopic studies and for plasma diagnostics- see for example, (Peskov, 1980a; Bogomolov, 1987; Francke 2004a).

We also made contributions in development and applications of TMAE- MWPCs in the CP-LEAR RICH (Rickenbach, 1989; Francke, 1991), CAPRICE 94 and CAPRICE 98 space experiments (Carlson 1994; Carlson 1994a; Carlson 1994b; Francke 1989; Francke 1989a; Francke 1989b, Francke 1992; Papini, 2004) and CsI-MWPCs in ALICE RICH (Hoedlmoser, 2006; Braem, 2007; Volpe, 2008). Note that above only a few of our publications are mentioned, the real list of references is much longer.

Some efforts were invested to upgrade the ALICE RICH detector, in particular the one based on a CsI-TGEMs and CsI-RETGEMs –see, for example (Di Mauro, 2011; Martinengo 2011; Peskov, 2012).

We also made applications of gaseous photomultipliers for the readout of various UV scintillators: gaseous (Peskov, 1980; Periale 2002; Periale, 2003), solid (Francke, 1989b; Charpak, 1991; Charpak, 1993) and even noble liquids (Periale, 2004; Periale, 2005; Periale, 2006; Periale, 2007; Peskov, 2007; Breskin, 2011).

One of the most exciting applications of the photosensitive gaseous detectors is recording sparks and open flames by their UV radiation (Carlson 2003; Carlson 2003a). Gaseous photomultiplier are 100-1000 times more sensitive than commercial devices and moreover can also determine the position of the flames. Rodionov, 2005; Bidault, 2006; Bidault, 2007; Periale, 2007; Di Mauro, 2007; Charpak, 2008; Charpak, 2009).

Finally, let's mention that the authors of this book performed developments of some other original types of gaseous detectors and methods based on photosensitive materials, for example detectors with CsI secondary electron emitters for the detection of X-rays (Crotty, 2003) and charged particles (Crotty, 1996; Ceron-Zebalos, 1997; Francke, 2003) or optical chambers based on the light emission of some photosensitive vapours (Charpak, 1988; Peskov, 1989).

REFERENCES

Anderson, D., Kwan, S., Honeisen, B., & Peskov, V. (1991). *Technical memo on new results on CsI photocathodes: enhancement and aging*. Fermilab technical Memo TM-1753. Retrieved from http://lss.fnal.gov/archive/test-tm/1000/fermilab-tm-1753.pdf

Anderson, D., Kwan, S., & Peskov, V. (1994). CsI and some new photocathodes. *Nuclear Instruments & Methods in Physics Research. Section A, Accelerators, Spectrometers, Detectors and Associated Equipment, 343*, 109–116.

Anderson, D. F., Kwan, S., & Peskov, V. (1993). CsI photocathode quantum efficiency and a simple production technique. *Nuclear Instruments & Methods in Physics Research. Section A, Accelerators, Spectrometers, Detectors and Associated Equipment, 326*, 611–612.

Anderson, D. F., Kwan, S., & Peskov, V., (1992). Properties of CsI and CsI-TMAE photocathodes. *Nuclear Instruments & Methods in Physics Research. Section A, Accelerators, Spectrometers, Detectors and Associated Equipment, 323*, 626–634.

Bidault, J. M., Crotty, I., & Di Mauro, A., (2007). The first applications of newly developed gaseous detectors with resistive electrodes for UV imaging in daylight conditions. *Nucl. Instrum. and Methods in Phys. Res., A580*, 1036-1041...., Bogomolov, G.D., Dubrovski, Y.V., & Peskov, V.D. (1978a). Multiwire gas counter for coordinate measurements in the VUV region. *Instruments and Experimental Techniques, 21*, 778–782.

Bidault, J. M., Fonte, P., & Francke, T., (2006). A novel UV photon detector with resistive electrodes. *Nuclear Physics B - Proceedings Supplements, 158*, 199–203.

Bogomolov, G. D., Dubrovski, I., Yu, V., Letunov, A. A., & Peskov, V., (1987). Experimental investigation of non-stationary streamer formation in a microwave plasma column at high pressure. *Soviet Physics, JETP, 66*, 295–302.

Bogomolov, G. D., Dubrovski, Y. V., & Peskov, V. D. (1978b). Photoelectrical readout for multiwire counters. *Instruments and Experimental Techniques, 21*, 639–642.

Borovik-Romanov, A., & Peskov, V. (1994). Cs based photocathodes for gaseous detectors. *Nuclear Instruments & Methods in Physics Research. Section A, Accelerators, Spectrometers, Detectors and Associated Equipment, 348*, 269–274.

Braem, A., De Cataldo, G., & Davenport, M., (2007). The current progress of the ALICE ring imaging Cherenkov detector. *Nuclear Instruments & Methods in Physics Research. Section A, Accelerators, Spectrometers, Detectors and Associated Equipment, 580*, 1144–1147.

Breskin, A., Peskov, V., & Cortesi, M., (2011). CsI-THGEM gaseous photomultipliers for RICH and noble-liquid detectors. *Nuclear Instruments & Methods in Physics Research. Section A, Accelerators, Spectrometers, Detectors and Associated Equipment, 639*, 117–120.

Carlson, P., (1994a). A RICH counter for identifying antiprotons and light isotopes in the cosmic radiation. *Nuclear Instruments & Methods in Physics Research. Section A, Accelerators, Spectrometers, Detectors and Associated Equipment, 343*, 19–23.

Carlson, P., (1994b). A RICH counter for identifying antimatter and light isotopes in the cosmic radiation. *IEEE Transactions on Nuclear Science, 41*, 1321–1324.

Carlson, P., Francke, T., & Löfgren, S., (1994). Results from beam tests of the CAPRICE RICH detector. *Nuclear Instruments & Methods in Physics Research. Section A, Accelerators, Spectrometers, Detectors and Associated Equipment, 349*, 577–591.

Carlson, P., Francke, T., & Lund-Jensen, B., (2003). Gaseous photomultipliers with solid photocathodes for the detection of sparks, flames and dangerous gases. *Nuclear Instruments & Methods in Physics Research. Section A, Accelerators, Spectrometers, Detectors and Associated Equipment*, *505*, 207–210.

Carlson, P., Iacobaeus, C., & Francke, T., (2003a). Beyond the RICH: Innovative photosensitive gaseous detectors for new fields of applications. *Nuclear Instruments & Methods in Physics Research. Section A, Accelerators, Spectrometers, Detectors and Associated Equipment*, *502*, 189–194.

Ceron-Zeballos, E., Crotty, I., & Hatzifoiadou, D., (1997). Resistive plate chambers with secondary electron emitters and microstrip readout. *Nuclear Instruments & Methods in Physics Research. Section A, Accelerators, Spectrometers, Detectors and Associated Equipment*, *392*, 150–154.

Charpak, G., Banaben, P., & Breuil, P., (2008). Development of new hole-type avalanche detectors and the firts resulsts of their applications. *IEEE Transactions on Nuclear Science*, *55*, 1657–1663.

Charpak, G., Benaben, P., & Breuil, P., (2009). Progress in development of a S-RETGEM-based detector for an early forest fire warning system. *JINST*, *4*, 1–16.

Charpak, G., Dominik, W., & Fabre, J. P., (1988). Some applications of the imaging proportional chamber. *IEEE Transactions on Nuclear Science*, *35*, 483–486.

Charpak, G., Fonte, P., & Peskov, V., (1991a). Investigation of operation of a parallel-plate avalanche chamber with a CsI photocathode under high gain conditions. *Nuclear Instruments & Methods in Physics Research. Section A, Accelerators, Spectrometers, Detectors and Associated Equipment*, *307*, 63–68.

Charpak, G., Gaudean, I., & Giomataris, I., (1993). Some studies of the applications of CsI photocathodes in gaseous detectors. *Nuclear Instruments & Methods in Physics Research. Section A, Accelerators, Spectrometers, Detectors and Associated Equipment*, *333*, 391–398.

Charpak, G., Lemenovski, D., & Peskov, V., (1991). New photocathodes for fast gaseous detectors. *Nuclear Instruments & Methods in Physics Research. Section A, Accelerators, Spectrometers, Detectors and Associated Equipment*, *310*, 128–132.

Charpak, G., Peskov, V., & Sauli, F. (1992a). Preliminary results of the study of gaseous detectors with solid photocathodes sensitive in the spectral region from 105 to 300 nm. *Nuclear Instruments & Methods in Physics Research. Section A, Accelerators, Spectrometers, Detectors and Associated Equipment*, *323*, 445–451.

Charpak, G., Peskov, V., & Sauli, F., (1989). Ethylferrocene in gas, condensed, or adsorbed phases: Three types of photosensitive elements for use in gaseous detectors. *Nuclear Instruments & Methods in Physics Research. Section A, Accelerators, Spectrometers, Detectors and Associated Equipment*, *277*, 537 546.

Charpak, G., Peskov, V., Scigoscki, D., Peskov, V., (1989). Mew scintillators for photosensitive gaseous detectors. In *Proceedings of the Sympos. on particle identification at high-luminosity hadron colliders*. Retrieved from http://lss.fnal.gov/conf/C890405/

Crotty, I., Fonte, P., & Francke, T., (2003). High-rate, high-position resolution microgap RPCs for X-ray imaging applications. *Nuclear Instruments & Methods in Physics Research. Section A, Accelerators, Spectrometers, Detectors and Associated Equipment*, *505*, 203–206.

Crotty, I., Fonte, P., & Lemenovski, D., (1996). A new resistive plate chamber with secondary electron emitters and two dimensional microstrip readout. *IEEE Nucl. Sci. Conf. Record, 1*, 362–366.

Di Mauro, A., Agocs, A., & Alfaro, R., (2011). The VHMPID RICH upgrade for ALICE at LHC. *Nuclear Instruments & Methods in Physics Research. Section A, Accelerators, Spectrometers, Detectors and Associated Equipment, 639*, 274–277.

Diatroptov, D. B., & Peskov, V. D. (1972). Vacuum ultraviolet spectra of a free plasma filament in a microwave high-pressure discharge. *Soviet Phys. JETF, 34*, 554–558.

Fonte, P., Francke, T., & Pavlopoulos, N., (2005). Novel single photon detectors for UV imaging. *Nuclear Instruments & Methods in Physics Research. Section A, Accelerators, Spectrometers, Detectors and Associated Equipment, 553*, 30–34.

Francke, T. (1991). *A ring imaging Cherenkov counter using a solid radiator*. (Ph.D. Thesis). Royal Institute of Technology, Stockholm, Sweden

Francke, T., (1989). Compact and fast RICH detector using a solid radiator: Principle and first experimental results. *Nuclear Instruments & Methods in Physics Research. Section A, Accelerators, Spectrometers, Detectors and Associated Equipment, 279*, 317–324.

Francke, T., (1991a). Particle identification by Cherenkov ring imaging using a neural network approach. *Nuclear Instruments & Methods in Physics Research. Section A, Accelerators, Spectrometers, Detectors and Associated Equipment, 307*, 47–51.

Francke, T., (1991b). A fast and compact solid radiator RICH counter. *Nuclear Instruments and Methods, A310*, 146–150.

Francke, T., Bergstrom, D., & Boezio, M., (1999). A gas- RICH detector for space. *Nuclear Instruments & Methods in Physics Research. Section A, Accelerators, Spectrometers, Detectors and Associated Equipment, 433*, 87–91.

Francke, T., Fonte, P., & Peskov, V., (2003a). Potential of RPCs for tracking. *Nuclear Instruments & Methods in Physics Research. Section A, Accelerators, Spectrometers, Detectors and Associated Equipment, 508*, 83–87.

Francke, T., & Peskov, V. (2003). Aging in gaseous photodetectors. *Nuclear Instruments & Methods in Physics Research. Section A, Accelerators, Spectrometers, Detectors and Associated Equipment, 515*, 292–301.

Francke, T., & Peskov, V. (2004a). Photosensitive gaseous detectors and their applications. *Nuclear Instruments & Methods in Physics Research. Section A, Accelerators, Spectrometers, Detectors and Associated Equipment, 525*, 1–5.

Francke, T., Peskov, V., & Rodionov, I., (2002). Novel position-sensitive gaseous detectors with solid photocathodes. *IEEE Transactions on Nuclear Science, 49*, 977–983.

Francke, T., Peskov, V., & Rodionov, I., (2004). High rate (up to 10^5 Hz/cm^2), high position resolution (30 μm) photosensitive RPCs. *Nuclear Instruments & Methods in Physics Research. Section A, Accelerators, Spectrometers, Detectors and Associated Equipment, 533*, 163–168.

Francke, T., & Suffert, M. (1989a). Rich counter using a solid radiator and a low pressure MWPC. *Nuclear Instruments & Methods in Physics Research. Section A, Accelerators, Spectrometers, Detectors and Associated Equipment, 283*, 60–64.

Francke, T., & Suffert, M. (1989b). Yield of photoelectrons produced in TMAE and solid TMPD by BaF2 scintillation light. *Nuclear Instruments & Methods in Physics Research. Section A, Accelerators, Spectrometers, Detectors and Associated Equipment, 276*, 263–267.

Francke, T., Suffert, M., & Carlson, P. (1992). Identification of antiprotons in space using a compact RICH counter. *IEEE Transactions on Nuclear Science, 39*, 971–979.

Giomataris, Y., Charpak, G., & Peskov, V., (1992). New generation of Cherenkov counters. *Nuclear Instruments & Methods in Physics Research. Section A, Accelerators, Spectrometers, Detectors and Associated Equipment, 323*, 431–438.

Giomataris, Y., Charpak, G., & Peskov, V., (1992). New generation of Cherenkov counter. *Nuclear Instruments & Methods in Physics Research. Section A, Accelerators, Spectrometers, Detectors and Associated Equipment, 323*, 431–438.

Giunji, S., Peskov, V., & Sakurai, H., (2002). Micro capillary glass plates as a gaseous detector of photons and particles. *Nuclear Instruments & Methods in Physics Research. Section A, Accelerators, Spectrometers, Detectors and Associated Equipment, 477*, 8–20.

Hoedlmoser, H., & Braem, A., De Cataldo, et al. (2006). Production technique and quality evaluation of CsI photocathodes for the ALICE/HMPD detector. *Nuclear Instruments & Methods in Physics Research. Section A, Accelerators, Spectrometers, Detectors and Associated Equipment, 566*, 338–350.

Iacobaeus, C., Breskin, A., & Danielsson, M., (2004). Advances in capillary-based gaseous UV imaging detectors. *Nuclear Instruments & Methods in Physics Research. Section A, Accelerators, Spectrometers, Detectors and Associated Equipment, 525*, 42–48.

Iacobaeus, C., Fonte, P., & Francke, T., (2003). The development and study of high-position resolution (50 μm) RPCs for imaging x-ray and UV photons. *Nuclear Instruments & Methods in Physics Research. Section A, Accelerators, Spectrometers, Detectors and Associated Equipment, 513*, 244–249.

Iacobaeus, C., Francke, T., & Danielsson, M., (2004a). Study of capillary-based gaseous detectors. *IEEE Transactions on Nuclear Science, 51*, 952–959.

Imrie, D.C., & Moghul, A.H., Wells, et al. (1991). Development of large-area, air-stable photocathodes for the ultraviolet. *Nuclear Instruments & Methods in Physics Research. Section A, Accelerators, Spectrometers, Detectors and Associated Equipment, 310*, 122–127.

Imrie, D. C., Peskov, V., & Tamadoni-Jahromi, R. (1992). Air-stable solid photocathodes with good ageing properties for fast gaseous detectors. *Nuclear Instruments & Methods in Physics Research. Section A, Accelerators, Spectrometers, Detectors and Associated Equipment, 317*, 92–96.

Martinenego, P., Peskov, V., & Breskin, A., (2011). R&D results on a CsI-coated triple thick GEM-based photodetector. *Nuclear Instruments & Methods in Physics Research. Section A, Accelerators, Spectrometers, Detectors and Associated Equipment, 639*, 126–129.

Papini, P., Piccardi, S., & Spillantini, P., (2004). High-energy deuteron measurement with the CAPRICE98 experiment. *The Astrophysical Journal, 615*, 259–274.

Periale, L., Peskov, V., & Braem, A., (2007). Development of new sealed UV sensitive gaseous detectors and their applications. *Nuclear Instruments & Methods in Physics Research. Section A, Accelerators, Spectrometers, Detectors and Associated Equipment, 572*, 189–192.

Periale, L., Peskov, V., & Carlson, P., (2002). Detection of the primary scintillation light from dense Ar, Kr and Xe with novel photosensitive gaseous detectors. *Nuclear Instruments & Methods in Physics Research. Section A, Accelerators, Spectrometers, Detectors and Associated Equipment, 497*, 242–248.

Periale, L., Peskov, V., & Carlson, P., (2003). Evaluation of various planar gaseous detectors with CsI photocathodes for the detection of primary scintillation light from noble gases. *Nuclear Instruments & Methods in Physics Research. Section A, Accelerators, Spectrometers, Detectors and Associated Equipment, 478*, 377–383.

Periale, L., Peskov, V., & Iacobaeus, C., (2004). The development of gaseous detectors with solid photocathodes for low-temperature applications. *Nuclear Instruments & Methods in Physics Research. Section A, Accelerators, Spectrometers, Detectors and Associated Equipment, 535*, 517–522.

Periale, L., Peskov, V., & Iacobaeus, C., (2005). The successful operation of hole-type gaseous detectors at cryogenic temperatures. *IEEE Transactions on Nuclear Science, 52*, 927–931.

Periale, L., Peskov, V., & Iacobaeus, C., (2006). A study of the operation of especially designed photosensitive gaseous detectors at cryogenic temperatures. *Nuclear Instruments & Methods in Physics Research. Section A, Accelerators, Spectrometers, Detectors and Associated Equipment, 567*, 381–385.

Peskov, V. (1980a). Coordinate gas scintillation x-ray and UV photon counters. *Instruments and Experimental Techniques, 23*, 507–510.

Peskov, V. (1980b). *Complex/multiparameter plasma diagnostics in UV and X-ray regions.* (Doctor of science thesis). Kapitza Institute for Physical problems, Moscow, Russia.

Peskov, V. (1988). Gaseous emitters and detectors of VUV radiation based on Cs cathodes. *Journal of Applied Spectroscopy, 48*(2), 316–317.

Peskov, V. (1995). Gaseous detectors with solid photocathodes. In *Proceedings of the First International Four Seas Conference.* Retrieved from http://cds.cern.ch/record/282759/files/CERN-97-06.pdf

Peskov, V., Baibussinov, B., & Centro, S., (2007). Development and first tests of GEM-like detectors with resistive electrodes. *IEEE Transactions on Nuclear Science, 54*, 1784–1791. doi:10.1109/TNS.2007.905160

Peskov, V., Bencze, G., & Di Mauro, A., (2012). First observation of Cherenkov rings with a large area CsI-TGEM-bases RICH prototype. *Nuclear Instruments & Methods in Physics Research. Section A, Accelerators, Spectrometers, Detectors and Associated Equipment, 695*, 154–158.

Peskov, V., Borovik-Romanov, A., & Sokolova, T., (1994). Gaseous detectors of ultraviolet and visible photons. *Nuclear Instruments & Methods in Physics Research. Section A, Accelerators, Spectrometers, Detectors and Associated Equipment, 353*, 184–188.

Peskov, V., Charpak, G., & Dominik, W., (1989). Investigation of light emission from a parallel-plate avalanche chamber filled with noble gases and with TEA, TMAE and H_2O vapors at atmospheric pressure. *Nuclear Instruments & Methods in Physics Research. Section A, Accelerators, Spectrometers, Detectors and Associated Equipment*, *277*, 547–556.

Peskov, V., Charpak, G., & Mine, P., (1988). Liquid and solid organic photocathodes. *Nuclear Instruments & Methods in Physics Research. Section A, Accelerators, Spectrometers, Detectors and Associated Equipment*, *269*, 149–160.

Peskov, V., Charpak, G., & Sauli, F., (1989). Organometallic photocathodes for parallel-plate and wire chambers. *Nuclear Instruments & Methods in Physics Research. Section A, Accelerators, Spectrometers, Detectors and Associated Equipment*, *283*, 786–791.

Peskov, V., & Silin, E. (1995). High efficiency photocathodes for gaseous detectors. *Nuclear Instruments & Methods in Physics Research. Section A, Accelerators, Spectrometers, Detectors and Associated Equipment*, *367*, 347–352.

Peskov, V., Silin, E., & Sokolova, T., (1999). *First attempts to combine capillary tubes with photocathodes*. Academic Press.

Peskov, V., & Zichichi, A. (1992). Fast (<1ns) gaseous detectors with solid photocathodes for frontier physics. *Nuclear Instruments & Methods in Physics Research. Section A, Accelerators, Spectrometers, Detectors and Associated Equipment*, *315*, 77–81.

Rickenbach, R., Montanet, F., & Backenstoss, G., (1989). The fast strangeness trigger of the CP-LEAR experiment. *Nuclear Instruments & Methods in Physics Research. Section A, Accelerators, Spectrometers, Detectors and Associated Equipment*, *279*, 305–309.

Rodionov, I., Bidault, J-M., Crotty, I., et al. (2005). Advanced gaseous photodetectors for hyperspectroscopy and other applications. *IEEE Nucl., Sci., Conf. Record, J04-2*, 3045-3049

Rodionov, I., Francke, T., & Peskov, V., (2002). Hybrid gaseous photomultipliers. *Nuclear Instruments & Methods in Physics Research. Section A, Accelerators, Spectrometers, Detectors and Associated Equipment*, *478*, 384–390.

Séguinot, J., & Charpak, G., Giomataris, et al. (1990). Reflective UV photocathodes with gas-phase electron extraction: Solid, Liquid and adsorbed films. *Nuclear Instruments & Methods in Physics Research. Section A, Accelerators, Spectrometers, Detectors and Associated Equipment*, *297*, 133–147.

Séguinot, J., & Ypsilantis, T. (1977). Photo-ionization and Cherenkov ring imaging. *Nuclear Instruments and Methods*, *142*, 377–391.

Volpe, G., Di Bari, D., & Di Mauro, A., (2008). Gas Cherenkov detectors for high momentum charged particles identification in the ALICE experiment at LHC. *Nuclear Instruments & Methods in Physics Research. Section A, Accelerators, Spectrometers, Detectors and Associated Equipment*, *595*, 40–43.

Glossary

1-D: One-dimensional.
2-D: Two-dimensional (e.g. position resolutions).
ADC: Analog to Digital Converter.
ALD: Atomic Layer Deposition.
ALICE: A large ion collider experiment.
Amplex: A monolithic analog signal processor fabricated in 3 μm n-well CMOS.
AMS: Alpha Magnetic Spectrometer.
APD: Avalanche Photodiode.
ASIC: Application Specific Integrated Circuit.
ATLAS: A toroidal LHC apparatus.
Belle-II: An upgrade particle physics experiment conducting by the Belle collaboration, Japan.
CALICE: Calorimeter for Linear Collider Experiment.
CAPRICE: Cosmic Anti Particle Ring Imaging Cherenkov Experiment.
CBM: Compressed Baryonic Matter.
CCD: Charged Coupled Devices.
CERN PS/T10: Beat test facility.
CERN: European Center for Nuclear research.
CMOS: Complimentary Metal-Oxide Semiconductor.
CMS: Compact Muon Solenoid.
CNC-Machine: Computerized Numerical Control drilling machine.
COBRA: A hybrid micropattern detector derived from gas electron multiplier and microhole plate, having zigzag shape electrodes.
COMPASS: Common Muon and Proton Apparatus for Structure and Spectroscopy.
CP LEAR: Charge Parity violation studies at Low Energy Antiproton Ring experiment at CERN.
CP Violation: Charge Parity Violation.
CPIHMB: Cyclopentadienyl-Iron-Hexamethyl-Benzene.
CPU: Central Processing Unit.
CRAMS: Card Random Access Memory.
CRID: Cherenkov Ring Imaging Detector.
CsI-MWPC: A multiwire proportional chamber combine with a CsI photocathode.
CsI-PPAC: Parallel-Plate Avalanche Chamber combined with a CsI photocathode.
CsI-TGEM: A thick gas electron multiplier coated with a CsI layer.
CsI-TIC: Threshold Imaging Cherenkov detector containing a CsI-MWPC.

DELPHI: DEtector with Lepton, Photon and Hadron Identification.
DEP: Delft Electronic Products.
ECAL: Electromagnetic Calorimeter.
EF: Ethylferrocene.
eV: Electron-Volt.
FC72: Fluorinert, C_6F_{14} (perfluorohexane) Flash ADC. The name comes from the very fast conversion from analog to digital.
F-R-MHSP: Flipped-Reversed-Micro-Hole & Strip Plate.
FWHM: Full width at half maximum.
GASSIPLEX: An integrated front-end analog processor developed for the ALICE RICH.
GEM: Gas Electron Multiplier.
HADES: High Acceptance Di-Electron Spectrometer.
HAPD: Hybrid Avalanche Photodetector.
HBD: Hadron Blind Detector.
HCAL: Hadronic Calorimeter.
Hera-B: A large aperture high rate spectrometer at a proton-electron accelerator at DESY, Germany.
HPD: Hybrid Photodetector.
HV: High Voltage.
ICARUS: Imaging of cosmic and rare underground signals.
JLAB: Hall A Jefferson Lab Experiment.
JLAB: Jefferson National Accelerator Facility.
KLOE-k: Long experiment for studying CP violation at French collider DAPHNE.
LAr: Liquid Ar.
LEAR: Low Energy Antiproton Ring.
LHC: Large Hadron Collider.
LHC-B: Large Hadron Collider Beauty experiment.
LYSO: Crystals of $Lu_{1.9}Y_{0.1}SiO_5$(Ce) 0.5 mol %.
MAPMT: Multianode Photomultiplier Tube.
MB-GP: Micro-Blasting Glass Plate.
MCP: Microchannel Plate.
MCPO-PMT: A vacuum photomultiplier based on microchannel plates.
Medipix: A family of photon counting pixel detector.
MGC: Micogap Chamber.
MHSP: Microhole Strip Plate.
MICROMEGAS: Micromesh Gas Chamber.
MIP: Minimum Ionizing Particle.
MSGC: Microstrip Gas Chamber.
MWPC: Multiwire-Proportional Chamber.
NA-48: Experiment at CERN super-proton synchrotron for search direct CP violation.
NA-49: Large acceptance hadron detector for an investigation of Pb-induced reactions at the CERN.
NE561: White reflective paint used in some Cherenkov detectors.

NMSU-WIZARD/CAPRICE94 Spectrometer (New Mexico State University): A balloon-borne magnet spectrometer equipped with a solid radiator Ring Imaging Cerenkov (RICH) detector, a time-of-flight system, a tracking device consisting of drift chambers and multiwire proportional chambers, and a silicon-tungsten calorimeter.
NP: Neopentane.
P10: Gas mixture of Ar+10%CH_4.
PANDA: anti-Proton ANnihiliation at Darmstadt.
PCB: Printed Circuit Board.
PDE: Photon Detection Efficiency.
PET: Positron Emission Tomography.
PHENIX: Pioneering High-Energy Nuclear Interaction Experiment.
PM: Photomultiplier.
PMMA: Polymethyl Methacrylate.
PMT: Photomuliplier Tube.
PMT: Photomultiplier Tube.
PPAC: Parallel-Plate Avalanche Chamber.
PPO: Diphenyloxazole.
PSI: *Paul Scherrer Institute* and *ETH* (Laboratory of Ion Beam Physics) *Zurich.*
Q-Bus: One of the several bus technologies used for example with MicroVAX computer systems.
QE: Quantum Efficiency.
R&D: Research and Developments.
RD26: Development of CsI base RICH.
RE-MHSP: Resistive Microhole-Microstrip detectors.
RETGEM: Resistive Electrode Gas Electron Multiplier.
RICH: Ring Imaging Cherenkov detector.
RMS: Root Mean Square.
RPC: Resistive Plate Chamber.
SiPM: Silicon Photomultiplier.
SLD: SLAC Large Detectors.
Sm: Samarium.
STAR: Solenoidal Tracker at RHIC.
SWC: Single-Wire Counter.
TEA: Triethylamine.
TFT: Thin-Film Transistors.
TGEM: Thick Gas Electron Multiplier.
TIC: Threshold Imaging Cherenkov detector.
Timepix: A multifunctional chip derived from the Medipix developments.
TMA: Trimethylamine.
TMAE: Tetrakis(dimethylamino)ethylene.
TMP: Tetramethyl Pentane.
TMPD: Tetramethyl-P Phenylenediamine.
TMS: Tetramethyl Silane.
TOF-PET: Time Of Flight Positron Emission Tomography.
TOP: Time-Of-Propagation.

TORCH: Time Of Internally Reflected Cherenkov Light.
TPC: Time Projection Chamber.
TV: Television tube.
UHF: Ultra High Friquency.
UV: Ultraviolet.
VERSA Module Eurocard (VME): A high speed bus system with a multiprocessor capability.
VLSI: Very Large Scale Integration electronics.
VUV: Vacuum Ultraviolet.
WIMP: Weakly Interacting Massive Particles.
XENON Experiment: Experiment attempting to detect dark matter in the form of weakly interacting massive particles (WIMPs) via recoils created by them in a liquid xenon.
ZEPLIN: A dark matter experiment aiming to detect Galactic WIMPs using a liquid Xe target.

Compilation of References

Abbon, P., Alekseev, M., Angerer, H., Apollonio, M., Birsa, R., Bordalo, P., & Wollny, H. et al. (2008). The COMPASS RICH-1 detector upgrade. *The European Physical Journal. Special Topics*, *162*(1), 251–257. doi:10.1140/epjst/e2008-00800-2

Adams, J., & Manley, B. W. (1966). The mechanism of channel electron multiplication. *IEEE Transactions on Nuclear Science*, *NS-13*(3), 88–99. doi:10.1109/TNS.1966.4324084

Adam, W., Albrecht, E., Ambec, I., Augustinus, A., Barnoux, C., Boštjančič, B., & Zevgolatakos, E. et al. (1994). The forward ring imaging Cherenkov detector of DEPHI. *Nuclear Instruments & Methods in Physics Research. Section A, Accelerators, Spectrometers, Detectors and Associated Equipment*, *338*(2-3), 284–309. doi:10.1016/0168-9002(94)91314-5

Agakichiev, G., Drees, A., Glässel, P., Irmscher, D., Ososkov, G. A., Panebrattsev, Y., & Ullrich, T. (1996). Cherenkov ring fitting techniques for the CERES RICH detectors. *Nuclear Instruments & Methods in Physics Research. Section A, Accelerators, Spectrometers, Detectors and Associated Equipment*, *371*(1-2), 243–247. doi:10.1016/0168-9002(95)01168-4

Agocs, G., Clark, B., Martinengo, P., Oliveira, R., Peskov, Pietropaolo, F., & Picchi, P. (2008). Developments and the preliminary tests of resistive GEMs manufactured by a screen printing technology. *JINST*, *3*, 1–10.

Agocs, A. G., Di Mauro, A., Ben David, A., Clark, B., Martinengo, P., Nappi, E., & Peskov, V. (2008). Study of GEM-like detectors with resistive electrodes for RICH applications. *Nuclear Instruments & Methods in Physics Research. Section A, Accelerators, Spectrometers, Detectors and Associated Equipment*, *595*(1), 128–135. doi:10.1016/j.nima.2008.07.031

Akkerman, A., Boutboul, T., Breskin, A., Chechik, R., & Gibrekhterman, A. (1994). Low-energy electron transport in alkali halides. *Journal of Applied Physics*, *76*(8), 4656–4662. doi:10.1063/1.357303

Alavarez, V., Aznar, F, & Borges, F.I.G.M. (2014). Description and commissioning of NEXT-MM prototype: first results from operation in a xenon-trimethylamine gas mixture. *JINST,* 9P030101, 1-20.

Albrecht, E., Baum, G., Birsa, R., Bosteels, M., Brandamante, F., & Braem, A. et al. (2002). COMPASS RICH-1. *Nuclear Instruments & Methods in Physics Research. Section A, Accelerators, Spectrometers, Detectors and Associated Equipment*, *478*(1-2), 340–343. doi:10.1016/S0168-9002(01)01823-X

Albrecht, E., van Apeldoorn, G., Augustinus, A., Baillon, P., Battaglia, M., Bloch, D., & van Vulpen, I. et al. (1999). Operation, optimization and performance of the DELPHI RICH detectors. *Nuclear Instruments & Methods in Physics Research. Section A, Accelerators, Spectrometers, Detectors and Associated Equipment*, *433*(1-2), 47–58. doi:10.1016/S0168-9002(99)00320-4

Aleksan, R., Besson, P., Bourgeois, Ph., Garganne, P., Hamel, G., & Monchenault, P. (1994). Measurements of CsI photocathode quantum efficiency in methane. *Nuclear Instruments & Methods in Physics Research. Section A, Accelerators, Spectrometers, Detectors and Associated Equipment*, *343*(1), 173–177. doi:10.1016/0168-9002(94)90547-9

Alexeev, M., Birsa, R., Bodlak, M., Bradamante, F., Bressan, Buchele, M., ... Veloso, J.F.C.A. (2014). MPGD-based counters of single photons developed for COMPASS RICH-1. *JINST, 9*, 1-9.

Alexeev, M., Birsa, R., Bradamante, F., Bressan, A., Chiosso, M., Ciliberti, P., ... Veloso J.F.C.A. (2013a). Ion backflow in thick GEM-based detectors of single photons. *JINST, 8*, 1-17.

Alexeev, M., Birsa, R., Bradamante, F., Bressan, Buchele, M., Chiosso, M., ... Veloso, J.F.C.A. (2013). Status and progress of novel photon detectors based on THGEM and hybrid MPGD architectures. *JINST,* 8, 1-18.

Alexeev, M., Barbosa, F., Birsa, R., Bradamante, F., Bressan, A., Chiosso, M., & Wollny, H. et al. (2012). Detection of single photons with Thick GEM-based counters. *Nuclear Instruments & Methods in Physics Research. Section A, Accelerators, Spectrometers, Detectors and Associated Equipment, 695*, 159–162. doi:10.1016/j.nima.2011.11.079

Alexeev, M., & Birsa, , R., Bradamante, F., Bressan, A., Chiosso, Ciliberti, P., ... Veloso, J.F.C.A. (2012a). Development of TGEM-based Photon Detectors for COMPASS RICH-1. *Physics Procedia, 37*, 781–788. doi:10.1016/j.phpro.2012.02.422

Alexeev, M., Birsa, R., Bradamante, F., Bressan, A., Büchele, M., Chiosso, M., & Veloso, J. F. C. A. et al. (2014). Status and progress of the novel photon detectors based on TGEM and hybrid MPGD architectures. *Nuclear Instruments & Methods in Physics Research. Section A, Accelerators, Spectrometers, Detectors and Associated Equipment, 766*, 133–137. doi:10.1016/j.nima.2014.07.030

Alexeev, M., Birsa, R., Bradamante, F., Bressan, A., Chiosso, M., Ciliberti, P., & Wollny, H. et al. (2009). The quest for a third generation of gaseous photon detectors for Cherenkov. *Nuclear Instruments & Methods in Physics Research. Section A, Accelerators, Spectrometers, Detectors and Associated Equipment, 610*(1), 174–177. doi:10.1016/j.nima.2009.05.069

Alkazov, G. D. (1970). Statistics of electron avalanches and ultimate resolution of proportional chambers. *Nucl. Instr. and Meth. in Phys. Res., A89*, 155.

Alkhazov, G. D. (1970). Statistics of electron avalanches and ultimate resolution of proportional counters. *Nuclear Instruments and Methods, 89*, 155–165. doi:10.1016/0029-554X(70)90818-9

Almedia, J., Amadon, A., Besson, P., Bourgeous, P., Braem, A., Breskin, A., & Williams, T. D. et al. (1995). Review of the development of cesium iodine photocathodes for applications to large RICH detectors. *Nuclear Instruments & Methods in Physics Research. Section A, Accelerators, Spectrometers, Detectors and Associated Equipment, 367*(1-3), 332–336. doi:10.1016/0168-9002(95)00571-4

Ambriola, M. L., Barbiellini, G., Bartalucci, S., Basini, G., Bellotti, R., & Bergstroem, D. (1998). *CAPRICE98: A balloon borne magnetic spectrometer to study cosmic ray at different atmospheric depths.* 26th International Cosmic Ray Conference, Salt Lake City, UT.

Ambriola, M., Barbiellini, G., Bartalucci, S., Basini, G., Bellotti, R., & Bergström, D. et al.. (1999). CAPRICE98: A balloon borne magnetic spectrometer to study cosmic ray antimatter and composition at different atmospheric depths. *Nuclear Physics B, 78*, 32.

Ambrosio, C. D. H., & Leutz, H. (2003). Hybrid photon detectors. *Nuclear Instruments & Methods in Physics Research. Section A, Accelerators, Spectrometers, Detectors and Associated Equipment, 501*(2-3), 463–498. doi:10.1016/S0168-9002(03)00431-5

Anassontzis, E. G., van Apeldoorn, G., & Aria, E. (1992). The barrel ring imaging Cherenkov counter of DELPHI. *Nuclear Instruments & Methods in Physics Research. Section A, Accelerators, Spectrometers, Detectors and Associated Equipment, 323*, 351–362.

Anassontzis, E., Resvanis, L., Voulgaris, G., Ypsilantis, T., Tripp, R., Bral, A. R. P., & Milotti, E. et al. (1992). Background-free detector for cosmic ray showers in the atmosphere. *Nuclear Instruments & Methods in Physics Research. Section A, Accelerators, Spectrometers, Detectors and Associated Equipment*, *315*(1-3), 267–271. doi:10.1016/0168-9002(92)90714-F

Anderson, D., Kwan, S., Honeisen, B., & Peskov, V. (1991). *Technical memo on new results on CsI photocathodes: enhancement and aging.* Fermilab technical Memo *TM-1753*. Retrieved from: http://lss.fnal.gov/archive/test-tm/1000/fermilab-tm-1753.pdf

Anderson, D. F. (1980). A xenon gas scintillation proportional counter coupled to a photoionization detector. *Nuclear Instruments and Methods*, *178*(1), 125–130. doi:10.1016/0029-554X(80)90866-6

Anderson, D. F. (1982). Extraction of electrons from a liquid photocathode into a low pressure wire chamber. *Physics Letters*, *118B*(1-3), 230–232. doi:10.1016/0370-2693(82)90634-7

Anderson, D. F. (1988). Measurement of TMAE and TEA vapour pressure. *Nuclear Instruments & Methods in Physics Research. Section A, Accelerators, Spectrometers, Detectors and Associated Equipment*, *270*(2-3), 416–418. doi:10.1016/0168-9002(88)90709-7

Anderson, D. F., Bouclier, R., & Charpak, G. (1983). Coupling of a BaF_2 scintillator to a TMAE photocathode and a low-pressure wire chamber. *Nuclear Instruments & Methods in Physics Research. Section A, Accelerators, Spectrometers, Detectors and Associated Equipment*, *217*, 217–223.

Anderson, D. F., Bouclier, R., Charpak, G., Majewski, S., & Kneller, G. (1983). Coupling of a BaF_2 scintillator to a TMAE photocathode and a low pressure wire chamber.D. *Nuclear Instruments and Methods*, *217*(1-2), 217–223. doi:10.1016/0167-5087(83)90137-0

Anderson, D. F., Kwan, S., & Peskov, V. (1993). CsI photocathode quantum efficiency and a simple production technique. *Nuclear Instruments & Methods in Physics Research. Section A, Accelerators, Spectrometers, Detectors and Associated Equipment*, *326*(3), 611–612. doi:10.1016/0168-9002(93)90866-G

Anderson, D. F., Kwan, S., & Peskov, V. (1994). High counting rate resistive–plate chamber. *Nuclear Instruments & Methods in Physics Research. Section A, Accelerators, Spectrometers, Detectors and Associated Equipment*, *348*(2-3), 324–328. doi:10.1016/0168-9002(94)90755-2

Anderson, D. F., Kwan, S., Peskov, V., & Hoeneisen, B. (1992). Properties of CsI and CsI-TMAE photocathodes. *Nuclear Instruments & Methods in Physics Research. Section A, Accelerators, Spectrometers, Detectors and Associated Equipment*, *323*(3), 626–634. doi:10.1016/0168-9002(92)90007-Q

Anderson, D. F., Kwan, S., & Salomon, M. (1994). A low-pressure, microstrip gas chamber operated with secondary-electron emission. *Nuclear Instruments & Methods in Physics Research. Section A, Accelerators, Spectrometers, Detectors and Associated Equipment*, *346*(1-2), 102–106. doi:10.1016/0168-9002(94)90692-0

Anderson, D., Kwan, S., & Peskov, V. (1994). CsI and some new photocathodes. *Nuclear Instruments & Methods in Physics Research. Section A, Accelerators, Spectrometers, Detectors and Associated Equipment*, *343*(1), 109–116. doi:10.1016/0168-9002(94)90539-8

Anderson, W., Azmoun, B., Cherlin, A., Chi, C. Y., Citron, Z., & Connors, M. et al. (2011). Design, construction, operation and performance of a Hadron Blind Detector for the PHENIX experiment. *Nuclear Instruments & Methods in Physics Research. Section A, Accelerators, Spectrometers, Detectors and Associated Equipment*, *646*(1), 35–58. doi:10.1016/j.nima.2011.04.015

Andres, Y., Barbosa, J., Braem, A., Carrone, E., Cozza, D., Davenport, M., ... Williams, T.D. (2002). R&D in ALICE: The CsI-based RICH high momentum particle identification detector. *EPJ Direct C, 4*(S1), 1-7.

Aoyama, T. (1985). Generalized gas gain formula for proportional counters. *Nuclear Instruments & Methods in Physics Research. Section A, Accelerators, Spectrometers, Detectors and Associated Equipment*, *234*(1), 125–131. doi:10.1016/0168-9002(85)90817-4

Aprile, E. (2001). XENON: a Liquid Xe Experiment for Dark Matter. Proposal # 0201740 for National Science Foundation. Columbia University.

Aprile, E., Bolotnikov, A. E., & Bolozdynya, A. I. (2006). Noble gas detectors. Wiley-VCH. doi:10.1002/9783527610020

Arino, I., Bastos, J., Broemmelsiek, D., Carvalho, J., Chmeissani, M., Conde, P., & Živko, T. et al. (2004). The HERA-B ring imaging Cherenkov counter. *Nuclear Instruments & Methods in Physics Research. Section A, Accelerators, Spectrometers, Detectors and Associated Equipment*, *516*(2-3), 445–461. doi:10.1016/j.nima.2003.08.173

Arnold, R., Baillon, P., Besch, H. J., Bosteels, M., Christophel, E., Dracos, M., & Ypsilantis, T. et al. (1988). A ring imaging Cherenkov detector, the DELPHI barrel RICH prototype. Part A: Experimental studies of the detection efficiency and the spatial resolution. *Nuclear Instruments & Methods in Physics Research. Section A, Accelerators, Spectrometers, Detectors and Associated Equipment*, *270*(2-3), 255–288. doi:10.1016/0168-9002(88)90695-X

Artuso, M., Ayad, R., Bukin, K., Efimov, A., Boulahouache, C., Dambasuren, E., & Smith, A. et al. (2003). Construction, pattern recognition and performance of the CLEOIII LiF-TEA RICH detector. *Nuclear Instruments & Methods in Physics Research. Section A, Accelerators, Spectrometers, Detectors and Associated Equipment*, *502*(1), 91–100. doi:10.1016/S0168-9002(02)02162-9

Astruc, D., Charpak, G., & Mine, P. (1988). Organometallic compounds as photocathodes in gaseous detectors. *Proceeding. Int. Workshop on liquid-sate electronics*. Berlin: Hahn-Meitner Inst.

Astruc, D. (1986). Organoiron electron-reservoir complexes. *Accounts of Chemical Research*, *19*(12), 377–383. doi:10.1021/ar00132a001

Azevedo, C. D. R., Cortesi, M., Lyashenko, A. V., Breskin, A., Chechik, R., Miyamoto, J., & dos Santos, J. M. F. et al. (2010). Towards TGEM UV-photon detectors for RICH: On single-photon detection efficiency in Ne/CH4 and Ne/CF4. *JINST*, *5*(01P01002), 1–12. doi:10.1088/1748-0221/5/01/P01002

Azevero, C. D. R., Cortesi, M., & Lyashenko, A. V. (2010). Towards THGEM UV-photons detectors for RICH: On single-photon detection efficiency in Ne/CH_4 and Ne/CF_4. *JINST*, *5*(P01002), 1–12.

Azmount, B., Anderson, W., Crary, D., Durham, J., Hemmik, T., Kamin, J...Woody, C (2006). A study of gain stability and charging effects in GEM foils. *IEEE Nucl. Sci. Conf. Record, MP3-4*, 3847-3854.

Balcerzyk, M., Mormann, D., Breskin, A., Sinh, B. K., Fritas, E. D. C., Chechik, R., & Rappaport, M. et al. (2003). Methods of preparation and performance of sealed gas photomultipliers for visible light. *IEEE Transactions on Nuclear Science*, *50*(4), 847–854. doi:10.1109/TNS.2003.815142

Barbiellini, G. (1997). A gas-RICH detector for cosmic ray studies. *Proceedings of the 25th International Cosmic Ray Conference*.

Barbiellini, G., Basini, G., Bellotti, R., Bocciolini, M., Boezio, M., Massimo Brancaccio, F., & Zampa, N. et al. (1996). Performance of the CAPRICE RICH detector during the 1994 balloon flight. *Nuclear Instruments & Methods in Physics Research. Section A, Accelerators, Spectrometers, Detectors and Associated Equipment*, *371*(1-2), 169–173. doi:10.1016/0168-9002(95)01152-8

Barnyakov, A. Yu., Barnyakov, M. Yu., Barutkin, V. V., Bobrovnikov, V. S., Buzykaev, A. R., Inami, K., & Yurikusa, Y. et al. (2009). Photomultiplier tubes with three MCPs. *Nuclear Instruments & Methods in Physics Research. Section A, Accelerators, Spectrometers, Detectors and Associated Equipment*, *598*(1), 160–162. doi:10.1016/j.nima.2008.08.084

Barouch, G. (1999). Development of a fast gaseous detector: Micromegas. *Nucl. Instr. and Meth. in Phys. Res.*, *A432*, 32–48.

Bartoli, B., Peruzzo, L., Sartori, G., Bedeschi, F., Betrolucci, E., & Mariotti, M. (1991). Detection of ultraviolet Cherenkov light from high energy cosmic ray atmospheric showers: A field test. *Nuclear Instruments & Methods in Physics Research. Section A, Accelerators, Spectrometers, Detectors and Associated Equipment*, *302*(3), 515–521. doi:10.1016/0168-9002(91)90369-2

Basini, G. (1999). *Proceedings of the 26th International Cosmic Ray Conference*. Salt Lake City, UT: Academic Press.

Bateman, J. E., Apsimon, R. J., & Barlow, F. E. (1976). A new photomultiplier tube utilizing channel plate electron multiplier as the gain producing element. *Nuclear Instruments and Methods*, *137*(1), 61–70. doi:10.1016/0029-554X(76)90249-4

Baur, R., Breskin, A., Chechik, R., Drees, A., Faschingbauer, U., Fischer, P., & Wurm, J. P. et al. (1994). The CERES RICH detector system. *Nuclear Instruments & Methods in Physics Research. Section A, Accelerators, Spectrometers, Detectors and Associated Equipment*, *343*(1), 87–98. doi:10.1016/0168-9002(94)90537-1

Beaver, E. A., Mc, C. E., & Ilwain, C. E. (1971). A Digital Multichannel Photometer. *The Review of Scientific Instruments*, *42*(9), 1321–1324. doi:10.1063/1.1685376

Bellazzini, R., & Spandre, G. (2003). MicroPattern Gas Detectors with pixel read-out. *Nuclear Instruments & Methods in Physics Research. Section A, Accelerators, Spectrometers, Detectors and Associated Equipment*, *513*(1-2), 231–238. doi:10.1016/j.nima.2003.08.038

Bellazzini, R., Spandre, G., Brez, A., Minuti, M., Baldini, L., Latronico, L., & Braem, A. et al. (2007). Imaging with the invisible light. *Nuclear Instruments & Methods in Physics Research. Section A, Accelerators, Spectrometers, Detectors and Associated Equipment*, *581*(1-2), 246–253. doi:10.1016/j.nima.2007.07.098

Bellazzini, R., Spandre, G., Minuti, M., Baldini, L., Brez, A., Cavalca, F., & de Oliveira, R. et al. (2006). Direct reading of charged multipliers with a self- triggered CMOS analogue chip with 105 k pixels and 50 μm pitch. *Nuclear Instruments & Methods in Physics Research. Section A, Accelerators, Spectrometers, Detectors and Associated Equipment*, *566*(2), 552–562. doi:10.1016/j.nima.2006.07.036

Bellazzini, R., Spandre, G., Minuti, M., Baldini, L., Latronico, L., Omodei, N., & Lempinen, E. et al. (2007). A sealed gas pixel detector for X-ray imaging. *Nuclear Instruments & Methods in Physics Research. Section A, Accelerators, Spectrometers, Detectors and Associated Equipment*, *579*(2), 853–858. doi:10.1016/j.nima.2007.05.304

Bellazzini, R., Spandre, G., Minuti, M., Brez, A., Baldini, L., Latronico, L., & Siegmund, O. et al. (2008). Single photon imaging at ultra-high resolution. *Nuclear Instruments & Methods in Physics Research. Section A, Accelerators, Spectrometers, Detectors and Associated Equipment*, *591*(1), 125–128. doi:10.1016/j.nima.2008.03.040

Bergström, D. (1999). *Performance of the CAPRICE98 gas-RICH detector*. (Lic. Thesis). Royal Inst. of Tech., Stockholm, Sweden.

Bergström, D. (1999). A gas RICH for cosmic ray studies. *Proceedings of the 26th International Cosmic Ray Conference*.

Bergström, D. (2001). Performance of the CAPRICE98 balloon-borne gas-RICH detector. *Nuclear Instruments & Methods in Physics Research. Section A, Accelerators, Spectrometers, Detectors and Associated Equipment*, *463*, 161.

Bergström, D., Boezio, M., Carlson, P., Francke, T., Grinstein, S., Khalchukov, F., & Stochaj, S. J. et al. (2000). First Mass-resolved Measurement of High-Energy Cosmic-Ray Antiprotons. *The Astrophysical Journal*, *534*(2), 177. doi:10.1086/312665 PMID:10813676

Bidault, J. M., Fonte, P., Francke, T., Galy, P., Peskov, V., & Rodionov, I. (2006). A novel UV photon detector with resistive electrodes. *Nuclear Physics B - Proceedings Supplements*, *158*, 199–203. doi:10.1016/j.nuclphysbps.2006.07.009

Bidault, J.-M., Crotty, I., Di Mauro, A., Martinengo, P., Fonte, P., Galy, F., & Zanette, O. et al. (2007). The first applications of newly developed gaseous detectors with resistive electrodes for UV imaging in daylight conditions. *Nuclear Instruments & Methods in Physics Research. Section A, Accelerators, Spectrometers, Detectors and Associated Equipment*, *580*(2), 1036–1041. doi:10.1016/j.nima.2007.06.061

Biteman, V., Guinji, S., Peskov, V., Skurai, H., Silin, E., Sokolva, T., & Rodionov, I. (2001). Position sensitive gaseous photomultipliers. *Nuclear Instruments & Methods in Physics Research. Section A, Accelerators, Spectrometers, Detectors and Associated Equipment*, *471*(1-2), 205–208. doi:10.1016/S0168-9002(01)00967-6

Black, J. K., Deines-Jones, P., & Readys, S. E. (2003). X-ray polarimetry with an active-matrix pixel proportional counter. *Nuclear Instruments & Methods in Physics Research. Section A, Accelerators, Spectrometers, Detectors and Associated Equipment*, *513*(3), 639–643. doi:10.1016/j.nima.2003.07.015

Bocciolini, M., Celletti, F., Finetti, N., Grandi, M., Papini, P., Perego, A., & Zampa, N. et al. (1996). N. Zampa, *The WiZard/CAPRICE silicon-tungsten calorimeter. Nuclear Instruments & Methods in Physics Research. Section A, Accelerators, Spectrometers, Detectors and Associated Equipment*, *370*(2-3), 403–412. doi:10.1016/0168-9002(95)00841-1

Bocciolini, M., Celletti, F., Grandi, M., Perego, A., Spillantini, P., & Basini, G. et al.. (1993). Silicon calorimeter for cosmic antimatter search. *Nuclear Physics*, *C32*, 77.

Boezio, M., Carlson, P., Francke, T., Weber, N., Suffert, M., Hof, M., & Stochaj, S. J. et al. (1999). Stochaj, S. J., *New Measurement of the Flux of Atmospheric Muons. Physical Review Letters*, *82*(24), 4757–4760. doi:10.1103/PhysRevLett.82.4757

Bogomolov, G.D., Dubrovski Yu., Peskov, V.D. (1978a). Multiwire Gas Counter for Coordinate Measurements in the VUV Region. *Instr. Exp. Tech*, *21*, 778-782.

Bogomolov, G.D., Dubrovski, Y., & Peskov, V.D. (1978a). Multiwire Gas Counter for Coordinate Measurements in the VUV Region. *Instr. Exp. Tech*, *21*, 778-782.

Bogomolov, G. D., Dubrovski, I. Yu., & Letunov, A. A. (1987). Experimental investigation of non-stationary streamer formation in a microwave plasma column at high pressure. *Soviet Physics, JETP*, *66*, 295–302.

Bogomolov, G. D., Dubrovski, Yu. V., Letunov, A. A., & Peskov, V. (1987). Experimental investigation of non-stationary streamer formations in a microwave plasma column at high pressures. *Soviet Physics, JETP*, *66*(2), 295–302.

Bogomolov, G. D., Dubrovski, Yu., & Peskov, V. D. (1978b). Photoelectrical Readout for Multiwire Counters. *Instruments and Experimental Techniques*, *21*, 639–642.

Bogomolov, G. D., Dybrovski, U. V., & Peskov, V. D. (1978). Position-sensitive gaseous counters for spectral investigations in vacuum ultraviolet. *Prib. Tekhn. Eksp.*, *4*, 210–213.

Bogomolov, G. D., Yu, V., Dubrovski, V., & Peskov, V. D. (1978). Multiwire Gas Counter for Coordinate Measurements in the VUV Region. *Instruments and Experimental Techniques*, *21*, 778–782.

Bolozdynya, A. I. (1999). Two-phase emission detectors and their applications. *Nuclear Instruments & Methods in Physics Research. Section A, Accelerators, Spectrometers, Detectors and Associated Equipment, 422*(1-3), 314–320. doi:10.1016/S0168-9002(98)00965-6

Bondar, A., Buzulutskov, A. L., Shekhtman, L., Snopkov, R., & Tikhonov, Y. (2004). Cryogenic avalanche detectors based on gas electron multipliers. *Nuclear Instruments & Methods in Physics Research. Section A, Accelerators, Spectrometers, Detectors and Associated Equipment, 524*(1-3), 130–141. doi:10.1016/j.nima.2004.01.060

Bondar, A., Buzulutskov, A., Grebenuk, A., Pavlyuchenko, D., Snopkov, R., & Tikhonov, Y. (2007). First results of the two-phase argon avalanche detector performance with CsI photocathode. *Nuclear Instruments & Methods in Physics Research. Section A, Accelerators, Spectrometers, Detectors and Associated Equipment, 581*(1-2), 241–245. doi:10.1016/j.nima.2007.07.086

Bondar, A., Buzulutskov, A., Grebenuk, A., Pavlyuchenko, D., Tikhonov, Y., & Breskin, A. (2008). Thick GEM versus thin GEM in two-phase argon avalanche detectors. *JINST, 3*(07P07001), 1–20. doi:10.1088/1748-0221/3/07/P07001

Borovik-Romanov, A., & Peskov, V. (1994). Cs based photocathodes for gaseous detectors. *Nuclear Instruments & Methods in Physics Research. Section A, Accelerators, Spectrometers, Detectors and Associated Equipment, 348*(2-3), 269–274. doi:10.1016/0168-9002(94)90745-5

Bottcher, E. H. (1984). *Experimentelle Untersuchung der photoelektrischen Leitung reiner und mit aromatischen moluculen dotierler organisher.* (PhD Thesis). Free University Berlin.

Boucher, R. G., Charpak, G. W., & Gao, W. (1988). Test of an electromagnetic calorimeter using BaF_2 scintillators and photosensitive wire chamber between 1 and 9 GeV. *Nuclear Instruments & Methods in Physics Research. Section A, Accelerators, Spectrometers, Detectors and Associated Equipment, 267*(1), 69–86. doi:10.1016/0168-9002(88)90631-6

Bradbury, N. E. (1932). Photoelectric current in gases between parallel plates as a function of the potential difference. *Physical Review, 40*(6), 980–987. doi:10.1103/PhysRev.40.980

Braem, A. (1995). Private communication

Braem, A., Chesi, E., Dulinski, W., Filthaut, F., Go, A., Joram, C., & Ypsilantis, T. et al. (2000). Highly segmented large-area hybrid photodiodes with bialkali photocathodes and enclosed VLSI readout electronics. *Nuclear Instruments & Methods in Physics Research. Section A, Accelerators, Spectrometers, Detectors and Associated Equipment, 442*(1-3), 128–135. doi:10.1016/S0168-9002(99)01210-3

Braem, A., Chesi, E., Joram, C., Séguinot, J., Weilhammer, P., Giunta, M., & Sartori, G. et al. (2003). Development of a 10-inch HPD with integratedreadout electronics. *Nuclear Instruments & Methods in Physics Research. Section A, Accelerators, Spectrometers, Detectors and Associated Equipment, 504*(1-3), 19–23. doi:10.1016/S0168-9002(03)00744-7

Braem, A., De Cataldo, G., Davenport, M., Di Mauro, A., Franco, A., Gallas, A., & Peskov, V. et al. (2007). The current progress of the ALICE Ring Imaging Cherenkov Detector. *Nuclear Instruments & Methods in Physics Research. Section A, Accelerators, Spectrometers, Detectors and Associated Equipment, 580*(2), 1144–1147. doi:10.1016/j.nima.2007.06.050

Braem, A., DiMauro, A., Nappi, E., Ljubicic, A. Jr, Paic, G., Piuz, F., & Williams, T. D. et al. (1994). Fast RICH detector with a cesium iodide photocathode at atmospheric pressure. *Nuclear Instruments & Methods in Physics Research. Section A, Accelerators, Spectrometers, Detectors and Associated Equipment, 343*(1), 163–172. doi:10.1016/0168-9002(94)90546-0

Braem, A., Fabjan, C. W., Franz, A., Kaneta, M., Paić, G., & Piuz, F. et al. (1998). A threshold imaging Cherenkov detector with CsI photocathodes. *Nuclear Instruments & Methods in Physics Research. Section A, Accelerators, Spectrometers, Detectors and Associated Equipment, 409*(1-3), 426–431. doi:10.1016/S0168-9002(98)00118-1

Brauning, H., Breskin, A., Chechik, R., Dangendorf, V., Demian, A., Ullmann, K., & Schmidt-Bocking, H. (1994). A large volume 3D imaging gas scintillation counter with CsI-based wire chamber readout. *Nuclear Instruments & Methods in Physics Research. Section A, Accelerators, Spectrometers, Detectors and Associated Equipment, 348*(2-3), 223–227. doi:10.1016/0168-9002(94)90734-X

Breskin, A. (1995). Secondary emission gaseous detectors: A new class of radiation imaging devices. *Nuclear Physics B, 44*(1-3), 351–363. doi:10.1016/S0920-5632(95)80054-9

Breskin, A. (1996). CsI UV photocathodes: History ad mystery. *Nuclear Instruments & Methods in Physics Research. Section A, Accelerators, Spectrometers, Detectors and Associated Equipment, 371*(1-2), 116–136. doi:10.1016/0168-9002(95)01145-5

Breskin, A., Alon, R., Cortesi, M., Chechik, R., Miyamoto, J., & Dangendorf, V. et al. (2009). A concise review on THGEM detectors. *Nuclear Instruments & Methods in Physics Research. Section A, Accelerators, Spectrometers, Detectors and Associated Equipment, 598*(1), 107–111. doi:10.1016/j.nima.2008.08.062

Breskin, A., Buzulutskov, A., & Chechik, R. (2002). GEM photomultiplier operation in CF_4. *Nuclear Instruments & Methods in Physics Research. Section A, Accelerators, Spectrometers, Detectors and Associated Equipment, 483*(3), 670–675. doi:10.1016/S0168-9002(01)01940-4

Breskin, A., Buzulutskov, A., Chechik, R., Sinh, B. K., Bondar, A., & Shekhtman, L. (2002). Sealed GEM photomutiplier with a CsI photocathode: Ion feedback and aging. *Nuclear Instruments & Methods in Physics Research. Section A, Accelerators, Spectrometers, Detectors and Associated Equipment, 478*(1-2), 225–229. doi:10.1016/S0168-9002(01)01752-1

Breskin, A., Buzulutskov, A., Shefer, E., Chechik, R., & Prager, M. (1998). Removable organic protective coating for alkali-antimonide photocathodes. *Nuclear Instruments & Methods in Physics Research. Section A, Accelerators, Spectrometers, Detectors and Associated Equipment, 413*(2-3), 275–280. doi:10.1016/S0168-9002(98)00547-6

Breskin, A., Chechik, R., & Frumkin, I. (1996). On atmospheric pressure operation of secondary electron emission gaseous X-ray imaging detectors. *Nuclear Instruments & Methods in Physics Research. Section A, Accelerators, Spectrometers, Detectors and Associated Equipment, 380*(3), 562–567. doi:10.1016/0168-9002(96)00579-7

Breskin, A., Chechik, R., Vartsky, D., Malamud, G., & Mine, P. (1994). A correction of the quantum efficiency of CsI and other photocathodes due to the recalibration of the reference photomultipliers. *Nuclear Instruments & Methods in Physics Research. Section A, Accelerators, Spectrometers, Detectors and Associated Equipment, 343*(1), 159–162. doi:10.1016/0168-9002(94)90545-2

Breskin, A., Peskov, V., Cortesi, M., Budnik, R., Chechik, R., & Duval, S. et al. (2011). CsI-THGEM gaseous photomultipliers for RICH and noble –liquid detectors. *Nuclear Instruments & Methods in Physics Research. Section A, Accelerators, Spectrometers, Detectors and Associated Equipment, 639*(1), 117–120. doi:10.1016/j.nima.2010.10.034

Breuil, P., Fonte, P., & Nappi, E. (2013). Advances in development of micropattern gaseous detectors with resistive electrodes. *Eur. Phys. J. Plus, 160*, 1–16.

Britting, A., Eyrich, W., & Lechmann, A. (2009). *Studies of MCP properties*. Retrieved from: http://www.google.fr/url?sa=t&rct=j&q=&esrc=s&source=web&cd=3&ved=0CDQQFjAC&url=http%3A%2F%2Fwww-hep.uta.edu%2F~ihowley%2FAFP%2FReadings%2Flehman-giessen.pdf&ei=ZT1_VM7HCNHrarOQgLAF&usg=AFQjCNEkTXwOt8a3gu-eW7ZR1woIU_A1mA&bvm=bv.80642063,d.d2s

Bruining, H. (1954). *Physics and Applications of Secondary Electron Emission*. London: Pergamon Press.

Bruyndonckx, P., Xuan, L., Rajeswaran, S., Smolik, W., Tavernier, S., & Shuping, Z. (1996). Design and physical characteristics of a small animal PET using BaF_2 crystals and a photosensitive wire chamber. *Nuclear Instruments & Methods in Physics Research. Section A, Accelerators, Spectrometers, Detectors and Associated Equipment, 382*(3), 589–600. doi:10.1016/S0168-9002(96)00804-2

Buckley, J., Dwyer, J., Müller, D., Swordy, S., & Tang, K. K. (1992). A ring imaging Cherenkov telescope for observations of high energy cosmic rays. *Nuclear Instruments & Methods in Physics Research. Section A, Accelerators, Spectrometers, Detectors and Associated Equipment, 323*(1-2), 380–385. doi:10.1016/0168-9002(92)90318-X

Busulutskov, A. F., Turchanovich, L. K., Vasilchenko, Y. L., Jansosn, Y. L., & Rachko, Z. A. (1990). Investigation of the crystal scintillators in the VUV region. *Nuclear Instruments & Methods in Physics Research. Section A, Accelerators, Spectrometers, Detectors and Associated Equipment, 288*(2-3), 659–661. doi:10.1016/0168-9002(90)90165-3

Buzulutskov, A. F. (2008). Gaseous photodetectors with solid photocathodes. *Physics of Particles and Nuclei, 39*(3), 424–453. doi:10.1134/S1063779608030052

Buzulutskov, A., Breskin, A., & Chechik, R. (1995). Heat enhancement of the photoefficiency from CsI, NaI and CuI photocathodes. *Nuclear Instruments & Methods in Physics Research. Section A, Accelerators, Spectrometers, Detectors and Associated Equipment, 366*(2-3), 410–412. doi:10.1016/0168-9002(95)00863-2

Buzulutskov, A., Breskin, A., Checik, R., Prager, M., & Shefer, E. (1997). Protection of caesium-antimony photocathodes. *Nuclear Instruments & Methods in Physics Research. Section A, Accelerators, Spectrometers, Detectors and Associated Equipment, 387*(1-2), 176–179. doi:10.1016/S0168-9002(96)00984-9

Byrne, J. (1962). Proc. Roy. Soc. Edinburgh. *Sect., A66*, 33.

Cabello, J., Barrillon, P., Barrio, J., Bisogni, M. G., Del Guerra, A., Lacasta, C., & Llosá, G. et al. (2013). High resolution detectors based on continuous crystals and SiPMs for small animal PET. *Nuclear Instruments & Methods in Physics Research. Section A, Accelerators, Spectrometers, Detectors and Associated Equipment, 718*, 148–150. doi:10.1016/j.nima.2012.08.094

Calvi, M., Giachero, A., Gotti, C. (2011). Aging and time resolution measurements for the Hamamatsu R 7600 multianode photomultiplier tube. *IEEE Nucl. Sci. Sympos. Conference Record.*

Calvi, M., Giachero, A., Gotti, C., Maino, M., Matteuzzi, C., & Pessina, G. (2013). Characterization of a Hamamatsu R7600 multi-anode photomultiplier tube with single photon signals. *JINST, 8*(02P02012), 1–11. doi:10.1088/1748-0221/8/02/P02012

Carlson, P., Iacobeaus, C., Francke, T., Lund-Jensen, B., Periale., Peskov, V., & Rodionov, I. (2003). Beyond the RICH: innovative photosensitive gaseous detectors for new fields of applications. *Nucl. Instrum. and Methods in Phys. Res., A502*, 189-194.

Carlson, P., Francke, T., Löfgren, S., Weber, N., & Suffert, M. (1994). Results from beam tests of the CAPRICE RICH detector. *Nuclear Instruments & Methods in Physics Research. Section A, Accelerators, Spectrometers, Detectors and Associated Equipment, 349*(2-3), 577–591. doi:10.1016/0168-9002(94)91229-7

Carlson, P., Francke, T., Lund-Jensen, B., & Peskov, V. (2003). Gaseous photomultipliers with solid photocathodes for the detection of sparks, flames and dangerous gases. *Nuclear Instruments & Methods in Physics Research. Section A, Accelerators, Spectrometers, Detectors and Associated Equipment, 505*(1-2), 207–210. doi:10.1016/S0168-9002(03)01053-2

Carruthers, G. R. (1969). Magnetically focused electronographic image convertors for space astronomy applications. *Applied Optics, 8*(3), 633–638. doi:10.1364/AO.8.000633 PMID:20072272

Carver, J. H., & Mitchell, P. (1964). Ionization chambers for the vacuum ultraviolet. *Journal of Scientific Instruments, 41*(9), 555–557. doi:10.1088/0950-7671/41/9/308

Cebrian, S., Dafni, T., Ferrer-Ribas, E. (2013). Micromegas-TPC operation at high pressure in xenon-trimethylamine mixtures. *JINST,* 8P01012, 1-16.

Ceron-Zeballos, E., Crotty, I., Fonte, P., Hatzifotiadou, D., Lamas Vlaverde, J., Peskov, V., … Zichichi, A. (1995). New developments of RPC: secondary electron emission and microstrip readout. In S.P. Ratti, & M. Merio (Eds.), *Proceedings of the International Workshop on RPCs and Related Detectors.* INFN.

Ceron-Zeballos, E., Crotty, I., Hatzifotiadou, D., Lamas, D., Valverde, J., Williams, M. C. S., & Peskov, V. et al. (1997). Resistive plate chambers with secondary electron emitters and microstrip readout. *Nuclear Instruments & Methods in Physics Research. Section A, Accelerators, Spectrometers, Detectors and Associated Equipment, 392*(1-3), 150–154. doi:10.1016/S0168-9002(97)00256-8

Chamberlain, O. (1960). The early antiproton work. In *Les Prix Nobel an 1959.* P.A. Norstedt & Söner.

Chamberlain, O., Segré, E., Wiegand, C., & Ypsilantis, T. (1955). Observation of antiprotons. *Physical Review, 100*(3), 947–950. doi:10.1103/PhysRev.100.947

Chandrasekharan, R., Messina, M., & Rubbia, A. (2006). Detection of VUV light at high quantum efficiency with large area avalanche photodiodes (LAAPDs). *Nuclear Instruments & Methods in Physics Research. Section A, Accelerators, Spectrometers, Detectors and Associated Equipment, 567*(1), 45–47. doi:10.1016/j.nima.2006.05.058

Charpak, G. (1999). *High resolution position detector for high-flux ionizing particles.* Patent WO1996FR01576.

Charpak, G., Dominik, W., Sauli, F. (1982).The gas photodiodes as a possible large-area photon detector. *Preprint* CERN-EP/82-169

Charpak, G., Peskov, V., & Schmidt, W. F. (1988). Liquid or solid photocathodes based on TMAE or TMPD: physics and application. *Proceeding. Int. Workshop on liquid-state electronics.* Berlin: Hahn-Meitner Inst.

Charpak, G., Peskov, V., & Scigocki, D. (1989). New scintillators for photosensitive gaseous detectors. Proceedings of Sympos. on particle identification at high-luminosity hadron colliders. Batavia, IL: CERN-LAA-EC-89-003.

Charpak, G., Peskov, V., & Scigocki, D. (1989b). *Study of BaF_2 calorimeters in future colliders.* Report at the 9th INFN Eloisotron project workshop: perspectives for new detectors in future supercolliders, Erice, Italy.

Charpak, G., Peskov, V., Breuil, P., Benaben, P. Patent: Electron multiplication device and systems for detection ionizing radiation, FR20070001202, WO2008129159, 2007.

Charpak, G., Peskov, V., Scigoscki, D., & Valbis, J. (1989). Mew scintillators for photosensitive gaseous detectors. *Proceedings of the Symposium on particle identification at high-luminosity hadron colliders.* Retrieved from: http://lss.fnal.gov/conf/C890405/

Charpak, G., Peskov, V., Scigoscki, D., & Valbis, J. (1989a). New scintillators for photosensitive gaseous detectors. *Proceedings of the Sympos. on particle identification at high-luminosity hadron colliders.* doi:10.1142/9789812795878_0057

Charpak, G., Peskov, V., Scigoscki, D., & Vlalbis, J. (1989). Mew scintillators for photosensitive gaseous detectors. *Proceedings of the Sympos. on particle identification at high-luminosity hadron colliders.* Retrieved from: http://lss.fnal.gov/conf/C890405/

Charpak, G. (1982). Some recent trends in the evolution of gaseous detectors. *Nuclear Instruments and Methods, 196*(1), 1–9. doi:10.1016/0029-554X(82)90608-5

Charpak, G. (1995). *Research on particle imaging detectors*. Singapore: World scientific.

Charpak, G. (2002). *Micromegas, a multipurpose gaseous detector*. Nucl. Instr. and Meth.in Phys. *Res., A478*, 26–36.

Charpak, G., Benaben, P., Breuil, P., Di Mauro, A., Martinenego, P., & Peskov, V. (2009). Development of new hole-type avalanche detectors and the first results of their applications. *IEEE Transactions on Nuclear Science*, *55*(3), 1657–1663. doi:10.1109/TNS.2008.924084

Charpak, G., Benaben, P., Breuil, P., Martinengo, P., Nappi, E., & Peskov, V. (2009). Progress in development of a S-RETGEM-based detector for an early forest fire warning system. *JINST*, *4*(12P12007), 1–16. doi:10.1088/1748-0221/4/12/P12007

Charpak, G., Bouclier, R., Bressani, T., Favier, J., & Zupančič, Č. (1968). The use of multiwire proportional counters to select and localize charged particles. *Nuclear Instruments and Methods in Physics Research*, *62*(3), 262–268. doi:10.1016/0029-554X(68)90371-6

Charpak, G., Dominik, W., Sauli, F., & Majewski, S. (1983). The gas photodiode as a possible large-area detector. *IEEE Transactions on Nuclear Science*, *NS-30*(1), 134–137. doi:10.1109/TNS.1983.4332236

Charpak, G., Domini, W., & Fabre, J. P. (1988). Some applications of the imaging proportional chamber. *IEEE Nucl. Sci*, *35*(1), 483–486. doi:10.1109/23.12770

Charpak, G., Fonte, P., Peskov, V., Sauli, F., Scigocki, D., & Stuart, D. (1991). Investigation of operation of a parallel-plate avalanche chamber with a CsI photocathode under high gain conditions. *Nuclear Instruments & Methods in Physics Research. Section A, Accelerators, Spectrometers, Detectors and Associated Equipment*, *307*(1), 63–68. doi:10.1016/0168-9002(91)90131-9

Charpak, G., Gauden, I., & Giomataris, Y. (1993). Some studies of the applications of CsI photocathodes in gaseous detectors. *Nuclear Instruments & Methods in Physics Research. Section A, Accelerators, Spectrometers, Detectors and Associated Equipment*, *333*(2-3), 391–398. doi:10.1016/0168-9002(93)91181-L

Charpak, G., Lamb, D., & Peskov, V. (1989a). New developments in calorimetry based on VUV scintillators coupled to photosensitive gaseous detectors.*Proceedings of the ECFA Study Week on Instrumentation Technology for High-luminosity Hadron Colliders*. Doi:10.5170/CERN-1989-010-V-2.593

Charpak, G., Lamb, D., Peskov, V., & Valbis, J. (1989). New developments in calorimetry based on VUV scintillators coupled to photosensitive gaseous detectors.*Proceedings of the ECFA Study Week on Instrumentation Technology for High-luminosity Hadron Colliders*.

Charpak, G., Lemenovski, D., Peskov, V., & Scigocki, D. (1991). New photocathodes for fast gaseous detectors. *Nuclear Instruments & Methods in Physics Research. Section A, Accelerators, Spectrometers, Detectors and Associated Equipment*, *310*(1-2), 128–132. doi:10.1016/0168-9002(91)91010-S

Charpak, G., Majewski, S., Melchart, G., Sauli, F., & Ypsilantis, T. (1979). Detection of far-ultraviolet photons with the multistep avalanche chamber. Application to Cherenkov light imaging and to some problems in high-energy physics. *Nuclear Instruments and Methods in Physics Research*, *164*(3), 419–433. doi:10.1016/0029-554X(79)90074-0

Charpak, G., Melchart, G., Petersen, G., & Sauli, F. (1979). High- accuracy localization of minimum ionizing particles using the cathode –induced charge centre-of-gravity read-out. *Nuclear Instruments and Methods*, *167*(3), 455–464. doi:10.1016/0029-554X(79)90227-1

Charpak, G., Peskov, V., & Sauli, F. (1992). Preliminary results of the study of gaseous detectors with solid photocathodes sensitive in the spectral region from 105 to 300 nm. *Nuclear Instruments & Methods in Physics Research. Section A, Accelerators, Spectrometers, Detectors and Associated Equipment*, *323*(1-2), 445–451. doi:10.1016/0168-9002(92)90330-7

Charpak, G., Peskov, V., Sauli, F., & Scigocki, D. (1989). Ethylferrocene in gas, condensed, or adsorbed phases: Three types of photosensitive elements for use in gaseous detectors. *Nuclear Instruments & Methods in Physics Research. Section A, Accelerators, Spectrometers, Detectors and Associated Equipment*, *277*(2-3), 537–546. doi:10.1016/0168-9002(89)90787-0

Charpak, G., Peskov, V., & Schmidt, W. F. (1988). Liquid and solid photocathodes based on TMAE or TMPD: physics and application.*Proceedings of the Internat. Workshop on Liquid-state electronics*.

Charpak, G., Peskov, V., & Scigocki, D. (1991a) New readout for BaF2 calorimeters: gaseous detectors with solid photocathodes. *Proc.Large Hadron Collider Workshop*.

Charpak, G., Peskov, V., Scigocki, D., & Lemenovski, D. (1991). Liquid and solid photocathodes and their applications for fast gaseous particle detectors. *IEEE Transactions on Electrical Insulation*, *26*(4), 623–630. doi:10.1109/14.83681

Charpak, G., Policarpo, A., & Sauli, F. (1980). The photo-ionization proportional scintillation chamber. *IEEE Transactions on Nuclear Science*, *NS-27*(1), 212–215. doi:10.1109/TNS.1980.4330830

Charpak, G., & Sauli, F. (1978). The multistep avalanche chamber: Anew high-rate, high accuracy gaseous detectors. *Physics Letters*, *78B*(4), 523–528. doi:10.1016/0370-2693(78)90502-6

Chechik, R., Cortesi, M., Breskin, A., Vartsky, D., Bar, D., & Dangendorf, V. (2006). *Thick GEM-like (TGEM) detectors and their possible applications*. Retrieved from: http://arxiv.org/ftp/physics/papers/0606/0606162.pdf

Chen, M., Mullins, M., & Pelly, , D., Shotkin, S., Sumorok, K., Akyuz, D., …Sugimoto, S. (1993). Homogeneous scintillating LKr/Xe calorimeters. *Nuclear Instruments & Methods in Physics Research. Section A, Accelerators, Spectrometers, Detectors and Associated Equipment*, *327*, 187–192.

Chepel, V., & Araujo, H. (2013). Liquid noble gas detectors for low energy particle physics. *JINST*, *8*(04), 1–114. doi:10.1088/1748-0221/8/04/R04001

Chepel, V., Lopes, M. I., Kuchenkov, A., Ferreira Marques, R., & Policarpo, A. J. P. L. (1997). Performance study of liquid xenon detector for PET. *Nuclear Instruments & Methods in Physics Research. Section A, Accelerators, Spectrometers, Detectors and Associated Equipment*, *392*(1-3), 427–432. doi:10.1016/S0168-9002(97)00196-4

Cherenkov, P. A. (1934). The visible glow of pure liquids under action of γ-rays. *Soviet Physics, Doklady*, *2*, 451.

Cherenkov, P. A. (1937). Izv. Akad. Nauk SSSR. *Ser. Fiz. OMEN*, *1*, 455.

Cherenkov, P. A. (1937). Visible radiation produced by electrons moving in a medium with velocities exceeding that of light. *Physical Review*, *51*(4), 378–379. doi:10.1103/PhysRev.52.378

Cherenkov, P. A. (1986). At the threshold of discovery. *Nuclear Instruments and Methods*, *S248*(1), 1–4. doi:10.1016/0168-9002(86)90487-0

Choi, H. T., Sethi, D. S., & Braun, C. L. (1982). Geminate charge recombination in the photoionization of *N,N,N′,N′*-tetramethyl-*p*-phenylenediamine (TMPD) in various solvents. *The Journal of Chemical Physics*, *77*(12), 6027–6039. doi:10.1063/1.443846

Chubb, T. A., & Friedman, H. (1955). Photon counters for far ultraviolet. *The Review of Scientific Instruments*, *26*(5), 493–498. doi:10.1063/1.1771334

Coelho, L. C. C., Veloso, J. F. C. A., Covita, D. S., Requicha Ferreira, L. F., & Dos Santos, J. M. F. (2006). High-pressure operation of a xenon-GPSC/MSGC hybrid detector for hard X-ray spectrometry. *Nuclear Instruments & Methods in Physics Research. Section A, Accelerators, Spectrometers, Detectors and Associated Equipment*, *569*(3), 796–802. doi:10.1016/j.nima.2006.09.061

Comby, G., Mangeot, Ph., Prugne, P., Chalot, J. F., Coulareau, E., Quidort, J., & Tichit, J. et al. (1980). Multi-needle detector with cathode focusing. *Nuclear Instruments and Methods*, *176*(1-2), 313–321. doi:10.1016/0029-554X(80)90722-3

Conneely, T. M., Milnes, J. S., & Howorth, J. (2013). Extended lifetime MCP-PMTs: Characterization and life time measurements of ALD coated microchannel plates, in a sealed photomultiplier tube. *Nuclear Instruments & Methods in Physics Research. Section A, Accelerators, Spectrometers, Detectors and Associated Equipment*, *732*, 388–391. doi:10.1016/j.nima.2013.07.023

Crotty, I., Fonte, P., Lemenovski, D., & Peskov, V. (1996). A new resistive plate chamber with secondary electron emitters and two dimensional microstrip readout. *IEEE Nucl Sci. Conf. Record*, *1*, 362–366.

Dangendorf, V., Breskin, A., Chechik, R., & Schmidt-Boecking, H. (1990a). An X-ray imaging gas scintillation detectors with CsI-wire chamber UV photon readout. *Proceedings SPIE 1235, Instrumentation in Astronomy VII.*

Dangendorf, V., Breskin, A., Chechik, R., & Schmidt-Boecking, H. (1990). A gas-filled UV –photon detector with CsI photocathode for the detection of Xe light. *Nuclear Instruments & Methods in Physics Research. Section A, Accelerators, Spectrometers, Detectors and Associated Equipment*, *289*(1-2), 322–324. doi:10.1016/0168-9002(90)90276-C

Danielsson, M., Fonte, P., Francke, T., Iacobaeus, C., Osling, J., & Peskov, V. (2004). Novel gaseous detectors for medical imaging. *Nuclear Instruments & Methods in Physics Research. Section A, Accelerators, Spectrometers, Detectors and Associated Equipment*, *518*(1-2), 406–410. doi:10.1016/j.nima.2003.11.038

de Beauville, E., & Borer, K. (1990). Amplex, a low-noise, low-power analog CMOS signal processor for multi-element silicon particle detectors. *Nuclear Instruments & Methods in Physics Research. Section A, Accelerators, Spectrometers, Detectors and Associated Equipment*, *288*(1), 157–167. doi:10.1016/0168-9002(90)90481-K

De Cataldo, G. (2014). The ALICE–HMPID performance during the LHC run period 2010–2013. *Nuclear Instruments & Methods in Physics Research. Section A, Accelerators, Spectrometers, Detectors and Associated Equipment*, *766*, 19–21. doi:10.1016/j.nima.2014.07.039

Dekker, A. J. (1962). *Solid State Physics*. Englewood Cliffs, NJ: Prentice-Hall.

Del Guerra, A., Belcari, N., & Bisogni, M. G. (2011). Silicon Photomultipliers (SiPM) as novel photodetectors for PET. *Nuclear Instruments & Methods in Physics Research. Section A, Accelerators, Spectrometers, Detectors and Associated Equipment*, *648*, S232–S35. doi:10.1016/j.nima.2010.11.128

Del Guerra, A., Conti, M., Gorini, G., Fujieda, I., Mulera, T. A., & Perez-Mendez, V. (1987). High resistance lead glass tubing for RICH counters and electromagnetic calorimeters. *Nuclear Instruments & Methods in Physics Research. Section A, Accelerators, Spectrometers, Detectors and Associated Equipment*, *257*(3), 609–613. doi:10.1016/0168-9002(87)90971-5

Di Mauro, A. (2014). Status and perspectives of gaseous photon detectors. *Nuclear Instruments & Methods in Physics Research. Section A, Accelerators, Spectrometers, Detectors and Associated Equipment*, *766*, 126–132. doi:10.1016/j.nima.2014.07.008

Di Mauro, A., Cozza, D., Davenport, M., Di Bari, D., Elia, D., Martinengo, P., & Williams, T. D. et al. (1999). Performance of large area CsI-RICH prototypes for ALICE at LHC: For the ALICE collaboration. *Nuclear Instruments & Methods in Physics Research. Section A, Accelerators, Spectrometers, Detectors and Associated Equipment*, *433*(1-2), 190–200. doi:10.1016/S0168-9002(99)00300-9

Di Mauro, A., Lund-Jensen, B., Martinengo, P., Nappi, E., Oliveira, R., Peskov, V., & Santiard, J. C. et al. (2007). Development of innovative micro-pattern gaseous detectors with resistive electrodes and first results of their applications. *Nuclear Instruments & Methods in Physics Research. Section A, Accelerators, Spectrometers, Detectors and Associated Equipment, 581*(1-2), 225–231. doi:10.1016/j.nima.2007.07.083

Di Mauro, A., Martinenego, P., & Nappi, E. (2009). GEM with double layered micropattern electrodes and their applications. *IEEE Transactions on Nuclear Science, 56*(3), 1550–1557. doi:10.1109/TNS.2009.2020864

Diatropto, D. B., & Peskov, V. (1972). Vacuum ultraviolet spectrum of a free plasma filament in a microwave high-pressure discharge. *Soviet Physics JETF, 34*, 554–558.

Diatroptov, D. B., & Peskov, V. D. (1972). Vacuum ultraviolet spectra of a free plasma filament in a microwave high-pressure discharge. *Soviet Phys. JETF, 34*, 554–558.

Dinu, N. (2005). *Silicon photomultiplier (SiPM) new technique for photon counting*. Presentation at College de France. Retrieved from https://www.google.ch/search?q=Silicon+photomultiplier+(SiPM)+new+technique+for+photon+counting&oq=Silicon+photomultiplier+(SiPM)+new+technique+for+photon+counting&aqs=chrome.69i57.4540j0j7&sourceid=chrome&es_sm=93&ie=UTF-8#q=Dinu%2C+Silicon+photomultiplier+(SiPM)+new+technique+for+photon+counting

Dinu, N. (2014). *Silicon Photomultiplier - characteristics and applications*. Presented at the Seminar in Geneva University. Retrieved from http://dpnc.unige.ch/seminaire/talks/dinu.pdf

Dutt, N. V. K., Kahol, A. P., & Jyothlraju, P. (1982). Vapour pressure of trimethyl phosphate and trimethylamine. *Journal of Chemical & Engineering Data, 27*(4), 369–370. doi:10.1021/je00030a001

Duval, S., Breskin, A., Budnik, R., Chen, W. T., Carduner, H., Cortesi, M., & Thers, D. et al. (2011). On the operation of a micropattern UV-photomultipliers in liquid Xe. *JINST, 6*(04), 1–12. doi:10.1088/1748-0221/6/04/P04007

Edmends, J. S., Miller, D. J., & Barlow, F. (1988). Investigation of prototype gas-filled photomultipliers. *Nuclear Instruments & Methods in Physics Research. Section A, Accelerators, Spectrometers, Detectors and Associated Equipment, 273*(1), 145–156. doi:10.1016/0168-9002(88)90809-1

Ekelöf, T., Séguinot, J., Tocqueville, J., & Ypsilantis, T. (1981). The Cherenkov ring-imaging detectors: Recent progress and future development. *Physica Scripta, 23*(4B), 718–724. doi:10.1088/0031-8949/23/4B/023

Ekspong, G. (1958). *Cherenkovstrålning,* Kosmos Fysiska Uppsatser, P.A. *Norstedt & Söner, 36*, 6.

Endo, I., Kawamoto, T., & Mizuno, Y. (1981). Systematic shifts of evaluated charge centroid for the cathode readout multiwire proportional chamber. *Nuclear Instruments & Methods in Physics Research. Section A, Accelerators, Spectrometers, Detectors and Associated Equipment, 188*, 51–58.

Enomoto, R., Sumiyoshi, T., Fujita, Y, (1994). Test of various photocathodes. *Nucl. Instrum. and Methods in Phys. Res., A343*, 117-120.

Enomoto, R., Sumiyoshi, T., & Fujita, Y. (1994). Test of various photocathodes. *Nuclear Instruments & Methods in Physics Research. Section A, Accelerators, Spectrometers, Detectors and Associated Equipment, 343*(1), 117–120. doi:10.1016/0168-9002(94)90540-1

Fano, U. (1947). Ionization yield of radiations. II. The fluctuations in the number of ions. *Physical Review, 72*(1), 26–29. doi:10.1103/PhysRev.72.26

Farnsworth, P. T. (1930). *Electron multiplier*. US Patent number 1 969 399.

Filippov, S. (2014). *First years of running for the LHCb calorimeter system.* Retrieved from https://cds.cern.ch/record/1667194/files/LHCb-TALK-2014-035.pdf

Fisher, P. (2012). *Amplifying Photo Detectors: Avalanche Photodiodes Silicon Photomultiplier.* Retrieved from: http://sus.ziti.uni-heidelberg.de/Lehre/Detectors1213/Part7_PhotoDetectors.pdf

Fonte, P., Francke, T., Pavlopoulos, N., Peskov, V., & Rodionov, I. (2005). Novel single photon detectors for UV imaging. *Nucl. Instrum. and Methods in Phys. Res., A553*, 30-34.

Fonte, P., Francke, T., Pavlopoulos, N., Peskov, V., & Rodionov, I. (2005a). *Novel single photon detectors for UV imaging.* Retrieved from http://www.arxiv.org/abs/physics/0502027

Fonte, P. T., Francke, T., Pavlopoulos, N., Peskov, V., & Rodionov, I. (2005). Novel single photon detectors for UV imaging. *Nuclear Instruments & Methods in Physics Research. Section A, Accelerators, Spectrometers, Detectors and Associated Equipment, 553*(1-2), 30–34. doi:10.1016/j.nima.2005.08.002

Fonte, P., Ferreira-Marques, R., Peskov, V., & Policarpo, A. (1999). Single-electron pulse-height spectra in thin-gap parallel plate chambers. *Nuclear Instruments & Methods in Physics Research. Section A, Accelerators, Spectrometers, Detectors and Associated Equipment, 433*(1-2), 513–517. doi:10.1016/S0168-9002(99)00309-5

Fonte, P., Martinengo, P., Nappi, E., Oliveira, R., & Peskov, V. (2009). Progress in developing hybrid RPC:GEM-like detectors with resistive electrodes. *Nuclear Instruments & Methods in Physics Research. Section A, Accelerators, Spectrometers, Detectors and Associated Equipment, 602*(3), 850–853. doi:10.1016/j.nima.2008.12.154

Fonte, P., & Peskov, V. (2000). Micro-gap parallel-plate chambers with porous secondary electron emitters. *Nuclear Instruments & Methods in Physics Research. Section A, Accelerators, Spectrometers, Detectors and Associated Equipment, 454*(1), 260–266. doi:10.1016/S0168-9002(00)00838-X

Fonte, P., Peskov, V., & Ramsey, B. (1998). Rate and gain limitations of MSGC and MGC combined with GEM and other preamplification structures. *Nucl. Instr. and Meth.in Phys. Res., A419*, 405–409.

Fonte, P., Peskov, V., & Ramsey, B. (1999a). A study of breakdown limits in microstrip gas counters with preamplification structure. *Nucl. Instr. and Meth.in Phys. Res., A416*, 23–31.

Fonte, P., Peskov, V., & Ramsey, B. (1999b). The fundamental limitations of high-rate gaseous detectors. *IEEE Transactions on Nuclear Science, 46*(3), 321–325. doi:10.1109/23.775537

Fonte, P., Peskov, V., & Sauli, F. (1991). UV emission and breakdown in parallel-plate chambers. *Nuclear Instruments & Methods in Physics Research. Section A, Accelerators, Spectrometers, Detectors and Associated Equipment, 310*(1-2), 140–145. doi:10.1016/0168-9002(91)91013-L

Fonte, P., Peskov, V., & Sauli, F. (1991). *VUV emission and breakdown in parallel plate chambers.* Nucl. Instr. and Meth. in Phys. *Res., A310*, 140–145.

Fonte, P., Peskov, V., & Sauli, F. (1991a). Feedback and breakdown in parallel-plate chambers. *Nuclear Instruments & Methods in Physics Research. Section A, Accelerators, Spectrometers, Detectors and Associated Equipment, 305*(1), 91–110. doi:10.1016/0168-9002(91)90522-R

Francke, T. (1991). *A Ring Imaging Cherenkov Counter using a Solid Radiator.* (Ph. D. Thesis). Royal Institute of Technology, Stockholm, Sweden.

Francke, T. (1991). *A ring imaging Cherenkov counter using a solid radiator.* (PhD Thesis). Royal Institute of Technology, Stockholm, Sweden.

Francke, T., & Peskov, V. (2002). *Patent: Flame and spark detector, automatic fire alarm and methods related thereto.* WO02097757, IPC: G08B17/12,H01J47/02.

Francke, T., & Peskov, V. (2014). *Innovative applications and developments of micro-pattern gaseous detectors.* Hershey, PA: Engineering Science Reference.

Francke, T., Peskov, V. (2005). *Patent: Detector assembly, WO2006135321 (A1).* EP1894179, IPC: G08B17/12H01J47/022006.

Francke, T., Bergstrom, D., Boezio, M., Carlson, P., & Suffert, M. (1999). A gas- RICH detector for space. *Nuclear Instruments & Methods in Physics Research. Section A, Accelerators, Spectrometers, Detectors and Associated Equipment, 433*(1-2), 87–91. doi:10.1016/S0168-9002(99)00311-3

Francke, T., Eklund, M., Ericsson, L., Kristofferson, T., Peskov, V., Rantanene, J., & Ullberg, C. et al. (2001). Dose reduction in medical X-ray imaging using noise free photon counting. *Nuclear Instruments & Methods in Physics Research. Section A, Accelerators, Spectrometers, Detectors and Associated Equipment, 471*(1-2), 85–87. doi:10.1016/S0168-9002(01)00920-2

Francke, T., & Peskov, V. (2003). Aging in gaseous photodetectors. *Nuclear Instruments & Methods in Physics Research. Section A, Accelerators, Spectrometers, Detectors and Associated Equipment, 515*(1-2), 292–301. doi:10.1016/j.nima.2003.09.013

Francke, T., & Peskov, V. (2004a). Photosensitive gaseous detectors and their applications. *Nuclear Instruments & Methods in Physics Research. Section A, Accelerators, Spectrometers, Detectors and Associated Equipment, 525*(1-2), 1–5. doi:10.1016/j.nima.2004.03.017

Francke, T., & Peskov, V. (2014). *Innovative applications and developments of micro-pattern gaseous detectors.* IGI Global. doi:10.4018/978-1-4666-6014-4

Francke, T., Peskov, V., Rodionov, I., & Fonte, P. (2004). High rate (up to 10^5 Hz/cm^2), high position resolution (30 μm) photosensitive RPCs. *Nuclear Instruments & Methods in Physics Research. Section A, Accelerators, Spectrometers, Detectors and Associated Equipment, 533*(1-2), 163–168. doi:10.1016/j.nima.2004.07.021

Francke, T., Peskov, V., Rodionov, I., & Sokolova, T. (2002). Novel position-sensitive gaseous detectors with solid photocathodes. *IEEE Transactions on Nuclear Science, 49*(3), 977–983. doi:10.1109/TNS.2002.1039600

Frank, I. M., & Tamm, I. J. (1937). *Coherent Visible Radiation of Fast Electrons Passing through Matter*, Dokl. *Academie des Sciences de l'URSS, 14*, 109.

Fraser, G. W. (2009). X-ray detector in astronomy. Cambridge University Press.

Funfer, E., & Neiert, H. (1965). *Detectors of radiation, Atomisdat.* Moscow. (in Russian)

Funfer, E., & Neiert, H. (1965). *Detectors of radiation.* Moscow: Atomizdat.

Garty, G., Breskin, A., Chechik, R., & Shefer, E. (1999). Single photoelectron detection with a low-pressure gas electron multiplier coupled to a CsI photocathode. *Instrum. and Methods in Phys. Res., A433*, 476–481.

Giomataris, Y., Rebourgeard, P., Robert, J. P., & Charpak, G. (1996). MICROMEGAS: A high-granularity position-sensitive gaseous detector for high-flux environment. *Nuclear Instruments & Methods in Physics Research. Section A, Accelerators, Spectrometers, Detectors and Associated Equipment, 376*(1), 29–35. doi:10.1016/0168-9002(96)00175-1

Giunji, S., Peskov, V., Sakurai, H., Silin, E., Sokolova, T., & Rodionov, I. (2002). Micro capillary glass plates as a gaseous detector of photons and particles. *Nuclear Instruments & Methods in Physics Research. Section A, Accelerators, Spectrometers, Detectors and Associated Equipment*, *477*(1-3), 8–20. doi:10.1016/S0168-9002(01)01906-4

Golden, R. L., Badhwar, G. D., Jeffrey, L., Lacy, J., & Zipse, J. (1978). A magnetic spectrometer for cosmic ray studies. *Nuclear Instruments and Methods*, *148*(1), 179–185. doi:10.1016/0029-554X(78)90351-8

Golden, R. L., Grimani, C., Hull, R., Kimbell, B. L., Park, R., Stephens, S. A., & Picozza, P. et al. (1991). Performance of a balloon-borne magnet spectrometer for cosmic ray studies. *Nuclear Instruments & Methods in Physics Research. Section A, Accelerators, Spectrometers, Detectors and Associated Equipment*, *306*(1-2), 366–377. doi:10.1016/0168-9002(91)90343-O

Golden, R. L., Horan, S., Mauger, B. G., Badhwar, G. D., Lacy, J. L., Stephens, S. A., & Zipse, J. E. et al. (1979). Evidence for the Existence of Cosmic-Ray Antiprotons. *Physical Review Letters*, *43*(16), 1196–1199. doi:10.1103/PhysRevLett.43.1196

Golovin, V. (1999). *Avalanche Photodetector*. Russian Agency for Patents and Trademarks, Patent No. RU 2142175.

Goodrich, G. W., & Wiley, W. C. (1962). Continuous channel electron multiplier. *The Review of Scientific Instruments*, *33*(7), 761–762. doi:10.1063/1.1717958

Grand, D., & Bemas, A. (1977). The ionization potential of a solute and the ground state energy of the excess electron. *Journal of Physical Chemistry*, *81*(12), 1209–1211. doi:10.1021/j100527a018

Grignon, C., Breskin, A., Carlier, T., Couturier, O., Cussonneau, J. P., & Ferrer, L. …Vasseur, A. (2005). Simulation of a High Performance γ-Camera Concept for PET based on LXe and Gaseous Photomultipliers. *Dielectric liquids.2005 IEEE International Conference on Dielectric Liquids*, (pp. 357-360). IEEE.

Guzhov, A. A., & Shuba, Yu. A. (1969). Measurement of photoelectron energy spectra with a secondary electron multiplier. *Journal of Applied Spectroscopy*, *10*(6), 692–693. doi:10.1007/BF00614424

Gys, T. (2005). *Particle Detectors – Principles and Techniques*. Retrieved from: http://ph-dep-dt2.web.cern.ch/ph-dep-dt2/CAT2005_3b.pdf

Gys, T. (2005). *Particle Detectors –Principles and Techniques(3/5) Lecture 3b – Photodetection*. Retrieved from http://ph-dep-dt2.web.cern.ch/ph-dep-dt2/CAT2005_3b.pdf

Gys, T. (2014). *Micro-channel plate photomultiplier tubes for fast photon detection*. Presentation at CERN detector seminar. Retrieved from https://indico.cern.ch/event/288433/

Haba, J. (2008). Status and perspectives of Pixelated Photon Detector (PPD). *Nuclear Instruments & Methods in Physics Research. Section A, Accelerators, Spectrometers, Detectors and Associated Equipment*, *595*(1), 154–160. doi:10.1016/j.nima.2008.07.061

Haitz, R. H. (1964). Model for the Electrical Behavior of a Microplasma. *Journal of Applied Physics*, *35*(5), 1370–1376. doi:10.1063/1.1713636

Hakamata, T., Kume, H., & Okano, K. (2007). *Photomultiplier tubes. Basics and applications. Edition 3a*. Retrieved from https://www.hamamatsu.com/resources/pdf/etd/PMT_handbook_v3aE.pdf

Heller, C., (1980). *Tetrakis(dimethylamino)ethylene (TMAE) purification report*. CERN.

Hempstead, M., Chauveau, J., Comby, G., Séguinot, J., & Ypsilantis, T. (1991). A measurement of the refractive index of sodium fluoride in the far ultraviolet. *Nuclear Instruments & Methods in Physics Research. Section A, Accelerators, Spectrometers, Detectors and Associated Equipment*, *306*(1-2), 207–211. doi:10.1016/0168-9002(91)90321-G

Heroux, L., & Hinteregger, H. E. (1960). Resistance strip magnetic electron multiplier for the extreme ultraviolet. *The Review of Scientific Instruments*, *31*(3), 280–286. doi:10.1063/1.1716957

Heroux, L., McMahon, W. J., & Hinteregger, H. E. (1966). The influence of cathode thickness and aging on the photoelectric efficiencies of LiF and CsI in XVUV. *Applied Optics*, *5*(8), 1338–1339. doi:10.1364/AO.5.001338 PMID:20057543

Hilke, H.G. (2011). *Gaseous Detectors for Particles and Radiation. Part 1: Principles and Methods*. Landolt-Börnstein - Group I Elementary Particles, Nuclei and Atoms Volume 21B1, 72-106.

Hoedlmoser, H., Braem, A., De Cataldo, G., Davenport, M., Di Mauro, A., Franco, A., & Peskov, V. (2006). Production technique and quality evaluation of CsI photocathode for the ALICE/HMPID detector. *Nuclear Instruments & Methods in Physics Research. Section A, Accelerators, Spectrometers, Detectors and Associated Equipment*, *566*(2), 338–350. doi:10.1016/j.nima.2006.07.047

Hoedlmoser, H., Braem, A., De Cataldo, G., Davenport, M., Di Mauro, A., Franco, A., & Schyns, E. et al. (2007). Long term performance and ageing of CsI photocathodes for the ALICE/HMPID detector. *Nuclear Instruments & Methods in Physics Research. Section A, Accelerators, Spectrometers, Detectors and Associated Equipment*, *574*(1), 28–38. doi:10.1016/j.nima.2007.01.101

Hof, M., Bremerich, M., Menn, W., Pfeiffer, C., Reimer, O., & Simon, M. et al.. (1994). Performance of drift chambers in a magnetic rigidity spectrometer for measuring the cosmic radiation. *Nuclear Instruments & Methods in Physics Research. Section A, Accelerators, Spectrometers, Detectors and Associated Equipment*, *345*(3), 561–569. doi:10.1016/0168-9002(94)90516-9

Holder, J. (2013). Atmospheric Cherenkov telescopes for high-energy γ-ray astronomy. *Particle Data Group Review of Particle Properties*. Retrieved from: http://pdg.lbl.gov/2014/reviews/rpp2014-rev-particle-detectors-non-accel.pdf

Holl, P., Andritschke, R., & Eckhardt, R. (2006). A New High-Speed, Single Photon Imaging CCD for the Optical. *2006 IEEE Nuclear Science Symposium Conference Record*. doi:10.1109/NSSMIC.2006.354202

Holroyd, R. A., Preses, J. M., & Woody, C. L. (1986). *Measurements of the quantum efficiency of TMAE and TEA from threshold to 120 nm*. Report at the, XXIII International Conference on High Energy Physics, Berkeley, CA. Retrieved from http://www.iaea.org/inis/collection/NCLCollectionStore/_Public/18/052/18052811.pdf

Holroyd, R. A., & Anderson, D. F. (1985a). The physics and chemistry of room –temperature liquid-filled ionization chambers. *Nuclear Instruments & Methods in Physics Research. Section A, Accelerators, Spectrometers, Detectors and Associated Equipment*, *236*(2), 294–299. doi:10.1016/0168-9002(85)90164-0

Holroyd, R. A., Ehrenson, S., & Preses, J. M. (1985b). Electron mobility, ion efficiencies, and photoconductivity in liquid tetrakis (dimethylamino) ethylene. *Journal of Physical Chemistry*, *89*(20), 4244–4249. doi:10.1021/j100266a019

Iacobaeus, C. A., Breskin, A., Danielsson, M., Francke, T., Mormann, D., Ostling, J., & Peskov, V. (2004). Advances in capillary-based gaseous UV imaging detectors. *Nuclear Instruments & Methods in Physics Research. Section A, Accelerators, Spectrometers, Detectors and Associated Equipment*, *525*(1-2), 42–48. doi:10.1016/j.nima.2004.03.022

Imrie, D. C., Moghul, A. K., & Wells, K. (1990). Decamethylferrocene-a new photocathode for the detection of BaF_2 scintillation light. *Nuclear Instruments & Methods in Physics Research. Section A, Accelerators, Spectrometers, Detectors and Associated Equipment*, *297*(1-2), 179–182. doi:10.1016/0168-9002(90)91364-H

Imrie, D. C., Peskov, V., & Tamadoni-Jahromi, R. (1992). Air-stable solid photocathodes with good ageing properties for fast gaseous detectors. *Nuclear Instruments & Methods in Physics Research. Section A, Accelerators, Spectrometers, Detectors and Associated Equipment*, *317*(1-2), 92–96. doi:10.1016/0168-9002(92)90595-U

Inami, K. (2014). TOP counter for particle identification at the Belle-II experiment. *Nuclear Instruments & Methods in Physics Research. Section A, Accelerators, Spectrometers, Detectors and Associated Equipment*, *766*, 5–8. doi:10.1016/j.nima.2014.07.006

Inami, K., Kishimoto, N., Enari, Y., Nagamine, M., & Ohshima, T. (2006). A 5ps TOF-counter with an MCP–PMT. *Nuclear Instruments & Methods in Physics Research. Section A, Accelerators, Spectrometers, Detectors and Associated Equipment*, *560*(2), 303–308. doi:10.1016/j.nima.2006.01.027

Jelley, J. V. (1958). *Cherenkov radiation and its applications*. Pergamon Press.

Jonsson, K. (1988). *Test of two PIDs with Cosmic Rays*. (PhD Thesis). KTH, Stockholm, Sweden.

Joram, C. (2003). *Hybrid photon detectors (HPD) and multi anode photomultiplier tubes (MAPMT)*. Retrieved from: http://www.google.fr/url?sa=t&rct=j&q=&esrc=s&source=web&cd=1&cad=rja&uact=8&ved=0CCYQFjAA&url=http%3A%2F%2Findico.cern.ch%2Fevent%2Fa035690%2Fsession%2Fs10%2Fcontribution%2Fs10t1%2Fmaterial%2F0%2F1.pdf&ei=ojh_VMnEL9LUasS6gugK&usg=AFQjCNG9rhdupwqOLC7A1Hphv23INhzMOw

Joram, C. (1999). Large area hybrid photodiodes. *Nuclear Physics B - Proceedings Supplements*, *78*(1-3), 407–415. doi:10.1016/S0920-5632(99)00578-2

Kadyk, J. (1991). Wire chamber aging. *Nuclear Instruments & Methods in Physics Research. Section A, Accelerators, Spectrometers, Detectors and Associated Equipment*, *300*(3), 436–479. doi:10.1016/0168-9002(91)90381-Y

Kaisser, T. R. (1974). Heaviside radiation. *Nature*, *247*(5440), 400–401. doi:10.1038/247400a0

Kamin, J. A. (2007). A hadron blind detector for the PHENIX experiment at RHIC. *The European Physical Journal C*, *49*(1), 177–180. doi:10.1140/epjc/s10052-006-0100-y

Karabadjak, G. F., & Peskov, V. D. (1981). Multiwire X-Ray Counter with High Energy Resolution. *Prib. Tekhn. Eksp.*, *6*, 174–176.

Karabadjak, G. F., Peskov, V. D., & Podoliak, E. R. (1982). Gaseous counters for detection plasma radiation in the region 3-200 nm with spatial and time resolutions. *Journal of Applied Spectroscopy*, *3*, 509–516.

Karabadjak, G. F., Peskov, V. D., & Podolyalk, E. R. (1983). Coordinate gas counters with super high energy resolution for plasma diagnostics. *Nuclear Instruments & Methods in Physics Research. Section A, Accelerators, Spectrometers, Detectors and Associated Equipment*, *217*, 56–64.

Karabadzhak, G. F., Peskov, V. D., & Podolyak, E. R. (1983). Coordinate gas counters with supehigh energy resolution for plasma diagnostics. *Nuclear Instruments & Methods in Physics Research. Section A, Accelerators, Spectrometers, Detectors and Associated Equipment*, *217*, 55–64.

Karabadzhak, G. F., Peskov, V. D., & Podolyak, E. R. (1983). Coordinate gas counters with super high energy resolution for plasma diagnostics. *Nuclear Instruments and Methods*, *A217*(1-2), 56–64. doi:10.1016/0167-5087(83)90109-6

Kononenko, W., Heinrich, J. G., Lockyer, N. S., Miller, J. M., Woody, C., & Kwan, S. (1997). Photon emission from 511 keV gamma rays incident on BaF_2 and LaF_3: Nd^{3+} crystals using a caesium iodide photocathode detector. *Nuclear Instruments & Methods in Physics Research. Section A, Accelerators, Spectrometers, Detectors and Associated Equipment*, *396*(1-2), 172–180. doi:10.1016/S0168-9002(97)00776-6

Korpar, S., Adachi, I., Hamada, N., Higuchi, M., Iijima, T., Iwata, S., & Yusa, Y. et al. (2014). A 144-channel HAPD for the Aerogel RICH at Belle II. *Nuclear Instruments & Methods in Physics Research. Section A, Accelerators, Spectrometers, Detectors and Associated Equipment*, *766*, 145–147. doi:10.1016/j.nima.2014.05.060

Korpar, S., Chagani, H., Dolenec, R., Križan, P., Pestotnik, R., & Stanovnik, A. (2010). A module of silicon photomultipliers for detection of Cherenkov radiation. *Nuclear Instruments & Methods in Physics Research. Section A, Accelerators, Spectrometers, Detectors and Associated Equipment*, *613*(2), 195–199. doi:10.1016/j.nima.2009.11.043

Korpar, S., Krizan, P., Stanovnik, A., Staric, M., & Skrk, D. (1999). Aging and rejuvenation of a TMAE + methane multiwire photon detector. *IEEE Transactions on Nuclear Science*, *46*(3), 317–320. doi:10.1109/23.775536

Korpar, S., Tahirovic, E., Krizan, P., & Pestotnik, R. (2014a). Test of the Hamamatsu MPPC module S11834 as a RICH photon detector. *Nuclear Instruments & Methods in Physics Research. Section A, Accelerators, Spectrometers, Detectors and Associated Equipment*, *766*, 107–109. doi:10.1016/j.nima.2014.05.074

Krizan, P., Cindro, M., Cindro, V., Korpar, S., Omahen, G., Stanovnik, A., & Starič, M. (1995). Measurements of the quantum efficiency of CsI photocathodes in a multiwire proportional chamber. *Nuclear Instruments & Methods in Physics Research. Section A, Accelerators, Spectrometers, Detectors and Associated Equipment*, *364*(2), 243–252. doi:10.1016/0168-9002(95)00471-8

Lanciart, A., & Morucci, J. P. (1962). Article. *Journal de Physique et le Radium*, *23*(suppl. 6), 102A. doi:10.1051/jphysap:01962002306010200

Lau, K., Atiya, M., Broemmelsiek, D., Hamacher, T., Ispiryan, M., Korpar, S., & Staric, M. et al. (2003). Test-beam aging studies of a TMAE prototype for the HERA-B RICH. *Nuclear Instruments & Methods in Physics Research. Section A, Accelerators, Spectrometers, Detectors and Associated Equipment*, *515*(1-2), 313–318. doi:10.1016/j.nima.2003.09.016

Lechmann, A., Britting, A., & Eyrich, W. (2013). Significantly improved life time of micro-channel plate PMTs. *Nuclear Instruments & Methods in Physics Research. Section A, Accelerators, Spectrometers, Detectors and Associated Equipment*, *718*, 535–540. doi:10.1016/j.nima.2012.11.109

Leith, D. (1989). A review of 4π Cherenkov Ring Imaging detectors. In *Proceedings ofSymposium on particle identification at high luminosity hadron colliders*. Fermi National Accelerator Laboratory.

Lennander, S. (1958). Nobelpriset i Fusik 1958, Kosmos Fysiska Uppsatser, P.A. *Norstedt & Söner*, *36*, 3.

Lenzen, G., Schyns, E., Thadome, J., & Werner, J. (1994). The use of fluorocarbon radiators in the DELPHI RICH detectors. *Nuclear Instruments & Methods in Physics Research. Section A, Accelerators, Spectrometers, Detectors and Associated Equipment*, *343*(1), 268–272. doi:10.1016/0168-9002(94)90562-2

Linka, A., Picek, J., Volf, P., & Ososskov, J. (1999). New solution to circle fitting problem in analysis of RICH detector data. *Czechoslovak Journal of Physics*, *49*(S2Suppl), 161–168. doi:10.1007/s10582-999-0099-z

Llopart, X., Ballabriga, R., Campbell, M., Tlustos, L., & Wong, W. (2007). Tmepix, a 65k programmable pixel readout chip for arrival time, energy and/or photon counting measurements. *Nuclear Instruments & Methods in Physics Research. Section A, Accelerators, Spectrometers, Detectors and Associated Equipment*, *581*(1-2), 485–494. doi:10.1016/j.nima.2007.08.079

Lopes, J. A. M., dos Santos, J. M. F., Conde, C. A. N., & Morgado, R. E. (1999). A new integrated photosensor for gas proportional scintillation counters based on the gas electron multiplier (GEM). *Nuclear Instruments & Methods in Physics Research. Section A, Accelerators, Spectrometers, Detectors and Associated Equipment*, *426*(2-3), 469–476. doi:10.1016/S0168-9002(98)01419-3

Lubsandorzhiev, B. K. (2006). *On the history of photomultiplier tube invention.* ArXiv:physics/0601159

Lu, C., & McDonald, K. T. (1994). Properties of reflective and semitransparent CsI photocathodes. *Nuclear Instruments & Methods in Physics Research. Section A, Accelerators, Spectrometers, Detectors and Associated Equipment, 343*(1), 135–151. doi:10.1016/0168-9002(94)90543-6

Lu, Y., Yang, K., Zhou, K., Zhang, Q., Pang, B., & Ren, Q. (2014). Development of a SiPM-based PET imaging system for small animals. *Nuclear Instruments & Methods in Physics Research. Section A, Accelerators, Spectrometers, Detectors and Associated Equipment, 743*, 30–38. doi:10.1016/j.nima.2014.01.010

Lyashenko, A. V., Breskin, A., Chechik, R., Velosos, J. F. C. A., Dos Santos, J. M. F., & Amaro, F. D. (2007). Further progress in ion back-flow reduction with patterned gaseous hole–multipliers. *JINST, 2*(08P08004), 1–18. doi:10.1088/1748-0221/2/08/P08004

Lyashenko, A. V., Breskin, A., Chechik, R., Velosos, J. F. C. A., Dos Santos, J. M. F., & Amaro, F. D. (2009). Development of high-gain gaseous photomultipliers for the visible spectral range. *JINST, 4*(07P07005), 1–22. doi:10.1088/1748-0221/4/07/P07005

Maier-Komor, P., Bauer, B. B., Friese, J., Gernhäuser, R., Kienle, P., Körner, H. J., & Zeitelhack, K. et al. (1995). Preparation of reflective CsI photocathodes with reproducibly high quantum efficiency. *Nuclear Instruments & Methods in Physics Research. Section A, Accelerators, Spectrometers, Detectors and Associated Equipment, 362*(1), 183–188. doi:10.1016/0168-9002(95)00314-2

Majewski, S., Kuzmiertz, W, Sauli, F. (1983). *Operation of gas photodiodes in strong magnetic fields.* Preprint CERN-EP/83-89.

Malamud, G., Mine, P., Vartsky, D., Equer, B., Breskin, A., & Chechik, R. (1994). The photo-efficiency of CsI, amorphous silicon and organometallic reflective photocathodes. *Nuclear Instruments & Methods in Physics Research. Section A, Accelerators, Spectrometers, Detectors and Associated Equipment, 343*, 121–128. doi:10.1016/0168-9002(94)90541-X

Malter, L. (1936). Thin film field emission. *Physical Review, 50*(1), 48–58. doi:10.1103/PhysRev.50.48

Martinengo, P., Nappi, E., Oliveira, R., Peskov, V., Pietropaolo, F., & Picchi, P. (2009). A new generation of GEM detectors and their applications. *Nuclear Instruments & Methods in Physics Research. Section A, Accelerators, Spectrometers, Detectors and Associated Equipment, 604*(1-2), 8–10. doi:10.1016/j.nima.2009.01.043

Martinengo, P., Peskov, V., Breskin, A., Di Mauro, A., Mayani Paras, D., Molnar, L., & van Hoorne, J. et al. (2011). R&D results on a CsI-coated triple thick GEM-based photodetector. *Nuclear Instruments & Methods in Physics Research. Section A, Accelerators, Spectrometers, Detectors and Associated Equipment, 639*(1), 126–129. doi:10.1016/j.nima.2010.10.033

Massam, T., Muller, T., & Zichichi, A. (1963). A telescope to identify electrons in the presence of pion background. Preprint CERN/63-25, Nucl. Phys Div., 1-10.

Mathieson, E., & Gordon, J. S. (1984). Cathode charge distribution in multiwire chambers, II. Approximate and empirical formulae. *Nuclear Instruments & Methods in Physics Research. Section A, Accelerators, Spectrometers, Detectors and Associated Equipment, 227*, 277–282. doi:10.1016/0168-9002(84)90133-5

Mazin, D., Bangale, P., & Sitarek, J. (2014). *Towards SiPM camera for current and future generations of Cherenkov telescopes.* ArXiv: 1410.5070

McDaniel, E.W. (1964). *Collision phenomena in ionized gases.* Wiley.

McDaniel, E. (1964). *Collision phenomena in ionized gases.* Wiley.

McIntyre, R. J. (1961). Theory of Microplasma Instability in Silicon. *Journal of Applied Physics*, *32*(6), 983–995. doi:10.1063/1.1736199

Meinschad, T., Ropelewski, L., & Sauli, F. (2004). GEM-based photon detector for RICH applications. *Nuclear Instruments & Methods in Physics Research. Section A, Accelerators, Spectrometers, Detectors and Associated Equipment*, *535*(1-2), 324–329. doi:10.1016/S0168-9002(04)01670-5

Melai, J., Breskin, A., Cortesi, M., Bilevych, Y., Fransen, M., van der Graaf, H., & Schmitz, J. et al. (2011). A UVsensitive integrated Micromegas with Timepix readout. *Nuclear Instruments & Methods in Physics Research. Section A, Accelerators, Spectrometers, Detectors and Associated Equipment*, *628*(1), 133–137. doi:10.1016/j.nima.2010.06.301

Melin, G. (1974). Handbuch der Anorganischen Chemie. Springer-Verlag.

Michau, J. C., Pichard, B., Soirat, J. P., Spiro, M., & Vignaud, D. (1986). The performance of a UV sensitive multiwire proportional chamber filled with TMAE. *Nuclear Instruments & Methods in Physics Research. Section A, Accelerators, Spectrometers, Detectors and Associated Equipment*, *244*(3), 565–570. doi:10.1016/0168-9002(86)91085-5

Milov, A. (2006). *Hadron Blind Detector for the PHENIX Experiment at RHIC.* ArXV: nucl-ex/0609014v1

Mine, Ph., Malamud, G., Vartsky, D., & Vasiliais, G. (1997). Organic UV photocathodes for gaseous particle detectors. *Nuclear Instruments & Methods in Physics Research. Section A, Accelerators, Spectrometers, Detectors and Associated Equipment*, *387*(1-2), 171–175. doi:10.1016/S0168-9002(96)00983-7

Moghal, A. (1991). *Ultraviolet photocathodes for use with solid scintillator proportional counter.* Brunel University.

Montanari, C. (2004). The ICARUS project. A 3000 ton detector for neutrino and matter stability search. *Nuclear Instruments & Methods in Physics Research. Section A, Accelerators, Spectrometers, Detectors and Associated Equipment*, *518*(1-2), 216–219. doi:10.1016/j.nima.2003.10.064

Moriya, T., Tokanau, F., Okazaki, K., Sakurai, H., Gunji, S., Kawabata, H., & Kishimito, S. et al. (2013). A concise quantum efficiency measurements system for gaseous photomultipliers. *Nuclear Instruments & Methods in Physics Research. Section A, Accelerators, Spectrometers, Detectors and Associated Equipment*, *732*, 269–272. doi:10.1016/j.nima.2013.08.054

Mormann, D., Breskin, A., Chechik, R., & Shalem, C. (2004). Operation principles and properties of the multi-GEM gaseous photomultiplier with reflective photocathode. *Nuclear Instruments & Methods in Physics Research. Section A, Accelerators, Spectrometers, Detectors and Associated Equipment*, *530*(3), 258–274. doi:10.1016/j.nima.2004.03.212

Musienko, Y. (2007). *Tests and performance of multipixel Geiger mode APD's and APD's for the CMSECAL.* Presented at the International workshop on new photon-detectors PD07 Kobe University, Kobe, Japan.

Nakato, Y., Ozaki, M., & Tsobomura, H. (1972). Photoionization and Rydberg States of Tetraaminoethylenes in Organic Solutions. *Journal of Physical Chemistry*, *76*(15), 2105–2112. doi:10.1021/j100659a008

Nappi, E., & Peskov, V. (2013). Imaging gaseous detectors and their applications. Wiley-VCH. doi:10.1002/9783527640294

Nappi, E., & Peskov, V. (2013). Imaging gaseous detectors and they applications. Wiley-VCH.

Nappi, E. (2001). CsI RICH detectors in high-energy physics experiments. *Nuclear Instruments & Methods in Physics Research. Section A, Accelerators, Spectrometers, Detectors and Associated Equipment*, *471*(1-2), 18–24. doi:10.1016/S0168-9002(01)00910-X

Nappi, E., Posa, F., Satalino, G., Scognetti, T., Tomasicchino, G., Braem, A., & Vranic, D. (1992). Development of a large area advanced fast RICH detectors for particle identification at the large hadron collider operated with heavy ions. Detector R&D proposal. CERN/DRDC 92-3. *DRDC, P35*, 1–19.

Nappi, E., & Séguinot, J. (2005). Ring Imaging Cherenkov Detectors: The state of the art and perspectives. *Rivista del Nuovo Cimento*, *28*, 1–130.

Neumann, W. (1980). *The ultimate energy resolution of the proportional counter and the analysis of unresolved low energy spectra*. 7th Symp. on Microdosimetry, Oxford, UK.

Ochi, A., Tanimori, T., Nishi, Yu., Nishi, Ya., Nagayoshi, T., Ohashi, Y., & Toyokawa, H. et al. (2002). Use of a Micro-Strip Gas Chamber conductive capillary plate for time-resolved X-ray area detection. *Nuclear Instruments & Methods in Physics Research. Section A, Accelerators, Spectrometers, Detectors and Associated Equipment*, *477*(1-3), 48–54. doi:10.1016/S0168-9002(01)01907-6

Ohring, M. (1992). *The Materials Science of Thin Films*. New York: Academic Press. doi:10.1016/B978-0-08-051118-4.50007-4

Oliveira, R., Peskov, V., Pietropaolo, F., & Picchi, P. (2007). First tests of thick GEMs with electrodes made of a resistive Kapton. *Nuclear Instruments & Methods in Physics Research. Section A, Accelerators, Spectrometers, Detectors and Associated Equipment*, *576*(2-3), 362–366. doi:10.1016/j.nima.2007.03.010

Orito, S., Maeno, T., Matsunaga, H., Abe, K., Anraku, K., Asaoka, Y., & Streitmatter, R. (2000). Precision Measurement of Cosmic-Ray Antiproton Spectrum. *Physical Review Letters*, *84*(6), 1078–1081. doi:10.1103/PhysRevLett.84.1078 PMID:11017448

Oschepkov, P. K., Skvortsov, B. N., & Osanov, B. A. (1960). Application of a continuous secondary electron multiplication for amplifying small currents (Translation). *Pribory Tekh Eksper.*, *4*, 89–91.

Östling, J., Brahme, A., Danielsson, M., Francke, T., Iacobaeus, C., & Peskov, V. (2003). Study of hole-types gas multiplication structures for portal imaging and other high count rate applications. *IEEE Transactions on Nuclear Science*, *50*(4), 809–819. doi:10.1109/TNS.2003.814562

Östling, J., Brahme, A., Danielsson, M., Iacobaeus, C., & Peskov, V. (2004). A radiation-tolerant electronic readout system for portal imaging. *Nuclear Instruments & Methods in Physics Research. Section A, Accelerators, Spectrometers, Detectors and Associated Equipment*, *525*(1-2), 308–312. doi:10.1016/j.nima.2004.03.079

Ott, R. J., Flower, M. A., Divoli, A., & Binnie, D. (2003). Performance of the PETRRA positron camera: Reduction of scatter and random events. *Nuclear Instruments & Methods in Physics Research. Section A, Accelerators, Spectrometers, Detectors and Associated Equipment*, *513*(1-2), 61–64. doi:10.1016/j.nima.2003.08.002

Paic, G., Piuz, F, Nappi, E., Manzari, V., Posa, F., Tomasicchio, G., ...Tustonic, T. (1996). *Status report of the CsI-RICH Collaboration 1995: development of a large area advanced fast RICH detector for particle identification at the LHC operated with heavy ions*. LDRB- Status report-RD-26., CERN-LHCC-96-020, 1-25

Papaevangelou, T. (2011). Micromegas technology. *Proc. 6th DITANET Topical Workshop on Detection Technique*.

Park, S., Hie-Hanng, C., & Ha, Y. (2007). Development of digital imaging device based on the gas electron multiplier. *Radiation Measurements*, *42*(10), 1661–1665. doi:10.1016/j.radmeas.2007.09.009

Particle Data Group. (1988). Review of particle properties. *Physics Letters. [Part B]*, *204*, 64.

Periale, L., Peskov, V., & Carlson, P. (2001). *Detection of the primary scintillation light from dense Ar, Kr and Xe with novel photosensitive gaseous detectors*. ArXive Physics/0106048

Periale, L., Peskov, V., Carlson, P., Francke, T., Pavlopoulos, P., Picchi, P., & Pietropaolo, F. (2001). *Detection of the primary scintillation light from dense Ar, Kr and Xe with novel photosensitive gaseous detectors*. ArXive, Physics/0106048.

Periale, L., Peskov, V., Carlson, P., Francke, T., Pavlopoulos, P., Picchi, P., & Pietropaolo, F. (2001). *Detection of the primary scintillation light from dense Ar, Kr and Xe with novel photosensitive gaseous detectors*. ArXive; Physics/0106048

Periale, L., Peskov, V., Braem, A., Di Mauro, A., Martinenego, P., Picchi, P., & Sipila, H. et al. (2007). Development of new sealed UV sensitive gaseous detectors and their applications. *Nuclear Instruments & Methods in Physics Research. Section A, Accelerators, Spectrometers, Detectors and Associated Equipment*, *572*(1), 189–192. doi:10.1016/j.nima.2006.10.272

Periale, L., Peskov, V., Carlson, P., Francke, T., Pavlopoulos, P., Picchi, P., & Pietropaolo, F. (2002). Detection of the primary scintillation light from dense Ar, Kr and Xe with novel photosensitive gaseous detectors. *Nuclear Instruments & Methods in Physics Research. Section A, Accelerators, Spectrometers, Detectors and Associated Equipment*, *478*(1-2), 377–383. doi:10.1016/S0168-9002(01)01779-X

Periale, L., Peskov, V., Carlson, P., Iacobaeus, C., Francke, T., Pavlopoulos, N., & Picchi, P. et al. (2003). Evaluation of various planar gaseous detectors with CsI photocathodes for the detection of primary scintillation light from noble gases. *Nuclear Instruments & Methods in Physics Research. Section A, Accelerators, Spectrometers, Detectors and Associated Equipment*, *497*(1), 242–248. doi:10.1016/S0168-9002(02)01918-6

Periale, L., Peskov, V., Iacobaeus, C., Francke, T., Lund-Jensen, B., Pavlopoulos, N., & Pietropaolo, F. et al. (2004). The development of gaseous detectors with solid photocathodes for low-temperature applications. *Nuclear Instruments & Methods in Physics Research. Section A, Accelerators, Spectrometers, Detectors and Associated Equipment*, *535*(1-2), 517–522. doi:10.1016/S0168-9002(04)01719-X

Periale, L., Peskov, V., Iacobaeus, C., Francke, T., Lund-Jensen, B., Pavlopoulos, P., & Tokanai, F. et al. (2005). The Successful Operation of Hole-Type Gaseous Detectors at Cryogenic Temperatures. *IEEE Transactions on Nuclear Science*, *52*, 927–931.

Periale, L., Peskov, V., Iacobaeus, C., Lund-Jensen, B., Picchi, P., Pietropaolo, F., & Rodionov, I. (2007a). Photosensitive gaseous detectors for cryogenic temperature applications. *Nuclear Instruments & Methods in Physics Research. Section A, Accelerators, Spectrometers, Detectors and Associated Equipment*, *573*(1-2), 302–305. doi:10.1016/j.nima.2006.10.264

Person, J. C. (1965). Isotope Effect in the Photoionization Efficiency for Benzene. *The Journal of Chemical Physics*, *43*(7), 2553–2555. doi:10.1063/1.1697162

Pertsev, A. N., & Ruzin, V. A. (1968). O*n a change in the spectral sensitivity of a photocathode during a change in the potential. Journal of Applied Spectroscopy*, *8*(5), 517–518. doi:10.1007/BF01103244

Peskov, V. (1980). *Complex/multiparamers plasma diagnostics in UV and X-ray regions*. (Doctor of Science Thesis). Kapitza Institute for Physical Problems, USSR Academy of Sciences.

Peskov, V. (1980). *Complex/multiparameter plasma diagnostics in UV and X-ray regions*. (Doctor of science thesis). Kapitza Institute for Physical Problems, Moscow, Russia.

Peskov, V. (1995). Gaseous detectors with solid photocathodes. *Proceedings of the first International four seas Conference*. Trieste, Italy: CERN. Retrieved from: http://cds.cern.ch/record/282759/files/CERN-97-06.pdf

Peskov, V. (2013). *Stability tests of GEMs with cylindrical holes fo the ALICE TPC upgrade*. Retrieved from https://indico.cern.ch/getFile.py/access?contribId=1&sessionId=1&resId=1&materialId=slides&confId=245535

Peskov, V., & Fonte, P. (2009). *Research on discharges in micropattern and small gap gaseous detectors*. ArXiv:0911.0463

Peskov, V., & Zichichi, A. (2007). *A New Supersensitive Flame Detector and it's Use for Early Forest Fire Detection* arXiv: 0709.2819

Peskov, V., (1988). Gaseous emitters and detectors of VUV radiation based on Cs cathodes. *Journal of Appl. Spectr., 48*(2), 316-317.

Peskov, V.D. (1979b). On the VUV Radiation of Townsend Avalanches. *J. Phys. (France), 40*, C7-331-C7333.

Peskov, V. (1976). On the Secondary Processes on Gas Counters-I. *Soviet Physics. Technical Physics, 20*, 791–794.

Peskov, V. (1977). On the Secondary Processes on Gas Counters-II. *Soviet Physics. Technical Physics, 22*, 335–338.

Peskov, V. (1980). Coordinate gas scintillation x-ray and UV photon counters. *Instruments and Experimental Techniques, 23*, 507–510.

Peskov, V. D. (1979a). VUV Spectrum of Townsend Avalanches. *Zh. Prikl. Spectr, 30*, 860–866.

Peskov, V. D. (1980). Coordinate Gas Scintillation Counters of VUV and X-Ray Photons. *Instruments and Experimental Techniques, 23*, 507–511.

Peskov, V. D. (1988). Gas detectors and sources of UV radiation based on Cs cathodes. *Journal of Applied Spectroscopy, 48*, 316–318.

Peskov, V., Baibussinov, B., Centro, S., Di Mauro, A., Lund-Jensen, B., Martinengo, P., & Ventura, S. et al. (2007a). Development and first tests of GEM-like detectors with resistive electrodes. *IEEE Transactions on Nuclear Science, 54*(5), 1784–1791. doi:10.1109/TNS.2007.905160

Peskov, V., Bencze, G., Di Mauro, A., Martinenego, P., Mayani, D., Molnar, L., & Shukla, I. et al. (2012). First observation of Cherenkov rings with a large area CsI-TGEM-based RICH prototype. *Nuclear Instruments & Methods in Physics Research. Section A, Accelerators, Spectrometers, Detectors and Associated Equipment, 695*, 154–158. doi:10.1016/j.nima.2011.11.045

Peskov, V., Borovik-Romanov, A., Sokolova, T., & Silin, E. (1994). Gaseous detectors of ultraviolet and visible photons. *Nuclear Instruments & Methods in Physics Research. Section A, Accelerators, Spectrometers, Detectors and Associated Equipment, 353*(1-3), 184–188. doi:10.1016/0168-9002(94)91634-9

Peskov, V., Charpak, G., Dominik, W., & Sauli, F. (1989). Investigation of light emission from a parallel-plate avalanche chamber filled with noble gases and with TEA, TMAE and H_2O vapours at atmospheric pressure. *Nuclear Instruments & Methods in Physics Research. Section A, Accelerators, Spectrometers, Detectors and Associated Equipment, 277*(2-3), 547–556. doi:10.1016/0168-9002(89)90788-2

Peskov, V., Charpak, G., Mine, P., Sauli, F., Scigocki, D., Séguinot, J., & Ypsilantis, T. et al. (1988). Liquid and solid organic photocathodes. *Nuclear Instruments & Methods in Physics Research. Section A, Accelerators, Spectrometers, Detectors and Associated Equipment, 269*(1), 149–160. doi:10.1016/0168-9002(88)90872-8

Peskov, V., Charpak, G., Sauli, F., Scigocki, D., Diep, V., & Janjic, D. (1989). Organometallic photocathodes for parallel-plate and wire chambers. *Nuclear Instruments & Methods in Physics Research. Section A, Accelerators, Spectrometers, Detectors and Associated Equipment, 283*(3), 786–791. doi:10.1016/0168-9002(89)91458-7

Peskov, V., Cortesi, M., Chechik, R., & Breskin, A. (2010). Further evaluation of a TGEM UV-photon detector for RICH – comparison with MWPC. *JINST, 5*(11P11004), 1–25. doi:10.1088/1748-0221/5/11/P11004

Peskov, V., Fonte, P., Danielsson, M., Iacobaeus, C., Ostling, J., & Wallmark, M. (2001). The study and optimization of new micropattern gaseous detectors for high-rate applications. *IEEE Transactions on Nuclear Science*, *45*(4), 1070–1074. doi:10.1109/23.958725

Peskov, V., Martinengo, P., Nappi, E., Oliveira, R., Paic, G., Pietropaolo, F., & Picchi, P. (2009). Progress in the development of photosensitive GEMs with resistive electrodes manufactured by a screen printing technology. *Nuclear Instruments & Methods in Physics Research. Section A, Accelerators, Spectrometers, Detectors and Associated Equipment*, *610*(1), 169–173. doi:10.1016/j.nima.2009.05.070

Peskov, V., & Silin, E. (1995). High efficiency photocathodes for gaseous detectors. *Nuclear Instruments & Methods in Physics Research. Section A, Accelerators, Spectrometers, Detectors and Associated Equipment*, *367*(1-3), 347–352. doi:10.1016/0168-9002(95)00654-0

Peskov, V., Silin, E., Sokolova, T., & Rodionov, I. (1999). First attempts to combine capillary tubes with photocathodes. *Nuclear Instruments & Methods in Physics Research. Section A, Accelerators, Spectrometers, Detectors and Associated Equipment*, *433*(1-2), 492–501. doi:10.1016/S0168-9002(99)00346-0

Peskov, V., & Zichichi, A. (1991). A new scintillator, $LiBaF_3$, and its basic properties. In *New technologies for supercolliders* (pp. 49–55). New York: Prenum Press. doi:10.1007/978-1-4684-1360-1_4

Peskov, V., & Zichichi, A. (1992). Fast (< 1 ns) gaseous detectors with solid photocathodes for frontier physics. *Nuclear Instruments & Methods in Physics Research. Section A, Accelerators, Spectrometers, Detectors and Associated Equipment*, *315*(1-3), 77–81. doi:10.1016/0168-9002(92)90684-V

Peyaud, A., Angelopoulos, A., Chemlis, C., Costopoulos, V., Chica, M., Giomataris, Y., & Quinlan, F. et al. (2015). The ForFire photodetector. *Nuclear Instruments & Methods in Physics Research. Section A, Accelerators, Spectrometers, Detectors and Associated Equipment*, *787*, 102–104. doi:10.1016/j.nima.2014.11.044

Philippov, P. (2014). *First year of running for the LHC-b calorimeter system*. Retrieved from https://www.google.ru/url?sa=t&rct=j&q=&esrc=s&source=web&cd=1&cad=rja&uact=8&ved=0CBwQFjAA&url=https%3A%2F%2Findico.cern.ch%2Fevent%2F301616%2Fcontribution%2F7%2Fmaterial%2Fslides%2F0.pptx&ei=dh-YVdPKN4m0ygPntJyIDg&usg=AFQjCNHALpPVepyHDolqTAJjJj2b3ZEqTg

Piuz, F. (1996). CsI-photocathode and RICH detectors. *Nuclear Instruments & Methods in Physics Research. Section A, Accelerators, Spectrometers, Detectors and Associated Equipment*, *371*(1-2), 96–115. doi:10.1016/0168-9002(95)01442-X

Piuz, F., Andres, Y., Braem, A., Davenport, M., Mauro, A. D., & Goret, B. et al. (1999). The CsI-based ring imaging detector for the ALICE experiment: Technical description of a large prototype. *Nuclear Instruments & Methods in Physics Research. Section A, Accelerators, Spectrometers, Detectors and Associated Equipment*, *433*(1-2), 222–234. doi:10.1016/S0168-9002(99)00293-4

Policarpo, A. J. P. L. (1978). Coupling the gas scintillation proportional counter to photoinization detectors. *Nuclear Instruments and Methods*, *153*(2-3), 389–391. doi:10.1016/0029-554X(78)90977-1

Pyrlik, J., Atiya, M., Broemmelsiek, D., Hamacher, T., Ispiryan, M., Korpar, S., & Starič, M. et al. (1998). Aging measurements of a TMAE based photon detector for the HERA-B RICH. *Nuclear Instruments & Methods in Physics Research. Section A, Accelerators, Spectrometers, Detectors and Associated Equipment*, *414*(2-3), 170–181. doi:10.1016/S0168-9002(98)00645-7

Qiang, Y., Zorn, C., & Fernando Barbosa, F. (2013). Radiation hardness tests of SiPMs for the JLab Hall D Barrel calorimeter. *Nuclear Instruments & Methods in Physics Research. Section A, Accelerators, Spectrometers, Detectors and Associated Equipment*, *698*, 234–241. doi:10.1016/j.nima.2012.10.015

Raether, H. (1964). *Electron avalanches and breakdown in gases*. London: Butterworths.

RCA Technical manual PT-60. (1963). Radio Corporation of America.

Reserva, R., Arogancia, D., Bacala, A., Boudjemline, K., Burke, D., Colas, P., & Zerguerras, T. et al. (2007). Studies on the drift properties and spatial resolution using a microMEGAS-equipped time projection chamber. *Pramana Jorn. of Phys.*, *69*(6), 989–994. doi:10.1007/s12043-007-0217-z

Ricci, M. (1999). *Proceedings of the 26th International Cosmic Ray Conference*. Salt Lake City, UT: Academic Press.

Rickenbach, R., Montanet, F., Backenstoss, G., Carlson, P., Chardalas, M., Franke, T., & Wildi, M. et al. (1989). Francke, T., ... Wildi, M., (The CPLEAR Collaboration), *The fast strangeness trigger of the CP-LEAR experiment. Nuclear Instruments & Methods in Physics Research. Section A, Accelerators, Spectrometers, Detectors and Associated Equipment*, *279*(1-2), 305–309. doi:10.1016/0168-9002(89)91098-X

Riegler, W., Schindler, H., & Veenhof, R. (2008). *Avalanche statistics*. Report at the RD-51 collaboration meeting. Retrieved from http://indico.cern.ch/event/35172/other-view?view=standard

Rodionov, I., Bidault, J.-M., Crotty, I., Fonte, P., Galy, F., Peskov, V., & Zanette, O. (2005). *Advanced gaseous photodetectors for hyperspectroscopy and other applications*. 2005 IEEE Nuclear Science Symposium Conference Record.

Rodionov, I., Bidault, J-M., Crotty, I., Fonte, P., Galy, F., Peskov, V., & Zanette, O. (2005). Advanced Gaseous Photodetectors for Hyperspectroscopy and Other Applications. *IEEE Nucl., Sci., Conf. Record.*

Rodionov, I., Francke, T., Peskov, V., & Sokolova, T. (2002). Hybrid gaseous photomultipliers. *Nuclear Instruments & Methods in Physics Research. Section A, Accelerators, Spectrometers, Detectors and Associated Equipment*, *478*(1-2), 384–390. doi:10.1016/S0168-9002(01)01778-8

Roncali, E., & Cherry, S. R. (2011). Application of Silicon Photomultipliers to Positron Emission Tomography. *Annals of Biomedical Engineering*, *39*(4), 1358–1377. doi:10.1007/s10439-011-0266-9 PMID:21321792

Rutherford, E., & Geiger, H. (1908). An electrical method of counting the number of alpha particles from radio-active substances. *Proc. R. Soc. Lond., Ser. A*, *81*, 141–161. doi:10.1098/rspa.1908.0065

Sadygov, Z. (1998). *Avalanche Detector*. Russian Agency for Patents and Trademarks, Patent No. RU 2102820.

Saito, K., Tawara, H., Sanami, T., Shibamura, E., & Sasaki, S. (2002). Absolute number of scintillation photons emitted by alpha particles in rare gases. *IEEE Transactions on Nuclear Science*, *49*(4), 1674–1680. doi:10.1109/TNS.2002.801700

Sakurai, H., Tamura, T., Gunji, S., & Noma, M. (1966). A new type of proportional counter using a capillary plate. *Nuclear Instruments & Methods in Physics Research. Section A, Accelerators, Spectrometers, Detectors and Associated Equipment*, *374*(3), 341–344. doi:10.1016/0168-9002(96)00226-4

Santiard, J. C. (2004). The ALICE HMPID on-detector front-end and readout electronics. *Nuclear Instruments & Methods in Physics Research. Section A, Accelerators, Spectrometers, Detectors and Associated Equipment*, *518*(1-2), 498–500. doi:10.1016/j.nima.2003.11.068

Sauli, F. (2006). *Gas electron multiplier detectors: performance and applications*. Presentation at the MPGD Workshop, CERN, January 2006. Retrieved from http://gdd.web.cern.ch/GDD/

Sauli, F. (2009). *Principles of gas detectors*. Talk given at KEK. Retrieved from http://view.officeapps.live.com/op/view.aspx?src=http%3A%2F%2Fcompassweb.ts.infn.it%2Frich1%2Fjarda%2FPresentations%2FTIPP09_japan%2FSAULI_PRINCIPLES.ppt

Sauli, F. (2011). Seminal Papers by Georges Charpak. Elsevier, Nuclear Instruments and Methods. *Physics Research Section A: Accelerators, Spectrometers, Detectors and Associated Equipment.*

Sauli, F. (2011). Seminal Papers by Georges Charpak. In *Nuclear Instruments and Methods in Physics Research Section A: Accelerators, Spectrometers, Detectors and Associated Equipment*. Elsevier.

Sauli, F., (1977). *Principles of operation of multiwire proportional and drift chambers*. CERN Yellow report 77-09.

Sauli, F. (1982). Rediscovering the gas-filled photodiode. *Nuclear Instruments and Methods*, *203*(1-3), 601–604. doi:10.1016/0167-5087(82)90681-0

Sauli, F. (1997). GEM: A new concept for electron amplification in gas detectors. *Nuclear Instruments & Methods in Physics Research. Section A, Accelerators, Spectrometers, Detectors and Associated Equipment*, *386*(2-3), 531–534. doi:10.1016/S0168-9002(96)01172-2

Sauli, F., Ropelewski, L., & Everaerts, P. (2006). Ion feedback suppression in time projection chambers. *Nuclear Instruments and Methods in Physics Research*, *560*(2), 269–277. doi:10.1016/j.nima.2005.12.239

Schindler, H., & Veenhof, R. (2008). *Avalanche statistics*. Retrieved from http://webcache.googleusercontent.com/search?q=cache:Oml9YMQffgAJ:https://indico.cern.ch/event/58953/session/0/contribution/1/material/slides/+&cd=1&hl=en&ct=clnk&gl=ch

Schlumbohm, H. (1958). Zur statistic der electronenlawinen im ebenen feld. *Zeitschrift fur Physik*, *151*(5), 563–572. doi:10.1007/BF01338427

Schmidt, W. (1997). Liquid state electronics of insulating liquids. CRC Press.

Schmidt, W. (1997). Liquid-state electronics of insulating liquids. SRS Press.

Schmidt, W.F. (1984). Electronic conduction processes in dielectric liquids. *IEEE Trans. Electric. Insul., E1-19*, 389-418.

Schmidt, W. F. (1974). Electron migration in liquids and glasses. In *Electron solvent and anion solvent interaction.* Amsterdam: Elsevier.

Scholze, F., Rabus, H., & Ulm, G. (1998). Mean energy required to produce an electron-hole pair in siliconfor photons of energies between 50 and 1500 eV. *Journal of Applied Physics*, *84*(5), 2926–2939. doi:10.1063/1.368398

Schotanus, P., Van Eijk, C. W. E., Hollander, R. W., & Pijpelink, J. (1987). Development study of a new gamma camera for positron emission tomography. *Nuclear Instruments & Methods in Physics Research. Section A, Accelerators, Spectrometers, Detectors and Associated Equipment*, *252*(2-3), 255–259. doi:10.1016/0168-9002(86)91186-1

Schukraft, J., Piuz, F., Klempt, W., Leistam, L., De Groot, J., Beoli, S., & Tustonic, T. (1998). Technical Design Report of the High Momentum Particle Identification Detector. *CERN / LHCC 98–19 ALICE TDR 1*. Retrieved from: http://www.google.co.uk/url?sa=t&rct=j&q=&esrc=s&source=web&cd=1&ved=0CCEQFjAA&url=http%3A%2F%2Fcern.ch%2Falice-hmpid%2Fpublicns%2FTDR%2FTDRFirstpage.pdf&ei=IW2IVJHNO4fuaK2kgvAD&usg=AFQjCNEs8miubRdegKxtCzkqSVDfDIgEJQ&bvm=bv.81456516,d.d2s

Séguinot, J (1989). *Lescompteurs Cherenkov: applications et limites pour l'identification des particules, developments et persepectives*. Preprint CERN-EP/89-92, LPC/89-25, 1-77

Séguinot, J., *Les compteurs Cherenkov: applications et limites pour lídentification des particules. Dévelopments at perspectives.*, CERN preprint, CERN EP/89-92 (1989).

Séguinot, J., Charpak, G., Giomataris, Y., Peskov, V., Tischhauser, J., & Ypsilantis, T. (1990). Reflective UV photocathodes with gas –phase extraction: Solid, liquid and absorbed thin films. *Nuclear Instruments & Methods in Physics Research. Section A, Accelerators, Spectrometers, Detectors and Associated Equipment*, *297*(1-2), 133–147. doi:10.1016/0168-9002(90)91359-J

Séguinot, J., Tocqueville, J., & Ypsilantis, T. (1980). Imaging Cherenkov detectors: Photoionization of triethylamine. *Nuclear Instruments and Methods*, *173*, 283–298.

Séguinot, J., & Ypsilantis, T. (1977). Photo-ionization and Cherenkov ring imaging. *Nuclear Instruments and Methods in Physics Research*, *142*(3), 377–391. doi:10.1016/0029-554X(77)90671-1

Senftleben, A., Pflüger, T., Ren, X., Al-Hagan, O., Najjari, B., Madison, D., & Ullrich, J. et al. (2010). Search for interference effects in electron impact ionization of aligned hydrogen molecules. *Journal of Physics. B, Atomic, Molecular, and Optical Physics*, *43*(8), 081002–081007. doi:10.1088/0953-4075/43/8/081002

Shalem, C., Chechik, R., Breskin, A., & Michaeli, K. (2006). Advances in Thick GEM-like gaseous electron multiplier-Part-1: Atmospheric pressure operation. *Nuclear Instruments & Methods in Physics Research. Section A, Accelerators, Spectrometers, Detectors and Associated Equipment*, *558*(2), 475–489. doi:10.1016/j.nima.2005.12.241

Shefer, E., Breskin, A., Buzulutskov, A., Chechik, R., Klin, M., & Prager, M. (1998). Laboratory production of efficient alkali-antimonide photocathodes. *Nuclear Instruments & Methods in Physics Research. Section A, Accelerators, Spectrometers, Detectors and Associated Equipment*, *411*(2-3), 383–388. doi:10.1016/S0168-9002(98)00350-7

Shefer, E., Breskin, A., Chechik, R., Buzulutskov, A., Singh, B. K., & Prager, M. (1999). Coated photocathodes for visible photon imaging with gaseous photomultipliers. *Nuclear Instruments & Methods in Physics Research. Section A, Accelerators, Spectrometers, Detectors and Associated Equipment*, *433*(1-2), 502–506. doi:10.1016/S0168-9002(99)00358-7

Siegmund, O. H. W., John, V., Vallerga, J. V., & Tremsin, A. S. (2011). Microchannel Plate Imaging Photon Counters for Ultraviolet through NIR Detection with High Time Resolution. *Proceedings of the Society for Photo-Instrumentation Engineers*, *8033*, 80330V-1, 80330V–12. doi:10.1117/12.884271 PMID:23833700

Siegmund, O. H. W., McPhate, J. B., Tremsin, A. S., Jelinsky, S. R., Hemphill, R., Frisch, H. J., & Mane, A et al.. (2012). Atomic layer deposited borosilicate glass microchannel plates for large area event counting detectors. *Nuclear Instruments & Methods in Physics Research. Section A, Accelerators, Spectrometers, Detectors and Associated Equipment*, *695*, 168–171. doi:10.1016/j.nima.2011.11.022

Siegmund, O. H. W., Tremsin, A. S., Vallerga, J. V., Abiad, R., & Hull, J. (2003). High resolution cross strip anodes for photon counting detectors. *Nuclear Instruments & Methods in Physics Research. Section A, Accelerators, Spectrometers, Detectors and Associated Equipment*, *504*(1-3), 177–181. doi:10.1016/S0168-9002(03)00816-7

Simon, F., Azmount, B., Becker, U., Burns, L., Crary, D., Kearney, K., & Woody, C. et al. (2007). Development of tracking detectors with industrially produced GEM foils. *IEEE Nucl. Sci*, *54*(6), 2646–2652. doi:10.1109/TNS.2007.909912

Simons, D. G., Fraser, G. W., De Korte, P. A. J., Pearson, J. F., & De Jong, L. (1987). UV and XUV quantum detection efficiencies of CsI-coated microchannel plates. *Nuclear Instruments & Methods in Physics Research. Section A, Accelerators, Spectrometers, Detectors and Associated Equipment*, *261*(3), 579–586. doi:10.1016/0168-9002(87)90371-8

Skovoroda, A. A., & Zvonkov, A. V. (2001). A stationary microwave discharge at high pressure of Argon. *Journal of Experimental and Theoretical Physics*, *92*(1), 78–85. doi:10.1134/1.1348463

Sommer, A. H. (1968). Photoemissive materials: preparation, properties, and uses. New York, NY: Wiley.

Sommer, A. H. (1968). *Photoemissive materials: preparation, properties and uses*. New York: Wiley.

Sommer, A. H. (1968). *Photo-emissive materials: preparation, properties, and uses*. New York, NY: Wiley.

Suffert, M., Charpak, G. (1986). *Some studies of a BaF_2 scintillator coupled to a photosensitive wire chamber: the influence of the temperature on the photocathode efficiency and the existence of a possible absorbed layer of TMAE*. Preprint CERN-EP-INT-86-03, 1-13.

Sumiyoshi, T., Tokanai, F., Sugiyama, H., Okada, T., Ohishi, N., Ohmura, T., & Gunji, S. et al. (2011). Development of gaseous PMT with micropattern gas detector. *Nuclear Instruments & Methods in Physics Research. Section A, Accelerators, Spectrometers, Detectors and Associated Equipment*, *639*(1), 121–125. doi:10.1016/j.nima.2010.10.032

Summer, W. (1957). *Photosensitors: a treatise on photo-electric devices and their application to industry*. London: Chapman and Hall Ltd.

Swift, E. Jr, & Hochanadel, H. P. (1945). The vapour pressure of trimethylamine from 0 to 40°. *Journal of the American Chemical Society*, *67*(5), 880–881. doi:10.1021/ja01221a508

Tamm, I. (1959). *General Characteristics of Radiation Emitted by Systems Moving with Superlight Velocities with some Applications to Plasma Physics. In Les Prix Nobel en 1958* (p. 122). P.A. Norstedt & Söner.

Teshima, M., Dolgoshein, B., & Mirzoyan, R. (2008). SiPM development for astroparticle physics applications. *Proceedings of the 30th International Cosmic Ray Conference*. Universidad Nacional Autónoma de México.

Tessarotto, F., Abbon, P., Alexeev, M., Birsa, R., Bordalo, P., Bradamante, F., …Takekawa, S. (2014). Long term experience and performance of COMPASS RICH-1. *JINST, 9*, 1-27.

Theobald, J. K. (1953). Investigation of back diffusion of photoelectrons in various standard gases as it affects secondary electron emission coefficients. *Journal of Applied Physics*, *24*(2), 123–128. doi:10.1063/1.1721223

Timothy, J. G. (2010). In M. C. E. Huber, A. Pauluhn, J. L. Culhane, J. G. Timothy, K. Wilhelm, & A. Zehnder (Eds.), Microchannel plates for photon detection and imaging in space. I n: Observing Photons in Space. ISSI Scientific Reports Series, ESA/ISSI (pp. 365–390). Academic Press.

Titov, M. (2003). *Radiation damage and long-term aging in gas detectors*. A talk given at the INFN Eloisotron project's 42nd Workshop on Innovative detectors for supercolliders, Erice, Italy.

Titov, M. (2012). Gaseous Detectors. In Handbook of Particle Detection and Imaging, (pp. 239-264). doi:10.1007/978-3-642-13271-1_11

Tokanai, F., Moriya, T., Takeyama, M., Sakurai, H., Gunji, S., Sumiyoshi, T., & Kishimoto, S. et al. (2014). Newly developed gaseous photomultiplier. *Nuclear Instruments & Methods in Physics Research. Section A, Accelerators, Spectrometers, Detectors and Associated Equipment*, *766*, 176–179. doi:10.1016/j.nima.2014.05.014

Tokanai, F., Sumiyoshi, T., Sugiyama, H., Okada, T., Ohishi, N., Ohmura, T., & Gunji, S. et al. (2009). Development of a gaseous PMT with micro-pattern gas detectors. *Nuclear Instruments & Methods in Physics Research. Section A, Accelerators, Spectrometers, Detectors and Associated Equipment*, *610*(1), 164–168. doi:10.1016/j.nima.2009.05.111

Tokanai, F., Sumiyoshi, T., Sugiyama, H., Okada, T., Ohishi, N., Sakurai, H., & Kishimoto, S. et al. (2011). Sealed gaseous photomultiplier with CsI photocathode. *Nuclear Instruments & Methods in Physics Research. Section A, Accelerators, Spectrometers, Detectors and Associated Equipment*, *628*(1), 190–193. doi:10.1016/j.nima.2010.06.314

Townsend, J. S. E. (1903 b). The genesis of ions by motion of positive ions in gas and a theory of the sparking potential. *Philosophical Magazine*, *6*(35), 598–618. doi:10.1080/14786440309463061

Townsend, J. S. E. (1903a). The conductivity produced in gases by the aid of ultraviolet light. *Philosophical Magazine*, *5*(28), 389–398. doi:10.1080/14786440309462937

Townsend, J. S. E., & Kirkby, M. A. (1901). Conductivity produced in Hydrogen and carbonic acid by motion of negatively charged ions. *Philosophical Magazine*, *1*(6), 630–642. doi:10.1080/14786440109462656

Tripathi, N., Bell, L. D., Nikzad, S. H. O. U. L. E. H., Tungare, M., Suvarna, P. H., & Sandvik, F. S. (2011). Novel Cs-Free GaN Photocathodes. *Journal of Electronic Materials*, *40*(4), 382–387. doi:10.1007/s11664-010-1507-7

Uhlig, F., Britting, A., Eyrich, W., Lehmann, A., Schwarz, C., & Schwiening, J. (2012). Performance studies of microchannel plate PMTs. *Nuclear Instruments & Methods in Physics Research. Section A, Accelerators, Spectrometers, Detectors and Associated Equipment*, *695*, 68–70. doi:10.1016/j.nima.2011.12.062

Ullrich, T. (1996). Cherenkov ring fitting techniques for the CERES RICH detectors. *Nucl. Instr. and Meth.in Phys. Res.*, *A371*, 243.

Va'vra, J. (1987). Wire aging of hydrocarbons with TMAE additions. *IEEE Transactions on Nuclear Science*, *NS-34*(1), 486–490. doi:10.1109/TNS.1987.4337390

Va'vra, J. (1992). Wire chamber gases. *Nuclear Instruments & Methods in Physics Research. Section A, Accelerators, Spectrometers, Detectors and Associated Equipment*, *323*(1-2), 34–47. doi:10.1016/0168-9002(92)90266-7

Va'vra, J. (1995). Can TMAE photocathode be used for high rate applications? *Nuclear Instruments & Methods in Physics Research. Section A, Accelerators, Spectrometers, Detectors and Associated Equipment*, *367*(1-3), 353–357. doi:10.1016/0168-9002(95)00536-6

Va'vra, J. (1997). Wire aging with TEA photocathode. *Nuclear Instruments & Methods in Physics Research. Section A, Accelerators, Spectrometers, Detectors and Associated Equipment*, *387*(1-2), 183–185. doi:10.1016/S0168-9002(96)00986-2

Va'vra, J., Kady, J., & Wise, J. (1996). Study of photosensitive mixtures of TMAE and helium, hydrocarbons or CF4-based carrier gases. *Nuclear Instruments & Methods in Physics Research. Section A, Accelerators, Spectrometers, Detectors and Associated Equipment*, *370*(2-3), 352–366. doi:10.1016/0168-9002(95)00842-X

Vasileiadis, G., Malamud, G., Miné, P., & Vartsky, D. (1996). Solid and vapour phase UV photocathodes for gaseous detectors. *Nuclear Instruments & Methods in Physics Research. Section A, Accelerators, Spectrometers, Detectors and Associated Equipment*, *372*(1-2), 31–34. doi:10.1016/0168-9002(95)01294-X

Vavilov, S. I. (1934). On the possible cause of the blue γ glow in liquids. *Soviet Physics, Doklady*, *2*, 457.

Veloso, J. F. C. A., dos Santos, J. M. F., & Conde, A. N. (2000). A proposed new microstructure for gas radiation detectors: The microhole and strip plate. *The Review of Scientific Instruments*, *A71*(6), 2371–2376. doi:10.1063/1.1150623

Volpe, G., (2009). *Results from cosmic rays and first LHC beam with the ALICE HMPID detector.* ArXV:0907.4579v2

Volpe, G. (2014). Pattern recognition and PID procedure with the ALICE-HMPID. *Nuclear Instruments & Methods in Physics Research. Section A, Accelerators, Spectrometers, Detectors and Associated Equipment*, *766*, 259–262. doi:10.1016/j.nima.2014.05.031

Von Halleer, B., Roukoutakis, F., Chapeland, S., Altini, V., Carena, F., & Carena, W. et al. (2010). The ALICE data quality monitoring. *Journal of Physics: Conference Series*, *219*(022023), 1–9.

Winn, D. R. (1989). Large photocathode detector for large calorimetry. *IEEE Transactions on Nuclear Science*, *NS-36*(1), 128–131. doi:10.1109/23.34418

Woody, C. L. (1988). Aging effects in low pressure wire chambers containing TMAE. *IEEE Transactions on Nuclear Science*, *35*(1), 498–497. doi:10.1109/23.12772

Xu, C., Klanner, R., Garutti, E., & Hellweg, W.-L. (2014). Influence of X-ray irradiation on the properties of the Hamamatsu silicon photomultiplier S10362-11-050C. *Nuclear Instruments & Methods in Physics Research. Section A, Accelerators, Spectrometers, Detectors and Associated Equipment*, *762*, 149–161. doi:10.1016/j.nima.2014.05.112

Yakovlev, B. S., & Lukin, L. V. (1985). Photoionization in Non-Polar Liquids. *Advances in Chemical Physics*, *60*, 99–160. doi:10.1002/9780470142844.ch3

Yamamoto, K., Yamammura, K., Sato, K. (2007). Development of multi-pixel photon counter (MPPC). *IEEE Nucl. Sci. Conf. Record*, *24*(292), 1511-1515.

Ypsilantis, T., & Séguinot, J. (1994). Theory of ring imaging Cherenkov counters. *Nuclear Instruments & Methods in Physics Research. Section A, Accelerators, Spectrometers, Detectors and Associated Equipment*, *343*(1), 30–51. doi:10.1016/0168-9002(94)90532-0

Zeitelhack, K., Elhardt, A., Friese, J., Gernhauser, R., Homolka, J., & Kastenmuller, A. et al. (1999). The HADES RICH detector. *Nuclear Instruments & Methods in Physics Research. Section A, Accelerators, Spectrometers, Detectors and Associated Equipment*, *433*(1-2), 201–206. doi:10.1016/S0168-9002(99)00371-X

About the Authors

Tom Francke is an internationally recognized scientist and entrepreneur, specializing in innovation and development of advanced radiation instruments and other novel technology. For 15 years, Dr. Francke served as a senior scientist in astroparticle physics developing radiation instruments for particle physics experiments and space research, mainly for CERN and NASA. During these years he served at CERN, Switzerland, CNRS, France, KTH, Sweden and New Mexico State University, USA. He has since then founded and/or contributed to grow a number of successful companies serving as CEO, CBO or chairman of the board, including XCounter (listed on AIM of the London Stock Exchange), Super Sonic Imagine (France), Tacton (Sweden), Conflux (Sweden) etc. All companies have invented and developed advanced new technology which has been commercialized internationally. Tom Francke is associate professor in particle physics from KTH, Sweden. He is the inventor of 61 patent families, has published more than 100 scientific articles and received a number of international awards.

Vladimir Peskov is a chief scientist at the Institute for Chemical Physics Russian Academy of Sciences (RAS) working currently at CERN for ALICE experiment. He obtained his academic degrees (Ph.D) in 1976 and Doctor of Sciences in 1982 from the Institute of Physical Problems, RAS, in Moscow. He worked there on plasma studies in the Physics Laboratory RAS led by P.L. Kapitza, and developed new methods and instruments for plasma diagnostic in far ultraviolet allowing to discover some new phenomena in plasma. From 1986 he mainly worked abroad: at CERN (in Charpak's group), Fermi National Laboratory, NASA and the Royal Institute of Technology, Sweden, on the instrumentation for high energy physics, astrophysics and medicine. He has developed photosensitive gaseous photomultipliers which have been used in several experiments, including ALICE. He is an author and co-author of more than 250 publications, two scientific books and twelve International patents, member of the International Scientific Advisory and Organizing Committees in several conferences and workshops. For many years served as an editor of Journal of Instrumentation.

Index

N

O

P

Q

R

S

T

U

V

W

X

www.ingramcontent.com/pod-product-compliance
Lightning Source LLC
LaVergne TN
LVHW080619300926
849162LV00005B/1208
* 9 7 8 1 5 2 2 5 0 2 4 2 5 *